TEXTBOOK OF PRODUC

TEXTBOOK OF PRODUCTION ENGINEERING

SECOND EDITION

K.C. JAIN

Formerly
Senior Professor
Prestige Institute of Engineering and Science, Indore
Director, Govindram Seksaria Institute of Management and Research, Indore
and
Dean, Faculty of Industrial Technology
Rajiv Gandhi University of Technology, Bhopal

A.K. CHITALE

Formerly
Director and Academic Advisor
Govindram Seksaria Institute of Management and Research, Indore

PHI Learning Private Limited
Delhi-110092
2014

₹ 695.00

TEXTBOOK OF PRODUCTION ENGINEERING, Second Edition
K.C. Jain and A.K. Chitale

© 2014 by PHI Learning Private Limited, Delhi. All rights reserved. No part of this book may be reproduced in any form, by mimeograph or any other means, without permission in writing from the publisher.

ISBN-978-81-203-4749-6

The export rights of this book are vested solely with the publisher.

Second Printing (Second Edition) ... ... **February, 2014**

Published by Asoke K. Ghosh, PHI Learning Private Limited, Rimjhim House, 111, Patparganj Industrial Estate, Delhi-110001 and Printed by Mohan Makhijani at Rekha Printers Private Limited, New Delhi-110020.

Contents

Preface .... *xxiii*
Preface to the First Edition .... *xxxi*

Part I METAL CUTTING

1 Introduction to Materials and Processes 3–23

1.1 Materials and Their Necessity for Engineering Application 3
1.2 Types of Materials 4
    1.2.1 Metallics 4
    1.2.2 Polymers 5
    1.2.3 Ceramics 6
    1.2.4 Composites 6
    1.2.5 Other Materials 7
1.3 Overview of Manufacturing Processes 9
    1.3.1 Basic or Conventional Processes 9
1.4 Constructional Features of Basic Machines 19
Review Questions 22

2 Metal Cutting Tools: Basic Concepts 24–49

2.1 Definition of Metal Cutting Tool 24
2.2 Classification of Metal Cutting Tools 24
    2.2.1 Single-point Cutting Tools 24
    2.2.2 Multi-point Cutting Tools 27
    2.2.3 Form Tools 27

2.3	Pre-requisites of Cutting Tools	27
2.4	Standard Angles of Cutting Tool	28
	2.4.1 Rake	28
	2.4.2 Side Relief	29
	2.4.3 End Relief	30
	2.4.4 Recommended Rake and Relief Angles	30
	2.4.5 Nose Radius	30
	2.4.6 Tool Holder Angle	31
	2.4.7 Flat or Drag	31
	2.4.8 Clearance Angle	31
	2.4.9 Side Cutting Edge	31
	2.4.10 End Cutting Edge	32
	2.4.11 Nose Angle and Angles in Normal Plane	33
2.5	Working Angles	33
	2.5.1 Setting Angle	33
	2.5.2 Entering Angle	33
	2.5.3 True Rake Angle	34
	2.5.4 Cutting Angle	34
	2.5.5 Lip Angle	34
	2.5.6 Working Relief Angle	34
2.6	Cutting Tool Nomenclature Systems	34
	2.6.1 British Maximum Rake System	34
	2.6.2 American System (ASA System)	35
	2.6.3 German System (DIN System)	35
	2.6.4 Normal Rake System	35
2.7	Reference Planes	36
	2.7.1 Coordinate System of Reference Planes	36
	2.7.2 Orthogonal System of Reference Planes or Orthogonal Rake System (ORS) or International Orthogonal System (ISO)	36
	2.7.3 Transformation from ASA System to ISO System	37
2.8	Cutting Tool Signature	39
2.9	Geometry of Cutting Tools	40
2.10	Factors Affecting Tool Geometry	40
2.11	Illustrative Examples on Tool Geometry	40
Review Questions		49

3 Cutting Tool Materials 50–66

3.1	Introduction	50
3.2	Requirements or Characteristics of a Tool Material	50
3.3	Types of Tool Materials	51
	3.3.1 Plain Carbon Steels	51
	3.3.2 Low Alloy Steels	52
	3.3.3 High Speed Steels	52
	3.3.4 Non-ferrous Cast Alloys (Super High Speed Tools)	52
	3.3.5 Cemented Carbides	53

	3.3.6	Ceramic Tool Materials	57
	3.3.7	Cermets	58
	3.3.8	Diamonds	58
	3.3.9	Abrasives	58
3.4	Characteristics and Uses of Diamond Tools		58
	3.4.1	General	58
	3.4.2	Classification	59
	3.4.3	Origin	59
	3.4.4	Special Characteristics	59
	3.4.5	Relation to Metal Cutting Process	59
	3.4.6	Machining Characteristics	60
3.5	Comparison of Cutting Tool Materials		61
3.6	Non-ferrous Cutting Tool Materials		61
3.7	Non-metallic Cutting Tool Materials		63
	3.7.1	Plastics	63
	3.7.2	Rubber	63
	3.7.3	Wood	63
	3.7.4	Hard Board	63

Review Questions 64

4 Design of Metal Cutting Tools 67–77

4.1	Introduction		67
4.2	Elements of Cutting Tool Design		67
4.3	Forces Acting on Cutting Tools		68
	4.3.1	Force of Resistance to Cutting	68
	4.3.2	Force of Drilling	69
	4.3.3	Design of Single-point Cutting Tool	70
4.4	Brazed Tool Seats		72
4.5	Cut-off Tools		72
4.6	Boring Tools		73
4.7	Brazed Tipped Tools		73
4.8	Mechanically Held Tipped Tools		74
4.9	Diamond Tools		74
	4.9.1	Brazed Diamond Tool	75
	4.9.2	Mechanically Clamped Diamond Tool	75
4.10	Form Tools and Their Design		75
	4.10.1	Circular and Flat Form Tool	75

Review Questions 77

5 Theory of Metal Cutting 78–121

5.1	Historical Development	78
5.2	Metal Cutting Defined	78
5.3	Characteristics of Metal Cutting	79
5.4	Representation of Metal Cutting Process	79
5.5	Orthogonal and Oblique Cutting	79

5.6	Difference between Orthogonal and Oblique Cutting	80
5.7	Mechanism of Chip Formation	81
5.8	Curling of Chip	82
5.9	Geometry of Chip Formation	82
5.10	Methods Used for Determining Chip Geometry	83
5.11	Classes of Chips	83
	5.11.1 Discontinuous Chips	83
	5.11.2 Inhomogeneous Chips	84
	5.11.3 Continuous Chips	84
	5.11.4 Fractured Chips	85
5.12	Effect of Various Factors on Chip Formation or Metal Cutting Characteristics	85
5.13	Methods of Reducing Friction	86
5.14	Physical Aspects of Chip Control and Chip Breakers	86
	5.14.1 Chip Control Through Tool Grinding	86
	5.14.2 Chip Control Through Chip Breakers	87
5.15	Velocity Relationships in Orthogonal Cutting—Merchant's Analysis of Metal Cutting Kinematics	91
5.16	Chip Thickness Ratio	92
5.17	Forces Acting on a Cutting Tool	94
5.18	Principle of Minimum Energy Applied to Metal Cutting—Merchant's Shear Angle Relation	96
5.19	Stress and Strain in Chip	99
5.20	Shear and Strain Rate	99
5.21	Energy Consideration in Metal Cutting	100
5.22	Stress and Strain Distributions in Plane Flow—Lee and Shaffer's Model	101
5.23	Different Theories of Shear Angle Relationship	104
Review Questions		119

6 Theory of Multipoint Machining 122–154

6.1	Introduction	122
6.2	Mechanism of Drilling	123
6.3	Torque and Thrust in Drilling Processes at Lips and Chisel	124
6.4	Mechanics of Metal Cutting in Chisel Edge Zone of Twist Drill	125
	6.4.1 Analysis for Cutting Zone	126
	6.4.2 Thrust Due to Chisel Edge Indentation	128
6.5	Use of Formulae for Torque and Thrust at Lips of Twist Drill	129
6.6	Milling	130
	6.6.1 Determination of Undeformed Chip Length	130
	6.6.2 Feed Rate and Cutter Wear	132
	6.6.3 Depth of Cut and Cutter Wear	132
	6.6.4 Chip Thickness	132
	6.6.5 Schlesinger's Formula	133
	6.6.6 Cutting Forces and Power	134
	6.6.7 Total Force Acting on a Cutter	135
	6.6.8 Work Done in Milling	136
	6.6.9 Relationship between Face and Peripheral Milling	137

	6.7	Mechanism of Grinding	137
		6.7.1 Undeformed Chip Length	138
		6.7.2 Maximum Chip Thickness in Cylindrical Grinding	141
		6.7.3 Grinding Kinetics	143
	6.8	Power in Broaching	144
		Review Questions	154

7. Heat in Metal Cutting and Temperature Measurement 155–192

7.1	Introduction		155
7.2	Computation of Temperatures in Orthogonal Cutting		156
	7.2.1	Stationary Heat Source—Friction Slider	156
	7.2.2	Shear Plane Temperature	160
	7.2.3	Tool Face Temperature	162
	7.2.4	Energy Balance	166
7.3	Cutting Tool Temperature by Dimensional Analysis		167
	7.3.1	Introduction	167
	7.3.2	Dimensional Analysis of Tool Temperature	167
	7.3.3	Comparison with Experimentally Obtained Expressions	171
7.4	Heat and Temperature in Milling		173
7.5	Heat and Temperature in Drilling		173
7.6	Heat and Temperature in Grinding		176
	7.6.1	Heat in Grinding	176
	7.6.2	Energy Considerations	178
	7.6.3	Thermodynamic Background for Thermal Aspect in Grinding	178
	7.6.4	Calculation of Grinding Temperature	180
7.7	Experimental Techniques of Temperature Measurement		182
	7.7.1	Orthogonal and Oblique Cutting	182
	7.7.2	Arrangement for Measurement by Thermocouple	182
	7.7.3	Calibration Procedure	184
	7.7.4	Radiation Method of Temperature Measurement	186
	7.7.5	Imbedded Thermocouple Method of Temperature Measurement	187
7.8	Measurement of Temperature in Milling		188
7.9	Measurement of Temperature in Drilling		189
7.10	Measurement of Temperature in Grinding		190
	Review Questions		191

8. Dynamometry 193–219

8.1	Introduction	193
8.2	Forces at the Cutting Edge	193
8.3	Desirable Characteristics of a Dynamometer	194
8.4	Strain and Strain Measurement	195
8.5	Need for Strain Measurement	195
8.6	Properties of an Ideal Strain Gauge	195
8.7	Mechanical Strain Measurement	196
8.8	Optical Strain Measurement	196

8.9	Electrical Strain Measurement	196
8.10	Uses of Strain Gauges	197
8.11	Survey of Various Types of Dynamometers	198
	8.11.1 Mechanical Type	199
	8.11.2 Hydraulic Type	199
	8.11.3 Pneumatic Type	200
	8.11.4 Optical Type	200
	8.11.5 Electrical Type	200
8.12	Use of Electrical and Electronic Transducers	203
	8.12.1 Electrical Transducer Tube	203
	8.12.2 Differential Transformer	203
	8.12.3 Unbounded Wire Resistance Strain Gauge	203
	8.12.4 Bonded Strain Gauges	204
8.13	Strain Gauge Force and Torque Transducer	204
	8.13.1 Strain Gauge Bridge Circuit	205
	8.13.2 Wheatstone Bridge	206
8.14	Practical Applications of Resistance Gauges	208
8.15	Bonding Techniques	211
8.16	Surface Preparation for Bonding Strain Gauges	211
8.17	Nitrocellulose Cement	212
8.18	Epoxy Cements	212
8.19	Strain Gauge Lathe Dynamometer	212
8.20	Turning Dynamometers	213
	8.20.1 Turning Dynamometer Based on Basic Principle	213
	8.20.2 Another Type of Turning Dynamometer	214
8.21	Three-component Lathe Tool Dynamometer	215
8.22	Milling and Grinding Dynamometers	216
	8.22.1 Milling Dynamometers	217
	8.22.2 Grinding Dynamometer	218
8.23	Response Curve of Dynamometer	218
Review Questions		219

9 Tool Failures and Tool Life 220–246

9.1	Tool Failure Defined	220
9.2	Criterion of Tool Failure	220
9.3	Types of Tool Failure	221
	9.3.1 Temperature Failure	221
	9.3.2 Mechanical Chipping	222
	9.3.3 Built-Up Edge (BUE)	222
	9.3.4 Spalling or Crumbling	223
9.4	Tool Wear (Microscopic Wear)	223
	9.4.1 Effects of Tool Wear	223
	9.4.2 Types of Tool Wear	223
9.5	Causes or Mechanisms of Wear	225
	9.5.1 Abrasion	225

	9.5.2	Adhesion	226
	9.5.3	Diffusion	227
	9.5.4	Electrochemical	228
9.6	Measurement of Tool Wear		228
9.7	Tool Life Definition		228
9.8	Tool Life Specifications		228
9.9	Measurement of Tool Life		229
	9.9.1	Expected Tool Life	229
	9.9.2	General Empirical Relationship between Cutting Speed, Tool Life, Feed, and Depth of Cut	229
	9.9.3	Formula Connecting Tool Life, Cutting Speed, Chip Thickness and Length of Tool Engagement	230
	9.9.4	Equation Incorporating the Effect of Size of Cut	230
9.10	Factors Affecting Tool Life		230
	9.10.1	Cutting Speed	231
	9.10.2	Effect of Feed Rate and Depth of Cut	231
	9.10.3	Microstructure of Workpiece	232
	9.10.4	Effect of Workpiece Hardness	232
	9.10.5	Effect of Tool Material	233
	9.10.6	Rigidity of Workpiece Machine Tool System	233
	9.10.7	Nature of Cutting	233
	9.10.8	Cutting Fluids and Tool Life	233
	9.10.9	Effect of Shape and Tool Angles on Tool Life	233
Review Questions			245

10 Machinability 247–252

10.1	Machinability Defined		247
10.2	Evaluation of Machinability		247
	10.2.1	Tool Life	248
	10.2.2	Chip Control	249
	10.2.3	Power Consumption	249
	10.2.4	Surface Finish	249
10.3	Factors Affecting Machinability		250
	10.3.1	Microstructure	250
	10.3.2	Strength	250
	10.3.3	Strength at Elevated Temperature	250
	10.3.4	Coefficient of Thermal Dispersion	251
	10.3.5	Built-up Edge Formation	251
	10.3.6	Work Hardening	251
	10.3.7	Effect of Alloying Elements	251
	10.3.8	Heat Treatment	252
	10.3.9	Tool Geometry	252
	10.3.10	Cutting Fluid	252
	10.3.11	Machine Tool, Tool and Work Factors	252
10.4	Machinability Index		252
Review Questions			252

11 Economics of Metal Machining 253–260

11.1 Cost Analysis and Economics 253
11.2 Indexable Insert System 257
11.3 Brazed Carbide Tool 258
11.4 Variation of Costs with Speed for Different Tools 258
Review Questions 260

12 Metal Cutting and Metal Working Fluids 261–276

12.1 Definition of Cutting Fluids 261
12.2 Functions of Cutting Fluids 261
12.3 Advantages of Cutting Fluid Applications 262
12.4 Characteristics of Good Cutting Fluid 262
12.5 Theory of Cutting Fluid 263
12.6 Action of Cutting Fluids as Lubricant 263
12.7 Cutting Fluids and Tool Life 264
12.8 Application of Metal Working Fluids 264
12.9 Type of Cutting Fluids 265
    12.9.1 Solid Cutting Fluids 265
    12.9.2 Liquid Cutting Fluids 265
12.10 Synthetic or Chemical Cutting Fluids 268
    12.10.1 Types of Synthetic Cutting Fluids 269
    12.10.2 Commonly Used Chemical Constituent Fluids 269
    12.10.3 Advantages of Synthetic Fluids 269
    12.10.4 Disadvantages of Synthetic Fluids 269
12.11 Gaseous Cutting Fluids 270
12.12 Extreme Pressure (EP) Lubrication Mechanism in Metal Cutting 270
    12.12.1 Advantages of EP 270
    12.12.2 Conditions of Use of EP 271
    12.12.3 Examples of EP 271
12.13 Criteria of Selection of Cutting Fluids 271
    12.13.1 Disadvantages of Using Cutting Fluids 271
Review Questions 275

Part II MACHINE TOOLS

13 Introduction to Machine Tools 279–283

13.1 Machine Tool—Definition 279
    13.1.1 Machine Tool Design Requirements 279
13.2 Present Trend of Design Optimisation 280
13.3 Elements of Machine Tools 280
13.4 Classification of Metal Cutting Machine Tools 280
13.5 Machine Tool Industry—Progress in India 281
13.6 Common Features of Machine Tools 282

13.7	Selection of Machine Tools	282
13.8	Systems for Control of Machine Tools	283
Review Questions		283

14 Design of Machine Tool Beds 284–293

14.1	Introduction		284
14.2	Factors in Design of Beds		285
	14.2.1	Strength	285
	14.2.2	Volume to Weight Ratio	286
	14.2.3	Vibration Response	289
	14.2.4	Damping	290
	14.2.5	Vibration and Chatter	290
14.3	Calculation for Design of Beds		292
	14.3.1	Material of Beds	292
	14.3.2	Shape of Beds	293
Review Questions			293

15 Design of Machine Tool Guides and Ways 294–309

15.1	Introduction		294
15.2	Working Surface of Guides		295
	15.2.1	Flat Guideways	295
	15.2.2	Inverted V-Shaped Guideways	295
	15.2.3	Combined Sliding Guideways (Flat and Inverted V)	296
	15.2.4	Dovetail Guideways	296
	15.2.5	Cylindrical or Round Guideways	297
	15.2.6	Roller Antifriction Guideways	297
	15.2.7	Mixed Guideways	298
15.3	Guide and Slideway Material		299
15.4	Pressure on Sliding Ways		299
	15.4.1	Pressure Calculation	299
	15.4.2	Considerations for Ballways	302
	15.4.3	Roller and Ball Guideways	303
15.5	Lubrication and Protection of Guideways		303
	15.5.1	Lubrication Theory	303
	15.5.2	Protection of Guideways	308
Review Questions			309

16 Design of Feed Power Mechanism and Screw 310–327

16.1	Translatory Motion Mechanisms		310
	16.1.1	Rack Gear and Rack-toothed Sector	310
	16.1.2	Worm and Worm Rack	310
	16.1.3	Nut and Screw	310
16.2	Design of a Screw		311
	16.2.1	Strength of a Screw	312
	16.2.2	Rigidity Checking	313

16.3	Kinematic Pair—Link Mechanisms	315
16.4	Kinematic Pair—Cam Follower and Link	315
16.5	Hinged Leverage Systems	316
16.6	Mechanisms for Intermittent Motions	317
16.7	Ratchet and Pawl	318
16.8	Safety Devices on Metal Cutting Machines	319
16.9	Spindles and Shafts	319
16.10	Edge Effect	320
16.11	Wear	321
16.12	Vibration of Spindles and Shafts	322
16.13	Spindle Nose	323
16.14	Supports of Spindles and Shafts	324
16.15	Antifriction Bearings	324
	16.15.1 Selection of Ball and Roller Bearings	324
	16.15.2 Mounting of Antifriction Bearings	325
Review Questions		327

17 Design of Machine Tool Gear Box 328–364

17.1	Introduction—Machine Tool System	328
	17.1.1 Drives and Regulation of Motion on Metal-Cutting Machines	328
	17.1.2 Various Motions of Machine Tool System	328
17.2	Fundamentals of Mechanical Regulation	329
17.3	Development of Series of Numbers	332
17.4	Ray Diagram for Overlapping Speeds	340
17.5	Ray Diagrams for Return Step of Speed	340
17.6	Kinematic Arrangement for Two or More Speeds at Input Shaft	341
17.7	Determination of Number of Teeth on Gears of Stepped Control Mechanisms	342
	17.7.1 Method of Least Common Multiple	342
	17.7.2 Method of Difference	345
	17.7.3 Constructive Method	346
17.8	Practical Aspects in the Design of Drives	352
17.9	Mechanical Regulation of Drives	352
	17.9.1 Belt and Cone Pulley Drive	353
	17.9.2 Belt Pulley Drive with Back Gear	355
	17.9.3 Gearbox Drives	358
Review Questions		364

18 Stepless Regulation of Speeds 365–419

18.1	Introduction	365
	18.1.1 Classification	365
	18.1.2 Reversing of Motion in Machine Tools	366
	18.1.3 Mechanical Regulation	366
	18.1.4 Methods of Increasing Range of Regulation	370
18.2	Hydraulic Drives for Speed Regulation	370
	18.2.1 Introduction	370

	18.2.2	Advantages and Disadvantages of Hydraulic Drives	372
	18.2.3	Requirement of Fluids Applied in Hydraulic Systems	372
	18.2.4	Mineral Oils as Fluids for Hydraulic Systems	372
	18.2.5	Properties of Hydraulic Oils	373
	18.2.6	Efficiency in Stages	374
	18.2.7	Cylinders for Hydraulic Drive Pumps	374
	18.2.8	Piston and Piston Rod Seals	377
	18.2.9	Pumps for Hydraulic Drives of Machine Tools	379
	18.2.10	Throttle Valves for Hydraulic Drives	382
	18.2.11	Fluid Control Valves	384
	18.2.12	Resistance to Flow Due to Obstruction	385
	18.2.13	Rotary Control Valves	386
	18.2.14	Hydraulic Piping and Its Joints	386
	18.2.15	Cross-section Calculation	387
	18.2.16	Speed Control	388
	18.2.17	Design of the Components of Grinding Machine Hydraulic Drive	393
	18.2.18	Hydraulic Drive for Rotary Motion	397
	18.2.19	Advantages and Disadvantages of Hydraulic Drive	397
18.3	Electrical and Electronic Regulation of Speeds		398
	18.3.1	Introduction	398
	18.3.2	Power Required by a Machine Tool	398
	18.3.3	Selection of Motor for Speed Regulation	400
	18.3.4	Classification of Drives	400
	18.3.5	Characteristics of Electric Motors	400
	18.3.6	Speed Regulation by Amplidyne	412
	18.3.7	Emotrol System of Speed Regulation (Electronic Motor Control)	412
	18.3.8	Selsyn System of Speed Regulation—The Word Sel-Syn Stands for Self-Synchronous Device	412
	18.3.9	Braking	413
	18.3.10	Analysis of Braking	413
	18.3.11	Starting and Stopping of Motors	417
	18.3.12	Clutch Control	417
	18.3.13	Relays	418
Review Questions			419

19 Machine Tool Vibrations...420–433

19.1	Introduction		420
19.2	Types of Machine Tool Vibration		420
	19.2.1	Forced Vibration	421
	19.2.2	Self-excited Vibration	422
19.3	Causes of Chatter		423
	19.3.1	Mathematical Analysis of Chatter Vibrations	425
	19.3.2	Chatter Vibrations	428
19.4	Closed Loop Representation of the Metal Cutting Process		430
19.5	Vibration Elimination		432
Review Questions			433

20 Mechanization and Automation 434–478

- 20.1 Machine Tools for Quantity Production 434
- 20.2 Semi-automatic Multi-tool Centre Lathes 434
- 20.3 Principal Parts of Capstan and Turret Lathes 436
- 20.4 Automation Mechanisms on Capstan and Turret Lathes 439
  - 20.4.1 Indexing Mechanism 439
  - 20.4.2 Bar Feeding Mechanism 440
  - 20.4.3 Bar Holding Mechanism 440
- 20.5 Tooling Layout for Capstan and Turret Lathes 442
- 20.6 Single-spindle Automatic Lathes 447
- 20.7 Hydraulic Copying Systems 457
- 20.8 Electric Copying System 459
- 20.9 Transfer Machines 460
  - 20.9.1 Transfer Machines and Automated Flow Lines 460
  - 20.9.2 Automated Flow Lines 460
  - 20.9.3 Classification of Transfer Lines 461
  - 20.9.4 Elements of Transfer Lines 462
  - 20.9.5 Transfer Mechanisms 472
  - 20.9.6 Selection of Transfer Devices 474
  - 20.9.7 Methods of Work Transfer 475
  - 20.9.8 Arrangement of Transfer Lines 477
- *Review Questions* 478

21 Numerical and Computer Numerical Controlled Machines 479–521

- 21.1 Introduction 479
- 21.2 History of NC 480
- 21.3 Working Principle of NC Machine 482
- 21.4 Basic Elements of NC System 482
- 21.5 Coordinate System in NC Machine Tools 488
- 21.6 Procedure in NC 490
- 21.7 Structure of NC Program 491
- 21.8 Tape Format 492
- 21.9 Types of Motion Control System in NC 496
- 21.10 Criteria for Classification of Numerical Controlled Systems 497
- 21.11 Advantages and Limitations of NC Machines 503
- 21.12 Computers and NC Machines 505
- 21.13 Computer Aided Part Programming (on CNC Lathe Machine Having FANUC Control) 508
- 21.14 Part Programming in APT Example 512
- 21.15 Miscellaneous and Preparatory Functons and their Codes for CNC Milling 513
- *Review Questions* 520
- *Objective Type Questions* 521

22 Gear Cutting, Broaching and Thread Cutting ... 522–541

- 22.1 Introduction ... 522
- 22.2 Various Methods of Gear Production ... 522
- 22.3 Various Kinds of Gears ... 523
- 22.4 Involute Gear Tooth Fundamentals ... 524
- 22.5 Gear Teeth Manufacturing ... 524
- 22.6 Broaching Operation ... 531
- 22.7 Production of Screw-Threads ... 535
  - 22.7.1 Screw Thread Geometry and Chasing of Screw Threads ... 536
  - 22.7.2 Internal Screw Threads ... 538
  - 22.7.3 Thread Rolling ... 538
- *Review Questions* ... 541

Part III PRECISION MEASUREMENT AND MANUFACTURING

23 Metrology and Precision Measurement ... 545–615

- 23.1 Linear Precision Measurement ... 545
  - 23.1.1 External Micrometer ... 545
  - 23.1.2 Vernier Calipers ... 547
- 23.2 Length Standards ... 548
- 23.3 Points of Support ... 550
- 23.4 Interferometry ... 551
- 23.5 Calibration of Length Standards ... 553
- 23.6 Slip Gauges—BS 888 and BS 4311 ... 555
- 23.7 Some Sources of Error in Linear Measurement ... 555
- 23.8 Angular Measurement ... 556
- 23.9 Measurement of Small Linear Displacements ... 557
- 23.10 Optical Magnification of Workpiece ... 563
- 23.11 Measurement of Small Angular Displacements ... 564
- 23.12 Limits, Fits and Tolerance ... 565
- 23.13 Geometrical Tolerances ... 571
- 23.14 Screw Threads ... 572
  - 23.14.1 Tolerance for ISO Metric Threads ... 573
  - 23.14.2 Magnitudes of Tolerance and Deviation ... 574
- 23.15 Limit Gauging ... 576
  - 23.15.1 Gauge Tolerances ... 577
- 23.16 Screw Gauge ... 579
  - 23.16.1 Gauging Principles ... 580
  - 23.16.2 Screw-thread Gauging ... 582
- 23.17 Surface Finish and Its Measurement ... 586
  - 23.17.1 Surface Finish Terminology ... 586
  - 23.17.2 Evaluation of Surface Roughness ... 588

	23.17.3	Representation of Surface Roughness	589
	23.17.4	Relationship of Surface Roughness to Manufacturing Process	589
	23.17.5	Measurement of Surface Roughness	590
23.18	Testing of Machine Tools		595
	23.18.1	Equipment Used for Testing	595
	23.18.2	Test Procedures	596
	23.18.3	Tests for Acceptance	596
Review Questions			602

24 Jigs and Fixtures 616–671

24.1	Definitions		616
24.2	Distinction between Jig and Fixture		617
24.3	Advantages		617
24.4	Principles of Design of Jigs and Fixtures		618
24.5	Design Procedure		619
24.6	Locations		620
	24.6.1	Factors Affecting Location	620
	24.6.2	Principles of Location	620
	24.6.3	Six-point Location of Rectangular Block	621
	24.6.4	Types of Locators	621
	24.6.5	Locating Devices and their Choice	621
24.7	Clamping		631
	24.7.1	Functions of Clamps	631
	24.7.2	Clamping Devices	631
	24.7.3	Types of Clamps	632
	24.7.4	Classification of Jig Bushes	645
24.8	Jig Base and Jig Feet		647
	28.8.1	Jig Base	647
	24.8.2	Jig Feet	647
24.9	Classification of Drill Jigs and Fixtures		648
	24.9.1	Classification of Jigs	648
24.10	Classification of Fixtures		655
	24.10.1	Milling Fixtures	656
	24.10.2	Turning Fixtures	662
	24.10.3	Grinding Fixtures	663
	24.10.4	Broaching Fixtures	664
	24.10.5	Assembly Fixtures	665
	24.10.6	Welding Fixtures	665
	24.10.7	Slotting Fixtures	666
	24.10.8	Boring Fixtures	666
	24.10.9	Miscellaneous Fixtures	666
24.11	Materials for Manufacturing of Jigs and Fixtures		666
24.12	Accuracy		668
24.13	Possible Ways of Avoiding Inaccuracies		668
24.14	Economic Aspects		669
Review Questions			670

Part IV METAL WORKING

25 Metal Working Processes 675–686
    25.1   Definition 675
    25.2   Types of Metal Working Processes 675
            25.2.1   Cold Working 675
            25.2.2   Hot Working 681
    Review Questions 686

26 Theory of Metal Working Processes 687–713
    26.1   Introduction 687
    26.2   Methods of Plasticity Analysis of Manufacturing Processes 687
    26.3   Forging Analysis Using Slab Method 689
    26.4   Flat Rolling Analysis Using Slab Method 692
    26.5   Deep Drawing Analysis Using Theory of Plasticity 695
    26.6   Extrusion Analysis Using Slab Method 698
            26.6.1   Particular Cases 699
    26.7   Rolling 701
    Review Questions 709

27 Press Tools and Their Design 714–770
    27.1   Introduction 714
    27.2   Advantages 714
    27.3   Types of Presses 715
            27.3.1   Classification Based on Source of Power 715
            27.3.2   Classification Based on Design of Frame 715
            27.3.3   Classification Based on Actuation of Ram 715
            27.3.4   Classification Based on Number of Slides 715
            27.3.5   Classification Based on Intended Use 716
    27.4   Design of Press Frame 717
            27.4.1   Power Press Driving Mechanism 718
    27.5   Methods of Punch Support 721
    27.6   Methods of Die Support 721
    27.7   Press Capacities 723
            27.7.1   Tonnage Capacity 723
            27.7.2   Catalogue Tonnage Rating 723
            27.7.3   Rating of Hydraulic Presses 725
    27.8   Press Operations 725
            27.8.1   Cutting Operations 725
            27.8.2   Non-Cutting Operations 726
    27.9   Choice of Press 729
    27.10 Design Fundamentals and Construction Features of Blanking, Piercing
           and Cropping Tools 729
            27.10.1 Design Analysis 729

27.10.2	General Notes on Press Tool Design		730
27.10.3	Press Tool and Its Parts		731
27.10.4	Important Design Consideration for Press Tool Clearances		734
27.10.5	Practical Die Clearance		736
27.10.6	Determination of Blanking Pressure		738
27.10.7	Determination of Press Size		738
27.10.8	Determination of Centre of Pressure		738
27.10.9	Design of Screws and Dowels		739
27.10.10	Design of Die Elements		744
27.10.11	Burr Height		750
27.10.12	Stripping Factors		750
27.10.13	Design of Punch		752
27.10.14	Shear Diagram for Punch Force Analysis		756
27.10.15	Punch Plate Design		757
27.10.16	Stripper Design		758
27.10.17	Knockout Design		762
27.10.18	Design of Bushes		764
27.10.19	Pneumatic Loading of Dies		764
27.10.20	Design of Press Tools with Ferrotic as Die Material		766

Review Questions 770

Part V MODERN METHODS OF MANUFACTURING

28 Unconventional Methods of Machining 773–815

28.1	Introduction	773
28.2	Definition of Unconventional Method	773
28.3	Major Unconventional Machining Processes	773
28.4	Process Capabilities of Unconventional Machining Processes	774
28.5	Ultrasonic Machining	776
28.6	USM Machine	777
28.7	Abrasive Jet Machining	779
28.8	Electrical Discharge Machining (EDM)	781
	28.8.1 Theory and Analysis of EDM	781
	28.8.2 Basics of EDM	782
	28.8.3 Construction Features of EDM Machine	783
	28.8.4 Wear Ratio	784
	28.8.5 Metal Removal Rate (MRR)	785
	28.8.6 Flushing the Electrode	786
28.9	Determination of Metal Removal Rate in Relaxation Circuit	786
28.10	Critical Resistance	788
28.11	Condition for Maximum Power	788
28.12	Wire Cut EDM	790
28.13	Electrochemical Machining (ECM) Process	792
28.14	Metal Removal Rate	796
28.15	Electrochemical or Electrolytic Grinding for Tools and Cutters	798

28.16	Electrochemical Deburring (ECD)		802
28.17	Laser Beam Machining Applications and Problems		803
28.18	Electron Beam Machining (EBM)		807
28.19	Electrolytic Sawing Mechanism and Machines		808
28.20	More Unconventional Processes		809
28.21	Hot Machining		809
Review Questions			811

29 Grinding and Other Abrasive Metal Removal Processes 816–869

29.1	Introduction		816
29.2	Grinding Wheel Variables		818
	29.2.1	Types of Abrasives	818
	29.2.2	Bond Material	819
	29.2.3	Grade	821
	29.2.4	Structure	821
29.3	Standard Codification of a Grinding Wheel		822
29.4	Selection of Grinding Wheels		822
29.5	Grinding Wheel Shapes		824
29.6	Wheel Dressing and Truing		824
	29.6.1	Dressing and Truing Procedures	826
	29.6.2	Rules for Using Diamond Tools	828
29.7	Balancing of Grinding Wheel		830
29.8	Recommended Wheel and Work Speeds for Grinding		831
	29.8.1	Recommended Wheel Speeds	831
	29.8.2	Work Speeds	832
29.9	Types of Grinding Machines		832
	29.9.1	Surface Grinding Machines	833
	29.9.2	Cylindrical Grinders	834
	29.9.3	Centreless Grinding	837
	29.9.4	Other Grinding Machines and Processes	844
29.10	Honing Operation		847
	29.10.1	Type of Honing Machines	849
	29.10.2	Lapping	850
	29.10.3	Superfinishing Operation	854
29.11	Snagging and Off-hand Grinding		855
	29.11.1	Types of Snagging	855
	29.11.2	Off-hand Grinding	857
29.12	Coated Abrasives for Other Industrial Applications		858
29.13	Centreless Belt Grinding		860
29.14	Tool Post Belt Grinding		862
29.15	Coated Abrasives		863
29.16	Trouble Shooting		864
29.17	Grinding Errors		865
Review Questions			868

References 871–878

Index 879–884

Preface

The second edition of this book is a revised and updated version of the first edition. Several additions and improvements have been done based on the feedback received from readers to make the contents more useful. In some chapters, there are very few changes while in others there are marked changes with additions which make the new edition in tune with the syllabi of proactive universities.

ORGANIZATION OF THE BOOK

Part I Metal Cutting

Chapter 1: Introduction to Materials and Processes, as in the first edition, covers various materials like, metallics polymers, ceramics, ferrous and non-ferrous metals. This is followed by an outline of several primary material working processes with neat diagrams to highlight their working. At the end of this chapter, the various machine tools are illustrated with their essential anatomy and constructional features.

Chapter 2: Metal Cutting Tools: Basic Concepts describes the geometrical features of various cutting tools highlighting their rake angle, clearance angle and other features covering the tool signature/geometry. While the British and German systems of specifying tool geometry are given, the major focus is upon the geometry of the tool in the ASA (American Standards Association System) and the ISO system. The geometrical relationship which converts back rake angle, side rake angle and plan approach angle (ϕ) is used to obtain the orthogonal rake and the angle of inclination of the cutting edge. Several numerical examples are taken up based on Eqs. (2.3) and (2.4). This is because the orthogonal rake angle in (ISO system) is most important and determines the basic wedge angle responsible for chip formation. Likewise, the angle of inclination (ISO system) is another fundamental angle which determines the direction of chip flow. Other important aspects of cutting, which governs the chip flow direction, is the tool plan shape (if the rake angle is small). In such cases, the Colwell's formula correctly determines the chip flow direction. In Example 2.15 of this chapter the authors have intentionally shown that the Colwell's formula does not correctly predict the chip flow direction because the rake angle of the tool is more in that example. In this chapter, the difference between nominal and working angles is also explained.

Chapter 3: Cutting Tool Materials discusses the various cutting tool materials (in the increasing order of hardness and decreasing order of toughness) like moving from plain carbon steel and high speed steel, to cast non-ferrous alloys such as Stellite, Tungsten and Titanium carbides with a Cobalt binder, to oxides and diamond as tool materials. Hardness, toughness, wear resistance, ease of manufacture are also discussed as desirable tool material properties.

Chapter 4: Design of Metal Cutting Tools discusses the Russian approach to design of cutting tools. Essential design of cutting tool has two aspects: Design of shank or body of tool and design of shape of the cutting tool point or edge. The designer should be adept in three-dimensional concepts of geometry.

In case of form tool, the reader should note that the profile to be obtained on the workpiece differs from profile of the form tool particularly in the case of circular form tools. This is owing to offset of the cutting tool centre from workpiece centre.

Chapter 5: Theory of Metal Cutting, initially discusses the action of chip brokers. Later on, this chapter focuses on two of the most important aspects of metal cutting theory. All through the chapter the pioneering work of Dr. M.E. Merchant has been discussed from the point of view of geometrical aspects, velocity relations, etc. His all important shear angle relationship to establish the effect of rake angle and chip tool friction angle on the shear angle have been covered.

The calculation of forces, shear strain and power in metal cutting based on Merchant's theory have been discussed. Another theory, due to Lee and Shaffer states that shearing occurs on a single shear plane and the chip is stress free. However, a plastically deformed stress field of triangular shape with shear plane as one side, shape tool control line as second side, is proposed by them. Using Mohr's circle for the plastically deformed, triangular field, a shear angle relationship, as determined by Lee and Shaffer, is also explained.

Chapter 6: Theory of Multipoint Machining analyses the cutting action of four types of multipoint tools: Drilling, Milling, Grinding and Brushing. The first section covers mechanics of drilling process highlighting the cutting action of drill occurs in two distinct modes: first one is at the two cutting tips where a merchant type analysis can be used. Second one is at chisel edge which acts as a rotating wedge indenter cutting and extruding the work material.

In the analysis of milling, the authors have assumed the path taken by a single tool as a trochoidal arc instead of a circular arc in the milling cutter. Later in Schlesinger's formula, a circular path is assumed.

In the present chapter, grinding is treated as a 'micro milling' and a relation and analysis is carried out to find comma shaped thickness of cut in term of D_W, D_G, V_W and V_G, to arrive at some useful results.

Chapter 7: Heat in Metal Cutting and Temperature Measurement deals with the thermal aspects in metal cutting. There are heat sources in metal cutting. The main heat source is due to shearing of metal at the shear plane. The heated chip slides as a friction slider along the chip tool interface and gets further heated due to chip tool interface friction. The third source is at the tool flank. In the presentation of thermal aspect, as a first step, a friction slider analysis is presented and the temperature rise is estimated using a stationary heat source concept. An illustrative example is taken up to clarify this concept. Next, the shear plane temperature is calculated considering the shearing velocity and shearing force. A numerical example is taken up to illustrate the thermal aspect of chip formation. The heating of the chip is owing to shearing of metal in shear plane as well as due to the heated chip rubbing at the rough chip tool interface (using Jaeger's slider formula).

This chapter ends with measurement of temperature by:

1. Use of tool-work thermocouple: The tool chip interface acts as hot junction while cold junction is created with copper wires joined to the tool end.
2. Embedding of thermocouples either in the workpiece or into the tool depending upon the spot where temperature measurement is needed.
3. Use of radiation intensity and using a photo heat sensitive call placed below a workpiece plate with drilled hole.

Chapter 8: Dynamometry discusses the topic of dynamometry. Because theoretical models are not able to accurately estimate the forces in metal cutting, experimental techniques using various forms of dynamometer have been developed. In this chapter, the discussion is confined to two-component and three-component half octagonal lathe tool dynamometer and three-component octagonal ring dynamometer suitable for milling and grinding.

Chapter 9: Tool failures and Tool life discusses various criteria resulting in its inability to hold the right dimensions of the job, go into chatter vibrations, increased power consumption, chipping, etc. Temperature is the most important factor giving rise to increased tendency of various modes which give rise to tool failure such as thermal cracks, chipping and B U E formation. Hard workpieces give abrasive flank wear, and ductile workpieces give rise to cratering due to diffusion, chemical wear, adhesive and abrasive wear. The Taylor's extended tool life equation is widely used. Of the various factors affecting tool life, the cutting speed and chip slenderness ratio, i.e. ratio of feed to depth of cut affect tool life adversely. The surface finish of the workpiece depends much on the tool plan shape. One highlight of the chapter is the discussion of Fick law of diffusion.

Chapter 10: Machinability focuses on the important aspect 'machinability' which explains the ease with which the work material can be cut. Apart from other aspects, machinability is influenced by strain hardening quality of the work material and its microstructure.

Higher carbon percentage makes steel harder. Cutting tool designers prefer a certain optimum hardness of steel because too soft steel causes formation of ferrite rich areas. On the other hand, too much carbon percentage results in formation of hard form flakes (cementite) making it difficult to machine and affects tool life adversely. Addition of lead to bronzes and brasses improves their machinability. Materials giving snarled continuous chips endanger the operator and suitable chip breaking is desirable.

Chapter 11: Economics of Metal Machining focuses on the development of equation of optimum cutting speed for minimizing the total cost. The total cost of producing a workpiece has three parts:

1. Costs which are independent of cutting speed. Examples of these costs are loading and unloading cost of workpiece. Cost of time when the tool is only approaching the work without taking a cut, etc.
2. Machine running cost during actual cutting. This cost component decreases with increase in cutting speed per finished workpiece produced.
3. Tooling cost per finished workpiece increases with increase in cutting speed.

In this chapter, the most economic cutting speed for machining have been derived for minimum cost.

Chapter 12: Metal Cutting and Metal Working Fluids deals with functions of cutting fluids, characteristics of good cutting fluids, and their advantages, theory of cutting fluid action, their effect on tool life, methods of application of cutting fluids, classification of cutting fluids such as soluble oils, mineral oils. Chemicals like carbon tetrachloride, gaseous lubricants, solid lubricants and their area of application to different metal cutting processes are also covered.

Part II Machine Tools

Chapter 13: Introduction to Machine Tools attempts to explain what is a machine tool. That includes its definition, its various elements, design requirements, different types of machine tools, main features and control for a machine tool. The status of machine tool industry in India is also explained.

Chapter 14: Design of Machine Tool Beds covers the various factors such as strength, volume to strength ratio, volume to stiffness ratio, etc. which enable the selection of the correct material for machine tool beds. Design criteria for machine tool structure such as strength and stiffness have been analyzed and shown graphically for decision making in selecting material for machine tool structure.

Chapter 15: Design of Machine Tool Guides and Ways discusses the various requirements for machine tool guideways. The importance of working surface of guideways of different designs has been explained. Additionally, the properties of different materials have been discussed. Guideways are subjected to cutting pressure during operation. Thus, pressure distribution calculation along the lubricated guideways have been explained as given by Siebel, Mansurov, Mehta, and others.

Chapter 16: Design of Feed Power Mechanism describes special analysis of the screw and nut as machine elements used in the machine tool design. Procedure for the calculation of torque and thrust acting on the feed screw of a machine tool has been discussed. The importance of vibrations and their damping in machine tools has been explained in reference to spindles and shafts. A special note on the selection of antifriction bearings with special reference to ball bearings has been discussed. Methods of mounting antifriction bearings in their housings has been discussed in detail in this chapter.

Chapter 17: Design of Machine Tool Gear Box explains design of gear box in simple manner. The flow of power from electric motor to the driven machinery is never direct. It is through intermediate shaft from where it goes to the driven shaft. To achieve minimum speed, deviation of the workpiece form its most economic speed (as decided by the diameter of the workpiece). A necessary condition is that the r.p.m. values of the spindle should follow a geometric series. There should be a common ratio ϕ (phi) between one r.p.m. and its nearest neighbouring speed. In addition the size of the gear box as determined by the centre distance between the driving shafts, intermediate shaft and the driven shaft should be prescribed. These factors make the work of designing a gear box a challenging one.

Chapter 18: Stepless Regulation of Speeds in Machine Tools discusses four approaches of mechanical regulation utilizing cone pulleys, rolling discs, ball variators, and positive infinite variable drive or PIV drive.

Hydraulic regulation techniques using positive displacement pumps of gear type, vane type and so on are reviewed. All these methods are based on achieving mechanical displacements through development of hydraulic pressure in servo cylinders using fluid pressure on either side of a piston through use of regulating valve. These applications are covered in the chapter using rarely available mathematical models.

Electric motors of induction type are widely used in machine tools. The chapter explains the Speed–Torque characteristics of induction motors as well DC shunt motors. The relationship between motor torque and external load torque decides the operating speed of the system. Speed control can be achieved by changing pole of the motor as well as resistance of shunt field winding in case of DC motors. Though current practice is to use induction motors for most machine tool applications, the Ward Leonard system is widely used for heavy duty planning machines.

The chapter also deals with electronic motor control and Selsyn devices. A machine tool engineer should be familiar at electrical technology as the chapter needs acquaintance with the same.

Chapter 19: Machine Tool Vibrations covers the theory of such vibrations in case of machine tools. Metal cutting machines are subjected to very large amount of forces and energy inputs due to nonhomogeneity in workpiece transmitted as natural vibratory movements of the machine elements.

Theory of chatter of machine tools is termed regenerative vibration which along with mathematical analysis using Laplace transform is also explained here.

Chapter 20: Mechanization and Automation discusses the constructional features of most of the automatic and semiautomatic machine tools from their constructional features. For mass production a centre lathe, which has only 2 stations, is unsuitable. Automatic and semiautomatic machine tools with 11 possible workstations are the right choice in such cases. The machine operator becomes merely a machine setter in these machine tools. The machine tools covered are capstan, turret, single spindle automat, hydraulic and electrical copying machines, transfer machines, flow lines and semiautomatic and automatic transfer lines.

Chapter 21: Numerical and Computer Controlled Machines reviews the history of the development of NC machines along with their principle of working and structural elements. Linear and rotary displacements in NC machines are defined through right hand co-ordinate system. The method of NC machining has been explained through a block diagram along with the structure of NC program and tape format. Control systems in NC machines are classified according to motion control, control circuit design and type of programming.

Installation of mini-computer on NC machines has enhanced the memory and flexibility of these machines, they are termed as CNC machines. When a single computer controls several NC/CNC machines, the arrangement is called DNC system. Finally, in this chapter several machine codes, generally termed as G/M codes, are utilized to illustrate the machining of various types of components.

Chapter 22: Gear Cutting, Broaching and Thread Cutting covers forming and generating methods of producing gear teeth. While gear forming methods are used for producing relatively inaccurate gears for jobbing production, low speed transmission of motion, the gear generating methods such as gear shaping and gear hobbing are used for the mass production of high speed gears such as used in automotive industry. This chapter also deals with the processes of gear shaving used for generating highly accurate gears.

The section on broaching deals with internal broaching as well as surface broaching methods. The section on screw thread manufacture deals with thread cutting using thread chasers as well as the various methods for the production of screw threads by thread rolling explaining the design of profile of thread rolling tools by cold working.

Part III Precision Measurement and Manufacturing

Chapter 23: Metrology and Precision Measurement begins with the explanation of the principles of linear measurement using the micrometer and the Vernier calipers. These being the most basic length measuring instruments. Later on, the topics of dimensional metrology like the international length standards are taken up. Then the theory of airy points of support for length standards is derived from first principles. This is followed by the topics of interferometry and a discussion on various types of comparators based on mechanical, optical electrical and electronic principles. Next, the important topics of limits, fits and tolerances are covered along with geometrical tolerencing.

Lastly the Taylor's principle of gauging, metrology of screw thread and design principles of gauges are taken up.

Section 23.17 of this chapter deals with the topic of surface finish and its measurement with the use of various surface finish measuring instruments based on electrical and electronic principle like the Talysurf.

Finally the surface finish obtained by various machining processes is discussed. Numerical examples on surface finish are taken up to make the concepts clear.

Section 23.18 of this chapter deals with the various alignment tests used in the metrology of machine tools to verify the accuracy with which these machine tools will perform in regard to perpendicularity and parallelism. The machine tools considered in the discussion are, the lathe, milling machine and the drilling machine.

Chapter 24: Jigs and Fixtures gives a detailed coverage of principles of location and clamping of a workpiece to restrict its 12 degrees of freedom. A jig or fixture dispenses with the tedious process of doing marking of every workpiece prior to machining to accurate and precise dimensions. The chapter discusses practically all types of jigs used in industry like pillar jig, box jig, angular jig and tumble jig. Various types of jig bushes such as fixed bushes, liner bushes, slip bushes are covered. Amongst fixtures, milling fixtures, turning fixtures, grinding fixtures, etc. are covered along with discussion of their design procedures.

Part IV Metal Working

Chapter 25: Metal Working Processes gives a complete description of cold working processes such as tube drawing, wire drawing, embossing, cold spinning, stretch forming, squeezing, cold rolling, thread rolling, rotary swaging, cold bending, roll forming, cold extrusion and cold pressing. All these processes are illustrated with neat diagrams.

Amongst the hot working processes the following processes are discussed in detail: hot rolling, hot forging, hot extrusion, hot spinning, hot cupping, roll piercing and pipe welding. The hot working processes are also illustrated diagrammatically.

Apart from describing the cold and hot working processes, their metallurgical differences and implications to the finished product are also explained.

Chapter 26: Theory of Metal Working Processes emphasizes on the theory of plasticity. Mathematical analysis of plastic deformation of several metal working processes is done in detail considering two basic equations: (1) equation of equilibrium and (2) equation of plasticity.

Chapter 27: Press Tools and Their Design classifies the various press machines on the basis of

1. Power source (e.g. manual, hydraulic, motorized, etc.)
2. Type of frame, e.g. C frame or portal frame type
3. Pneumatic

The essential parts of a press, e.g. ram, punch holder, punch, lower plate, bolster, die holder, die, etc. are clearly explained. It is emphasized that in piercing or making a hole, the punch size must correspond to the hole dimension on the drawing. On irregular shaped punches and dies, correction should be provided in dimensions. The kind of drive, e.g. crank type, eccentric type knuckle joint, rack and pinion are employed for double action toggle mechanism.

Double action press is suitable for drawing of cup shaped part. A blank holding ring becomes essential in this case. It is essential to have the ram axis to coincide with the centre of pressure in case of irregular shaped pieces to be blanked or pierced.

It has been shown how in a compound die the upper portion acts as a punch as well as a die and vice versa. In progressive dies such a provision is not necessary. In this case a channel type stripper is required and a fixed or adjustable pin helps to locate the sheet after every stroke.

In case of very high volume production as in case of laminations for electric motor rotors, a special die and punch material ferrotic has been developed. It is much more wear resistant than even carbide material.

Part V Modern Methods of Manufacturing

Chapter 28: Unconventional Methods of Machining covers the various processes. Ultrasonic machining process utilizes a tool made of tough and ductile material vibrating with a high frequency due to magnetostrictures effect. Abrasive slurry flows continuously in the small gap between the tool and work surface. Workpiece material is removed due to the hammering action of the tool. The chapter discusses the abrasive get machining next. In this process, the abrasive particles move with a high speed aided by air discharge stream. Next, the electric discharge machining process is discussed. When discharge takes place between two points with workpiece as anode and tool as cathode, the sparking takes place because of the dielectric medium separating the two breaks down and thermal mode of metal removal takes place. An RLC circuit is responsible for charging and discharging. The process has been extended to wire cut EDM.

The chapter then deals with electrochemical machining. This process is the reverse of electroplating because workpiece is positive and the tool is negative. The positive ions from the workpiece get it eroded and, try to deposit on the tool. This deposition is prevented by the flow of electrotype. The process is based on Faraday's law of electrolysis. Electrochemical grinding (ECG) is an extension of the process which contain abrasive coated copper wheel. The chapter then discusses laser beam machining. In LBM, a ruby rod is usually used as a lasing material. The electrons in the ruby are excited to move into higher orbit under the influence of high voltage applied by xenon flashlight. Material removal is possible due to thermal effect of photons emitted by the ruby rod. Another process discussed in the chapter is electron beam machining in which electrons emitted by a hot tungsten electrode are focused by electromagnetize lens to melt the workpiece. Finally the process of hot machining has been discussed.

Chapter 29: Grinding and other Abrasive Metal Removal Processes states that grinding is done on shapes and materials of all types. It is carried out to produce accurate and precise surfaces. Grinding wheel have large number of cutting tools. Due to this fact, grinding wheel has large number of variables as compared with conventional machining processes. An exhaustive list of variables is documented in the beginning of the chapter. These multiple variables make grinding a most complex material removal process. In the present chapter, different wheel variables, standardization and specification of grinding wheels, their prudent selection, and their dressing has been described in detail. Balancing and mounting of grinding wheels has also been dealt in detail. Different types of grinding processes and the machines which carry out these processes such as surface grinding machines, cylindrical grinding machines and centreless grinding machines and their application have been dealt with in detail. Honing, lapping and super finishing processes are micro finishing processes to correct geometrical errors, and polish the surface respectively. Special details are provided on these processes. Snagging, a large scale metal removal process applied for cleaning castings and forgings is dealt with in detail. Coated abrasives are having wide application in industry and coated abrasive wheels nowadays compete with bonded abrasive wheels. These are also covered in the chapter. In the end the topic of troubleshooting has been dealt with.

During the preparation of this revised Second Edition the authors received continual guidance from Mr. Darshan Kumar, former Executive Editor of PHI Learning. We express our heartfelt thanks to him for the guidance provided by him. We now feel confident that this new edition which is thoroughly revised should be of much more utility to students as well as teachers in various institutes of engineering and technology including the national institutes and the IITs. The authors wish to express their thanks to editorial and production staff members of PHI Learning. They would specifically acknowledge the support provided by Mr. V. Balamurugan, General Manager (Sales), Ms. Babita Misra, Editorial Coordinator, Mr. Ajai Kumar Lal Das, Assistant Manager (Production), Mr. Arvind Pahwa, General Manager (Production).

K.C. JAIN
A.K. CHITALE

Preface to the First Edition

If you haven't the strength to impose your own terms upon life, you must accept the terms it offers you.

—T.S. Eliot

In the past few decades, manufacturing has undergone notable transformation and has oriented itself to market demands, frequently assuming radical changes in contrast to what its pioneers had envisioned. At the same time, the hitherto unparalleled economic prosperity of the developed nations, resulting in extraordinary opportunities of economic growth, has recently declined; as a result, the developed nations are struggling to maintain a growth rate of even 5%.

The recent trend of conservative consumer behaviour has compelled the manufacturing sector to engage in strategic planning, and the production function has therefore undergone revolutionary changes. The manufacturing sector faces the turbulence of changing times and yet needs to maintain sustainable growth rates. The future times are going to be even more difficult, as the developing nations may also face the crisis that the developed nations are facing. The only way out is value addition through manufacturing. The modern manufacturing industries therefore need to have high production rates, enhanced flexibility in operations, and exceptional standards of product quality.

All mechanical and production engineers have to face the problem of developing innovative designs and manufacturing methods during such times. Competence in metal cutting analysis and machine tool design helps them in facing such problems confidently. Recently, corporate thinking has changed to "we will make it only when we can sell it" from "we can sell it if we can make it". Due to the latter strategy, many enterprises have either shut down or are on the brink of closure. Innovative and competitive manufacturing strategies must be adopted by all companies engaged in manufacturing. Therefore, strong emphasis must be given on effective manufacturing management.

In developed nations like the USA, value additions by the manufacturer amount to about one-third of all developing nations' combined GDP. This type of manufacturing capability is the foundation of national wealth and strength, high levels of employment, and the only way to fulfil the aspirations of the people of a nation.

The purpose of this book is to present in an integrated manner the available knowledge in diverse areas of production engineering such as machine tool design, metal cutting analysis, and unconventional methods of metal machining. These three areas have been covered previously by many authors, and quite a few books on these topics are available in the market. However, as teachers and specialists of this subject, we have attempted to explain certain theories and principles of design and metal cutting that other authors have ignored. The success of a strategy in manufacturing depends on one's level of knowledge of the subject. Therefore, a course in production engineering provides a knowledge base which places mechanical engineering graduates in a position of competitive advantage over those who have not undergone such a course.

The text is divided into twenty-eight chapters. Of these, twelve are devoted to metal cutting analysis, while another seven are dedicated to machine tool design, analysis, and operation. The remaining nine chapters are devoted to miscellaneous topics such as NC machines, gear cutting, precision measurement, jigs and fixtures, metal working processes, and unconventional machining processes.

The authors have attempted the presentation of production engineering theory, methodology, and analysis in a manner that makes the subject easy to understand for the readers. Wherever possible, a number of solved numerical examples have been provided to ensure that the students can easily comprehend the practical applications of the subject matter. The aim throughout has been to present the subject in the most comprehensive form. However, the scope of the subject is immense, and therefore certain topics have been treated in brief.

The authors welcome all constructive suggestions for the improvement of the contents of the book. They would also like to express their sincere appreciation of all efforts made by the staff of PHI Learning, New Delhi, towards bringing out this book within a reasonable period of time.

It is fervently hoped that the book proves useful to the students as well as the teachers of the subject.

K.C. JAIN
A.K. CHITALE

PART I
METAL CUTTING

CHAPTER 1

Introduction to Materials and Processes

1.1 MATERIALS AND THEIR NECESSITY FOR ENGINEERING APPLICATION

Almost everything that people have ever done has involved materials. Materials are directly related to our very existence. The major epochs of our history have been labelled after materials. Stone Age, Bronze Age and Iron Age—all convey the intimate relationship of people to materials. It is interesting to ponder the "Age" that future historians may label the short period of years from the end of the 20th century, to the early 21st century—perhaps the "Age of Semiconductors", or the "Age of Composite Materials", or better still the "Age of Materials". It can be said that materials form a basic resource for humanity. They play a crucial role not only in our way of life but also in the well-being and security of the nations across the world.

Whatever people build, materials are always used. There are various types of construction materials, each possessing different types of properties to fit a particular application. From dried mud and straw for huts, flint for tools and arrowheads, wood for ships and petroleum, coal and uranium for energy to composite materials for space vehicles, materials have been and will continue to be used by humanity to meet its needs for construction, energy and weapons. Regardless of the age, humans have been directly involved with materials. They have shaped cultures and civilisations. The peoples possessing an abundance of materials combined with an ability to use them have prospered. To the present day, the rapidity with which many of these sources of materials are being depleted, as well as the accessibility of materials in the light of political upheavals and wars, is a cause for concern.

It is possible to group much of the technological support of our society into three systems—communication, transportation and production. Each system relies heavily on engineering materials. The scientists, craftsmen, technologists, engineers, technicians and other workers involved in these three systems make decisions about materials throughout their career. Often these decisions relate to the selection of materials to meet specific needs. The interrelation of materials and the methods of processing the materials always exist; so whenever a design requires a material, a consideration for processing it must accompany the selection.

Both the communication and transportation industries have made significant gains due to rapid developments in miniaturised circuitry. Through ingenious selection of materials, appropriate designing of circuitry and highly sophisticated and clean manufacturing processes, the composites of ceramic and metallic memory chips have not only reduced sizes and weights but also lowered cost and energy demands. The chips allow the devices such as computers, calculators, radios, televisions and watches to offer greater capabilities, with only minute electrical energy needed to power them. The field of telephone and telegraph communications is undergoing a revolutionary change brought about by breakthroughs in materials science, especially in fibre optic technology.

The transportation industry has long been dependent on aluminium alloys because of their favourable strength to weight ratios and good corrosion resistance. Although the cost per kilogram of these alloys is more than that of steel, their weight per volume makes them cheaper. Composite materials are nowadays continuously being developed and widely used in road transport, aerospace and aircraft industries, and sports and recreation industries.

The high cost of many of the new materials, such as graphite fibre-reinforced composites, results from the production system required to generate these and fabricate parts, products, and systems that use these materials. Plastic production technology is rapidly developing due to plentiful supply of raw materials and ease of production processes.

The need for strong and lightweight materials to meet the demands for fuel-efficient ground, air, and sea vehicles led to the development of radically different materials, such as magnesium, virgin aluminium, and graphite fibres. Advances in the production of materials used with electronic circuitry made rapid strides because of tremendous advantages provided through miniaturization.

The future challenges in the areas of materials posed by computer-aided design (CAD), computer-aided manufacturing (CAM), computer-integrated manufacturing (CIM) and flexible manufacturing systems (FMS) include elimination of threaded fasteners, use of minimum number of components, continuous processing, reduced material handling and intense use of robots.

A modern vacuum cleaner is an example of a complex product that requires a multitude of different materials for its manufacture and successful use by a consumer. It has various metals such as those in the motor housing, various plastics for hand gears, hose and cleaner shell, as well as rubber wheels and an electrical plug. Such materials make up material systems. Each separate material component must be compatible with every other component, and at the same time contribute its distinct properties to the overall characteristics of the system of which it is a part.

1.2 TYPES OF MATERIALS

Materials can be divided into five major groups:
1. Metallics
2. Polymers
3. Ceramics
4. Composites
5. Other materials.

1.2.1 Metallics

Metallics or metallic materials include metal alloys. A *metal* in strict definition only refers to an element, such as iron, gold, aluminium and lead. Alloys consist of metal elements combined with other elements. Steel is an iron alloy made by combining iron, carbon, and some other elements.

Metallic materials are divided into three categories—ferrous, nonferrous, and powdered.

Ferrous metals

Ferrous is a Latin word meaning *iron like*. Ferrous metals include iron and alloys of at least 50% iron, such as cast iron, wrought iron, steel, and stainless steel. Each of these alloys is highly dependent on the presence of the key element of carbon. Steel is the most widely used alloy. Sheet steel is used in car bodies, desk bodies, cabinets for refrigerators, stoves, washing machines, doors, tin, cans, shelves and thousands of other things. Heavier steel, such as plates, beams, angle irons, pipes and bars, is used in making the structural frames of buildings, bridges, ships, automobiles, roadways, and many other structures.

Nonferrous metals

Metal elements that consist of more than 50% non-iron element are called *nonferrous alloys*. The nonferrous subgroup includes both common light-weight metals, such as titanium and beryllium, and common heavier metals, such as copper, lead, tin, zinc, and alloys of brass and bronze. Among the heavier metals is a group of white metals, including tin, lead, and cadmium. These white metals have low melting points of around 230 to 330°C. The nonferrous metals that have high melting points are chromium, nickel, tantalum, and tungsten. The possible combinations of nonferrous alloys is practically infinite.

Powdered metals

Alloying of metals involves melting the main ingredients so that on cooling, the metal alloy made is generally a nonporous solid. Powdered metal is often used because it is undesirable or impractical to combine elements through alloying, or produce parts by casting or other forming processes. Powdered metal is also called *sintered metal*. The process of forming a powdered metal consists of producing small particles of metal, squeezing them together and then sintering (applying heat below the melting point of the main alloy). The squeezing pressure with added heat converts the metal powder into a strong solid. Powdered metals can be ferrous, nonferrous, or a combination of both ferrous and nonferrous elements with non-metallic elements.

1.2.2 Polymers

Polymeric materials are basically the materials that contain many parts—*poly* means many and *mer* stands for unit. A polymer is a chainlike molecule made up of smaller molecular units (monomers). The atoms of the monomers (mostly carbon) bond together covalently, and then the chains of polymers covalently bond together to form solids.

Polymers can be classified as plastics, wood, elastomers, and other natural polymers.

Plastics

The term *plastic* is used to define fabricated polymers containing carbon atoms covalently bonded with other elements. It also stands for mouldable or workable things such as dough and wet clay. Plastic materials are either liquid or mouldable during the processing stage, after which they turn into solids. After processing, some plastics cannot be converted back into the plastic or mouldable state. Such plastics are called *thermosetting plastics* or *thermosets*. Common thermosetting plastics include epoxy, phenolic, and polyurethane. Other plastics can be repeatedly heated and converted back to the plastic stage. Such plastics are called *thermoplastics*. Examples of thermoplastic materials are acrylics (e.g. plexiglass, Lucite), nylon, and polyethylene. Though today most plastics are produced from oil, they can also be made from other organic (carbon) materials such as coal or agricultural crops, including wood and soyabean.

Wood

Of all the materials used in industry, wood is the most familiar and most used. Wood is a natural polymer. In the same manner in which polymers of ethylene are joined to form polyethylene, glucose monomers polymerize in wood to form cellulose polymers ($C_6H_{10}O_5$). Glucose is a kind of sugar made up of carbon (C), hydrogen (H) and oxygen (O). Cellulose polymers are joined in the layers with the glue-like substance *lignin*, which is another polymer.

Elastomers

Prior to the Second World War, most rubbers (elastomers) were natural rubbers. Today, synthetic or fabricated rubbers far exceed natural rubber in use. An *elastomer* is defined as any polymeric material that can be stretched at room temperature to at least twice its original length and can return to its original length after the stretching force has been removed. Elastomers have a molecular, amorphous structure similar to other polymeric materials. To further increase the strength of the elastomers, the process of vulcanisation is used to form the necessary cross-links (strong bonds) between the adjacent polymers. *Vulcanisation* is a chemical process that produces covalent bonding between the adjacent polymers with the help of a small amount of sulphur.

Other natural polymers

Human skin is most amazing natural polymer. Animal skin or hide in the form of fur and leather has limited industrial use. Medical science continues to study such natural polymers as bones, nails and tissues of humans and animals in order to synthesize these materials for their replacement when they are damaged due to injury or illness.

1.2.3 Ceramics

Ceramics are crystalline compounds of metallic and non-metallic elements. Glass is grouped with ceramics because it has similar properties, but most glass is amorphous. Included in ceramics are porcelain such as pottery, abrasives such as emery used on sandpaper, refractories such as tantalum carbide with a melting temperature around 3870°C, and structural clay, such as brick ceramics, including glass. These are hard, brittle and stiff, and have high melting points. Ceramics primarily have ionic bonds, but covalent bonding is also present. Their structure usually consists of two main ingredients. One major ingredient contains the atoms that form the crystalline structure. The second ingredient is a glassy substance, which acts as cement to bond together the crystalline structure. Silica is a basic unit in many ceramics, when it is combined with such metals as aluminium, magnesium and other elements.

1.2.4 Composites

A composite is a material containing two or more integrated materials (constituents), with each material keeping its own identity. Some of the most familiar composites are fibre glass, plywood and laminated coins of silver and copper.

The subgroup of composites include wood-based, plastic-based, metallic-based, concrete and cermets. It is also possible to classify composites by their structure. Composite structures include layers, fibres, particles and any combination of the three. Examples of layer composites include plywood, laminated boards covered with plastic sheets, safety glass, card board and Alclad. Particle composites include concrete, powdered metal and particle board. Fibre composites include fibreglass, hard board, boron fibres in aluminium matrix, reinforced glass, and graphite epoxy.

INTRODUCTION TO MATERIALS AND PROCESSES 7

1.2.5 Other Materials

Among these are the various types of cast iron, as discussed below.

Cast Iron (C.I.)

Cast iron is produced by melting and refining pig iron in a cupola furnace together with a definite amount of lime stone, steel scrap, and spoiled casting. It contains about 2 to 4% carbon, small percentages of silicon, sulphur, phosphorus and manganese, and certain amounts of alloying elements, e.g. nickel, chromium, molybdenum, copper and vanadium.

Properties of C.I.

C.I. has the following properties:

1. It can be economically cast into complicated shapes.
2. It has the capacity of damping out vibrations in supports and frames of moving machinery.
3. It has resistance to wear, abrasion and corrosion.
4. It can be machined easily.
5. It has high compressive strength.
6. It is brittle, and has no plasticity and low resistance to tension.
7. It can be forged, magnetised or tempered.

Types of C.I.

The varieties of C.I. in common use are:

1. Grey C.I.
2. White C.I.
3. Malleable C.I.
4. Nodular C.I.
5. Chilled C.I.
6. Alloy C.I.
7. Mechanite C.I.

Grey C.I.: This is obtained by allowing the molten metal to cool and solidify slowly. It contains carbon in free form. It contains C = 2.5 to 3.5%, Si = 1 to 2.5%, Mn = 0.4 to 1%, S = 0.02 to 0.15%, and P = 0.15 to 1%. It is soft and dark grey in colour. It is easily machinable. It is quite cheap. It is widely used for cylinders, car wheels, columns, automobiles, locomotives, and pipes.

White C.I.: Sudden cooling of molten C.I. produces white C.I. It contains carbon in combined form. It is very hard, brittle and unmachinable. It has high compressive strength but a low resistance to impact. It has very limited applications. It is used for making wrought iron. Its composition include C = 1.75 to 2.3%, Si = 0.85 to 1.2%, Mn = 0.1 to 0.5%, S = 0.1 to 0.35%, and P = 0.05 to 0.2%.

Malleable C.I.: It is obtained by annealing white C.I. at 950 to 1000°C with the object of separating carbon in a finely divided form. It is tough, machinable and malleable. It is of two types:

(a) White heart malleable C.I.
(b) Black heart malleable C.I.

White heart malleable C.I. on breaking shows a brightly crystalline fracture, whereas black heart malleable shows a dark grey or black fracture. Malleable C.I. is less brittle and more ductile. It is used where small intricate jobs are needed. It is used in differential and steering gear housings, axles, break pedals, structural components for railway carriages, and agricultural and textile machinery. Spanned cranks, a typical composition of white heart malleable C.I., includes C = 3.2 to 3.6%, P = 0.1%, Si = 0.4 to 1.1%, S = 0.1 to 0.3%, Mn = 0.1 to 0.4%.

Nodular C.I. (or Spheroid Graphite Iron or Ductile Iron): It is obtained by the addition of magnesium into molten cast iron. This causes the graphite to take a spheroidal or nodular form. It behaves like steel and gives a better machined surface. It has high strength, ductility, and impact strength. It is used for the manufacture of hydraulic cylinders, cylinder heads, gears, cams, gear boxes, liners of I.C. engines, and heat-resisting of furnaces. It can be welded and galvanised easily. A typical composition includes $C = 3.3\%$, $Mn = 0.4\%$, $P = 0.03\%$, $Ni = 0.75\%$, $Mg = 0.06\%$, and $Si = 0.6\%$.

Chilled C.I.: Quick cooling is called *chilling* and the iron produced by quick cooling is called *chilled iron*. Castings are chilled at their outer surface by bringing them in contact with the cool sand in mould. Chilling gives a hard outer surface, while the interior of the casting remains grey. It is used in making hard the running surface of the wheels of railway carriages and also used in casting of rolls.

Alloy C.I.: Alloys are obtained by adding alloying elements like nickel, chromium, and silicon to cost iron. Nickel C.I. is resistant to caustic corrosion and is therefore used for caustic pots, pipes, etc. Addition of chromium leads to grain refinement, and increased strength and wear resistance. Nickel C.I. is used in cylinder or cylinder liners, steam and hydraulic machinery compressors, and I.C. engines.

Mechanite C.I.: C.I. in which metal has been treated with calcium silicate is known as *mechanite cast iron*. It possesses high strength, toughness, ductility and easy machineability. They can be heat-treated. It is ideally suited for machine tool castings.

Wrought Iron (W.I.)

It is a pure form of iron and contains about 99.8% iron. It is obtained by remitting pig iron through the puddling process. The process is carried out in a puddling furnace, which is a reverberatory furnace (see Fig. 1.1). It consists of a hearth and is lined with fire bricks. It is an oxidation process and is carried out at 100–1300°C. The pig iron is first melted and a blast of air is passed over the surface of molten metal, so that carbon is oxidised and white iron is obtained. White iron is put into the puddling furnace and W.I. is obtained in the spongy form called *puddled balls*.

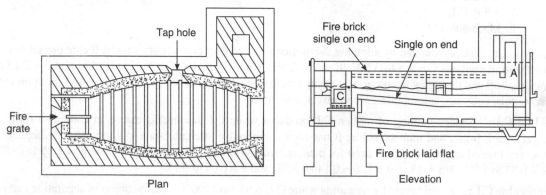

Fig. 1.1 Reverberatory furnace.

Properties and uses

1. It can be rolled, drawn, forged and welded.
2. It is ductile and malleable, and can be used for wire drawing and sheets.
3. It has resistance to shock.
4. It can be hardened.

It is used for railway couplings, chains, crane hooks, anchors, and steam, water and oil pipes.

INTRODUCTION TO MATERIALS AND PROCESSES

Effects of alloying elements on C.I.: The following alloying elements create significant effects on C.I.:

Carbon. If it is present in free form, it makes the C.I. easily machinable but C.I. will be weak and brittle. If it is present in combined form, it makes C.I. hard. Carbon percentage ranges between 2 and 4%.

Silicon. If it is less than 2.5%, it makes C.I. soft, while more than 2.5% silicon presence makes C.I. hard.

Sulphur. It makes the iron hard and brittle, accelerates the rate of solidification, and promotes defects like blow holes. Its presence should not exceed more than 0.2%.

Manganese. It minimises the effects of sulphur. Its amount usually varies from 0.5 to 1%.

Phosphorus. It aids feasibility and fluidity but induces brittleness. It is rarely allowed to exceed 1%.

Nickel. It makes iron machinable and corrosion resistant. It varies from 0.25 to 2%.

Chromium. It increases the hardness and tensile strength of C.I.

Steel. It is an alloy of iron and carbon. It also contains very small amounts of sulphur, phosphorus, manganese and silicon. Steel basically differs from C.I. in the amount of carbon contained. The maximum amount of carbon allowed is 2%.

1.3 OVERVIEW OF MANUFACTURING PROCESSES

A large number of processes with their limited characteristics are available as manufacturing resources. Each process has its own operating procedure and capability, in terms of the tolerance and surface finish that each can produce. Depending upon the characteristics required, a process is selected to produce a component of desired quality.

Broadly, these processes can be put into two basic categories. viz. *primary* and *secondary* processes.

Primary processes are those which can be used to shape the material (raw material blank), and later a secondary process is used as the finishing operation if dimensional accuracy of the component is required.

In case the primary process is not followed by the secondary process, and the component as the output of the so-called primary process is accepted for assembly, then such primary processes are not categorised as primary. It can be concluded that there is hardly a line of demarcation. Primary processes may include metal rolling, metal casting, etc. while the list of secondary processes may include metal machining, metal grinding, investment casting, die casting, etc.

This categorisation is broadly to identify where a secondary operation is desired because of dimensional accuracy requirement. Thus, application of a process decides whether it is primary or secondary.

1.3.1 Basic or Conventional Processes

The conventionally used production processes can be categorised into four groups, namely:

Metal casting processes: Used as primary as well as secondary processes.

Mechanical working processes: Used as primary as well as secondary processes.

Machining processes: Used as primary as well as secondary processes.

Auxiliary operations: Metal joining processes (welding processes), heat treatment operations, polishing, assembly operation, etc.

During the manufacturing process of the product(s), the assembly operations cannot be overlooked since they consume time and money.

A brief outline of the manufacturing processes is presented here to emphasise (highlight) the basic scientific principles involved in making products out of different materials. The importance of manufacturing can be realised from the fact that the manufacturing processes are responsible for converting raw materials of discrete properties into final products (end-products).

It is evident that there is a continuous struggle for cheaper production processes (cheaper converting technology) that can produce the product(s) of desired quality. The involvement of production processes in the manufacturing of products is shown in Fig. 1.2.

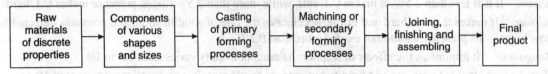

Fig. 1.2 Production processes involved in achieving final the product.

All the preceding production processes have undergone tremendous change in terms of sophistication to achieve the desired quality of the final product. In order to have proper understanding of these discrete processes, three engineering research activities—designing, production and development of new techniques— are to be understood, i.e. at the design stage, manufacturing conditions are to be taken into account, not only to produce the components in the most economic manner but also to impart the required mechanical properties, e.g. hardness and finish.

Obviously, a production engineer must have a good understanding of the processes, so that he can select and implement the appropriate process. Further, the development of new manufacturing processes and modification of the existing technology must be taken up as a continuous effort. Thus, the production engineer would develop capability to face challenges and to solve problems efficiently.

Here in the context of the book, we would take up in detail—(a) metal working processes and (b) machine processes.

Metal working processes (Forming processes)

These are as follows:

(i) Forging (ii) Rolling
(iii) Drawing (iv) Bending
(v) Extrusion processes.

A brief description of the abovementioned processes is given below; however, it is pertinent to mention here that all the above categories of processes are broadly covered in two classes—(i) processes that shape the object after it undergoes *plastic deformation*, and (ii) processes which shape the desired object after it undergoes *elastoplastic formation* (e.g. sheet metal working).

In plastic deformation, the applied stress in the process is greater than the yield strength of blank-materials but less than their fracture strength. The type of loading may include tensile, compressive, bending or shearing, or a combination of these. It is an economic process, since the desired shape and size and finished parts can be obtained without any significant loss of work material. Further, a part of energy is also effectively utilised in improving the mechanical properties of the product-formed material, owing to having undergone strain-hardening. This category of metal forming involves the transformation of raw stock (blank or workpiece) into useful parts by deforming the work material plastically (*permanent deformation*).

Plastic deformation (bulk deformation) can be carried out under two categories of temperature. One process is termed *cold forming*, wherein the working temperature is kept below the recrystallisation temperature, while the other process is termed *hot forming*, which is carried out at the temperature above the recrystallisation temperature. Basically, the flow stress behaviour of a material is entirely different both above and below its recrystallisation temperature. Strain hardening is higher in cold forming, while it is very low in hot working. Cold forming or working processes thus require a low coefficient of friction (μ_i) of about 0.1, while hot forming requires as high as 0.6 approximately. It is also a fact that hot forming or working requires less energy even for bulk deformation, while cold forming requires more energy even for lesser degrees of deformation.

Elastoplastic deformation processes, i.e. drawing, deep drawing, and bending processes work with materials below the recrystallisation temperature and are included in the calculation of elastic spring-back deflections.

Typical forming processes are schematically presented in Fig. 1.3. Of these, while rolling, forging (open and closed die forging), extrusion, squeeze forging, wire drawing, etc. are bulk deformation processes, punching, blanking, and shearing are cutting processes, and pressing, deep-drawing, etc. are elastoplastic deformation processes.

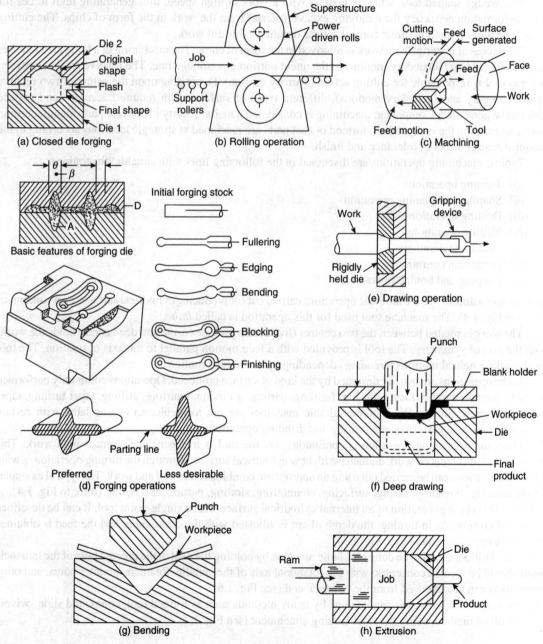

Fig. 1.3 Schematic view of various forming processes.

12 TEXTBOOK OF PRODUCTION ENGINEERING

Machining processes

Forming processes, as detailed earlier, work on the principle of flow of materials and generally result in metallographic structural changes. Also, such operations require very large forces handled by very huge machines, particularly for large jobs. Sometimes, a forming process followed by a machining process is utilised for achieving the desired level of geometric complexity in the final shape. These problems can be overcome by adopting the process of machining, where the excess of material is removed in the form of chips by using a wedge shaped tool, while in contact with a work at high speed, thus generating high forces for plastic deformation necessary for removing excess material from the work in the form of chips. The cutting tool and the work remain in direct contact for the duration of useful work.

The excess of material from work is removed in the form of chips. The machining of a finite area requires a continuous feeding (secondary motion) of the uncut portion at a suitable rate. The relative motion between work and tool is responsible for cutting action (primary motion). Depending upon the nature of two relative motions (primary and secondary motions), different types of surfaces with required accuracy and surface finish can be generated. Component machining is considered a major activity in discrete parts manufacturing. Prior to assembling the end-product, formed or cast parts are machined at strategic locations according to the functional requirements of tolerance and finish.

Typical machining operations are discussed in the following lines with suitable illustrations:

  (i) Turning operations
 (ii) Shaping and planing operation
(iii) Drilling operations
 (iv) Milling operations
  (v) Grinding operations
 (vi) Broaching operations
(vii) Lapping and honing operations.

Turning operations: These are basic operations carried out for producing cylindrical surfaces on a continuous basis (see Fig. 1.4). The machine tool used for this operation is called *lathe*.

The work is rotated between the two centres (live centre provides drive and dead centre holds the work) about the axis of symmetry. The tool is provided with a feed motion parallel to the axis of rotation. The tool can also take a helical motion (*threading*) depending upon the feed motion.

Lathe operations are usually designed by the kind of surface produced. Operations commonly performed on a lathe include turning, facing, boring, forming, parting, grooving, knurling, drilling, taper turning, taper boring and screw cutting. When more suitable machines are not available, an engine lathe with certain accessories may also perform some milling and grinding operations.

Here, the machining operation is continuous, i.e. the tool is in mechanical contact with work. This operation provides reduced work diameter with new cylindrical surface. Through the turning operation, a wide variety of operations can be carried out using an appropriate combination of tool and work. The typical examples include tapering, threading, facing, surfacing, chamfering, slotting, parting, and boring (refer to Fig. 1.4).

Boring. This is the generation of an internal cylindrical surface with a single-point tool. It can be described as *internal turning*. As in turning, the depth of cut is adjusted with the cross-slide, and the feed is obtained with the carriage movement (see Fig. 1.5).

Drilling. Drilling can also be done on a lathe machine by holding the drill in the tapered hole of the tailstock, which should be located concentric with the rotational axis of the spindle. Reamers, counter bores, and other cutting tools can also be used from the tailstock end (see Fig. 1.6).

Taper turning. On lathe, this can be done by many methods such as formed tool, compound slide swivell, tailstock offset method, using the taper turning attachment (see Fig. 1.7).

INTRODUCTION TO MATERIALS AND PROCESSES

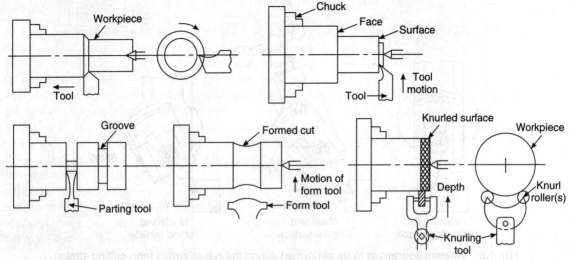

Fig. 1.4 Turning operations.

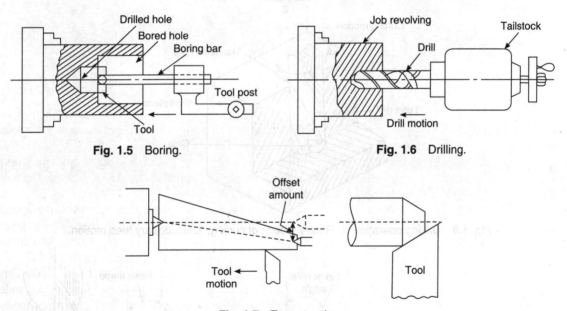

Fig. 1.5 Boring. **Fig. 1.6** Drilling.

Fig. 1.7 Taper turning.

Shaping and planing operations: The surface obtained is plain. The cutting tool gets a reciprocating motion and after every cutting stroke, the work is fed perpendicularly (see Fig. 1.8).

Drilling operations: This operation is used for making a hole into a solid body (see Fig. 1.9). The tool used is called *drill* or *drilling machine* which gets cutting motion by rotating the two cutting edges, and the feed is obtained through a rectilinear motion of the drill in the axial direction. The final surface obtained is an internal cylindrical surface. Dotted lines in the shank portion in Fig. 1.10 indicate the straight shank, and the solid line indicates the taper shank.

14 TEXTBOOK OF PRODUCTION ENGINEERING

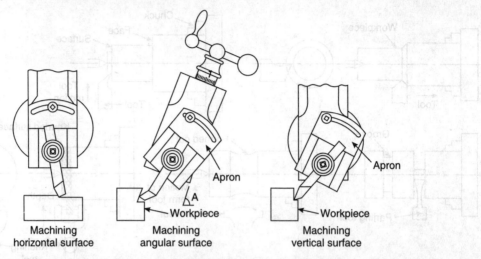

Fig. 1.8 Different operations to be performed during the return stroke (non-cutting stroke).

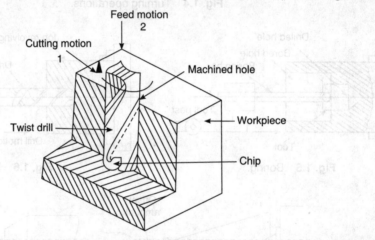

Fig. 1.9 Drilling operation: 1. Rotary motion of cutting, 2. Secondary feed motion.

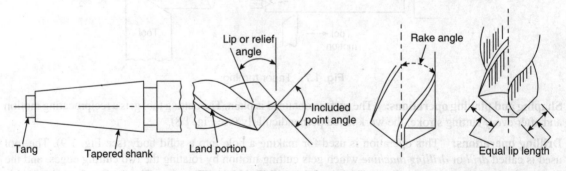

Fig. 1.10 Elements of twist drill.

Milling operation: It is a versatile machining operation capable of producing surfaces of different shapes. This process uses a multipoint tool called *milling cutter*. The cutter is provided with a rotary motion and the work is gradually fed. The excess material is removed as small chips through each cutting edge during its revolution, and finally a flat surface is generated. The various types of milling operations are shown in Figs. 1.11 to 1.18.

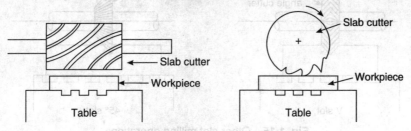

Fig. 1.11   Helical slab milling.

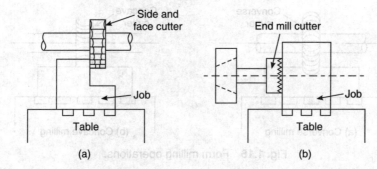

Fig. 1.12   (a) Arbour-mounted surface and (b) shank-mounted surface.

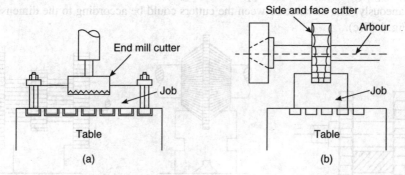

Fig. 1.13   (a) Channel slot shank-mounted milling and (b) channel slot arbour-mounted milling.

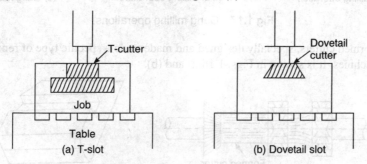

Fig. 1.14   Slot milling operations.

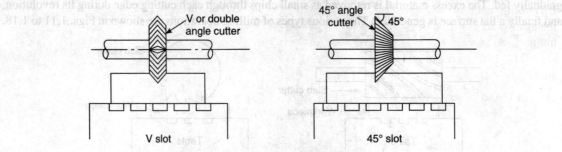

Fig. 1.15 Other slot milling operations.

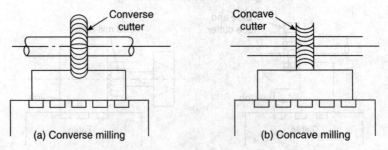

(a) Converse milling (b) Concave milling

Fig. 1.16 Form milling operations.

Straddle milling or gang milling. In this type of milling, more than one cutter is used to mill two or more surfaces simultaneously. The distance between the cutters could be according to the dimensions of the job. It is shown in Fig. 1.17(c).

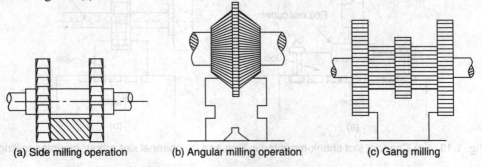

(a) Side milling operation   (b) Angular milling operation   (c) Gang milling

Fig 1.17 Gang milling operations.

Form milling. Formed cutters, specially designed and made to do a specific type of repetitive job, are also used in milling machines. It is shown in Figs. 1.18(a) and (b).

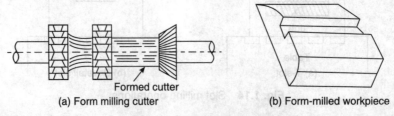

(a) Form milling cutter   (b) Form-milled workpiece

Fig. 1.18 Form milling.

Up milling (conventional milling). In this type of milling, the job is fed against the direction of rotation of the cutter. Up milling is shown in Fig. 1.19.

This is the conventional way of milling and has the following advantages:

(a) The cutter teeth generate the finished surface of the job. Hence due to less wear on the cutter and cutting edges, longer life is achieved.
(b) Due to vertical force F_v which tries to separate the cutter and the workpiece, the operations are safe.
(c) No special arrangements for backlash error are needed as the cutter tries to take the job in the backward direction due to force F_h.

The only disadvantage is that due to the vertical component of the force (F_v) there is a tendency of the job to lift upwards, therefore rigid clamping fixtures are needed.

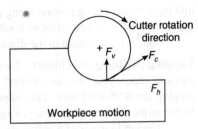

Fig. 1.19 Up milling.

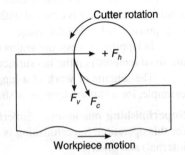

Fig. 1.20 Down milling.

Down milling (climb-cut). In this, the job is fed in the direction of the after rotation. Cutting edges take their cut in the downward direction (see Fig. 1.20).

Grinding operations: These operations are carried out with a grinding wheel, which contains sharp cutting edges of abrasive bonded grains. These cutting edges are available on the cutting surface in very large numbers and are randomly distributed. The common grinding methods used for finishing are shown in Fig. 1.21.

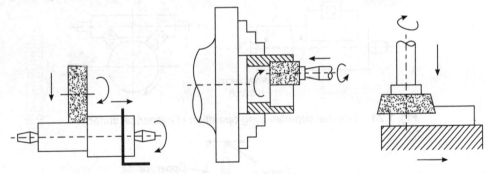

Fig. 1.21 Commonly employed grinding methods as finishing processes.

Diamond machines: These machines consist of small diamonds fixed on a special tool-holder and serve as cutting tools. As a rule, these are boring machines with vertical spindles. Their special features are:

1. Great rigidity of machine construction
2. Cutting speed from 600 to 1000 m/min
3. Depth of cut 0.1–0.3 mm
4. Feed 0.01–0.03 mm revolution.

Honing machines: Honing machines usually have vertical spindles and are used for finishing holes (e.g. of hydraulic cylinders).

On the spindle is mounted a honing-head, which holds abrasive stones (hones) of small grains. The stones, situated at a certain angle to each other, touch a generated surface. They rotate together with the spindle and have reciprocating motion along its axis. The correlation between the number of spindle revolutions and the number of its double strokes is odd.

Feed is the increase in the diameter of a honing head, measured by abrasive stones, per double stroke.

Lapping (Refining) machines: Lapping machines are designed to remove small amounts of metal from the surfaces (flat, cylindrical and others). Allowances for lapping are 0.05–0.02 mm. The lapping process also improves the geometrical accuracy of a generated surface.

The lap is made of a softer metal than that of a workpiece. Often, grey iron is used as the material for a lap. The lap surface is covered with an abrasive paste (e.g. powder, oils, paraffin, etc.).

In lapping process, the grains of abrasive power are fixed (entered) on the lap surface.

The scheme of work of a lapping machine, for example, for a flat workpieces is shown in Fig. 1.22.

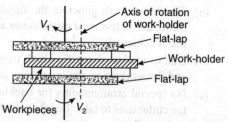

Fig. 1.22 Scheme of work of lapping machine.

Superfinishing machines: Superfinishing produces surfaces of the highest quality. No allowance is left for this operation. Superfinishing is applied for generating the surfaces of both rotative bodies (external and internal) and planes.

The cutting tool is a soft, fine abrasive stone which is pressed (either with springs or hydraulically) against a manufactured surface. Superfinishing is carried out with the application of lubricating and cooling fluids (e.g. kerosene with oil).

The schemes of various superfinishing operations are shown in Figs. 1.23 and 1.24.

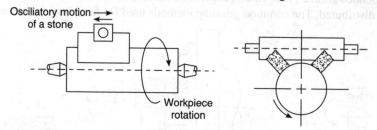

Fig. 1.23 External superfinishing operation of cylindrical surface.

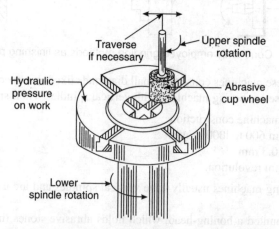

Fig. 1.24 Superfinishing operation of plain surface.

1.4 CONSTRUCTIONAL FEATURES OF BASIC MACHINES

The constructional features of a lathe are shown in Fig. 1.25. It shows the two-dimensional diagram of a lathe machine with details of various components of such type of machines.

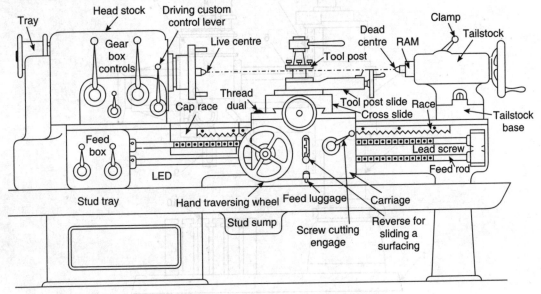

Fig. 1.25 Two-dimensional diagram of lathe.

The constructional features of a drilling machine are shown in Fig. 1.26.

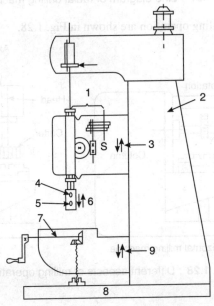

1. Transmission of motion from the spindle to the feed box 2. Upright 3. Manual adjustment 4. Hole for tool removal 5. Hole for fastening a tool 6. Automatic feed 7. Table 8. Base 9. Manual adjustment

Fig. 1.26 Block diagram of drilling machine.

The line diagram of a radial drilling machine is shown in Fig. 1.27.

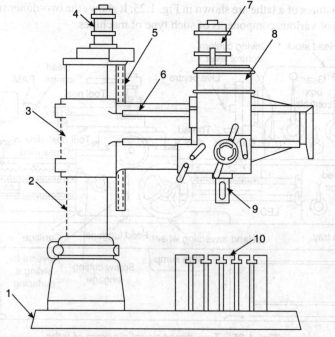

1. Base 2. Column 3. Radial arm 4. Motor for elevating the arm 5. Elevating screw 6. Guide ways
7. Motor for driving the drill spindle 8. Drill head 9. Drill spindle 10. Table

Fig. 1.27 Line diagram of radial drilling machine.

The different aspects of milling operation are shown in Fig. 1.28.

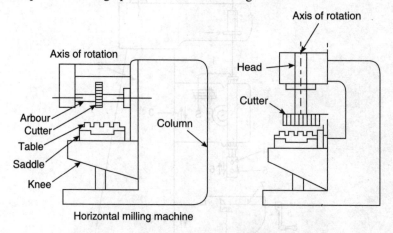

Fig. 1.28 Different aspects of milling operation.

INTRODUCTION TO MATERIALS AND PROCESSES

The block diagram of a shaper is shown in Fig. 1.29.

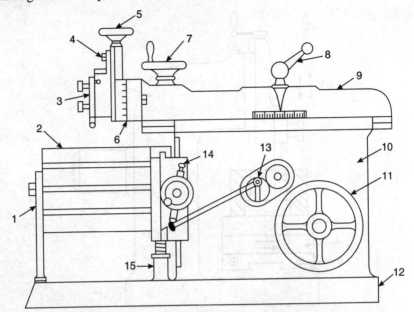

1. Table support 2. Table 3. Clapper box 4. Apron bolt 5. Downfeed hand wheel
6. Swivel base graduations 7. Stroke adjusting hand wheel 8. Ram locking handle 9. Ram 10. Column
11. Drive pulley 12. Base 13. Feed disc 14. Pawl mechanism 15. Elevating screw

Fig. 1.29 Block diagram of shaper.

The line diagram of a shaper is shown in Fig. 1.30.

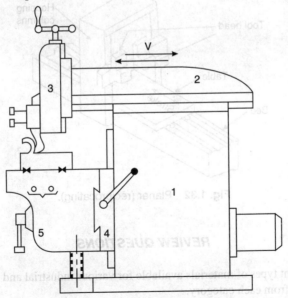

1. Column 2. Ram 3. Toolhead assembly 4. Crossrail 5. Table

Fig. 1.30 Line diagram of shaper.

The slotter is shown in Fig. 1.31.

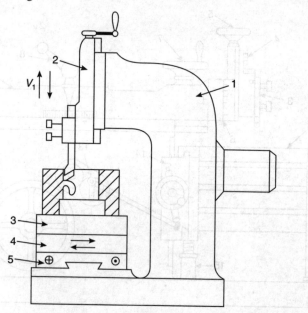

1. Base and column 2. Ram 3. Rotary table 4. Cross saddle 5. Longitudinal saddle

Fig. 1.31 Slotter.

The reciprocating planer is shown in Fig. 1.32.

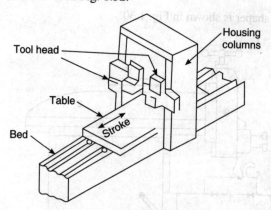

Fig. 1.32 Planer (reciprocating).

REVIEW QUESTIONS

1. What are the different types of materials available for various industrial and domestic purposes? Name at least one product from each category.
2. What are the different types of plastics? What are their manufacturing applications?
3. What is meant by 'Ferrous Metals'?

4. What is the effect of alloying elements on the parent material? What are the various alloying elements and their uses?
5. What are the various areas of application of iron and steel?
6. What is the metallurgical difference between low carbon steel, mild steel and high carbon steel?
7. Describe the applications of ceramics and glasses.
8. Explain why several stages are needed in the forging of a connecting rod type component?
9. Explain how the transformation from a raw material to the final product takes place. Describe in detail the transformation process for the following:
    (i) A pulley
    (ii) A gear wheel
    (iii) A plastic bucket
    (iv) A toothpaste tube.
10. Which alloying elements provide the following properties in steel?
    (i) Hot hardness
    (ii) Wear resistance
    (iii) Thermal conductivity
    (iv) Flow ability.
11. Name a product for which casting and welding are used exclusively as manufacturing processes.
12. What are the different secondary production processes?
13. Brazing and soldering are processes which have specific uses in transforming raw materials to finished products. State whether or not these are used as sole processes or as intermediate secondary transformation processes.
14. Describe the process of manufacturing a part by powder metallurgy.
15. List and describe in detail the following:
    (i) Bulk material forming processes
    (ii) Finishing processes
    (iii) Superfinishing processes.

CHAPTER 2

Metal Cutting Tools
Basic Concepts

2.1 DEFINITION OF METAL CUTTING TOOL

It is a device used in metal cutting processes on machine tools, for removing layers of material from the blank, to obtain a product of specified shape and size with specified accuracy and surface finish.

2.2 CLASSIFICATION OF METAL CUTTING TOOLS

Depending upon the number of cutting edges, the metal cutting tools are classified as follows:

    (a) Single-point cutting tools
    (b) Multi-point cutting tools
    (c) Form tools.

2.2.1 Single-point Cutting Tools

This type of tool has an effective cutting edge and removes excess material from the workpiece along the cutting edge. Single-point cutting tools are used on lathes, shapers, planers, etc.

Types of single-point cutting tools

Single-point cutting tools are classified as follows:

    (a) According to the method of holding the tool:
        (i) Forged tool

(ii) Tipped tool fastened mechanically to the carbon steel shank
(iii) Tipped tool brazed to the carbon steel shank.
(b) According to the method of holding the tool:
  (i) Solid tool
  (ii) Tool bit inserted in the tool holder.
(c) According to the method of applying feed:
  (i) Right hand
  (ii) Left hand
  (iii) Round nose.

Forged tool: Forged tools are manufactured from high carbon steel or high speed steel. The required shape of the tool is given by forging the end of a solid tool shank. The cutting edges are then ground to shape to provide the necessary tool angles. Figure 2.1 shows a forged tool.

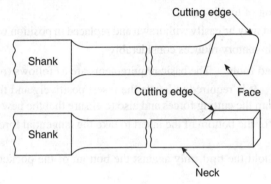

Fig. 2.1 Forged tool.

Mechanically fastened tipped tool: To ensure rigidity, tips are sometimes clamped at the end of a tool shank by means of a clamp and bolt as shown in Fig. 2.2(a).

Brazed tipped tool: In a tipped type cutting tool, the cutting edge is in the form of a small tip made of stellite and cemented carbide tool materials, which is welded to the end of a carbon steel shank by a brazing operation. Figure 2.2(b) shows a brazed tipped tool.

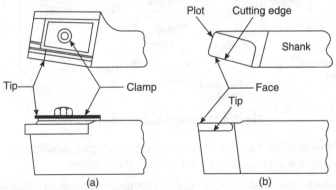

Fig. 2.2 (a) Mechanically fastened tipped tool and (b) brazed tipped tool.

Solid tool: Solid tools are made of high carbon steel, forged and ground to the required shape. They are directly mounted on the tool post of a lathe.

Tool bit inserted in tool holder: A tool bit is a small piece of cutting material, having a very short shank which is inserted in a forged carbon steel tool holder, and clamped in a position by a bolt or screw as shown in Fig. 2.3.

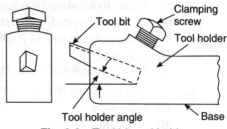

Throwaway carbide bits are made in a wide variety and different shapes (thickness 3 to 12 mm, size 10–25 mm^2). Positive rake inserts have 3 or 4 cutting edges with 5–8° relief angle. In negative rake inserts, 6 to 8 edges are incorporated.

Fig. 2.3 Tool bit and holder.

Advantages of tool bit over solid tool: These include:

Adjustment facility. The tool can be adjusted to the correct height easily by simply adjusting the position of the tool bit in the slot.

Reduced cost. It is cheaper than the solid tool.

Easy regrinding. Regrinding of tool is easier.

Easy replacement. A tool bit may be easily withdrawn and replaced in position without disturbing the setting.

Reduced tool inventories. Inventory reduces considerably.

Basic requirements of tipped tools: The basic requirements are as follows (refer to Fig. 2.3):

A pocket on the tool holder. It is required to locate the insert positively and thus to take the longitudinal side and end radial thrusts from the cutting forces and also to ensure that the new tip will cut to the same size.

A solid seat. It is required for the bottom of the insert to take the tangential force and not allow the tip to be subjected to a bending force.

A clamp. It is required to hold the tip firmly against the bottom of the pocket and prevent it from being pulled out.

Right/Left hand tool: A tool is said to be right/left hand type if its cutting edge is on the right/left hand side when viewing the tool from the point end.

Parts of a single-point cutting tool

The various important parts of a single-point cutting tool are shown in Fig. 2.4(a). These are described below.

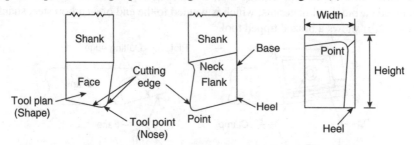

Fig. 2.4(a) Single-point cutting tool.

Shank: It is the main body of the tool, at one end of which the cutting portion is formed, or the tip or bit is supported. The shank in turn is supported on the tool post of the machine.

Neck: The portion which is reduced in section to form the necessary cutting edges and angles is called the neck. A relatively small point as required in the boring is sometimes attached to the shank by a neck.

Face: The face is that surface on which the chip apparently impinges as it is separated from the work. It may be provided with an edge or a narrow band ground along the cutting edge to support the built-up edge.

Base: It is the surface on which the tool rests.

Heel or lower face: It is the horizontal surface at the end of the base in the neck portion which does not participate in the cutting process.

Cutting edge or lip: It is that part of the face edge along which the chip is separated from the work. The cutting edges consist of:
  (a) Nose
  (b) Side cutting edge or major cutting edge
  (c) End cutting or minor cutting edge.

The nose is the corner, arc or chamfer at the junction of major and minor cutting edges.

Flank: It is the surface below the cutting edge.

Tool plan (shape): It is the contour of the face when viewed in a direction at right angle to the base (from the top).

Tool point: The tool point or nose is that part of the tool which is shaped to produce the cutting edge and the face.

2.2.2 Multi-point Cutting Tools

Multi-point cutting tools used on drilling, milling, and broaching machines, etc. have two or more cutting edges. In such tools, all the cutting edges may either operate one by one or engage simultaneously. Each edge of a multi-point cutting tool can be regarded as a separate single-point tool, having almost all the single-point features.

2.2.3 Form Tools

Form tools are generally ground on cutting tool elements as a replica of the work profile. These tools are used to shape short jobs for capstan or turret lathes. These are fed radially into the job. Form tools are classified as given in Table 2.1.

Table 2.1 Classification of form tools

Type of form tool	Description	Use
Circular	Such tools have zero rake, positive rake and with a swivelled axis. Its complex production method limits its use in special cases only.	When the contour is such that tools with parallel axis cannot be used. Used in automatic bar machines.
Helical circular form tool	May have zero or positive rake.	Used in special purpose machines.
Tangential (skewing) form tool	This is a modified design of the flat form tool, and it has a longer tool life. It is held in an inclined tool holder to get a front clearance angle.	Can be used on capstan, turret or heavy duty lathes.
Flat form tool	This is the simplest type of form tool. The tool profile in the plane normal to the front face shall be the same as that of the workpiece.	Reground by removing material on top face; can be used with heavy duty lathes.

2.3 PRE-REQUISITES OF CUTTING TOOLS

The following are the main pre-requisites of a good cutting tool to give maximum production with minimum maintenance and trouble:

28 TEXTBOOK OF PRODUCTION ENGINEERING

Sufficient strength: It should possess sufficient strength to maintain a sharp cutting edge.

Resistance to wear: It should possess sufficient resistance to withstand wear of the cutting edge.

Sufficient hardness: It must have sufficient hardness to prevent picking up chips.

Proper material: It should be made of a proper material for the given cutting material.

Proper heat treatment: It should be properly heat treated to obtain good mechanical and technological properties.

Accurate design: It should be accurately designed to meet the basic requirements.

2.4 STANDARD ANGLES OF CUTTING TOOL

The standard angles of a cutting tool are the angles which depend upon the shape of the tool. These are described in Fig. 2.4(b) to (d), and explained below.

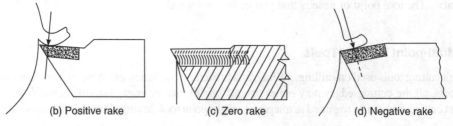

(b) Positive rake   (c) Zero rake   (d) Negative rake

Fig. 2.4(b) to (d) Standard angles of cutting tool.

2.4.1 Rake

It is the slope of the top, away from the cutting edge. It is the slope of the tool face.

Functions of rake

The main functions of rake are as follows:

Guidance to chips: It allows the chip to flow in a convenient direction.

Increased tool life: It reduces the cutting force required to shear the metal and thus increases the tool life.

Reduced power consumption: Due to the reduced force, the power consumption gets reduced too.

Improved surface finish: It improves the surface finish of the workpiece.

Improved heat dissipation: It improves the heat conduction to environment, making the tool cool.

Factors affecting the rake

The amount of rake angle to be given on a tool depends on the following factors:

Type of material being cut: A harder material like cast iron requires a smaller rake angle than that required by soft metals like mild steel or aluminium.

Type of tool material: It has been observed that at a very high cutting speed, the rake angle has a little influence on cutting pressure. Tool materials like cemented carbide, ceramics, etc. permit very high cutting speeds. Under such conditions the rake angle is kept minimum or even negative to increase the tool strength.

Depth of cut: Tools may have larger rake angle for small depth of cuts.

Rigidity of tool holder and condition of machine: For an improperly supported tool on an old and worn-out machine, a larger rake is provided.

Rakes are of the following two types:

Front rake (back rake)

It measures the downward slope of the top surface of the tool from the nose to the rear along the longitudinal axis. It is the angle by which the face of the tool is inclined towards the back side.

Functions of back rake: This angle helps in turning the chip away from the workpiece. It determines the thickness of the tool behind the cutting edge.

Types of back rake: Back rake angles are of three types:

Positive back rake. The angle is positive if the face slopes downward from the point towards the shank, tending to reduce the lip angle (refer to Fig. 2.4(b)).

Neutral back rake or zero rake. When the face is parallel to shank, it is called zero rake. It increases the strength of the tool and prevents the cutting edge from digging into the work. In brittle materials, cutting tools such as brass tools are provided with zero rake (refer to Fig 2.4(c)).

Negative back rake. The back rake angle is negative if the face slopes upward to the shank. Cemented carbide tipped or ceramic tipped tools are provided with negative rake (refer to Fig. 2.4(d)).

Use of negative rake provides the following advantages:

 (a) Higher cutting speed can be used for metal removal.
 (b) It allows taking heavier depth of cuts.
 (c) It gives better finish.
 (d) Increased strength of the cutting tool can be obtained.
 (e) It reduces the tendency for a built-up edge.
 (f) It allows better heat conduction from the tool.

However, there are various limitations of negative rake. The use of an increased negative rake angle leads to increased cutting force during machining. This causes vibration, reduced machining accuracy, and increased power consumption. Tools with negative back rake should be used, except for carbide and ceramic tools, only when absolutely necessary.

The size of the back rake angle depends upon the material to be machined, tool material, depth of cut, etc. the softer the material, the greater should be the positive back rake angle. Aluminium requires a larger back rake angle than C.I. or steel. Increased back rake angle leads to better tool life, reduced power consumption, improved surface finish, etc. But extremely high values weaken the tool. Table 2.2 shows the different values for various materials.

Side rake

It is the angle between the face of a tool and a line parallel to the base. It is measured in a plane at right angle to the base and at right angle to the centre line of point.

Functions: It guides the direction of chip away from the job. The amount of chip bending depends upon the tilt angle. The larger the side rake angle, the less will be the chip bending, leading to reduced power consumption. The larger the side rake angle, the better will be the surface finish.

2.4.2 Side Relief

It is the angle between the portion of the shank immediately below the cutting edge and a line drawn through this cutting edge perpendicular to the base. It is measured in a plane at right angles to the centre line of the point. The normal side relief angle is measured in a plane perpendicular to the base of the shank and the cutting edge.

Function

This angle permits the tool to be fed sideways into the job, so that it can cut without rubbing.

Effect of size of side relief angle

If the side relief angle is made very large, the cutting edge of the tool will break because of insufficient support, whereas if the side relief angle is either very small, zero or negative, the tool cannot be fed into the job, it will rub against the job, get overheated and blunt, and result in poor surface finish of the workpiece.

Side relief angle on carbide tools

Since carbide is brittle and carbide tips are relatively thin (3 mm to 6 mm), a large side relief angle would weaken the tip and it may be damaged or may break easily. For this reason, relief angles on carbide tips are kept as small as possible.

2.4.3 End Relief

It is the angle between the portion of the end-flank immediately below the cutting edge perpendicular to the base. It is measured in a plane parallel to the centre line of the point. The normal end relief angle is measured in a plane perpendicular to the base of the shank and the end cutting edge.

Function

This angle prevents the cutting tool from rubbing against the job. If the angle is very large the cutting edge of the tool will be unsupported and tool will break off, whereas if the angle is very small the tool will rub on the job. Its value varies from 6° to 10°.

2.4.4 Recommended Rake and Relief Angles

Table 2.2 shows the recommended tool angles for HSS and carbide single-point tools for some of the most commonly used materials.

Table 2.2 Recommended tool angles

Material	HSS single-point tool				Carbide single-point tool			
	Relief angle (degrees)		Rake angle (degrees)		Relief angle (degrees)		Rake angle (degrees)	
	Side	End	Back	Side	Side	End	Back	Side
Aluminium	12–14	8–10	30–35	14–16	6–10	0–10	10–20	10–20
Copper	10–12	8–10	0	1–3	6–8	6–8	0–4	15–20
Brass	8–10	6–8	3–5	10–12	6–8	6–8	0–(–5)	8–(–5)
C.I.	12–14	12–14	14–16	18–20	5–8	5–8	0–(–7)	6–(–7)
Steel	7–9	6–8	5–7	8–10	5–10	5–10	0–(–7)	6–(–7)

2.4.5 Nose Radius

The nose of a tool is slightly rounded.

Main functions

The main functions of nose radius are as follows:

To improve surface finish: A greater nose radius clears up the feed mark caused by the previous shearing action and provides better surface finish.

To reduce wear: It increases the strength of the cutting edge and tends to minimise the wear.

To increase tool life: It enhances the tool life.

To improve heat conduction: Accumulation of heat is less.

To reduce cutting force: It reduces the cutting force.

To reduce power consumption: It helps in minimising the power consumption.

Value of nose radius

A very large nose radius may cause chatter, noise, or vibration. For rough turning its value is 0.4 mm and for finish turning it varies from 0.8 to 1.6 mm.

2.4.6 Tool Holder Angle

It is the angle between the bottom of the bit slot and the base of the tool holder shank.

2.4.7 Flat or Drag

It is the straight portion of the end cutting edge at zero degree angle (refer to Fig. 2.5). It improves surface finish.

2.4.8 Clearance Angle

The clearance angle is greater than the relief angle. It is the angle between the perpendicular to the base of a tool and that portion of the shank immediately below the relieved flank. The side clearance angle is measured in the plane of the side rake angle. The end clearance angle is measured in the plane of the back rake angle.

The surface of the tool below the cutting edge is sometimes forged or rough-ground before hardening, so as to reduce the amount of grinding on the flank.

The ground flank sometime extends to the heel, in which case the clearance equals the relief. The clearance angle may be measured in a plane perpendicular to the base of the shank and the cutting edge, in which case it is known as *normal clearance*. A forged tool is shown in Fig. 2.5.

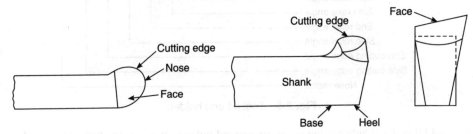

Fig. 2.5 Forged tool.

2.4.9 Side Cutting Edge

The side cutting edge angle is the angle between the straight side cutting edge and the side of the tool shank. In the case of a bent tool, this angle is measured from the straight portion of the shank.

Effect of increased angle of side cutting

It has the following advantages:
    (i) Increased tool life
    (ii) Better surface finish
    (iii) Greater feed, depth of cut and speed
    (iv) Reduced power consumption.

2.4.10 End Cutting Edge

The end cutting edge angle is the angle between the cutting edge on the end of the tool and a line at right angles to the side edge of the straight portion of the tool shank. Its value varies from 0° to 90°, and the usual value is 15°.

Ground chip breaker type B is located from side cutting edge of carbide-tipped, right-cut, single-point tool.

Figure 2.6 shows a typical tool holder with a 30° left-bent shank and a 15° tool holder angle. A right-cut tool bit suitable for steel is shown. The rake and relief angles are shown turning soft, based on the centre line of the tool bit.

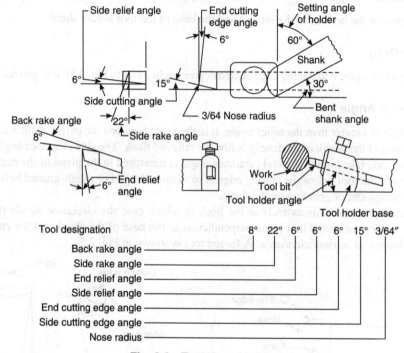

Fig. 2.6 Tool bit and holder.

As the tool bit end and side cutting edges are ground independently of the bent shank holder, they are so indicated with respect to the centre line of the tool holder head in Fig. 2.7.

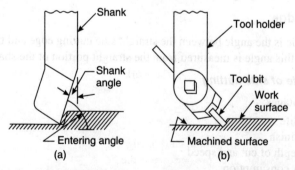

Fig. 2.7 Bent tools—(a) A forged, left-bent shank, right-cut tool and (b) a right-cut tool bit in a right-bent tool holder.

As the setting angle is changed, the relation of the end cutting edge to the work is also changed (see Section 2.6).

2.4.11 Nose Angle and Angles in Normal Plane

The nose angle is the angle included between the side cutting edge and the end cutting edge, as shown in Fig. 2.10. Whenever a relief and a clearance below the cutting edges are used, the tool designation may be written as 8, 22, 6, 6, 6, 15, 1.5 mm as illustrated in Fig. 2.6. It may also be written as 8, 22, 6(10), 6(10), 6, 15, 3/64, in which the value of (10) would indicate the clearance relief angle as shown in Fig. 2.8. In case of relief and clearance angles below 6 degrees, given as being normal to the cutting edges, the tool designation may be written as indicated in Fig. 2.8. However, the 'normal' angle system is rarely used.

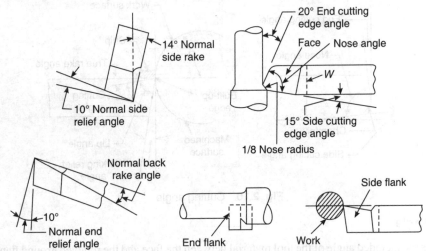

Fig. 2.8 Typical solid tool of high speed steel for turning steel of 250 to 300 Brinell (the rake and relief angles shown as normal to cutting edges).

2.5 WORKING ANGLES

The working angles are those angles between the tool and the work which depend not only on the shape of the tool, but also on the position of the toll with respect to the work, as shown in Fig. 2.9.

2.5.1 Setting Angle

The setting angle is the angle made by the straight portion of the shank of a tool with the machined surface of the work, as shown in Fig. 2.9.

2.5.2 Entering Angle

The entering angle is the angle which the side cutting edge of a tool makes with the extended machine surface of the work, measured on the cutting-edge side of the tool point, i.e. the right side of a right-cut tool. See Figs. 2.7(a) and 2.9.

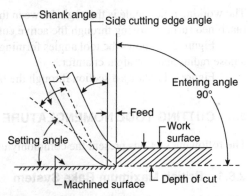

Fig. 2.9 Working, setting and entering angles of a right-bent shank, a right-cut tool, showing the depth of cut and the feed.

2.5.3 True Rake Angle

The true rake angle (or top rake), under actual cutting conditions, is the slope of the tool face towards the tool base from the active cutting edge in the direction of chip flow. It is the combination of the back rake and the side rake angles, and varies with the setting of the tool and with the feed and depth of cut. Rakes normal to the cutting edges are illustrated in Figs. 2.8 and 2.10.

2.5.4 Cutting Angle

The cutting angle is the angle between the face of the tool and a tangent to the machined surface at the point of action. It equals 90° minus the true rake angle. See Fig. 2.10.

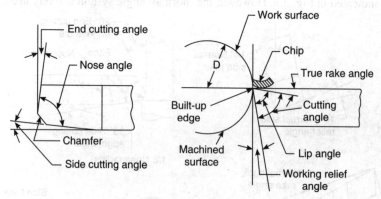

Fig. 2.10 Cutting angle.

2.5.5 Lip Angle

The lip angle is the included angle of the tool material between the face and the relieved ground flank, measured in a plane at right angles to the cutting edge. When measured in a plane perpendicular to the cutting edge at the end of the tool, it is called the *end lip angle*. When measured at the point of chip flow, it is called the *true lip angle*. See Fig. 2.10.

2.5.6 Working Relief Angle

The working relief angle is the angle between the relieved ground flank of the tool and the line tangent to the machined surface passing through the active cutting edge, as shown in Fig. 2.10.

Figure 2.10 shows the tool angles forming the cutting edges on a ground tool. The loci is shown within a nose radius and a straight chamfer.

Figure 2.11 shows a section through the plane of chip flow, illustrating working angles.

2.6 CUTTING TOOL NOMENCLATURE SYSTEMS

The following sections present the commonly used systems of cutting tool nomenclature.

2.6.1 British Maximum Rake System

The British system, where the maximum slope on the cutting tool face is measured in a plane perpendicular to the base, and the direction of the maximum slope with the side of the shank is measured in a plane parallel to the base, is specified and illustrated in Fig. 2.11.

Additionally the various other rake systems are also shown in Fig. 2.11. These are—American (A), British (B), German (G).

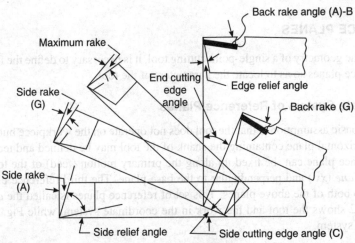

Fig. 2.11 Rake systems.

It is claimed that this method has the advantage of the specified angles being set on a universal vice and the face ground to specified angles.

The difficulty with the system is that the angles specified are quite independent of the position of the cutting edge, and therefore give no indication of the behaviour of the cutting tool in operation. At one time it was claimed that chip flow is in the direction of maximum rake. This is not borne out by facts.

2.6.2 American System (ASA System)

In this system the cutting tool face is specified by its slope in the orthogonal plane, parallel and perpendicular to the side of the shank. The two angles thus specified are known as American back rake and American side rake. In this system, like the British system, the angles specified are quite independent of the position of the cutting edge end and therefore give no indication of the behaviour of the cutting tool vis-a-vis chip flow in operation.

The American system has, however, the advantage that it is convenient to set up the universal vice for sharpening. However, the tool cannot be accurately sharpened without using equations and curves. The error for small angles is, however, negligible.

2.6.3 German System (DIN System)

In this system, the two rake angles are specified and referred to as *German back rake* and *German side rake*; unlike in the American system, the German back rake is the slope of the cutting edge measured in a plane containing this edge and the perpendicular to the base, and the German side rake is the slope of the cutting tool face in a plane perpendicular to the plane in which the back rake is measured and also perpendicular to the base.

In the German system, difficulties arise when this system is used for sharpening. The problems of this system are similar to those of the American system.

2.6.4 Normal Rake System

This system is similar to the German system, with the exception that the side rake is measured in a plane normal to the cutting edge, and is named *normal rake*. It will be noted that this system, like the German system, has some physical meaning—the tool angles are specified in relation to the cutting edge, but unlike the German system the specified angles may be set directly on the universal vice.

2.7 REFERENCE PLANES

In order to define the geometry of a single-point cutting tool, it is necessary to define the following convenient systems of reference planes so as to locate the parameters of the tool.

2.7.1 Coordinate System of Reference Planes

In this system the basic assumption is that the tool does not operate on the workpiece but on the held position in the space. A horizontal plane containing the shank of the tool may be defined and termed the *base plane*. The second reference plane can be fixed up along the primary motion (feed) of the tool. It can be named the *longitudinal plane* (xx') and perpendicular to the base plane. The third reference plane can be one that is perpendicular to both of the above planes. This set of reference planes is called the *coordinate system of planes*. Figure 2.12 shows the tool and the work in the coordinate system, while Fig. 2.13 shows the tool angles in the same system.

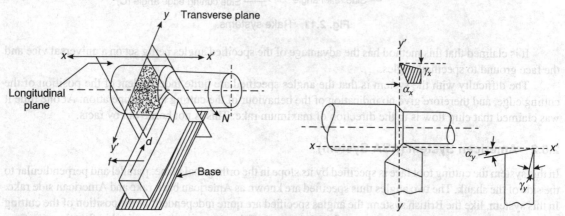

Fig. 2.12 Tool and workpiece in the coordinate system. **Fig. 2.13** Tool angles in the coordinate system.

2.7.2 Orthogonal System of Reference Planes or Orthogonal Rake System (ORS) or International Orthogonal System (ISO)

This system is also called the Orthogonal Rake System (ORS) or the international orthogonal system (ISO). In this system a the basic assumption is that the tool operates on the workpiece on the machine tool. Referring to Figs. 2.14 and 2.15, a horizontal plane containing the shank of the tool may be defined as the *base plane* (m). The plane which contains the principal cutting edge is called the *cutting plane* (c) and is the second plane. The third plane perpendicular to the above two planes can be termed the *orthogonal plane* (n). This set of reference planes is called the *orthogonal system*.

Figure 2.14 shows the tool and the workpiece in the orthogonal system, while Fig. 2.15 shows the tool angles γ_0 and λ in the same system.

METAL CUTTING TOOLS: BASIC CONCEPTS **37**

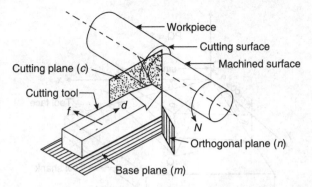

Fig. 2.14 Tool and workpiece in the international orthogonal system.

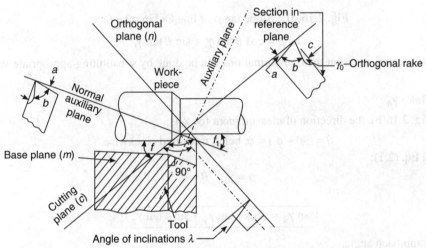

Fig. 2.15 Tool angles in the international orthogonal system (ISO).

2.7.3 Transformation from ASA System to ISO System

In transformation from ASA system to ISO system the objective is to derive equations for γ_o and λ in terms of γ_y, γ_x and ϕ_s (or its complement ϕ). In the present section a simple three dimensional approach is taken up as explained below.

General equation of slope of a line in the rake face

The earlier Sections 2.7.1 and 2.7.2 described the coordinate system or the American Standards Association System (ASA) and the ISO system respectively. Though the ASA system is simple to understand, it does not specify the Fundamental angles Orthogonal Rake γ_0 and angle of inclination λ of the main cutting edge. It becomes essential to carry out transformation from ASA system to the ISO system to have a better understanding of the cutting process. Referring to Fig. 2.16(a) the general equation of the slope tan ρ of a line OQ in the rake face can be written as:

$$\tan \rho = \frac{PQ}{OP} = (PR + RQ), \quad \text{adopting } OP = 1 \text{ for *developing equation of slope it is taken up}$$

This approach is different from the one adopted by show, Bhattacharyya, Kronenberg and others.

---

*It was developed by Chitale and published as: "A simplified approach to cutting tool geometry" by Dr. A.K. Chitale in *Indian Journal of Technical Education*, September 1972 (Ed. Dr. L.S. Srinath and others).

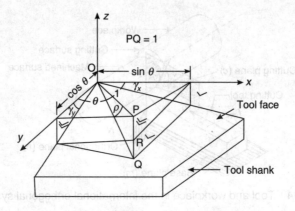

Fig. 2.16(a) Slope tan ρ of line OQ in rake face.

$$\tan \rho = \cos \theta \tan \gamma_y + \sin \theta \tan \gamma_x \quad (2.1)$$

From the above general equation the transformations can be done by substituting appropriate values of θ in Eq. (2.1).

Orthogonal Rake γ_0

Referring to Fig. 2.16(b), the direction of measurement for γ_0 is:

$$\theta = (90 - \phi_s) = \phi, \text{ being measured clockwise}$$

Substituting in Eq. (2.1):

$$\rho = \gamma_0, \quad \theta = \phi$$

We have,

$$\boxed{\tan \gamma_0 = (\cos \phi \tan \gamma_y + \sin \phi \tan \gamma_x)} \quad (2.2)$$

where ϕ is the approach angle.

Angle of inclination λ

Referring to Fig. 2.16(c), the direction of measurement of λ is:

$$\theta = -(\phi_s), \text{ being measured clockwise}$$

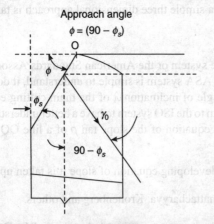

Fig. 2.16(b) $\theta = (90 - \phi_s)$ for determining angle λ_0

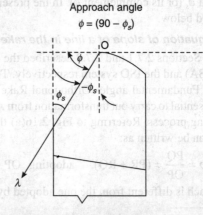

Fig. 2.16(c) $\theta = -(\phi_s)$ for determining angle λ.

Substituting in Eq. (2.1):

$$\rho = \lambda, \; \theta = -(\phi_s)$$

We have,

$$\tan \lambda = \cos(-\phi_s) \tan \gamma_y + \sin(-\phi_s) \tan \gamma_x$$

$$\therefore \quad \tan \lambda = \cos \phi_s \tan \gamma_y - \sin \phi_s \tan \gamma_x$$

Putting $\phi_s = (90 - \phi)$, where ϕ is approach angle. We have

$$\boxed{\tan \lambda = \sin \phi \tan \gamma_y - \cos \phi \tan \gamma_x} \tag{2.3}$$

Equations (2.2) and (2.3) are the transformation equations which find extensive application in cutting tool geometry.

Note: The reader should note that (2.2) and (2.3) are equivalent to the orthogonal 'Kronenberg Relations' in terms of side cutting edge angle ϕ_s adopted by some authors.

$$\tan \gamma_0 = (\sin \phi_s \tan \gamma_y + \cos \phi_s \tan \gamma_x)$$

and $\quad \tan \lambda = (\cos \phi_s \tan \gamma_y - \sin \phi_s \tan \gamma_x)$

Equation (2.2) which enables the calculation of angle γ_0 is of importance. Angle (γ_0) is the orthogonal rake angle which determines the basic wedge shape responsible for chip formation and also determines the cutting forces. Equation (2.3) which enables the calculation of the angle of inclination (λ) is of importance because (λ) determines the obliquity of cutting and the direction of chip-flow in metal cutting. The chip-flow should be directed away from the machined surface to maintain its surface finish.

The shape of the tool may be specified in a sequence as given in the two widely used rake systems as indicated below.

1. American System (AS stated in Section 2.8) in detail:

$$\gamma_y, \gamma_x, \alpha_y, \alpha_x, \phi_e, \phi_s, r''$$

2. Orthogonal Rake System (ORS) or ISO system:

$$\lambda, \gamma_o, \alpha_o, \alpha_o', \phi_1, \phi, r \text{ (mm)}$$

where
$\quad \lambda$ = Inclination angle
$\quad \gamma_0$ = Orthogonal rake
$\quad \alpha_0$ = Orthogonal clearance
$\quad \alpha_0'$ = Auxiliary edge clearance
$\quad \phi_1$ = End cutting edge angle
$\quad \phi$ = Principal cutting edge angle or Plan approach angle
r(mm) = Nose radius mm

2.8 CUTTING TOOL SIGNATURE

The seven elements of a single-point tool, which describe the tool completely in the American system and are known as the *signature* of the cutting tool, are always stated in the following order:

(a) Back rake angle γ_y
(b) Side rake angle γ_x
(c) End relief angle ϕ_0, α_y
(d) Side relief angle ϕ_x

(e) End cutting edge angle ϕ_e
(f) Side cutting edge angle ϕ_s
(g) Nose radius r'' (mm).

Other than this, sometimes the effective rake angle and the normal rake angles are also determined and represented respectively as γ_e and γ_n, or sometimes even as α_e and α_n, and are mutually related as

$$\sin \gamma_e = \cos \lambda \cos \rho \sin \gamma_n + \sin \lambda \sin \rho$$

If $\rho = \lambda$, as per Stabler's Rule we have,

$$\sin \gamma_e = \cos^2 \lambda \sin \gamma_n + \sin^2 \gamma$$

where ρ is the chip flow angle and λ is the inclination angle.

Generally, the symbols for degrees and mm are omitted, and the numerical value of each component is simply enlisted for a cutting tool signature, e.g. 5 5 3 3 10 30 1.

2.9 GEOMETRY OF CUTTING TOOLS

Cutting tool geometry refers to the angles provided on the cutting tool face and flanks, and the shape of the cutting edge. These angles, except the side cutting edge angles, are provided to reduce, to a minimum, the rubbing of the cutting tool against the chips and the work material, and to cause the chips to impinge on wider areas. With these angles, a cutting tool acquires a chisel-like shape, which facilitates shearing action.

2.10 FACTORS AFFECTING TOOL GEOMETRY

The optimum cutting tool geometry in any specific case depends on the following major factors:

(a) Material of work
(b) Material of cutting tool
(c) Type of cutting tool
(d) Cutting variables, i.e. feed, speed, and depth of cut
(e) Nature and quality of coolant
(f) Area of cut.

2.11 ILLUSTRATIVE EXAMPLES ON TOOL GEOMETRY

The objective of this section is to expose the reader to the wide range and variety of problems find their solution in cutting tool geometry.

EXAMPLE 2.1 In an orthogonal turning operation the cutting tool has the following angles:

(a) Side rake $\gamma_x = 8°$
(b) Principal cutting edge angle $\phi = 75°$

Calculate the back rake angle γ_y.

Solution We know that in orthogonal cutting the angle of inclination, $\lambda = 0$
We also know from Eq. (2.3) that:

$$\tan \lambda = \sin \phi \tan \gamma_y - \cos \phi \tan \gamma_x \qquad (i)$$

Also it is given that

$$\lambda = 0$$

$$\phi = 75°$$
$$\gamma_x = 8°$$

Substituting the values, we get

$$\tan 0° = \sin 75° \tan \gamma_y - \cos 75° \tan 8°$$
$$= 0.966 \tan \gamma_y - 0.259 \times 0.141$$

∴ $\gamma_y = 2°7'$

EXAMPLE 2.2 For a turning operation, a tool of the geometry with side rake angle 9° and back rake angle 9° and principal cutting edge angle 60° was used. Calculate the angle of inclination λ and the orthogonal rake angle γ.

Solution We are given that

Principal cutting edge angle, $\phi = 60°$
Side rake angle, $\gamma_x = 9°$
Back rake angle, $\gamma_y = 9°$

We know from Eq. (2.3)

$$\tan \lambda = \sin \phi \tan \gamma_y - \cos \phi \tan \gamma_x \qquad (i)$$

where λ is the inclination angle.

Substituting the values, we get

$$\tan \lambda = \sin 60° \tan 9° - \cos 60° \tan 9°$$
$$= 0.867 \times 0.158 - 0.50 \times 0.158 = 0.0579$$

∴ $\lambda = 3°19'$

Also, we know from Eq. (2.2) that

$$\tan \gamma_0 = \cos \phi \tan \gamma_y + \sin \phi \tan \gamma_x$$
$$= \cos 60° \tan 9° + \sin 60° \tan 9° = 0.50 \times 0.158 + 0.867 \times 0.158$$
$$= 0.2159$$

∴ $\gamma_0 = 12°11'$

EXAMPLE 2.3 A single-point cutting tool has a back rake angle of 10° and a side rake angle of 12°. Calculate the orthogonal rake angle γ and the inclination angle λ, if the principal cutting edge angle is 60°.

Solution We are given that

$\gamma_y = 10°$ (Back rake angle)
$\gamma_x = 12°$ (Side rake angle)
$\phi = 60°$ (Principal cutting angle or approach angle)

We know from Eq. (2.2) that

$$\tan \gamma_0 = \tan \gamma_y \cos \phi + \tan \gamma_x \sin \phi \qquad (i)$$

where γ is the orthogonal rake angle.

Substituting the values, we get

$$\tan \gamma_0 = \tan 10° \cos 60° + \tan 12° \sin 60°$$
$$= 0.176 \times 0.50 + 0.213 \times 0.866 = 0.272$$

∴ $\gamma_0 = 15°48'$

We also know from Eq. (2.3) that

$$\tan \lambda = \tan \gamma_y \sin \phi - \tan \gamma_x \cos \phi \qquad (ii)$$

where λ is the inclination angle.

Substituting the values, we get

$$\tan \lambda = \tan 10° \sin 60° - \tan 12° \cos 60°$$
$$= 0.176 \times 0.866 - 0.213 \times 0.5 = 0.0459$$
$$\therefore \quad \lambda = 2°38'$$

EXAMPLE 2.4 The geometry of a cutting tool used for an operation is given below:

$$7 \quad 10 \quad 6 \quad 6 \quad 8 \quad 15 \quad 1 \text{ mm (ASA)}$$

Calculate the orthogonal rake angle and the inclination angle.

Solution We are given that

$$\text{Back rake angle, } \gamma_y = 7°$$
$$\text{Side rake angle, } \gamma_x = 10°$$
$$\text{Side cutting edge angle, } \phi_s = 15°$$

Principal cutting edge angle, $\phi = 90° - 15° = 75°$

We know from Eq. (2.2) that

$$\tan \gamma_0 = \tan \gamma_y \cos \phi + \tan \gamma_x \sin \phi$$

where γ_0 is the orthogonal rake angle.

Substituting the values, we get

$$\tan \gamma_0 = \tan 7° \cos 75° + \tan 10° \sin 75°$$
$$= 0.123 \times 0.259 + 0.176 \times 0.966$$
$$= 0.202$$
$$\therefore \quad \gamma_0 = 11°25'$$

The inclination angle is calculated from Eq. (2.3):

$$\tan \lambda = \tan \gamma_y \sin \phi - \tan \gamma_x \cos \phi$$

Substituting the values, we get

$$\tan \lambda = \tan 7° \sin 75° - \tan 10° \cos 75°$$
$$= 0.123 \times 0.966 - 0.176 \times 0.259 = 0.0732$$
$$\therefore \quad \lambda = 4°12'$$

EXAMPLE 2.5 In a certain cutting operation the back rake angle is 8° and the side rake angle is 0°. Calculate the true rake angle γ and the inclination angle λ when the principal angle is 70°, 90° or 0°.

Solution

(a) We are given that:

$$\text{Principal angle, } \phi = 70°$$
$$\text{Side rake angle, } \gamma_x = 0°$$
$$\text{Back rake angle, } \gamma_y = 8°$$

For calculating the orthogonal rake γ_0, we use Eq. (2.2):

$$\tan \gamma_0 = \cos \phi \tan \gamma_y + \sin \phi \tan \gamma_x$$

Substituting the values, we get

$$\tan \gamma = \cos 70° \tan 8° + \sin 70° \tan 0°$$
$$= 0.342 \times 0.141 + 0.940 \times 0 = 0.0482$$
$$\therefore \quad \gamma = 2°46'$$

For calculating the inclination angle λ, we use Eq. (2.3):

$$\tan \lambda = \sin \phi \tan \gamma_y - \cos \phi \tan \gamma_x$$
$$= \sin 70° \tan 8° - \cos 70° \tan 0°$$
$$= 0.940 \times 0.141 - 0.342 \times 0.0 = 0.1325$$
$$\therefore \quad \lambda = 7°33'$$

(b) When $\phi = 90°$

$$\tan \gamma_0 = \cos 90° \tan 8° + \sin 90° \tan 0°$$
$$= 0.0 \times 0.141 + 1.0 \times 0.0$$
$$\therefore \quad \gamma_0 = 0$$
$$\tan \lambda = \sin 90° \tan 8° - \cos 90° \tan 0°$$
$$= 1 \times 0.141 - 0.0 \times 0.0$$
$$\therefore \quad \lambda = 8°2'$$

(c) When $\phi = 0°$

$$\tan \gamma_0 = \cos 0° \tan 8° + \sin 0° \tan 0°$$
$$= 1 \times 0.141 + 0.0 \times 0.0$$
$$\therefore \quad \gamma_0 = 8°2'$$
$$\tan \lambda = \sin 0° \tan 0° - \cos 0° \tan 0°$$
$$= 0 \times 0.0 - 1.0 \times 0.0$$
$$\therefore \quad \lambda = 0$$

EXAMPLE 2.6 A tool bit has a rake angle α of 10° and a front relief angle of 8°. If the diameter of the shaft is 25.00 mm and the offset is above the centreline by 1.00 mm, calculate the effect on the front relief angle with both 10° positive rake angle and 10° negative rake angle.

Solution From geometry shown in Fig. 2.17, the correction angle is

$$\sin \phi = \frac{h}{R} = \frac{h}{D/2} = \frac{2h}{D}$$

where ϕ is the correction angle
 D is the diameter of the shaft
 and h is the offset.

Substituting the values of D and h, we get

$$\sin \phi = \frac{2 \times 1.00}{25} = 0.08$$
$$\therefore \quad \phi = 4°34'$$

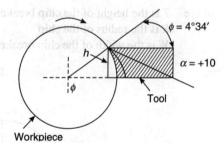

Fig. 2.17 Effect of tool offset (h) on working rake angle.

(a) For positive rake angle, $\alpha = 10°$ and front relief angle, $b = 8°$

Assume that α' = effective rake angle
 b' = effective front relief angle

Then
$$\alpha' = \alpha + \phi$$
$$= 10° + 4°34'$$
$$= 14°34'$$
$$b' = b - \phi$$
$$= 8' - 4°34' = 3°26'$$

(b) For negative rake angle, $\alpha = -10°$

$$\alpha' = \alpha + \phi$$
$$= -10° + 4°34' = -5°26'$$
$$b' = b - \phi = 8° - 4°34' = 3°26'$$

EXAMPLE 2.7 Assuming the same data as in Example 2.6, except that the tool bit is held below the centreline of the workpiece by 1.00 mm, and calculate the effect on the front relief angle.

Solution The correction angle ϕ is the same as in Example 2.6, except that it is below the centreline and hence taken negative.

$$\therefore \qquad \phi = -4°34'$$

(a) For positive rake angle,
$$\alpha' = \alpha + \phi = 10° + (-4°34') = 5°26'$$
$$b' = b - \phi = 8 - (-4°34') = 12°34'$$

(b) For negative rake angle,
$$\alpha' = \alpha + \phi = -10° + (-4°34') = -14°34'$$
$$b' = b - \phi = 8° - (-4°34') = 12°34'$$

EXAMPLE 2.8 Calculate the width (W) of a chip breaker if its height (T) is 0.5 mm and the radius of curvature (R) of the chip removed is 19 mm.

Solution Referring to the Fig. 2.18 of a step type chip breaker with radius R and width W and depth T we have:

$$W \cdot W = T(2R - T)$$

$$\therefore \qquad W = \sqrt{2TR - T^2}$$

Substituting the values, $T = 0.5$ mm and $R = 19$ mm

where T is the height of the chip breaker in mm
R is the radius of the chip
W is the width of the chip breaker.

$$\therefore \qquad W = \sqrt{2(0.5)(19) - (0.5)^2}$$
$$\cong 4.33 \text{ mm}$$

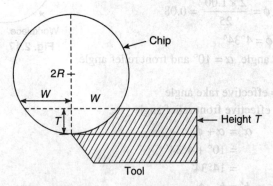

Fig. 2.18 Geometry for calculating chip breaker width.

EXAMPLE 2.9 Given a 63.5 mm bar diameter and a back rake angle α of 6°, calculate the height above the centre of the tool to maintain a 0° effective rake.

Solution We know that:
Referring to Fig. 2.19.

$$h = \frac{D}{2} \sin \phi$$

where h is the height above the centre of the tool
D is the diameter of bar
ϕ is the correction angle.

Substituting the given values, we get

$$h = \frac{63.5}{2} \sin 6°$$
$$= 31.75 \times 0.1045 \cong 3.30 \text{ mm}$$

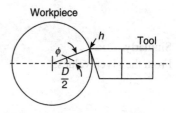

Fig. 2.19 Calculation of tool offset h.

EXAMPLE 2.10 A tool used in a relieving lathe has its rake face in the plane of the axis of work rotation. The work rotates at 1 rev/min and the tool is displaced towards the work axis by the relieving cam at a velocity of 101.6 mm/min. Find at the point where the work diameter is 101.6 mm, the cutting speed and the effective rake angle in a plane perpendicular to the work axis. What front clearance angle relative to the tool shank is required to give an effective cutting clearance angle of 8° in this plane?

Solution Figure 2.20 shows the relationships between the various velocities upon which the solution depends.

The velocity of work relative to the tool edge is represented by PR. Therefore, $\angle SPR = 90°$ and hence $\angle QPR = \gamma$ (effective rake angle).

The velocity of the work relative to the tool edge is the vector sum of PQ and QR.

$$PQ = \pi \times 15 \times \frac{101.6}{1000} = 4.78 \text{ m/min}$$

$$PR = \sqrt{(4.78)^2 + \left(\frac{101 \cdot 6}{1000}\right)^2} = 4.78 \text{ m/min}$$

i.e. the true cutting velocity is 4.78 m/min.

$$\tan \gamma = \frac{QR}{PQ} = \frac{101.6}{15 \times 1000 \times \pi} = 0.212$$

or
$$\gamma = 11°3' \text{ (nearly)}$$

i.e. the effective rake angle $\gamma = 11°3'$.

The clearance angle relative to tool shank = $11°3' + 8° = 19°3'$.

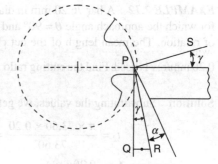

Fig. 2.20 Example 2.11.

EXAMPLE 2.11 A tool has an approach angle $\phi = 55°$, effective rake angle $\gamma = 22°$ and a cutting edge which lies in a plane parallel to the base of the tool. Find the side rake angle γ_x and the back rake angle γ_y (American System) for these conditions.

Solution The solution is based upon the geometry shown in Fig. 2.21(a).
From the diagram,

$$\angle RTU = \angle TSU = \phi$$

Point T in the rake face of the tool lies at a distance UT tan γ below the horizontal plane containing the edge PQ [see Fig. 2.21(b)].

Hence,
$$\tan \gamma_y = \frac{UT \tan \gamma}{RT} = \frac{UT \tan \gamma}{UT/\cos \phi}$$

$$= \tan \gamma \cos \phi$$

Similarly,
$$\tan \gamma_x = \tan \gamma \sin \phi$$

From the given values,

$$\tan \gamma_y = 0.4040 \times 0.5736 = 0.2317$$

$$\gamma_y = 13°3'$$

$$\tan \gamma_x = 0.4040 \times 0.8192 = 0.3309$$

$$\gamma_x = 18°19'$$

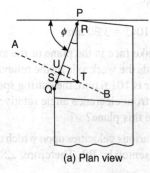
(a) Plan view

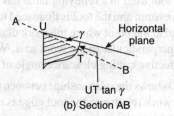
(b) Section AB

Fig. 2.21 Example 2.11.

EXAMPLE 2.12 A bar 76.20 mm in diameter is reduced to 73.60 mm diameter by means of a cutting tool for which the approach angle $\theta = 90°$ and for which the cutting edge lies in a plane containing the work axis of rotation. The mean length of the cut chip $l2 = 73.91$ mm, the effective rake angle $= 15°$, and a feed of 0.20 mm/rev is used. Find the cutting ratio $\frac{t_1}{t_2}$ and the value of the shear plane angle. Use the graphical method.

Solution Substituting the values, we get

$$t_2 = \frac{\pi \times 73.60 \times 0.20}{73.60} = 0.628 \text{ mm/rev}$$

$$\therefore \quad \frac{t_1}{t_2} = \frac{0.20}{0.628} = 0.32$$

The shear plane angle ϕ can then be found by a scale drawing as shown in Fig. 2.22. Set off TP and PS, such that $\angle TPS = 90° + 15°$, and PS represents the rake face of the tool. To some convenient scale, set off a line parallel to TP at a distance 0.20 mm and a line parallel to PS at a distance 0.625 mm to intersect at Q.

Join PQ and measure the $\angle TPQ$. By measurement, the shear plane angle $\phi = 19°$.

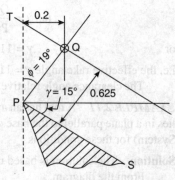

Fig. 2.22 Example 2.12.

EXAMPLE 2.13 Calculate the side rake angle γ_x and the back rake angle γ_y when the inclination angle is $5°9'$, the orthogonal rake angle γ_0 is $6°38'$ and the side cutting edge angle is $15°$ in an ASA system.

Solution Substituting the values, we get

$$\tan \gamma_x = \tan \gamma_0 \sin(90° - \phi) + \tan \lambda \cos(90° - \phi)$$
$$= (-0.11629 \times 0.96593) + (0.9600 \times 0.925882)$$
$$= -0.11133 - 0.02485 = -0.08748$$

∴ $\gamma_x = -5°$

and $\tan \gamma_y = \tan \gamma_0 \sin \phi - \tan \lambda \cos \phi$
$$= (0.11629 \times 0.25882) - (0.09600 \times 0.96593)$$
$$= -0.03010 - 0.09273 = -0.12283$$

∴ $\gamma_y = -7°$

EXAMPLE 2.14 A −10, 10, 5, 5, 5, 0, 0.8 ASA designation HSS tool is set for a turning cut of 9.5 mm and 0.4 mm per revolution feed on an MS forging (200 BHN). Determine the following: (a) the angle of obliquity, (b) the approximate direction of chip flow, and (c) the effect if the back rake angle is −10°.

Solution We have

−10° is the back rake angle (γ_y)
10° is the side rake angle (γ_x)
5° is the end relief angle (α_y)
5° is the side relief angle (α_x)
5° is the end cutting angle (ϕ_e)
0° is the side cutting angle (ϕ_s)
0.8 mm is the nose radius (r)

In the equation given below θ is the complement of the chip flow angle given by the relation is due to Colwell (Fig. 2.23). He assumed the rake angles to be very small:

$$\tan \theta = \frac{d}{[NR + (d - NR) \tan SCEA]}$$

where d is the depth of cut
NR is the nose radius
SCEA is the side cutting edge angle

Putting the values of the above variables, we get

$$\tan \theta = \frac{9.5}{0.8 + (9.5 - 0.8) \tan 0°}$$
$$= 11.8$$

∴ $\theta = 85°$

The Colwell formula therefore gives the chip flow angle,

$$\eta = 90° - 85° = 5°$$

We have the following data:

$\gamma_y = -10°$
$\gamma_x = 10°$
$\phi_s = 0°$

Therefore, from Eq. (2.3),

$$\tan \lambda = \tan \gamma_y \sin \phi_s - \tan \gamma_x \cos \phi_s$$

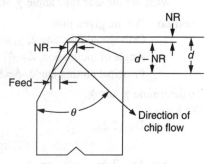

Fig. 2.23 Example 2.14.

48 TEXTBOOK OF PRODUCTION ENGINEERING

$$= \tan(-10°) \sin 0° - \tan 10° \cos 0°$$
$$= -\tan 10°$$
$$\therefore \quad \lambda = -10°$$

Thus, the cutting is oblique, and hence the direction of chip flow will not be at 5° (as calculated above) with respect to perpendicular to the cutting edge, Stabler's rule gives the chip flow angle $\eta = 0.9\lambda = -9°$. There is a difference of $9° + 5° = 14°$ in the two estimates of the chip flow angle. This deviation is because the Colwell's formula applies to only very small rake angles.

EXAMPLE 2.15 The principal cutting edge angle is 75°, the side rake angle is –5°, and the back rake is 10°. Calculate λ and γ.

Solution From Eq. (2.3), we know that
$$\tan \lambda = \sin \phi \tan \gamma_y - \cos \phi \tan \gamma_x$$
where λ is the angle of inclination
   ϕ is the principal cutting angle, i.e. 75°
   γ_x is side rake angle, i.e. –5°
   γ_y is back rake angle, i.e. 10°
Putting the values,
$$\tan \lambda = \sin 75° \tan 10° - \cos 75° \tan (-5°)$$
$$= 0.965 \times 0.176 - 0.258 \times (-0.08)$$
$$\therefore \quad \lambda = 10°55'18''$$

We also calculate γ_0, i.e. orthogonal rake angle:
$$\tan \gamma_0 = \cos \phi \tan \gamma_y + \sin \phi \tan \gamma_x$$
$$= \cos 75° \tan 10° + \sin 75° \tan (-5°)$$
$$= 0.258 \times 0.176 + 0.965 \times (-0.08)$$
$$\therefore \quad \gamma_0 = -2°13'33''$$

EXAMPLE 2.16 In a single-point cutting tool used for turning, the geometry is specified in ISO system of nomenclature as follows:
   $\gamma_0 = 8°$
   $\lambda = 4°$
   ϕ_s is side cutting edge angle = 15°
What are the side rake angle γ_x and the back rake angle γ_y in the American system?

Solution We are given that:
   Orthogonal rake angle, $\gamma_0 = 8°$
   Angle of inclination, $\lambda = 4°$
   Principal cutting angle, $\phi = 90° - \phi_s = 90° - 15° = 75°$.
To determine γ_x and γ_y.
We know
   from Eq. (2.2),
$$\tan \gamma_0 = \cos \phi \tan \gamma_y + \sin \phi \tan \gamma_x \qquad \text{(i)}$$
   from Eq. (2.3),
$$\tan \lambda = \sin \phi \tan \gamma_y - \cos \phi \tan \gamma_x \qquad \text{(ii)}$$

Substituting the values, we get

$$\tan 8° = \cos 75° \tan \gamma_y + \sin 75° \tan \gamma_x \quad \text{(iii)}$$
$$\tan 4° = \sin 75° \tan \gamma_y - \cos 75° \tan \gamma_x \quad \text{(iv)}$$

or
$$0.1405 = 0.2588 \tan \gamma_y + 0.966 \tan \gamma_x \quad \text{(v)}$$

and
$$0.0699 = 0.966 \tan \gamma_y - 0.2588 \tan \gamma_x \quad \text{(vi)}$$

Solving Eqs. (v) and (vi), we get

$$\gamma_x = 6°42'$$
$$\gamma_y = 5°56'$$

REVIEW QUESTIONS

1. Define a tool. What are the major kinds of tools?
2. (i) How do you define the cutting ability of a cutting tool?
   (ii) What is the essential criterion for a cutting tool to give maximum production with minimum maintenance and trouble?
3. State the advantages of mechanically held inserted tools.
4. Describe the back rake and side rake angles and their uses.
5. What is the difference between true rake, orthogonal rake and effective rake of a cutting tool?
6. Define the various tool angles of a single-point cutting tool. What are the standard angles of a cutting tool? What is the difference between standard angles and working angles?
7. Sketch three view diagrams of a 25 mm square tool bit with the following tool signatures. Also, show the various parts and angles on the diagrams.
   (i) 15, 15, 10, 10, 15, 10, 3 mm
   (ii) –8, +8, 10, 10, 10, 6, 6.0 mm
   (iii) 10, 10, 6, 6, 8, 0, 0.8 mm
8. What factors are influenced by the shape of a cutting tool?
9. Explain the effect of varying the side angle from positive to negative.
10. What is meant by tool signature?
11. What is the importance of describing the tool geometry in XYZ-planes?
12. Discuss the advantages of throw-away cutting tips.
13. What are the advantages of increasing the nose radius?
14. What is the rake angle requirement for ductile work materials and brittle materials?
15. What are the advantages of providing a side cutting edge angle (lead angle or principal edge angle) on cutting tools?
16. Why are relief angles on carbide tools kept as small as possible?
17. Why are relief or clearance angles never zero or negative?

CHAPTER 3

Cutting Tool Materials

3.1 INTRODUCTION

The technology of metal removal by chip formation made tremendous strides in the 20th century. Machining operations which required 105 minutes in the year 1900 can be done today within 1 minute. There are reasons to believe that this time could be reduced still further in the future. This improvement has been possible because of the immense sophistication in machine tools and remarkable progress in cutting tool materials.

Generally, the progress of cutting tool materials has been a step ahead of machine tool development. But the aerospace industry has been developing high-strength and exotic materials which are rather difficult to machine. This situation has necessitated the development of still better cutting tool materials. In brief, the most important factors responsible for the development of tool materials are:

(a) Economic competition
(b) Shortage of raw materials at critical times
(c) Need to machine materials of higher strength
(d) Military necessities.

3.2 REQUIREMENTS OR CHARACTERISTICS OF A TOOL MATERIAL

The most essential requirements for all the types of cutting tool materials are described in the following lines:

Hot hardness or red hardness: It is the ability of the cutting tool to withstand high temperatures without losing its cutting edge. The red hardness of the tool materials can be increased by adding chromium, molybdenum, tungsten and vanadium, all of which form hard carbides.

CUTTING TOOL MATERIALS 51

Abrasion resistance: It is the ability to resist wear. Abrasion resistance not only depends on hardness but also on the extent of hard, undissolved carbides present. This abrasion resistance increases as the carbon and alloy contents increase.

Toughness: Toughness is the ability to resist shock and/or impact forces and also to resist a high unit pressure against the cutting edge. The term actually implies a combination of strength and ductility.

Frictional coefficient: In order to have low tool wear and better surface finish, the coefficient of friction between the chip and tool should be as low as possible in the operating range of speed and feed.

Thermal conductivity and specific heat: It is very much desired that the tool material should possess high thermal conductivity and specific heat, so that the materials may conduct away the heat generated at the cutting edge.

Machinability: This is the property of a material which defines the ease with which a material would machine. The tool material should be comparatively easier to machine.

Cost and ease of fabrication: The cost and ease of fabrication should be within reasonable limits.

Resistance to deformation: The tool steel material should retain shape and size during the heat treatment process.

Resistance to decarburisation: Decarburisation causes soft spots on the tool surface, which may get cracked due to quenching by the application of cutting fluid.

Quality: The tool material must produce acceptable quality parts in terms of surface finish.

Ease of grinding: The tool material should be easy to form, grind and sharpen to the desired geometry.

The above functional and property requirements are often contradictory in nature; therefore, as with all engineering problems, compromises are sought.

3.3 TYPES OF TOOL MATERIALS

As a result of research, the following types of tool materials, each suitable for specific ranges of application, have been evolved:

1. Plain carbon steels
2. Low alloy steels
3. High speed steels
4. Non-ferrous cast alloys, also referred to as *super high speed tools*
5. Cemented carbides
6. Ceramics
7. Cermets
8. Diamonds
9. Abrasives.

3.3.1 Plain Carbon Steels

Plain carbon steel, generally called *carbon tool steel*, possesses 0.85 to 1.5% carbon. Hardness of these materials ranges from Rockwell C 55 to C 64. The material begins to lose hardness at about 150°C. At about 200°C, it loses its hardness too rapidly to be effective as a cutting tool material beyond this temperature. Carbon tools are cheap on weight basis. They also have low fabricating cost. The plain carbon steel can be sharpened to a keen cutting edge. Their application is limited to cases where operations are done at low speeds.

The use of plain carbon steels is recommended for form tools, including thread cutting tools and parting tools for small-quantity manufactures. Apart from these uses, plain carbon steel cutting tools have little application in modern production.

3.3.2 Low Alloy Steels

The properties of plain carbon steels can be improved by adding small alloying elements such as chromium (Cr) up to 0.25%, vanadium (V) up to 0.25%, manganese (Mn) up to 1.2% and silicon (Si) up to 0.25%. With all or some of the above additions, the temperature at which the material loses its hardness is increased to 300°C.

These steels are generally placed between plain carbon steels and high speed steels in hardness and have much the same application as plain carbon steels. The costs both for material and for grinding are higher by 25%. The speeds used are also higher by about 30%. They have a hardness in the same range as plain carbon steels.

3.3.3 High Speed Steels

High speed steel as a cutting tool material is the most widely used cutting tool material in the world. When properly heat-treated, high speed steel has a hardness in the range of Rockwell C 62 to C 65, and it retains the hardness at high temperatures up to 500°C. Consequently, high speed steel tools can be worked at substantially higher cutting speeds.

High speed steel is essentially a high carbon tool steel to which elements like tungsten, chromium, vanadium, etc. in large quantities have been added. The most common type of high speed steel used for cutting contains tungsten 18%, chromium 4% and vanadium 1%, and is referred to as **18–4–1 high speed steel**. During the Second World War, owing to shortage of tungsten, molybdenum was used as a substitute. The most common type was molybdenum 8%, chromium 4.1% and vanadium 2%, and was referred to as **8–4–2 high speed steel**.

Molybdenum steels are tough but less heat resistant; in some cases cobalt was added. It increased the hot hardness.

The typical composition of a high speed steel cutting tool is carbon 0.75–1.5%, chromium 3.15–42%, vanadium 1.78–2% and tungsten 18% (alternatively, molybdenum 8%).

Carbon tool steel and high speed steel can be equally hard and have the same strength and nearly the same toughness. High speed steel has a greater wear resistance due to the presence of tungsten, chromium and vanadium carbides in matrix. The cost of material and its fabrication is, however, higher. Its use is widespread in all shapes and sizes. They are available as solid tools used directly in the tool post and as bit tools used in cutting tool holders of various shapes. When used in cutting tool holders, the feeds and depths of cut should be reduced by about 20% each.

3.3.4 Non-ferrous Cast Alloys (Super High Speed Tools)

These are cobalt-based materials whose hardness is inherent and which require no heat treatment. Because of the inherent hardness, they are cast to size and then ground to final shape. These materials typically contain cobalt 40%, chromium 35% and tungsten 20%. Some alloys contain a few per cent of carbon, vanadium, iron or nickel. Usually these materials can be operated at interface temperature above 540°C. As cast, they are relatively brittle and cannot withstand severe impact or vibration to the same extent as the high-speed steels can. Speeds used are appreciably higher—one and a half to twice those used on high speed steel cutting tools. These tools are however costly and are available as butt welded tools. Considering the cost, their advantages do not appear to be many and they have not gained much popularity. Their use for casting is however recommended.

CUTTING TOOL MATERIALS

3.3.5 Cemented Carbides

Modern technology demands a wide choice of various materials having well defined properties. In fact, there has been a tremendous progress in the field of engineering materials during the last century, specially during and after the Second World War. Cemented carbide has thus found its important place in the domain of design and engineering materials. A wide range and a combination of properties make cemented carbide a versatile material for wear resistant surfaces and metal forming tools. Cemented carbides are used for parts which should withstand extremes of any or several destructive conditions like wear, deflection, corrosion, high pressure impact, and temperature. In many applications they provide long life where other materials quickly fail.

Manufacturing process

Cemented carbide is a product of powder metallurgy. It consists of a hard principal ingredient (usually tungsten carbide) and a binding metal (usually cobalt).

Chemical action

Crushed concentrated tungsten ore, with scheelite as the tungsten bearing mineral, is the normal starting material. The first stage of manufacture is the chemical process where the ore containing calcium tungstate is ground in a ball mill with hydrochloric acid. Calcium tungstate reacts with hydrochloric acid to form tungstic acid H_2WO_4 and calcium chloride $CaCl_2$. Washed tungstic acid is treated with ammonium hydroxide, NH_4OH. The ammonium tungstate so formed is filtered and again treated with hydrochloric acid to obtain tungstic acid, H_2WO_4. After washing and drying it is calcined in air at about 800°C to give tungstic oxide WO_3.

Tungsten powder is produced from WO_3 by reduction with hydrogen. By varying the conditions of reduction, tungsten powder of various grain sizes can be manufactured.

Carbide powder

Carefully weighed quantities of tungsten powder and carbon black are thoroughly mixed. This mixture is charged in carburising furnaces using high frequency heating. Carburising takes place at a high temperature under a protective atmosphere of hydrogen to form WC, tungsten carbide.

Using various types of tungsten powder obtained from the reduction process, it is possible to produce a series of tungsten carbide powders required for different grades of cemented carbides.

Instead of pure titanium carbide, TiC, a solid solution of tungsten carbide in titanium carbide is normally used. Equal parts of TiC and WC are ground together and the mixture is heated to about 2000°C. At this temperature, TiC dissolves WC, forming (Ti, W)C. Tantalum and niobium carbides are produced in a similar manner of forming (Ta, Nb)C.

Blending

Carefully weighed quantities of carbides of tungsten, titanium, tantalum and niobium, and binder metal (cobalt) are thoroughly mixed by grinding in ball mills. Alcohol is generally used as a grinding medium. Grinding balls are made of cemented carbide. During grinding, an adhesive contact develops between the grains of binding metal and the carbide. The speed of rotation and duration of rotation of a ball mill vary from grade to grade. The blended slurry is taken to a drying furnace for removal of alcohol. Mixed blended powder, in the form of a cake, is further crushed, sieved and kept ready for use. A small amount of wax is added during ball milling to act as a lubricant during compacting.

Compacting

Pressing of cemented carbide powder to compacts is usually done in hydraulic presses, using powder pressing dies made of cemented carbide. In pressing, the compacts assume shape but not size. This is because the green briquette is porous. The porosity disappears during pre-sintering, corresponding to a linear shrinkage of 20%. Pressing calls for a pressure of about 500–1000 kg/cm^2.

Pre-sintering

Pre-sintering has the dual function of driving off wax which was added to act as a lubricant and giving some strength and machineability to the green briquette.

Shaping

By reason of the flow properties of cemented carbide powder and the limitations imposed by tooling considerations, many briquette shapes cannot be produced by direct pressing. Such items are given their final form after pre-sintering. This is effected by using diamond and carbide tooling in grinding, turning, drilling and boring operations.

Sintering

Sintering is the last step which converts the pre-sintered blanks to cemented carbide. At temperatures between 1400°C and 1600°C, the binding metal melts and dissolves a considerable amount of carbides. Also, carbides dissolve in one another. Concurrently with the structure changing reactions, the cemented carbide shrinks and becomes non-porous. On cooling, the bulk of the dissolved carbides will be precipitated out of the binding metal.

The sintering of carbides is done either in a vacuum or under an atmosphere of hydrogen in high frequency furnaces.

Quality control

Though there is quality control at every stage, the final and pre-delivery tests are carried out on samples with respect to porosity, structure, grain size, coercive force, specific gravity and hardness. After the dimensions have been checked, the material is ready for use.

Outstanding properties

The outstanding properties of cemented carbides are as follows:

Hardness and wear resistance: High wear resistance is the property associated with cemented carbides. Wear resistance varies considerably from one grade to another. But, as a general guideline, cemented carbides may be taken to be 10–100 times more wear resistant than steel. The relative wear resistance value of various grades of carbide, and that of several other common wear resistant materials, is presented in Fig. 3.1.

The hardness range of cemented carbides begins at the level of hardest steel (about 80 Vickers) and goes up to about 2000 Vickers, sufficient to cut glass (see Fig. 3.2).

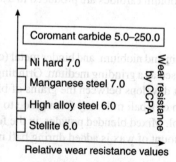

Fig. 3.1 Relative wear resistance values for different materials vis-a-vis cemented carbides.

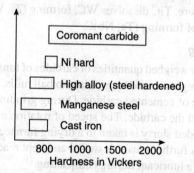

Fig. 3.2 Hardness range of different materials vis-a-vis cemented carbides.

Even at high temperatures (about 1000°C), the hardness of cemented carbides is fairly high. See Fig. 3.3. Hardness and wear resistance are linked together, as shown in Fig. 3.4. A small change in the value of hardness can make an appreciable change in the wear resistance and hence the performance of the material.

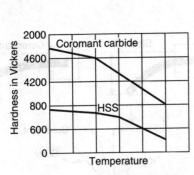

Fig. 3.3 Hardness of cemented carbides compared to HSS at elevated temperatures.

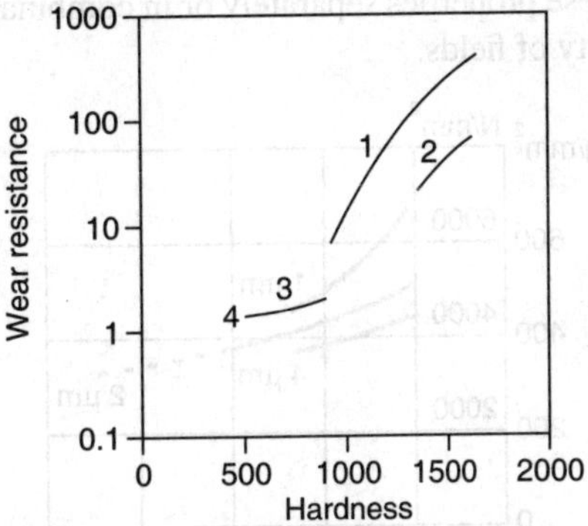

Fig. 3.4 Relative wear resistance as function of hardness.

Besides, the hardness of cemented carbides is related to grain size, as shown in Fig. 3.5.

1. Straight carbides (WC–Co).
2. Alloyed carbides (TiC content results reduced Crater wear while Tantalum carbide controls flank wear (E.W. Trent)].
3. Case hardened alloy steel.
4. Hardened and tempered steel.

Compressive strength: Next to wear resistance, high compressive strength is the most outstanding property of cemented carbides.

The compressive strength of cemented carbides is higher than virtually all melted and cast (or forged) metals and alloys. The carbide grades which exhibit the greatest wear resistance also possess the highest compressive strength—up to $700/mm^2$ (see Fig. 3.6). Also, the cobalt content of the carbides is a strong determinant of their compressive strengths (see Fig. 3.7).

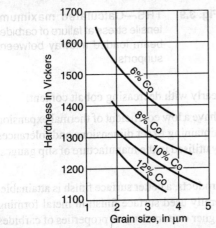

Fig. 3.5 Hardness as function of grain size at various cobalt contents of cemented carbides. (grain size is in microns)

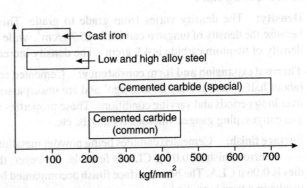

Fig. 3.6 Compressive strength of carbides in relation to other materials.

Rigidity: High rigidity is another unique property of cemented carbides. The modulus of elasticity is 2–3 times that of steel. In Fig. 3.8, steel is compared to cemented carbides. Using the same force, steel bends three times as much as cemented carbides.

These properties separately or in combination are exploited in a multitude of different applications and in a variety of fields.

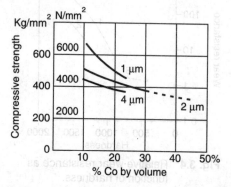

Fig. 3.7 Compressive strength of cemented carbides as a function of their cobalt content for different grain sizes.

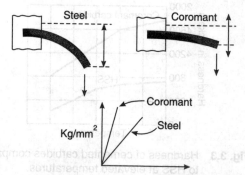

Fig. 3.8 Steel bends three times as much as cemented carbides under identical load.

Other properties

Toughness: Toughness is interpreted as the ability to withstand fracture. The toughness level of carbides is considered low when compared with most other metallic materials. The closest comparison, to an extent, is of steel hardened to a high level.

Transverse rupture strength: Transverse Rupture Strength (TRS) is an ultimate strength property which is used as a standard in the carbide industry. It is the calculated maximum tensile stress at failure of a carbide beam loaded midway between supports (see Fig. 3.9).

TRS is a combination of shear strength, compressive strength and tensile strength. The value of TRS is in the range of 100–300 kg/mm^2.

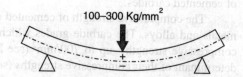

Fig. 3.9 TRS—Calculated maximum tensile stress at failure of carbide beam loaded midway between supports.

Density: The density varies from grade to grade. This is because the density of tungsten carbide is 15.7 g/cm^3, while the density of titanium carbide is 4.9 g/cm^3. The density increases linearly with decreasing cobalt content.

Thermal expansion and form consistency: Cemented carbides have a low coefficient of thermal expansion (about half of steel). They do not 'age' and are unsurpassed in maintaining exact dimensions and tolerances over long periods and varying conditions. These properties are fully utilised in the manufacture of slip gauges, gap gauges, plug gauges, micrometer anvils, etc.

Surface finish: Cemented carbides being powder metallurgical products, a better surface finish is attainable, and a mirror finish of 0.01 μ CLA is feasible. However, the generally used surface finish in metal forming dies is 0.06 μ CLA. The better surface finish accompanied by the higher wear resistance properties of carbides results in a very long die life.

Friction characteristics: Cemented carbides are an excellent bearing materials. Two different grades of carbide can work satisfactorily against each other even without conventional lubrication. Therefore, the pharmaceutical and dairy industries use carbide bearings in some applications.

The same property is utilised in the manufacture of seal rings for chemical pumps.

Thermal conductivity: The thermal conductivity of WC–Co carbides is approximately twice that of the unalloyed steels. The presence of TiC lowers the thermal conductivity.

Corrosion resistance: Corrosion resistance of carbides is dependent not only on their chemical composition, but also on the structure and micro-composition of the carbide grain and binder metal. Carbide is more resistant to corrosion than the binder metal and, therefore, cavities and passages may be formed upon corrosion. In these passages, new chemical compounds may develop and these may offer protection against further corrosion attacks. Resistance to corrosion and wear can, therefore, often be traced to interaction between the strength of the carbide skeleton and the chemical compounds formed in the carbide.

If resistance to corrosion is the only factor, then stainless steel may outperform cemented carbides. However, when both corrosion resistance and wear resistance properties are desired, certain grades of cemented carbides may be outstandingly superior.

Typical applications

Owing to increased mechanisation and the trend to automation in industry, more complicated and consequently more expensive equipment is employed for production. Breakdowns or inadequate performance of the equipment can be caused by the wearing out of vital parts. A critical examination of process equipment employed in different branches of industry will often reveal a justified demand for increased resistance to wear. By manufacturing the vital parts in cemented carbides, one can increase their service life, improve the performance function and impart overall working economy.

Some of the important applications are tooling for bar, tube and wire drawing; extrusion, cold heading, hot forging, cold rolling, hot rolling, swaging, deep drawing, powder pressing and blanking operations.

Wear resistant surfaces like work rest blades, lathe centres, seal rings, nozzles, measuring gauges, scribing tools, etc. are manufactured using cemented carbides.

The application list can be made a great deal longer. However, the examples given here illustrate the great demand for hard, wear resistant cemented carbides and show how their good qualities are turned to a good account in numerous fields.

3.3.6 Ceramic Tool Materials

The latest development in the metal cutting tools is the use of aluminium oxide, generally referred to as *ceramics*. However, the following cutting tool materials are also known as ceramics:

1. Silicon carbide
2. Boron carbide
3. Titanium carbide
4. Titanium boride.

The name *ceramic* has been derived from the fact that these materials are made by sintering the elements at extremely high temperatures, approaching those of pottery ceramics.

The materials are extremely brittle, but retain their hardness even at temperature above 1200°C. Being brittle like tungsten carbides, they are available in the form of tips which are clamped on to very rigid shanks.

The properties of ceramics are: compressive strength of 500 kgf/mm^2, hardness of 90–95 Ra, retention of hardness up to 1200°C, and bending strength of 45 kgf/mm^2. Ceramics poor thermal conductivity, but corrosion resistance to the strongest acid is outstanding. Ceramics has no problem of built-up edge.

Ceramic cutting tools have the advantages of higher cutting speeds, longer tool life, superior surface finish, lower coefficient of friction, and greater machining flexibility. However, owing to their extreme brittleness, they have not found much application as yet in our country. The ceramic material has a great advantage in its very high red hardness. With the development of high speed machine tools, its use in India is likely to increase substantially.

3.3.7 Cermets

Cermets may be described as cutting tool materials whose properties lie somewhere between tungsten carbide and ceramics. The bending strength of cermets is greater than that of ceramics, but less than that of tungsten carbide. On the other hand, its red hardness is less than that of ceramics, but greater than that of tungsten carbide.

The hard element in cermets is either aluminium oxide or beryllium oxide. Various metals such as tungsten, molybdenum, boron and titanium, in amounts up to about 10%, are used as binders.

3.3.8 Diamonds

Diamonds are the hardest of all materials (hardness 10 on Mho scale). Diamonds have low chemical activity, hardly any tendency for adhesion to metals, a low coefficient of friction, and a very high wear resistance. Diamonds have a very high red hardness of the order of 1500°C, and can be sharpened to a keen cutting edge. Thus, diamond appears to be an ideal cutting tool material.

3.3.9 Abrasives

Abrasive grains in various forms—loose, bonded into wheels and stone, and embedded in papers and cloths—find numerous applications in industry. They are mainly used for grinding harder materials and where a superior finish is desired on hardened or unhardened materials.

The abrasives commonly used may be either natural or artificial (manufactured). Natural abrasives include corundum, emery, quartz, garnet and diamond. Manufactured abrasives include aluminium oxide, silicon carbide and boron carbide. The aluminium oxide abrasives are used for grinding all high tensile materials, whereas silicon carbide abrasives are more suitable for low tensile materials and non-ferrous metals.

3.4 CHARACTERISTICS AND USES OF DIAMOND TOOLS

3.4.1 General

The diamond has special characteristics which adapt it to industrial processes of various types. Of interest here is the application of the stone to single-point cutting operations, such as boring, turning, chamfering, grooving and facing. Usage has increased, in this respect, generally with the development of operational data, and the stimulus of war time exigencies has not only broadened the industry's acceptance of this cutting tool material but also produced savings in post-war competition.

The fundamental consideration of any "cost per piece" analysis includes the set-up and tool regrinding time as well as the surface finish and size control requirements. Within its scope, the diamond produces more work of superior finish and to close tolerances with less set-up and tool regrinding time than any other cutting medium known. However, in the application of diamond cutting tools, the high first cost and re-lapping cost, together with the need of special machine tool characteristics and operating skills, must be carefully balanced against the cited advantage of diamond.

3.4.2 Classification

Diamonds with definite planes of cleavage have great refractive power, and when of suitable colour and comparatively free from flaws and inclusions, are valuable for decorative purposes (gems). Diamonds which are more likely to be used for industrial purposes are those without cleavage planes or are crystalline diamonds with the aforementioned imperfections. For industrial purposes, diamonds can be divided into two groups as follows:

Bort

Crystalline diamonds or bort have distinct cleavage planes and are most commonly found as octahedra (eight-sided units) in the cubic system. Octahedral bort has six hard corners and is inherently adaptable to single-point tools.

Carbon and ballas

The crystals of these types are closely distributed in all directions, and so offer greater resistance against breaking forces when compared to stones having a regular crystalline structure. Although carbons can be lapped easily, they are now not commonly used for shaped-diamond tools because they vary greatly in hardness and are so porous that a keen cutting edge is hard to establish. Ballas diamonds can be lapped only with difficulty and are infrequently used in single-point machining operations.

3.4.3 Origin

The diamond industry may also classify a stone by using a geographical system. In this respect, African bort is highly regarded for the making of shaped-diamond tools. Brazilian bort, though harder in some cases than the African bort, is more susceptible to chipping and thus not so desirable for matching operations. Congo stones, though crystalline in structure, are of a lower grade and rarely used for shape-cutting tools.

3.4.4 Special Characteristics

A diamond is hard, incompressible, of large grain structure, conducts heat readily, and has a low coefficient of friction. These properties are among those required in a cutting tool material if long tool life and good work finish are to be realised. Note that these properties permit the establishment of a durable and keen cutting edge practically free from burrs. These properties with the permissible high cutting speeds help minimise ragged finish and dimensional distortion, so undesirable in finishing operations.

3.4.5 Relation to Metal Cutting Process

What are the peculiar physical properties of diamond that promote its ability to cut metal under certain conditions, is still a matter for scientific explanation. There is unquestionably a relation between that and the known facts of what happens in the general process of metal cutting.

With ductile materials, work finish appears to be partially a function of elimination of the built-up edge. Conditions which help eliminate this built-up edge are:

(i) Small chip thickness
(ii) Keen and smooth cutting edge
(iii) High cutting speed
(iv) Maximum temperature of material at cutting edge
(v) Minimum opposition to chip flow over tool face
    (a) by providing high polish on tool face
    (b) by use of tool material having inherent low coefficient of friction
    (c) by use of positive rake on tool face.

Since high cutting speed is desirable to produce finish under conditions outlined above, and since the main stress and wear on a tool face approach the direct cutting edge under high speed and small chip conditions, it becomes clear that a tool can quickly become blunted at its most sensitive point unless it possesses great strength and wear resistance.

It is obvious then that diamond possesses those characteristics which contribute to the production of high quality and size control over a great length of cutter travel. By the same token, it becomes clear that these characteristics satisfy the requirements of a finishing tool, while they do not especially help to adapt it to roughing cuts.

3.4.6 Machining Characteristics

The machining characteristics of diamonds with reference to materials to be machined and the machining operations are as follows:

Machine materials

Many materials can be cut successfully with diamond tools. Notable in this list are aluminium, magnesium, babbitt, brass, copper, silver, bronze, lead-bronze, soft and hard rubber, and plastics. Noticeably absent in this list are the ferrous materials, though the technical literature has dealt with this latter application to some extent.

Machining operations

Turning, boring and facing operations are probably the most widely used applications of diamond to single-point tool machining. This is due partly to the existence of acceptable machine tools for that work and partly due to the difficulty in shaping diamonds to forms other than the simple ones.

Tool forms

A typical boring or turning tool is shown in Fig. 3.10. This tool has the advantage of conforming to the natural cleavage planes of certain types of diamonds, minimising the time for initial tool shaping, and increasing the resistance of the stone to cutting pressures. Other advantages are listed below in the discussion under *Adjustment*.

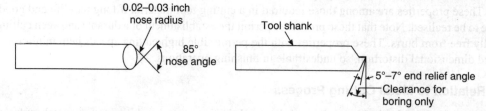

Fig. 3.10 A typical diamond boring or turning tool used to machine various non-ferrous materials. (Dimensions shown will vary depending on material, finish required and dimensional tolerances of workpiece.)

Specialised types of diamond cutting edges have been evolved, and though commonly used in Europe, are rarely found in American industry. For instance, tools with large circular or multi-facet edges can be adjusted in the tool post to offer new and sharp portions of the edge to the workpiece, as the old edge becomes dull through use. Another design incorporates the zero degree or a very small end cutting edge angle to produce a flat, as often used in sintered carbide finishing tools on cast iron to produce smooth surfaces with no feed marks. These specialised types and some similar types also have certain disadvantages. These disadvantages frequently include the need for larger and thus more costly stones, greater difficulty in holding or setting the stone into the tool shank, and an occasional need for auxiliary mechanical or optional means of adjusting the cutting-edge to the workpiece. Actually, the use of the tool with a small end cutting edge angle might be undesirable, as in the case of boring silver or lead-bronze borings.

Tool angles

The commonly used tool angles are shown in Fig. 3.10. Note that the tool has zero degree back and side rake. While it can be shown that for ductile materials a positive back rake or side rake on any tool material is conducive to smooth chip flow, and although these rake angles satisfy one of the tool requirements for good finish (see Section 3.4.5 entitled *Relation to Metal Cutting Process*), they at the same time reduce the strength of the diamond cutting edge beyond the point of usefulness on most machinable materials.

Sometimes the tool shank is adjusted in the tool holding block, so that the cutting edge presents a negative side rake to the workpiece for the purpose of chip control. For plain turning and boring, a nose angle of 85° is desirable as are end relief angles of 5° to 7°. They may be slightly greater for boring, depending upon the size of the hole.

Nose radii values of 0.02 to 0.03 inch should be used. The size of the nose radius with its appropriate feed is of critical importance in a diamond tool application, since it is related directly to micro-inch values of finish and tool life, and to the types of spindle bearings in the machine tool on which the diamond cutter is used.

Adjustment

Due to its shape, the diamond tool shown in Fig. 3.10 can be adjusted readily to the workpiece by means of a simple tool post or tool holder. For 'through boring' or turning cylindrical workpieces, the axis of the tool can be set at right angles to the axis of the workpiece, thus permitting a desirably simple relation of the diamond relief angles to the tool shank. Considering that in the initial shaping operations, the stone is lapped and polished in a lead pad separate from the tool shank and then set by hand in the shank, the advantages of the right angle relationship should be obvious. However, when step boring or turning to a shoulder is necessary, the tool shank must be set at some practical angle to the axis of the workpiece. Under these conditions, special precautions are necessary to maintain proper working relief angles at the cutting edge if unusual wear on the edge is to be avoided. This is accomplished by specifying to the tool manufacturer, the relief angles in a plane perpendicular to the axis of the workpiece and the setting angle of the tool shank with the workpiece.

Normally, the diamond cutting edge should be set at the centre, although a slight adjustment above the centre may be permitted for turning operations. This setting will create the effect of a slight top rake without reducing the strength of the cutting edge. It also reduces the relief angle, which may be detrimental. In no case should the cutting edge be allowed to drop below the centre, as this may cause chipping of the stone.

3.5 COMPARISON OF CUTTING TOOL MATERIALS

Table 3.1 depicts the general comparison between the various types of cutting tool materials.

3.6 NON-FERROUS CUTTING TOOL MATERIALS

There are many non-ferrous materials which are used in tool, jig and fixture design. Such materials are aluminium, magnesium, zinc base alloys, bismuth alloys, lead base alloys and beryllium. These materials may be used where the lightness of weight is a factor, e.g. for temporary dies, limited production runs, or for some special purposes.

Aluminium-bronze and duralumin are two of the more widely used aluminium alloys. The former are lightweight and have strengths which permit their use for lightweight fixture bodies.

Magnesium is another of the lightweight (two-thirds of aluminium's weight) materials which may be used in large fixtures. Like aluminium, it is machinable and may be welded. It is corrosion resistant in most atmospheres, but needs surface protection when in saltwater atmospheres. It is also used as a facing material on forming blocks for short-run productions.

Table 3.1 Types of cutting tool materials

Sl. no.	Tool material	Typical composition	Relative material cost	Cost of grinding	Hardness	Hot hardness temperature	Cutting speed of grey cast iron	Toughness	Wear resistance	Typical uses
1.	Plain carbon steels	C 1.2% Balance Fe	Low	Low	R_c 55–60	200°C	20 m/min	Good	Poor	Form tools
2.	Low alloy steels	Si 0.25% Cr 0.25% Va 0.25% Mn 1.2% Balance Fe	Slightly higher	Low	R_c 60–64	300°C	25 m/min	Good	Poor	Form tools
3.	High speed steels	C 0.75–1.5% W 18% Cr 3.15–4.2% Va 1.78–2% Balance Fe	High	High	R_c 62–65	500°C	35 m/min	Fair	Fair	Wide use everywhere
4.	Cast non-ferrous alloys	Mn 0.3% Cr 20% W 14% C 1.7% Cv 53%	Higher	High	R_c 62–65	600°C	70 m/min	Fair	Fair	Single-point cutting tools
5.	Cemented carbides	W and other carbides such as Ta, Li, Co used as binder	Very high	Very high	R_c 87–92	800°C	150 m/min	Poor	Very good	Wide use (particularly for cast iron)
6.	Ceramics	Al₂O₃ 95% Co 5%	Very high	Not known	R_c 90–95	1200°C	300–400 m/min	Very poor	Very good	Use will increase
7.	Cermets	CrC 89% Ni 11%	Not known	Not known	R_c 88–90	1000°C (Exactly not	250 m/min (Enough data not available)	Better than ceramics	Very good	May prove very useful
8.	Diamonds	C 100% (six atoms equivalently placed on a plane forming intertwined layers)	Prohibitive	Not known	10 (Mho's scale)	1500°C	500 m/min (Enough data not available)	Poorest	Excellent	Hardly any

Zinc-based materials may be cast into shapes quickly for the purpose of use as short-run punches and dies either for short-run productions or for experimental short runs. When punching aluminium sheet parts, these cast tools require very little further machining. Since the material is comparatively cheap and easy to cast, it is an ideal material for short, quick production runs.

One of the better-known zinc-based die materials is *kirksite*. It is cast into complicated shapes, polished and used in blanking or drawing dies. It is the sole material in a die, as the hard component in combination with the matting component made from a softer material or as a part of a die in combination with the steel components. Another advantage of kirksite is the fact that the material may be re-melted and used again.

Bismuth alloys are low-melting alloys which expand upon solidification. This is especially true of the high bismuth alloys. These characteristics of bismuth make it possible to use this material for duplicating mould configurations which would otherwise require many hours to reproduce. The material is suitable for making tracing or engraving patterns that have intricate shapes and for making forms used in stretch forming. Bismuth has been used effectively as another component in punches and dies.

Lead-based alloys, when one of the alloys is antimony, have been used with kirksite dies.

Beryllium, when alloyed with copper, produces a material which has characteristics similar to aluminium bronze. When only 2% beryllium is alloyed with copper and heat-treated, it develops properties which make it suitable for chisels and other cutting tools.

3.7 NON-METALLIC CUTTING TOOL MATERIALS

3.7.1 Plastics

The thermosetting plastics such as phenolic, epoxy, urethane and polyester plastics are used widely as tooling materials. Where the base plastic is not completely suitable as a tooling material, it may be impregnated with metallic or other abrasive powders. In some instances, steel wear plates are inserted in the plastic. They are currently used in many operations such as drawing and forming dies, stretching dies, drill jigs, assembly, machining, and inspection of fixtures. These materials have advantages over many of the other materials in that they are resistant to moisture, chemicals and temperature. They are generally easy to machine, thereby, saving labour costs. Related to this are the advantages of ease of machine repairs, rework and design changes.

3.7.2 Rubber

A rubber pad confined in a container is used in several processes as the female die. The desired form is contained on the solid punch. The formed punch pushes the metal into the rubber which takes on the shape of the punch and in so doing causes the metal to be formed. It should be noted that if the punch has sharp edges, and the rubber pad has the appropriate hardness, the metal workpiece will be cut or blanked. Operations such as forming, blanking, bulging and drawing may be done with rubber as one of the components of the die set.

3.7.3 Wood

Densified wood is made by impregnating wood with phenolic resin and compressing it to about half its original thickness. This wood may then be used in dies for drawing or forming soft materials.

3.7.4 Hard Board

Hard board materials such as masonite, made from compressed wood material, is used to form or draw thin gauge metals. It may also be used as the solid form with rubber-forming materials or with stretching materials. When used for punching or blanking, steel inserts are used as the cutting edges. Special high-tensile-strength masonite of desired thickness may be purchased.

REVIEW QUESTIONS

1. Briefly explain the essential characteristics of tool materials.
2. What are the different types of tool materials? Give details of at least four most commonly used tool materials.
3. What is meant by 18 : 4 : 1 high speed steel? What do the numbers 18, 4, and 1 indicate?
4. Cutting tool materials can be classified as forged/cast, sintered and synthetic. Give two examples in each category.
5. What do you understand by whisker-reinforced tool materials?
6. Briefly explain the manufacturing process of cement carbide tools.
7. What are the outstanding properties of cement carbide tools, which are not present in other tools? Give details.
8. Tungsten carbide is a hard metal but possesses less toughness. How can its toughness be improved? What is the difference between straight tungsten carbides, double carbides and triple carbides?
9. What is the purpose of depositing Golden Titanium Nitride on carbide-tipped tools?
10. What are ceramic tools? List them and also give their characteristics and uses.
11. What is the area of application of stellite as a tool material? Why is stellite called cast non-ferrous alloy?
12. HSS tools should be used for machining at a speed of around
    (i) 20 m/min
    (ii) 40 m/min
    (iii) 60 m/min
    (iv) 80 m/min.
13. The composition of HSS tools can be
    (i) 18% W, 4% Cr, 1% V
    (ii) 4% W, 18% Cr, 1% V
    (iii) 1% W, 4% Cr, 18% V
    (iv) 10% W, 10% Cr, 10% V.
14. To increase the cutting efficiency of HSS specially at high temperatures, it is desirable to add
    (i) vanadium
    (ii) cobalt
    (iii) chromium
    (iv) nickel.
15. HSS tools retain cutting ability up to the temperature of
    (i) 200°C
    (ii) 400°C
    (iii) 600°C
    (iv) 800°C.
16. In carbon steel tools, the cutting ability remains up to around
    (i) 250°C
    (ii) 450°C
    (iii) 600°C
    (iv) 800°C.
17. Carbon steel tool machines should work at speeds around
    (i) 80 m/min
    (ii) 60 m/min
    (iii) 40 m/min
    (iv) 25 m/min.
18. Super HSS tools, in addition to tungsten, chromium and vanadium, contain
    (i) nickel
    (ii) cobalt
    (iii) manganese
    (iv) sulphur.
19. Vanadium in HSS tools is added to increase
    (i) hardness
    (ii) corrosion resistance
    (iii) wear resistance
    (iv) cutting ability at very high temperatures.
20. HSS can be used to make
    (i) drills, reamers and broaches
    (ii) milling cutters
    (iii) gear cutters
    (iv) all of these.

21. Cast alloys known as stellites are
    - (i) non-ferrous alloys
    - (ii) high carbon alloys
    - (iii) high tungsten alloys
    - (iv) high chromium alloys.
22. Stellite is harder than HSS at
    - (i) all temperatures
    - (ii) over 1000°F
    - (iii) room temperature only.
23. Hot hardness of stellite compared to high speed is
    - (i) lower
    - (ii) equal
    - (iii) higher
    - (iv) incomparable.
24. Stellite is not widely used for making cutting tools because it
    - (i) is very brittle
    - (ii) is very ductile
    - (iii) has very hot hardness.
25. The maximum permissible cutting speed for stellite is
    - (i) equal to that for HSS tools
    - (ii) twice that of HSS tools
    - (iii) half that of HSS tools
    - (iv) three times that of HSS tools.
26. The basic ingredient of most cemented carbide tools is
    - (i) iron carbide
    - (ii) chromium carbide
    - (iii) tungsten carbide
    - (iv) vanadium carbide.
27. The cemented carbide tools can retain cutting ability up to
    - (i) 200°C
    - (ii) 400°C
    - (iii) 600°C
    - (iv) 900°C.
28. The optimum cutting speed of cemented carbide tools for machining HSS is around
    - (i) 60 m/min
    - (ii) 80 m/min
    - (iii) 100 m/min
    - (iv) 150 m/min.
29. The carbide tips on steel shanks are
    - (i) brazed
    - (ii) soldered
    - (iii) welded.
30. Throwaway inserts compared to brazed tip tools are
    - (i) costlier
    - (ii) cheaper
    - (iii) equally costly.
31. The main constituent of ceramic tools is
    - (i) tungsten oxide
    - (ii) magnesium oxide
    - (iii) aluminium oxide
    - (iv) silicon oxide.
32. Cemented carbide tips are manufactured by
    - (i) powder metallurgy technique
    - (ii) casting
    - (iii) machining
    - (iv) brazing.
33. Ceramic tools are given
    - (i) zero rake
    - (ii) negative rake
    - (iii) positive rake.
34. The negative back rake on ceramic tools is in the range of
    - (i) 20–25°
    - (ii) 10–15°
    - (iii) 5–7°
    - (iv) 0–5°.
35. Negative rake on ceramic tools is given to
    - (i) strengthen the cutting edge
    - (ii) make machining easy
    - (iii) reduce cutting forces
    - (iv) all of these.

36. To make cutting easy on tools, the rake angle should be
    (i) negative
    (ii) positive
    (iii) zero.
37. Ceramic tools can maintain high hot hardness at temperatures over
    (i) 260°C
    (ii) 540°C
    (iii) 1500°C
    (iv) 980°C.
38. The cutting speed with ceramic tools could be as high as
    (i) 100 m/min
    (ii) 200 m/min
    (iii) 300 m/min
    (iv) 500 m/min.
39. Cermet tools compared to ceramic tools have
    (i) lower brittleness
    (ii) equal brittleness
    (iii) higher brittleness.
40. The cermet tools are
    (i) like sterlite tools
    (ii) like ceramic metal tools
    (iii) like cemented carbide tools
    (iv) like none of these.
41. Cermet tools contain mainly
    (i) tungsten, chromium and vanadium
    (ii) iron, carbon and cobalt
    (iii) aluminium oxide, iron, chromium and titanium
    (iv) none of these.
42. A diamond tool's brittleness is
    (i) lower than that of ceramic tools
    (ii) lower than that of cermet tools
    (iii) lower than that of stellite tools
    (iv) higher than that of ceramic, cermet and stellite tools.
43. Cutting speeds with diamond tools can be as high as
    (i) 500 m/min
    (ii) 1000 m/min
    (iii) 2500 m/min
    (iv) 5000 m/min.
44. The chance of built-up edge formation on diamond tools is
    (i) nil
    (ii) little
    (iii) nigh.

CHAPTER 4

Design of Metal Cutting Tools

4.1 INTRODUCTION

The principle of metal removal for varied jobs is almost the same irrespective of the type of tool and machine used. However, the general characteristics of the cutting process affect the tool design; hence the shape and size of all the parameters of a cutting tool must be determined either graphically or analytically.

4.2 ELEMENTS OF CUTTING TOOL DESIGN

The design of a cutting tool can be divided into the following four main categories:

Design for cutting forces: On the basis of the work material properties, the tool forces that might be developed during machining with a tool having an optimum tool geometry are estimated theoretically. After this, a most suitable material is selected for an effective and economic material removal rate. The selection of a set of tool angles is a tricky job as there are many parameters to be taken into account. However, the design can be easily finalised by an experienced engineer.

Design of job profile: Work material and its shape and accuracy determine the tool angles, tolerances and tool material. The shape of the tool has to be such that it can be easily reproduced after it is worn out.

Design of tool holding part: In cases where the tool holding portion is separate from the cutting element, the strength and rigidity of the cutting element holder is to be separately calculated, and a suitable design of such a holder is to be obtained.

Preparing the drawings: A working drawing of the cutting element and the holder portion providing full details has to be prepared, showing how to proceed for manufacturing. The drawing should indicate all the necessary information regarding tool angles, shape, size and tolerances of the holder.

The design of a cutting tool depends on the cutting mechanics. Hence it is necessary to analyse the process of cutting. This has already been discussed in the previous chapters. The cutting tool removes material from the work, only when there exists a relative motion between the tool and the work. Furthermore, proper angle or angles must be maintained between the work and the cutting edge of the tool to remove material at an optimum rate.

It is to be understood that a knife starts cutting the pencil only when a proper angle between the cylindrical surface of the pencil and the cutting edge of the knife is maintained. Not only this, the angle has to be adjusted during every cut for efficient cutting.

Therefore the tool designer should know the kinematics involved in the cutting process of a given machining operation, in order to determine the exact values of the tool angles, as different kinds of cutting tools have different schemes of load distribution.

4.3 FORCES ACTING ON CUTTING TOOLS

A force of magnitude F_c is necessary to be employed in the direction of the cutting velocity of the tool, so that the cutting takes place. This force should overcome the resistance of the metal for penetration and removal of the material in the form of chips.

During the cutting operation, both elastic and plastic deformations occur for a layer of metal to be cut. Hence both type of forces F_{ce} (elastic) and F_{cp} (plastic) are exerted on the flank and rake faces occur. There is a system of friction forces, too, acting on both the faces of the tool.

To overcome the resultant effect of all these forces on the tool, tending to move the tool away from its position, the tool is clamped in the tool holder.

If it were possible to accurately calculate all the forces acting on the tool for the layer of work material being removed, then it would have been quite simple to find out all the forces, i.e. F_c (vertical cutting force), F_h (feed force), and F_t (thrust force) acting on the tool in mutually perpendicular directions. But it is as much difficult to calculate these forces as it is to measure them.

Because of these difficulties encountered in calculating and measuring these forces, the method proposed here in this chapter for calculating these forces is not yet widely accepted by designers and research workers. However, it may give a good approximation for the purpose of design.

4.3.1 Force of Resistance to Cutting

The resultant of all the forces acting on the tool is not employed in a major way for practical calculations, but the components of this force resolved in three mutually perpendicular directions as shown in Fig. 4.1 are used. Their magnitudes can be readily measured by using a lathe tool dynamometer.

The resultant force R can then be found as follows:

$$R = \sqrt{F_c^2 + F_h^2 + F_t^2}$$

In view of the complexity that may arise during force measurement, it may be quite convenient for the purpose of design to estimate the magnitude of forces by empirical relations. These empirical relations as hereinafter used, are based on experiments and on logical decisions.

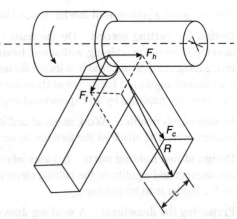

Fig. 4.1 Resultant force R to F_c, F_h and F_t.

The three forces, F_c, F_h and F_t, depend on five variables while turning, viz.,

  (i) Tool geometry
  (ii) Depth of cut
  (iii) Feed rate
  (iv) Work material and
  (v) Cutting conditions.

Considering all these factors, the force F_c can be expressed as follows:

$$F_c = C_p d^m f^n$$

where C_p is a constant dependent on the physical properties of the work material, d is the depth of cut in mm, and f is the feed rate in mm/rev.

Sometimes the above expression for F_c is multiplied by a correlation coefficient which accounts for the actual working conditions and tool angles. This factor varies from 0.9 to 1.00.

Table 4.1 gives the values of the constant C_p and exponents m and n.

Table 4.1   Constant C_p and exponents m and n

Work material	f_t (Ultimate strength of the material) (in kg/mm^2)	Hardness (in BHN)	C_p	m	n	Method of working
Steel	75	215	225	1.0	0.75	Turning
Steel	75	215	264	1.00	1.00	Facing and cutting off
C.I. (Grey)	75	190	100	1.00	0.75	Turning and boring
C.I. (Grey)	75	111	135	1.00	1.00	Parting and facing

4.3.2 Force of Drilling

The drill force can be approximately estimated by the expression given as follows:

$$F_{th} = KDf^n$$

where F_{th} is the drill thrust in kg
  K is a constant dependent on the work material and other parameters of the drill
  D is the drill diameter in mm (hole that can be drilled)
  f is the feed rate in mm/rev.

$$F_{th} = 0.60 \times Df^{0.8} \text{ kg} \quad \text{for steel}$$

and
$$F_{th} = 85 \times Df^{0.7} \text{ kg} \quad \text{for C.I.}$$

The turning moment T in drilling can be obtained by

$$T = K_t D^m F^n \text{ kg mm}$$

where
  $K_t = 38.8$ for steel
  $= 23.3$ for C.I.
  $m = 1.9$
  $n = 0.8$

∴
$$T = 38.8\, D^{1.9} f^{0.8} \text{ kg mm} \quad \text{for steel}$$

and
$$T = 23.3\, D^{1.9} f^{0.8} \text{ kg mm} \quad \text{for C.I.}$$

4.3.3 Design of Single-point Cutting Tool

The tool holder of a single-point tool is most often made rectangular or square in section. However, sometimes it is made round in section too. The cross-sectional view of a tool holder as shown in Fig. 4.2 is quite often used. Thus, rectangular cross-sections with varying h to b ratios are used. Mostly, the ratio of h to b varies from 1.125 to 1.6, where the magnitude of b varies from 10 mm to 40 mm. The ratio h/b equal to 1.6 is recommended for tools to be used in finishing and semi-finishing operations. In contrast, h/b equal to 1.25 is generally recommended for roughing operations.

A square cross-section of a tool shank is used for boring, capstan and current lathes, and screw machine tools. In situations where the distance from the tool base to the line of centres of the machine tool is not sufficient enough to accommodate the height of a rectangular tool, a square cross-section for the tool holders is used.

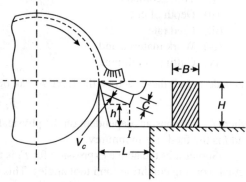

Fig. 4.2 Cross-sectional view of tool holder.

Sometimes, round tool holders are also used, e.g. in boring and thread cutting operations. This is done to adjust the tool in any direction in the tool holder.

Some of the common sizes of the shank are as shown in Table 4.2.

Table 4.2 Shank sizes

Sr. No.	B (in mm)	H (in mm)
1	10	15
2	12	15
3	15	20
4	20	25
5	25	30
6	25	40
7	30	40
8	40	50

To determine the cross-section of the shank, it is necessary to know the cutting forces. From these, the bending moment on the section due to cutting forces can be compared with the resisting moment of the tool.

Like in the case of a single-point turning tool, the three components of forces on the tool acting in mutually perpendicular directions at a point are the cutting force F_c and the feed forces F_h and F_t. Normally, 95% of the energy in metal cutting operations is derived from the cutting force component F_c. The cutting the force depends mostly upon the material to be cut and the cross-sectional area of the cut.

Therefore, as stated above, the maximum moment on the cutting tool section is due to the cutting component F_c of the force system, which can be estimated by the empirical relationship as explained earlier. This moment on the tool can be found out as explained below.

$$M_c = F_c \times l = Z \times f_b$$

where M_c is the bending moment due to cutting
l is the tool overhang in cm

f_b is the permissible bending stress for the shank material in kg/cm^2.
Z is the sectional modulus of the shank of the tool.

$$Z = \frac{bh^2}{6} \text{ cm}^3$$

$$F_c \times l = \frac{bh^2}{6} f_b$$

or

$$bh^2 = \frac{6 F_c l}{f_b}$$

Assuming

$$\frac{h}{b} = 1.6$$

i.e.

$$h = 1.6b$$

$$b \times (1.6b)^2 = \frac{6 F_c l}{f_b}$$

or

$$b = \sqrt[3]{\frac{6 F_c l}{f_b \times (1.6)^2}}$$

$$= \sqrt[3]{\frac{6}{2.56} \frac{F_c l}{f_b}} \text{ cm}$$

For square shanks, $h = b$

Therefore,

$$b = \sqrt[3]{\frac{6 F_c l}{f_b}}$$

For round shanks, the section modulus is changed and we have

$$F_c \cdot l = \frac{\pi d^3}{32} \cdot f_b$$

where $\frac{\pi d^3}{32}$ is the section modulus.

Therefore,

$$d = \sqrt[3]{\frac{32 \cdot F_c \cdot l}{\pi f_b}} \text{ cm}$$

The permissible bending stress of the shank of the tool varies with its material. This is shown in Table 4.3.

Table 4.3 Permissible bending stress versus shank material

Shank material	Permissible bending stress, f_b (kg/mm^2)
Hardened structural steel	20
Hardened steel	25
Carbon steel (not heat treated)	30
Carbon steel (heat treated)	40

In spite of the significant role of the cutting force F_c in calculating the power consumption, which amounts to a magnitude of 90% of the total energy expended, the effect of the other two force components viz. F_h and F_t acting on the tool cannot be ignored while determining the energy requirements and stresses on the tool shank. Thus, F_h and F_t are taken into consideration together with the cutting force F_c for finding out the bending stresses on the shank; the combined bending stresses are higher (in comparison to the bending stresses in plane bending due to F_c) to an extent of 100%. Such stresses are dependent on the plan approach angle (or the side cutting edge angle) and other variables of the cutting tool.

It is essential to check the above calculations for the rigidity of the tool shank. The maximum permissible load per unit deflection can be determined from the given formula for the tool. This varies from 100 mm to 500 mm. It is generally decided after considering (i) the size of the tool point (ii) the tool overhang from the tool holder (iii) the dimensions of the tool holder (iv) the number of clamping screws (at least two screws should be employed) and (v) the distance between these screws.

4.4 BRAZED TOOL SEATS

In brazed tools, as shown in Fig. 4.3, the seat or the recess in the holder for the tip may be either open or closed.

The open seat is easy to make and is used for mostly all types of single-point tools. Semi-closed seats are used to accommodate tips having one corner radius closed, and slot type seats are for small tips and to ensure a more reliable joint between the tip and the shank.

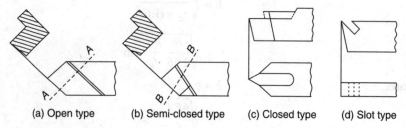

(a) Open type   (b) Semi-closed type   (c) Closed type   (d) Slot type

Fig. 4.3 Various types of seats in tool holder for tool tip mounting.

4.5 CUT-OFF TOOLS

Cut-off tools as shown in Fig. 4.4 are standard carbide-tipped cut-off tools having a cutting edge of length approximately $0.6D^{0.5}$, where D is the workpiece diameter in mm.

$$F_c = \frac{3 \times \delta \times E \times l}{l^3}$$

where δ is the permissible deflection of the tool (in cm; 0.01 cm for rough turning, 0.005 cm for finish turning);

E is Young's modulus of the tool shank or holder material. Generally, for carbon structural steel, $E = 20{,}000$ to $22{,}000$ kg/cm^2; l is the moment of the shank cross-section; for rectangular cross-section, $l = BH^3/12$ and for round cross-section, $l = 0.05d^4$, where d is the diameter of the tool holder.

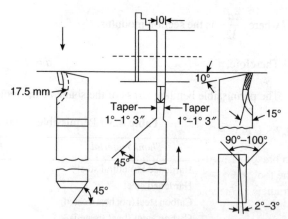

Fig. 4.4 Cut-off tool design.

While designing a single-point cutting tool for a particular machine tool, the height of the tool tip or shank should be checked from the line of centres of the machine to the surface of the tool.

Generally, tool overhang is heavy [$l = 1$ to $1.5\,d$]. Furthermore, design and checking should be carried out for the weakest section in the given construction of the tool. Other than the cross-section ($b \times h$), the overall length is also a design factor that depends on the size of the point and the size of the stock to be cut off.

4.6 BORING TOOLS

Boring tools should possess high vibration absorption properties. Generally, the cutting edge is located on the neutral surface of the tool shank. These tools generally operate with a large overhang from the tool holder. This fact does not allow a shank type boring tool to take a heavy cut. Therefore, deep holes are commonly bored by bar type tools, called *bits*, held in a bar.

Boring tools are used in turret lathes and screw machines and have a round cross-section held in special tool holders. A device for minor change in size is shown in Fig. 4.5. For an adjustment of the tool (2) for the required size 'x', screw (3) is released and the adjusting screw (4) is taken out of the bar (1). The locking screw (5) is loosened and the adjusting screw (4) is screwed in or out to obtain the dimension 'x'. After this, the locking screw (5) is replaced in its position for final use. The two types of tool holders are shown in Fig. 4.6.

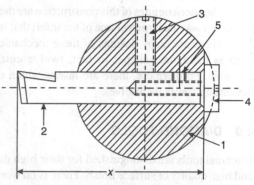

Fig. 4.5 Tool adjustment for boring bar tool.

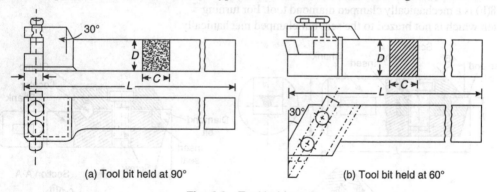

(a) Tool bit held at 90°   (b) Tool bit held at 60°

Fig. 4.6 Tool holders.

4.7 BRAZED TIPPED TOOLS

In brazed cemented carbide tipped tools, cracks are often found after some use or when the tool is brazed on the tool shank. These cracks are developed because of the brazing stress developed owing to a temperature gradient in the shank. This shortens the tool life.

Another shortcoming in smooth operation of brazed tools is the grinding of steps in tool shank or employing chip breakers and loss of tool shank steel while sharpening the tool. Finally, the loss of tool shank steel can also occur when the carbide is reduced to scrap; here the tool holder is also reduced to scrap.

4.8 MECHANICALLY HELD TIPPED TOOLS

The drawbacks of brazed tipped tools have led to the development of mechanically-held tipped tools as shown in Fig. 4.7. Here, the tool tip rests on a hardened steel shim, which is welded to a shank. The shim can be firmly secured mechanicaly. It has 60° serration at the right hand. The tip or insert is held to the shim by a screw, which fastens through a washer and a clamp, which works also as a chip breaker.

The shortcomings of this construction are the complexity of the design and the large volume of the insert that is thrown away.

There are many designs of these mechanically held tips, such as hexagonal shape; in fact, twelve cutting edges are available. Furthermore, there are inserts which can be held by the cutting forces themselves.

4.9 DIAMOND TOOLS

Diamond tools are distinguished for their high dimensional life and high quality of surface finish. There is no work hardening of this tool material; it is most widely used for precision turning and burnishing of non-ferrous and non-metallic materials.

In Fig. 4.8, two types of single-point diamond tools are shown. Figure 4.8(a) shows a brazed diamond tool for boring and Fig. 4.8(b) is a mechanically clamped diamond tool. For turning operation which is not brazed to the seat but clamped mechanically.

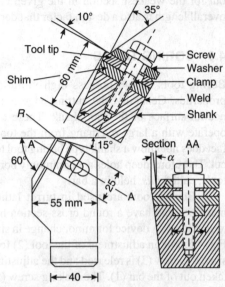

Fig. 4.7 Single-point cutting tool holder for throwaway tips.

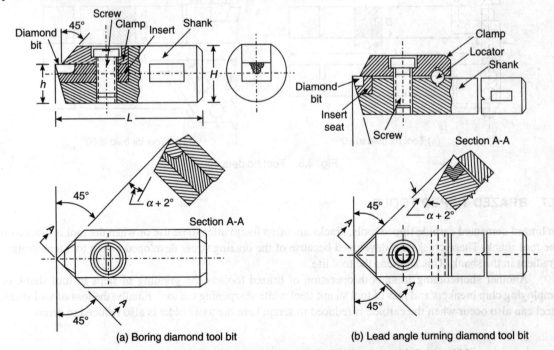

(a) Boring diamond tool bit

(b) Lead angle turning diamond tool bit

Fig. 4.8 Diamond tool holders having a mechanically clamped diamond bit.

The weight of diamonds used in single-point tools ranges from 0.5 to 0.8 carats (one carat equals 0.2 gram). The cutting edge of the diamond tool should be located in such a manner that the cutting forces must not pass along the cleavage planes of the diamond tool material.

4.9.1 Brazed Diamond Tool

In the manufacture of brazed diamond tools, a slot is milled in the tool bit. The brazed diamond is fitted in the slot and brazed with the silver metal. The brazing of these tools enables diamonds of small size to be used.

Increased temperature lowers the strength of diamonds; hence brazing should be limited to small sizes of diamonds. Brazed diamond tool bits are much smaller and create problems in redressing but there is saving of bit material.

4.9.2 Mechanically Clamped Diamond Tool (Refer Fig. 4.8)

Mechanically clamped diamond tools can be easily removed, reground and replaced for further operation.

Diamond tools are sharpened and lapped in special machines using cast iron discs charged with diamond powder and olive oil and at a speed of 30 to 40 metres per minute.

Table 4.4 shows the speeds, feeds and depths of cut for various diamond tools.

Table 4.4 Speeds, feeds and depths of cut for single-point diamond tools

Material to be machined	V (cutting speed) (m/min)	Feed (mm/rev)	Depth of cut (mm)
Aluminium	400–500	0.03–0.08	0.1–0.3
Aluminium alloys	600	0.02–0.04	0.05–0.1
Brass	400–500	0.02–0.07	0.03–0.06
Babbitt	400–500	0.02–0.05	0.05–0.15
Tin–Bronze	300–400	0.03–0.06	0.05–0.25
Lead–Bronze	800	0.02–0.04	0.025–0.05
Copper	350–400	0.02–0.1	0.01–0.4
Titanium	200	0.03–0.05	0.03–0.5
Magnesium and its alloys	800–1000	0.02–0.1	0.01–0.4
Plastics	100	0.0–0.03	0.05–0.15

4.10 FORM TOOLS AND THEIR DESIGN

A form tool can be defined as "a tool having a cutting edge of shape that produces the desired contour on the workpiece in a forming operation". The use of a form tool ensures high output, uniform contour and uniform dimensions. It is a tool adopted for mass production. Some of the common types of form tools and the various aspects involving them are shown in Figures 4.9, 4.10 and 4.11.

4.10.1 Circular and Flat Form Tool

Form tools (circular and flat) with the axis or mounting surface at an angle to the workpiece axis are seldom used, owing to their complex manufacture, and are only used where the shape of the workpiece contour is such that tools with a parallel axis cannot be employed.

Most form tools are made of high-speed steel, but cemented carbides are also extensively used.

A graphical construction is recommended for determining the outside diameter of a form tool as shown in Fig. 4.10 having a positive rake angle.

76 TEXTBOOK OF PRODUCTION ENGINEERING

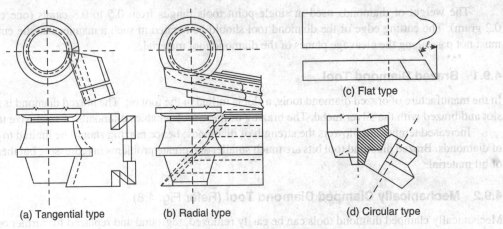

(a) Tangential type   (b) Radial type   (c) Flat type   (d) Circular type

Fig. 4.9 Various types of form tools.

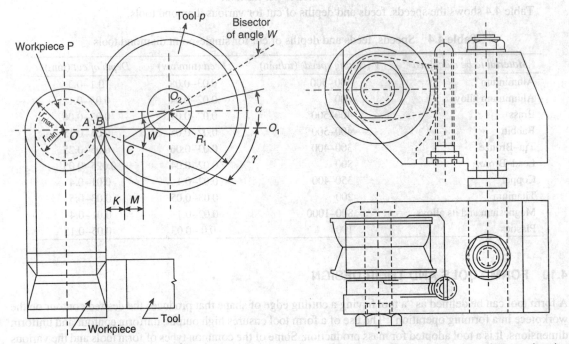

Fig. 4.10 Graphical method for determining the outside diameter of positive rake circular form tool.

Fig. 4.11 Method of clamping a circular form tool in tool holder for lathes and automatics.

About the axis O of the work, the following types of form tools can be drawn:
  (a) Circular form tool with zero rake angle
  (b) Circular form tool with positive rake angle

(c) Helical circular form tool
(d) Radially fed flat form tool
(e) Skewing form tool
(f) Circular form tool with swindled axis
(g) End form tool.

The graphical construction can be obtained as follows. Concentric circles with radii equal to the maximum and minimum radii of the contour to be turned are drawn. Through point A (refer to Fig. 4.10) at an angle γ, a line representing the trace of the plane containing the face is drawn. From the point A a second line at a relief angle α is drawn. At the distance K from the point of contact B, a line perpendicular to OO_1 is drawn. The distance K is the minimum distance that will permit chip disposal from the tool face. From the point of intersection C of the vertical line and the line of the tool face, a line bisecting the angle W is drawn. The point of intersection of this bisector and the line drawn at angle α is the point being sought O_2, the centre of the circular tool. Distance K is taken as between 3 and 12 mm, depending upon the amount of chip thickness and its volume. After getting O_2 as the tool centre, it is possible to draw a circle of radius R and then to determine all other dimensions graphically. For determining the diameter of the mounting hole, the thickness M can be taken as 6 mm to 10 mm.

The method of mounting a form tool on a tool holder is shown in Fig. 4.11.

If the tool is too small to be mounted in the tool holder, it is made in one piece, together with the holder or shank, from a solid piece of metal. When the shank is not very large, it is butt welded to another shank.

The profile of a form tool does not coincide with the profile of the workpiece. The profile determined from the graphical method is simple and straightforward but not very accurate. The analytical method enables the dimensions of the tool to be determined to a great degree of accuracy. But for a curvilinear profile, calculations are quite complex.

REVIEW QUESTIONS

1. Explain the elements of cutting tool design.
2. Discuss the forces acting on a cutting tool.
3. How will you design a single-point cutting tool?
4. What is a form tool? What are the various types of form tools and where are they used?
5. How are form tools designed?

CHAPTER 5

Theory of Metal Cutting

5.1 HISTORICAL DEVELOPMENT

Research on the mechanism of chip formation was first carried out by the French scientist Tresca, as far back as 1873. However, fundamental theories have been developed only during the past three or four decades. The compression of the chip, however, was recognised correctly by Tresca, who wrote that the material of the workpiece starts to flow over the tool face in an upward direction, shearing along an oblique plane as the planer tool advances. He measured the length of the chip and came to the conclusion that chip length is only between one-third and one half of the distance travelled by the tool.

Many other earlier scientists were interested in the problem of the shearing of metals under cut; among them were Thime (1877), Hausser (1892), Reauteaux, Tallner, and others. However, they did not arrive at any important conclusion. Cokers' research on plastic models provided that a shear plane extends from the point of the tool to the surface of the workpiece. Piispanen, Kronenberg, and Merchant contributed considerably to the development in metal shearing theories. New aspects of heat produced in metal cutting, microstructure of the metal, etc. have also been investigated.

5.2 METAL CUTTING DEFINED

The metal cutting process may be defined as removing a layer of metal from the blank to obtain a product of specified shape and size with specified accuracy and surface finish. The cutting process is carried out on metal cutting machine tools with the help of metal cutting tools.

5.3 CHARACTERISTICS OF METAL CUTTING

1. Metal is cut by way of removal of chips which are thicker than the depth of cut and are correspondingly short.
2. There is no flow on the material at right angles to the direction of chip flow.
3. Metal cutting involves shearing mechanism, a fact which is proved from the presence of new lines on the side and back of a chip.
4. Examination of flow lines on the surface of a chip reveals that chips are formed by block-wise slip of the metal. The front surface usually becomes smooth due to burnishing action.
5. There are three important areas in any cutting process. The first one is along the shear plane, the second is the interface between the chip and the tool face, and the third is the finished or machined surface and the material of the tool adjacent to that surface.
6. Generally, no crack is observed in front of the cutting tool point. Due to strain hardening, the hardness of the metal in the chip and the built-up edge, and near the finished surface, is usually greater than that for the metal.
7. Sometimes a built-up edge is formed at the tip of the tool and it significantly alters the cutting process. It deteriorates the surface finish and the rate of tool wear is increased.
8. A lot of heat is generated in the process of cutting due to the friction between the chip and the tool.

5.4 REPRESENTATION OF METAL CUTTING PROCESS

Figure 5.1 represents the elements of the mechanism of the metal cutting process. In any cutting process, the following are the main elements:

(i) Workpiece
(ii) Tool (including holding devices)
(iii) Chip
(iv) Cutting fluid.

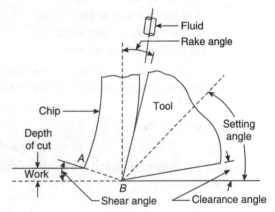

Fig. 5.1 Elements of metal cutting process.

In Fig. 5.1, the line *AB* is the dividing line between the work and the chip. The material above this line is deformed by an internal shearing process and comes out in the form of a chip. The metal below this line is underformed. The *shear plane* is the plane along the line *AB* and perpendicular to the plane of paper. The inclination of the plane *AB* with respect to the surface of the work is known as the *shear angle*. When the shear angle increases, the thickness of the chip as well as the plastic deformation of the chip get reduced.

5.5 ORTHOGONAL AND OBLIQUE CUTTING

The two basic methods of metal cutting are the orthogonal and the oblique, or the two-dimensional and three-dimensional methods of metal removal respectively.

The *orthogonal* or two-dimensional process of cutting takes place when the cutting edge is perpendicular to the cutting velocity vector, *V*, or to the direction of movement of the tool [see Fig. 5.2(a)]. If the cutting edge is inclined at an angle, which is either less or more than 90° to the direction of travel of the tool, the process is known as *oblique cutting*.

There are two important advantages of the three-dimensional cutting process. As we see, the depth of cut and feed are the same in both the cases, but the force which is used to shear the metal acts on a longer area in this case of oblique cutting as shown in Fig. 5.2(b).

In Fig. 5.2(b), we see that when the cutting edge of the tool is inclined at an angle i, the chip does not flow in a direction perpendicular to the movement of the tool but flows up along V_c inclined at an angle η_c. This is further illustrated in Fig. 5.2(c). Thus we can say that this process slightly deviates from the three-dimensional one.

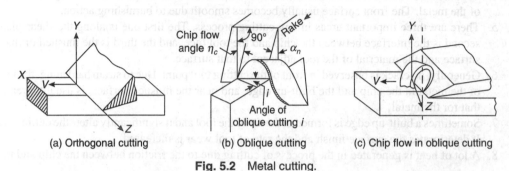

Fig. 5.2   Metal cutting.

From the geometry given in the figure,

α_n is the normal rake angle

η_c is the chip flow angle

α_e is the effective rake angle

i is the tool inclination angle,

The effective rake angle, α_e, determines the strain; the chip flow angle, η_c, is a dependent variable and a function of friction generally equal to the inclination i. From the geometry,

$$\sin \alpha_e = \sin \eta_c \sin i + \cos \eta_c \cos i \sin \alpha_n \quad (5.1)$$

If we assume $\eta_c = i$, then

$$\alpha_e = \sin^2 i + \cos^2 i \sin \alpha_n$$

Thus if we can determine α_n and i, we can estimate the value of α_e and then treat the cutting mechanism as two-dimensional.

5.6   DIFFERENCE BETWEEN ORTHOGONAL AND OBLIQUE CUTTING

Orthogonal cutting	Oblique cutting
Direction of cutting edge	
The cutting edge of the tool is perpendicular to the direction of tool travel.	The cutting edge is inclined at an angle to the normal to the direction of tool travel.
Chip flow direction	
The chip flows over the tool face, and the direction of chip flow velocity is normal to the cutting edge.	Chip flows on the tool face and the direction of chip flow velocity makes an angle with the normal on the cutting edge.
Nature of chip generated	
Chip coils in a light flat spinal.	The chip flows sideways in a long cut.

(Contd.)

Orthogonal cutting	Oblique cutting
Clearing action of cutting edge	
The cutting edge clears the width of the workpiece on either end.	The cutting edge may or may not clear the width of the workpiece.
Point of maximum chip thickness	
Maximum chip thickness occurs at its middle.	There may or may not be a maximum thickness.
Heat developed per unit area	
For the same feed and depth of cut, the heat developed per unit area is less.	For the same feed and depth of cut, the heat generated is more.
Tool life	
Tool life is less because the shear force acts on a smaller area.	Tool life is more because the shear force acts on a larger area.
Cutting force components	
Only two components of the cutting force act on the tool. These are perpendicular to each other and can be represented in a plane.	Three components of the cutting force, which are mutually perpendicular to each other, act at the cutting edge.

5.7 MECHANISM OF CHIP FORMATION

Irrespective of the basic nature of the chips obtained during machining of metals, the main factor governing the formation of chips is the plastic deformation of the metal by shear process.

Merchant (1945) used an idealised concept of chip formation, for which a precise geometry may be taken to be the basis for his studies of the mechanism of metal cutting. In developing the geometry, the following assumptions are made:

(i) The process can be adequately represented by any two-dimensional cross-sections of the cut.
(ii) The tool is perfectly sharp and contacts only the chip on its front or rake face.
(iii) The primary deformation takes place in a very thin zone adjacent to the shear plane AB.
(iv) The cutting edge is perpendicular to the cutting velocity vector V.
(v) The chip does not flow to the sides.

Figure 5.3 illustrates the concept. As the tool advances into the workpiece, the metal ahead of the tool is severely stressed. The cutting tool causes internal shearing action in the metal, and so the metal below the cutting edge yields and flows plastically in the form of chip. First of all, compression of the metal under the tool edge takes place. Then, the plastic flow takes place, followed by the separation of metal when the compression limit of that metal is exceeded. Plastic flow takes place in a localised region called shear plane, which extends from the cutting edge obliquely up to the uncut surface ahead of the tool. Here ϕ is the shear angle and ψ is the grain elongation angle.

It may be mentioned that the deformation of metal, in the process of separation of chip, does not occur sharply across the shear plane. The grains of the metal ahead of the cutting edge of the tool start elongating along the line AB in Fig. 5.3 and continue to do so until they are completely deformed along the line CD. The region between AB and CD is called the *shear zone*.

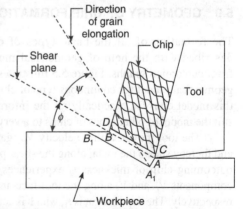

Fig. 5.3 Illustrating the basic mechanism of chip formation.

After passing out of the shear zone, the deformed metal slides along the tool face due to velocity of the cutting tool. For all mathematical analysis, this shear zone is treated as a plane and called *shear plane* (see Fig. 5.4).

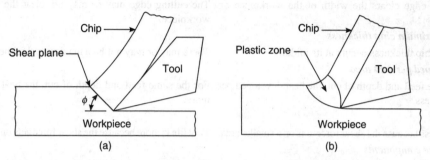

Fig. 5.4 Shear plane and shear zone in cutting action.

The size of the shear zone is thick if the metal is machined at low cutting speeds and thin if the metal is machined at high cutting speeds. Some investigations have proposed two shearing zones instead of one. The additional zone has been described at chip tool interface. Kececioglu (1958) proposed that the shear plane is not the only means by which deformation takes place in shear zone; there may also be slip on the cleavage planes of the grains, adjustment of grain boundaries for accommodating the elongated grain movement in the grain, boundary material accompanied by the auxiliary movements within the grains, rotation of slip plane, and sub-division of existing grains into smaller units at the microscopic level.

5.8 CURLING OF CHIP

As shown in Fig. 5.3, the two lines AB and CD confining the shear zone may not be parallel. They may often produce a wedge-shaped zone which is comparatively thicker near the tool face at the right side than at the left. This is probably one of the main causes of curling of chip. Also, due to non-uniform distribution of cutting forces at the tool–chip interface and on the shear plane, the shear can then be slightly curved concave downwards, thereby causing the chip to curl away from the cutting face of the tool.

5.9 GEOMETRY OF CHIP FORMATION

The formation of all the basic types of chips can be described with the help of geometrical models derived from photomicrographs. Figure 5.5 illustrates the simplest geometrical model for continuous type of chips. However, this model conveys practically all the information which suit the model of other types of chips to a very great extent.

The tool moves with a velocity V_c against the work and thereby shears the metal along the shear plane AB. The outcoming chip of thickness t_2 experiences two velocity components V_f and V_s along the tool face and shear plane respectively. The depth of cut is t_1, which is actually the feed in the machining operation as shown in Fig. 5.5. From the

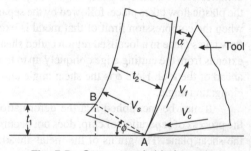

Fig. 5.5 Geometry of chip formation.

geometry of chip, it is possible to compute the value of shear angle ϕ in terms of measurable parameters t_1, t_2 and α, which in turn fully defines the model of continuous chips.

5.10 METHODS USED FOR DETERMINING CHIP GEOMETRY

Two experimental methods have been used in order to discover the way in which chips are formed. In one of these, the cutting process is stopped suddenly to leave the chip attached to the workpiece and in contact with the tool. This gives a still picture of chip formation, which probably suffers some distortion due to elastic recovery when the cutting stress is released.

The second method is that of high speed photography, which enables a 'slow motion' film of chip formation to be made. The method has been employed by Loxham et al. at the Department of Aircraft Economics and Production of the College of Aeronautics at Cranfield, UK. In both methods, the region around the cutting edge of the tool is viewed through a microscope so that chip formation may be examined in detail.

5.11 CLASSES OF CHIPS

Depending on the properties of the work material and the conditions of cutting, the chips are formed in different shapes and types. These are:

(i) Discontinuous chips
(ii) Inhomogeneous chips
(iii) Continuous chips with and without built-up edge (BUE), and
(iv) Fractured chips.

5.11.1 Discontinuous Chips

These chips are small individual segments which may adhere loosely to each other. The chips are produced as the tool advances in the direction of the feed, due to plastic deformation of the material ahead of the tool nose and in the vicinity of the cutting edge. The reason for generation of such chips is that as the material gets ahead, due to advancement of the tool, it ruptures intermittently, thus producing segmented or discontinuous chips (refer to Fig. 5.6).

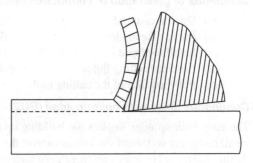

Fig. 5.6 Discontinuous or segmented chip.

Conditions favouring discontinuous chip formation

(i) Brittle and non-ductile work materials such as cast iron, brass castings, etc.
(ii) Small or negative rake angle
(iii) Low cutting speed
(iv) Dry cutting (cutting without application of cutting fluid)
(v) Large chip thickness, i.e. large depth of cut and high feed rate.

Characteristics

(i) Easy handling and quick disposal due to the small size of chips
(ii) Good degree of surface finish as the chips do not interfere with the work surface
(iii) More tool life
(iv) Less power consumption.

5.11.2 Inhomogeneous Chips

These are the chips with inhomogeneous strain obtained while machining steels at medium cutting speeds. In this type of chip the side adjoining the tool face is smooth and the opposite side has notches showing the orientation of the separately bonded segments (see Fig. 5.7).

5.11.3 Continuous Chips

Such chips are in the form of long coils having the same thickness throughout. The chips are produced due to plastic deformation of metal without rupture. Continuous chips without the built-up edge are difficult to obtain at normal cutting speeds. However they can be obtained at very high speeds when the surface finish and the tool life improve and the power consumption reduces (see Fig. 5.8).

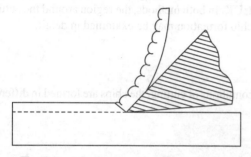

Fig. 5.7 Inhomogeneous chip formation.

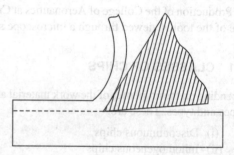

Fig. 5.8 Continuous chip without the built-up edge.

Conditions of generation of continuous chips

(i) Ductile material
(ii) High cutting speeds
(iii) Large rake angles
(iv) Small depth of cut
(v) Small feed rate
(vi) Sharp cutting edge
(vii) Efficient cutting fluids
(viii) Low coefficient of friction at the tool–chip interface
(ix) Polished face of the cutting tool.

Continuous chips with built-up edge (BUE)

The term *built-up edge* implies the building up of a ridge of metal on the top surface of the tool and above the cutting edge. As the chip moves over the tool face due to high normal load on the tool face, high temperature and high coefficient of friction between the chip and the tool interface, a portion of the chip gets welded on the tool face, forming the embryo of the built-up edge. The strain-hardened chip is so hard that now it becomes the practical cutting edge and starts cutting the material. Since this built-up edge is irregular in shape, the surface produced becomes quite rough. As the machining continues, more and more chip material gets welded on the embryo of the built-up edge; this increases its size and ultimately it becomes unstable and gets sheared off. This cycle is repeated. These chips are also in the form of coil (see Fig. 5.9).

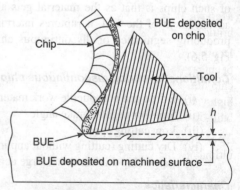

Fig. 5.9 Continuous chips of coil form with built-up edge.

Characteristics of continuous chips with built-up edge:

(i) Difficult to handle and dispose off (ii) More frequent tool failures
(iii) More power consumption (iv) Poor surface finish.

Conditions of generation of continuous chips with built-up edge:

(i) Ductile material (ii) Coarse feed
(iii) Small rake angle (iv) Low cutting speed
(v) Dull cutting edge (vi) Insufficient cutting fluid
(vii) High friction at the tool surface interface.

Methods of avoiding built-up edge:

(i) Increasing the cutting speed (ii) Decreasing the feed
(iii) Heating the work (iv) Increasing the rake angle
(v) Improving the cutting fluid system.

5.11.4 Fractured Chips

Such chips are obtained in machining metals of low plasticity like hard C.I. and hard bronze. The chip consists of separate specimens that seem to be broken out from the chip in continuity. In such chips, the leading crack spreads immediately over the whole shear surface from which the chip is formed. Such fractured chips are not subjected to appreciable plastic deformations, but an instantaneous non-uniform load is applied to the machine–work tool system. The surface produced is rough and there is no relative movement of the chip over the tool (see Fig. 5.10).

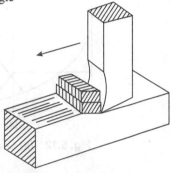

Fig. 5.10 Different fractured types of chip.

5.12 EFFECT OF VARIOUS FACTORS ON CHIP FORMATION OR METAL CUTTING CHARACTERISTICS

Velocity: It directly affects the temperature at the tool point. It does not affect the direction of chip flow. Low cutting speed causes built-up edge and discontinuous chips. On ductile materials, high velocity forms continuous chips.

Material of workpiece: Ductile materials produce continuous chips (normally with built-up edge), whereas brittle materials produce segmented or discontinuous chips.

Depth of cut and feed: Increasing the feed and depth of cut results in greater distortion of chip. The direction of chip flow changes with the change in the size of cut. The high values of feed and depth of cut lead to formation of built-up edge and poor surface finish.

Tool geometry: This changes the shear angle and ultimately the chip thickness. When $\phi_1 < \phi_2$, we have $t_{c_2} < t_{c_1}$ and $t_{c_1} > t_{c_2}$. A large shear plane angle ϕ_2 results in a smaller shear plane area and better cutting conditions.

Figure 5.11 shows the effect of large and small shear angle on the chip thickness and length of shear plane for a given tool and depth of cut.

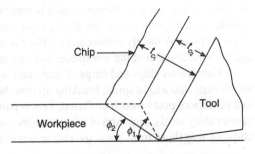

Fig. 5.11 Effect of large and small shear angles on chip thickness and length of shear plane for a given tool and depth of cut.

The temperature of the cutting tool may reach a high value, particularly when a heavy cut is taken at a high speed. This is evident when the work or tool is touched by the presence of temper colours ranging from red to blue of the chip, work or tool; it may even be evident due to the loss of hardness of the tool point, with an attendant loss of tool geometry and failure by excessive flow occurs.

Friction in metal cutting: Friction between the chip and the tool plays a significant role in the cutting process. It is found that the value of the coefficient of friction at the tool–chip interface is very high. Zorev in 1948 established that the sliding of a chip on the tool face is different in different zones. Two zones called *sticking region* and *sliding region* are shown in Fig. 5.12(b). The texture of the sticking zone is different from that of the sliding zone, which is composed of longitudinal scratches. Due to the two different types of contacts shown in Fig. 5.12(a), the laws for them for the variation of the coefficient of friction differ from each other very much. The measured value of the coefficient of friction is the average of the value of the two zones.

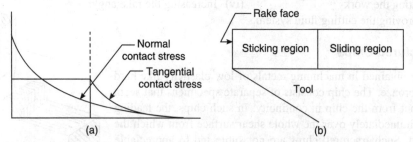

Fig. 5.12 (a) Contact stresses and (b) zones of contact of chip on tool face.

5.13 METHODS OF REDUCING FRICTION

The coefficient of friction between the chip and the tool can be reduced by either using an effective cutting fluid, taking a thicker chip (feed per revolution), decreasing the effective rake angle, or using a higher cutting speed. It can also be reduced by using certain chemical additives in the work material, e.g., the presence of lead or graphite in steel will reduce the friction of the chip on the rake face of the tool. Thus, the performance of the tools can be improved by additives. Increase in the coefficient of friction is associated with increase in rake angle on the positive side.

5.14 PHYSICAL ASPECTS OF CHIP CONTROL AND CHIP BREAKERS

Effective chip control is essential in the machining of steels of high tensile strength and of most metals when machined at high speeds. Proper control is particularly necessary when machining with sintered carbide tool material, which operates at a relatively high speed and produces continuous, sharp, blue chips at high temperatures. These chips are definitely detrimental to effective machining from the standpoint of the serviceability of the tool and the surface finish of the workpiece, and they are hazardous to the safety of the machine operator.

Continuous chips and chips of large coil are difficult to handle and dangerous to the operator. A chip which curls into a tight spiral, breaking up into short sections against the unfinished surface of the work, rigid tool or the tool post is much preferred. This is particularly true when tough materials such as some of the low carbon alloy steels, high nickel alloy stainless steels and aluminium are machined. Discontinuous or short chips such as those from brass and cast iron require no special tool treatment.

5.14.1 Chip Control through Tool Grinding

A tool can be ground in two ways to control chips—(i) by getting the right combination of back and side rake angles for a given feed and speed, and (ii) by grinding a groove or shelf in the face of the tool.

Method (i) is usually preferred where a strict control of the chip is not required, since it is simpler to grind and more likely to preserve the strength of the cutting edge. However, method (ii) involving the use of a ground groove as chip breaker is less dependent on the precision of rake angles, and since it provides a better control of the chip it is more universal in application.

5.14.2 Chip Control through Chip Breakers

Chip breakers are of two general types—those made of an additional piece clamped to the tool to deflect the chip and those ground into the face of the tool. The latter are of the shelf type and the groove type.

Clamp-type chip breaker

A chip breaker, separate from the cutting tool, consists of some form of block fastened to the cutting tool. As the chip flows over the tool face, it hits the shoulder of the chip breaker, which is so curved as to force the chip into the desired spiral or helix. The chip exerts considerable force against the shoulder and also has an abrasive effect on its face.

The chip breaker, therefore, must be rigidly clamped in position and the active face should be of a material at least as hard as that used for the cutting tool. A sintered carbide tip clamped to the shank by a screw-controlled clamp, the beveled edge of which is nearly parallel to but slightly behind the cutting edge (see Fig. 5.13), serves as a chip breaker. This tool was used to finish turn 155 mm shells and gun barrels.

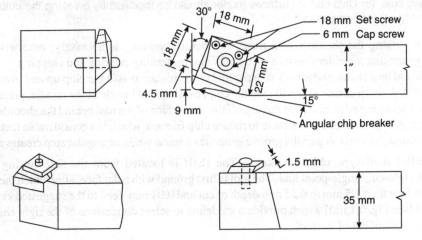

Fig. 5.13 Clamp-type chip breaker.

Ground-in type chip breakers

It is a common practice to grind a small groove in the face of a tool, nearly perpendicular to the direction of chip flow, and immediately behind the cutting edge, as shown for the tangential-type bar turning tools in Fig. 5.14. Such a groove turns the chip into a spiral or helix, which breaks up as it is forced against the shoulder, the work, or tool, or comes off with a small diameter.

Instead of grinding a groove in the tool face as shown in Fig. 5.14, the active portion may be ground somewhat lower than the balance of the face, with a small radius connecting the two levels as shown in Fig. 5.15. This type of chip breaker is more satisfactory than that shown in Fig. 5.16 in that it does not weaken the tool.

The parallel type breaker is a simple step or shelf, ground in the top surface of the tool along the cutting edge and parallel to it. The shoulder thus left behind the cutting edge will have a cooling or crimping action on the chip. The chip has a natural tendency to coil upwards as it strikes the face of the tool, and the parallel type breaker should be wide enough to ensure that the chip has already left the tool surface and started up again. Figure 5.13 shows the angular chip breaker ground in sintered carbide tip. This tip is not brazed to the shank, but is held by the mechanical clamp. The two set screws under the clamp raise one end of the clamp plate to concentrate pressure on the tip. Vibration will not loosen the tip.

Figure 5.14 shows three types of ground-in chip breakers for the tangential-type bar turning tool as recommended by the American Rolling Mill Company (also see Fig. 5.16).

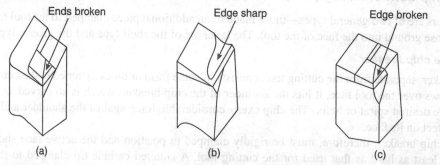

Fig. 5.14 (a) Chip breaker ground-in for the carbide box tool. May also be ground-in for the other types of tools for light or medium cuts. (b) Chip-curl groove as usually applied to high speed cobalt type tools for heavy cuts. (c) Chip-curl in carbides (curler should be modified by breaking the cutting edge with hand hone) for heavy cuts.

The chip striking the shoulder of a fixed stop before breaking is necessary; otherwise the chip is uncontrolled, generating a greater pressure on the tool and promoting catering and chipping.

There should be a radius at the back of a parallel chip breaker to skid the chip up and over the shoulder [see Fig. 5.15(b)]. A sharp corner at D will stop the chip, causing it to buckle in the middle, greatly increasing the pressure on the cutting edge and often flaking off the top surface of the tool behind the shoulder or chipping the cutting edge. A large side rake is desirable to reduce chip friction, which is a considerable factor, especially when cutting the tough metals. A parallel groove generates a spiral while an angular step creates a helical chip.

Ground parallel shelf-type chip breaker: The shelf is located from the side-cutting edge of the carbide-tipped, right-cut, single-point tool. The tool is first ground with a flat face, after which the chip breaker is ground in the face for a 1.5 mm to 6.25 mm depth of cut and 0.04 mm, feed to the suggested key with values from Table 5.1 [see Fig. 5.15(a)] which provide a guideline to select dimensions of the right chip breaker for given feed and depth of cut of the tool.

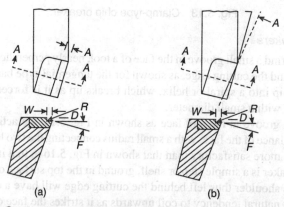

Fig. 5.15 (a) Groove-type parallel chip-breaker and (b) angular step chip breaker.

Ground-in angular chip breaker for sintered carbide-tipped tool: The dimensions of the breakers shown in Table 5.1 are for a variable depth of cut of 3.30 mm in steel with a feed of 0.50 m.

Table 5.1 Recommended dimensions for both parallel and angular chip breakers for all common cuts and feeds

Feed (mm/R)	0.20–0.30 (1)	0.33–0.43 (2)	0.45–0.55 (3)	0.58–0.68 (4)	0.70–0.80 (5)
Depth of cut (mm)		Width of breaker shelf W (in mm)			
0.4–1.20	1.50	2.00	2.40	2.75	3.00
1.5–6.20	2.40	3.00	4.00	4.36	4.75
8.00–12.70	3.00	4.00	4.75	5.10	5.50
14.30–19.00	4.00	4.75	5.50	6.00	6.00

Requirements of the ground-in type chip breakers: The depth of a parallel chip-breaker is also of utmost importance. In regrinding a dull tool, the surface of the tip should be ground down, so as to permit regrinding of the chip breaker without deepening it too much. A deep chip breaker crimps the chip too tightly, causing unnecessary pressure on the cutting edge and frequently fracturing the tip.

The second type is the angular chip breaker type as shown in Fig. 5.16. It is similar to the parallel type in cross-section, but is ground with the shoulder at an angle to the cutting edge, being narrower at the back of the tip than at the front. An angular chip breaker ground-in at about five to ten degrees with the side cutting edge runs out nearer the back of the tip, making it unnecessary to grind the steel shank with the diamond wheel. This not only increases the life of the diamond wheel, but also prevents loading it up and thus permits faster grinding of the carbide. The tip is usually mounted above the surface of the shank.

A greater angle on the chip-breaker (10° or more) curls the chip forward towards the upturned shoulder of the workpiece, and on shallow cuts will break the chip up into shorter curls unlike a parallel chip-breaker.

Figures 5.16(a) and (b) show the influence of width, depth, and shoulder radius of the chip breaker on the coil of continuous chips of tough metal. Figure 5.16(c) shows the modified shelf-type chip breaker. The negative rake of the ridge adjacent to the cutting edge with positive rake ground in itself gives a strong cutting edge with a free cutting action.

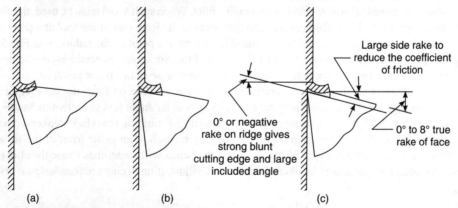

Fig. 5.16 (a) and (b) Influence of width, depth, and shoulder radius of the chip breaker on the coil of continuous chips of tough metal and (c) modified shelf-type chip breaker.

The angular chip breaker must be used on a fairly constant depth of cut, as the width varies with the distance from the cutting point, and a breaker used on a greater depth of cut than that for which it is designed will be too narrow at the back end and will cause trouble.

Both the parallel and angular type chip breakers may be modified to give a normal side rake angle different from that of the face. One of the most common modifications is to grind the shelf in, at zero or a slightly negative side rake, for use on rough work requiring a very strong cutting edge. Usually between 0 and 3 degrees negative is sufficient to give a blunt included angle of the carbide along the cutting edge. The tip on such a tool should be set in at zero side rake, in order to have a sufficient shoulder behind the chip breaker shelf.

Another modification is to mount the carbide tip at zero side rake and then grind a positive side rake on the chip breaker. This requires less grinding to get the same height of the shoulder or effective depth of the chip breaker than in the case of the standard angular or parallel type, although it has a tendency to round the corner of the diamond wheel.

To strengthen the cutting edge of carbide tipped tools as shown in Fig. 5.17, a ridge may be ground or honed on, at zero or slightly negative side rake, since the main chip action is behind the cutting edge—a distance approximately equal to the feed (assuming that the proper speed is used). The fragile cutting edge is thus removed without destroying the positive true rake at the point of contact between the chip and the carbide (see Fig. 5.17).

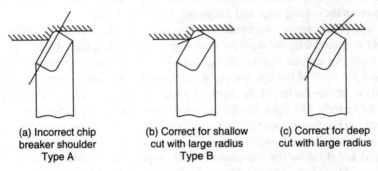

(a) Incorrect chip breaker shoulder Type A

(b) Correct for shallow cut with large radius Type B

(c) Correct for deep cut with large radius

Fig. 5.17 Chip breaker angles for shallow and deep cuts.

Most carbide tools are used with a small nose radius due to greater life between the grinds, but occasionally a large tool radius is required in order to finish with a radius fillet. When such a tool must be used, the chip breaker becomes more of a problem. If the chip breaker shoulder runs to the front end of the tool at a point on the end cutting edge, as in Fig. 5.13, there is no difficulty. But if it runs out at a point on the radius, as at Fig. 5.17(a), the cutting edge at that point is irregular and will mar the finish of the workpiece, generate excessive heat and dull rapidly. Widening the chip breaker enough to include a very large nose radius is not practical.

If the depth of cut is small and fairly constant, as is often the case on finishing cuts requiring a fillet of larger radius, an angular chip breaker may be ground across at an angle just sufficient to handle the depth of cut, as shown in Fig. 5.17(b). On a deeper cut it is necessary to combine two chip breakers, a parallel one to accomplish the chip curling and an angular one to prevent the high point at the front of the shoulder from dragging on the work, as shown at Fig. 5.17(c). A large nose radius will sometimes cause the chip to curl and break without the need for a ground-in breaker of any kind. An illustration of chip breaker design is given below.

Type	C
Radius of fillet	1/16
Shoulder angle	0°
Normal rake angle of shelf	γ
Width of shelter breaker	1/8

THEORY OF METAL CUTTING **91**

Depth of shelter groove	0.017
Width of ridge	0.02
Depth of ridge	0
Normal rake angle of ridge	8°

The third general type of chip breaker has the groove and ridge as shown in Fig. 5.18 (types C, D and E). The design has a normal rake angle ground in the breaker, but it is distinctly different in that a narrow flat (ridge) is grouped or honed along the top of the cutting edge.

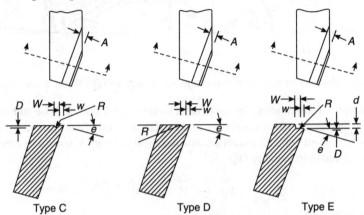

Fig. 5.18 Chip breakers ground for HSS tools.

While the chip breaker types A and B in Fig. 5.17(a) and (b) are restricted to tools with sintered carbide tips since they do not seem to reduce the tool life of this material as they do for high speed steel, types C, D and E are ground often in high speed steel tools. Generally speaking, the width of the ridge w and the width of the shelf W for types C and E depend on the feed, the depth of cut and the hardness of the material cut. The heavier the feed and the harder the material, the wider the ridge and the groove should be. Satisfactory results are usually obtained when the ridge is about 1 to 1½ times the feed and when the shelf is from 3 to 4 times the feed.

5.15 VELOCITY RELATIONSHIPS IN ORTHOGONAL CUTTING—MERCHANT'S ANALYSIS OF METAL CUTTING KINEMATICS

M. Eugene Merchant of US is, rightly called the father of the 'Science of Metal Cutting'. He proposed kinematic and kinetic relations including his famous shear angle relation.

As a starting point, the authors begin with the kinematic or velocity relationships proposed by Merchant as applied to metal cutting. The kinetics or Force Analysis will be taken up in Figure 5.17.

Three velocities come into play when the tool cuts the material. These are as follows:

Cutting velocity (V_c): It is the velocity of the tool relative to the work and directed parallel to the compressive force on the shear plane. This is equal to the effective cutting speed.

Chip velocity (V_f): It is the velocity of the chip relative to the tool and directed along the tool face.

Shear velocity (V_s): It is the velocity of the chip relative to the workpiece and directed along the shear plane.

Figure 5.19(a) shows the velocity relationships in orthogonal cutting, and Fig. 5.19(b) represents the shear zone in cutting.

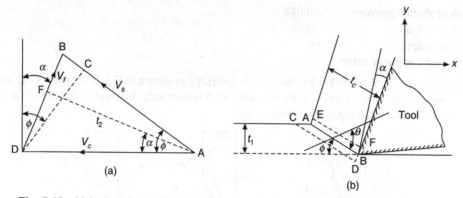

Fig. 5.19 Velocity relationship in orthogonal cutting and (b) shear zones in cutting.

From the principle of kinematics, the relative velocity of two bodies (tool and chip) is equal to the vector difference between their velocities relative to the reference body (workpiece). So the vectors of these velocities should form a closed velocities diagram, and the shear velocity V_s should be equal to the vector sum of cutting velocity (V_c) and chip velocity (V_f).

Therefore,
$$\vec{V_s} = \vec{V_c} + \vec{V_f}$$

Using trigonometric principles
$$\frac{V_c}{\sin(90-\phi+\alpha)} = \frac{V_f}{\sin\phi} = \frac{V_s}{\sin(90-\alpha)}$$

or
$$\frac{V_c}{\cos(\phi-\alpha)} = \frac{V_f}{\sin\phi} = \frac{V_s}{\cos\alpha}$$

∴
$$V_s = V_c \frac{\cos\alpha}{\cos(\phi-\alpha)}$$

and
$$V_f = V_c \frac{\sin\alpha}{\cos(\phi-\alpha)}$$

5.16 CHIP THICKNESS RATIO

The outward flow of the metal causes the chip to be thicker after separation from the parent metal. That is, the chip produced is thicker than the depth of cut as shown in Fig. 5.19(c).

The chip thickness ratio (r_c) is defined as the ratio of the depth of cut (t_1) (feed in case of turning) to the chip thickness (t_2). It may be noted that in Fig. 5.19(c), AF which is perpendicular to the tool-chip interface represents t_2, i.e. chip thickness.

From the right-angled triangle ABD in Fig. 5.19(c),
$$AB = \frac{t_1}{\sin\phi}$$

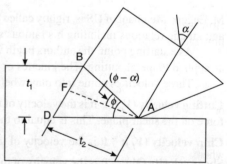

Fig. 5.19(c) Geometry of chip formation.

and from the right-angled triangle ABF in Fig. 5.19(c),

$$AB = \frac{t_2}{\cos(\phi - \alpha)}$$

Comparing these two equations, we have the chip thickness ratio,

$$r_c = \frac{t_1}{t_2} = \frac{\sin\phi}{\cos(\phi - \alpha)}$$

$$= \frac{\sin\phi}{\cos\phi\cos\alpha + \sin\phi\sin\alpha}$$

or

$$1 = \frac{r_c \cos\phi \cos\alpha}{\sin\phi} + \frac{r_c \sin\phi \sin\alpha}{\sin\phi}$$

or

$$r_c \cos\alpha = (1 - r_c \sin\alpha)\tan\phi$$

or

$$\tan\phi = \frac{r_c \cos\alpha}{1 - r_c \sin\alpha} \tag{5.2}$$

Note that $\dfrac{1}{r_c}$ is termed the *chip reduction coefficient* and denoted by K.

However, the chip thickness ratio can be expressed in a different manner. Let l_2 be the length of cut chip in one revolution of the workpiece. As the volume of uncut and cut chip should remain constant, it may be written that

$$l_2 \times t_2 \times b_2 = \pi D \times t_1 \times b_1$$

where b_1 and b_2 are the width of cut and the width of chip respectively, and D is the diameter of workpiece. If there is no side flow, then

$$b_1 = b_2$$

∴

$$l_2 \times t_2 = \pi D \times t_1$$

Let

$$\pi D = l_1$$

then

$$l_2 \times t_2 = l_1 \times t_1$$

or

$$\frac{t_1}{t_2} = \frac{l_2}{l_1}$$

However, if side flow is to be considered, the chip thickness ratio is to be multiplied by a factor λ, the side flow factor. Thus,

$$\frac{l_2}{l_1} \times \frac{b_1}{b_2} = \frac{t_1}{t_2}$$

or

$$\frac{l_2}{l_1} = \lambda \frac{t_1}{t_2}$$

where

$$\lambda = \frac{b_1}{b_2}$$

5.17 FORCES ACTING ON A CUTTING TOOL

The study of the forces acting on a tool during the process of metal cutting is necessary for a quantitative analysis of the process. Figure 5.20 shows a typical diagram of the force system in a case of orthogonal cutting. In the actual practice of machining, the force system acting on a cutting tool is always three-dimensional. The relationship among the various forces was determined by Merchant with a large number of assumptions, noted as follows.

(i) The cutting edge of the tool is sharp and it does not make any flank contact with the workpiece.
(ii) Only the continuous chip without the built-up edge is produced.
(iii) The shear surface is a plane extending upwards from the cutting edge.
(iv) The cutting edge is a straight line, extending perpendicular to the direction of motion, and generates a plane surface as the work moves past it.
(v) The chip does not flow to either side.
(vi) The cutting velocity remains constant.
(vii) The depth of cut is constant.
(viii) The width of the tool is greater than that of the workpiece.
(ix) The inertia force of the chip is entirely neglected.
(x) The work moves relative to the tool with uniform velocity.
(xi) The chip behaves as a free body in stable equilibrium under the action of two equal, opposite and almost collinear resultant forces.

Merchant developed a very useful concept of treating the chip as a 'free body' held in equillibrium due to two equal and opposite forces R' and R.

The force R' is the resultant of two forces exerted by the chip on the tool as a result of chip tool-friction.

1. Force of friction (F) acting at the chip-tool interface.
2. Normal force (N) exerted by the chip on the tool.

These two forces F and N at the chip-tool interface have a 'braking action' at the interface.

The second set (R) of forces during cutting are

1. F_s which is the internal shear force on the shear plane which is resisted by the work material along a shear plane.
2. F_n which is the internal normal force at the shear plane.

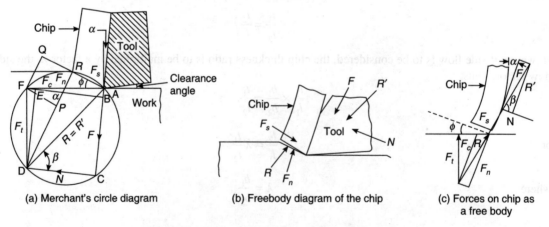

(a) Merchant's circle diagram (b) Freebody diagram of the chip (c) Forces on chip as a free body

Fig. 5.20 Force components acting on the chip.

THEORY OF METAL CUTTING

While the (R) set tries to severe the work material, the set (R') tries to prevent shear of the metal at the shear plane.

Figures 5.20(b) and (c) indicate Merchants insight into the metal cutting mechanism. Since $R = R'$, Merchant developed his famous 'Circle diagram' which is shown in Fig. 5.20(a).

The forces F, N, F_s and F_n cannot be measured. However, it is possible to draw the circle diagram if the cutting power component F_c exerted by the moving workpiece against the tool and the thrust component F_t can be measured by Dynamometry. This makes Merchant's model of metal cutting a very useful practical technique apart from its theoretical utility.

In orthogonal cutting, the resultant force that acts on a tool is R, having two basic components F_c and F_t.

F_c: It is the horizontal or longitudinal tool force that sets the direction of tool travel. It determines the amount of work needed to move the tool a given distance.

F_t: It is the vertical or tangential tool force that sets in a direction tangential to the rotating workpiece. It does no work but aids in deflecting the tool relative to the workpiece.

As the chip slides over the tool face under pressure, the kinetic coefficient of friction may be expressed as [see Fig. 5.20(a)]

$$\mu = \frac{F}{N} = \tan \beta$$

where μ is the kinetic coefficient of friction
F is the frictional force or frictional resistance met by the chip as it slides over the tool
N is the normal force
β is the frictional angle.

From Fig. 5.20(a), we have

$$F = AB + BC$$
$$= AB + DE$$
$$= F_c \sin \alpha + F_t \cos \alpha$$
$$F_s = AQ - QR$$
$$= AQ - FP = F_c \cos \phi - F_t \sin \phi$$
$$F_n = DR = DP + PR$$
$$= DP + FQ$$
$$\frac{F_c}{F_s} = \frac{R \cos(\beta - \alpha)}{R \cos(\phi + \beta - \alpha)}$$
$$= \frac{\cos(\beta - \alpha)}{\cos(\phi + \beta - \alpha)}$$
$$\frac{F}{N} = \left[\frac{F_c \sin \alpha + F_t \cos \alpha}{F_c \cos \alpha - F_t \sin \alpha}\right]$$
$$= \left[\frac{F_t + F_c \tan \alpha}{F_c - F_t \tan \alpha}\right]$$

Utility of the above formulae

As stated in the beginning, the above formulae link the measurable forces F_c and F_t with the other forces in the Merchant circles. This facilitates analysis.

Plastic deformation will occur along that plane, thus forming chips. The cutting force required to cause shear deformation along the plane identified by ϕ as in Fig. 5.21 will then be the lowest cutting force to produce a chip; shear deformation along any other plane would require a greater cutting force; but after shear deformations begin on one plane, the cutting force cannot exceed the minimum value. Also, there are two circles—

(i) Circle I, representing the force for shear zone and friction zone on the face of the tool, and
(ii) Circle II, representing the force for the tool top tip and flank friction. For ideal case with sharp tool this circle is not present. However, in practice the sharpness of the tool is the 1st basic assumption of Merchant's model. When this is not present, there is cutting, rubbing and thrust at the flank shown by small circle of radius R_1 shown in Fig. 5.21.

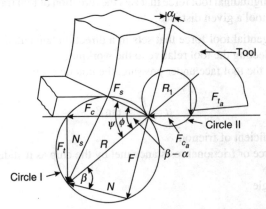

Fig. 5.21 Merchant's circle and circle due to blunt flank.

From Fig. 5.21,

$$F_c = R \cos(\beta - \alpha)$$
$$F_t = R \sin(\beta - \alpha)$$
$$F = R \sin \beta$$
$$N = R \cos \beta$$
$$F_s = R \cos \phi = \phi_s + A_s - \frac{\phi_2 A_0}{\sin \phi}$$
$$N_s = R \sin \phi = \sigma_s A_s = \frac{\sigma_s A_0}{\sin \phi}$$

5.18 PRINCIPLE OF MINIMUM ENERGY APPLIED TO METAL CUTTING—MERCHANT'S SHEAR ANGLE RELATION

This process is an application of the principle of minimum energy, that is, the cutting force F_c is responsible for the work done in metal cutting such that, for a given rake angle α and friction angle β, the shear plane angle ϕ will assume such a value as to make the energy required or the work done by F_c minimum. To determine the shear plane angle ϕ, we may thus write an equation for the cutting force in terms of ϕ.

From the geometry of Fig. 5.21,

$$F_c = R \cos(\beta - \alpha)$$
$$F_s = R \cos(\phi + \beta - \alpha)$$

∴
$$F_c = F_s \frac{\cos(\beta - \alpha)}{\cos(\phi + \beta - \alpha)} \qquad (5.3)$$

Let A be the cross-sectional area of the undeformed chip, that is, the depth of the cut multiplied by its width (feed). Then the area of the shear plane will be $A/\sin\phi$.

The force F_s on the shear plane can be replaced by the stress on the plane multiplied by the area,

or
$$F_s = f_s \cdot A_s = \frac{f_s A}{\sin\phi}$$

Equation (5.3) becomes
$$F_c = \frac{f_s A \cos(\beta - \alpha)}{\sin\phi \cos(\phi + \beta - \alpha)} \qquad (5.4)$$

In order to obtain an expression relating the shear plane angle ϕ to the friction angle β and the rake angle α, the principle of minimum energy (indirectly minimum F_c) is applied.

$$F_c = \frac{f_s A}{\sin\phi} \frac{\cos(\beta - \alpha)}{\cos(\phi + \beta - \alpha)}$$

For minimum F_c, we get
$$\frac{\partial F_c}{\partial \phi} = 0$$

Since the numerator of Eq. (5.3) is constant, the equation can be differentiated and written as
$$\frac{dF_c}{d\phi} = \frac{-\cos\phi \cos(\phi + \beta - \alpha) - \sin\phi \sin(\phi + \beta - \alpha)}{\sin^2\phi \cos^2(\phi + \beta - \alpha)}$$

Equating this to zero, we get
$$\cos\phi \cos(\phi + \beta - \alpha) - \sin\phi \sin(\phi + \beta + \alpha) = 0$$

From trigonometry, we get
$$\cos(\phi + \phi + \beta - \alpha) = 0$$
or
$$\cos(2\phi + \beta - \alpha) = 0$$
∴
$$2\phi + \beta - \alpha = \frac{\pi}{2}$$
∴
$$\phi = \frac{\pi}{4} - \frac{\pi}{4} + \frac{\pi}{4}$$
∴
$$\phi = 45 - \frac{\beta}{2} + \frac{\alpha}{2} \quad \text{(This is Merchant's first shear angle relation)} \qquad (5.5)$$

In 1945, Merchant found that, this theory fitted well when turning certain non-metallic materials, but not steel. He concluded that for metals, shear assumption was unreliable, and that shear strength was in some way dependent on normal stress acting at shear plane. He, therefore, modified his theory by assuming that shear and normal stresses are related in the form (see Fig. 5.22).

$f_s = f_s o + K f_n$ (referred to as the Bridgeman Effect) (5.6)

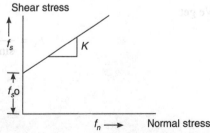

Fig. 5.22 Bridgeman effect.

where f_s is the shear stress
f_n is the normal stress
and $f_s o$ and K are constants for the material.

From Fig. 5.22.

$$f_s = f_n \cot(\phi + \beta - \alpha) \qquad (5.7)$$

and from Eqs. (5.2) and (5.3)

$$f_s = \frac{f_s o}{1 - K \tan(\phi + \beta - \alpha)} \qquad (5.8)$$

$$= R \cos(\beta - \alpha)$$

$$= \frac{f_s \cos(\beta - \alpha) A}{\sin \phi \cos(\phi + \beta - \alpha)} \qquad (5.9)$$

Combining Eqs. (5.3) and (5.5)

$$F_c = \frac{f_s o \, A \cos(\beta - \alpha)}{\sin \phi \cos(\phi + \beta - \alpha)[1 - K \tan(\phi + \beta - \alpha)]}$$

$$= \frac{f_s o \, A \cos(\beta - \alpha)}{\sin \phi [\cos(\phi + \beta - \alpha) - K \sin(\phi + \beta - \alpha)]} \qquad (5.10)$$

Differentiating Eq. (5.10) with reference to ϕ

$$\frac{dF_c}{d\phi} = \frac{\begin{bmatrix} \{(f_s o \cos(\beta - \alpha))(\cos \phi \cos(\phi + \beta - \alpha)) - K \sin(\phi + \beta - \alpha)\} \\ \{\sin \phi \sin(\phi + \beta - \alpha) - K \cos(\phi + \beta - \alpha)\} \end{bmatrix}}{[\sin \phi \cos(\phi + \beta - \alpha)1 - K \ldots]^2} = 0$$

$\therefore \quad \cos \phi \cos(\phi + \beta - \alpha) + \cos \phi \sin(\phi + \beta - \alpha) - K[\sin \phi \cos(\phi + \beta - \alpha) + \cos \phi \sin(\phi + \beta - \alpha)] = 0$

From trigonometric relationship

$$\cos(2\phi + \beta - \alpha) = K \sin(2\phi + \beta - \alpha)$$

or

$$\frac{\cos(2\phi + \beta - \alpha)}{\sin(2\phi + \beta - \alpha)} = K \qquad (5.11)$$

or

$$\cot C = K$$

whereby substituting

$C = (2\phi + \beta - \alpha)$ (This is Merchant's second shear angle relation.)

We get

$$F_C = \frac{f_s o A \cos(C - 2\phi)}{\sin \phi \cos(C - \phi)[1 - K \log(C - \phi)]} \qquad (5.12)$$

$$\boxed{r_c = \frac{\sin \phi}{\cos(\phi - \alpha)} = \frac{t_1}{t_2}} \qquad (5.13)$$

and values of t_1 and t_2 are determined by experiment. Constants K, f_so and C may be estimated by carrying out a series of tests and applying results in the above expressions. For steel, Merchant obtained a value of $C = 80$ for expression $2\phi + \beta - \alpha = C$.

5.19 STRESS AND STRAIN IN CHIP

The average stress on the shear plane is easily found by dividing F_s and N_s by the shear plane area $A_s = \dfrac{f_s t_1}{\sin \phi}$, in which b is the width of cut and t is the depth of cut, as shown in Fig. 5.15(b). Thus we have mean shear stress

$$f_s = \frac{F_s}{A_s}$$

Mean shear stress on the shear plane

$$f_n = \frac{N_s}{A_s} = \frac{F_s}{A_s}$$

or

$$f_s = \frac{(F_c \cos\phi - F_t \sin\phi)}{b \times t} \cdot \sin\phi \qquad (5.14)$$

Mean normal stress on the shear plane

$$f_n = \frac{(F_c \sin\phi + F_t \cos\phi)}{b \times t} \cdot \sin\phi \qquad (5.15)$$

The average shear stresses are difficult to estimate because of many reasons, and one of them is the chip tool contact length. This must be measured during or after the cut, and is rather difficult to measure with accuracy so as to give a correct value of stresses. The chip–tool contact varies during a cut, depending on the geometry of the chip curl.

5.20 SHEAR AND STRAIN RATE

The shearing process in metal cutting can be represented by a simple model of blockwise slip of stack of cards originally conceived by V. Pispaanen. It is shown in Figure 5.23. The shear strain can be written as

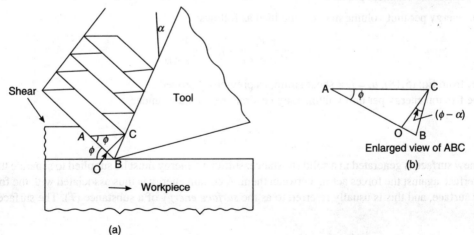

Fig. 5.23 (a) Card analogy and (b) geometry of shearing.

$$\gamma = \frac{\text{Distance sheared}}{\text{Thickness of card}}$$

$$= \frac{AB}{OC} = \frac{OA + OB}{OC}$$

Thus,
$$\gamma = \frac{OA}{OC} + \frac{OB}{OC} = \{\cot \phi + \tan (\phi - \alpha)\}$$

Strain rate: During shearing of metal in cutting, strain rate $= \frac{\partial s}{\partial t} \cdot \frac{1}{\partial y}$ where $\frac{\partial s}{\partial t}$ is velocity of shearing V_s and ∂y is Δy or thickness of lamellar block. For example if the value of velocity of shearing is assumed as $V_s = 100$ m.p.m. and thickness of block $= 10^{-3}$ mm.

We have $\gamma = \frac{100 \times 10^3}{60} = 1667$ metres per second which is a very high strain rate.

5.21 ENERGY CONSIDERATION IN METAL CUTTING

When metal is cut in a two-dimensional cutting operation, the total energy consumed per unit time is
$$U = F_s V \tag{5.16}$$
The total energy per unit volume of metal removed will therefore be
$$u = \frac{U}{V \times b \times t} = \frac{F_c}{b \times t} \tag{5.17}$$
where b and t are the width and depth of cut respectively. This total energy per unit volume will be consumed in the following ways:

  (i) As shear energy per unit volume (u_s) on the shear plane
  (ii) As friction energy per unit volume (u_f) on the tool face
  (iii) As the surface energy per unit volume (u_a) due to the formation of new surface area in cutting
  (iv) As momentum energy per unit volume (u_m) due to momentum change associated with the metal as it crosses the shear plane.

The shear energy per unit volume may be obtained as follows:
$$u_s = \frac{F_s V_s}{V \times b \times t} = f_s \times \frac{V_s}{V \sin \phi} \tag{5.18}$$

However, from Eq. (5.18), $u_s = f_s \times$ shear strain, as previously proved.

The friction energy per unit volume may be similarly found as under,
$$u_f = \frac{F \times V_c}{V \times b \times t} \tag{5.19}$$

When a new surface is generated in a solid substance, sufficient energy must be supplied to separate the atoms at the interface against the forces acting between them. A certain energy is thus associated with the formation of a new surface, and this is usually referred to as the *surface energy* of a substance (T). The surface energy of a solid is analogous to the surface tension of a liquid. The unit of surface energy is cm kg per sq. cm. The surface energy per unit volume in cutting will be

$$u_a = \frac{\lambda(2v \times b)}{V_t} = \frac{2\lambda}{t} \quad (5.20)$$

In Eq. (5.20), the factor 2 appears from the fact that two surfaces are generated simultaneously when a cut is made. The value of λ for most metals is about 0.0075 cm/kg.

The resultant momentum force associated with the change in momentum of the metal as it moves ahead of the shear plane will obviously be in the direction of slip or along the shear plane. This resultant momentum force (F_m) may therefore be obtained by applying the linear momentum law of applied mechanics along the shear plane from the fundamental law, i.e. $F_m = mV$.

$$\therefore \quad F_m = \rho \times V \times b \times t\,[V_t \sin(\phi - \alpha) + V_c \cos \phi] \quad (5.21)$$

where ρ is the mean density of the metal. By use of shear angle and velocity relationships, this may be written as

$$F_m = \rho V_c^2\, bt\, r_c \sin \phi \quad (5.22)$$

Therefore, the momentum energy per unit volume will be

$$u_m = \frac{F_m V_s}{V_c \times b \times t} = \rho V_c r_c \sin \phi\, V_s$$

After substituting the value of b and t, we get

$$u_m = \rho V_c^2 r_c^2 \sin^2 \phi \quad (5.23)$$

As it is established that the surface energy per unit volume (u_a) and momentum energy per unit volume (u_m) are negligible relative to the other two components; hence to first approximation, we can consider the total energy per unit volume u to be made up of u_s and u_f.

$$u = u_s + u_f$$

Practically all of the energy associated with a cutting operation is consumed in either plastic deformation or friction, and in a typical operation effective U_e can be approximated to a value of 75% of the total energy expended. As already derived,

$$u_s = \frac{F_s V_s}{V \times b \times t} = \frac{f_s}{\sin \phi} \times \frac{V_s}{V}$$

$$u_f = \frac{F V_c}{V \times b \times t}$$

The energy balance, i.e., $u = u_s + u_f$ can be verified.

5.22 STRESS AND STRAIN DISTRIBUTIONS IN PLANE FLOW—LEE AND SHAFFER'S MODEL

Lee and Shaffer (1951) used the methods of studying stress and strain distributions in the plane flow of a perfectly plastic material, which is free from work-hardening effects.

The field of stress is shown in Fig. 5.24(a), with the two systems of slip lines being parallel to AC and DB, AC being shear plane, and AB a plane free of stress, is inclined at 45° to AB. Angle ϕ of shear plane AC is chosen so that the chip above AB is stress-free, there is no stress across AB and the slip lines (showing the direction of maximum shear stress) meet AB at 45°. ABC therefore represents an area of uniform stress distribution which can be represented by the Mohr circle as shown in Fig. 4.24(b). The maximum shear stress condition (say, σ_s) fixes the radius of circle at this value, and since there is no force across AB, the circle passes through the origin. The points a to e marked on the circle represent stresses on correspondingly marked elements of chip (normal horizontal and shear stresses). For point e on tool face the shear is represented by ordinate and normal stress by abscissa so that $\angle$ebo will be the friction angle P. Also, from Fig. 5.24(b), $\angle$ebd = η = 45° – β. The perpendicular BD to the shear plane AC is inclined at $\eta + \alpha$ to the vertical, so that $\phi = \eta + \alpha$, and since

$$\eta = 45° - \beta$$
$$\phi = 45° - \beta + \alpha$$

indicating that the shear angle increases with the rake and decreases with increasing tool face friction. By considering velocity components along slip lines and showing them to be constant. Lee and Shaffer postulated that ABC moves as a rigid body with no plastic strain in it across AC there is a discontinuity in tangential component of velocity; and this, corresponding to a line of infinite strain rate, produces the deformation in the chip.

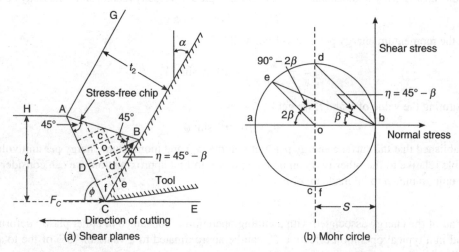

Fig. 5.24 Lee and Shaffer's model of shear angle relation.

An expression for tool force F_c can be derived by considering the equilibrium of the element ABC and since AB is stress-free, the forces on AC must be resolved.

Shear and normal stresses on AC are both equal to f; also, since t_1 = AC sin ϕ, the components of the total force along and normal to AC are both equal to $\dfrac{f_s t_1}{\sin \phi}$. The component of these along CE may be written as

$$F_c = \frac{f_s t_1}{\sin \phi} \cdot \cos \phi \,(\text{for shear}) + \frac{f_s t_1}{\sin \phi} \cdot \sin \phi \,(\text{for normal})$$
$$= f_s t_1 (1 + \cot \phi)$$
$$= f_s t_1 [1 + \cot(45° - \beta + \alpha)] \tag{5.24}$$

From this, it can be seen that F_c increases with friction and decreases with increase in rake. When $\beta = 45°$, it can be seen from Mohr circle that friction shear at tool face = S, the maximum which the chip material can transmit. For larger values of β, shear flow rather than frictional slip occurs at the tool face. Chip thickness t_2 = AC cos η (since AC is perpendicular to BD), and since

$$AC = \frac{t_1}{\sin\phi}$$

So
$$t_2 = \frac{t_1 \cos\eta}{\sin\phi} = \frac{t_1 \cos(45° - \beta)}{\sin(45° - \beta + \alpha)} \qquad (5.25)$$

The ratio $\frac{t_2}{t_1}$ is called the *chip thickness ratio* (r_c). This ratio increases with increasing friction up to $\beta = 45°$ and then decreases with increasing rake.

The solution given by Lee and Shaffer cannot apply for all possible values of β and α since $\sin(45° - \beta + \alpha)$, can be zero for reasonable values of these angles. For example, if $\beta = 45°$, $\alpha = 0$ we get $\phi = 0$, which is impractical. Therefore, for another condition, e.g. small and negative rake or high friction, a certain solution must be sought and such a solution is associated with occurrence of built-up edge on tool, which ensures a positive value of α.

Kronenberg's shear angle relation

Kronenberg disagreed with Merchant as well as Lee and Shaffer. He contended that during chip formation, the layer of metal cut undergoes a momentum change. The cutting process, according to Kronenberg is a dynamic process instead of the static process assumed by his predecessors.

Model

Notations: v_1 = Work valocity
v_2 = Chip velocity
m = Mass of particle
F = Friction
N = Normal reaction
α = Rake angle.

Kronenberg proposed that a particle moving up the rake face of a tool experienced two forces (1) Force F of retardation (2) A centrifugal force against the rake face resulting in a normal reaction N as shown in Fig. 5.25. Thus, consider the motion of a chip particle shown in the figure and applying principle of momentum change, we have the two equations

$$m\frac{dv}{dt} = -F = -\mu N \qquad (1)$$

and
$$m\frac{d\theta}{dt} = N \qquad (2)$$

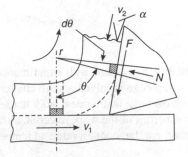

Fig. 5.25

Combining the above Eqs. (1) and (2) we have,

$$\left(\frac{dv}{v}\right) = -\mu d\theta$$

As the velocity changes from v_1 to v_2 we can write:

$$\int_{v_1}^{v_2} \frac{dv}{v} = -\mu \int_0^{(\pi/2)-\alpha} d\theta \qquad (3)$$

$$\therefore \quad \frac{v_1}{v_2} = e^{\mu(\pi/2)-\alpha} = \left(\frac{1}{r_t}\right) \qquad (4)$$

$$\tan\phi = \frac{r\cos\alpha}{1 - r\sin\alpha} = \frac{\cos\alpha}{\left(\frac{1}{r} - \sin\alpha\right)} \qquad (5)$$

$$\therefore \quad \tan\phi = \frac{\cos\alpha}{e^{\mu(\pi/2)-\alpha} - \sin\alpha} \qquad (6)$$

which approximates to

$$\phi = e^{\mu(\pi/2)-\alpha} \qquad (7)$$

5.23 DIFFERENT THEORIES OF SHEAR ANGLE RELATIONSHIP

The shear angle relationship formulae suggested by various researchers are as follows:

Researchers	Shear angle formulae
Merchant and Ernst, 1st formula	$\phi = \dfrac{\pi}{4} + \dfrac{\alpha}{2} - \dfrac{\beta}{2}$
Merchant and Ernst, 2nd formula	$\phi = \dfrac{C}{4} + \dfrac{\alpha}{2} - \dfrac{\beta}{2}$
Lee and Shaffer	$\phi = \dfrac{\pi}{4} + \alpha - \beta$
Kronenberg, 1st formula	$\tan\phi = \dfrac{\cos\alpha}{e^{\mu(\pi/2-\alpha)} - \sin\alpha}$
Kronenberg, 2nd formula	$\phi = e^{\mu(\pi/2-\alpha)}$

Summary

No theory yet produced is in quantitative agreement with experiments even for a limited range of cutting conditions, i.e. where there is no built-up edge involved and no chatter occurs. This is not surprising, since none of the theories takes into account the large strain rates, work hardening and temperatures which exist in the process and their effects on the properties of the work material. However, some of these factors produce effects which tend to cancel each other, e.g. large strain rate and temperature, and again the effects of large strain makes the behaviour of the material tend towards that of a perfectly plastic solid, i.e. its flow stress tends towards a substantially constant value over a wide range of conditions.

Despite the failure of current theories, understanding of the mechanism of the cutting process has been improved. Thus, experiments to check suggestive relations between shear angle, friction angle and rake angle, while not agreeing quantitatively, have proved the link between them with various models. For each of the several materials the existence of linear relation is unaffected by wide changes in the rake angle, cutting speed and use of lubricants.

Thus, for example, the effect of lubricants can be assessed by this effect on the chip friction angle; any reduction in this (or any increase in the rake angle) causes an increase in the shear angle and hence a decrease in chip thickness. For a given rake angle, this causes a decrease in plastic shear strain, and since the estimated shear stress remains approximately constant, the work done in plastic deformation in the cutting process is reduced. Because of decrease in chip friction, work done against friction can be expected to decrease, but this is offset by the increase in chip speed due to smaller thickness.

EXAMPLE 5.1 To investigate the tooling problems faced in a heavy machine shop, an orthogonal cutting test on an alloy tool was carried out, and the following values were observed during the experiment:

Rake angle	= 18°
Depth of cut	= 0.10 mm
Width of cut	= 3.5 mm
Width of chip	= 3.8 mm
Shear stress, f_s	= 20 kg/mm²
Length of chip	= 40 mm
Length of uncut chip	= 120 mm
Coefficient of friction	= 0.75
Velocity of cutting	= 25 m/min

The tool engineer wants to switch over to another tool material. As a consultant to the tool engineer, you are required to provide the necessary data pertaining to forces and the power consumption.

Solution We know
$$l_1 \cdot b_1 \cdot t_1 = l_2 \cdot b_2 \cdot t_2$$
or
$$r_c = \frac{t_1}{t_2} = \frac{l_2 b_2}{l_1 b_1} \qquad (i)$$

where l_1 = length of uncut chip = 120 mm
l_2 = length of chip = 40 mm
b_1 = width of cut = 3.5 mm
b_2 = width of chip = 3.8 mm
t_1 = thickness of uncut chip or depth of cut chip
t_2 = thickness of cut chip
r_c = chip thickness ratio.

Substituting the values, we have
$$\frac{t_1}{t_2} = r_c = \frac{40 \times 3.8}{120 \times 3.5} = \frac{3.8}{10.5} = 0.3619$$

106 TEXTBOOK OF PRODUCTION ENGINEERING

Now
$$\tan\phi = \frac{r_c \cos\alpha}{1 - r_c \sin\alpha} \quad \text{(ii)}$$

where ϕ = shear angle and α = rake angle = 18°. Therefore,

$$\tan\phi = \frac{0.3619 \cos 18°}{1 - 0.3619 \sin 18°}$$

$$= \frac{0.3619 \times 0.951}{1 - 0.3619 \times 0.309}$$

$$= \frac{0.3433}{1 - 0.1115} = \frac{0.3433}{0.8885}$$

$$= 0.38638$$

or $\phi = 21.12°$

$\sin\phi = 0.359$

$\cos\phi = 0.932$

$\mu = \tan\beta = 0.75$

∴ $\beta = 36.86°$

where β is the friction angle.

$\beta - \alpha = 36.86° - 18°$

$= 18.86°$

$\cos(\beta - \alpha) = \cos 18.86°$

$= 0.945$

$\cos(\phi + \beta - \alpha) = \cos(21.12° + 36.86° - 18°)$

$= \cos 39.98°$

$= 0.766$

We know that

$$F_c = \frac{f_s \, t_1 \cdot b_1}{\sin\phi} \times \frac{\cos(\beta - \alpha)}{\cos(\phi + \beta - \alpha)}$$

$$= \frac{20 \times 0.15 \times 3.0}{0.359} \times \frac{0.945}{0.766}$$

or $F_c = 24.0554$ kg

Also,
$$\frac{F_t}{F_c} = \tan(\beta - \alpha) \quad \text{(iii)}$$

or $F_t = F_c \tan(\beta - \alpha)$

$= F_c \tan 18.86°$

$= 24.0555 \times 0.34432$

$= 8.2815$ kg

Power consumption = $\dfrac{F_c \times V_c}{60 \times 75}$; where $F_c = 24.055$ kg and $V_c = 25$ m/min

$$= \dfrac{24.055 \times 25}{60 \times 75} = \dfrac{24.055}{180}$$

$$= 0.1336 \text{ HP}$$

EXAMPLE 5.2 The following observations were made during an orthogonal cutting with 10° positive rake HSS tool on a tubular workpiece with 3 mm thickness:
Feed = 0.20 mm/rev
$F_c = 150$ kg
Cutting speed = 100 m/min
$F_t = 50$ kg
Chip thickness = 0.45 mm
Calculate the chip thickness ratio, shear angle, coefficient of friction and length of shear plane, $F, N, F_s, F_n, \sigma, \beta$.

Solution Given:
$$f = 0.2 \text{ mm/rev}$$
$$V_c = 100 \text{ m/min}$$
Chip thickness = 0.45 mm
$$F_c = 150 \text{ kg}$$
$$F_t = 50 \text{ kg}$$
$$\alpha = 10°$$
$$b_1 = 3 \text{ mm}$$

From the given data

$$r_c = \dfrac{t_1}{t_2} = \dfrac{0.20}{0.45} = 0.44$$

$$\tan \phi = \dfrac{r_c \cos \alpha}{1 - r_c \sin \alpha}$$

$$= \dfrac{0.44 \times 0.985}{1 - 0.44 \times 0.174}$$

$$= \dfrac{0.434}{1 - 0.0765}$$

$$= 0.468$$

$$\phi = \tan^{-1}(0.468) = 25°$$

$$F = F_t \cos \alpha + F_c \sin \alpha$$
$$= 50 \times 0.985 + 150 \times 0.174$$
$$= 49.7 + 26.2$$
$$= 75.9 \text{ kgf}$$

$$N = F_c \cos \alpha - F_t \sin \alpha$$
$$= 150 \times 0.985 - 50 \times 0.174$$

$$= 148 - 8.7$$
$$= 139.3 \text{ kgf}$$
$$\mu = \frac{F}{N} = \frac{75.9}{139.3} = 0.545$$
$$\frac{t_1}{AB} = \sin\phi; \quad \text{where AB is length of shear plane.}$$
$$AB = \frac{t_1}{\sin\phi} = \frac{0.2}{\sin 25}$$
$$= \frac{0.2}{0.4225} = 0.473 \text{ mm}$$
$$F_s = F_c \cos\phi - F_t \sin\phi$$
$$= 150 \times 0.906 - 50 \times 0.4225$$
$$= 136 - 21.1$$
$$= 114.9 \text{ kgf}$$
$$F_n = F_t \cos\phi + F_c \sin\phi$$
$$= 50 \times 0.906 + 150 \times 0.4225$$
$$= 45.4 + 63.5$$
$$= 108.9 \text{ kgf}$$
$$\sigma = \frac{F_c \sin\phi + F_t \cos\phi}{b_1 \times t_1}$$
$$= \frac{63.5 + 45.4}{3 \times 0.2}$$
$$= \frac{108.9}{0.6}$$
$$= 181.5 \text{ kg/mm}$$
$$f_s = \frac{F_c \cos\phi - F_t \sin\phi}{b_1 \times t_1} \cdot \sin\phi$$
$$= \frac{114.9}{0.6} \times 0.4225$$
$$= 80.91 \text{ kg/mm}^2$$

EXAMPLE 5.3 Estimate by employing a plasticity relationship

$$\phi + \beta - \alpha = \frac{\pi}{4}$$

the cutting force involved in cutting SAE 1020 steel of ultimate strength $\sigma_u = 35$ kg/mm^2 and % elongation $\delta = 33\%$ with 0, 10, 7, 7, 8, 0–0.5 mm shape cemented carbide tool, under the following cutting conditions: feed = 0.3 mm/rev, depth of cut = 3.00 mm, kinetic coefficient of friction = 0.68.

Solution The given information can be presented as follows:

0	10	7	7	8	0	0.5 mm
Back rake	Side rake	End relief	End relief	ECE angle	SCE angle	Nose rod

$$f_s = 0.74\sigma_u \, 6^{0.6\delta}$$
$$= 0.74 \times 35 \times 6^{0.6 \times 0.33}$$
$$= 0.74 \times 35 \times 6^{0.2}$$
$$= 0.74 \times 35 \times 1.43$$
$$f_c = 37.2 \text{ kg/m}^2$$

This formula used estimate dynamic shear stress on the shear plane f_s.

σ_u = ultimate tensile strength of work material
δ = elongation fraction of a test specimen.

For the given tool
$$\alpha = 10°$$
$$\beta = \tan^{-1} \mu = \tan^{-1} 0.68 = 34.2°$$
$$\phi = \frac{\pi}{4} - \beta + \alpha$$
$$= 45 - 34.2 + 10 = 20.8°$$

Now we have
$$\tan(\beta - \alpha) = \frac{F_t}{F_c} = \tan 24.4$$

or
$$F_t = F_c \times 0.45$$
$$f_s = \frac{F_c \cos\phi - F_t \sin\phi}{b_1 \times t_1}$$
$$f_s = \frac{F_c \times 0.93 - 0.45 \times 0.36 \, F_c}{3 \times 0.3}$$
$$f_s = F_c \left(\frac{0.93 - 0.45 \times 0.36}{0.9} \right)$$
$$f_s = F_c \times 0.853 = 37.2$$
$$F_c = \frac{37.2}{0.853}$$
$$= 43.61 \text{ kg}$$

EXAMPLE 5.4 In an orthogonal cutting set-up, the depth of cut is 12 mm, feed 1 mm/revolution, cutting speed 70 mpm, back rake angle 8°, chip thickness ratio 0.30, and shear stress of material at zero compressive stress 100 kg/cm². Assume that the value of the constant k in the equation
$$2\phi + \beta - \alpha = \cot^{-1} k = C$$
is 0.22.

Calculate the resultant force, rate of metal removed, shear strain, and h.p. at the tool per cubic centimetre of metal removed per minute.

Solution We are given that

$$\text{Rake angle} = \alpha = 8°$$
$$\text{Chip thickness ratio} = r_c = 0.3$$

We know

$$\tan\phi = \frac{r_c \cos\alpha}{1 - r_c \sin\alpha}$$

Substituting the values, we have

$$= \frac{0.3 \cos 8°}{1 - 0.3 \sin 8°} = \frac{0.3 \times 0.9902}{1 - 0.3 \times 0.139}$$

$$= \frac{0.29706}{1 - 0.0417} = \frac{0.29706}{0.9583}$$

$$= 0.3099$$

or $\quad \phi = 17.218°$

Now we are given

$$\cot^{-1} k = 0.22 = 77.29 = C$$

Substituting the values of ϕ and α, we have

$$2 \times 17.218 + \beta - 8 = 77.29$$

$\therefore \quad \beta = 50.99°$

We also know

$$f_s = f_s o + k\sigma$$

where f_s is the shear stress;
$f_s o$ is the shear stress at zero compressive stress;
k is a constant.

$$\frac{f_s}{\sigma} = \cot(\phi + \beta - \alpha)$$

$$\sigma = f_s \tan(\phi + \beta - \alpha)$$
$$f_s = f_s o + k f_s \tan(\phi + \beta - \alpha)$$

But $\quad f_s o = 1000 \text{ kg/cm}^2 \text{ or } 10 \text{ kg/mm}^2$

Now

$$f_s = \frac{\tau \cdot \tau \, bt \cos(\beta - \alpha)}{\sin\phi \cos(\phi + \beta - \alpha)} \qquad \text{(i)}$$

Substituting the values of f_s, b, t, ϕ, β, α in Eq. (i), we have

$$F_c = \frac{f_s o \, bt \cos(\beta - \alpha)}{\{1 - k \tan(\phi + \beta - \alpha)\}\{\sin\phi \cos\theta(\phi + \beta - \alpha)\}}$$

$$= \frac{10 \times 12 \times 1 \times \cos(50.99 - 8)}{\{1 - 0.22 \tan(17.2 + 50.99 - 8)\} \times \{\sin 17.2 \cos(17.2 + 50.99 - 8)\}}$$

$$= \frac{120 \cos 42.99}{\{1 - 0.22 \tan 60.19\} \times \{\sin 17.2 \cos 60.19\}}$$

$$= \frac{120 \times 0.7325}{(1 - 0.22 \times 1.74) \times (0.295 \times 0.4971)}$$

$$= \frac{87.9051}{(1 - 0.383) \times (0.14700)}$$

$$= \frac{87.9051}{0.617 \times 0.147} = \frac{87.9051}{0.09069}$$

$$F_c = 969.29 \text{ kg}$$
$$F_t = f_c \tan(\beta - \alpha)$$
$$= f_c \tan(50.99 - 8)$$
$$= f_c \tan 42.99$$
$$= 969.29 \times 0.9322$$
$$= 903.57 \text{ kg}$$

$$\text{Resultant force } (R) = \sqrt{F_c^2 + F_t^2}$$
$$= \sqrt{(969.29)^2 + (903.57)^2}$$
$$= \sqrt{939523 + 816493}$$
$$= \sqrt{1776581.615}$$
$$= 1325.15 \text{ kg}$$

Rate of metal removed $= bt \times V$
$$= 12 \times 1 \times 70 \times 100 \text{ mm}^3/\text{min}$$
$$= 840 \times 100 \text{ mm}^3/\text{min}$$
$$= 84 \text{ cm}^3/\text{min}$$

$$\text{HP} = \frac{F_c \times V_c}{60 \times 75} = \frac{969.29 \times 70}{4500}$$
$$= 15.08$$

$$\frac{\text{HP}}{\text{Rate of metal removed}} = \frac{15.08}{84} = 0.18$$

Shear strain $\gamma = \cot\phi + \tan(\phi - \alpha)$
$$= \cot 17.2 + \tan(17.2 - 8)$$
$$= \cot 17.2 + \tan 9.2 = 0.16 + 3.23$$
$$= 3.3942$$

EXAMPLE 5.5 ABC Ltd., engaged in the manufacture of tubes and pipes for power projects, has a modern tool room. Suddenly the production targets on the shop floor for condenser tubes are increased to cope with the increased demand. The tool engineer, after investigation, found that the present tool material needs to be replaced by better tool material. To find out the same, he conducted an experiment on a M.S. tube of 250 mm diameter and 3.3 mm thickness. An orthogonal cut was taken with a cutting speed of 109 mpm and 0.15 mm feed per revolution with a cutting tool having back rake angle of –10° and cutting force 200 kg, and feed force of 40 kg is provided. The net power consumed for cutting is 3.5 h.p. and the chip thickness generated is 0.3 mm. Compute the data required for replacement of tool material.

Solution   We know:

Chip thickness ratio,
$$r_c = \frac{\text{Feed}}{\text{Chip thickness}}$$

for orthogonal cutting.

We are given that feed = 0.15 and chip thickness = 0.30.

$\therefore$
$$r_c = \frac{0.15}{0.30} = 0.5$$

We also know for negative rake angle
$$\tan \phi = \frac{r_c \cos \alpha}{1 + r_c \sin \alpha}$$

We are given that
$$\alpha = -10°$$

Substituting the values, we have
$$\tan \phi = \frac{0.5 \cos 10°}{1 + 0.5 \sin 10°}$$

$$= \frac{0.4925}{1.0865} = 0.4532$$

$$= 24.38°$$

Shear strain = $\tan (\phi - \alpha) + \cot \phi$

$\qquad\qquad\quad = \tan (24.38° - (-10°)) + \cot 24.38°$

$\qquad\qquad\quad = \tan 34.38° + \dfrac{1}{\tan 24.38°}$

$\qquad\qquad\quad = 0.679 + \dfrac{1}{0.4532}$

$\qquad\qquad\quad = 2.8855$

Shear energy per unit volume = Shear stress × Shear strain

$$\text{Shear stress } f_s = \frac{F_s}{A_s} = \frac{F_c \cos \phi - F_t \sin \phi}{b \cdot t} \times \sin \phi$$

Substituting the values, we have

$$f_s = \frac{200 \cos 24.38° - 50 \sin 24.38°}{0.15 \times 3.5} \times \sin 24.38°$$

$$= \frac{200 \times 0.91 - 50 \times 0.413}{0.15 \times 3.5} \times 0.413$$

$$= \frac{182 - 20.65}{0.525} \times 0.413$$

$$= \frac{161.35 \times 0.413}{0.525}$$

$$= 126.93 \text{ kg/mm}^2$$

Shear energy per unit volume = 126.93×2.8855

$$= 366.3 \text{ kg/mm}^2$$

EXAMPLE 5.6 A small tool manufacturer is interested in the development of a single-point tool to be used in the machining of small servomotor cylinders, which is required in the hydraulic control system of the loop line of the automatic NC machine tool. The basic requirements established by tool engineer for tool development are as follows:

(a) Coefficient of friction
(b) Shear plane angle
(c) Velocity of chip along tool face
(d) Chip thickness.

To determine above factors, a seemless tubing 40 mm outside diameter is turned on a lathe. The rake angle used on tool is 32°. Cutting speed 15 mpm and feed 0.8 mm/rev is used. The length of continuous chip in one revolution is 60 mm. The cutting force and feed force are 250 kg and 100 kg respectively. Compute the required data for tool engineer.

Solution We are given:
Rake angle (α) = 32°
Outside diameter of tube = 40 mm
Cutting speed V_c = 12 m/min
Feed = 0.8 mm/rev
l_c = 60 mm
Feed force = F_t = 100 kg
Cutting force = F_c = 250 kg.

We know that the coefficient of friction is given by

$$\mu = \frac{F_t + F_c \tan \alpha}{F_c - F_t \tan \alpha}$$

By substituting the values, we get

$$\mu = \frac{100 + 250 \tan 32°}{250 - 100 \tan 32°}$$

$$= \frac{100 + 250 \times 0.624}{250 - 100 \times 0.624}$$

$$= \frac{100 + 156}{250 - 62.4}$$

$$= \frac{256.0}{187.6} = 1.354$$

Shear plane angle is given by

$$\tan\phi = \frac{r_c \cos\alpha}{1 - r_c \sin\alpha}$$

where r_c = chip thickness ratio.

But

$$r_c = \frac{t_1}{t_2} = \frac{l_2}{l_1}$$

where t_1 = depth of cut (feed in this case)
t_2 = thickness of chip
l_1 = length of uncut chip
l_2 = length of chip = 60 mm
   $= \pi D$

$$r_c = \frac{60}{3.14 \times 40} = 0.477$$

$$\tan\phi = \frac{r_c \cos\alpha}{1 - r_c \sin\alpha}$$

$$= \frac{0.47 \times \cos 32°}{1 - 0.477 \sin 32°}$$

$$= \frac{0.47 \times 0.8480}{1 - 0.477 \times 0.5299}$$

$$= \frac{0.404}{1 - 0.248} = \frac{0.404}{0.7552} = 0.529$$

$$\phi = 27.5°$$

Chip velocity

$$V_f = V_c \times r_c$$
$$= 15 \times 0.477$$
$$= 7.155 \text{ m/min}$$

Feed chip thickness

$$t_2 = \frac{\text{Feed}}{r_c}$$

$$= \frac{0.8}{0.477} = 1.677 \text{ mm}$$

EXAMPLE 5.7 In an orthogonal cutting experiment conducted in a tool and gauge department of a company, the tangential force was 150 kg and feed force was 200 kg. In the experiment the tool used had a rake angle of 10° and it was established that chip thickness ratio was 0.4. The tool engineer is interested to find out the compressive and shear forces on the shear plane and also the coefficient of friction of chip on the tool face, for further investigation of tool life pattern and improvement in tool signature. Compute the data required by the tool engineer.

Solution Compressive force on shear plane (N_s) is given by the formula:
$$N_s = F_c \sin\phi + F_t \cos\phi \qquad \text{(i)}$$
where F_c = Cutting force in kg
F_t = Tangential force in kg
ϕ = Shear angle.
Also, the shear force on shear plane is given by
$$F_s = F_c \cos\phi - F_t \sin\phi \qquad \text{(ii)}$$
But
$$\tan\phi = \frac{r_c \cos\alpha}{1 - r_c \sin\alpha} \qquad \text{(iii)}$$
where r_c = chip thickness ratio
α = rake angle.
In the problem, we are given:
$$\alpha = 10° \qquad F_t = 150 \text{ kg}$$
$$r_c = 0.4 \qquad F_c = 200 \text{ kg}$$

Substituting the values in Eq. (iii), we get
$$\tan\phi = \frac{0.4 \cos 10°}{1 - 0.4 \sin 10°}$$
or
$$\tan\phi = \frac{0.4 \times 0.984}{1 - (0.4 \times 0.173)}$$
$$= \frac{0.3936}{0.9308} = 0.4228$$
or
$$\phi = 22.91°$$

Substituting the values of ϕ, F_c and F_t in Eq. (i) and (ii), we get
$$N_s = F_c \sin\phi + F_t \cos\phi$$
$$= 200 \times 0.3892 + 150 \times 0.921$$
$$= 77.85 + 138.16$$
$$= 216.01 \text{ kg}$$
$$F_s = F_c \cos\phi - F_t \sin\phi$$
$$= 200 \times 0.921 - 150 \times 0.389$$
$$= 184.20 - 58.450$$
$$= 125.75$$

Coefficient of friction
$$\mu = \frac{F_t + F_c \tan\alpha}{F_c - F_t \tan\alpha}$$
$$= \frac{150 + 200 \tan 10°}{200 - 150 \tan 10°}$$
$$= \frac{150 + 200 \times 0.176}{200 - 150 \times 0.176}$$

$$= \frac{150 + 35.2}{200 - 26.4}$$

$$= \frac{185.2}{173.6}$$

$$= 1.067$$

EXAMPLE 5.8 Show that the shearing strain γ in orthogonal cutting is given by

$$\gamma = \frac{r_c^2 - 2r_c \sin\alpha + 1}{r_c \cos\alpha}$$

where α = rake angle

r_c = chip thickness ratio = $\dfrac{t_1}{t_2}$; where t_2 is chip thickness and t_1 is depth of cut.

Solution We know that

$$\tan\phi = \frac{r_c \cos\alpha}{1 - r_c \sin\alpha}$$

where ϕ = shear angle (see Fig. 5.26)

$$AC = \sqrt{(r_c \cos\alpha)^2 + (1 - r_c \sin\alpha)^2}$$

$$= \sqrt{r_c^2 \cos^2\alpha + 1 + r_c^2 \sin^2\alpha - 2r_c \sin\alpha}$$

$$AC = \sqrt{1 + r_c^2 - 2r_c \sin\alpha}$$

$$\therefore \quad \sin\phi = \frac{r_c \cos\alpha}{\sqrt{1 + r_c^2 - 2r_c \sin\alpha}}$$

$$\cos\phi = \frac{1 - r_c \sin\alpha}{\sqrt{1 + r_c^2 - 2r_c \sin\alpha}}$$

Now

$$\gamma = \frac{\cos\alpha}{\sin\phi \cos(\phi - \alpha)}$$

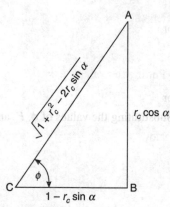

Fig. 5.26 Example 5.8.

Putting the values, we get

$$\gamma = \frac{\cos\alpha}{\dfrac{r_c \cos\alpha}{\sqrt{1 + r_c^2 - 2r_c \sin\alpha}} \left(\dfrac{1 - r_c \sin\alpha}{\sqrt{1 + r_c^2 - 2r_c \sin\alpha}} \cos\alpha + \dfrac{r_c \sin\alpha \cos\alpha}{\sqrt{1 + r_c^2 - 2r_c \sin\alpha}} \right)}$$

$$= \frac{\cos\alpha (1 + r_c^2 - 2r_c \sin\alpha)}{r_c(1 - r_c \sin\alpha)\cos^2\alpha + r_c^2 \cos^2\alpha \sin\alpha}$$

$$= \frac{r_c^2 - 2r_c \sin\alpha + 1}{r_c \cos\alpha}$$

Hence proved.

EXAMPLE 5.9 Calculate by employing plasticity relationship, $\phi + \beta - \alpha = \pi/4$, the cutting force F_c involved in cutting SAE 1020 steel (ultimate strength $\sigma = 55$ kg/mm^2 and % elongation = 33%) employing a 0, 10, 7, 7, 8, 0, 0.5 mm shaped cemented carbide tool under the following cutting conditions: feed = 0.3 mm/rev, depth of cut = 3 mm, approximate kinetic coefficient of friction = 0.68. Assume that the values of relationship for dynamic stress are

$$f_s = 0.74 \sigma_u\, 6^{0.6\delta}$$

Solution Coefficient of friction

$$\tan \beta = \mu = 0.68$$

$$\therefore \quad \beta = \tan^{-1} 0.68$$
$$= 34.2$$

Now, from the given relation we can determine the angle

$$\phi + \beta - \alpha = \frac{\pi}{4}$$

or
$$\phi + 34.2° - 10° = 45°$$
$$\phi = 55° - 34.2°$$
$$= 20.8°$$

Again, given that

$$f_s = 0.74 \times \sigma_u \times 6^{0.6} \times \delta$$

where f_s = shear stress
δ (percentage elongation) = 0.33
σ_u (ultimate strength) = 35 kg/mm^2.

Putting the values, we get

$$f_s = 0.74 \times 35 \times 6^{0.6} \times 0.33$$
$$= 25.9 \times 1.425 = 36.9$$

Now F_S (shear force) $= f_s \times A_s = \dfrac{f_s \times A}{\sin \phi}$

$$A = b \times t$$
$$\sin \phi = \sin 20.8$$
$$F_s = \frac{36.9 \times 3 \times 0.3}{0.355}$$
$$= 93.6 \text{ kg}$$

Now, from Fig. 5.27, we determine the different angles

$$\angle AOB = 90° - \beta$$
$$\angle HOC = 90° - \alpha$$
$$\angle BOC = 90° - \alpha - 90° + \beta$$
$$= \beta - \alpha$$

Therefore in $\triangle BOD$

$$\angle BOD = \phi + \beta - \alpha$$
$$= 20.8° + 34.2° - 10°$$
$$= 20.8° + 24.2°$$
$$= 45°$$

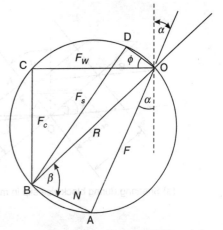

Fig. 5.27 Example 5.9.

$$R = \frac{F_s}{\cos 45} = 93.6 \times \sqrt{2} \text{ kg}$$

Now considering $\triangle BOC$

$$F_c = R \cos BOC$$
$$= 93.6 \times \sqrt{2} \times \cos 24.2°$$

∴ Cutting force $F_c = 121$ kg

EXAMPLE 5.10 Show that the condition of minimum shearing strain in metal cutting is $\phi = \frac{\pi}{4} + \frac{\alpha}{2}$.

Solution Figure 5.28(a) indicates the overall geometry of shearing causing blockwise slip along the shear plane during metal cutting. From the local configuration given in Fig. 5.28(b)

Shearing strain
$$\gamma = \frac{\Delta s}{\Delta y} = \frac{BC + CO}{AC}$$

Since AO is assumed equal to AC

$$\gamma = \frac{BC}{AC} + \frac{OC}{AC}$$
$$= \cot \phi + \tan(\phi - \alpha)$$
$$= \frac{\cos \phi}{\sin \phi} + \frac{\sin(\phi - \alpha)}{\cos(\phi - \alpha)}$$
$$= \frac{\cos \phi \cos(\phi - \alpha) + \sin \phi \sin(\phi - \alpha)}{\sin \phi \cos(\phi - \alpha)}$$
$$= \frac{\cos(\phi - \phi + \alpha)}{\sin \phi \cos(\phi - \alpha)}$$
$$= \frac{\cos \alpha}{\sin \phi \cos(\phi - \alpha)}$$

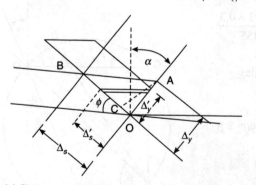

(a) Shearing during blockwise slip in metal cutting

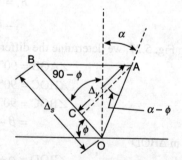

(b) Enlarged view for geometrical derivation

Fig. 5.28 Example 5.10.

For minimum shearing force ϕ_{max}

$$\frac{d\gamma}{d\phi} = 0$$

or

$$\frac{0 - \cos\phi \cos(\phi - \alpha) + \sin\phi \sin(\phi - \alpha)}{\sin\phi \cos(\phi - \alpha)^2} = 0$$

or

$$\cos(\phi + \phi - \alpha) = 0$$

$$2\phi - \alpha = \frac{\pi}{2}$$

Therefore, for minimum shear strain

$$\boxed{\phi = \frac{\pi}{4} + \frac{\alpha}{2}}$$

REVIEW QUESTIONS

1. What is metal cutting? Explain the characteristics of metals.
2. What are the elements of the metal cutting process? How is metal cutting process presented?
3. Differentiate between orthogonal and oblique cutting.
4. What is the effect of cutting speed, depth of cut and feed rate on the forces on cutting tool?
5. Explain the concept of chip formation.
6. What do you understand by the term *geometry of chip*? What are the different methods used for determining chip geometry?
7. Explain the various types of chips.
8. Why is a built-up edge on a tool undesirable?
9. Explain the effects of the following parameters on chip formation:
    (i) Velocity
    (ii) Material of workpiece
    (iii) Depth of cut and feed
    (iv) Tool geometry.
10. What is the need of chip control?
11. Discuss the methods of chip control.
12. Draw Merchant's force diagram. State the assumptions made in the development of such a diagram.
13. In an orthogonal cutting process, following data were observed—chip length of 80 mm was obtained with an uncut chip length of 200 mm and the rake angle used was 20° and depth of cut 0.5 mm. The horizontal and vertical components of cutting force F_H and F_V were 2000 N and 200 N respectively. Determine the shear plane angle, chip thickness, friction angle and resultant cutting force.
14. A mild steel tubing of 50 mm outside diameter is turned on a lathe with cutting speed of 20 mpm with a tool having tool rake angle of 35°. The tool is given a feed of 0.10 mm/rev. and it is found by dynamometer that the cutting force is 250 kg and feed force is 100 kg. Length of continuous chip in one revolution is 80 mm. Calculate the coefficient of friction, shear plane angle, velocity of chip along tool face and chip thickness.

15. A.M.S. bar of 12 cm diameter is turned with a HSS tool at a cutting speed of 0.2 mpm with feed of 0.2 mm per rev. and depth of cut 5 mm. It is found that machine consumes 0.5 kW when running ideal and 2.5 kW when cutting. Determine the tangential force on the tool, the normal pressure on the chip, and shear resistance assuming the shear angle as 20°.

16. An experiment was conducted on a M.S. tube of 200 mm diameter and 3 mm thick. An orthogonal cut was taken with a cutting speed of 80 mm and 0.15 mm per rev. feed with a cutting tool having back rake angle of –10°. It was determined that cutting force was 150 kg, and feed force was 40 kg, net horse power for cutting was 3 h.p. and chip thickness was 0.25 mm. Calculate the shear strain and strain energy per unit volume.

17. What do you understand by shear strain? Derive an expression for it.

18. A lathe is used to machine a steel bar of 100 mm diameter at 900 rpm. The cutting force applied by the tool to the work is 600 N.
    Determine
    (i) Initial cutting velocity
    (ii) Cutting velocity when bar is reduced to half the size
    (iii) Power consumed in both the cases.

19. A steel specimen requires 4 W/mm³/s of power to turn it. If maximum power available at a machine spindle is 5 kW, determine the maximum material removal rate.
    Also determine the cutting force and depth of cut at maximum material removal rate if cutting speed is 30 m/min and feed rate is 0.2 mm/rev.

20. A lathe while running consumes 500 W and 2500 W when cutting a steel specimen at 30 m/min. Determine the cutting force and torque at the spindle at 120 rpm. Also determine the specific power consumption if the depth of cut is 4 mm and feed is 0.25 mm/rev.

21. A bar 75 mm diameter is required to 70 mm diameter by means of a cutting tool for which $x = 90°$ and for which the cutting edge lies in the plane containing the work axis of rotation. The mean length of the cut chip of 73 mm, the rake angle α of 15° and a feed of 0.2 mm/rev. is used. Find the cutting ratio and the value of shear plane angle.

22. The following data were obtained from a cutting test—$\alpha = 20°$, $k = 90°$, depth of cut = 6.4 mm, feed = 0.25 mm/rev., chip length before cutting = 29.4 mm, chip length after cutting = 12.9 mm. The cutting forces were—axial force = 427 N, vertical force = 1050 N. Use Merchant's analysis to calculate
    (i) the direction and magnitude of the resultant force
    (ii) the shear plane angle
    (iii) the frictional force
    (iv) the friction angle.

23. The following values relate to a cutting test under orthogonal cutting conditions for machining aluminium. Forces as determined by dynamometer F_c and F_t are:
    $$F_c = 1500 \text{ N}, F_t = 1000 \text{ N}, \alpha = 10°, r = t_1/t_2 = 0.37$$
    Determine as per Merchant's theory, the cutting forces N_s, F_s, N and F. Also determine the coefficient of friction at the chip-tool interface.

24. During orthogonal cutting of a MS tube at 15 m/min with a HSS tool having 15° rake angle, the chip thickness ratio was 0.35 and the friction force on the tool chip interface measured by means of a special setup was 48 kg with coefficient of friction 0.6. Estimate the component of the cutting force's shear angle, shear strain and work done in deformation.

25. Using Merchant's circle diagram, derive the questions to find
   (i) specific energy of cutting and power consumption, shear and friction
   (ii) kinetic coefficient or friction
   (iii) normal and shear stresses on rake and shear planes.

26. (i) Show with a neat sketch the forces acting on a chip in orthogonal machining. Derive an expression to calculate the coefficient of friction between tool chip interface.

   (ii) During an orthogonal machining operation on mild steel, the results obtained are:

   Uncut chip thickness = 0.25 mm, chip thickness = 0.75 mm, width of the cut = 2.5 mm, rake angle = 0°, horizontal cutting force = 900 N, thrust force = 400 N.

   Compute the coefficient of friction between the tool and chip interface. Determine also the ultimate shear stress of the work material.

CHAPTER 6

Theory of Multipoint Machining

6.1 INTRODUCTION

Cutters which have more than one edge to work usually have a higher metal removal rate than single-edged tools, and the life of the cutter between regrinds is also raised by increasing the number of cutting edges. The mechanism of multi-point cutting tools is described below.

Drilling consists of originating holes in workpieces by rotary and axial movement of either the tool (drill) or the workpiece. The cutting action of the cutting lips of a drill is more or less an inclined and oblique cutting process. Shaw, Cook and Smith (1952) linked the drilling process to the basic three-dimensional cutting process and demonstrated the essential similarities between the drill point and the conventional turning tool. As shown in Fig. 6.1, a drill is superimposed on the basic turning tool.

The twist drill is unique in a way that it involves two different metal deformation processes. The main cutting action is along the lip of the drill, which is like a single-point tool. Apart from the curvature of the finished surface, the chip is quite similar to the chip produced by a single-point tool. The metal deformation under the chisel edge is much more complex. At the exact centre of the hole the only motion of the drill is axial; so the deformation resembles that produced by an indenting punch. As the radius increases, the rotation of the drill becomes important and the chisel edge wedge appears to both cut and extrude the metal. An analysis of this region of the drill based upon such an assumption fits experimental data. An attempt was also made by Oxford to analyse the basic mechanics of the drilling process.

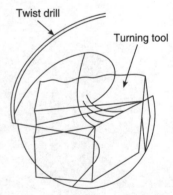

Fig. 6.1 Drill superimposed on the basic turning tool.

6.2 MECHANISM OF DRILLING

The twist drill having two flutes shown in Fig. 6.2 is the type used in job-shop and production work, unless special conditions require the use of another type.

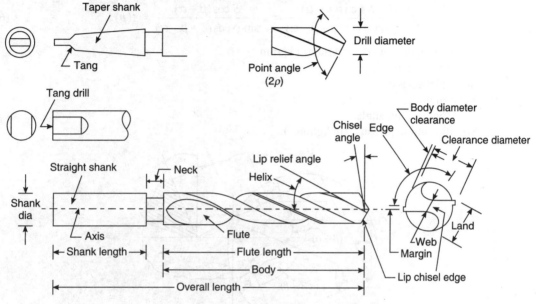

Fig. 6.2 Nomenclature for twist drills.

Factors affecting the drill, which fix the magnitude of the cutting force, depend on:

(i) Properties of the material to be drilled
(ii) Shape of drill (diameter, rake angles, length and position of cross edge, etc.)
(iii) Chip section (drill diameter, feed rate, etc.)
(iv) Cutting conditions (depth of drilled hole, coolant, etc.).

Consider that the effective rake angle varies over the length of the cutting edge. It is difficult to state approximate specific cutting resistance for different materials without knowing the drill diameter, because the specific cutting resistance depends upon the rake angle.

The chip section is determined by the drill diameter and feed rate. Increase in diameter and feed rate increases the drill torque and drill thrust. Use of a pilot hole (of web thickness diameter) reduces the drill thrust by 50%, but has insignificant influence on drill torque. The forces decrease with growing helix angle. The axial thrust decreases with decreasing point angle whilst the torque increases, because at equal diameter the length of the cutting edge grows, and the chip are thinner at equal feed rate. The smallest forces occur if the cross edge lies at an angle between 55° and 60°. The influence of relief angle upon the forces is negligible.

Finally, consider the cutting conditions—lubrication, coolant, depth of drilled hole, friction, etc. The friction at the outside diameter of the bore as well as the requirements of chip removal add to the cutting resistance; consequently the magnitudes of the forces and torques are affected by the efficiency of cooling and lubrication during cutting.

6.3 TORQUE AND THRUST IN DRILLING PROCESSES AT LIPS AND CHISEL

It is convenient to calculate the torque and thrust in metal drilling due to lips by using Merchant formulae.
From Merchant's circle (refer to Fig. 6.3)

$$F_c = \frac{SA\cos(\beta-\alpha)}{\sin\phi\cos(\phi+\beta-\alpha)} = A\frac{S\cos(\beta-\alpha)}{\sin\phi\cos(\phi+\beta-\alpha)} = A(p) \qquad (6.1)$$

where S = Shear strength of work material (in kg/mm^2)
 A = Uncut area (in mm^2)
 ϕ = Shear plane angle
 α = Rake angle
 β = Chip tool friction angle
 p = Specific cutting pressure (in kg/mm^2)

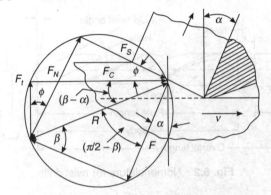

Fig. 6.3 Merchant's circle diagram.

If the specific cutting pressure is known and uncut area is known, it is possible to estimate the cutting force F_c.

From Merchant's circle diagram, thrust

$$F_t = F_c \tan(\beta-\alpha) \qquad (6.2)$$

The formulae for F_c and F_t can be used to estimate torque and thrust in drilling. If D is the drill diameter and c is the chisel edge length, the figure is given by (refer to Fig. 6.4)

$$T = F_c\left(\frac{D+c}{2}\right)$$

Substituting $F_c = A \cdot p$ and $A = \left(\frac{D-c}{2}\right)\frac{f}{2}$, we have

$$T = p\left(\frac{D-c}{2}\right)\frac{f}{2} \times \left(\frac{D+c}{2}\right)$$

$$= pf\left(\frac{D^2-c^2}{8}\right) \qquad (6.3)$$

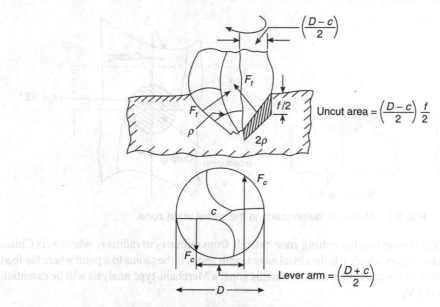

Fig. 6.4 Cutting action of two-fluted twist drill—force system.

Drill thrust per lip = $F_t \sin \rho$

Thrust on both the lips

$$THL = 2F_t \sin \rho$$

where 2ρ is the point angle of the drill.

While the chisel edge does not contribute much to the total torque, it contributes 50% to the total thrust. Thus

$$\text{Total thrust} = 4F_t \sin \rho$$

Noting that from Merchant's circle diagram

$$F_t = F_c \tan(\beta - \alpha)$$

the final equation for drill thrust is total thrust = $4F_c \tan(\beta - \alpha) \sin \rho$

Substituting $F_c = \left(\dfrac{D-c}{2}\right)\dfrac{f}{2} \times p$, we get total thrust, i.e.

$$4p\left[\left(\frac{D-c}{2}\right)\frac{f}{2}\tan(\beta-\alpha)\right]\sin \rho \qquad (6.4)$$

6.4 MECHANICS OF METAL CUTTING IN CHISEL EDGE ZONE OF TWIST DRILL

The chisel edge of a twist drill cut, metal in two modes. These are as follows:

1. In the first mode there is metal cutting at a high negative rake from radius r_1 to radius r_2 as shown in Fig. 6.5.

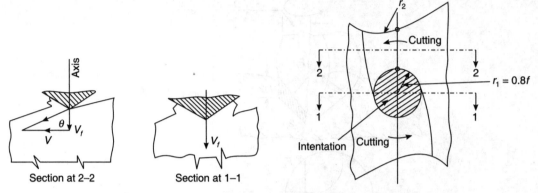

Fig. 6.5 Modes of deformation in the chisel edge zone.

Oxford (1955) found that the cutting zone extends from radius r_2 to radius r_1 where r_2 is Chisel edge corner radius $c/2$, in which, c is the chisel edge length and r_1 is the radius to a point where the feed velocity V_f is 20% of the rotational velocity. In this zone, a Merchant-type analysis will be essential.

At r_1: $V_f = 0.2\ V_1$

Thus
$$\tan \theta_1 = \frac{V_1}{V_f} = \frac{2\pi r_1 N}{fN} = \frac{1}{0.2} = 5 \tag{6.5}$$

∴
$$r_1 = \frac{5}{2}\frac{f}{\pi} = 0.8f$$

Helix angle of drill feed at $r_1 = \tan^{-1} 5 = (78.7°) \simeq 78°$

Helix angle of drill feed at $r_2 \times 90°$, and at r_m it is $\dfrac{90° + 78°}{2} = 84°$.

2. In the second mode, the chisel edge acts as a wedge-type indentor. In this zone, the plasticity theory of wedge indentation is essential. This zone extends from radius zero to radius r_1.

6.4.1 Analysis for Cutting Zone

Torque
$$T = 2r_m F_T = 2r_m (F_c \sin \theta_m - F_t \cos \theta_m) \tag{6.6}$$

Thrust
$$P = 2F_a = 2 (F_t \sin \theta_m + F_c \cos \theta_m) \tag{6.7}$$

To estimate torque and thrust in the cutting zone, F_c and F_t should be calculated from Merchant's analysis. The formulae for F_c and F_t are, if K_c is shear strength in the chisel edge zone (Fig. 6.6).

$$F_c = \frac{K_c (r_2 - r_1)\left(\dfrac{f}{2}\right) \cos (\beta - \alpha)}{\sin \phi \cos (\phi + \beta - \alpha)} \tag{6.8}$$

$$F_t = F_c \tan (\beta - \alpha) \tag{6.9}$$

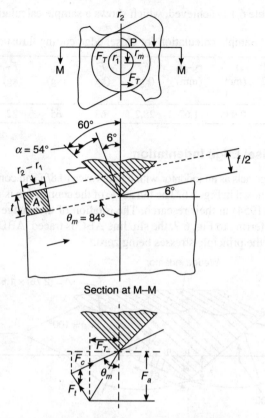

Fig. 6.6 Chip formation and force system of chisel edge in cutting zone.

The following data typical for high negative rake cutting is utilized to derive various forces, torque and thrust:

Rake angle, $\alpha = 60° - 6° = 54°$

$\theta_m = 84°$

$\beta = 12°$

and $\phi = 9°$

A value of shear plane angle of about 9° has also been observed in photomicrographs of Oxford's research paper.

Other data:

Point angle, $2\rho = 120°$

Drill diameter = 25 mm

Chisel edge length = 3.66 mm

Feed rate, $f = 0.2$ mm

Shear stress factor, $K_c = 17$ kg/mm^2

The result shown in Table 6.1 is achieved, which shows a sample calculation for a feed if 0.2 mm/rev.

Table 6.1 Sample calculation for chisel edge cutting thrust and torque

Feed (f) (mm/rev)	$r_1 = 0.8f$ (mm)	$r_2 = c/2$ = 3.66/2 = 1.83	r_m (mm)	$r_2 - r_1$ (mm)	F_c (kg)	F_t (kg)	F_a (kg)	F_T (kg)	P (kg)	T (kg) (mm)	P_c (kg)
0.2	0.16	1.83	0.995	1.67	28.2	63	65	22	130	42	26.6

6.4.2 Thrust due to Chisel Edge Indentation

It is assumed that the chisel edge acts as an indentor with length twice $(0.8f)$ and contact length $h = 2.76c = 2.76f$. The slipline configuration is shown in Fig. 6.6. The fan angle of the centred fan is 100°. These are experimental findings of Grunzweg et al. (1954) in their research. The rotation of the wedge is ignored and only vertical indentation is considered. Referring to Fig. 6.7, the slip line ABC is traced. ABD is an isosceles triangle with free surface at A with one of the principle stresses being zero.

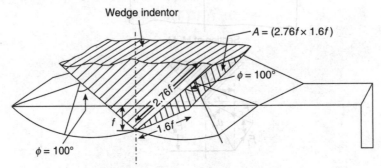

Fig. 6.7 Chisel edge acting as wedge indentor.

Mohr's circle touches the origin as shown in Fig. 6.8. Between points B and C, the hydrostatic pressure increases by $2\phi K_c$, where ϕ is the angle of turn of the slip line BC. Thus, from Hencky's rule, the centre of Mohr's circle shifts to left by $2K_c\phi$. It will be seen that the pressure on the tool face

$$p = K_c + 2K_c\phi + K_c = 2K_c(1 + \phi)$$

Contact area

$$A = (1.6f \cdot 2.76f) = 4.416f^2 \tag{6.10}$$

Refer to Fig. 6.8.

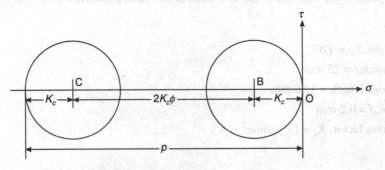

Fig. 6.8 Mohr's circle to calculate p.

THEORY OF MULTIPOINT MACHINING 129

Pressure, $p = 2K_c(1 + \phi)$

$\phi = 100° = 0.56\pi$ rad $= 1.75$ rad $\quad \left(\because 1° = \dfrac{\pi}{180} \text{ rad}\right)$

Therefore, $\quad p = 2K_c(1 + 1.75) = 5.5K_c$ \hfill (6.11)

Load per face $= 5.5K_c \times 4.416 f^2 = 24.29 K_c f^2$

Vertical indentation force per face component $= 24.29 K_c f^2 \cos 30° = 21.03 K_c f^2$

Vertical indentation force due to both faces $P_c = 42.06 K_c f^2$ \hfill (6.12)

As an illustration, for a feed rate of 0.2 mm/rev and $K_c = 17$ kg/mm², the chisel edge thrust

$$P_c = 42.06 \times 17 \times 0.2^2 = 28.60 \text{ kg}$$

The chisel edge does not contribute much torque in the indentation zone and can be neglected.

Total thrust due to chisel edge

$$P + P_c = 130 + 28.60 = 158.60 \text{ kg} \tag{6.13}$$

6.5 USE OF FORMULAE FOR TORQUE AND THRUST AT LIPS OF TWIST DRILL

The use of the relevant formulae can be illustrated with the help of the following example:

EXAMPLE 6.1 Calculate the torque and the thrust in drilling from a solid block of steel from the following data:

Diameter of drill $D = 25$ mm
Chisel edge length $c = 3$ mm
Feed $= 0.2$ mm/rev
Specific cutting pressure, $p = 300$ kg/mm²
Friction angle $\beta = 35°$
Mean true rake angle $\alpha = 18°$
Point angle $2\rho = 120°$.

Solution Applying formula for drilling torque, we get

$$\text{Torque} = pf\left(\dfrac{D^2 - c^2}{8}\right) \text{kg-mm}$$

Substituting $p = 300$ kg/mm², $D = 25$ mm, $c = 3$ mm, $f = 0.2$, we get

$$\text{Torque} = 300 \times 0.2 \left(\dfrac{25^2 - 3^2}{8}\right) = \dfrac{60(625 - 9)}{8} = \dfrac{60 \times 616}{8} = 4620 \text{ kg-mm}$$

Now,

$$\text{Thrust} = 4p\left[\dfrac{D-c}{2} \times \dfrac{f}{2} \tan(\beta - \alpha) \sin \rho\right]$$

$$= 4 \times 300 \left[\left(\dfrac{25-3}{2}\right)\dfrac{0.2}{2} \tan(35° - 18°) \sin 60°\right]$$

$$= 1200 \ (1.1 \tan 17° \times 0.866)$$

$$= 349.48 \text{ kg}$$

6.6 MILLING

As discussed earlier in peripheral milling, the teeth traverse a circular arc with respect to the workpiece.

However, in actual practice the path traversed by the tip of the tooth is a trochoid. This is when the feed of the table containing the workpiece is considered apart from the circular path of the cutter tooth. To simulate the true kinematics of the tooth tip of a milling cutter, it is useful to think of a disc of radius r and rolling without slipping on a straight edge. The tool point Q linked to the straight edge at the instantaneous centre U during the rolling of the disc describes a trochoidal arc AB as shown in Fig. 6.9. The chip formed is the shaded portion between two trochoidal paths AC and AB which are generated by two successive teeth of a milling cutter. Figure 6.9 shows the up-cut milling operation in which the tool tip moves in a direction opposite to the table feed, whereas Fig. 6.10 shows down-cut milling where the tool tip and the table move in the same direction.

It is to be noted that in Fig. 6.9, ST is the straight edge, R is the cutter radius, d is the depth of cut, t_{max} is the maximum undeformed chip thickness, F is the feed rate in mm/rev, and r is the radius of the disc such that $F = 2\pi r$. A rigorous analysis of milling treats the path of the milling cutter teeth as a trochoidal arc. It is presented comprehensively in books such as *A Treatise on Milling and Milling Machines*, published by Cincinnati Milling Machine Company (1951) and *Principles of Engineering Production* by Lissaman and Martin (1983), cited in the references at the end of this book.

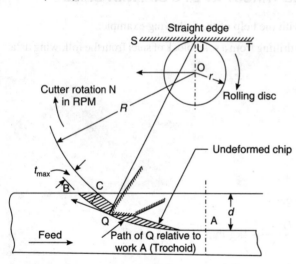

Fig. 6.9 Geometry of chip formation—up-cut milling.

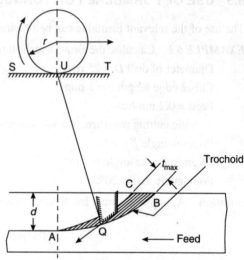

Fig. 6.10 Geometry of chip formation—down-cut milling.

6.6.1 Determination of Undeformed Chip Length

Consider a small element of the trochoid AB (see Fig. 6.11) generated by rotation of the cutter through a small angle $\delta\theta$. The length of the element PQ depends upon the two displacements

(i) $PT = R\,\delta\theta$ due to rotation of cutter
(ii) $TQ = r\,\delta\theta$ due to the feed motion of the work.

The vector sum of these displacements is given by

$$\delta\theta[(R\cos\theta + r)^2 + R^2 \sin^2\theta]^{1/2}$$

and does lead to a solution in a convenient form. Extend PT to the point U such that TUQ is a right angle. Now, since r is very small in relation to R, length PQ is very nearly the same as length PU.

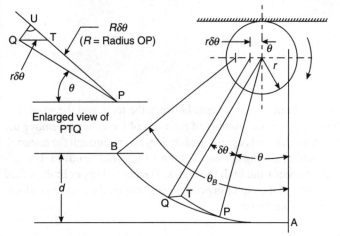

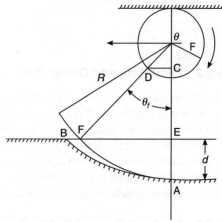

Fig. 6.11 Determination of undeformed chip length.

Fig. 6.12 Geometry for determination of CD and r.

$$PQ \cong PU$$
$$PQ \cong R\delta\theta + r\delta\theta \cos\theta$$
$$= (R + r\cos\theta)\,\delta\theta$$

Hence
$$AB = -\int_0^{\theta_B}(R + r\cos\theta)\,\delta\theta$$
$$= R\,\theta_B + r\sin\theta_B \text{ (up-cut milling)} \quad (6.14)$$

Similar reasoning gives
$$BA = R\,\theta_B - r\sin\theta_B \text{ (down-cut milling)} \quad (6.15)$$

From Fig. 6.12, it can be seen that
$$CD = r\sin\theta_F \quad (6.16)$$

which is very nearly the same as $r\sin\theta_f$.
Also, by similar triangles

$$CD = \frac{r}{R}(EF) \quad (6.17)$$

and by Pythagoras' theorem
$$EF = [(R^2 - (R - d)^2]^{1/2}$$
$$= (2Rd - d^2)^{1/2} \quad (6.18)$$

Let f be the feed/rev of the cutter, i.e.
$$f = 2\pi r$$

Hence
$$r = \frac{f}{2\pi} \quad (6.19)$$

By substituting Eqs. (6.16), (6.17), (6.18) and (6.19) in Eqs. (6.14) and (6.15) for up-cut milling

$$AB = R\theta_B + \frac{f}{2\pi r}(2Rd - d^2)^{1/2} \quad (6.20)$$

and for down-cut milling

$$\text{BA} = R\theta_B - \frac{f}{2\pi r}(2Rd - d^2)^{1/2} \qquad (6.21)$$

6.6.2 Feed Rate and Cutter Wear

The life of a cutter is influenced by the amount of sliding which occurs between the teeth and the work. If a workpiece of length t cm is cut at a feed of f mm/rev, then the number of rotations of the cutter to remove the metal is t/f (approach distance neglected) and each tooth travels approximately $R\theta_B \, x_{t/f}$ cm through the material. It is obvious that high values of f reduce the amount of sliding per unit volume of material removed. For this reason, milling should be done at moderate cutting speeds and high feed rates. The normal upper limit of feed rate will depend either upon the mechanical strength of the cutter or workpiece, or upon the power available at the machine spindle. Vibration may also be a limiting factor.

6.6.3 Depth of Cut and Cutter Wear

The depth of cut (d) is also significant in relation to cutter wear. The volume of metal removal per pass is proportional to d, and the cutter wear is proportional to $R\theta_B$. It can be seen from Fig. 6.14 that θ increases according to the relationship

$$\cos\theta = \frac{R-d}{R}$$

The best relationship between cutter life and volume of metal removed is achieved by taking one pass only at the full depth of cut; moreover, as d increases, the torque at the arbor gets smoother.

6.6.4 Chip Thickness

As seen from Figs. 6.9 and 6.10, the chip thickness varies during cutting. The maximum thickness t_{max} occurs almost at the end of the cut for up-cut milling and at the start of cut for down-cut milling.

A close approximation for the value of t_{max} can be obtained by reference to Fig. 6.13.

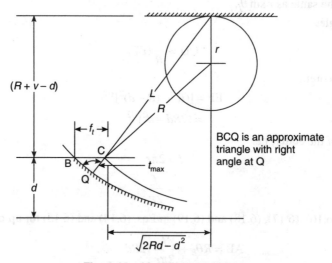

Fig. 6.13 Maxi-chip thickness.

By similar triangles

$$\frac{f_t}{t_{max}} = \frac{L}{(2Rd - d^2)^{1/2}} \qquad (6.22)$$

where f_t is the feed per tooth.

Also

$$L^2 = (R + r - d)^2 + 2Rd - d^2$$
$$= (R + r)^2 - 2rd \qquad (6.23)$$

By substituting Eq. (6.23) in Eq. (6.22)

$$t_{max} = f_t \left[\frac{(2Rd - d^2)}{(R + r)^2 - 2rd} \right]^{\frac{1}{2}}$$

for up-cut milling.

The same can be shown that for down-cut milling. However,

(a) r is generally small and can be neglected; then

$$t_{max} = f_t \left[\frac{(2Rd - d^2)}{R^2} \right]^{\frac{1}{2}} \qquad (6.24)$$

(b) If d is small relative to R (shallow cuts), d^2 can be neglected

$$t_{max} = f_t \left[\frac{2Rd}{R^2} \right]^{\frac{1}{2}} = t_{max} = 2 f_t \left[\frac{d}{D} \right]^{\frac{1}{2}} \qquad (6.25)$$

where D is the diameter of cutter. If d can not be neglected

$$t_{max} = f_t \left[\frac{d}{R} \left(2 - \frac{d}{R} \right) \right]^{\frac{1}{\sqrt{2}}}$$

by rearrangement of Eq. (6.23) above, and substituting the value of cutter diameter

$$= \frac{2}{D} f_t [d(D - d)]^{\frac{1}{\sqrt{2}}} \qquad (6.26)$$

6.6.5 Schlesinger's Formula

The geometry of Schlesinger's formula is derived from Fig. 6.14. It is assumed that t_{mean} occurs at angular position $\frac{\theta}{2}$. From the diagram

$$t_{mean} = f_t \sin \frac{\theta}{2} \qquad (6.27)$$

but

$$\sin \frac{\theta}{2} = \left(\frac{1 - \cos \theta}{2} \right)^{\frac{1}{2}}$$

and

$$\cos\theta = \frac{R-d}{R}$$

hence

$$\sin\frac{\theta}{2} = \left(\frac{d}{2R}\right)^{\frac{1}{2}}$$

or

$$\sin\frac{\theta}{2} = \sqrt{\frac{d}{D}} \qquad (6.28)$$

where D is the diameter of cutter.

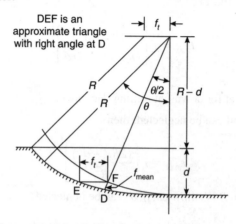

Fig. 6.14 Geometry of Schlesinger's formula for mean chip thickness.

Substituting Eq. (6.28) in Eq. (6.27), we have

$$t_{mean} = f_t \sqrt{\frac{d}{D}} \qquad (6.29)$$

This is called *Schlesinger's formula*.

6.6.6 Cutting Forces and Power

Figure 6.15 shows those components of the force, exerted by the work on a cutting tooth, which act in a plane perpendicular to the cutter axis. T, the tangential force, determines the torque on the cutter; F_r, the radial force, may be regarded as the rubbing force between the workpiece and the tooth. The value of T will depend upon the chip area being cut and on the specific cutting pressure.

Work done in cutting a chip = $T \times$ undeformed chip length

Direct measurement of force

T and F_r are difficult to measure because they are oscillating rapidly during the cutting. A suitable dynamometer capable of recording the fluctuations on a time base must be used.

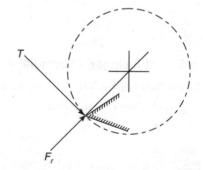

Fig. 6.15 Force acting on cutter tooth.

Figure 6.16 shows the trend of these forces in relation to the undeformed chip section for up-cut milling, while Fig. 6.17 shows the forces in relation to the undeformed chip section for down-cut milling.

THEORY OF MULTIPOINT MACHINING   **135**

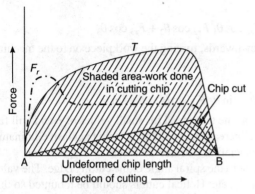
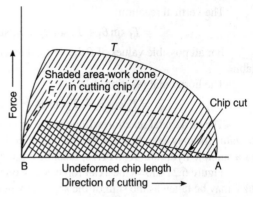

Fig. 6.16 Relationship between forces acting on cutter tooth and undeformed chip thickness (up-cut milling).

Fig. 6.17 Relationship between forces acting on cutter tooth and undeformed chip thickness (down-cut milling).

The high initial value F_r causes rapid wear of cutter edges, and this has the effect of work-hardening the surface which the following teeth must penetrate. In the milling of materials which work hardened readily, the effect as shown in Fig. 6.18 is quite serious; the cutting edges on the clearance side of the teeth rapidly develop a polished land. If dull cutters are kept in service, F_r rises to a very high value and pieces of metal may fracture from the rake face, as shown in Fig. 6.19.

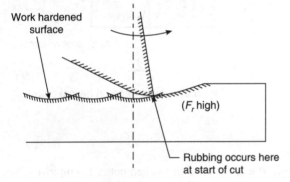
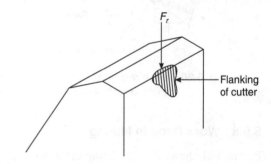

Fig. 6.18 Rubbing action caused by radial force.

Fig. 6.19 Compression fracture at cutting edge caused by very high radial force.

It is obvious that such conditions are unfavourable for application of carbide teeth cutters because of their brittleness.

Since down cut milling avoids the conditions described in the above paragraph, an improved cutter life should generally result from a change to this method. It is normally employed in peripheral milling with carbide tooth cutters (e.g. deep slot milling).

6.6.7 Total Force Acting on a Cutter

Suppose three teeth are in engagement with the workpiece and the down cut method of milling is employed. Figure 6.20 illustrates these conditions and shows how the forces acting may be resolved to give the magnitude of the reaction of the cutter on the workpiece.

The vertical reaction
$$= T_1 \sin\theta_1 + T_2 \sin\theta_2 + T_3 \sin\theta_3 + F_{r1} \cos\theta_1 \, F_{r2} \cos\theta_2 + F_{r3} \cos\theta_3$$

For all possible values of θ the vertical reaction is downwards, forcing the workpiece on to the machine table.

The horizontal reaction
$$= T_1 \cos\theta_1 + T_2 \cos\theta_2 + T_3 \cos\theta_3 - F_{r1} \sin\theta_1 \, F_{r2} \sin\theta_2 - F_{r3} \sin\theta_3$$

The horizontal reaction will be in the same direction as the table feed for small values of θ and will fall in value (or may even become reversed in direction) as θ is increased. A feed drive with backlash eliminator, or a suitable hydraulic feed drive, must be employed in association with the down-cut milling method.

Figure 6.21 shows the axial force acting on a cutter due to the spiral angle of the cutting edge. The value of T may be taken as the sum of the separate values T_1 and T_2, etc. Helical cutters should be mounted so that force A pushes the arbour into the spindle nose.

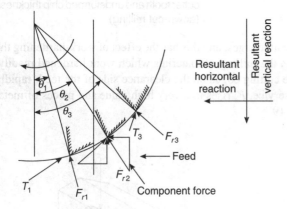

Fig. 6.20 Total reactions of cutter on workpiece down-cutting milling.

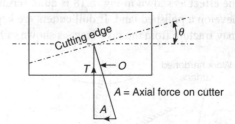

Fig. 6.21 Axial force on helical cutter.

6.6.8 Work Done in Milling

Figure 6.21 shows the area being cut at rotation angle θ when a straight toothed cutter is employed. The value of T for this position is given by $T = Pt$, where P is the specific cutting pressure. Figure 6.22 shows the variation which occurs with variation of chip thickness; but since t is a function of θ, P also is a function of θ.

The work done in a cutting chip
$$W_c = \text{Torque} \times \text{Angle turned}$$
$$W_c = R_b \int_0^{\theta_B} Pt \, d\theta$$

By Schlesinger's formula, $t = f_t \sqrt{\dfrac{d}{D}}$

then
$$W_c = R\theta b f_t \sqrt{\dfrac{d}{D}} \, P_{\text{mean}} \qquad (6.30)$$

If a cutter has N teeth and rotates at S revolutions per minute, the total work done per minute in cutting is equal to $W_c NS$, and from this expression a value for the power required at the cutter can be estimated.

Figure 6.23 shows the range of specific cutting pressure involved in peripheral milling.

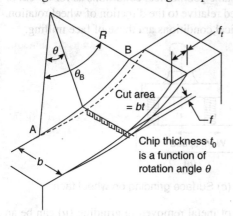

Fig. 6.22 Chip thickness as a function of angle of rotation.

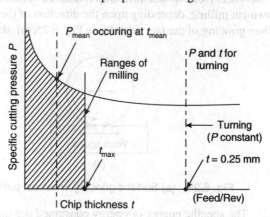

Fig. 6.23 Range of specific cutting pressure involved in peripheral milling.

6.6.9 Relationship between Face and Peripheral Milling

In face milling the chips are cut on the periphery of the cutter and are bound by trochoids; as for peripheral milling, a different portion of the trochoid is involved as shown in Figs. 6.24(a) and (b). The undeformed chip shapes for up-cut milling, face milling and down-cut milling lie within the crescent bounded by the trochoids. The machined surface is left by the face of the cutter.

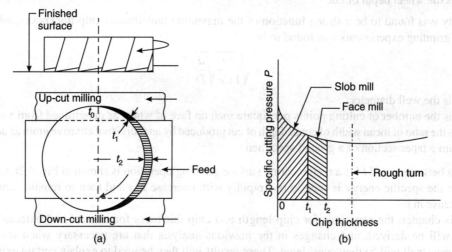

Fig. 6.24 (a) Comparison of undeformed chip thickness and specific cutting pressures involved in peripheral and face milling and (b) shaded portion $(t_0 - t_1)$ up-cut milling, portion $(t_1 - t_2)$—face milling, remaining—down-cut milling.

6.7 MECHANISM OF GRINDING

Grinding is a cutting process having geometry of chip formation similar to that of milling. The grit edges which project from the surfaces of the wheel act as very small cutting teeth. When cutting on the periphery of the wheel [see Fig. 6.25(a)], chip formation occurs under the same geometrical conditions as for up-cut or down-cut milling, depending upon the direction of the work speed relative to the direction of wheel rotation. When grinding of the face wheel [see Fig. 6.25(b)], the geometrical conditions are those of face milling.

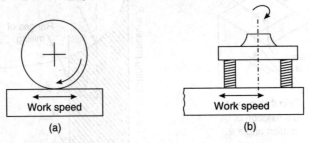

Fig. 6.25 (a) Surface grinding on wheel periphery and (b) Surface grinding on wheel face.

The specific energy or energy consumed per unit volume of metal removed in grinding (u) can be an important quantity which is defined as follows:

$$u = \frac{VF_p}{vb \cdot d} \quad (6.31)$$

where V is the wheel speed
v is the work speed
F_p is the tangential component of force on grinding wheel
b is the width of cut
d is the wheel depth of cut

This quantity was found to be a strong function of the maximum undeformed chip-thickness t, which from the surface-grinding experiments was found to be

$$t = \sqrt{\frac{4v}{VCr}} \sqrt{\frac{d}{D}} \quad (6.32)$$

where D is the well diameter
C is the number of cutting points per square inch on face of wheel as determined from a soot track
r is the ratio of mean width of cut to depth of cut produced by an individual abrasive grain as determined from a taper section of a ground specimen.

The relation between u and t for a reciprocating surface grinding operation is shown in Fig. 6.26 (Acherkan, 1968). Here the specific energy is seen to rise rapidly with decrease in t and then to remain constant with further decrease in t.

In this chapter, the equations for chip length and chip thickness for internal and external grinding operations will be derived, and changes in the previous analysis that are necessary when v/V becomes large or d very small will be then considered. These results will then be used to explain certain experimental grinding data.

6.7.1 Undeformed Chip Length

In Fig. 6.27(a), the surface grinding case is shown. In this case the undeformed chip length l is found to be

$$l_e = \sqrt{Dd} \quad (6.33)$$

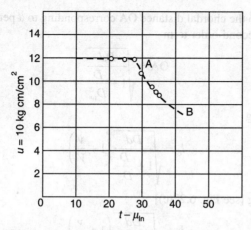

Fig. 6.26 Variation of specific grinding energy (u) with maximum chip thickness (t).

Figure 6.27(b) is for external grinding (where $v \le V$). If the work were the stationary arc of contact l_c would be OA, but since the work also moves it will extend to the additional distance AB.

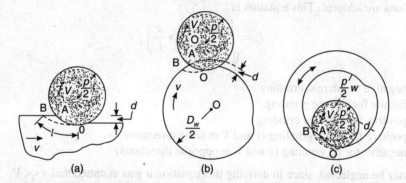

Fig. 6.27 Chips produced during (a) surface grinding, (b) external grinding and (c) internal grinding operations, when $v < V$.

Hence
$$l_c = OA + AB \tag{6.34}$$

If t is the time it takes the wheel to travel distance OA, then
$$OA = Vt \tag{6.35}$$
and
$$AB = vt \tag{6.36}$$

Thus eliminating t
$$AB = \frac{v}{V}(OA) \tag{6.37}$$

or, from Eqs. (6.34) and (6.37)
$$l_c = \left(1 + \frac{v}{V}\right) OA \tag{6.38}$$

Arc OA is very nearly equal to the chordal distance OA corresponding to a penetration of the two circles by an amount d. Neglecting the second-order term

$$\text{OA} = \sqrt{\frac{Dd}{1 + \dfrac{D}{D_w}}} \tag{6.39}$$

and hence for external grinding

$$l_e = \sqrt{\frac{Dd}{1 + \dfrac{D}{D_w}}}\left(1 + \frac{v}{V}\right) \tag{6.40}$$

Similarly, for internal grinding [see Fig. 6.27(c)]

$$l_e = \sqrt{\frac{Dd}{1 - \dfrac{D}{D_w}}}\left(1 + \frac{v}{V}\right) \tag{6.41}$$

It is worth noting that a general expression that holds for all types of peripheral grinding can be written, if certain conventions are adopted. This equation is

$$l_c = \sqrt{\frac{Dd}{1 + \dfrac{D}{D_w}}}\left(1 + \frac{v}{V}\right) \tag{6.42}$$

where D_w is negative or internal grinding
  D_w is infinite for surface grinding
  D_w is positive for external grinding
   v is positive for down grinding (v and V in same direction)
   v is negative for up grinding (v and V in opposite directions)

Of course v/V may be neglected, since in deriving this equation it was assumed that $v \ll V$.

Equation (6.42) is for Case I of Fig. 6.28, where distance AB $\ll$ OA. In this instance, chord OA is given by Eq. (6.39) and the value of l_c by Eq. (6.42). The other cases of Fig. 6.28 are for progressing greater values of v relative to V. If we consider the last Case V where AB $>$ OA, it is evident that chord OA is twice as great as that for Case I, and hence for Case V

$$l = \text{OA} + \text{AB} = 2\sqrt{\frac{Dd}{1 + \dfrac{D}{D_w}}}\left(1 + \frac{v}{V}\right) \tag{6.43}$$

This equation also holds for Case IV, but for other intermediate cases the coefficient on the right side of the equation will be less than 2. In the general case which covers all conditions of Fig. 6.29.

$$l = K\sqrt{\frac{Dd}{1 + \dfrac{D}{D_w}}}\left(1 + \frac{v}{V}\right) \tag{6.44}$$

where K is a constant that varies from 1 (for Case I where AB $\ll$ OA) to 2 (for Cases IV or V where AB $>$ OA).

Equation (6.44) is the most general expression for chip length l, and holds for chips of all sizes and for internal, external, or surface grinding, if the conventions listed under Eq. (6.42) are followed.

THEORY OF MULTIPOINT MACHINING **141**

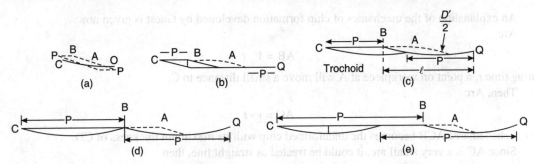

Fig. 6.28 Types of chips produced when work speed (*v*) is increased progressively relative to wheel speed (*V*) (*P* is the distance that work moves between passage of successive grains).

6.7.2 Maximum Chip Thickness in Cylindrical Grinding

The present section deals with a formula developed by J.J. Guest in 1953. Referring to Fig. 6.29, it will be seen that the comma type shape of the undeformed chip denoted by the arcial triangle CAB, the following relationship can be written.

Maximum undeformed chip thickness

$$CD = AC \sin (a + b)$$

$$t_m = f_g \sin (a + b)$$

Therefore,
$$t_m = \frac{v}{NV} \sin (a + b)$$

where v = surface velocity of work
  V = surface velocity of wheel
  t = time taken by a grit to move from A to B
  N = number of grits per unit length of wheel

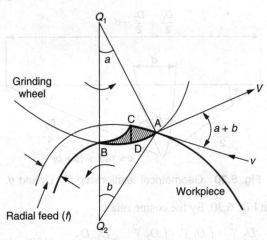

Fig. 6.29 Determining maximum chip thickness.

An explanation of the mechanics of chip formation developed by Guest is given now.

Arc
$$AB = V \cdot t$$

During time t, a point on workpiece at A will move a small distance to C.

Then, Arc
$$AC = v \cdot t$$

Hence, shaded area ACB becomes the undeformed chip with its maximum thickness of CD.

Since AC is a very small arc, it could be treated as straight line, then
$$CD = AC \sin(a+b) = v \cdot t \sin(a+b)$$

Here, a and b are the angles subtended by the arc of the contact at the centre of wheel and work.

Let N be the number of grits per unit length of the wheel circumference, then maximum chip thickness per grit is

$$t_m = \frac{CD}{N \times \text{Arc AB}} = \frac{v \cdot t \cdot \sin(a+b)}{N \cdot V \cdot t}$$

$$= \frac{1}{N} \times \frac{v}{V} \times \sin(a+b)$$

$$= \frac{v}{NV} \sin(a+b) \qquad (6.45)$$

A useful relation for maximum chip thickness (t_m) can be obtained after an equation for $\sin(a+b)$ is derived as follows:

Determination of sin (a + b)

Figure 6.30 shows the geometrical relationships between D_1 (diameter of workpiece), D_2 (diameter of wheel) and the depth of cut d.

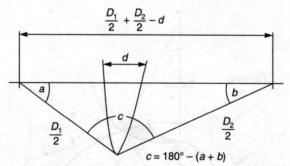

Fig. 6.30 Geometrical relationship D_1, D_2 and d.

Consider the triangle in Fig. 6.30. By the cosine rule,

$$\left(\frac{D_1}{2}+\frac{D_2}{2}\right)^2 = \left(\frac{D_1}{2}\right)^2 + \left(\frac{D_2}{2}\right)^2 - 2\frac{D_1}{2}\frac{D_2}{2}\cos[180°-(a+b)] \qquad (6.46)$$

since d is very small, as compared with D_1 and D_2 terms involving d^2 are negligible and omitted from the expansion below. Hence,

$$\frac{D_1 D_2}{2} \cos(a+b) = \left(\frac{D_1 D_2}{2} - dD_1 - dD_2\right) \tag{6.47}$$

or
$$\cos(a+b) = 1 - 2d\left(\frac{D_1 + D_2}{D_1 D_2}\right)$$

Also,
$$1 - \sin^2(a+b) = \cos^2(a+b) = 1 - 4d\left(\frac{D_1 + D_2}{D_1 D_2}\right) + \text{terms containing } d^2.$$

Hence
$$\sin^2(a+b) \cong 4d\left(\frac{D_1 + D_2}{D_1 D_2}\right)$$

or
$$\sin(a+b) \cong 2\left[d\left(\frac{1}{D_1} + \frac{1}{D_2}\right)\right]^{1/2} \tag{6.48}$$

Substituting this value of $\sin(a+b)$ in Eq. (6.45), we get

$$\frac{v}{NV}\sin(a+b) \cong \frac{2v}{(NV)}\left[d\left(\frac{1}{D_1} + \frac{1}{D_2}\right)\right]^{1/2} \tag{6.49}$$

By the substitution of this result for t_m, we have

$$t_m = \frac{2v}{NV}\sqrt{\left(d\frac{D_1 + D_2}{D_1 D_2}\right)} \tag{6.50}$$

This is the form generally stated for the result of Guest's theory, but it may also be written as:

$$t_m = \frac{2v}{NV}\sqrt{d\left(\frac{1}{D_1} + \frac{1}{D_2}\right)} \tag{6.51}$$

Also, feed per grit,

$$f_g = \frac{v}{NV} \tag{6.52}$$

Considering the main variables in external grinding, since undeformed chip cross-sectional area assuming a triangular cut is proportional to $(t_m)^2$. From Eqs. (6.51) and (6.52), we have

$$\text{Force per grit } (f_g) \propto (t_m)^2 \quad \text{and} \quad (t_m)^2 \propto \frac{v^2 d}{V^2}\left(\frac{1}{D_1} + \frac{1}{D_2}\right)$$

The pitch of the grits (N) being a constant for the grade of wheel selected. Clearly, raising of the work speed (v) is the most effective way of increasing self-sharpening action should dull grits not be broken from the bond.

From the above equation, it is obvious that decrease in chip thickness is possible by increase in wheel speed V. Decrease in chip thickness leads to better finish and tighter geometrical tolerances due to lower grinding forces. It is with this reason that higher speeds are tried for precision grinding applications.

6.7.3 Grinding Kinetics

Figure 6.31 shows the schematic diagram of a section through an abrasive grit as it removes a chip. The radial and tangential forces that the grit exerts on the metal are balanced by the normal and frictional forces that the chip places on the grit. For a given effective rake angle, the ratio of radial to tangential force is indicative of the coefficient of friction on the grit face. The magnitude of the tangential force itself can be related to the specific energy of metal removal as shown:

$$T_e = \mu A \tag{6.53}$$

Since the mean effective rake angle is unknown, the specific energy and the ratio of radial to tangential force may be considered the two fundamental parameters of grinding kinetics. In metal cutting operations the specific energy is found to vary very little with the width of the cut, but it does increase with decreasing chip thickness.

$$u = \frac{k}{t_0} \tag{6.54}$$

where u is instantaneous specific energy
       k is a constant of proportionality.

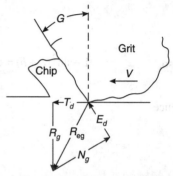

Fig. 6.31 Force system on mean chip.

The value of a in metal cutting is about 0.2. The specific energy also tends to decrease with increasing cutting speed, as the result primarily of a reduction in tool-chip interface friction at higher speeds.

The thickness of a chip in grinding starts at zero and increases with the distance along the grit-workpiece contact path. The effective specific energy for the overall chip must thus be found by integration as follows:

$$u_e = \frac{\int_0^{t_m} u \, A \, dl}{\int_0^{t_m} A \, dl} \tag{6.55}$$

Further, the frequency f of chips of any given size s in a sample of swarf will be related to α by some statistical distribution equation. The true mean effective energy can only be found after another integration

$$u_c = \frac{\int_0^\alpha f u \, ds}{\int_0^\alpha f \, ds} \tag{6.56}$$

Since the relationship between f and s is unknown, it must be assumed that the active grits are uniformly spaced on the wheel face, so that u_v equals v.

6.8 POWER IN BROACHING

The pull or the cutting force acting on the broach whilst broaching round holes in steel may be approximated by the formula

$$T = c f^{0.35} \, d \cdot z \text{ kg} \tag{6.57}$$

where c is a constant
       d is a broach diameter in mm
       f is the feed per tooth in mm
       z is the number of broach teeth cutting at a time

Power required

$$P = \frac{T \times V}{60 \times 120} \text{ kW} \qquad (6.58)$$

where T is the broach pull in kg
 V is the broaching speed in m/min, which is further given by

$$V = \frac{C}{T' f^y} K \text{ m/min}$$

where C is a constant depending upon the working conditions
 T' is the broach life in min
 K is a coefficient depending upon the broach material.

C	T	X	Y	K	Values of constants BHN
12	100	0.62	0.62	1–1.55	100–180

EXAMPLE 6.2 The feed of an 8 teeth face mill is 0.33 mm per tooth at 200 rpm. The material cut is 300 BHN steel. If the depth of cut is 3.17 mm and the width 100 mm, calculate (a) the HP at the cutter, (b) the HP at the motor.

Solution

$$HP_c = \frac{yxF_t nN}{k}$$

$$= \frac{100 \times 3.17 \times 0.33 \times 8 \times 200}{9.6} = 17.43$$

$$HP_m = \frac{HP_c}{E} = \frac{17.43}{0.60} = 29.05$$

where HP_c is the HP at cutter
 y (the width of chip) = 3.17 mm
 x (the depth of cut) = 100 mm
 k a constant = 9.6 hp/cm³.min
 F_t = 0.33 mm (feed)
 n = 8 teeth
 N = 200 rpm
 E = 60%

EXAMPLE 6.3 What would be the HP in Example 8.1 if the cutter tooth has 30° rake angle?

Solution

$$HP_c = \frac{y(x)(F_t \cos\alpha)nN}{k}$$

$$= \frac{3.17 \times 100 \times (0.33 \cos 30°) \times 8 \times 200}{9.6 \times 1000} = 15$$

$$HP_m = \frac{15}{0.69} = 2.5$$

EXAMPLE 6.4 The following data was observed while measuring power in drilling operation by calorimetric process:

Weight of calorimeter (steel) = 30 g
Weight of holder for test bar = 40 g

Weight of screens (steel) = 8.2 g
Weight of drill pan in water (steel) = 19 g
Weight of thermometer in water (steel) = 4.2 g
Weight of rubber gasket in water (steel) = 8.63 g
Weight of test bar = 4.25 g
Specific heat of steel = 0.11
Specific heat of steel thermometer = 0.2
Specific heat of steel rubber gasket = 0.48
Specific heat of steel test bar = 0.249
Weight of water = 50 g
Temperature difference = 13°C
Length of test bar = 25 mm
Spindle speed = 700 rpm
Feed = 0.0625 mm/rev.

Calculate the total heat and HP in cutting.

Solution Total heat developed

$$H = \text{Heat taken by water} + \text{Heat taken by accessories}$$
$$= (W_a + W_w)\theta$$

where H is the heat generated or taken by water and other appliances
 θ is the rise in temperature
 W_a is the water equivalent of accessories ($W_a = W_{steel} + W_{rubber} + W_{thermometer} + W_{test\ bar}$)
 W_w is the weight of water

Now,
$$W_a + W_w = 16.732 + 50$$
$$= 66.732 \text{ g}$$
$$\text{Heat generated} = 66.732 \times 13$$
$$= \frac{867.5}{1000} \text{ kcal}$$

But feed = 0.0625 mm/rev

$$\text{Work cut per min} = 0.0625 \times 700$$
$$= 43.75 \text{ mm}$$

Therefore, $\text{Heat generated per min} = \dfrac{867.5 \times 42.3}{1000} = 36.674$

Therefore, $\text{HP} = \dfrac{36.674}{175.5} = 0.209$

EXAMPLE 6.5 A 205 mm diameter face mill has 6 inserted teeth and is used to cut material 140 mm wide. The cutter operates at a speed of 30.0 m/min and table feed of 114 mm/min. The centre of the cutter is 57.00 mm from the edge of the work approached by the teeth. Find

(a) the variation of undeformed chip thickness
(b) the influence of the cutting conditions on the clearance angle

(c) the angles at which the teeth approach and leave the workpiece
(d) the cutting time to mill a 457 mm length
(e) the maximum number of teeth in contact with the work.

Solution (a) Spindle speed = $\dfrac{1000 \times 30.0}{\pi \times 205}$ = 45 rev/min

Feed/tooth, $f_t = \dfrac{114}{45 \times 6}$ = 0.42 mm

(b) Undeformed chip thickness are denoted by t_1 at entry, t_2 at exit; then we have

$$t_1 = \dfrac{2}{D} f_t \, [d_1(D-d)_1]^{1/2}$$

$$= \dfrac{2 \times 0.42}{205} [44.5 \times (205 - 44.5)]^{1/2}$$

$$= 0.34 \text{ mm}$$

$$t_{max} = f_t = 0.42 \text{ mm}$$

and now
$$t_2 = \dfrac{2 \times 0.42}{205} [25 \times (205-25)]^{1/2} = \dfrac{2 \times 0.42}{205} (25 \times 180)^{1/2}$$

$$t_2 = 0.27 \text{ mm}$$

Loss of clearance angle due to table feed, $\delta\phi$, from the vector triangle of velocities, is given by

$$\sin \delta\phi = \dfrac{\text{Table feed}}{\text{Cutting speed}}$$

$$\sin \delta\phi = \dfrac{114}{30.5 \times 1000}$$

$$\delta\phi = 0.21°$$

(c) Angles at entry and leaving (see Fig. 6.32):

$$\cos E = \dfrac{57}{102}$$

$$\cos E = 0.5625$$

$$E = 55.8°$$

$$\cos L = \dfrac{78.5}{102}$$

$$\cos L = 0.77$$

$$L = 39.7°$$

(d) The approach distance can be found with sufficient accuracy from a scale drawing. Alternatively, approach distance is calculated.

$$\text{Approach distance} = 102.5(1 - \sin 39.7°)$$

$$= 102.5 \times 0.639$$

$$= 65.50$$

Therefore, $\quad$ Cutting time $= \dfrac{457 + 65.50}{114} = \dfrac{522.50}{114} = 4.58$ min

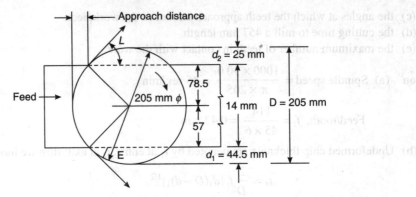

Fig. 6.32 Comparatively small change in thickness as compared with slab milling.

(e) Maximum number of teeth in contact:

Arc of engagement of cutter = 180° – (55.8° + 39.7°) × 84.5°

Angular spacing of teeth = $\dfrac{360°}{6}$ = 60°

Therefore, maximum number of teeth in contact = 2

EXAMPLE 6.6 A 203 mm diameter face mill has 6 inserted carbide teeth of length 0.25 mm corner angle 45° and takes a cut 6 mm deep across a steel slab 114 mm wide of 300 Brinell hardness. Assume cutting speed as 106.5 m/min. Find the power required at the spindle. Find also the minimum chip thickness measured perpendicular to corner edge.

Solution

$$\text{Spindle speed} = \dfrac{1000 \times 106.5}{\pi \times 203}$$

$$= 167 \text{ rev/min feed/tooth}$$

Table feed required, $F = 0.25 \times 6 \times 167$

$$= 250.5 \text{ mm/min}$$

Metal removal rate, $w = \dfrac{250.5 \times 6 \times 114}{60} = 2855.7 \text{ mm}^3/\text{s}$

Power consumed, $k = 0.19/\text{mm}^3$

$$\text{HP at the cutter} = \dfrac{2855.7}{0.19 \times 746} = 20.15$$

Minimum value of underformed chip thickness perpendicular to cutting edge (assume that cutter is positioned over centre of work and $\phi = 45°$)

$$t_{\max} = \dfrac{2}{D} f_t \, [d(D - d)]^{1/2}$$

$$= \dfrac{2 \times 0.25}{203} [44.5 \times (203 - 44.5)]^{1/2}$$

$$= 0.00246 \times 83.98 = 0.206 \text{ mm}$$

and
$$t_{min} = \frac{0.206}{\sqrt{2}} = 0.147 \text{ mm}$$

EXAMPLE 6.7 A carbide face mill has inserted teeth to provide radial rake $-12°$, axial rake $+5°$ and corner angle $45°$. Find the angle of inclination of the cutting edge and the approximate value of the effective rake angle.

Solution We know
$$\tan \psi = \cos \phi \tan \alpha_a - \sin \phi \tan \alpha_r$$

where ψ is the angle of inclination;
α_a is the axial rake angle;
α_r is the radial rake angle.

Substituting the values, we have
$$\tan \psi = \cos 45° \tan 5° - \sin 45° \tan (-12°)$$
$$= 0.7071 \times 0.0875 + (0.7071 \times 0.2126)$$
$$= 0.0619 + 0.1503 = 0.2122 = +12°$$

Approximate value of effective rake angle (α_e)
$$\tan \alpha_e = \tan \alpha_r \cos \phi + \tan \alpha_a \sin \phi$$
$$= \tan(-12°) \cos 45° + \tan 5° \sin 45°$$
$$= (-0.2126 \times 0.7071) + (0.0875 \times 0.7071)$$
$$= -0.1503 + 0.0619 = -0.0884$$
$$\alpha_e = 5° \text{ (app.)}$$

EXAMPLE 6.8 The following conditions relate to a surface grinding operation carried out with a 46 grit wheel:

$V = 1672$ m/min
$v = 9.02$ m/min
$C = 294.50$
$d = 0.025$ mm
$\gamma = 17$
$D = 203$ mm

Estimate the maximum chip thickness.

Solution We know
$$t = \left(\frac{4v}{V \times C \times \gamma} \sqrt{\frac{d}{D}} \right)^{1/2}$$

where t is the chip thickness;
v is the work speed in ft/min;
V is the cutting speed in ft/min;
C is the number of effective grits per cm^2 of grinding wheel surface;
γ is the ratio of width to depth of groove cut by a grit;
d is the depth of cut in mm;
D is the wheel diameter in mm.

Now substituting the values we have

$$t = \left(\frac{4 \times 9.00}{1672 \times 294.50 \times 17} \times \sqrt{\frac{0.025}{203}}\right)^{1/2}$$

$$= \frac{36 \times 10^{-6}}{0.4924} \times \frac{1.109}{100}$$

$$= 73 \times 10^{-6} \times \frac{1.109}{100}$$

$$= 81.07 \times 10^{-8} \text{ mm}$$

EXAMPLE 6.9 Determine the power required for the following face milling operation:

Work material = MS
Tool material = HSS
Cutter dia = 100 mm
Number of teeth = 10
Cutting speed = 25 m/min
Feed (f) = 75 mm/min
Width of cut (w) = 50 mm
Depth of cut (d) = 5 mm
Cut = Symmetrical
Cutting pressure = 300 kg/mm^2

Solution RPM of cutter, $N = \dfrac{1000\,V}{\pi D} = \dfrac{1000 \times 25}{\pi \times 100} = \dfrac{250}{\pi}$

$$f_t = \frac{75}{100} \times \frac{\pi}{250}$$

$$= 0.0942 \text{ mm/rev}$$

Mean chip area of cross section in mm^2, $A_m = \dfrac{w \times f \times d}{1000\,V}$

$$= \frac{50 \times 75 \times 5}{1000 \times 25} = 0.75 \text{ mm}^2$$

Peripherical force = $300 \times 0.75 = 225$ kg

HP required = $\dfrac{225 \times 25}{75 \times 60} = 1.25$

EXAMPLE 6.10 A side and face cutter of 127 mm diameter has 10 teeth. It operates at a cutting speed of 13.5 m/min and table feed of 108 mm/min. Find the maximum chip thickness for (a) a cutting depth of 5 mm; (b) a cutting speed of 25 mm.

Solution Spindle speed = $\dfrac{13.5 \times 1000}{\pi \times 127} = 34$ rev/mm

$$f_t = \frac{13.5 \times 1000}{34} \times \frac{1}{10} = 3.30 \text{ mm}$$

We also know that

$$t_{max} = 2 \times f_t \sqrt{\frac{d}{D}}$$

where t_{max} is the maximum thickness
D is the diameter of cutter
d is the cutting depth

(a) $t_{max} = 2 \times 3.3 \sqrt{\dfrac{5}{127}}$

$= 1.32$ mm

(b) $t_{max} = 2 \times 3.3 \sqrt{\dfrac{25}{127}}$

$= 2.86$ mm

EXAMPLE 6.11 A slot 32 mm deep is milled with a cutter 115 mm diameter operating at a feed/rev of 5.00 mm. Find the maximum reduction of the nominal clearance angle caused by the feed motion.

Solution df at 2 mm depth $= \dfrac{90 \times 5}{\pi^2 \times 58.5^2} \sqrt{32.90(11.5 - 32)}$

$= 0.72°$

EXAMPLE 6.12 A slab milling cutter has a radial rake of 7° and a spiral angle of 35°. Find the effective rake angle. Also give comments on the result obtained.

Solution From the geometry of general form of cutting edge, we have (see Fig. 6.33)

$$\tan \gamma_p = \tan \gamma_r \sec \sigma$$

where γ_r is the radial rake angle
γ_e is the effective rake angle
γ_p is the primary rake angle
σ is the spiral angle

Substituting the values, we have

$$\tan \gamma_p = \tan 7° \sec 35°$$

$= 0.1228 \times 1.221$

$= 0.1499$

$\gamma_p = 8.5°$

$\sin \gamma_e = \sin^2 35° + \cos 35° \sin 8.5°$

$= 0.3291 + (0.671 \times 0.1478)$

$= 0.4283$

$\gamma_e = 25.4°$

Comments: The results show that the introduction of the spiral angle has the effect of raising the value of the rake angle from $\gamma_c = 7°$ (when $\sigma = 0$) to $\gamma_c = 25.4°$. Since this large effective rake angle is achieved without producing a weak tooth section, this principle is of importance in cutter design. Refer to Fig. 6.34 for an illustration of the spiral angle in oblique cutting.

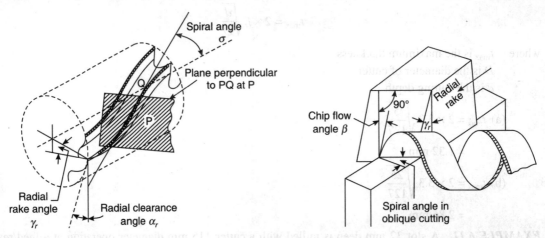

Fig. 6.33 Example 6.11—Cutting angle of a helical slab mill.   **Fig. 6.34** Oblique cutting process.

EXAMPLE 6.13 Estimate the power required to take a cut 100 mm wide and 3 mm deep at 76 mm/min feed, in alloy steel for which $k = 7.6$, if the cutter diameter is 90 mm and a cutting speed of 135 m/min is employed, find the mean torque at the arbour and estimate the force required to drive the machine table (up-cut milling). If the cutter has a spiral angle of 40°, estimate the axial thrust.

Solution

Power at the spindle, $HP = \dfrac{W}{k} = 3$

Mean force at the periphery of the cutter, $F = \dfrac{3 \times 4500}{13.5} = 1000$ kg

Mean torque at the arbour $= \dfrac{90 \times 1000}{1000 \times 2} = 45$ kg

Since for the cut being considered, Q_E is small, the estimated force to derive the machine table (neglecting friction at the slide) is approximately equal to T (1000 kgf).

Estimated axial thrust, $A = 1000 \sin 40° = 630$ kgf

Note that the feed force of 1000 kgf estimated above is about 8 times the feed force required to achieve an equal metal removal rate on lathe.

EXAMPLE 6.14 A 100 mm diameter cutter having 8 teeth cuts at 25 m/min. The depth of cut is 0.4 mm and the table feed 152.5 mm/min. Find the percentage reduction in sliding between the cutter edges and the material cut, which results from a change from up cut to down cut milling.

Solution   Spindle speed $= \dfrac{25 \times 1000}{\pi \times 100} = 80$ rev/mm

$$\text{Feed/rev of cutter} = \frac{152.5}{80} = 1.90 \text{ mm/rev}$$

Let θ is angle of engagement between cutter and work, then

$$\cos \theta = \frac{50 - 4.00}{50} = 0.92$$

$$\theta = 22.3° = 0.3892 \text{ rad}$$

$$y = \frac{1.90}{2\pi} = 0.3 \text{ mm}$$

Path length, $AB = 50 \times 0.3892 + 0.3 \times \sin 22.3°$

$= 19.4 + 0.11 = 19.5$ mm

Reduction of path length resulting from change of method $= 2 \times 0.11 = 0.22$ mm

Percent reduction in sliding $= \dfrac{0.22}{19.5} \times 100 = 1.12\%$

EXAMPLE 6.15 Show that the effective cutting clearance angle of a twist drill is lower towards the centre than at periphery. A 2554 mm/diameter drill, chiesel edge 3.2 mm in width, is used with a feed of 0.64 mm/rev. If nominal clearance angle is 7°, find the greatest and least effective cutting clearance angles.

Solution Due to the feed motion of the drill, each point on the cutting edge travels in a helical path, with the nominal clearance angle being reduced by the helix angle at the point being considered, as indicated in Fig. 6.35.

Since radius r is larger at P than at Q, the helix angle (δ) at the path travelled is given by $\tan \delta = f/2\pi r$, smaller at P than at Q. The effective clearance angle α is reduced due to the influence of the feed helix angle δ; hence the effective clearance angle is lower towards the centre than at the periphery.

Angle δ lies in plane CD; its effect in the plane AB, which contains the effective clearance angle Ψ, depends upon the point angle of the drill. Assuming this to be 120°, we have $\tan \delta AB = \tan \delta \cos 30°$, with AB indicating the magnitude in plane AB of the angle δ lying in plane CD

At P

$$\tan \delta\, AB = \frac{0.64}{25\pi} \times 0.866 = 0.0069$$

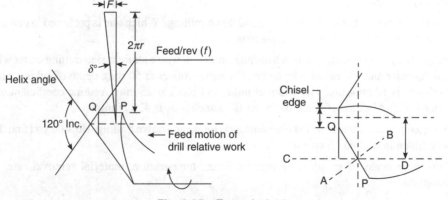

Fig. 6.35 Example 6.15.

AB is 24′ approximately and the effect on the cutting clearance of the periphery is negligible.

At Q

$$\tan \delta AB = \frac{0.64}{\pi \times 3.2} \times 0.866 = 0.0588$$

$$\delta AB = 3°22' \text{ approximately}$$

The least cutting clearance angle (effective) = $7° - 3°22' = 3°38'$

REVIEW QUESTIONS

1. (a) What are the commonly used values of point angle, lip relief angle, helix angle, margin material for a standard twist drill? How are these selected for various workpiece materials?
   (b) What is the purpose of lip relief angle?
   (c) Comparing lathe tool with twist drill, which angles in drill are equivalent to rake angle and side cutting edge angle on a lathe tool?
   (d) When will you adopt 90° point angle on drills, and when 135°–150° point angle?
2. Differentiate between drilling and single-point cutting.
3. Explain the function of chisel edge. How does the cutting action take place by chisel edge? Explain.
4. Draw shear plane model diagram for a drilling process. Indicate normal rake, friction angle, cutting velocity, and of chip velocity vectors, with the help of a diagram.
5. What is the function of drill flutes? Why is the web of a drill increased in thickness from a drill point towards the shank? What is meant by point thinning of a drill point?
6. Explain the following with reference to milling cutter geometry:
   (a) Axial rake angle, Radial rake angle and Corner angle for a face milling cutter.
   (b) Helix angle and Radial rake angle for a slab milling cutter.
7. How will you proceed for prediction of torque and thrust in drilling? Derive an equation giving all assumptions made therein.
8. Why is milling called an intermittent cutting operation? How smoothness of arbor torque is achieved in a milling operation?
9. What is the difference between up milling and down milling? Which one is preferred in certain situations and which one is preferred in other situations?
10. Determine the power consumption in a milling operation with a straight slab milling cutter with 24 teeth, 50 mm diameter and 10° radial rake angle. The cutter rotates at 50 rpm, depth of cut is 1 mm and table feed velocity is 10 mm/min. The width of mild steel block is 25 mm. Assume coefficient of friction at rake face is 0.5. Also, shear stress of the work materials, s_s is 420 N/mm^2.
11. Derive an equation for undeformed chip thickness for up and down milling. How do you find theoretically the work done in milling? Explain.
12. Explain the process of grinding as regards force, temperature, material removal rate, and power consumption.
13. Derive a general expression from first principles for undeformed chip length in grinding.

CHAPTER 7

Heat in Metal Cutting and Temperature Measurement

7.1 INTRODUCTION

The role of the temperature of the cutting tool in metal cutting processes was first studied as early as 1907 by F.W. Taylor, who observed that increased cutting speeds resulted in increased tool temperatures and hence decreased tool life. Such observations led to the development of the first high speed steels by Taylor and White. However, quantitative studies of tool temperatures were made by Shore in America and almost simultaneously by Gottwein in Germany and Herbert in England for measuring the mean temperature along the face of a cutting tool by the tool work thermocouple technique. This method of determining tool temperatures utilizes the fact that an emf is generated at the interface changes. The tool-work thermocouple method has been used in many of the studies which have increased our knowledge of the thermal behaviour of cutting tools.

There are several variables which influence the tool temperature, and it is desirable that the relative contribution of each of these be understood. However, it is difficult to obtain such information experimentally, although the radiation studies of Schwerd and the use of thermal sensitive paints by Pahlitzsch, Helmerdig Bickel and Widmer have sighted many factors affecting tool temperatures.

The heat developed at the point of a cutting tool has two principal sources, i.e. (a) the work of plastic deformation which is developed primarily along the plane AB in Fig. 7.1, and (b) the frictional work that is generated along the plane AC in Fig. 7.1. These sources of heat make the resulting average temperature along the interface AC much higher than that at B or the mean temperature of chip, tool, or workpiece. While a wide variety of methods have been used to estimate the chip–tool interface temperature, including the complicated radiation pyrometers, embedded thermocouples, temperature-sensitive paints, the development of temper colours, and indirect calorimetric techniques, all of these methods suffer from slow speed of response, adverse geometric considerations, or the indirectness of measurement.

156 TEXTBOOK OF PRODUCTION ENGINEERING

The most successful approach to this problem is the tool-work thermocouple, apparently first used by Shore in 1924, and since then used by many others to study not only cutting problems but also abstract studies of friction phenomena. In this method, the tool-work contact area serves as the hot junction in a thermoelectric circuit and the emf generated is proportional to its temperature. Actually the maximum tool-tip temperature is not measured, but rather the average temperature over the area of contact. Since the tool face on a microscopic scale is not a plane surface but rather a series of peaks and valleys, the hot junction is essentially a series of small thermocouples in parallel, all of which contribute to the observed emf, analytically estimated; the maximum temperature at the points of contact is approximately 1.27 times the observed mean surface temperature. Another point of interest in connection with chip–tool interface–temperature measurements is that a steady-state condition is not established when a built-up edge is present. The cutting forces, power consumed, chip thickness, and even the depth of cut vary appreciably with time as well the temperature at the tool tip.

According to thermoelectric theory, if two dissimilar metals are joined to form a closed loop and the two resulting junctions are maintained at temperatures T_1 and T_2 respectively, an emf will be generated, which is proportional to the quantity $(T_2 - T_1)$. Some important features of thermoelectric circuits include the following:

(i) If a junction of two metals is at a uniform temperature, the emf generated is not affected by the introduction of a third metal (solder, weld metal, etc.).
(ii) The emf generated is independent of temperature gradients along the wire constituting the circuit, but depends upon the difference between the hot and cold junctions temperature $(T_2 - T_1)$.
(iii) The emf generated is independent of the size or resistance of conductors.

7.2 COMPUTATION OF TEMPERATURES IN ORTHOGONAL CUTTING

Since practically all of the energy involved in metal cutting is expended in heating the workpiece, tool and chips, the temperatures near the cutting region are quite high.

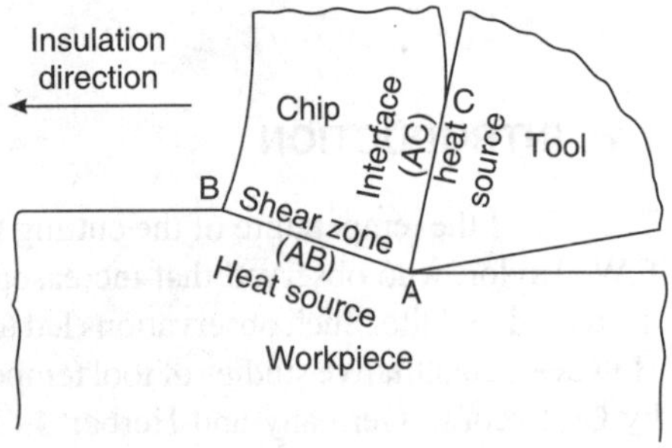

In orthogonal metal cutting where a continuous chip is produced, there are two sources of heat generation (see Fig. 7.1). At the shear zone (AB), the metal is deformed plastically and practically all of the strain energy involved in the deformation appears in the form of heat. The second source is at the chip–tool interface (AC) and is due to the friction of the chip rubbing against the face of the tool.

Undeformed and unheated workpiece material passes through AB, where its temperature is raised by the heat of deformation at the shear zone. After passing through AB, this same material becomes deformed and the heated chip material acquires more heat at AC. Therefore the chip at AC is quite hot and may even attain a temperature as high as the melting temperature of the chip or tool.

The energy expended along the shear zone AB is that which is almost entirely available as thermal energy to be distributed partly into the workpiece and partly into the chip. If the fraction of this energy which goes into the workpiece is known, the shear zone temperature can be found.

Fig. 7.1 Sources of heat generation.

7.2.1 Stationary Heat Source—Friction Slider

In order to compute the temperature rise on the tool face resulting from frictional heat, it is necessary to be able to solve the problem of Fig. 7.2 for the case where the non-conducting member is stationary. This may be done by using Kelvin's integration of the Fourier heat-transfer equation for the special case in which a quantity of heat Q is liberated instantaneously at any point (x', y', z') within an infinite, conducting body at time zero.

HEAT IN METAL CUTTING AND TEMPERATURE MEASUREMENT 157

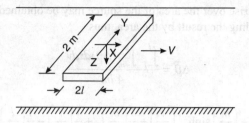

Fig. 7.2 Insulated slider on conducting surface.

The partial differential equation which relates the temperature and energy input is

$$\frac{k}{\rho c}\left(\frac{\partial^2 \theta}{\partial x^2} + \frac{\partial^2 \theta}{\partial y^2} + \frac{\partial^2 \theta}{\partial z^2}\right) + \frac{q}{\rho c} = \frac{d\theta}{dt}$$

[In this section, a simpler application considering a friction slider of dimensions $(2l \times 2m)$ is taken up before considering the case of shear plane and tool face temperatures in Sections 7.2.2 and 7.2.3].

Kelvin's result for the temperature θ at point (x, y, z) at time t is well known and available in texts on heat transfer.

$$\theta(x,y,z,t) = \frac{QK}{8k(\pi Kt)^{3/2}} \cdot e^{-\frac{r^2}{4Kt}} \tag{7.1}$$

where Q is heat released at point x', y', z'
x, y, z are coordinates of a point at distance (r) from point of heat release
k is thermal conductivity,
K is diffusivity,
also, $r^2 = (x - x')^2 + (y - y')^2 + (z - z')^2$.

For the case of a continuous heat source extending over a finite area, both a time and area integration are required. If the time integration is performed first for the semi-infinite body, and if we concentrate our interest on the steady-state solution (i.e. let $t \to \infty$), then the equation for the steady-state temperature rise anywhere in the semi-infinite body with a uniform heat source, of strength extending over $-l < x' < l$ and $-m < y' < m$ is given by

$$\Delta\theta = \frac{q}{2\pi k}\int_{-l}^{+l} dx' \int_{-m}^{+m} \frac{dy'}{\sqrt{(x-x')^2 + (y-y')^2 + (z)^2}} \tag{7.2}$$

The main interest lies in the temperature rise in the plane containing the source. Then, upon integrating Eq. (7.2) and letting $z = 0$, we have

$$\Delta\theta = \frac{q}{2\pi k}\left[(x+1)\left(\sinh^{-1}\frac{y+m}{x+1} - \sinh^{-1}\frac{y-m}{x+1}\right) + |x-1|\left(\sinh^{-1}\frac{y-m}{x-1} - \sinh^{-1}\frac{q+m}{x-1}\right)\right.$$

$$\left. + (y+m)\left(\sinh^{-1}\frac{x+1}{y+m} - \sinh^{-1}\frac{x-1}{y+m}\right) + |y-m|\left(\sinh^{-1}\frac{x-1}{y-m} - \sinh^{-1}\frac{x+1}{y-m}\right)\right] \tag{7.3}$$

The mean surface temperature rise over the area of the source may be obtained by integrating Eq. (7.3) over the area of the source and dividing the result by the area, thus

$$\Delta\bar{\theta} = \frac{\int_{-l}^{l} \int_{-m}^{m} (\Delta\theta) dx dy}{4\,lm}$$

$$\Delta\bar{\theta} = \frac{q \cdot l}{k} \cdot \frac{2}{\pi} \left\{ \sinh^{-1}\left(\frac{m}{l}\right) + \left(\frac{m}{l}\right)\sinh^{-1}\left(\frac{l}{m}\right) - \frac{1}{3}\left(\frac{m}{l}\right)^2 + \frac{1}{3}\left(\frac{l}{m}\right) - \frac{1}{3}\left(\frac{l}{m}\right) + \left(\frac{m}{l}\right)\right] \sqrt{1 + \left(\frac{m}{l}\right)^2} \right\} \quad (7.4)$$

This equation may be written as

$$\Delta\bar{\theta} = \frac{q \cdot l}{k} \bar{A} \quad (7.5)$$

where $\bar{A}$ is the area factor (a function of the aspect ratio m/l of the surface area only), plotted in Fig. 7.3. The function A_m also has a similar significance for the maximum temperature rise in the surface ($\Delta\theta_m$) as shown in Fig. 7.3, and is defined as

$$\Delta\theta_m = \frac{q \cdot l}{k} A_m \quad (7.6)$$

For values of m/l greater than 20, the quantities $\bar{A}$ and A_m may be found to a good approximation from the equations

$$\bar{A} = \frac{2}{\pi}\left(\ln\frac{2m}{l} + \frac{1}{3}\frac{l}{m} + \frac{1}{2} \right) \quad (7.7)$$

$$A_m = \frac{\pi}{2}\left(\ln\frac{2m}{l} + 1 \right) \quad (7.8)$$

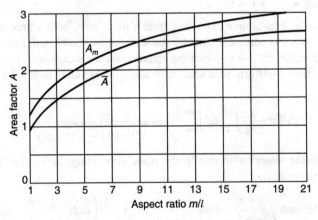

Fig. 7.3 Variation of area factors $\bar{A}$ and A_m with aspect ratio m/l.

Illustrative example of friction slider analysis

An illustrative example shall be taken up to explain a physical situation when a slider is moving on an extensive surface. A part (Rq) of the total heat (q) generated will flow in the extensive member and a part $(1 - R)q$ will flow in the slider member relative to the heat source. The quantity R which is a proportion of the total heat developed going with the moving member is derived by Shaw according to an equation here and derived in the end of this discussion as

$$R = \cfrac{1}{1 + \cfrac{0.754 k_2 / k_1}{\overline{A}\sqrt{L}}} \qquad (7.9)$$

where A is area factor
k_1 and k_2 are the conductivities of the two members of the sliding pair and
L is a non-dimensional velocity-diffusivity factor

Once R is known, the rise in temperature delta theta ($\Delta\theta$) may be obtained by substituting q, l and L into any of Eqs. (7.10) (7.11) or (7.12). At the physical level since a fraction R of the heat generated goes to the slider and a fraction $(1 - R)$ flows in the stationary body, multipliers R and $(1 - R)$ should be used in the original temperature rise expressions. The modified expressions are shown below in Eqs. (7.10), (7.11) and (7.12), where q is heat generated per unit area, and l is half the length of the slider as shown in Fig. 7.2.

$$\Delta\theta = 0.754 \frac{Rql}{k_1 \sqrt{L}} \qquad (7.10)$$

$$\Delta\theta = 0.754(1 - R) \frac{ql}{k_2 \sqrt{L}} \qquad (7.11)$$

$$\Delta\theta = \frac{(1 - R)ql}{k_2} \overline{A} \qquad (7.12)$$

Shaw's approach to finding R

Shaw equated the $\Delta\theta$ values given by Eqs. (7.10) and (7.12) since they represent the interface at which the temperature rise is the same and solved for R to get the formula to calculate the proportion of heat going with the slider (it has been found that 90 % of the heat generated at the interface is carried by the slider, i.e. the chip and $R = 0.9$)

$$R = \cfrac{1}{1 + \cfrac{0.754 k_2 / k_1}{\overline{A}\sqrt{L}}}$$

An application of preceding formulae to estimate mean temperature rise for a stainless steel slider on a carbon steel surface to the following data is given to fix the ideas.

Data: $l = 0.05$ inch (1.27 mm)
$m = 0.25$ inch (6.35 mm)
$V = 20$ inch/s (508 mm/s)
$P = $ Normal load at the interface $= 100$ lbs (445 N)
$\mu = $ Coefficient of friction $= 0.5$

Conductivities, $k_1 = 2.7 \times 10^{-4}$ BTU/inch^2/s (°F/inch)
$k_2 = 7.1 \times 10^{-4}$ BTU/inch^2/s (°F/inch)
Diffusivity, $K = 0.02$ inch^2/s (0.3 cm/s)

Calculations:

$$\text{Heat}, q = \frac{\text{Friction force} \times \text{Velocity}}{J \times \text{Area}} = \frac{\mu P V}{J(4lm)}$$

$$= \frac{0.5(100)(20)}{(9340)4(0.05)(0.25)} = 2.14 \text{ BTU/inch}^2\text{/s } (350 \text{ J/cm}^2\text{/s})$$

From Fig. 7.3, the value of $\bar{A}$ for $m/l = 5$ is 1.80. The final results are:

$$L = \left[\frac{Vl}{2K}\right] = 25, \; R = 0.82 \text{ and } \Delta\theta = 49 \, °\text{F}$$

7.2.2 Shear Plane Temperature

The rate at which shear energy is expended along the shear plane will be

$$U' = F_s V_s \tag{7.13}$$

where F_s is the component of force directed along the shear plane, and V_s is the velocity of the chip relative to the workpiece, which is also directed along the shear plane. In accordance the rate at which energy is expended per unit area on the shear plane will be for orthogonal cutting conditions, which assume

$$U' = \frac{F_s V_s}{tb \, \text{cosec} \, \phi} \tag{7.14}$$

where t is the thickness of the layer removed (see Fig. 7.2), b is the width of the workpiece, and ϕ is the shear angle. There is considerable experimental evidence which shows that the energy involved in plastic deformation largely goes into thermal energy, the extent of completion of the transformation depending upon the strain energy involved in the process. When the plastic-strain energy is as large as that involved in cutting, all but 1% or so of the strain energy appears as thermal energy, the small residum being associated with permanent lattice deformation. Thus to a good approximation, it may be assumed that all of the mechanical energy associated with the shearing process is converted into thermal energy, and the heat which flows from the shear zone per unit time per unit area will be

$$q_1 = \frac{F_s V_s}{Jtb \, \text{cosec} \, \phi} = \frac{u_s V_c \sin \phi}{J} \tag{7.15}$$

where J is the mechanical equivalent of heat, u_s the shear energy per unit volume of metal cut, and V_c the cutting speed.

Part of this heat will travel with the chip and the remainder will flow into the workpiece. If $R_1 q_1$ is the heat per unit time per unit area which leaves the shear zone with the chip, then $(1 - R_1)q_1$ is the heat per unit time per unit area that flows into the workpiece. The mean temperature of the metal in the chip in the vicinity of the shear plane will be

$$\theta_s = \frac{R_1 q_1 (bt \, \text{cosec} \, \phi)}{c_1 \rho_1 (Vbt)} + \theta_0 = \frac{R_1 u_s}{J c_1 \rho_1} + \theta_0 \tag{7.16}$$

where θ_0 is the ambient workpiece temperature and $c_1 \rho_1$ is the volumetric specific heat at the mean temperature between θ_s and θ_0.

Equation (7.16) gives the mean shear plane temperature in terms of unknown R_1, from the point of view of the chip. A value of θ may be computed from the point of view of the workpiece in the vicinity of the shear plane. This latter value will involve the quantity $(1 - R_1)$. According to the procedure of Block, these two values of θ may be equated to obtain the value of the single unknown R_1 and hence the value of θ_s.

The calculation of θ_s from the standpoint of the workpiece may be made by using Jaeger's (1942) solution for the mean temperature at the interface of a perfectly insulated slider moving across a conducting surface with velocity V, as heat is continuously and uniformly supplied to the interface. When the dimensions of the slider are as shown in Fig. 7.2 and $m/l > 2$, as in most practical cases, the mean temperature at the interface ϕ may be written as

$$\theta = 0.754 \frac{ql}{k\sqrt{L}} + \theta_0 \quad L > 0.2 \qquad (7.17)$$

where k is the thermal conductivity of the conducting stationary surface in cal/cm^2/s/°C and q is the rate at which thermal energy leaves the interface per unit area in cal/cm^2/s; and

$$L = \frac{Vl}{2K}$$

where K is the thermal diffusivity[1] of the conductor in cm^2/s. The parameter L is non-dimensional if velocity V is in cm/s and l is measured in cm. It can be shown that the mean temperature, given by Eq. (7.16), undergoes little change when q is not distributed exactly uniformly.

Equation (7.17) may be used to compute the shear-plane temperature as shown in Fig. 7.4. Here the chip may be considered a perfect insulator if the total heat flowing from the interface is $(1 - R_1)q_1$. The velocity of sliding is taken as V_s and not V in accordance with the generally adopted Piispanen picture of metal cutting shown in Fig. 7.5. By this picture, metal when cut behaves as a deck of cards, one card at a time sliding a finite distance across its neighbour. Heat is generated only when sliding occurs in direction V_s, and we might consider that one plate at a time is always in motion. An approximation involved in the use of Fig. 7.4 is that the metal which is assumed present in wedge A is not there, while that in wedge B is ignored. Since the heat flows associated with wedges A and B will tend to cancel each other, it is considered preferable that the idealised picture of Fig. 7.4 provides a good approximation.

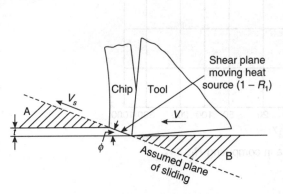

Fig. 7.4 Idealized diagram of shear plane moving heat source.

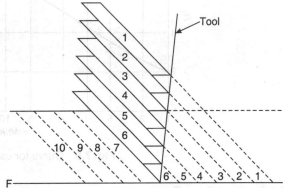

Fig. 7.5 Piispanen's mechanism of chip formation.

---

[1] The thermal diffusivity K is notation for the ratio of thermal conductivity k and volumetric specific heat $c\rho$, i.e. $K = k/c\rho$.

When Eq. (7.17) is used to compute the shear-plane temperature, it is found that

$$\theta_s = 0.754 \frac{(1-R_1)q_1 \left(\dfrac{t \operatorname{cosec}\phi}{2}\right)}{k_1 \sqrt{L_1}} + \theta_0 \qquad (7.18)$$

$$L_1 = \frac{V_s \left(\dfrac{t \operatorname{cosec}\phi}{2}\right)}{2k_1} = \frac{V\gamma t}{4K_1} \qquad (7.19)$$

where k_1 is the conductivity, K_1 the diffusivity of the workpiece material at temperature θ_s, and γ the strain in the chip. Then, equating Eqs. (7.16) and (7.18), and solving for R_1, it is found that

$$R_1 = \frac{1}{1 + \dfrac{0.664\,\gamma}{\sqrt{L_1}}} = \frac{1}{1 + 1.328\sqrt{\dfrac{K_1\gamma}{Vt}}} \qquad (7.20)$$

Once R_1 is known, θ_s may be calculated from Eq. (7.16), and R_1 can be computed from Fig. 7.6. It is evident that the percentage of energy going to the chip R_1 does not increase with increased cutting speed V alone, but rather with the non-dimensional quantity $(Vt/K_1\gamma)$. From Eq. (7.16), it may be seen that the temperature rise at the shear plane varies directly with the shear energy per unit volume going into the chip $R_1 u_s$, and inversely with the volume specific heat of the workpiece $c\rho$.

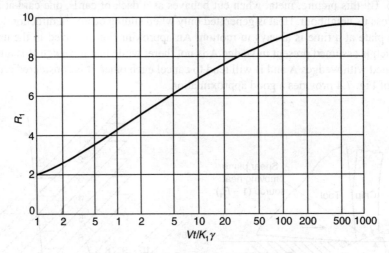

Fig. 7.6 Curve for use in computing R_1.

7.2.3 Tool Face Temperature

From an analytical viewpoint, the friction between chip and tool can be regarded as a heat source that is moving in relation to the chip and, at the same time, one that is stationary in relation to the tool. As before, it is this double treatment that allows calculation of the partition of friction energy between chip and tool, and hence the resulting temperature.

The total friction energy q_2 in kcal/cm²/s that will be dissipated at the chip-tool interface per unit time per unit area will be

$$q_2 = \frac{F_c V_c}{Jab} = \frac{u_f V t}{Ja} \tag{7.21}$$

where F_c is the friction force along the tool face, V_c is the velocity of the chip relative to the tool, a is the length of contact between chip and tool in the direction of motion, b is the width of the chip, and u_f is the energy per unit volume of metal cut that is consumed by friction on the tool face.

If R_2 is the fraction of q_2 that flows into the chip, then the average temperature rise in the surface of the chip due to friction $\Delta \bar{\theta}_f$ is, from Eq. (7.17)

$$\Delta \bar{\theta}_f = \frac{0.754 (R_2 q_2) \frac{a}{2}}{k_2 \sqrt{L_2}} \tag{7.22}$$

where k_2 is the thermal conductivity of the chip at its final temperature. The parameter L_2 is defined as

$$L_2 = \frac{V_c \frac{a}{2}}{2 k_2} \tag{7.23}$$

The mean temperature $\bar{\theta}_t$ of the chip surface along the tool face ($\bar{\theta}_f$) will be the sum of the mean shear-plane temperature ($\bar{\theta}_s$) and the mean temperature rise due to friction ($\Delta \bar{\theta}_f$).

$$\bar{\theta}_t = \bar{\theta}_s + \Delta \bar{\theta}_f = \bar{\theta}_s + \frac{0.754 (R_2 q_2) \frac{a}{2}}{k_2 \sqrt{L_2}} \tag{7.24}$$

The application of Eq. (7.24) is explained through an illustrative example given at the end of this section.

To find R_2 the stationary heat-source solution must be applied to the tool. The shaded area in Fig. 7.7(a) represents the area of contact between a chip and a two-dimensional cutting tool, the cutting edge being along the Y-axis and the tool face in the XY-plane. The tool represented by the solid lines may be considered a quarter-infinite body relative to the shaded area of contact. This is seen to represent a good approximation of an orthogonal cutting tool, provided the rake and clearance angles are not too large. By symmetry it is evident that the temperature at any point in the surface of the quarter-infinite body, subjected to the uniform heat source represented by the shaded area, would be the same as the temperature of the corresponding point in the semi-infinite body when subjected to a uniform heat source extending over A–B–C–D, provided that the YZ-plane is a perfect insulator, just as the XY-plane, which is a valid assumption. Thus the aspect ratio to be used in obtaining the mean tool face temperature is

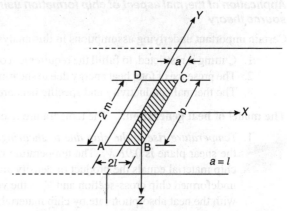

Fig. 7.7(a) Relationship between a two-dimensional tool and a semi-infinite body.

$$\frac{m}{l} = \frac{b}{2a} \quad \text{(orthogonal tool)} \tag{7.25}$$

By similar reasoning, the aspect ratio for a lathe tool which cuts at the corner of the tool would be

$$\frac{m}{l} = \frac{b}{a} \quad \text{(lathe tool)} \tag{7.26}$$

With the aspect ratio thus defined for any tool, the shape factor $\overline{A}$ may be determined from Fig. 7.3 and the temperature $\overline{\theta}_t$ found out for a point in the tool from

$$\overline{\theta}_t = \frac{(1-R_2)q_2 a}{k_3}\overline{A} + \theta_0 \tag{7.27}$$

where k_3 is the thermal conductivity of the tool material at the final tool temperature $\overline{\theta}_t$, and θ_0 is the ambient temperature of the tool.

By equating the two values of $\overline{\theta}_t$ given in Eqs. (7.24) and (7.27), it is possible to solve for R_2.

$$R_2 = \frac{q_2 \dfrac{a\overline{A}}{k_3} - \theta_s + \theta_0}{q_2 \dfrac{a\overline{A}}{k_3} + q_2 \dfrac{0.377a}{k_2\sqrt{L_1}}} \tag{7.28}$$

Or, substituting from Eqs. (7.21) and (7.23), and noting that $V_c = V_r$ where r is the chip thickness ratio (i.e. the ratio of the depth of cut t to chip thickness when taking an orthogonal cut), we have

$$R_2 = \frac{\dfrac{u_f V t \overline{A}}{J k_3} - \theta_s + \theta_0}{\dfrac{u_f V t \overline{A}}{J k_3} + \dfrac{0.754 u_f}{J \rho_2 c_2}\sqrt{\dfrac{V t^2}{ar k_2}}} \tag{7.29}$$

When R_2 is determined, θ_t may be found readily from Eq. (7.27).

As in all analytical heat transfer calculations, it has been necessary in the foregoing derivations to consider all thermal quantities to be constants.

Application of thermal aspect of chip formation using Merchant's analysis and Jaeger's moving heat source theory

Certain important underlying assumptions in this analysis are:

1. Cutting is restricted, to fulfill the requirement of Jaeger's heat source model at the chip-tool interface.
2. The presence of total heat energy due to shearing of the metal in the shear zone is 90%. Thus, $R_1 = 0.9$.
3. The thermal conductivity and specific heat are assumed to be independent of temperature.

The model of heat generation: The temperature rise at the chip-tool interface is owing to two processes:

1. *Temperature rise in the chip due to shearing of metal in the shear zone:* The heat generated in the shear plane is $0.9 F_s V_s$. The temperature rise in the work material is $(\theta_s - \theta_o)$; the mass of the chip material equals the product of density and volume, i.e. $\rho(t_1 V_c w)$ where ρ is density, $(t_1 \cdot w)$ is undeformed chip cross-section and V_c is the velocity of cutting tool. Equating the energy input rate with the heat absorption rate by chip material, we get

$$\rho(t_1 V_c w) C (\theta_s - \theta_o) = (0.9 F_s \times V_s)$$

$$\theta_s = \theta_o + \frac{0.9 F_s V_s}{(\rho t_1 V_c w) C} \tag{7.30}$$

where C is the specific heat of work material.

2. *Temperature rise of the hot chip further due to rubbing between hot chip and the rough chip-tool interface:* The Jaeger's moving heat source with chip as a slider rubbing at the chip-tool interface is given by

$$\Delta\theta = \frac{0.754 R_1 ql}{k\sqrt{L}} = \frac{0.754 \times 0.9 \times q \times 0.5a}{k\sqrt{L}} = \frac{0.339 qa}{k\sqrt{L}} \quad (7.31)$$

where $R_1 = 0.9$
Length of chip-tool contact $a = 2l$ [see Fig. 7.7(b) for turning]
k is thermal conductivity,

L (Non-dimensional velocity–diffusivity number) $= \left(\dfrac{V_c l}{2k}\right) = \dfrac{V_c a}{4k}$ with the length of chip-tool contact, $a = 2l$, for case of restricted turning.

q is heat due to rubbing $= \dfrac{FV_f}{aw}$, with w is the width of cut,

F is chip-tool friction force and (aw) is the area of contact.

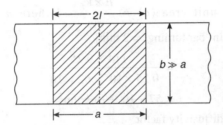

Fig. 7.7(b) Heat source for restricted cutting.

In this model, F_s and V_s can be estimated by using Merchant's analysis of cutting. The values of cutting force F_c and thrust force F_t are measured by a dynamometer. This concept was proposed by **Pande and Singh**.

Illustrative example
During turning of a steel bar with restricted cutting (w) 2 mm depth of cut, a tool with side rake angle (α) 10° is used; feed rate (t_1) 0.1 mm/rev, cutting speed (V_c) is 1 m/s, chip-thickness ratio (r_c) is 0.25; F_c is 187 N, F_t is 83 N; and chip-tool contact length a is 0.25 mm.

Work material properties: Thermal conductivity $k = 50$ w/mk; density $\rho = 6000$ kg/m^3, specific heat $C = 500$ J/kg.K

Solution
1. Calculating temperature rise due to shearing in shear zone by applying Merchant's model: From geometrical relationship, if ϕ is shear plane angle, α is rake angle and r_c is chip thickness ratio, then

$$\tan\phi = \frac{r_c \cos\alpha}{1 - r_c \sin\alpha} = \frac{0.25 \cos 10°}{1 - 0.25 \sin 10°} = 0.2572$$

∴ $\phi = 14°$

$$F_s = (F_c \cos\phi - F_t \sin\phi) = [(187 \times 97) - (83 \times 24)] = 160 \text{ kg}$$

$$V_s = \frac{V_c \cos\alpha}{\cos(\phi - \alpha)} = \frac{0.984}{0.997} = 0.987 \text{ mm/s}$$

where F_s is shear force on shear plane and
V_s is shear velocity

$$\therefore \quad \theta_s = \frac{0.9 F_s V_s}{\rho \times t_1 \times V_c \times w \times C} + \theta_0$$

$$= \frac{0.9 \times 160 \times 0.987}{6000 \times 0.1 \times 10^{-3} \times 1 \times 2 \times 10^{-3} \times 500} + 20$$

$= 257°C$ if ambient temperature θ_o is assumed 20°C

2. Calculation of friction heating effect:

$$V_f = V_c r_c = 1 \times 0.25 = 0.25$$

where V_f = Velocity of chip-flow.

Friction force $(F) = F_c \sin \alpha + F_t \cos \alpha$
$= 187 \sin 10° + 83 \cos 10°$
$= 114$ N

Heat generated per unit area $(q) = \dfrac{F \times V_f}{aw}$ [where $a = 0.25$ mm and w = depth of cut i.e. width of cut in 'bar turning' = 2 mm]

$$= \frac{114 \times 0.25}{0.25 \times 10^{-3} \times 2 \times 10^{-3}}$$

$= 57 \times 10^6$ N(m/s)/m²

Calculation of velocity diffusivity factor

$$L = \left(\frac{V_c a}{4k}\right) = \frac{0.25 \times 0.25 \times 10^{-3}}{4 \times (1.6 \times 10^{-5})} = 0.976$$

The diffusivity K is calculated as

$$K = \frac{k}{\rho C} = \left(\frac{50}{6000 \times 500}\right) = 1.6 \times 10^{-5} \text{ m/s}$$

To calculate temperature rise, $\Delta\theta$, Jaeger's formula is used as applied by Milton C. Shaw and Pandey and Singh

$$\Delta\theta = \frac{0.339 qa}{k\sqrt{L}} = \frac{0.339 \times 57 \times 10^6 \times 0.25}{50\sqrt{0.976}} = 97.8°C$$

Thus, adding the two effects, i.e. shearing in shear zone and chip as a slider we have, average chip-tool interface temperature = 257°C + 97.8°C = 354.8°C.

7.2.4 Energy Balance

The energies going to the chip u_c, tool u_t and workpiece u_w per unit volume of metal removed are functions of R_1 and R_2 and the specific energies of the cutting process (u_s, the shear energy per unit volume, and u_f, the friction energy per unit volume). Thus

$$u_c = R_1 u_s + R_2 u_f \qquad (7.32)$$
$$u_t = (1 - R_2) u_f \qquad (7.33)$$
$$u_w = (1 - R_1) u_s \qquad (7.34)$$

When the appropriate values are substituted into these equations for the tests, the distribution of energy shown in Fig. 7.8 is obtained. The percentage of the total energy going to the chip increases with increased speed, although the percentages going to the tool and workpiece decrease. At very high cutting speeds, practically all of the energy is carried away in the chip, with a small amount going into the workpiece and a still smaller amount to the tool.

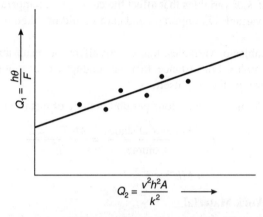

Fig. 7.8 Log-Log plot of dimensionless groups for interface temperature.

7.3 CUTTING TOOL TEMPERATURE BY DIMENSIONAL ANALYSIS

7.3.1 Introduction

Dimensional analysis as applied to metal cutting was used by Kronenberg first in 1949. In 1952, Chao and Trigger suggested a dimensionless quantity, the thermal number R_t, given as follows:

$$R_t = \frac{vhf}{k} \tag{7.35}$$

where, v is the cutting speed
h is the volumetric specific heat
f is the feed per revolution
k is the thermal conductivity of the work material

It is also found that the tool-chip interface temperature is not solely dependent on this thermal number, but there exists a functional relationship between this number and the energy expended in cutting per unit volume of metal removed.

The principle governing dimensional analysis is simple and is known as the *Principle of Dimensional Homogeneity*. It states that the terms on both sides of a correct physical equation must have the same dimensions.

We know that the dimensional formula for force is MLT^{-2}. Here M, L, T are fundamental quantities—Mass, Distance and Time, and exponents 1, 1 and –2 are the dimensions of the various quantities. A dimensionless number is one in which all these exponents are zero.

7.3.2 Dimensional Analysis of Tool Temperature

The treatment for the dimensional analysis for tool temperature can be best understood in five steps.

Selection of fundamental quantities

Firstly select basic units or fundamental quantities. The most commonly used units are M, L, T and θ, where M = Mass, L = Length, T = Time, θ = Temperature.

Selection of physical quantities

Select those physical quantities or variables that affect the cutting tool temperature most.

This step of choosing variables is important and care should be taken that no variable likely to affect results is omitted.

Inclusion of extra variables or variables, that do not affect the result appreciably, will unnecessarily make the final expression complex. For instance, thermal conductivity of tool is omitted as most of the heat dissipation (about 85%) is through the chip itself.

Unit Cutting Force: It is defined as work done per unit volume of metal removed.

$$F = \frac{\text{Force} \times \text{Distance}}{\text{Volume}} = \frac{MLT^{-2} \times L}{L^3}$$

$$= ML^{-1} \times T^{-2} \tag{7.36}$$

Thermal Conductivity of Work Material

$$\text{Heat transfer rate, } q = \frac{kA(t_1 - t_2)}{\Delta x}$$

$$k = \frac{q \Delta x}{A \cdot (t_1 - t_2)}$$

$$= \frac{\frac{MLT^{-2} \times L}{T} L}{L^2 \cdot \theta} = ML\, T^{-3} \times \theta^{-1} \tag{7.37}$$

Volumetric Specific Heat: It is defined as specific heat per unit volume of metal removed and is the product of density ρ and specific heat C.

$$h = \rho \times C$$

$$= \frac{M}{L^3} \times \frac{MLT^{-2} \times L}{M\theta}$$

$$= ML^{-1} T^{-2} \theta^{-1} \tag{7.38}$$

All these variables are shown here.

Physical quantity	Symbol	Dimensional formula
Temperature	θ	θ
Chip cross-sectional area	A	L^2
Cutting speed	v	LT^{-1}
Unit cutting force	F	$ML^{-1} T^{-2}$
Thermal conductivity of work	k	$MLT^{-3} \theta^{-1}$
Volumetric specific heat	h	$ML^{-1} T^{-2} \theta^{-1}$

Selection of basic variables

Number of fundamental quantities = 4
Number of physical quantities = 6

Then, according to Buckingham P_i Theorem, the number of dimensionless products required to correlate all these quantities would be equal to $6 - 4 = 2$.

Now choose 4 basic variables out of 6 physical quantities such that they do not make any dimensionless group in themselves. (Number of basic variables must be equal to the number of fundamental quantities.)

It is not difficult to find out that such basic variables are v, h, F and k.

Formation of dimensionless groups

One non-basic quantity is grouped with all the four basic variables to give one dimensionless number. So

$$Q_1 = v^a \cdot F^b \cdot k^c \cdot h^d \cdot \theta$$

$$Q_2 = v^e \cdot F^f \cdot k^g \cdot h^i \cdot A$$

Substituting dimensional formulae for all in the expression Q_1 $(M^0 L^0 T^0 \theta^0) = (LT^{-1})^a$ $(ML^{-1} T^{-2})^b$ $(MLT^{-3} \theta^{-1})^c (ML^{-1} T^{-2} \theta^{-1})^d$. Equating exponents of $M, L, T,$ and θ separately on both the sides of the equation, we get

$$b + c + d = 0$$
$$a - b + c - d = 0$$
$$-a - 2b - 3c - 2d = 0$$
$$-c - d + 1 = 0$$

Solving for values of a, b, c and d, which are—$a = 0, b = -1, c = 0, d = 1$ and substituting in equation for Q_1 we get,

$$Q_1 = v^0 \cdot F^{-1} \cdot k^0 \cdot h^1 \cdot \theta^1 = \frac{h\theta}{F} \quad (7.39)$$

Working on similar lines we get,

$$Q_2 = \frac{v^2 h^2 A}{k^2} \quad (7.40)$$

It is noted that the thermal number given by Chao and Trigger is nothing but square root of Q_2. Also, energy expended in cutting per unit volume of metal removed is nothing but the unit cutting force F.

A functional relationship between these two can be established as it is done in the last step of the analysis.

Relationship between dimensionless products Q_1 and Q_2

The two dimensionless groups Q_1 and Q_2 which include dependent variable θ also should have a functional relationship like

$$Q_1 = f(Q_2)$$

The value of f can be ascertained by either conducting experiments or by experimental data available. Data where all these quantities are given is available from Gotwein's Tests. From these, numerous values of Q_1 and Q_2 were calculated and plotted on a bi-log paper. Such a plot is called the *Master Curve for Cutting Temperature* (see Fig. 7.8).

All the points fit fairly well along a straight line. But if in an analysis they do not lie on a curve, it can be concluded that some essential parameter has been overlooked while selecting physical quantities. Writing the equation for the curve

$$\log Q_1 = n \log Q_2 + \log C_0$$

It can be seen that the slope is n and C_0 is a constant; Q_1 is equal to C_0 when $Q_2 = 1$.
This is given by the equation

$$Q_1 = C_0 \cdot Q_2^n$$

On measurement from the master curve, the exponent n is found to be 0.22.

$$\frac{h\theta}{F} = C_0 \left(\frac{v^2 h^2 A}{k^2}\right)^n$$

$$\theta = \frac{C_0 F V^{2n} A^n}{k^{-1} h^{1-2n}} \tag{7.41}$$

For $n = 0.22$, temperature θ would be given by expression

$$\theta = \frac{C_0 F V^{0.44} A^{0.22}}{k^{0.44} h^{0.56}} \tag{7.42}$$

The following conclusions can be drawn from the expression derived above. They are displayed in Fig. 7.9.

(i) Cutting temperature is affected most by the unit cutting force and it is directly proportional to it.

$$\theta \propto F$$

The two thermal quantities k and h have the next largest effect on temperature, and temperature is nearly proportional to the inverse of the square root of these.

$$\theta \propto \frac{1}{\sqrt{k}}$$

$$\theta \propto \frac{1}{\sqrt{h}}$$

According to the fact that specific heat varies with temperature, we should not expect an expression for θ containing h to be an independent one.

But this effect of temperature could be overlooked as thermal conductivity k decreases with temperature and their product $k \times h$ can be assumed to be constant over a wide range of temperatures.

Thus it can be argued that the temperature effect doesn't alter the denomination of the expression for θ, though this reasoning will not hold in cases where specific heat is substantially changed with temperature.

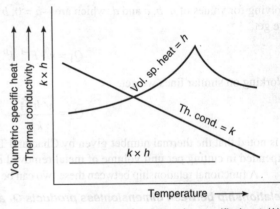

Fig. 7.9 Variation of volumetric specific heat (h), thermal conductivity (k) and their product ($k \times h$) with temperature.

(ii) Temperature varies nearly as square root of cutting speed, while it is proportional to only the fourth root of chip cross-sectional area.

$$\theta \propto \sqrt{v}$$

and

$$\theta \propto \sqrt[4]{A}$$

7.3.3 Comparison with Experimentally Obtained Expressions

Loewen and Shaw (1954) conducted tests and derived complicated formulae which can be simplified as

$$\theta = \frac{S_0 v^{0.50} t^{0.50} \theta^{0.50}}{k^{0.50} h^{0.50} J} \qquad (7.43)$$

where S_0 is the shear strength of the material
 t is the depth of cut or feed per revolution

A comparison reveals that Eqs. (7.42) and (7.43) agree in many respects, except in exponent of feed.

Here the exponents of both cutting speed and feed are same, while in Kronenberg's expression derived by dimensional analysis, exponent of feed is half of cutting speed. Kronenberg has justified his expression.

Also, J is the mechanical equivalent of heat
 S_s is the dynamic shear stress of work material (dynamic means shear occurring when tool and work are moving)
 R_t is the thermal number
 ξ, η are coordinate axes moving with heat source
 e is half of the length of two shear planes
 $\Delta\theta$ is the rise in temperature of chip.

S_s is constant for work material, while θ shows a functional relationship. Therefore we can say that for a certain location on the shear plane, the temperature rise above initial temperature depends on thermal number. In other words, the temperature distribution along the shear plane is grounded entirely by this dimensionless number under the ideal conditions.

The fraction θ (fraction of heat generated at shear zone conducted back into workpiece) depends on thermal number for orthogonal cutting. The larger the number, the smaller the proportion of heat conducted to the workpiece. This means that the proportion of heat at shear zone which goes into the workpiece will be less and less with increase in cutting speed, or feed, or both.

The practical importance of thermal number in metal machining is the fact that there exists a functional relationship between the number and energy expended in cutting per unit volume of metal removed. This latter quantity has been used to define the efficiency of machining operation, and the variation is shown in Figs. 7.10 and 7.11 herein.

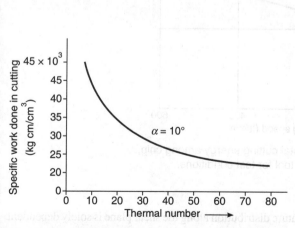

Fig. 7.10 Variation in thermal number with respect to specific work done for $\alpha = 10°$.

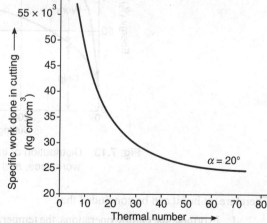

Fig. 7.11 Variation in thermal number with respect to specific work done for $\alpha = 20°$.

It appears from various experiments performed that better understanding of metal machining is possible when full significance of thermal number is appreciated. The same degree of correlation does not exist between the thermal number and the tool chip friction behaviour. The lack of correlation is supposedly due to change in the friction behaviour as influenced by temperature and in the temperature-dependent properties of the chip material with steep temperature gradient.

It is clear from Fig. 7.12 that for lower value of specific energy requirements and rate of metal removal, a high thermal number is desirable. Figure 7.13 shows the distribution of heat developed in metal cutting in the chip, workpiece and tool.

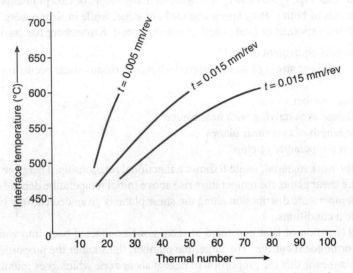

Fig. 7.12 Variation in interface temperature with change in thermal number.

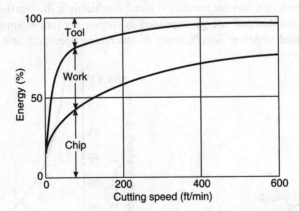

Fig. 7.13 Distribution of total cutting energy among chip, workpiece, and tool for test conditions.

Therefore, at last it can be concluded that:

1. In orthogonal cutting operations, the temperature distribution along the shear plane is solely dependent upon the thermal number.

2. Under usual cutting conditions, a larger thermal number results in more uniform temperature distribution along the shear frame.
3. Within the range investigated, the effect of cutting speed and feed on chip thickness ratio, shear strain, shear angle, and specific energy of metal removal can be expressed in terms of thermal number (constant rake angle).
4. Tool chip surface temperature is not solely dependent upon thermal number.

7.4 HEAT AND TEMPERATURE IN MILLING

Milling is a high-rate metal removal process, incorporating the phenomenon of either orthogonal or oblique cutting, characterised by the engagement of more than one tool simultaneously. Hence more energy dissipation and existence of very high instantaneous surface temperatures is found in milling.

Work done in a metal-cutting operation can be measured in the form of heat in the workpiece and tools. Most of the heat is in the chips, but usually from 5 to 30% of the total heat goes into the workpiece with the greater percentage occurring at lower speeds. It has been proved in many tests, and often can be observed visually, that tool tips frequently fail because of the very high temperatures attained near the cutting edge. However, excessive temperatures in the workpiece are seldom noticed in casual or general observations, but special tests confirm that high instantaneous temperatures occur in the machined surface of a workpiece during the machining process.

In X-ray diffraction investigations, the depth of penetration of plastic deformation beneath the surface of a milled workpiece is found to be several microns, varying with the rake angle.

Approximately 10% of the entire energy is confined to the workpiece in the form of heat. All of this machining heat in the workpiece must be contained initially in the thin deformed surface layer for a short period of time, before dispersing itself to the cooler metal beneath. Thus it is possible to compute for this layer a maximum average temperature which, although lower than the instantaneous surface temperature, at least provides an indication of the magnitude of that temperature, e.g., higher than tool-tip or surface temperatures measured during the cut. If the heat existing for a short period of time in a very small segment of the completely insulated cold-worked surface layer is considered, the results of the temperature computation will be the same.

If the surface layer is taken as a body of weight W having a mean heat capacity C_m, an initial temperature t_1 and a maximum average temperature t_2 after introduction of a quantity of heat Q, then

$$Q = WC_m(t_2 - t_1)$$

or

$$Q = WC_m \, \delta t$$

where

$$\delta t = t_2 - t_1$$

Hence

$$\delta t = \frac{Q}{WC_m}$$

The computations of δt involve C_m, which is either predetermined at high temperature, i.e. t_2, or may be available from tables.

7.5 HEAT AND TEMPERATURE IN DRILLING

The knowledge of the temperature distributions in non-orthogonal operations is also desirable to have for both academic and practical use. To investigate temperature distribution in a drill, the energy entering the drill is to be calculated analytically as given here.

174 TEXTBOOK OF PRODUCTION ENGINEERING

The temperature at any point along the cutting edge is determined by taking the average temperature in the tool-chip interface zone by the equation given below.

$$\theta = \frac{q_1}{2\pi\alpha pc}[2b_1(3 - \gamma + \log_e \alpha t - 2\log_e b_1)] + \theta_0 \tag{7.44}$$

where, q_1 is the amount of heat entering the rake face of the drill;
b_1 is the chip contact length;
γ is the Euler's constant;
α is the thermal diffusivity of the drill material;
t is the cutting time;
p is the density of the drill material;
c is the specific heat of the drill material;
θ_0 is the initial temperature of the drill.

However, it is thought that the average temperature in the tool-chip interface zone is a better representation of the true cutting edge temperature, even though it is found that the difference between these two values is insignificant.

Tseuda determined the intensity of the heat source q_1 by experimentally measuring the temperature θ_p at a point on the drill flank, but the intensity of heat source q_1 may be calculated using the approach of Loewen and Shaw (1954) after necessary modifications.

The energy dissipated in cutting with single-point tools is confined in two regions—(a) in the shear zone due to the plastic deformation associated with chip formation, and (b) along the tool face as a result of the frictional resistance between chip and tool. Thermal energy associated with plastic deformation ahead of the tool is dissipated partly into the chip and partly into the workpiece. The total frictional energy Q, dissipated at the tool–chip interface, is given by

$$Q = \frac{u_f V t}{J b_1} \tag{7.45}$$

where u_f is the friction energy in mkg/cm^3;
V is the cutting velocity in m/s;
t is the depth of cut in mm;
J is the mechanical equivalent of heat in mkg/kcal;
b_1 is the chip contact length.

A portion of the total frictional energy leaves with the chip, while the remaining portion enters the drill. The amount of heat that enters the rack face of the drill is given by

$$q_1 = (1 - R_2)Q \tag{7.46}$$

where R_2 is a partition factor (i.e. the fraction of Q which flows away with the chip).

Now,
$$u_f = \frac{F_H \sin\alpha_e + F_t \cos\alpha_e}{bt}\left[\frac{\sin\phi}{\cos(\phi - \alpha_e)}\right] \tag{7.47}$$

where
$$f = 48.0 - 0.81(\lambda - \alpha_e) \tag{7.48}$$

Also, partition factor R_2 is given by

$$R_2 = \frac{C' - \bar{\theta}_s + \theta_0}{C' + B'} \tag{7.49}$$

where F_H is the cutting force;
 F_t is the thrust force;
 λ is the friction angle;
 r_c is the chip thickness ratio;
 α_a is the effective rake angle;
 u is the drill torque;
 T is the drill thrust.
 $C' = f(\mu_f, V, t, B)$
 B is the area factor
 $B' = f(\mu_f, V, t, b)$
 $\theta_x = (U_s, R_1)$

In order to calculate the partition of the frictional energy, the chip-tool friction is regarded as a moving heat source with respect to the chip and, at the same time, a stationary heat source with respect to the drill. This treatment of the chip tool friction allows the partition of the frictional energy between chip and drill to be calculated.

Since drilling is a non-orthogonal cutting operation, the cutting edge can be divided into 60 divisions, so that each cutting edge is small enough to be assumed an orthogonal cutting. Here 60 divisions are chosen because it is found that the cutting edge temperature rises considerably with the number of segments into which the drill face is divided, as is illustrated in Fig. 7.14.

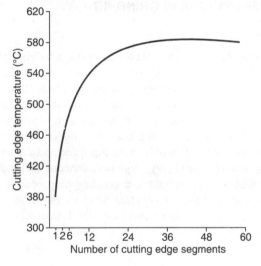

Fig. 7.14 Effect of the number of divisions into which the cutting edge is divided on the cutting edge temperature.

In Fig. 7.14, the temperature at a point 15 mm away from the chisel edge and on the cutting edge is plotted as a function of the number of segments into which the drill face is divided. The temperature at this point is observed to increase with the number of segments. The increase in the analytical temperature is large when the number of segments is increased from 1 to 12 and tends to level off as the number of segments is increased from 12 to 60. The analytical temperature at the above described point is 190°C when the drill edge is considered one segment. When the drill is divided into 3 segments, the analytical temperature increases by more than 70% to 330°C, while for 6 segments the increase was 35% to 365°C. As the number of segments increased from 12 (375°C) to 60 (390°C), the rate of increase in temperature with increasing number of segments decreases and approaches a constant temperature. The reason for this increase in cutting temperature is the increase in the mean radius of the segments. The mean radius of any segment, for example, the segment closest to the drill periphery, is larger when the drill face is divided into a large number of segments. With the increase in mean radius, the amount of heat entering the segment also increases, causing a higher temperature. The coordinate axes selected for the drill are (see Fig. 7.15)— (a) positive y corresponds to an increasing distance along the cutting edge form the chisel edge outwards, and (b) positive z corresponds to increasing distances on the flank face beginning at and perpendicular to the cutting edge. The sample calculation

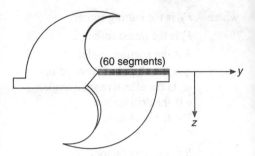

Fig. 7.15 Division of cutting edge and coordinate axes.

not given here is divided in three sections. The first section consists of given values, including—(a) the drill geometry parameters and the cutting conditions, (b) the properties of the drill material and the workpiece hardness, and (c) several values which are calculated using empirical equations. The references used for calculating these values are also given. The second section consists of—(a) the calculations of the intensity of the heat generated during cutting, (b) the partition factor R_2, and (c) the amount of heat source, including the amount of heat entering the rake face, the chip contact length, is also determined.

7.6 HEAT AND TEMPERATURE IN GRINDING

7.6.1 Heat in Grinding

Similar to previous analysis, the heat transfer theory involving a moving heat source is applied to the grinding process to obtain an equation for the mean surface temperature during grinding operation. This predicts temperatures as high as 1650°C for a representative fine grinding. It is found during the course of analysis that grinding variables such as depth of cut, feed rate, etc. influence surface temperatures.

The energy per unit volume is found to be unusually high in grinding as a result of the relatively small depth of cut associated with this process. At the same time, the grinding operation is normally carried out at a high cutting speed. From this combination of high cutting speed and a relatively large dissipation of energy per unit volume of metal removed (specific energy), high instantaneous values of surface temperatures are expected.

It can be that, to a good approximation, a grinding chip could be represented as a long slender wedge of constant width as shown in Fig. 7.16 and it has been found that the undeformed length l and maximum height t of such a chip could be obtained from the expressions

$$l = \sqrt{Dd} \quad (7.50)$$

$$t = \left(\frac{100 \times 4}{\pi DNCr}\sqrt{\frac{d}{D}}\right)^{1/2}\left(\frac{4v}{VCr}\sqrt{\frac{d}{D}}\right)^{1/2} \quad (7.51)$$

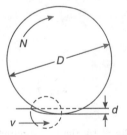

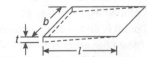

Fig. 7.16 Chip formation in surface grinding.

where D is the wheel diameter in mm;
 d is the wheel depth of cut in mm;
 v is the work speed in m/min;
 V is the wheel speed in m/min;
 N is the spindle speed in rpm;
 C is the number of cutting grits per sq. cm;
 r is the ratio of the width to the depth of the groove produced by the mean grit.

It was further observed that the energy per unit volume could be obtained from the equation

$$u = \frac{\pi D N F_H}{100 \, vbd} = \frac{V F_H}{vVd} \qquad (7.52)$$

where F_H is the mean tangential force on the wheel in kg;
 b is the width of the workpiece in mm.

This equation may be readily verified in the following way:

The number of chips produced per revolution is $\pi D\, b\, C$, where b is the width of the workpiece. At the same time the volume of metal removed per revolution is $1000\, vdb/N$. Hence the volume per chip will be $1000\, vdb/N\pi DbC$. From Fig. 7.16, the volume per chip is also $(1/2)tb'l$, where b is the width of the chip and equal to $(1/2)r$. Thus

$$\frac{1}{2} t \left(\frac{1}{2} r \right) l = \frac{1000 \, v b^{1/2}}{\pi N D C}$$

Upon substitution from Eq. (7.51) and solving for t, we get

$$t = \left(\frac{4000 \, v}{\pi D N C r} \sqrt{\frac{d}{D}} \right)^{1/2}$$

A surface grinding cut in which the wheel width is greater than the work width and the entire width of the specimen is ground in a single pass is assumed.

As in the case of milling, two types of grinding may be distinguished: up grinding in which the work and wheel travel in opposite directions and down grinding in which the wheel and workpiece move in the same direction (see Fig. 7.17). It was found in Eq. (7.52) that the horizontal force is essentially the same for up and down surface grinding. A significant difference is observed in the nature of the vertical component of force in the two types of grinding and the two types of milling. In up milling the vertical component of force tends to lift the workpiece from its support, thus making it more difficult to hold the part, whereas in down milling the vertical force is directed into the surface of the workpiece. However, in both types of grinding, the vertical force on the workpiece is directed downward into the finished surface. The reason for this difference between milling and grinding may be found in the difference in the geometry of the chips produced in the two processes. In grinding, the maximum chip thickness t is so small, compared with the undeformed chip length l, that the circumferential component of force on a grit remains very nearly horizontal throughout a cut. This is not the case with milling, and indeed here all the reversal in vertical force with up and down milling is due to the change in direction of the circumferential force which occurs during a cut. It is thus evident that without alteration, the dynamometer of grinding cannot be used to study the ordinary milling process.

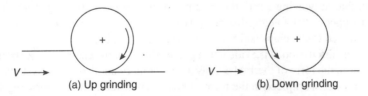

Fig. 7.17 Up and down surface grinding.

Actually, Eq. (7.52) should be written as follows for up (+ sign) and down (− sign) grinding:

$$u = \frac{F_H(V + v)}{vbd} \qquad (7.53)$$

However, the second term on the right side of this equation which was omitted in Eq. (7.53) is very small compared with the first. Inasmuch as v/V is of the order of 0.001, the error in neglecting the last term of Eq. (7.54) is but 0.1%, and hence Eq. (7.53) may be used for both up and down grinding.

7.6.2 Energy Considerations

The total specific energy involved in a fine grinding operation is of the order of 1.7×10^3 m kg/cm^3. This energy will appear in several forms which include:

  (i) Heating of the workpiece
  (ii) Heating of the wheel
  (iii) Kinetic energy of the chips
  (iv) Kinetic energy of moving air layer
  (v) Heating of the chips
  (vi) Radiation to the surroundings
  (vii) Generation of new surface
  (viii) Residual energy remaining in the lattice of the ground, surface and chips.

All but three of these are heat quantities and we must consider these three first. The chips leaving the wheel will move with a velocity equal to that of the periphery of the wheel and hence the kinetic energy involved will be

$$KE = \frac{1}{2}\left(\frac{1000\gamma}{g}\right)\left(\frac{V}{60}\right)^2 \qquad (7.54)$$

where γ is the specific weight of the chips in g/cm^3
        g is the acceleration due to gravity in m/s^2
        V is the wheel speed in m/min.

When a metal is plastically deformed, all of the strain energy involved in the deformation does not appear in the form of heat, but a certain portion of this energy is retained in the specimen in the form of the increased internal energy associated with residual lattice distortion. Taylor and Quininey investigated this matter and found that 5 to 15% of the strain energy does not appear in the form of heat when bars are twisted or elongated. When these results are further studied, it is found that smaller amounts of residual strain energy on a percentage basis is associated with those tests involving the greatest amounts of distortion energy. Thus their appears to be a saturation value with regard to the amount of energy that a given metal can retain in the form of lattice distortion, and hence when metals are subjected to increasing amounts of distortion the retained energy decreases on a percentage basis. It is recently found that the retained energy during the cut is 1 to 3% in an experiment similar to that of the drilling. Now, even as the specific shear energy involved in drilling is large compared with that in twisting a rod, the specific shear energy in grinding is much larger than that in drilling. From this, it appears that a negligible quantity (probably less than 1%) of the shear energy involved in grinding will be retained in the chips and finished surface in grinding.

From the foregoing discussion, it is evident that practically all of the energy (1.7×10^3 m kg/cm^3) involved in grinding will appear in the form of thermal energy at one point or another of the system. The portion of this energy which flows into the workpiece in the form of heat is considered in the following section.

7.6.3 Thermodynamic Background for Thermal Aspect in Grinding

The first law of thermodynamics may be written as

$$dQ - dW = dE \qquad (7.55)$$

and states that the net heat flowing into the matter contained within a rigid control volume in space (dQ) minus the work done by this contained matter on its surroundings (dW) is equal to the change in internal energy of the enclosed matter (dE). It is found empirically that the net heat flowing in through one surface of a control volume and out through another parallel surface a distance dS away is given by Fourier's equation

$$dQ = \frac{k d\theta A dT}{dS} \tag{7.56}$$

where $d\theta$ is the difference in temperature of the two surfaces;
A is the flow area;
dT is the time involved;
k is a constant of proportionality known as the *coefficient of thermal conductivity*.

The units of k are kcal/cm-K, and k for metals actually decreases with increased temperature. The change in the internal energy of a solid body is conveniently expressed by the change in temperature of the body as follows:

$$dE = c d\theta \gamma dV_0 \tag{7.57}$$

where γ is the specific weight of the body of volume V_0, and c is a proportionality constant known as *specific heat*. The unit of c is cal/g°C and c for metals actually increases with increased temperature. If now we apply Eqs. (7.46), (7.47), and (7.48) to a control volume of infinitesimal magnitude as shown in Fig. 7.20, the following differential equation is obtained:

$$\frac{k}{c\gamma}\left(\frac{\delta^2\theta}{\delta x^2} + \frac{\delta^2\theta}{\delta y^2} + \frac{\delta^2\theta}{\delta z^2}\right) + \frac{q}{c\gamma} = \frac{d\theta}{dT} \tag{7.58}$$

where q is the rate at which work or energy is added to the system per unit volume. The quantity q may be due to the absorption or radiant energy, radioactive decay, friction, plastic deformation, or other causes. In our application q arises from plastic deformation during the grinding process; however, it is convenient to consider the case of a plane friction slider first. The quantity $k/\gamma c$ which appears frequently in thermal calculations has been called *thermal diffusivity* by Kelvin and is represented by the symbol K. The unit of K is cm^2/s and this quantity decreases with increased temperature. Figure 7.18 shows a material particle P and increments $\Delta x, \Delta y, \Delta z$ in three dimensions.

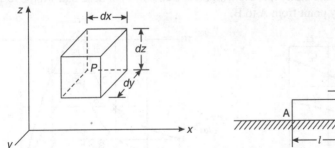

Fig. 7.18 Material particle.

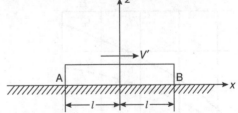

Fig. 7.19 Plane slider.

Equation (7.58) may be applied to a plane slider (see Fig. 7.19) which moves with uniform velocity V(m/s) and which is very long in the y-direction compared with its length in the x-direction. The slider is assumed to be a perfect insulator ($k = 0$), so that all of the frictional energy that is dissipated at the sliding interface at the rate of q cal/cm^2/s will flow into the lower stationary surface and cause its temperature to rise. The details of the solution for the temperature θ at any point may be found in Carslaw and Jaeger (1959). The mean θ and maximum θ_m temperatures at the interface are of most practical interest, and the following results have been taken from Jaeger's (1942) paper:

For $L > 5$,
$$\theta = 0.752 \frac{ql}{K\sqrt{L}} \tag{7.59}$$

$$\theta_m = 1.128 \frac{ql}{K\sqrt{L}} = 1.5\beta \tag{7.60}$$

For $L < 5$,
$$\theta = 0.636 \frac{K \cdot qf}{kV'} \tag{7.61}$$

$$\theta_m = 0.636 \frac{K \cdot ql_m}{kV'} = \frac{f_m}{2K'}\theta \tag{7.62}$$

where
$$L = \frac{V'l}{2K'} \cdot f$$

and f_m are given in Fig. 7.20, and all quantities are in British system units.

The manner in which the temperature varies along the stationary surface is shown non-dimensionally in Fig. 7.21 for several values of L. The ordinate of this plot is proportional to the temperature for a given metal and condition of friction. The position of the maximum temperature is seen to approach the trailing edge of the slider as the velocity of the slider increases (i.e. for large values of the velocity parameter L) and to approach the centre of the slider as the velocity approaches zero.

7.6.4 Calculation of Grinding Temperature

Using Eqs. (7.59) to (7.62) to compute the temperature that is reached in a ground surface at the tip of a grinding grit. The mean grinding point may be considered to cut with a zero degree rake angle, as shown in Fig. 7.19 (Jaeger, 1942) for reasons expressed in technical literature. The grit depth of cut is t. The line AB represents the shear plane which makes angle ϕ with the direction of motion. In any cutting process all of the strain occurs in a very narrow band extending along shear plane AB and hence as in the friction process the energy will be dissipated along a distinct line AB. If we assume the shear stress to be constant along shear plane AB, then the specific energy that is converted into thermal energy during the shearing process will be constant from A to B and equal to u m-kg/cm³. This will be known as the *specific shear energy*. This thermal energy will flow partly into the chip and partly into the workpiece in the form of heat, and will be responsible for the temperature that exists at any point from A to B.

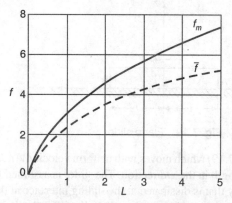

Fig. 7.20 Mean and maximum temperature characteristics for plane slider of infinite width sliding on a perfect insulation.

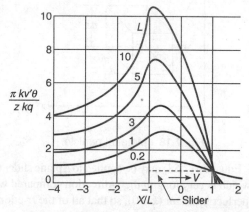

Fig. 7.21 Temperature distribution (curves for plane slider of infinite width sliding on a perfect insulation.

In Fig. 7.21, it is evident that relatively little heat has a chance to escape at the front of the slider. This is particularly the case at high speeds (large values of L). Hence, we may ignore the inclination of slider AB in Fig. 7.22 and replace it by slider AC. Then the temperature at any point in surface ABE will be given to a very good approximation by the temperature at the vertical projection of this point on plane surface ACD. CE and D points are in the plane perpendicular to the plane of the paper.

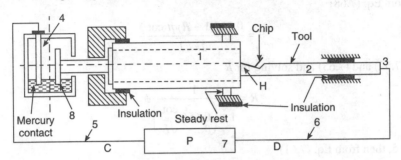

Fig. 7.22 Test apparatus.

While Eqs. (7.59) to (7.62) cannot be applied directly because our slider (chip) is not a perfect insulator, we may take care of this difficulty by applying the ingenious method which Block employed in his solution of the temperature distribution. If R is the fraction of the total shear energy which leaves with the chip, the fraction which passes into the workpiece will be $(1-R)$. Thus, Eq. (7.59) or (7.62) may be applied to find the mean temperature across AB if q is taken equal to $(1-R)$ times the actual energy dissipated per unit area per unit time. An expression for the mean temperature from A to B may also be obtained by assuming the surface below AB to be a perfect heat insulator, the quantity q this time being taken equal to R times the actual energy dissipated per unit area per unit time. By equating these two values of mean temperature, we obtain an equation in which R is the only unknown and the distribution of energy between chip and work is thus established. This procedure will now be followed in detail. The quantity q will be obtained first. By definition, we have

$$q = \frac{W_s}{t \cos\phi \, (b' \, 100 \, J)} \tag{7.63}$$

where W_s is the total shear work involved in cutting in cm-kg;
       J is the mechanical equivalent of heat (427 kg m/kcal);
       b' is the width of the cut in cm;
       t is the time.

This equation may be written as

$$q = \left(\frac{W_s}{V'b't}\right)\left(\frac{V't}{t\cos\phi}\right)\left(\frac{1}{100\,J}\right) = \frac{u_s V'}{100 \, J \cos\phi} \tag{7.64}$$

where u_s is specific shear energy.

For a condition of given data,

$$q = \frac{(5.25 \times 10^6)(1200)}{100(427) \cot 19.9°} = 11000 \text{ kcal/cm}^2\text{/s}$$

Assuming all of the energy R_q passes off with the chip, the mean temperature across AB will be

$$\theta = \frac{(Rq)b't \cot\phi}{c\gamma V'tb'} = \frac{Rq \cot\phi}{c\gamma V'} \tag{7.65}$$

The expression for $\bar{\theta}$ when all of the energy $(1 - R) q$ flows into the lower surface may be obtained from Eq. (7.59) or (7.60), depending on whether L is greater or less than 5, where

$$L = \frac{V't \cos\phi}{4K} \qquad (7.66)$$

If $L > 5$, then from Eq. (7.68)

$$\theta = \frac{0.752(1 - R)qt \cot\phi}{2k\sqrt{L}} \qquad (7.67)$$

Equating Eqs. (7.59) and (7.65) and solving for R

$$R = \frac{1}{1 + \left(0.665 \dfrac{\cot\phi}{\sqrt{L}}\right)} \qquad (7.68)$$

Similarly if $L < 5$, then from Eq. (7.61)

$$\bar{\theta} = \frac{4.636\, K(1 - R)q\, f}{kV'} \qquad (7.69)$$

Equating Eqs. (7.61) and (7.69) and solving for R

$$R = \frac{1}{1 + \dfrac{1.57 \cot\phi}{f}} \qquad (7.70)$$

7.7 EXPERIMENTAL TECHNIQUES OF TEMPERATURE MEASUREMENT

7.7.1 Orthogonal and Oblique Cutting

Orthogonal or oblique machining operation is a most widely used process, so much so that 70% to 80% of the total metal machined is carried out by this process. Temperature distribution around and within a metal cutting tool is of much interest for increasing productivity by improved tool materials.

The most common cause of tool wear and failure is the high temperature which is inherent in the cutting operation. A knowledge of the ways in which the cutting conditions, tool material, and work material affect this temperature is helpful in choosing correct combinations and ferreting out troubles. Many have studied this problem of temperature distributions from both the experimental and analytical aspects with varying degrees of success. Some investigators like Herbert Gottwein and Shore first developed the chip-tool thermocouple concept; Schwerd with a radiation technique obtained some information on temperature fields, Schallbroch and Lang applied temperature-sensitive paints to the tool, Tsueda, Hasegawa and Nisina (1961) have buried thermocouples within the tool to obtain temperature fields therein, and Schmidt used calorimetric methods to obtain average temperatures.

7.7.2 Arrangement for Measurement by Thermocouple

The test arrangement is illustrated in Fig. 7.22. The standard carbide tool has a hole drilled in the shank through which a rod of carbide (3), insulated from the shank, is placed in contact with the carbide insert (2), the assembly being electrically insulated from the tool post (see Fig. 7.22). The test apparatus is shown in Fig. 7.22 on page 181 while the details of tool tip temperature set-up have been shown in Fig. 7.23.

HEAT IN METAL CUTTING AND TEMPERATURE MEASUREMENT 183

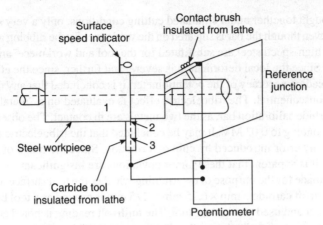

Fig. 7.23 Details of tool tip temperature set-up.

The brush (4), insulated from the machine frame, is made from a section of the workpiece. Iron-wire leads (5 and 6), with suitable connectors were attached to the potentiometer (7). The completed arrangement thus consists of several thermocouples, namely 1–2, 3–6, and 4–5. Couples 6–7 and 5–7 are of no consequence since any parasitic emf's developed would cancel one another. Since the lead wires are at constant temperature during any test, the reference junction is at couples 3–6 and 4–5, and the temperature at these points are maintained at an essentially constant level during any calibration or cutting test.

No thermocouple exists at 2–3 since the carbide insert, contact rod, and calibration strip are made from the same batch of carbide mixture. Similarly, junction 1–4 is of no consequence, since both metals in contact are from the same bar.

Thus the hot junction of the thermocouple circuit is at 1–2, and the reference junction are at 3–6 and 4–5.

The validity of the operating principle of the tool-work thermocouple may be affected by several factors: (a) the effect of chip force upon the tool face; (b) the electrical effect of chip deformation; (c) the triboelectric effect due to friction of the steel rubbing on the carbide; (d) the temperature gradient at the contacting surface of the chip and the tool; and possibly by (e), the dissimilarity of temperature gradients during calibration (see Fig. 7.24) as compared with those during cutting.

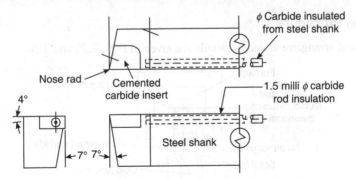

Fig. 7.24 Diagram of calibration arrangement.

The effect of chip force is investigated by slowly forcing a carbide strip into a steel strip under a bearing stress of over 6000 kg/cm². No electrical effect was observed at the potentiometer. The thermoelectric effect of chip deformation was investigated by substituting a bar of the workpiece for the tool. When this bar and the workpiece were brought together under simulated cutting conditions, only a very slight (0.04 to 0.05 MV negative) emf is noted, even though the bar is smeared on the workpiece by the rubbing action. When two pieces of chemically identical high-speed steel are substituted for the tool and workpiece, an indication of the same magnitude is observed. Since the local deformation is severe, and further, since the effect is within the limits of reproducibility and reading accuracy of the potentiometer, it is concluded that any deviation introduced by chip deformation is inconsequential. The triboelectric effect is evaluated on a "metal-sorter", using a bar of the test stock and the carbide calibration bar, as the two metals are in contact. The observed deflection is about 27 mm positive, corresponding to 0.03 MV. It may be concluded that the triboelectric effect is negligible and, in fact, tends to cancel any error introduced by chip deformation. Since the range of readings is 10–30 MV during the cutting tests, it is apparent that these minor deviations are insignificant.

Observations are made for the purpose of determining whether the tool surface and chip surface attain a stable temperature. A strip of carbide 5 mm × 6.25 mm × 275 mm is ground as a tool bit on one end, insulated from a supporting steel bar and used as a cutting tool. The millivolt reading is noted at the beginning of a test and virtually the same reading prevails during the duration of cut. This may be concluded that the contact surface of the tool and chip reached, at least a state of quasi-equilibrium after a cutting test is started. Therefore it was assumed that the separating surface of the chip and the face of the tool reached essentially that same temperature during the cutting test.

Generally the temperature gradients along the calibration strips during calibration are not the same as those during a cutting test, but this is of no consequence in the light of the so-called *law of the homogeneous circuit* which states as a corollary: "If one junction of two dissimilar homogeneous materials is maintained at a temperature T_1 and the other junction at a temperature T_2, the thermal emf developed is independent of the temperature gradient and distribution along the wires". While steel and cemented carbide cannot be considered homogeneous materials, they may be thought of as being uniformly unhomogeneous and the law of the homogeneous circuit would apply. As a result of these preliminary investigations, it may be concluded that the possible errors are not of sufficient magnitude to outlaw the tool work thermocouple as a means—perhaps the only means—of evaluating the temperature where tool failure is investigated.

In considering the experimental results, it must be remembered that the tool-work thermocouple consists, in reality, of a number of thermocouples in parallel, so that the emf observed is an average emf of all points of contact (thermocouples) of the tool and the chip. Cutting temperatures so determined would represent average temperature at the tool-chip interface.

7.7.3 Calibration Procedure

The general calibration arrangement and its details are given in Figs. 7.25 and 7.26.

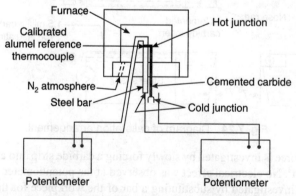

Fig. 7.25 Diagram of calibration thermocouple.

The steel member is prepared from a portion of the workpiece with a rounded end, so as to provide line contact with the carbide calibration strip. A 1.5 mm hole is drilled within 0.25 mm of the contact area for the reference of chromel-alumel thermocouple. Contact between the carbide and steel is maintained by means of a small clamp and blocks of ceramic insulation.

The assembly is placed in an electric furnace through a hole in the door, so that the opening could be parked with asbestos yarn and aluminium cement. The cold ends of both steel and carbide members project about 50 mm beyond the door, so that cooling water could be applied to keep the cold junction at the desired reference temperature. The lead wires connecting the cold junction to the potentiometer are the pair used in the cutting tests.

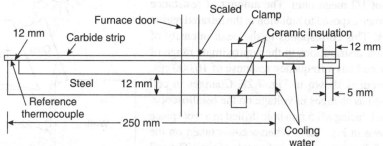

Fig. 7.26 Experimental set-up for measuring cutting temperatures using a PbS cell to measure radiation.

During the calibration, a protective furnace atmosphere of nitrogen gas is used for temperature above 150°C to prevent oxidation or decarburisation. The cold junction temperature was kept constant by cooling water, the temperature of which was the same as that prevailing at the potentiometer binding posts. Thus each calibration curve had a reference temperature, to which all results from the cutting tests are corrected.

After establishing uniformity of temperature, simultaneous readings of reference thermocouple temperatures, tool-work thermocouple temperatures, and tool–work thermocouple millivolts are taken as per the following increments—3°C intervals up to 40°C; 30°C intervals from 350°C to 700°C; 15°C intervals during the allotropic transformation of the steel, and 30°C intervals up to the maximum temperature attained during a particular calibration. It may be noted that the temperature millivolt relationships are straight lines in the ranges up to 700°C. The change of slope at 300°C is evidence of the allotropic change in the steel either as it affects the thermocouple proper as it causes variation in the structure along the steel member. A few points are checked on cooling to determine whether any decarburisation had occurred in the steel at the contact point. Generally, the points are found to be on the curve, though in some calibrations the points were slightly higher than the corresponding ones obtained during heating.

In applying the calibration curve for cutting temperatures the extended portion of the curve is used, since by the time 15 MV is indicated the cutting speeds were such that there is not sufficient time for the allotropic transformation to occur while the chip is in contact with the tool. A metallographic examination of the separating surface of the chip revealed the same type of structure as that in the annealed bar. Thus it is concluded that transformation does not occur, and the extended portion of the curve must be used for temperatures in excess of 650°C.

Numerous calibration tests are conducted with the steel member of the tool-work thermocouples in various conditions such as mill-annealing, normalising, heat-treatment to 300 Bhn, and cold-working to 50% cross-section. The temperature millivolt relationships are affected very little (less than 0.10 MV) by change in the steel member, probably because the same two phases, ferrite and carbide, exist in the steel and are altered only in their distribution. Thus the same calibration curve can be used for all conditions of a particular steel, and although slight errors may result, these are considered to be negligible.

7.7.4 Radiation Method of Temperature Measurement

A new technique of measuring temperature distributions in the cutting region is being presented here. This technique is shown schematically in Fig. 7.27 which shows the experimental setup. The actual cutting is done on a 600 mm stroke shaper. The stroke is set at 250 mm and cutting is done on a 150 mm long specimen. The cutting velocity is checked stroboscopically and kept essentially constant during the cut. The shaper is powerful enough to ensure that it does not slow down appreciably in taking the rather heavy cuts necessary.

The temperature-sensing element is a PbS (lead sulphide) cell, a fairly recent development. A PbS cell is essentially a resistance of about 1/2 mega ohm. The amount of resistance changes slightly when exposed to radiation in the infrared region of 1 to 3 microns. This corresponds to the peak intensity of blackbody radiation distributions in the approximate range of 250–1050°C. The cell has a frequency response of 10,000 cps. The electrical output is shown in Fig. 7.28. Changes in cell resistance are noted as changes in voltage on the oscilloscope. The cell is mounted, facing a 0.5 mm hole drilled in a workpiece of 16 mm as shown in Fig. 7.27. When a cut is taken on the opposite side, the following sequence of events occurs: The cell sights out through the hole upon the spotlight which is placed at a considerable distance to ensure parallel light rays. There exists sufficient infrared radiation in the light to activate the cell. As the tool advances, the shear plane arrives at the hole and closes over it, as shown in Fig. 7.27, cutting off the light and producing a voltage change in the PbS cell. This signal is used

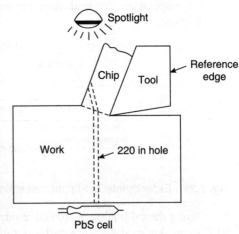

Fig. 7.27 Experimental setup.

to trigger the sweep on the oscilloscope. As the cut proceeds, the PbS cell sights on various points of the shear plane and then on the clearance face of the tool. Finally the flat reference edge of the tool passes, re-exposing the spotlight. Figure 7.28 shows the trace obtained during such a cut. The shaper is then run without cutting and a trace as shown in Fig. 7.28 is obtained. The traces in Fig. 7.28 represent the cutting and return stroke of the shaper. The cutting stroke, has longer trace, is scaled off and the same distance measured on the trace is obtained during cutting. Since this square wave represents the width of the tool, this procedure locates the cutting edge of the tool. Also, if the width of the tool from cutting edge to reference edge is known, it gives a convenient relation between shaper cutting speed and oscilloscope sweep rate.

Fig. 7.28 (a) Output signal from PbS cell without cutting and (b) Output signal from PbS cell during cut.

A calibration of the cell is needed before the trace of Fig. 7.28 can be interpreted. The cell can be calibrated in place by removing the cutting tool from the shaper and placing in the tool holder a small electrically heated furnace. The shaper is then set in operation reciprocating the furnace back and forth. Each time the heated surface passes the 0.5 mm hole, a signal is recorded on the oscilloscope. A standard chromel–alumel thermocouple monitors the furnace temperature. Various surfaces are placed on the furnace to check the effect of changes in emissivity. If one starts with a clean surface of 1020 steel, the calibration curve on rising temperature is not the same as that with falling temperature as a result of oxidation (see Fig. 7.30). Since it is felt that the hot surface in the actual cutting test would not be oxidized owing to the very short time available, the initial portion of the calibration curve is extrapolated (dotted line) to higher temperatures parallel to the descending temperature curve. Figure 7.29 shows electrical circuit used for PbS cell.

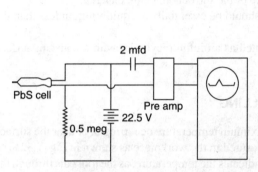

Fig. 7.29 Electrical circuit used for PbS cell.   **Fig. 7.30** Schematic calibration curve for PbS cell.

7.7.5 Imbedded Thermocouple Method of Temperature Measurement

In this technique the measuring element is a small single wire thermocouple imbedded in the side of the workpiece, a small distance down from the top. If a planing cut is made with sufficient depth of cut so that the tool point passes below the thermocouple, a complete history of temperature at the thermocouple is obtained. As the tool approaches, the temperature rises ahead of the shear plane by conduction; as the shear plane passes, there is a sharp rise in temperature caused by the plastic deformation. The chip still containing the thermocouple now slides up the face of the tool and additional heating due to friction is recorded. Figure 7.31 shows such an arrangement with the output of the thermocouple fed into an amplifier–oscilloscope combination.

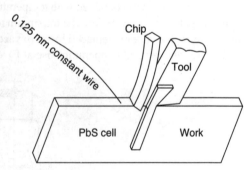

Fig. 7.31 Test arrangement for measurement of workpiece temperatures.

The wire used is 0.12 mm constantan which stakes into a hole drilled about 0.5 mm deep with a 0.125 mm spade drill. The staking procedure imbedded the wire firmly enough so that it is possible actually to lift the specimen by means of the wire. The shearing operation tends to deform the round hole into an ellipse, gripping the wire all the more firmly. All cuts using this procedure can be at a 1.25 mm depth of cut, since it is felt that with any lighter cut the diameter of the wire would be too great a percentage of the chip thickness and would interfere with the cutting process.

The point at which the temperature is measured is by necessity close to the side surface, and the temperatures here may be less than average. Radiation losses should be small, but there are losses due to conduction along the wire. The most serious objection is that the cutting operation is not completely two-dimensional, there being some side flow which is not uniformly distributed. Most of the side flow takes place near the side surfaces and hence this is a region of less severe plastic flow than average. However, in spite of these difficulties, the readings are repeatable and form a temperature field similar to what is expected.

There is some doubt whether the thermocouple wire would come to equilibrium as it goes through the shear process. To arrive at an approximation of the time constant, the following assumptions are made:

(a) The thermocouple wire does not deform itself and hence does not generate any energy within its boundaries.
(b) The wire surface is maintained at the temperature of the surrounding bulk material.
(c) For the readings to be useful, the wire surface should be essentially at equilibrium in less than the time it takes to move 0.125 mm.
(d) The thermocouple junction may be approximated as an infinite cylinder with a constant surface temperature.

7.8 MEASUREMENT OF TEMPERATURE IN MILLING

To determine experimentally as closely as possible the maximum temperature occurring in or near the surface of a workpiece while it is being milled, temperature is measured in the workpiece as shown in Fig. 7.32. The test bar has a slot in the centre where the thermocouple P indicates the temperature, as the tool cuts through the workpiece material. As expected, the temperatures recorded are highest when the surface being cut is closest to the thermocouple. The general distribution of maximum temperatures in a workpiece during and shortly after machining is shown in Fig. 7.34. For the last cut the thickness of the remaining layer is only 625 mm (see Fig. 7.33), and the potentiometer used does not respond fast enough to indicate the peak temperatures, which are consequently obtained with temperature-indicating crayons. When the distance S is 3 mm or greater, a time lag always occurs before the maximum temperature is attained. (In determination of maximum surface temperature, the cutter is replaced by an oxyacetylene flame moving over the workpiece at the same feed rate and causes an identical temperature rise at P).

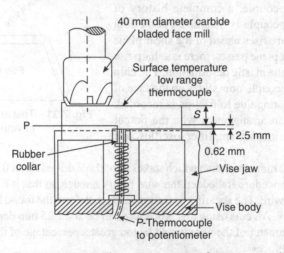

Fig. 7.32 Test arrangement for measurement of temperature in milling.

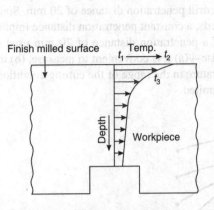
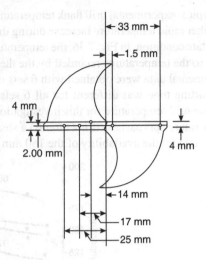

Fig. 7.33 Temperature as a function of depth of finish milling.

Fig. 7.34 Thermocouple locations along drill flank.

7.9 MEASUREMENT OF TEMPERATURE IN DRILLING

The drill flank temperature distribution can be measured experimentally using iron-constantan thermocouples attached to various locations on the flanks of a 60 mm-diameter drill. The geometrical characteristics of the drill employed must be listed. The thermocouple positions on the drill flanks are indicated in Fig. 7.36. Thermocouples 3 and 4 are placed 5 mm back from the cutting edge, while Thermocouples 1, 2, and 5 are placed 2 mm back from the cutting edge. Thermocouple 1 is nearest to the chisel edge while Thermocouple 5 is nearest to the drill periphery.

Temperatures are obtained with the drill held stationary in the turret of a 25 hp Gisholt 1L turret lathe. The use of a stationary drill, as opposed to a rotating one, is assumed not to affect the drill temperature distribution and eliminates the need for a rotating pick up assembly for the thermocouple signals. The thermocouple junctions are tungsten inert gas (TIG) welded on the drill flanks in the positions shown in Fig. 7.35. Shallow semicircular grooves are ground into the drill flanks and up the drill lands to provide space for the thermocouple wires, which are held in place using an epoxy cement. The temperatures can be recorded on temperature recorders having the capacity to measure all the signals simultaneously.

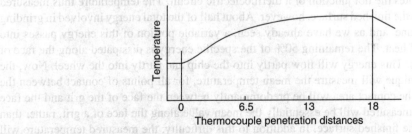

Fig. 7.35 Temperature as a function of drill penetration.

In investigating the effect of the chisel edge energy source, several workpieces containing a pilot hole are to be employed. The pilot hole diameter must be somewhat larger than web thickness.

A typical experimental drill flank temperature versus penetration distance curve is illustrated in Fig. 7.32. After a rather rapid temperature increase during the initial stages of cutting, the temperature tends to approach a steady-state condition. In Fig. 7.36, the temperature drops rapidly as the drill feed is disengaged. Temperature data refer to the temperatures recorded by the thermocouples at a drill penetration distance of 20 mm. Since the experimental data were obtained with 6 sets of feeds and speeds, a constant penetration distance implies that the cutting time was different for all 6 sets. The choice of a penetration distance of 20 mm used in defining the drill temperature for this investigation is made because—(a) it is convenient to measure, (b) the drill temperature approaches (if not attains) a steady-state temperature in the range of the cutting conditions employed, and (c) the availability of the 150 mm workpieces is limited.

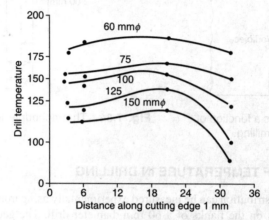

Fig. 7.36 Drill flank temperature as a function of Workpiece diameters 60 mm to 100 mm.

7.10 MEASUREMENT OF TEMPERATURE IN GRINDING

A little overview reveals that it is not easy to measure the temperatures which have been computed in the foregoing section; in fact, it is not at all evident that such temperatures can be measured. Inasmuch as the temperature decreases extremely rapidly from the surface downward into the metal, it is obviously hopeless to bury thermocouples below the surface at known distances and attempt to extrapolate the measured temperatures up to the surface of the metal. The extreme rate at which the heat leaves the surface also rules out the use of a sensitive bolometer, to measure the radiation from the surface, after the tool (wheel) has passed. However, grinding temperature can be measured by use of the tool work thermocouple technique, in which the wheelwork interface constitutes the hot junction of a thermoelectric circuit. The temperature thus measured will not correspond to that on the finished surface, however. About half of the total energy involved in grinding is dissipated on the shear plane, and as we have already seen, a variable portion of this energy passes into the workpiece in the form of heat. The remaining 50% of the specific energy is dissipated along the face of the grit as a result of friction. This energy will flow partly into the chip and partly into the wheel. Now, the chip-grit thermocouple technique will measure the mean temperature, for all points of contact between the metal and the wheel. Since the contact area will be predominantly between the face of the grit and the face of the chip, the temperature measured will be essentially the mean value along the face of a grit, rather than the desired temperature of the finished surface. In addition to this difficulty, the measured temperature will be the time mean value for a number of simultaneous cuts that are in varying stages of completion. Despite this difficulty of interpreting the results, it would seem worth noting the order or magnitude of the mean chip grit interface temperature.

To do this, a vitreous-bonded silicon-carbide wheel having a relatively low-contact resistance is found. While the thermoelectric power of a silicon-carbide-steel combination is very much higher than that for ordinary metal combinations, the high impedance of the grinding wheel compared with that of ordinary metals makes it very difficult to measure the thermoelectric emf. The advantage which exists in the high thermometric power is more than offset by the difficulty introduced by the high impedance. By use of a vacuum-tube voltmeter in the manner indicated in Fig. 7.37, it is possible, to obtain some approximate data.

The wheelwork couple is calibrated in a furnace against a chromel-alumel couple to a temperature of 900°C. The straight calibration line was extrapolated linearly beyond 900°C, and some investigations showed the thermal emf of silicon carbide against metals to be linear over a wide range of temperatures. The slope of the curve corresponded to 0.01 volts per 38°C.

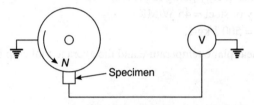

Fig. 7.37 View of Chip-grit thermocouple.

REVIEW QUESTIONS

1. What are the principal sources of heat at the point of cutting during metal machining? Explain.
2. What is the source of energy that heats the chip during a metal cutting operation? Explain.
3. How will you estimate analytically the temperatures at a mar shear plane? State the equation governing the temperature from five principles.
4. How will you proceed to derive the following equation for mean tool face temperature:

$$\theta_t = \frac{(1 - R_2)q_2 a}{k_s} \bar{A} + \theta_0$$

5. What is thermal number? How does this number control the temperature distribution along the shear plane?
6. How do you compare grinding and milling processes with respect to:
   (i) Energy expended per unit volume
   (ii) Depth at cut
   (iii) Velocity of cutting
   (iv) Metal removal rate.
   Explain why and where each one is preferable.
7. What is the purpose of 'helix' in a drill? How does this not affect the strength of the drill?
8. What are the various methods of measuring average and absolute temperatures at the cutting edge in orthogonally oblique cutting? Explain any one of them in detail.
9. How would you obtain the amount of heat in chip, tool and work by using the calorimeter method?
10. "A machine tool is the reverse of a heat engine." Explain.

11. The following data pertains to the orthogonal cutting test on MS tool carbide:
   Cutting speed = 2 m/s
   Feed = 0.5 mm/rev
   Depth = 2.0 mm
   Tool chip contact length = 1.2 mm
   Tangential component of the cutting force = 500 N
   Chip thickness ratio = 0.5
   Ambient temperature = 30°C
   Work = 0.2
   Proportion of frictional energy going to the tool = 0.15
   Thermal conductivity of steel = 45 W/MK
   Specific heat of MS = 500 J/kg-K.

Calculate the average shear plane temperature and the average tool chip interface temperature.

Chapter 8

Dynamometry

8.1 INTRODUCTION

The experimental determination of the cutting forces is necessary not only to analyse the force relationship during the process of chip removal but also for the investigation of tool wear and temperature at chip tool interface. Further knowledge of cutting force behaviour is required for:

(i) Determining the power requirements;
(ii) Designing the work tool configuration;
(iii) Analysing the static and dynamic behaviour of the machine tool; and
(iv) Investigating tool wear, tool design and machineability problems.

In the past, attempts have been made to find the forces acting at the cutting edge theoretically. The results obtained from various theoretical equations have also been verified by suitable experimental techniques and have been found quite comparable.

8.2 FORCES AT THE CUTTING EDGE

An oblique conventional cutting tool has pre-chosen coordinates x, y, and z, with the respective forces being depicted by P_x, the force acting in direction of feed, P_y, the thrust force acting in direction perpendicular to the surface generated, and P_z, acting along tool axis, the main cutting force acting in the tangential direction to the generated surface.

The resultant cutting force can be measured by measuring P_x, P_y, and P_z individually and then determining the resultant force R by the equation

$$R = \sqrt{(P_x^2 + P_y^2 + P_z^2)}$$

or

$$R = \sqrt{(P_t^2 + P_y^2)}$$

in case of orthogonal cutting.

In order to analyse the metal cutting operation on a qualitative basis, certain observations must be made before, during and after the cut. The number of observations that can be made during that cutting process is rather limited. One of the most important observations of this type is the determination of cutting forces. These forces can be measured by dynamometers, operated either mechanically, electrically, or electronically. However, in recent years the developments of metal cutting tool dynamometers have been so much so that the results obtained are quite reliable. The characteristics that are desirable for a dynamometer are discussed in the following section.

8.3 DESIRABLE CHARACTERISTICS OF A DYNAMOMETER

It is desirable for a dynamometer to possess the following characteristics:
  (i) It should be simple in design as far as possible.
  (ii) It should be easy to assemble and vice versa.
  (iii) It should be rigid enough so as not to give rise to vibrations at least within the operating range.
  (iv) It should be sufficiently elastic (in case of sensitive displacement type) so as to give appreciable deformation on small loads.
  (v) It should not obstruct the chip flow and hence should not in any way disturb the process.
  (vi) It should be free from cross effect as far as possible. This means that if a force is applied in either direction of measurement, the other component should not show any reading.
  (vii) A dynamometer must be stable with respect to time, temperature and humidity, thus requiring only occasional checks once it is calibrated.

The two requirements that are always in opposition in dynamometer design are sensitivity and rigidity. The sensitivity of a good research dynamometer should be such that the determinations are accurate to within ± 1%. That is, if a dynamometer is designed for a mean force of 40 kg, 400 g increment should be easily detectable.

Some deformation is associated with the operation of every dynamometer. However, a dynamometer should be rigid enough so that the cutting operation is not influenced by the accompanying deflections. Frequently, the dominating stillness criterion is the natural frequency of the dynamometer. All machine tools operate with some vibration, and in certain cutting operations (i.e. milling, grinding, and shaping) these vibrations may have large amplitudes. In order that the recorded force is not influenced by any vibrating motion of the dynamometer, its natural frequency must be large (at least 4 times) compared to the frequency of the existing vibrations.

For the purpose of analysis, any dynamometer can be reduced to a mass supported by a spring. This natural frequency (ω_n) of such an n-mass spring system is given by

$$\omega_n = \frac{1}{2\pi}\sqrt{\frac{k}{m}} \text{ in cps}$$

where k is the spring constant in kg/cm and m is mass in kg-s²/cm.

If for example, a grinding machine is being run at 3600 r.p.m., machine vibrations are apt to be present with an exciting frequency of 60 cps; in this case the natural frequency of a dynamometer should be at least 240 cps.

In general, a dynamometer must measure at least two force components in order to determine a two-dimensional resultant cutting force. In a three-dimensional cutting operation, three force components are necessary to be measured, while in drilling or taping, only a torque measurement, and a thrust measurement are required.

The dynamometer must be capable of indicating the individual force component without any cross effects, where such forces are measured simultaneously. The simplest and reliable out of all the dynamometers is the strain gauge dynamometer.

8.4 STRAIN AND STRAIN MEASUREMENT

Strain is a fundamental engineering phenomenon. It exists in all matters at all times due to external loads or due to the weight of all the matter itself. Strain varies in magnitude from atomic dimensions; this change in dimension can be seen by the naked eye depending upon the material and loads involved. The terms *strain* and *uniform deformation* are synonymous and refer to the change in any linear dimensions of a body, usually due to the application of external force. Average unit strain is the deformation of the body in a given direction, divided by the original length in that direction. Strain gauges are used to determine unit strain.

8.5 NEED FOR STRAIN MEASUREMENT

The ability of a material to support applied load or force is usually expressed in terms of stress rather than in terms of strain. For economic reasons, material cost, transportation and handling cost, and for general convenience, it is desired to keep the functional components of any machine as small and light as possible. This means that the parts should be stressed in service to the highest permissible value. Prior to the evidence of accurate strain determination, it was necessary to design complex mechanical parts, principally on cut and try basis. This involved making some calculations based on a theory and approximately multiplying this by a "safety factor" of 3 to 5 times, building and testing of pieces, and in the event of failure, adding material in the critical section until a suitable component is evolved. Designing by this method is often extremely wasteful of both time and material. The cut and try process became increasingly dissatisfactory as the demand for higher performance from very complex parts grew towards its present state. The need of aircrafts is minimum weight and maximum performance for every part. It is desirable to determine accurately the local stress, so that the least amount of material would be distributed to the greatest advantage in new design or in modifications of old design.

A great deal of effort has been done to develop a universal strain gauge. The ideal strain gauge properties are discussed in Section 8.6.

8.6 PROPERTIES OF AN IDEAL STRAIN GAUGE

  (i) Extremely small size
 (ii) Insignificant mass
(iii) Easy to attach to the member being gauged
(iv) Highly sensitive to strain
 (v) Unaffected by temperature, vibration, humidity or other ambient conditions, which are likely to be encountered in testing machine parts under service loads

(vi) Capable of indicating both static and dynamic strains
   (vii) Capable of remote indication and recording
   (viii) Least expensive
   (ix) Characterised by an infinitesimal gauge length.

The technique of measurement of strain has been developed step by step, with the mechanical method as the first one.

8.7 MECHANICAL STRAIN MEASUREMENT

This was the earliest method of measuring strain and involved the use of a screw micrometer to measure the overall change in length of a body under load (i.e. total strain). This actually gives the average strain over the entire gauge length and gives no indication of what the local strain may be in neighbour of a discontinuity. In stress analysis it is common to find that a piece which fails under load does not need more material but merely better distribution of the material already present, so that all sections are stressed approximately to the same level. Stress analysis will show infact that the piece can be made stronger simply by removing material.

8.8 OPTICAL STRAIN MEASUREMENT

Since the beam of light is easier to manipulate than a mechanical device, and is weightless and free of friction, hence efforts were made towards applying light-beam amplification to the problem of measurement. This type of measurement has high sensitivity indicating strains as low as 2×10^{-6} units. Optical interference phenomena have also been employed in strain measurement. This interference method is extremely sensitive, accurate as well as delicate.

8.9 ELECTRICAL STRAIN MEASUREMENT

Electrical strain gauges are instruments so constructed that any strain of the body to which they are attached is accompanied by a proportional change in some electrical characteristic of the gauge. The electrical variables commonly used are resistance, inductance and capacitance.

The capacitance strain gauge is composed of a condenser, the capacity of which can be made to vary as strain. The inductance strain gauge is essentially an iron core foil, the inductance of which can be employed for strain measurement. The piezoelectric effect of certain types of crystals is also used to the measurement of strain. The voltage difference across the faces is of a relatively high magnitude. The crystal gauge are bulky compared to the resistance type gauge.

The carbon resistor gauge is one of the resistance wire strain gauge which operates on the principle that any lengthening or shortening of a carbon resistor is accompanied by a change in the electrical resistance of the carbon.

The ultimate in strain measuring devices to date is the bonded resistance gauge. The more commonly used gauge consists of a short length of very fine wire or thin conductive foil which is attached to the piece being tested, so that the filament of the gauge is strained equally with the surface of the test piece. The electrical resistance of the filament material used for these gauges change with strain. The change in resistance when detected by proper instruments is an accurate measure of the strain in the filament and hence the strain in the underlying material being tested.

Reviewing the specifications for ideal strain gauge as applied to the bonded resistance strain gauge, the final results can be presented as follows:

Category	Gauge factor
Size	Very small
Weight	Insignificant
Ease of attachment to the test piece	Relatively simple
Sensitivity to strain	Fair; higher output would be advantageous
Static and dynamic strain indication	Will indicate both with equal ease
Sensitivity to ambient variable	Slightly affected, but the gauges can be protected or the variables compensated for
Remote indication and recording	Easily accomplished
Expense	Comparatively expensive
Gauge length	Shortest length presently available—0.4 mm

8.10 USES OF STRAIN GAUGES

It is usual in most experimental set-ups to show the use of strain gauges to have one pair of gauges only, mounted on a cantilever, or more commonly, in the centre of a simply supported beam with equal overhangs. Sometimes the compensating gauge is omitted to show the temperature sensitivity as well. These experiments are extremely satisfactory, as the practical result usually compares very well with the theoretical. However, during the course of advance engineering education and later, the stress systems to be met with are apt to be far more complicated; therefore it seems essential to introduce to the student as soon as possible to complex stress analysis. The following lines describe some further uses of strain gauges.

The problem is to set up a complex stress system, with the stresses amenable to fairly simple theoretical calculations, but at the same time showing the use of these gauges and some of the difficulties and inaccuracies that may arise with strain gauge rosettes. A fairly obvious way of setting up such a stress system is to have a cantilever with an offset end load. If the cantilever is tubular and has been designed with reasonably good rigidity with this system, the maximum stress occurs at the built-in end of the cantilever at the top of the tube. The main difficulty is that if the tube has a small internal diameter, the gauge rosette covers too great an area and may extend down to the neutral axis where there is no bending stress at all: on the other hand, if the tube is very large it may ordinarily take large end loads to produce satisfactory stress.

The idea seems sufficiently attractive and the necessary materials are inexpensive. The apparatus can be set up as shown in Figure 8.1. This can be mounted on a lathe bed. The tube is of 400 mm length, 60 mm dia and 3 mm wall thickness. One end of the tube is welded to an upright plate and to the other is welded a lever arm 300 mm long. Provision is made for loading the tube at its end or at the end of the lever arm to set up either a simple or a complex system at the fixed end. The strain gauge rosette is stuck on the top of the tube about 25 mm from the vertical plate to avoid end effects. The compensating rosette is struck on to a plate nearby, and both are connected to a strain gauge bridge to the usual way explained later.

If the three pairs of gauges are balanced with no applied load, and strain readings are taken for the load in both the central and offset positions, then the principal stresses can be calculated from the strain readings using the following formulae for strain gauges rosettes:

Delta: type setup of strain gauges

$$p_1, p_2 = \frac{E}{3}\left[\frac{e_1 + e_2 + e_3}{(1-\mu)} \pm \frac{\sqrt{2}}{1+\mu}\{(e_1-e_2)^2 + (e_2-e_3)^2 + (e_3-e_1)^2\}^{1/2}\right]^{*}$$

---

[*] Perry C.C., Data Reduction Algorithms for Strain Gauge Rosette Measurements, Experimental Techniques, pp. 13–18, May 1989.

where p_1, p_2 are principal stresses;
 e_1, e_2 and e_3 are strains in the three directions recorded by rosette;
 μ is Poisson's ratio;
 E is the modulus of elasticity.

Or for rectangular rosette set up of strain gauges, the formula is:

$$p_1, p_2 = \frac{E}{2}\left[\frac{e_1 + e_3}{1-\mu} \pm \frac{\sqrt{2}}{1+\mu}\sqrt{(e_1-e_2)^2 + (e_2-e_3)^2}\right]$$

Thus, the principal stresses can be measured, and the Mohr circle diagram can be drawn for practical verification of complex stress system in a metallic body when subjected to loads, as shown in Fig. 8.1.

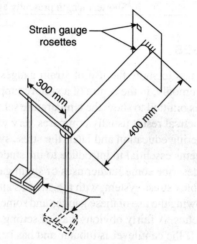

Fig. 8.1 Alternative loading positions in bending pressure and torsion.

8.11 SURVEY OF VARIOUS TYPES OF DYNAMOMETERS

The measurement of cutting forces is performed either by measuring directly the deformation due to the cutting force or by measuring the transformed deformation by a transducing element. Direct force indicating dynamometers are of mechanical type, with the deformation being picked up by dial indicators. The basic scheme of such a dynamometer is shown in Fig. 8.2. However, the sensitivity of the deformation is often magnified by the lever system.

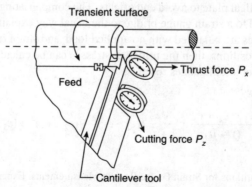

Fig. 8.2 Basic scheme of a mechanical dynamometer.

8.11.1 Mechanical Type

A one-dimensional mechanical dynamometer based on this repositioning of measuring forces by dial indicator has been developed and is shown in Fig. 8.3. In this arrangement, the tool holder is depressed under the action of vertical force, thus actuating a lever mechanism which deflects the dial indicator fastened to the shank. However, for correlating and investigating metal data, two- or three-dimensional cutting force dynamometers are needed. A two-dimensional mechanical type dynamometer is shown in Fig. 8.4.

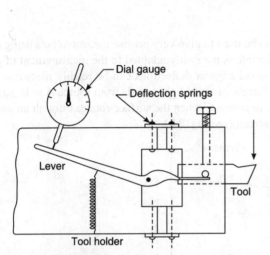

Fig. 8.3 Scheme of one-dimensional tool force dynamometer.

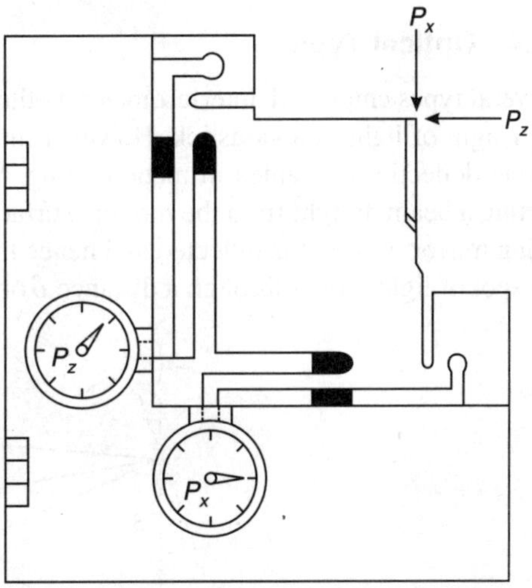

Fig. 8.4 Two-dimensional mechanical type dynamometer used by Merchant.

8.11.2 Hydraulic Type

Hydraulic pressure cells have also been used in conjunction with pressure gauges to measure and record the force on a tool. The manner in which this may be done is illustrated diagrammatically in Fig. 8.5. When this method is used, the force may be read at a distance from the pressure cell.

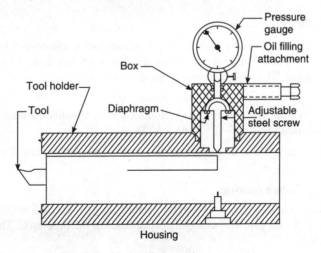

Fig. 8.5 Hydraulic type dynamometer.

8.11.3 Pneumatic Type

These devices, such as the Solen micrometer in which change in back pressure occurs when a flat surface is brought into closer contact with a sharp edged orifice, are sometimes used to measure deflections in tool dynamometer. Such systems are simple and reliable, if carefully applied with clean construction and compressed air. There is, however, a limited region over which they are linear. Besides, dynamometers of this type tend to be bulky.

8.11.4 Optical Type

Of several types employed, interferometer methods can be used to give very precise measurements using the wave length of light as a yardstick. However, this principle is not easily adapted to the measurement of the dynamic deflections obtained in metal cutting. Very small angular deflections can be readily measured by reflecting a beam of light from the moving surface. A narrow beam of light travels from source A to B, strain bending mirror, where it is reflected and hence travels to screen C when the surface rotates through an angle α the spot of light moves through a distance δ on the screen (see Fig. 8.6).

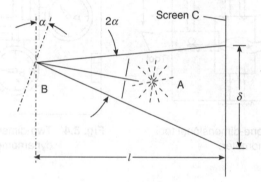

Fig. 8.6 Optical lever.

This device acts as an optical level and may be used to measure small angular displacements. However, even though such a device is simple in principle, it is difficult in practice.

8.11.5 Electrical Type

This type of dynamometer is most widely used. It employs transducing elements which convert mechanical deformations into electrical signals. The device that is used to transform mechanical deformation into electrical output is known as *transducer*. The various forms of these devices are:

(i) Capacitive pick-up (ii) Inductance
(iii) Piezo-electric pick-up (iv) Bounded and unbounded wire strain pick-up
(v) Electronic tube transducer (vi) LVDT.

While the first three have been discussed in this section, the others are discussed in Section 8.12 onwards.

Dynamometer using capacitive pick-up

A capacitive pick-up consists of two plates with an intervening air gap. The deformation in the object subject to loads causes a change in the air gap, thus changing capacitance of the pick-up C. This change can be found by the formula given below:

$$Q = \frac{\varepsilon A V}{\mu \delta}$$

where Q is the electrical charge of the condenser;
 ε is the dielectric constant;
 A is the area of plates in cm^2;
 δ is the gap between the plates;
 V is the voltage;
 μ is the magnetic permeability.

Thus, change in capacitance due to change in air gap is converted into an electric current and thus measured. This type of dynamometer used in CMTI in the former USSR is shown in Fig. 8.7.

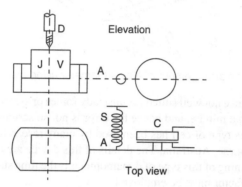

Fig. 8.7 Capacitive type pick-up used for torque measurement.

Dynamometer using inductive pick-up

The dynamometers using inductance pick-ups employ unbalancing of a circuit due to the change in mutual inductance caused by mechanical deformation. This type of dynamometer was used by Optiz.

Dynamometer using piezo-electric pick-up

Piezo-electric pick-ups, when subjected to mechanical deformations, produce change in electrical output and vice versa. Due to this property, they are used for the measurement of loads that produce deformations on the tool. However, this pick-up is most widely used for the measurement of frequency acceleration and velocities. It has the greatest sensitivity of all the electrical methods. This pick-up has a minimum detectable displacement, approximately less than that of resistance strain gauge. Hence the best tool dynamometer available today uses piezo-electric crystal transducers. The schematic arrangement of a dynamometer using piezo-electric pick-up is shown in Fig. 8.8.

A piezo-electric crystal dynamometer has a capability of measuring force from 1 g to 1000 kg, which is quite higher than that of strain gauge dynamometers.

The disadvantage of piezo-electric crystal when used as force transducer is its property of producing electric charge rather than voltage. Hence it requires a charge amplifier to convert the signal into usable form. As this is expensive, dynamometers using piezo-electric crystal transducers are obviously costlier than the strain gauge type. Furthermore, they are sensitive to temperature variations.

It is possible to make a cheaper device using piezo-electric cell by incorporating a small integrated circuit amplifier into the cell. This device gives an output in voltage form and may thus be measured directly; such devices are also available commercially. On the contrary, the stability of this type of load cell is as good as

202 TEXTBOOK OF PRODUCTION ENGINEERING

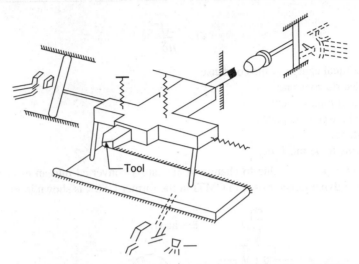

Fig. 8.8 Schematic arrangement of a dynamometer using piezo-electric pick-up.

that of charge amplifiers, and hence not well suited for unsteady forces or systems. But in cutting, the unsteady state does not last for more than a minute, and hence the error is not so serious.

A dynamometer using this type of cell has been used by many research workers and recently used by the CSIRO laboratory at Melbourne, Australia. The details of this device are as shown in Fig. 8.9.

The most serious shortcoming of this type of dynamometer is the time stability; for cutting time of more than one minute, a conversion factor must be employed.

Temperature sensitivity is another shortcoming, but this is adequate for short duration tests. Besides, the cost is quite low—it is of the same order as that of the resistance strain gauge dynamometer.

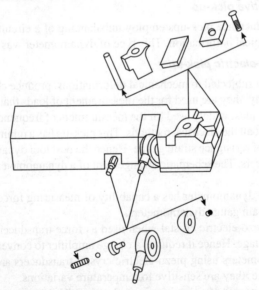

Fig. 8.9 Detailed view of the piezo-electric lathe tool dynamometers.

8.12 USE OF ELECTRICAL AND ELECTRONIC TRANSDUCERS

8.12.1 Electrical Transducer Tube

An electronic transducer tube is essentially a very small 6×12 mm triode vacuum tube with a movable plate, with the tube characteristics being changed as the plate moves. The pin at A can be rotated approximately $\pm 1/2°$ about the diaphragm, and the minimum motion of the end of the pin that can be accurately measured is of the order of one millionth of an inch (25 microns). The electrical system associated with this tube is relatively simple and there are no serious frequency limitations. A dynamometer employing this device will be described later.

8.12.2 Differential Transformer

Differential transformer consists essentially of three transformer axis with a common movable core. AC current is supplied to coils on a common the central primary coil which induces an e.m.f. in the two secondary coils. The outputs of the two secondary coils are wired to oppose each other, so that when the core is in the centre there is zero net output. When the core is displaced, an output is obtained which is proportional to the displacement, with a phase depending upon the direction of motion. The electrical system associated with this instrument is rather complex because of the necessary high amplification and phase sensitivity. However, ac wave type strain gauge amplifiers and re-carders are commercially available and capable of quite satisfactory sensitivity.

8.12.3 Unbounded Wire Resistance Strain Gauge (Refer to Fig. 8.10)

The fact that the electrical resistance of a wire is changed when it is stretched has led to the development of extremely useful wire resistance strain gauges. These are of two types. In the unbounded type of gauge shown in Fig. 8.10, a movable plate P is hung from two spring members. Four coils R_1, R_2, R_3 and R_4 preloaded in tension, are supported by small pins projecting from base B and plate P. When the plate is moved relative to the bases of two of the coils, they will undergo an increase in tension, while the other two coils undergo a decrease in tension. The maximum deflection of an unbounded gauge is usually limited by stops to about 0.375 mm.

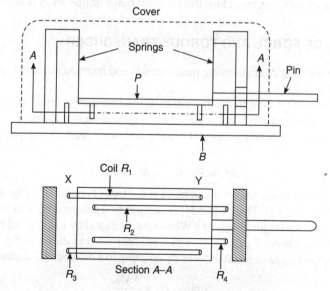

Fig. 8.10 Unbounded resistance strain gauge.

When this type of gauge is connected in the form of a Wheatstone bridge, it can measure deflections of the order of 0.05 μm. One of the chief advantages of this type of gauge over that of bonded type strain gauge is that it can be detached from the part under test and used again.

8.12.4 Bonded Strain Gauges

Two types of bonded wire resistance strain gauges are commercially available as shown schematically in Fig. 8.10. In each case the necessary length of gauge wire, which is usually about one metre is present in the form of a flat coil which is cemented between two thin insulating sheets of paper. Such a gauge cannot be used directly to measure deflection, but must be first cemented to the member to be strained. A strain gauge is applied to a structure by first cleaning the surface thoroughly and then cementing the gauge in place. Either Duco or Bakelite cement is used corresponding to the type of cement that is used in the manufacture of the gauge. After cementing, the unit is allowed to dry thoroughly and then coated with wax to provide some mechanical protection and also prevent atmospheric moisture from affecting the paper. When the resistance between the metal part under test and the gauge itself is tested, it should be at least 50 mega ohms, if the gauge has been properly applied.

Since several of the dynamometers to be described in detail later employ bonded wire gauges, it is justifiable to discuss their use in detail. The manufacture of a strain gauge supplies two important pieces of information with every gauge—the resistance R and the gauge factor F. The gauge factor is a measure of the sensitivity of the gauge, as detailed here:

$$F = \frac{\frac{\Delta R}{R}}{\frac{\Delta L}{L}} = \frac{\Delta R}{e \cdot R} \qquad (8.1)$$

where e is the unit strain. The gauge factor F varies from about 1.75 to 3.5 for most gauges. Bonded gauges can measure strains as low as 2.5×10^{-6} mm. However, commercial strain gauges supplied limit the practical range to about 2.5×10^{-5} mm. Due to fatigue, the maximum strain which it can undergo for a large number of cycles is of the order of 2000 cm/cm. Thus, the practical range of operation is about 2000 to 1.

8.13 STRAIN GAUGE FORCE AND TORQUE TRANSDUCER

In order to assist in the teaching of engineering measurement and instrumentation, a specially designed strain gauge transducer can be made.

It is frequently convenient to reduce lathe operation to a two-dimensional process (orthogonal cutting). In this case, only two force components are required F_V and F_H as shown in Fig. 8.11.

The force F_V will cause a bending moment M_V at a distance r from the cutting edge and F_H will cause a moment M_H where:

$$M_V = F_V \cdot r; \qquad M_H = F_H \cdot r$$

In order to measure M_V, two strain gauges are applied at the top, (T_1 and T_2) in Fig. 8.11(b) and two gauges are applied directly below (C_1, C_2). Thus, when M_V is applied, two gauges are put in tension and two others in compression. These satisfy the requirements of a Wheatstone bridge shown in Fig. 8.11(c). In a similar manner M_H is measured by the four gauges T_3, T_4, C_3, C_4 which are connected as shown in Fig. 8.11(d). Figure 8.11(a) indicates the strains e_{V_1} (and e_{V_2}) as a result of force F_V resulting in a moment M_V. It can be calculated as below:

$$\text{Strain, } e_V = \frac{6M_V}{ab^2 E}$$

DYNAMOMETRY

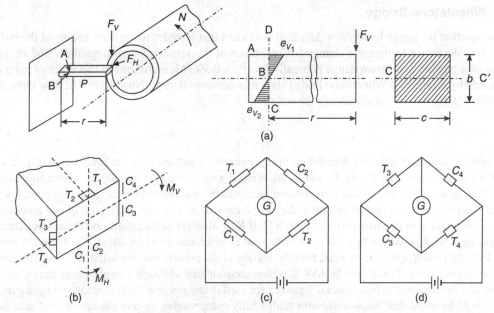

Fig. 8.11 Arrangement of strain gauges for force measurement.

Since for a rectangular cross-section,

$$I = \frac{ab^3}{12} \quad \text{and} \quad M_V = \frac{\sigma}{b/2}$$

which are well known relations in applied mechanics.

The forgoing discussion explains how forces in Turning operation can be measured using the Wheatstone's bridge principle. It of course has to be calibrated in advance.

8.13.1 Strain Gauge Bridge Circuit

Figure 8.12(a) shows the bridge circuit used. The bridge is supplied with a constant voltage V_1 (a 60 volt battery is suitable) and differential shunt balancing is used. The balancing resistor is a high resistance (from 10 kΩ to 50 kΩ) 10-turn potentiometer. The output V_2 from the bridge is measured by a dc micro-voltmeter with a high impedance, so that this supplies negligible current to the meter.

If all four gauges are originally of equal resistance R and the gauges are strained, so that their resistances become $R_1 + \Delta R_1$, $R_2 + \Delta R_2$, $R_3 + \Delta R_3$ and $R_4 + \Delta R_4$ respectively, the output voltage V_2 between points C and D of the bridge is given by

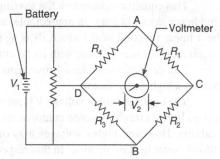

Fig. 8.12(a) Strain gauge bridge circuit.

$$\frac{V_2}{V_1} = \frac{1}{4R}[(\Delta R_1 - \Delta R_2) + (\Delta R_3 - \Delta R_4)] \tag{8.2}$$

8.13.2 Wheatstone Bridge

From the equation for gauge factor, $F = \Delta R/eR$, it is evident that in order to measure strains of the order of $e = 10^{-6}$, it is necessary to measure changes of resistance of the same order of magnitude. The change in resistance of a bonded wire strain gauge is usually less than 0.5%. Measurement of this order of magnitude may be made by means of a Wheatstone bridge shown in previous figures. No current will flow through the galvanometer (G) if the resistances satisfy the equation

$$\frac{R_1}{R_4} = \frac{R_2}{R_3} \tag{8.3}$$

In order to demonstrate how a Wheatstone bridge operates, a voltage scale has been drawn at B, A, C and D of Fig. 8.12(a). Assume that R_1, R_2, R_3 and R_4 are bonded gauges and that initially the above equations are satisfied. If R_1 is now stretched so that its resistance measures by one unit ($+\Delta R$), the voltage at point P will be measured from 0 to plus one unit of voltage ($+\Delta V$) and there will be a voltage difference of one unit between C and D, which will give rise to a current through G. If R_4 is also a bonded gauge, and at the same time that R_1 changes by $+\Delta R$, R_4 changes by $(-\Delta R)$, the voltage at D will move to $+2\Delta V$. Also, if at the same time, R_2 changes by $(-\Delta R)$ and R_3 changes by $+\Delta R$, then the voltage of the point C will move to $(-2\Delta V)$ and the voltage difference between C and D will now be $4\Delta V$. It is then apparent that although a single gauge can be used, the sensitivity can be increased to four times if 2 gauges are used in tension and 2 others are used in compression.

It should be noted that the 4-active-arm bridge fully compensates for any change in resistance due to temperature, provided all gauges experience the same temperature change. This is important since a single gauge is frequently sensitive to temperature changes as to make it useless for strain measurements in a region of steep temperature gradients. The output voltage E_0 from a strain gauge bridge is given by the following expression:

$$E_0 = \frac{a}{4}(F)(V) e \cdot l \tag{8.4}$$

where a is the number of active arms in the bridge (from 1 to 4);
F is the gauge factor of the gauge;
V is the input voltage of the bridge;
$e \cdot l$ is the longitudinal strain on the gauges.

This equation is based on the assumption that the instrument used to measure V has a high impedance and hence does not draw an appreciable current from the bridge. If this is not so, the output e will no longer be a linear function of the strain. It is again evident from Eq. (8.4) that when maximum sensitivity e/E is required. (This is usual case with metal cutting dynamometers.)

It is desirable to make all four arms of the bridge active. It is even more advantageous to use gauges having a high gauge factor.

The imposed bridge voltage V is another quantity which may be varied to alter the sensitivity of a strain gauge bridge circuit. In some commercial instruments, this voltage is fixed from 4 to 6 volts for reasons of stability. However, higher voltages may often be used, especially with the larger gauges whose greater area allows better heat dissipation. In this respect, aluminium makes a better material to cement gauges than steel because of its higher rate of heat conductivity. The use of two strain gauges in each arm of the bridge, in place of one, enables the imposed voltage to be doubled without overheating the gauges. The maximum safe voltage in each case is determined by the desired stability. Too high a voltage will cause zero difference. When a very stable zero setting is required over a long time, the voltage must be reduced. The following expression may be used as a rough guide in determining the maximum permissible applied voltage per gauge, when the gauge is cemented to steel.

$$V = 2\frac{(WR)^{1/2}}{2} \tag{8.5}$$

where W is the limiting wattage which a gauge may dissipate as heat, an average value being in the order of 0–0.75 watts. R is the resistance of a single gauge in ohms. As previously mentioned, the strain E should be kept below 2000 microns/cm to ensure an adequate fatigue life for the strain gauge making up the bridge. The disbalance of a Wheatstone bridge can be determined using several methods. A variable resistance can be used in parallel with one arm of the bridge and adjusted to give zero current at balance position. This is the usual potentiometer set-up. Again, a null instrument can be provided by use of a calibrated "backing" voltage applied across CD to give zero current. Through G, on the other hand, the voltage across CD can be measured directly.

The following analysis, however, shows that if the bridge circuit is used in an unbalanced condition, a microammeter without amplification cannot give the sensitivity that is directly available from a good potentiometer.

If all four bridge arms consist of identical strain gauges [Fig. 8.12(a)], then for small change in resistance R the galvanometer current I_g is represented in Fig. 8.12(b), as th thevenin equivalent.

The current through the galvanometer can be found by finding the thevenin-equivalent circuit. The thevenin or open circuit voltage between terminals C and D of Fig. 8.12(a) will open-circuited galvanometer is

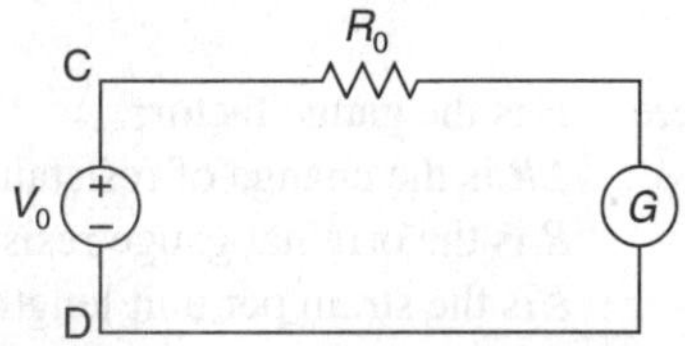

Fig. 8.12(b) Thevenin-equivalent circuit of Wheatstone Bridge.

$$V_0 = V_1 \frac{\Delta R}{4R} \quad \text{for a bridge with equal arms}$$

where V_0 is the thevenin-equivalent voltage
V_1 is the applied voltage to the bridge (emf of the battery)
R is the internal resistance.

The thevenin-equivalent resistance of the bridge is $-R_0 = R$ for a bridge with equal arms. Therefore, by the thevenin-equivalent circuit, the current in the galvanometer circuit is given by

$$I_g = \frac{V_0}{R_0 + R_g} \tag{8.6}$$

where I_g is the galvanometer current
R_0 is the thevenin-equivalent ressitance and
R_g is the resistance of the galvanometer circuit.

Further, for the bridge with equal arms the galvanometer current is given as

$$I_g = \frac{V(\Delta R/4R)}{R + R_g} \tag{8.7}$$

The galvanometer current may be determined from Eq. (8.7) by substituting the following representative values:

$V_1 = 6$ V
$\Delta R = 0.5\ \Omega$
$R = 120\ \Omega$
$R_g = 100\ \Omega$

$$I_g = \frac{6 \times (0.5/4 \times 120)}{120 + 100} = 0.284\ \text{mA}$$

8.14 PRACTICAL APPLICATIONS OF RESISTANCE GAUGES

Transducer sensitivities: These are
 Mean diameter $d = 29.7$ mm
 Wall thickness $t = 0.7$ mm
 Outside diameter $t + d = 30.4$ mm

Moduli of steel: These are
 Young's modulus $E = 21 \times 10^{10}$ N/M^2
 Shear modulus $G = 8 \times 10^{10}$ N/M^2
 Poisson's ratio $\gamma = 0.3$

Gauge factor definition: This is given by

$$F = \frac{\Delta R}{R/\varepsilon} \tag{8.8}$$

where F is the gauge factor;
 ΔR is the change of resistance of the gauge;
 R is the original gauge resistance;
 ε is the strain per unit length.

The four axially mounted gauges and the 2 circumferentially mounted gauges have a gauge length of 10 mm and a gauge factor of 2.03.

The four gauges mounted at 45° to the tube axis have a gauge length of 5 mm and a gauge factor 2.03.

Bending moment (B.M.): Figure 8.13 shows how four axially mounted gauges are connected into a bridge circuit to measure B.M.

If a B.M. (M) is applied, gauges R_1 and R_3 will be subjected to a tensile strain and will increase in resistance by ΔR. Gauges R_2 and R_4 will be subjected to an equal and opposite compressive strain and will decrease in resistance by ΔR.

From Eq. (8.4), the output from the bridge is given by the average strain under the gauge is given by

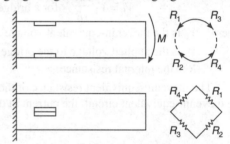

Fig. 8.13 Strain gauges mounted on cantilever and their connection in Wheatstone bridge for force measurement.

$$\frac{V_2}{V_1} = \frac{1}{4R}[+\Delta R - (-\Delta R) + \Delta R - (-\Delta R)]$$

$$= \frac{\Delta R}{R} \tag{8.9}$$

$$= \frac{M}{EI}$$

For a thin-walled tube of thickness t and dia. d

$$l = \frac{\pi d^3 E}{8} - 3(5) \cdot y = \frac{d+t}{2} \tag{8.10}$$

Therefore, transducer sensitivity is given by

$$\frac{V_2}{V_1} = \frac{4FM(d+t)}{E\pi d^3 t} \tag{8.11}$$

$$\mu = \left(\frac{2(d+t)}{E\pi d^3 t}\right)$$

$$\frac{V_2}{V_1 M} = \frac{2}{\mu} \quad \text{(in volts per volt/unit moment applied)}$$

Axial force: Figure 8.14 shows how two diametrically opposed axially mounted gauges and the two circumferentially mounted gauges are connected in bridge circuit to measure axial force.

If an axial force is applied, gauges R_1 and R_2 will be subjected to a tensile strain and will increase in resistance by ΔR. Gauges R_2 and R_4 will be subjected to a compressive strain and will decrease in resistance by ΔR.

From Eq. (8.9), the output from the bridges is given by

$$\frac{V_2}{V_1} = \frac{1}{4R}[(+\Delta R) - (-\Delta R) - (-\Delta R) + (\Delta R)]$$

[see Eq. (8.9)]

$$= \frac{\Delta R}{R} \qquad (8.12)$$

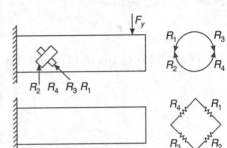

Fig. 8.14 Bridge circuit for measuring axial foce.

The strain in the axial gauges is given by

$$\varepsilon = \frac{F_x}{AE} \qquad (8.13)$$

where E is Young's modulus.

For a thin-walled tube

$$A = 2\pi dt$$

Therefore, from Eqs. (8.5), (8.11), (8.12) and (8.13), the transducer sensitivity is given by

$$\frac{V_2}{V_1 F_x} = \frac{F_x(1+V)}{2E\pi dt} \qquad (8.14)$$

$$\frac{V_2}{V_1 F_x} = 0.99 \ \mu\text{V per V/unit force}$$

Shear force: Figure 8.14 shows how the two pairs of 45° gauges mounted at the neutral axis on either side of the tube are connected into a bridge circuit to measure axial force.

If a force F_y is applied and the strains in the two gauges R_1 and R_3 are E_1 and E_2, the shear stress at the location of the gauges is given by

$$\gamma = G(E_1 - E_2) \qquad (8.15)$$

where G is the shear modulus of the material of the beam. The approximate shear stress at the neutral axis of a thin-walled tube is given by

$$\gamma = \frac{2F_x}{A} \qquad (8.16)$$

For a thin-walled tube

$$A = 3\pi dt \qquad (8.17)$$

If the force F_y, causes gauges R_1 and R_3 to change resistance by ΔR_1, gauges R_2 and R_4 change resistance by ΔR as seen from Eq. (8.8). The output from the bridge is given by

$$\frac{V_2}{V_1} = \frac{1}{4R}\{(\Delta R_1 - \Delta R_2 + \Delta R_1 - \Delta R_2)\} = \frac{\Delta R_1 - \Delta R_2}{2R} \tag{8.18}$$

From Eq. (8.8), we have

$$\frac{\Delta R_1 - \Delta R_2}{R} = (\varepsilon_1 - \varepsilon_2)k$$

From Eqs. (8.15), (8.16), (8.17) and (8.18), the transducer sensitivity is given by

$$\frac{V_2}{V_1 F_y} = \frac{K}{G_1 \pi d t} \tag{8.19}$$

$$\frac{V_2}{V_1 F_y} = 0.39 \text{ µV per mV}$$

Axial torque: Figure 8.15 shows that when a tube is subjected to an axial torque, the principal stress $+\sigma$ and $-\sigma$ are at 45° to the tube axis and equal in magnitude to the shear stress r.

Figure 8.16 shows how the four gauges mounted at 45° to the tube axis are connected to bridge circuit to measure axial torque.

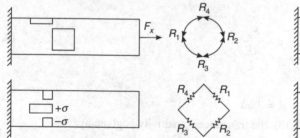

Fig. 8.15 Mounting of strain gauges for torque measurement.

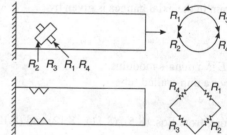

Fig. 8.16 Connection of strain gauges to bridge circuit for torque measurement.

If an axial torque T is applied, gauges R_1 and R_3 will be subjected to a tensile strain and will increase in resistance by ΔR. Gauges R_2 and R_4 will be subjected to an equal and opposite compressive strain and will decrease in resistance by ΔR.

From Eq. (8.18), the output from the bridge is given by

$$\frac{V_2}{V_1} = \frac{1}{4R}\{(+\Delta R) - (-\Delta R) + (+\Delta R) - (-\Delta R)\}$$

$$= \frac{\Delta R}{R} \tag{8.20}$$

The shear stress in the thin-walled tube and the axial torque are related by

$$\gamma = \frac{\pi d^2 t \pi}{2}$$

The strain ε at 45° and the stress σ are related by

$$\varepsilon = \frac{(1+v)\sigma}{\varepsilon} \qquad (8.21)$$

Therefore, from Eqs. (8.18), (8.19), (8.20) and (8.21), the transducer sensitivity is given by

$$\frac{V_2}{V_1} \times T = \frac{2k(1+v)}{E\pi d^2 t} \qquad (8.22)$$

8.15 BONDING TECHNIQUES

The resistance strain gauge is extremely versatile, accurate and sensitive device. Fundamentally though, its performance is absolutely dependent on the bond by which it is attached to the test piece. It is readily apparent that if a precise measuring instrument is used, the gauge must have this strain transmitted undiminished by the bonding cement. This can be accomplished only through a perfect bonding job.

The characteristics of cement for bonding the strain gauge to set on the test piece are dictated by three principal considerations: the material used in the gauge construction, the environmental consideration within which the gauge must perform satisfactorily, and the time available for making the gauge installation. For example, some strain gauges have paper backing and a nitrocellulose cement in their construction. It is obvious that the adhesive for bonding the gauge to a test surface must be selected for compactability within these materials, and the bonding procedure required for the adhesive should not subject the gauge to high temperature or other extreme conditions.

Commercial strain gauges for operations at 180°C and below fall into three linear categories according to the backing material: (1) paper with nitrocellulose adhesive ($T < 70°C$), (2) bakelite with cellulose fibre ($T < 180°C$) or glass fibre ($T < 250°C$), and (3) epoxy ($T < 100°C$).

The most commonly used cement for application at temperature below 180°C include Duco (nitrocellulose), epoxy, bakelite and acrylic.

The epoxy and acrylic cements have come into use as the results of efforts obtained from room temperature curing adhesives with short curing times and superior mechanical properties. These epoxy cements generally cure in less time than nitrocellulose cements and produce a bond which is satisfactory for static strain measurement at temperature 16°C to 28°C, greater than for the nitrocellulose cements.

8.16 SURFACE PREPARATION FOR BONDING STRAIN GAUGES

Before cementing any strain gauge to a test part, certain surface requirements should be met in order to ensure a strong bond between the gauge and surface. Preparations of the test surface prior to cementing the gauge are nominally the same for all gauges and cements. The surface on which the gauge is to be mounted should be smooth in a gross sense but not rigidly polished. Since the lathe conditions do not promote adhesion naturally, all scale, rust and paint should be removed from metal surfaces. This can usually be accomplished by successive application of two or three grades of abrasive paper finishing with fine paper (say 1000 grit) in a random motion pattern. If the material is apt for casting or forging, it may be necessary to use light grinding. Initially a highly polished surface should be roughened slightly with a medium fine grit abrasive paper or light sand blast to improve adhesion. Some volatile solvent such as acetone, tricolour-ethylene, toluene, or methyl ethylketone will ordinarily perform this function adequately. To secure maximum bond strength it is imperative that cleanliness be maintained until the gauge is firmly in place. A clean bottle of solvent and clean clothes should be used, and neither the gauge nor test surface should be touched by the finger after cleaning. If gauges are handled before the use the backs of the gauges can be cleaned by wiping with a clean cotton or swabbed dampened slightly with solvent. This would be done with extreme care in the case of paper gauges to avoid damaging the gauges with solvent. Wipe the fingers as frequently as necessary with solvent damped clean cotton gauge. In order to orient the strain gauge in the desired direction it is advisable to scribe guide lines on the prepared surfaces. The scribed lines should not pass under the strain gauge, burrs-raised by the scriber may pierce the gauge and rosette, resulting in a short or open circuit.

8.17 NITROCELLULOSE CEMENT

Nitrocellulose cement is ordinarily used only with paper-backed strain gauge, since the cement cures by the evaporation of a solvent is not suitable for use with the impermeable epoxy or bakelite strain gauges. For paper gauges, the manufacturer commonly supplies a cellulose cement, but ordinarily Duco household cement which is available in "limestone" form will work very successfully for this purpose. After preparing the test surface, the actual cementing process is about the same as would be used in repairing a broken piece of china clay. A fairly linear layer of cement is spread on both the gauge and the prepared surface, and the gauge is set in place at once. In the first few seconds after the gauge has been applied to the surface, it will be possible to slide the gauge around slightly, in order to align it with the guidelines and orient it to the lineslightly with a finger and to squeeze most of the excess cement. It is not necessary to squeeze out all the excess cement, since this will take care of the damping pressure. While the gauge is drying, it should be held in place with a 1/2 kg weight. Greater pressure may result in grounding of the gauge wire. Scrapping of the excess cement will shorten the required time of drying and curing.

8.18 EPOXY CEMENTS

The epoxy cements can be used with paper epoxy or bakelite strain gauges; these cements are available in a number of different formulations, requiring different curing procedures and exhibiting somewhat different characteristics. The room temperature curing epoxy cement will withstand temperature of 1400°C. It is significantly less hygroscopic than nitrocellulose cement, unaffected by oils and most common solvents, and can be cured by heating to less than an hour after cementing.

This adhesive is solid commercially and has two components, a resin and an activator packed separately or in a two-compartment, elastic mixing bag. The components are mixed in recommended proportions to get the full potential strength.

Some of the epoxy cements are self-polymerising at room temperature and begin to cure immediately after mixing. These cements will develop half their strength in 16 hours at room temperature and full strength after about a week. Floating to 70°C will drastically shorten the curing time to approximately an hour.

The gauge application procedure when using cement is similar to that for nitrocellulose adhesive.

8.19 STRAIN GAUGE LATHE DYNAMOMETER

For simplicity of analysis, it is frequently convenient to reduce the lathe operations to a two-dimensional process. In this case, the resultant forces will act in a known plane, and only two forces are required to be measured. Two force components may be measured, the axial components (F_t) and the tangential force (F_c).

It is apparent that the force F_c will cause a bending moment M_c at a distance r from the cutting edge and that F_t will cause a corresponding moment M_t where

$$M_c = F_c \times r \qquad (8.23)$$
$$M_t = F_t \times r \qquad (8.24)$$

The strain at a section through A and B is caused by bending moment M_c. At the surface the strain is maximum and equal to

$$e_{(c1-2)} = \frac{M_c \cdot \frac{b}{2}}{E \times Z} \quad (8.25)$$

where e_{c1} is a tensile strain, e_{c2} is a compressive strain, E is Young's modulus of elasticity and Z is the area moment of inertia of the section about the axis. For a rectangular section

$$Z = \frac{ab^2}{12} \quad (8.26)$$

∴ Eq. (8.25) becomes

$$e_{c(1-2)} = \frac{6M_c}{abE} \quad (8.27)$$

$$l_{t(1-2)} = \frac{6M_t}{abE} \quad (8.28)$$

The foregoing discussion indicates how lathe forces can be measured, and we will now consider practical designs employing such principles. Since it is inconvenient to place the gauges directly on the tool, a tool holder with a built-in measuring section is employed. A dynamometer of this type is shown in Fig. 8.17(a).

It is important that the cutting edge be kept on line axis of the measuring section at a known distance from the gauges. When the particular unit is connected to commercial strain recording equipment, a force gives 1 mm of deflection per unit of force, and this represents adequate sensitivity for turning studies. The stiffness of the unit shown is actually greater than the customary tool holder for the 16 mm square tool bit that is used.

8.20 TURNING DYNAMOMETERS

8.20.1 Turning Dynamometer Based on Basic Principle

This can be illustrated by the design, shown in Figures 8.17(a), (b) and (c), where the disbalance of the Wheatstone flow bridge determines the cutting forces.

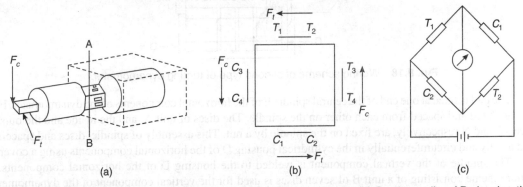

Fig. 8.17 (a) Working scheme of first type of turning dynamometer, (b) Section at line AB showing the details of gauge arrangement and (c) Arrangement of gauges connected in Wheatstone bridge.

Four gauges are mounted for measuring each force, i.e. F_t and F_c, and these are connected to two sets of Wheatstone bridge circuits for their measurement.

The section along the line AB shows the details where strain gauges are stuck.

This is of cantilever design and is made of alloy steel. The rectangular machined surface is obtained such that where gauges are stuck, they are at a distance from neutral axis in order to achieve maximum moment of inertia and to reduce torque sensitivity by increasing polar moment of inertia. The web is such that concentration of loads is in this section, but this is insensitive to loads in the direction of the dynamometer axis.

Working principle

As shown in the bridge, the gauges are connected in such a way that the output is increased 4 times. The thin section where the gauges are mounted is machined such that it is sufficient for mounting gauges only. The gauges fixed in this manner ensure that there is no cross-sensitivity.

This type of force measuring dynamometer is quite simple in design and fabrication.

8.20.2 Another Type of Turning Dynamometer

In Fig. 8.18 the working scheme of this type of turning dynamometer, in which R is the resultant force acting on the tool is shown. This force is resolved into two components, one vertical (in the direction of cutting speed) and the other horizontal (normal to the cutting speed). The dynamometer consists of two housings in which two sets of discs are fixed circumferentially and axially.

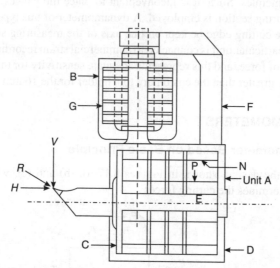

Fig. 8.18 Woring scheme of second type of turning dynamometer.

The tool is fixed at one end of the central spindle E of the horizontal components of the dynamometer. Five discs are fixed and spaced from each other on the spindle. The discs of unit A, and the inside and the outside spacers P and N respectively, are fixed on the spindle by a nut. This assembly of spindle, discs and spacers is fixed axially and circumferentially in the cylindrical housing D of the horizontal components using a cover C.

The spindle of the vertical component is welded to the housing D of the horizontal components. A similar scheme consisting of a unit B of seven discs is used for the vertical component of the dynamometer. The housing G of the vertical component is rigidly held on the toolpost by a strong structure.

An L-shaped fixture is put between the upper end of the spindle E and the housing D of the horizontal component. This support completes the rectangular frame and gives an additional rigidity to the welded joint between E and D. This reduces the turning moment acting on the welded joint and avoids the tilting of the horizontal component of loading—the other component should not show any reading.

Working principle

The horizontal component F_t of the resultant cutting force R acts axially on spindle E. The discs of group A are fixed to this spindle at their centre between a collar at the tool end of the spindle and the nut at the other end, while the discs are circumferentially and axially clamped at the boundary in the housing D. The force component F_v gives a central bulge of the discs relative to their circumferential plane, i.e., the disc takes the shape of a shallow saucer. This bulge and hence the displacement of spindle E along its axis is proportional to the load, and is measured relative to the circumference of the housing.

The working of the vertical component is similar under the vertical load component F_c. The spindle of the vertical component is welded to the housing D of the horizontal component. The housing D comes down, remaining horizontal by an amount equal to the bulge of the discs of group B. This displacement is proportional to the load component V.

A turning moment due to the distance between the tool edge and the welded joint acts upon the weld. However, this does not play any part in the deformation of the plates of either component, as the number of plates in group B are large enough to give a radial stability of spindle I. Moreover, this turning moment is reduced to a large extent by completing the rectangular frame I–D–F through an additional support E.

The displacement of spindle I, due to the vertical force V, is measured relative to the circumference of the housing G. The housing G is fixed with reference to the tool post.

Cross effect is observed to be completely absent in this scheme of the dynamometer. The horizontal component F_t trying to push unit B discs radially, along its own direction, cannot do so because the discs are clamped circumferentially. The same is true for vertical force component F_c and the discs of group A.

The lathe dynamometer used in the laboratory consists of a unit machined from one piece of steel, to provide maximum stiffness for the required sensitivity, and to achieve the high degree of linearity and freedom from hysteretic effects. The deflection taking place under load should be purely elastic and free of friction. The deflections are measured with suitably placed resistance strain gauges, that provide a means of convening forces into a conveniently measured electrical quantity.

8.21 THREE-COMPONENT LATHE TOOL DYNAMOMETER

A lathe dynamometer can be designed to consist of a unit machined from one piece of steel to provide maximum stiffness to the required sensitivity and to achieve the high degree of linearity and freedom from hysteresis effects. In Fig. 8.18 the deflections taking place under load should be purely elastic and free of friction. The deflections are measured with suitably placed resistance strain gauges, that provide a means of converting forces into a conveniently measured electrical quantity. Figure 8.19 shows the diagrammatic sketch of a three-component lathe tool dynamometer. Here, gauges 1 to 4 are for radial force, gauges 5 to 8 for tangential force, and

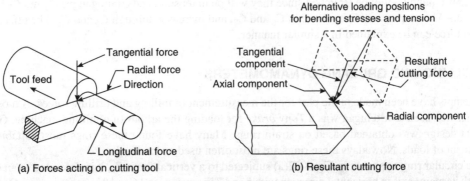

Fig. 8.19 Forces F_P, F_Q and F_R acting on three-component lathe tool dynamometer.

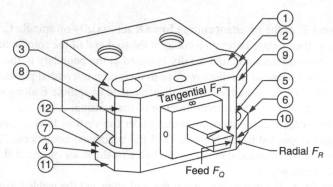

Fig. 8.20 Diagrammatic sketch of three-component lathe tool dynamometer.

gauges 9 to 12 for axial or feed force (see Fig. 8.20). In the octagonal ring dynamometer used for turning octagonal half rings are cut in the block. The basic octagonal ring dynamometer used for milling and grinding is explained in next Section 8.22. In this dynamometer F_R puts all octagonal half rings in compression. F_Q subjects all half rings to tangential force. F_P puts upper elements in tension and lower in compression due to bending moment.

There are 4 octagonal half ring in 3 component lathe dynamometer providing 12 positions for slicking strain gages.

The signal from the wheatstone bridge is channelled through a strain gauge amplifier to a recorder or oscilloscope (Fig. 8.21). The voltage input is between C and D, and the voltage output is between A and B.

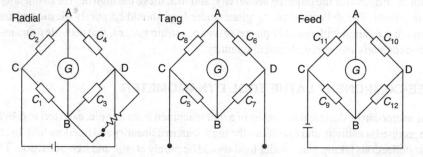

Fig. 8.21 Three Wheatstone bridge circuits used to measure each respective force encountered in lathe work.

When the tangential force is applied to the tool tip, the octagonal ring of the dynamometer will bend, strain gauges C_5 and C_8 will be stretched since they will be in tension, and strain gauges C_6 and C_7 will be in compression. The current will be reduced in C_6 and C_7, and increased through C_5 and C_8. The effect of axial and radial force can be explained in a similar manner.

8.22 MILLING AND GRINDING DYNAMOMETERS

Many attempts have been made in the past for the measurement of milling and drilling forces. A most simple device that was used at Michigan was a *Tony brake* for loading the arbour of a milling machine. Gradually a compact design was obtained based on strain rings. Many have found these rings most suitable for the measurement of loads. Nowadays these rings are most often used as load cells.

If a circular ring as shown in Fig. 8.22(a) subjected to a vertical force F_V, it is deformed to an elliptical shape with its major axis in horizontal direction with 5 and 7 in tension, and 6 and 8 in compression. On the other hand, a horizontal force F_H deforms the circular ring into an ellipse with its major axis at 39° (approximately) with vertical. The stressed points 1 and 2 in tension and those at 3 and 4 in compression. From a practical point of view, circular rings are not used but octagonal rings are used to simulate the circular ring as shown in Fig. 8.22(b). Referring to Fig. 8.22, strain gauges for vertical force F_V are at 5, 6, 7, 8 and strain gauges for F_H are at 1, 2, 3, 4. There is no cross sensitivity between F_V and F_H.

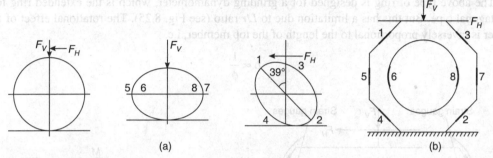

Fig. 8.22   (a) Deformation of circular ring under vertical and horizontal load (b) octagonal ring for milling dynamometer.

Regarding the design of rings for loads in horizontal and vertical direction, any strength of materials book may be consulted.

As circular rings have the disadvantage that they may roll under the action of cutting forces, octagonal rings are often used to avoid rolling tendencies, and these rings have nodes at 50°, 90°, and so on for all 360°.

8.22.1 Milling Dynamometers

These are designed so that octagonal rings are placed between two plates as shown in Fig. 8.23. In this dynamometer, the vertical load P_z is taken by all four rings, P_x is measured by (A, D) and P_y is measured by (B, C).

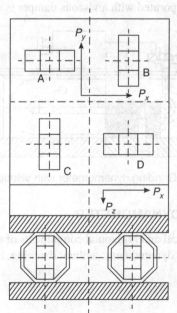

Fig. 8.23   Layout of octagonal rings for three-dimensional milling force dynamometer.

8.22.2 Grinding Dynamometer

A similar principle is extended to be used in the design of grinding dynamometers. An octagonal ring, with its horizontal faces much longer than the other faces, is designed. Horizontal top face carries the workpiece, where the forces arising out of machining mechanism are exerted. The arrangement is shown in Fig. 8.24.

The above type of ring is designed for a grinding dynamometer, which is the extended ring form of the octagonal type. But this has a limitation due to L/r ratio (see Fig. 8.25). The rotational effect of the top member is inversely proportional to the length of the top member, i.e.

$$L \propto \frac{1}{\phi}$$

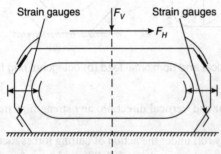

Fig. 8.24 Extended octagonal ring and strain gauge mounting for grinding dynamometer.

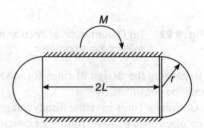

Fig. 8.25 Relationship for horizontal surface—L/r ratio.

Therefore, L should be limited to the extent of $4r$ when ϕ, the rotation of the top member shown in Fig. 8.25 by M reduces by 30 times.

A grinding dynamometer incorporated with a viscous damper is shown in Fig. 8.26.

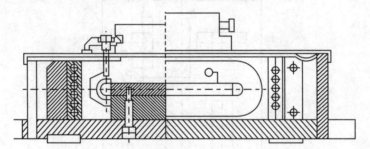

Fig. 8.26 Grinding dynamometer with viscous damping.

8.23 RESPONSE CURVE OF DYNAMOMETER

The performance curves for the vertical and horizontal components of a grinding dynamometer with viscous damping, compounded with that of a dynamometer without damping, is shown in Figs. 8.27(a) and (b).

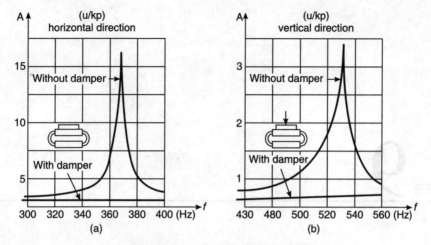

Fig. 8.27 Response curve of dynamometer.

REVIEW QUESTIONS

1. What are the functions of dynamometers in metal cutting?
2. What are the desirable characteristics of a dynamometer?
3. What is the principle of working of a tool dynamometer? Explain.
4. Give the history of development of dynamometers. What is the latest trend in the design of dynamometers? Explain.
5. What is the difference in working of two-dimensional and three-dimensional dynamometers?
6. What is the importance of strain gauge in the development of tool dynamometers?

CHAPTER 9

Tool Failures and Tool Life

9.1 TOOL FAILURE DEFINED

Tool failure may be defined as the state of the tool when it ceases to function satisfactory. Cutting tool failure is rather difficult to gauge, particularly when brittle materials like cast iron are machined.

9.2 CRITERION OF TOOL FAILURE

The following criteria are used in determining cutting tool failure, i.e. the time when the cutting tool requires sharpening.

Total tool destruction or breakdown: The cutting tool is unable to cut as required. It produces highly burnished surface of work, the nose completely worn-out and excessive parking.

Temperature failure or overheating: The tool gets overheated, and cutting edge gets softened, and stops functioning.

Increased feed: Increase in the need of feed of the tool by certain amount indicative of shank wear.

Increased power consumption: Tool failure leads to increased cutting forces therefore power requirements will be more.

Poor surface finish: Occurrence of a sudden, pronounced change in the finish of the work surface, either in the direction of improvement or deterioration.

Change in dimensions: Occurrence of the change in dimension(s) of the finished part. This occurs when the final cut is taken on a large workpiece surface like cutting threads in a lead screw.

Increased noise level: Sudden change in the cutting noise level. This is indicative of excessive rubbing between tool workpieces.

Chipping of tool: In this case, fine pieces of tool material break off along the cutting edge.

Formation of crack: Formation of cracks may occur at face or flank of cutting edge. This may be due to excessive heat.

9.3 TYPES OF TOOL FAILURE

Tool failures are mainly classified as follows:
1. Temperature failure. This can be due to either:
    (i) Plastic deformation of cutting edge due to high temperature, or
    (ii) Cracking at the cutting edge due to thermal stress.
2. Chipping of the edge or fracture due to mechanical impact.
3. Built-up edge.
4. Chemical decomposition of contact surface.
5. Gradual microscopic wear.

9.3.1 Temperature Failure

The temperature failure at cutting edge may occur at high cutting speeds when the tool becomes too soft due to high temperature developed, and stop functioning as a result of low strength; and it may also shear off a part of the cutting edge.

Plastic deformation

As the cutting speed is increased, the tool close to the cutting edge becomes hotter. As the feed rate is increased, both the temperature of tool tip and the cutting force are increased. With many work materials, a limit is reached above which tool begins to deform under the influence of temperature and pressure. Carbide tools have much more resistance to deformation than that of high speed steel tools, which enables them to be used at much higher speeds.

The tool deformation can be measured by putting the carbide tool after cutting for a certain time on optical glass and then examining under monochromatic light.

Under less severe conditions, deformation may lead to premature failure for another reason. The bulge on the clearance face may be of such a form that the clearance angle is greatly reduced and flank wear thereby much increased. This leads to rapid failure, usually concentrated at the nose of the tool.

Apart from speed and feed, the amount of deformation depends upon depth of cut and tool geometry among other factors. Increasing depth of cut gives increased deformation if other conditions remain the same. As regards tool geometry, the deformation appears to be particularly related to the nose radius of the tool, since this is the position from which the heat generated at the friction surfaces is least rapidly conducted away. The smaller the nose radius, the hotter is the tool at the nose and greater the amount of deformation. The grade of carbide is of importance in resistance to deformation. The harder grades of carbide are more resistant than the softer, tougher alloys.

If coolants could be applied sufficiently close to the cutting edge, they might greatly reduce deformation. As normally applied, it is doubtful whether they have any appreciable effect. This is a subject of further investigation.

Thermal cracks

Figure 9.1(a) shows tool failure due to thermal stresses. Thermal cracks are called so because they appear to arise from the stresses caused by local expansion and contraction of the tool material, where it is subjected to rapid changes of temperature and severe temperature gradients. Thermal cracks start from cutting edge and form on planes normal to the edge. These cracks are usually quite short and often cause loss of tool life, but they are undesirable because they may give rise to stress concentration leading to tool failure.

Thermal cracks occur most frequently where the depth of cut is large (greater than 9.5 mm) or where relatively long cutting edge is engaged. They also appear to be most severe in milling cutters, where very frequent interruption of cut results in fluctuating temperature at the tool edge.

9.3.2 Mechanical Chipping

Mechanical chipping of the cutting edge, also called rupture or tool of mechanical breakage, is a frequent cause of tool failure. While the tool actually cuts the portion of the cutting edge which is embedded in the metal, it is unlikely to be damaged. Damage is most likely to occur when engaging or disengaging the tool, or by swarf on the part of the tool edge not engaged in the cut, or by careless handling.

Where the cutting edge is chipped, subsequent cutting on this portion of the edge may act as a tool with excessively large negative rake, and cause severe rubbing rather than cutting. It is not subject to any general laws such as speed, feeds and depth of cut, etc. Its general effect is to contribute to the scatter in performance figures which is such a pronounced feature of the life of the cutting tools in service in the machine shop. Figure 9.1(b) shows tool failure due to mechanical impact.

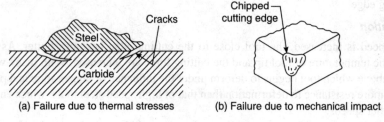

Fig. 9.1 Tool failure.

9.3.3 Built-Up Edge (BUE)

The built-up edge consists of work material wielded to the rake face and cutting edge of the tool under some condition of cutting. The presence or absence of the BUE is important in relation to tool life and surface finish. It may be either harmful or beneficial to the tool, depending upon the condition. When cutting cast iron, it is usually beneficial to cut cast iron under conditions when the BUE is present. This largely protects the rake face of the tool from wear and rate of flank wear is low.

There are cases in cutting steels when the presence of built-up edge may be beneficial, but more often it reduces tool life. If the conditions of cutting are such that the built-up edge is apt to be broken away at the end of cut, it frequently occurs. This can lead to a very rapid breakdown of the tool.

While the BUE may protect from wear the tool material immediately beneath it, under some conditions there may be a crater wear in the form of a groove at its outer edge, as shown in Fig. 9.2.

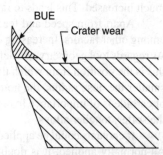

Fig. 9.2 Crater wear at outer edge of tool material.

This type of wear is similar in appearance to high speed cratering, but it is important to distinguish between the two since this type of wear cannot be reduced or prevent by the use of Ti-WC-CO tool.

The BUE can also cause severe damage in another way. The coefficient of thermal expansion of carbide tool alloys is considerably lower than that of steel. When a large BUE is welded to the rake face of the carbide tool under such conditions that the strength of weld is high, the greater contraction of the steel during cooling, when the tool is disengaged, may cause cracks to form in the carbide from the edge of the build-up as shown in Fig. 9.2. Such cracks are responsible to removal of large "flakes" of carbides from the surface of the tool. This type of damage is more severe where the weld is strong and coefficient of expansion of work material is high. These conditions are at their worst with austenitic steels and alloys of Nimonic type.

Unlike flank wear, which occurs under almost all conditions of cutting, the built-up edge is a factor affecting tool life only under a restricted set of conditions, mainly at low cutting speeds and feeds.

9.3.4 Spalling or Crumbling

This occurs when tool cuts a hard material. This results due to improperly ground relief angles or because of excessive clearance angles.

9.4 TOOL WEAR (MICROSCOPIC WEAR)

Tool wear may be defined as loss of weight or mass of tool.

9.4.1 Effects of Tool Wear

These are:
   (i) Increased cutting force
   (ii) Increased surface roughness
   (iii) Reduced dimensional accuracy
   (iv) Generation of vibration in tool-work system
   (v) Breaking of tool
   (vi) Damaging of workpiece.

9.4.2 Types of Tool Wear

Tool wear occurs at two places on a cutting tool. These are called: (a) flank wear; (b) crater wear.

Flank wear

The wear occurring at the cutting edge and principal flank of the tool is called *flank wear*. It is a flat portion worn behind the cutting edge which eliminates some clearance or relief. Flank wear takes place when machining brittle materials like C.I., or when feed is less than 0.15 mm/rev. The worn region at the flank is called *wear land*. The wear land width is measured accurately with a Brinell microscope. Wear land is not of uniform width. It is widest at a point farthest from the nose. This is due to the fact that the material cut by this part of cutting edge had been work-hardened during the previous cut. The frictional stress and maximum temperature at the flanks also go on increasing with time. A stage is reached when diffusion becomes the predominant wear mode on the flank. After a critical wear land has formed, further wear takes place at an accelerating rate [see Fig. 9.3(a)]. It is advisable to change the tool at this point. During the steady wear phase, flank wear [see Fig. 9.3(b)] is caused mainly through abrasion, whereas during rapid wear phase, it is caused by diffusion. This increases the number of cutting forces, ruins the surface finish, and dimensional accuracy.

The extreme rigidity at the tool-work contact on the flank surface results in much higher normal stress and proportionally greater wear, by plowing away micro-constituents on the tool flank. Figure 9.4 illustrates the flank wear characteristics of two grades of carbides.

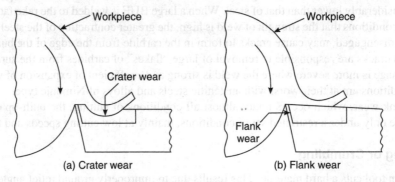

Fig. 9.3 Trends of wear representative of performance of different grades of carbide.

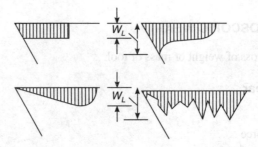

Fig. 9.4 Flank wear characteristics of two grades of carbides.

Table 9.1 shows the standard flank wear ranges recommended.

Table 9.1 Standard recommended flank wear ranges

Work material	Tool material	Cutting condition	Predominant wear	Recommended maximum flank wear (mm)
Steel	HSS	Semi-rough	Flank	0.6–1.00
CI	HSS	Semi-rough	Flank	1.5–2
Steel	Cemented carbide	Feed greater than 0.3 mm/rev	Flank	1.5–1.7
CI	Cemented carbide	Feed less than 0.3 mm/rev	Flank	0.75–1
CI/steel	Ceramic	—	Flank	0.6

Crater wear

It occurs on the rake face of a tool in the form of a depression called the *crater*. The crater is formed at some distance from the cutting edge. Experiments have shown that locations of maximum cratering and maximum chip-tool interfacial temperature coincide with each other. The crater significantly reduces the strength of tool and may lead to its total failure. The mechanism of crater wear is mainly adhesion and diffusion. This is caused by the pressure of the chip as it slides up the face of cutting tool. Both flank and crater wear take place when feed is greater than 0.15 mm/rev at low or moderate speeds. This type of failure occurs when high speed steel satellite or sintered carbide turn ductile materials.

Crater wear is defined by a ratio H_c as under [Fig. 9.5(a)]

$$H_c = \frac{d}{\frac{w}{2} + f}$$

where d is the crater height;
w is the width of crater;
f is the distance from cutting edge.

Figure 9.5(b) shows the development of crater wear with respect to time, cutting temperature which depends on speed, feed, etc.

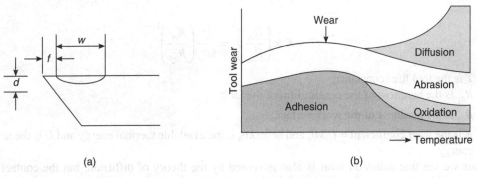

Fig. 9.5 (a) Geometry of wear ratio H_c, (b) wear mechanisms.

The limiting value of ratio H_c is 0.4 for carbide and 0.6 for HSS tools for machining steel. It does not develop at low speeds while machining at speeds greater than 100 m/min it becomes predominant. At high feeds (greater than 0.3 mm/rev) both flank and crater wear take place. Wear takes place only at flank when machining brittle material like C.I. For soft material, flank wear is high.

9.5 CAUSES OR MECHANISMS OF WEAR

The wear between two surfaces rubbing over each other occurs due to the following reasons:

1. Abrasion (mechanical wear)
2. Adhesion (mechanical wear)
3. Diffusion (thermochemical wear)
4. Electrochemical (chemical wear).

9.5.1 Abrasion

It is a mechanical type of wear. Abrasive particles in the work material which approach the hardness of the tool or are harder than the tool material plough into the softer matrix of tool (due to high temperatures, the tool loses its hardness) and remove tool materials in particles. This type of tool wear called *abrasion*. In the surface of C.I., entrapped sand particles will plough or gauge the tool materials off the cutting edge. Similarly, tightly adhered heat treat scale carbides in the microstructure of the work material or aluminium oxide particles in the structure of aluminium killed steels may bring about wear of the cutting edge by abrasion. Also, the fragment of built-up edge carried along by the underside of the chip remove tool material from the rake surface by abrasion. Such wear is predominant in the case where the prevailing strain rate may result in hard spots traversing through the asperitis in the point of contact of the interface. This is relatively more severe on the tool flank due to nature of the contact and hard backing provided by the work shoulder, as compared to the ribbon-like chip on the rake surface.

9.5.2 Adhesion

Theoretical analysis shows that in the process of adhesive wear over the medium (of homogeneous work materials), the path travelled by the tool before it becomes effective is given by

$$T \propto \left(\frac{R_{c_1}}{R_{c_2}}\right)^2$$

or

$$T = K \times \left(\frac{R_{c_1}}{R_{c_2}}\right)^2 = \frac{U}{k\theta_a}\left(\frac{R_{c_1}}{R_{c_2}}\right)^2$$

where T is the tool life (constant);
   R_{c_1} is the hardness of the contact film of the tool;
   R_{c_2} is the hardness of the work surface;
   K is the wear coefficient $= U/k\theta_a$ and here, $k\theta_a$ is the available thermal energy and U is the activation energy.

Thus we see that adhesion wear is also governed by the theory of diffusion, but the contact is only between the work and the tool flank face and not between the chip and the tool.

It has been shown that the diffusion at flank after time T is given by

$$W_f = k_b \sqrt{V_w v} \int_0^T C\sqrt{Dh_f}\, dt \qquad (9.1)$$

where h_f is the flank height;
   k_b is a constant;
   V_w is the work-tool surface velocity;
   D is the diffusion constant proportional to temperature;
   C is the concentration factor;
   v is the chip–tool interface velocity.

If $\int_0^T C\sqrt{Dh_f}\, dt = A$, then $W_f = k_b \sqrt{V_w v}\, A$.

The chips are very rough. When chips move over the tool surface (rake face), it makes point contact. The chip moves under high pressure, causing extremely high localized temperature, which brings about metallic bond between the chip and the tool material. This phenomenon is called *adhesion*.

The strength of the bond at the points of adhesion is often so great that when the chip moves, separations do not occur along the interface, but some of the tool material is also sheared off and carried along by the chip. Thus the chip lifts small particles of the tool material from the cutting edge. The quantity of metal transferred is proportional to the area of contact and the hardness of the mating pair.

9.5.3 Diffusion

Figure 9.6 shows an experimental set up which utilizes thermoelectric effect to measure effect of temperature θ and intermetallic diffusivity D. Figure 9.7 shows the chip formation under conditions of flank contact length h_f and face contact length l.

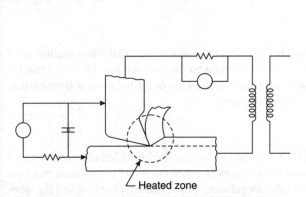

Fig. 9.6 Intermetallic and thermoelectric diffusion.

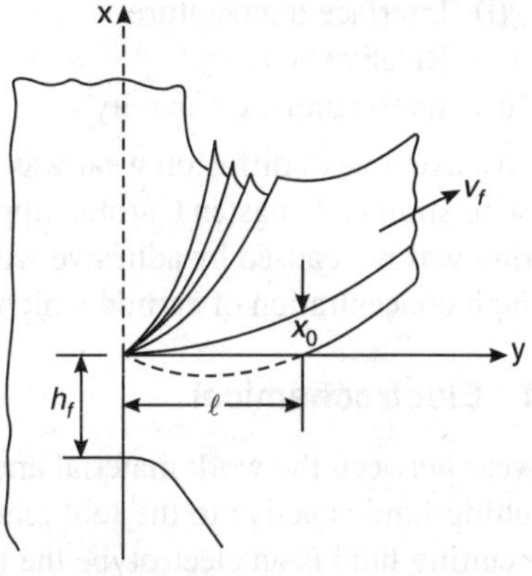

Fig. 9.7 Schematic diagram of tool and chip.

If the mechanical process involved in adhesion is capable of increasing the localised temperature of the real area in contact, surface or interstitial diffusion shall occur. Diffusion is the mechanism by which atoms in a metallic crystal shift from one lattice point to another, causing a transfer of the element in the direction of the concentration gradient. The rate of diffusion for any substance follows a hyperbolic relationship. Initially the rate of diffusion is very high and then decreases asymmetrically to zero. The process of diffusion during chip formation is dynamic. If the chip and tool are kept in the similar condition under static contact, the total amount diffused will be much less. Though all elements of the tool material diffuse into the chip to different extents, experiments have shown that it is in the case of tungsten in a monocarbide tool material when transfer causes the loss of form stability of the cutting edge.

What promotes diffusion: Take two cases. In the first case let the interface temperature be about 850°C obtained by increasing the cutting speed. In the second case, the interface is artificially heated to 850°C.

It is seen that the order of tool life is same. Therefore, it may be concluded that it is not the fraction that is responsible for the mean, but it is something dependent on temperature or activation energy to make a start of diffusion wear (crater).

How diffusion takes place: Atoms in a metallic crystal lattice move from a region of high concentration to that of low concentration. The rate of diffusion depends on concentration gradient and the chip-tool interface temperature. For example, the atoms of tool material diffuse into the adjacent chip material, thereby softening the tool. This causes diffusion wear of the tool along the chip-tool contact length (l). From Fick's first law flux of atoms

$$J = -D \frac{\Delta C}{\Delta x}$$

in units of atoms per m²/s where ΔC is the concentration gradient over a distance Δx. D is diffusivity which increases with temperature θ. Figure 9.8 indicates diffusion D as a function of temperature θ.

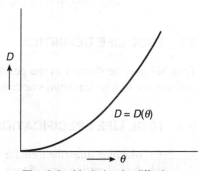

Fig. 9.8 Variation in diffusion.

Also, $D = D_0 \exp\left(-\dfrac{Q}{RT}\right)$ where D_0 is a constant for a material pair, Q is activation energy needed to move the atoms through the lattice, R is gas constant (7.937 J/mol.K) and T is the absolute temperature.

Diffusion characteristics: These are as follows:
  (i) Interface temperature
  (ii) Relative velocity
  (iii) Intermetallic diffusivity.

An example of diffusion wear was identified by Dr. E.M. Trent who discovered that while machining of steel with straight Tungsten Carbide tipped tools, a crater was formed on the rake surface. He found that the cratering was not caused by adhesive wear or abrasive wear but it was a case of diffusion wear from carbide with high concentration of carbon which diffuses to the steel chips.

9.5.4 Electrochemical

The wear between the work material and the tool in the presence of cutting fluid is called *chemical wear*. If the cutting fluid is active to the tool material, the tool wear is greatly accelerated due to chemical reaction. If the cutting fluid is an electrotype the tool wears out due to galvanic action between the tool and the work material. HSS tools are anodic to some stainless steels such as 16-25-6, S-816, etc. So, while cutting these steels with HSS tools, the tool wears out due to galvanic action.

9.6 MEASUREMENT OF TOOL WEAR

The tool wear could be measured by the following methods:

  (i) Flank wear (H_f)—It is measured by low magnification microscope using filar eye piece.

  (ii) Crater wear (H_c)—It is characterised by three variables viz. width, depth and location of maximum depth of crater. The craters are usually evaluated by calculating volume or maximum depth. It is measured by the following methods:

  (a) By series of low magnification photographs, with each one taken on a surface slightly below the previous one.
  (b) By the direct measurement of a sectioned tool, taking several cross-sections.
  (c) By spectrographic analysis of the chips to know total volume of crater.
  (d) By measuring the loss of weight of either the tool, tip or inserts, the total wear volume is estimated. It is simple and quick method.

9.7 TOOL LIFE DEFINITION

Tool life may be defined as the period during which tool cuts satisfactorily. It is the period between two consecutive tool resharpenings or replacements. Tool life can be specified in the following ways.

9.8 TOOL LIFE SPECIFICATIONS

  (i) Machine time (the time for which machine was operated or run).
  (ii) Actual cutting time of tool.
  (iii) Volume of metal removed.

TOOL FAILURES AND TOOL LIFE 229

   (iv) Number of jobs machined.
   (v) Equivalent cutting speed (Taylor speed), e.g. V_{60} cutting speed, at which a standard value of machine time or actual cutting time, such as 60 minutes, is obtained under a given set of cutting conditions from the Taylor equation.

It should be particularly noted that tool life may be, and even is, expressed in terms of cutting speed. The reason is that if all oilier machine and material variables are held constant while cutting speed is varied, the time required to dull the tool will be a direct function of the cutting speed.

The other criteria to specify tool life are:
(a) Failure of tool
(b) Presence of chatter
(c) Poor surface finish
(d) Sudden increase in power and cutting forces
(e) Overheating and fuming
(f) Dimensional unstability.

9.9 MEASUREMENT OF TOOL LIFE

Taylor gave the following relationship between tool life and cutting speed:
$$VT^n = C$$
where V is the cutting speed in m/min
   T is the actual cutting time between two resharpenings in min
   C is a constant whose value depends upon other machine variables and work material variables
   n is the exponent whose value depends to some extent upon machine variables and work material variables.

Tool life excludes the following periods:
(a) Removing
(b) Regrinding
(c) Resetting.

9.9.1 Expected Tool Life

Cast steel tool—22 min
HSS—60 to 120 min
Cemented carbide—240 to 480 min

9.9.2 General Empirical Relationship between Cutting Speed, Tool Life, Feed, and Depth of Cut

This is given by
$$V = \frac{C}{d^x f^y}$$
where V is the equivalent cutting speed in surface metre per min
   C is a constant whose value depends on other machine variables and work material variables
   f is the feed per revolution in mm/rev.

d is depth of cut in mm.

x, y are constant whose values depend on the workpiece material; $x = 0.14$ for steel, 0.10 for cast iron and $y = 0.42$ for steel and 0.30 for C.I.

9.9.3 Formula Connecting Tool Life, Cutting Speed, Chip Thickness and Length of Tool Engagement

This is given by

$$V = \frac{C}{t^{n_1} f^{n_2} T^n}$$

where *V* is the cutting speed surface metre per minute;
t is the chip thickness in mm;
f is the tool length engagement in mm;
T is the tool life (minute); and
n_1, n_2, n are experimental indices.

9.9.4 Equation Incorporating the Effect of Size of Cut

This is given by

$$C = VT^n f^{n_1} d^{n_2}$$

where *C* is the constant of proportionality
f is the feed in mm/rev
d is the depth of cut in mm
n_1 is the exponent of feed (varies from 0.8 to 2.2)
n_2 is the exponent of depth of cut (varies from 0.2 to 0.4)
n is the function of cutting tool material: 0.1 to 0.15 for HSS, 0.2 to 0.25 for carbides, and 0.6 to 1.00 for ceramics.

9.10 FACTORS AFFECTING TOOL LIFE

The following are the major factors which affect tool life:

1. Cutting speed
2. Feed rate
3. Depth of cut
4. Microstructure of workpiece
5. Hardness of workpiece
6. Tool material
7. Rigidity of workpiece (machine tool systems)
8. Nature of cutting
9. Type of cutting fluid and its application
10. Tool geometry.

9.10.1 Cutting Speed

Cutting speed has the maximum influence on tool life. Tool life decreases as the cutting speed increases. The tool wear is a function of cutting speed. It has been found that for turning of steel, the crater wear become more predominant above a cutting speed of 100 m/min. The relationship between cutting speed and tool life is shown in Fig. 9.9.

Cutting speeds varying from 30 to 200 m/min have been successfully used with carbide tools over different types of work materials. It is necessary to know exactly what material is to be machined before cutting conditions are selected. The cutting speeds and other parameters can be taken from the data book of their manufacturers.

Tool life vs. cutting speed for various tool materials is shown in Fig. 9.10, while Fig. 9.10 shows tool life versus cutting speed when machining cast steel with HSS tools and carbide tools.

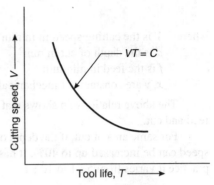

Fig. 9.9 Plot of tool life vs. cutting speed.

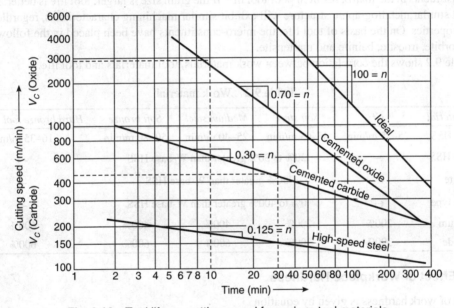

Fig. 9.10 Tool life vs. cutting speed for various tool materials.

9.10.2 Effect of Feed Rate and Depth of Cut

As the feed rate increases, the tool life decreases; so is the case with depth of cut. A relationship between the variables is given below for cemented carbide tool and low carbon steel combination.

$$V \cdot T^{0.20} = \frac{0.260}{f^{0.35} d^{0.08}}$$

where V is the cutting speed in m/min;
        f is the feed in m/min;
        d is the depth of cut in mm.

Another relation among variables is as noted below.

$$V = \frac{C}{d^x f^y}$$

where V is the cutting speed in m/min
   d is the depth of cut in mm
   f is the feed in mm/min
   x, y are constants of mechanical properties of material.

The above relationship shows that for a constant tool life, cutting speed decreases with the increase of feed and cut.

For same area of cut, if the depth of cut is increased twice and feed is decreased to one half, the cutting speed can be increased up to 40%. It has been found that the desirable ratio depth to the feed is 8 : 1, but in practice it varies from 5 : 1 to 10 : 1.

9.10.3 Microstructure of Workpiece

The tool life has a definite correlation with the microstructure of the workpiece. In general, hard micro-constituents in the matrix result in poor tool life. If the grain size is larger, tool life is better. It has been found that similar metallographic structure will exhibit similar machining characteristics, regardless to their relative properties. On the basis of tool life, the micro-constituents have been placed in the following order: pearlite, sorbite, troosite, bainite and martensite.

Table 9.2 shows the correlation between work materials, tool materials and tool life.

Table 9.2   Work material

Work material	CI	Soft steel	Medium steel	Soft bronze	Hard bronze tool material
W-base HSS	25–35 m/min	35–65 m/min	25–40 m/min	50–80 m/min	16–35 m/min
Co-base HSS	←	20% to 30% greater than W-base HSS			→
Stellite	←	50% greater than W-base HSS			→
Carbide type	←	0% to 400% greater than W-base HSS			→
Tantalum	200%	200%	400%	400%	600%
Carbide	300%	300%	600%	600%	600%

9.10.4 Effect of Workpiece Hardness

The effect of work hardness is given by equation

$$V = \frac{\text{Constant}}{BHN^{1.68} \, x\% \text{ reduction}}$$

As the hardness increases the velocity decreases. Harder the material, lesser will be tool life for the equivalent amount of material removal.

Area of cut

Cutting speed V is inversely proportional to the area of cut. For a given metal removal rate as the area of cut increases, the cutting speed has to be decreased for a given tool life.

The relationship between velocity and area of cut is expressed by an empirical relation:

$$V = \frac{K}{(A+b)} + c$$

where V is the cutting speed
  A is the area of cut
  K, b, c are constants.

9.10.5 Effect of Tool Material

Figure 9.10 shows tool life variation against cutting speeds for different tool materials. At any given cutting speed, the tool life is greatest for ceramic tools and lowest for high speed steel.

9.10.6 Rigidity of Workpiece Machine Tool System

Lower the rigidity of system, higher is the chance of tool failure. If rigidity is more, it will absorb more vibration and thus tool life will increase.

9.10.7 Nature of Cutting

If the cutting is intermittent, the tool experiences impact loading and there is more chance of its quick failure.

9.10.8 Cutting Fluids and Tool Life

The application of cutting fluid during the machining operation results in increase in tool life. Carbon steel tools that have less heat resistance have maximum increase in tool life (nearly 50%). The increase in tool life is nearly 25% in case of high speed steel tools. The effect of cutting fluid on carbide tool is not known very well. Carbide tools may show slight improvement in tool life, but may also be damaged by thermal chipping. Figure 9.11 shows the relationship between feed rate and cutting speed for particular types of tool material and work material.

9.10.9 Effect of Shape and Tool Angles on Tool Life

These are as follows:

Effect of rake angle

The rake angle has a complex influence on the temperature, forces and tool life of cutting tool. An increase in the rake angle reduces the deformation of chips, and hence a decrease in the amount of heat generated. Cutting forces also decrease, since the wedging action is more concentrated and this further reduces the temperature rise (see Fig. 9.12). Consequently, there is an increase in cutting life and better surface finish of the work. Increase in the rake angle, on the other hand, decreases the mass of the metal in the immediate vicinity of the cutting edge (region of maximum temperature is very close to it) and, therefore, the effective heat conductivity of the cutting tool is reduced. Loss of conductivity causes greater temperature rise and consequently a decrease in cutting tool life. Keeping in view the above contradictory influences the rake angles used in practice are shown in Table 9.3.

The rake angle can be either positive or negative. It is positive when the face slopes downwards from the plane perpendicular to the plane of the cut, zero when the face is in the same plane and negative when the face slopes upwards.

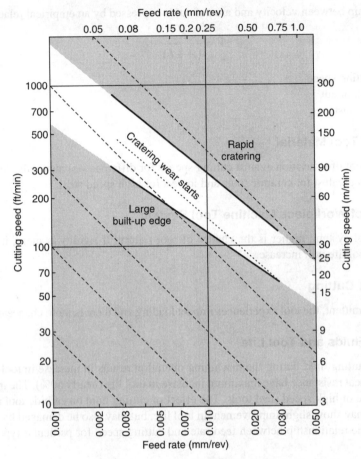

Fig. 9.11 Machining chart for WC + 6% Co tool cutting 0.4% C steel with hardness 200 HV.

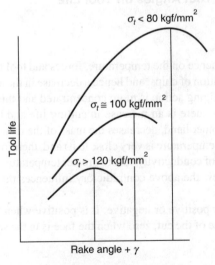

Fig. 9.12 Relationship between tool life and rake angle θ.

The back and side rakes combine to form the slope of the cutting tool face, which governs the cutting forces and the cutting temperature. The two rake angles have, however, their individual effects.

The back rake angle controls, to an extent, the direction of the chip flow. With a zero back rake, the chip would flow parallel to work surface and have a tendency to entangle with the work and cause problem of removal. A proper back rake angle will cause the chip to flow, at an angle to the axis of the work, and strike the tool holder or a suitable chip breaker curl and break into small fragments, and can be easily disposed off.

Table 9.3 Recommended rake angles for different materials

Material to be cut	HSS and nonferrous alloys		Cemented carbides		Clearance angle
	Back	Side	Back*	Side*	
Low carbon steel	12	15	5	8	6–8 for HSS and cast nonferrous alloys and 5–6 for tungsten carbide
Medium carbon steel	6	10	5	8	
High carbon steel	3	5	0	0	
12–15% manganese steel	3	5	0	0	
Grey cast iron	0	3	0	0	
White cast iron	0	0	0	0	
Aluminium	12	25	5	15	
Aluminium alloys	6	15	5	8	
Brass (soft)	6	10	0	0	
Brass (hard)	3	5	0	0	
Bronze (soft)	6	10	0	0	
Bronze (hard)	3	5	0	0	
Copper	6	15	5	15	
Zinc	3	5	5	8	

*Tips are generally polyhedral and tool holders are provided with required slopes at the lip seals.

The side angle has considerable effect on cutting forces. Increase in the side rake angle decreases the wedge angle, and therefore concentrates the forces on the primary deformation zone, thereby reducing the applied force considerably. At the same time, with the decrease in wedge angle, the effective thermal conductivity is reduced and there is a tendency for the edge to crumble due to a decrease in the supporting metal.

Figure 9.13 shows a graph for both positive and negative rakes between cutting force and cutting speed for a carbide tool when 20% nickel steel is machined.

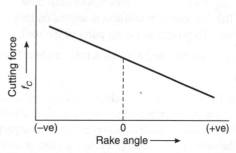

Fig. 9.13 Graph for both positive and negative rakes between cutting force and cutting speed for a carbide tool when 20% nickel steel is machined.

Machining with negative rake angles

Since the development of cemented carbides, machining with negative rakes have been widely used. Benefits of machining with negatives rakes have been recognized since the last 20 years. Its chief function is to strengthen end and side cutting edge and the finishing point of the tool.

Machining operations can be carried out with either side rake negative and back rake having any value, or back rake negative and side rake any value, or both negative.

The main purpose of negative side and back rake angles

These are respectively for:

Negative side rake angle:
- (i) It strengthens the side cutting edge;
- (ii) Tool life is improved by negative side rake angle;
- (iii) Negative side rake makes machining of tough material possible.

Negative back rake angle:
- (i) The negative back rake protects the finishing point of the tool because:
  - (a) It directs the cutting force in such a way as to place the front cutting edge under compression, rather than in shear or in tension (cemented carbide is strong in compression and weak in tension or shear).
  - (b) It increases the included angle or the end lip angle between the top and end relief faces.
  - (c) It creates a back pressure against the tool and reduces vibration in the machine tool.
- (ii) It uses the end cutting edge to be in compression rather than in shear or tension.

Amount of negative back rake angle

We cannot strictly define the limits of side and back rake angles which must be used. One can only state the principle that controls the size of this angle. This is, that, cemented carbides are good in compression rather than in tension or in shear. Hence the rake angles must be such that cutting edge can best cope with stresses and shocks that are subjected to, this is particularly important as regards cemented carbides.

To put side cutting edge area in compression, a negative side rake must be equal to or more than the side relief angles.

Amount of back rake angle

The amount of back rake angle is determined from the type of use. These are as follows:

- (i) To control the direction of chip flow.
- (ii) To maintain uniform diameter on long, slender shafts and strength surfaces on thin, fragile parts.
- (iii) To protect finishing point of the tool.

The recommended negative back rake angle is from 5° to 8° for general use.

Land

Studies have shown that the chips impinges on the face of cutting tool, not at the edge but some distance away from it. The distance from the cutting edge at which the chips impinge, however, varies according to the work material (in ductile materials this distance is larger) and the cutting variables. The magnitude of this distance is of the order of the feed given to the cutting tool. This observation has led to the development and use of cutting tools having a land on the face near the cutting edge.

The land in common tools used for cutting ductile materials like low carbon steel, is of the order of 0.25 mm. The provision of land increases in the mass of metal in the immediate vicinity of the cutting edge and this improves the cooling. The slope of the land varies from 0° to 5° negative.

Nose radius

The nose radius plays an important part in the machining of metals. It smoothens the feed marks on the machined surface, strengthens the cutting edge, and increases the effective conductivity of the cutting tool.

The effective feed is reduced in the curvilinear portion of the chip, and the length of the chip is increased. This increases effective heat conductivity. The longitudinal and the radial forces increase, making the use of cutting tools difficult on slender work. Chatter is also increased. In general, a nose radius of 1.5 mm is provided. On rigid work and on cast iron, a generous nose radius is used to advantage. Nose radius increases abrasion to an extent. In general, increase in nose radius improves tool life. Small nose radius results in excessive stress concentration and greater heat generation. In brief, large radius ensures high strength at the weakest zone of the tool (i.e. nose) and secondly, considerable improvement in surface finish is achieved. However, larger nose radius creates chatter.

Figure 9.14(a) indicates that there is an appreciable improvement in surface finish up to 0.3 mm nose radius. An increase beyond this does not bring about notable improvement on the finish; on the other hand it may lead to chatter in the machining process. The value of h which is a measure of surface finish alongwith feed rate f can be derived by referring to Fig. 9.14(b). Derivation for h:

From the geometry:

$$(2r - h)h = \frac{f}{2} \cdot \frac{f}{2}$$

$\Rightarrow \qquad 2rh - h^2 = \frac{f^2}{4} \qquad$ (ignoring h^2 as it is very small)

or
$$2rh = \frac{f^2}{4}$$

$$h = \frac{f^2}{8r}$$

where r is nose radius
        f is feed rate (mm)
        h is maximum depth of surface roughness.

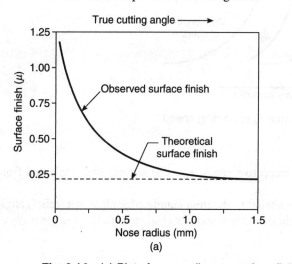
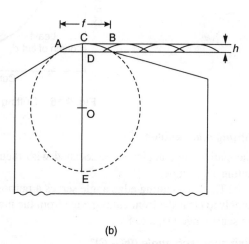

Fig. 9.14 (a) Plot of nose radius vs. surface finish and (b) Nose radius and surface finish.

Relief angles

The relief angle, also called *clearance angle*, is provided to keep the flanks from rubbing against the shoulder and machined surface, thereby eliminating unnecessary heat generation. The side relief angle is necessary to enable the cutting tool to be forced into the work.

The provision of relief angles reduce the effective mass of metal and thereby the effective conductivity of the cutting tool point. Hence relief angles should be as small as possible.

In normal practice this angle ranges from 5 to 8°. In some cases like tungsten carbide tipped tools, a secondary clearance angle of the order of about 10° is provided to reduce unnecessary grinding of the shank.

Cutting force and cutting speed

Refer to Fig. 9.15. It has been found that increase in cutting speed tends to decrease the cutting force. Further, lower feed rates result in higher values of cutting force for a fixed area of cut. With heavy feeds and lower depth of cut, the cutting force is lower.

Now,
$$\text{Area of cut} = fd$$
$$\text{Cutting force} = F_c$$
and
$$F_c = p(fd)$$
where p is specific cutting pressure which is high at low feed.

Symbolically,
$$\boxed{\begin{array}{c} f_1 < f_2 < f_3 \\ d_1 > d_2 > d_3 \end{array}}$$

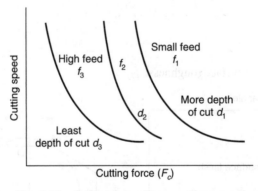

Fig. 9.15 Cutting force F_c vs. cutting speed.

Cutting edge angles

The cutting edge angles are associated with their respective edges, e.g. side cutting edge angle and front cutting edge angle.

The front cutting edge angle serves a purpose similar to the front cutting edge clearance (relief) angle namely to clear the front cutting edge from the machined surface and like clearance angles, it is generally of the same range viz. 5 to 8.

Plan approach angle (90 – C)°

In Fig. 9.16, *C* is side cutting edge angle, *E* is end cutting edge angle, *f* is feed in mm/revolution and H_{max} is maximum depth of groove. The side cutting edge angle, also called *plan approach angle*, serves an important

TOOL FAILURES AND TOOL LIFE

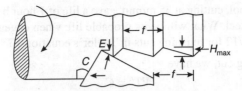

Fig. 9.16 Plan angles and surface finish.

purpose in cutting. Its influence, like rake angle, is complex. The effect of the plan approach angle on tool life is felt mainly due to direct influence on the undeformed chip thickness. Tools with smaller plan approach angle permit higher cutting speed. An increase in the side cutting edge angle decreases the chip thickness (uncut chip thickness is equal to feed × cos C, where C is the side cutting edge angle and the heat generated in the cutting tool. Simultaneously, it increases the chip width (uncut chip width is equal to depth of cut × l/cos a) and, therefore, increases the effective conductivity of the cutting tool. Consequently, either the cutting tool life increases or the cutting tool can be worked at higher speeds and or feeds.

A cutting tool with a large cutting edge angle introduces chatter at the cutting edge. Increased cutting edge angle also increases radial force which give rise to bending. Particularly in slender work, with cutting tools having a cutting edge angle up to 25°, chatter does not present any serious problem in rigid work.

End cutting angle

Change in end cutting angle has little effect on tool life. However, it is found that larger this angle, longer is the tool life. Similarly, larger is the side cutting angle, more is the tool life. But an angle more than 15° produces chipping and the tool life decreases.

From the geometry:

$$f = (H_{max} \tan C + H_{max} \cot E)$$

$$\therefore \quad H_{max} = \frac{f}{(\tan C + \cot E)}$$

True cutting angle

True cutting angle has a great influence on cutting speed. The relationship between true cutting angle and cutting speed is shown in Fig. 9.17 for different depths of cut and feeds for the same tool life.

As the cutting angle increases, the horse power required to machine a metal increases.

Effect of inclination angle

A positive value of the inclination angle of the main cutting edge strengthens the tool, and thus increases the tool life.

Effect of tool shank cross section

Tools with larger cross section often give higher speeds due to better heat conduction and rigidity and hence more tool life.

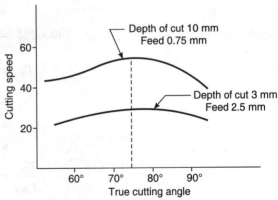

Fig. 9.17 True cutting angle vs. cutting speed.

EXAMPLE 9.1 A cutting tool, cutting at 35 m/min, gave a life of 1 hour between re-grinds when operating on roughing cuts with mild steel. What will be its probable life when engaged on light finishing cuts? Take n as 1/8 for roughening and 1/12 for finishing cuts in Taylor's equation $VT^n = C$.

Solution For the roughening cut, we have
$$VT^n = C$$
$$35 \times 60^{1/8} = C$$

For finishing cut, again
$$VT^n = C$$
$$35 \times T^{1/12} = C$$
$$35 \times T^{1/12} = 35 \times 60^{1/8}$$
$$T = 60^{12/8} = 60^{1.5} = 464.758 = 465 \text{ min}$$

EXAMPLE 9.2 If the relationship for HSS tools is $VT^n = C_1$ and for tungsten carbide tools $VT^{1/5} = C_2$, and assuming that at a speed of 30 m/min the tool life was 4 hour in each case, compare their cutting lives at 40 m/min.

Solution 4 hours = 240 min or $T = 240$ min

For HSS $30 \times T^{1/8} = C_1$
$$30 \times 240^{1/8} = C_1$$

and for WC $= 30 \times 240^{1/5} = C$

Now at cutting speed of $V = 40$ m/min
$$40 \times T^{1/8} = 30 \times 240^{1/8}$$
$$T_{\text{HSS}} = \left(\frac{30}{40}\right)^8 \times 240^{1/8 \times 8}$$
$$= (0.75)^8 \times 240$$
$$= 0.101 \times 240$$
$$= 24.214 \text{ min}$$

And
$$40 \times T_{\text{WC}}^{1/8} = 30 \times 240^{1/8}$$
$$T_{\text{WC}} = \left(\frac{30}{40}\right)^5 \times 240$$
$$= (0.75)^5 \times 240$$
$$= 0.23 \times 240$$
$$= 56.953 \text{ min}$$

$$\frac{T_{\text{WC}}}{T_{\text{HSS}}} = \frac{56.953}{24.214} = 2.352$$

EXAMPLE 9.3 A carbide tool with mild steel workpiece was found to have life of 3 hours while cutting at 60 m/min. Compute the tool life if the same tool is used at a speed 30% higher than the previous one. Also determine the value of cutting speed if the tool is required to have tool life of 4 hours. Assume Taylor's exponent $n = 0.3$.

Solution

$$VT^n = C$$

For initial values,

$$60 \times 180^{0.3} = C$$

When the speed is increased by 30%, new value of V is

$$V = 60 + 60 \times 0.3 = 60 + 18 = 78$$

Now put the value of $V_{in}VT^n = C$,

$$78 \times T^{0.3} = C$$

$$\left(\frac{T}{78}\right)^{0.3} = \left(\frac{60}{78}\right)$$

$$\left(\frac{T}{180}\right) = \left(\frac{60}{78}\right)^{3.33}$$

$$T = 180 \times \left(\frac{60}{78}\right)^{3.33}$$

$$= 180 \times (0.4769)^{3.33}$$
$$= 180 \times 0.4174 = 75.1348 \text{ min}$$

For tool life of 240 min

$$V \times (240)^{0.3} = 60 \times (180)^{0.3}$$

$$\left(\frac{240}{180}\right)^{0.3} = \left(\frac{60}{V}\right)$$

$$V = 60 \times \left(\frac{180}{240}\right)^{0.3}$$

$$= 60 \times \left(\frac{3}{4}\right)^{0.3}$$

$$= 60 \times 0.917 = 55.038 \text{ m/min}$$

EXAMPLE 9.4 When turning a 19 mm diameter bar on an automatic lathe employing tungsten carbide tools, the value of n is 1/5 and the value of V_{60} is 105.00 m/min. At what speed should the spindle run to give a tool life of 6 hours? If a length of 50 mm per component is machined and the feed used is 0.15 mm/rev, what is the cutting time per piece and how many pieces can be produced between one tool change?

Solution

$$V \times 360^{1.5} = 105 \times 60^{1/5}$$

$$V = 105 \times \left(\frac{60}{360}\right)^{1/5}$$

$$= \frac{105}{(180)^{0.2}} = 80 \text{ m/min}$$

$$\text{Spindle speed} = \frac{1000 \times 80}{\pi \times 19} = 1340 \text{ rpm}$$

$$\text{Cutting time per piece} = \frac{50 \times 60}{1340 \times 0.15} = 15 \text{ s}$$

$$\text{Numbers of components per tool change} = \frac{6 \times 60 \times 60}{15} = 1440$$

EXAMPLE 9.5 Find the percentage change in cutting speed required to give an 80% reduction in tool life (i.e. to reduce tool life to one-fifth of its former value), when the value of $n = 0.12$.

Solution
$$V_1 T_1^{0.12} = V_2 T_2^{0.12}$$

$$\frac{V_2}{V_1} = \left(\frac{T_1}{T_2}\right)^{0.12}$$

$$= \left(\frac{1}{1/5}\right)^{0.12}$$

$$= 5^{0.12} = 1.214$$

∴ Increase in cutting speed = 21.4%.

EXAMPLE 9.6 Under certain machining conditions the tool life equation is $VT^{0.2} = 600$. The time taken to change the tool is 10 minutes. Show that operating at a cutting speed of 90 m/min gives higher output than operating at either 120 m/min or 60 m/min, other cutting conditions remaining constant.

Solution One way of demonstrating the superiority of a cutting speed of 90 m/min is to determine the average cutting speed where the time spent in tool changing (during which no cutting occurs) is taken into account. The output will be highest where the average cutting is highest, because the remainder of the cutting conditions are constant.

$$\text{Tool life, } T = \left(\frac{600}{V}\right)^5$$

When $V = 60$, $T = 243$
When $V = 90$, $T = 32$
When $V = 120$, $T = 8$

When $V = 60$ m/min, average cutting speed $= \dfrac{243 \times 60}{243 + 10} = 57.63$ m/min

When $V = 90$ m/min, average cutting speed $= \dfrac{32 \times 90}{32 + 10} = 68.57$ m/min

When $V = 120$ m/min, average cutting speed $= \dfrac{8 \times 120}{8 + 10} = 53.33$ m/min.

On the basis of above discussion, a speed of 90 m/min is the best of the three cutting speeds quoted. However, on a cost basis the servicing of the tool every 32 minutes might be too expensive, and hence a speed somewhere between 60 and 90 m/min would probably be more profitable.

EXAMPLE 9.7 A 50 mm bar of steel was turned at 284 rpm and tool failure occurred in 10 minutes. The speed was changed to 232 rpm and the tool failed after 60 minutes of cutting time. What cutting speed should be used to obtain a 30-minute tool life.

Solution We know
$$VT^n = C \qquad (i)$$
where V is the speed in m/min;
T is the tool life in min;
n is the index whose value varies with machine variable;
C is a constant whose value depends upon machine variables and work material.

In the first case,
$$V_1 = \frac{\pi DN}{1000} = \frac{284 \times \pi D}{1000} = \frac{284 \times \pi \times 50}{1000} = 44.63 \text{ m/min}$$
$$T = 10 \text{ min}$$

Substituting these values in Eq. (i), we have
$$44.63 \times (10)^n = C \qquad (ii)$$

In second case,
$$V_2 = \frac{232 \times \pi D}{1000} = \frac{232 \times \pi \times 50}{1000} = 36.46 \text{ m/min}$$
$$T = 60 \text{ min}$$

Substituting these values in Eq. (i) again, we have
$$36.46(60)^n = C \qquad (iii)$$

Equating Eqs. (ii) and (iii), we have
$$44.63 \times 10^n = 36.46 \times 60^n$$
$$\left(\frac{60}{10}\right)^n = \frac{44.63}{36.46}$$
$$6^n = 1.22$$

or
$$n = 0.133$$

Now to obtain a life of 30 minutes, we have
$$VT^n = C \quad \text{or} \quad \frac{\pi DN}{1100}(30)^{0.133} = C$$

Comparing with Eq. (ii), we have
$$\frac{\pi DN}{1100} \times (30)^{0.133} = 44.63 \, (10)^{0.133} = \frac{\pi \times 50 \times N \times 1.47}{1100} = 44.63 \times 1.29$$
$$N = \frac{44.63 \times 1.29 \times 1100}{\pi \times 50 \times 1.47} = 273.85 \text{ rpm}$$

EXAMPLE 9.8 The following equation has been obtained when machining AISI 2340 steel with H.S.S cutting tools having a 8, 22, 6, 6, 15, 6, 0.117 mm tool signature. The work-tool system is governed by the following equation:
$$26.035 = VT^{0.13} f^{0.77} d^{0.37}$$

A 100 min tool life was obtained under cutting conditions—Velocity = 25 m/min; Feed = 0.3125 mm/rev; and Dia of bar = 2.50 mm.

Calculate the effect on the tool life for a 20% increase in the cutting speed, feed and depth of cut, taking each separately. Calculate the effect of a 20% increase in each of the above parameters taken together.

Solution We are given
Velocity $(V) = 25$ min
Feed $(f) = 0.3125$ mm/rev
Dia of bar $(d) = 2.50$ mm
For a tool life of 100 min.

Case I. V is increased by 20%; feed and diameter of the bar remain constant. Substituting the values, we have

$$26.035 = V(100)^{0.13} \qquad \text{(i)}$$

But
$$26.035 = 1.2V(T)^{0.13} \qquad \text{(ii)}$$

Rewriting Eqs. (i) and (ii), we have

$$\left(\frac{V}{1.2V}\right) = \left(\frac{100}{T}\right)^{0.13} = 1$$

or
$$T^{0.13} = \frac{(100)^{0.13}}{1.2}$$

$$T^{0.13} = \frac{1.82}{1.2}$$

$$T^{0.13} = 1.51$$

$$T = 24 \text{ min}$$

Case II. When feed is increased by 20%, V and d are constant; we have

$$26.035 = (100)^{0.13} \times (f)^{0.77}$$
$$26.035 = (T)^{0.13} \times (1.2 f)^{0.77}$$

or
$$1 = \frac{(100)^{0.13}}{T^{0.3}} \left(\frac{f}{1.2f}\right)^{0.77}$$

or
$$T^{0.13} = \frac{(100)^{0.13}}{(1.2)^{0.77}} = \frac{1.82}{1.15}$$

$$T^{0.13} = 1.58$$

or
$$T = 33.5 \text{ min}$$

Case III. When depth of cut is increased by 20%, V and feed rate are constant.

$$26.035 = (100)^{0.13} \times (d)^{0.37}$$
$$26.035 = (T)^{0.13} \times (1.2 \, d)^{0.37}$$

or
$$(T)^{0.13} = \frac{(100)^{0.13}}{(1.2)^{0.37}} = \frac{1.82}{1.011}$$

or
$$T^{0.13} = 1.7$$

or
$$T = 54.5 \text{ min}$$

Case IV. When V, f, d are all increased by 20%,

$$26.035 = V(100)^{0.13} f^{0.77} d^{0.37} \qquad \text{(iii)}$$
$$26.035 = 1.2V (T)^{0.13} (1.2f)^{0.77} (1.2 d)^{0.37} \qquad \text{(iv)}$$

Rewriting Eqs. (iii) and (iv), we have

$$(T)^{0.13} = \frac{(100)^{0.13}}{1.2 \times (1.2)^{0.77} \times (1.2)^{0.37}}$$

or

$$(T)^{0.13} = \frac{1.82}{1.2 \times 1.5 \times 1.07}$$

$$(T)^{0.13} = 1.23$$

$$T = 4.9 \text{ min}$$

REVIEW QUESTIONS

1. Define the term 'tool life'.
2. Discuss the effect of tool geometry and microstructure on tool life.
3. Explain how tool life curves are established.
4. What is Taylor's equation of tool life?
5. Discuss the factors affecting tool life.
6. Define the term tool failure. How will you find out whether a tool has failed?
7. Discuss the criterion of tool failure.
8. What are the types of tool failure?
9. Explain the mechanism of flank wear and crater wear.
10. Explain the following types of tool damage:
    (a) Flank wear
    (b) Crater wear
    (c) Chipping.
11. What are the main causes of tool wear?
12. How will you measure tool wear?
13. What is fretting corrosion and how it can be minimised?
14. Define adhesive wear. How it behaves? What is the function of specific energy in adhesive wear?
15. A tool life of 110 minutes is obtained at 25 mpm and 10 minutes at 65 mpm. What is the tool equation? Determine a cutting speed for tool life of 1 minute and 200 minutes. Also determine the tool life for a speed of 50 mpm and 80 mpm.
16. In an experiment on turning, the following data were obtained:

Cutting speed (mpm)	Feed (mm/rev)	Depth (mm)	Tool life (minutes)
90	0.10	2	100
90	0.10	2	40
120	0.10	2	60

Determine the tool life equation by computing the values of x, y and z in the relationship $C = VT^x f^y D^z$. Also compute the tool life at a cutting speed of 100 mpm and at a feed of 0.15 mm/rev. Take $C = 2.22$.

17. A carbide tool with mild steel workpiece was found to give tool life of 2 hours while cutting at 0.50 m/min. Compute the tool life when the same tool is used at a speed of 25% higher than the previous one. Also determine the value of the cutting speed if the tool is required to have a tool life of 3 hours. Assume Taylor's exponent n to be 0.27.
18. Wear life of a tool is reached when the component loses surface finish or size. Before the wear life, there could be several failures. Identify possible causes for following problems on carbide tooling:
    (a) Rapid flank wear
    (b) Rapid crater failure
    (c) Mechanical failure
    (d) Built-up edge
    (e) Comb cracks
    (f) Deformation.
19. A carbide tool with mild steel workpiece was found to give life of 2 hours while cutting at 0.50 mpm. Compute the tool life if the same tool is used at a speed of 25% higher than previous one. Also determine the value of cutting speed if the tool is required to have tool life of 3 hours. Assume Taylor's exponent $n = 0.27$.
20. A 50 mm dia-bar of steel was tuned at 300 rpm and tool failure occurred in 10 minutes. The speed was changed to 240 rpm and the tool failed on 60 minutes of cutting time. Assuming a straight line relationship exists, what cutting speed should be used to obtain a 30-minute tool life?
21. The following equation has been obtained when maching particular steel with HSS cutting tools having a 8, 22, 6, 6, 6, 15, 6, tool signature $20 = VT^{0.13} f^{0.77} d^{0.37}$. A 90-minute tool life was obtained using the cutting conditions—$V = 25$ mpm, $f = 0.25$ mm, $d = 2.5$ mm.

    Calculate the effect upon the tool life for 20% increase in the cutting speed, feed and depth of cut, taking each separately. Calculate the effect of a 20% increase in each of the above parameters taken together.

CHAPTER 10

Machinability

10.1 MACHINABILITY DEFINED

It may be defined as the property of a material that indicates the ease with which the material can be cut or marked. The most machinable material is one which permits the removal of material with a satisfactory finish at lowest cost. In other words, the most machinable material is one which will permit the fastest removal of the largest amount of material, per grind of tool, with satisfactory finish.

10.2 EVALUATION OF MACHINABILITY

Machinability can be evaluated by the following criterion:

Tool life/Tool wear or the cutting speed for a given tool: More the tool life, better will be machinability.

Specific power consumption: Lower the power consumption, better will be machinability.

Cutting force generated during machining: Lower the cutting force generated during machining, better will be machinability.

Quality of surface finish: Better the surface finish, better will be machinability.

Dimensional stability of the finished work: Harder the material, poorer will be machinability.

Shear angle: Greater the shear angle permitted, better will be machinability.

Heat generated during cutting: Lesser the heat generated, better will be machinability.

However from practical point of view, machinability is evaluated on the following criteria:

1. Tool life
2. Chip control
3. Power consumption
4. Surface finish.

10.2.1 Tool Life

The factor most commonly used is tool life. Cutting speed will give the predetermined tool life. A material with better machinability will give longer tool life or will allow higher cutting speed for the same tool life. Figure 10.1 shows the relative machinability with reference to cutting speed (tool life).

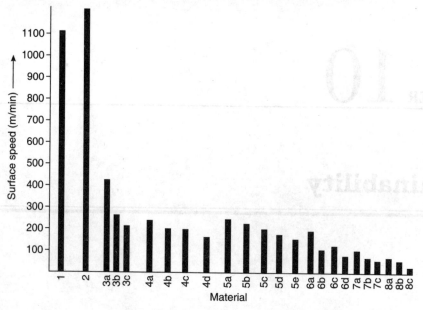

Fig. 10.1 Relative machinability.

The relative machinability of the following materials is shown in Fig. 10.1:

1. Aluminium alloys
2. Magnesium alloys
3. Copper alloys
   (a) Free machining
   (b) Brasses
   (c) Bronzes
4. Steel carbon
   (a) Free machining (Sulphur)
   (b) Flow carbon
   (c) Medium carbon
   (d) High carbon
5. Steel alloys
   (a) Free machining (Loaded)
   (b) Free machining (Sulphurised)
   (c) Goup 1
   (d) Group 2
   (e) Group 3
6. Stainless steel
   (a) Free machining
   (b) Ferrite
   (c) Martensitic
   (d) Austenitic
7. Titanium
   (a) Unalloyed
   (b) Alloy 5Al
   (c) Alloy 6Al; 4V
8. Super alloys
   (a) Nickel iron
   (b) Cobalt
   (c) Nickel

10.2.2 Chip Control

In certain materials and operations, chip control assumes greater importance than tool life. Ineffective chip control leads to various machining problems. For example, in copying or internal turning operations in low carbon steels, fully annealed non-ferrous material presents a serious problem of chip control. In such cases, machinability with respect to chip control is improved by cold working and/or heat treatment.

10.2.3 Power Consumption

Power consumption per cubic centimetre is another indicator of machinability. This factor assumes importance while machining very difficult materials, specially with lower-powered machines. Figure 10.2 illustrates relative power consumption for different materials and hardness values.

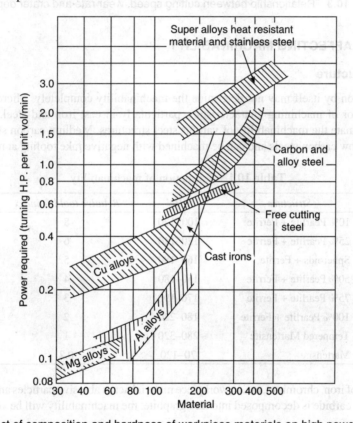

Fig. 10.2 Effect of composition and hardness of workpiece materials on high power requirements.

10.2.4 Surface Finish

Better surface finish, reduced wear rate and higher cutting speeds leading to more removal of material results in better machinability. Figure 10.3 shows the relationship between cutting speed, wear rate and crater depth.

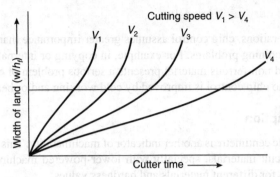

Fig. 10.3 Relationship between cutting speed, wear rate and crater depth.

10.3 FACTORS AFFECTING MACHINABILITY

10.3.1 Microstructure

Chemical composition by itself may not determine the machinability completely. Microstructure is a more fundamental indicator of machining characteristics, particularly in cast iron and steel. Table 10.1 can be used to roughly estimate the machinability of various steel structures. Medium carbon steels have the better machinability than low carbon steels and can be machined with negative rake tooling at moderate chip loads.

Table 10.1 Estimation of machinability

Structure	BHN	Relative tool life index
10% Pearlite + Ferrite	100–120	8
25% Pearlite + Ferrite	120–140	6
Spheroids + Ferrite	160–180	5
50% Pearlite + Ferrite	150–180	4
75% Pearlite + Ferrite	170–190	3
100% Pearlite + Ferrite	180–220	2
Tempered Martensite	280–320	1
Martensite	70–120	0.2

Free carbides of iron, chromium, etc. in workpiece material act as abrasive particles and reduce the tool life substantially. If iron carbide is decomposed into free graphite, the machinability will be very much improved.

10.3.2 Strength

Tensile strength and hardness of materials have generally been considered the index of machinability. Higher the strength of the material being machined, lower will be the machinability.

10.3.3 Strength at Elevated Temperature

If the workpiece material is such that it does not lose strength even at elevated temperature, tremendous cutting force is required to remove such material which shows its poor machinability characteristics.

10.3.4 Coefficient of Thermal Dispersion

Materials with good conductivity or high coefficient of thermal dispersion have good machinability, e.g. magnesium and aluminium alloys.

10.3.5 Built-up Edge Formation

The phenomenon of built-up edge is predominant in materials having large spread between yield strength and tensile strength. If an alloy has relatively low yield strength but high tensile strength or, in other words, it is of good ductility, it will be difficult to machine. Heat-resistant materials and stainless steel have a tendency for built-up edge formation; hence they have poor machinability.

10.3.6 Work Hardening

Certain materials get work-hardened when machined or drawn. Subsequent working becomes difficult unless they are annealed in between the operations. Stainless steels, super alloys and heat resistant materials fall in this category. However, this property of work-hardening is made use of in improving the machinability of low carbon steel and some nonferrous materials, which may not give effective chip control without work-hardening by cold drawing, and other methods.

10.3.7 Effect of Alloying Elements

All the alloying elements that are added to steel affect the tool life adversely. Table 10.2 shows the effect of various alloying elements on properties of material and also on the machinability. All the iron base alloys are comparatively easily machinable than the cobalt and nickel alloys.

Table 10.2 Effect of alloying elements

Alloying element	Hardness	Strength	Yield point	Elongation	Elasticity	High temp. stability	Machinability Y
Silicon	+	+	++	−	+++	+	−
Manganese in pearlitic steels	+	+	+	+	+	=	−
Manganese in Austenitic steels	−	+	−	+++	?	?	−
Chromium	+	++	++	−	+	+	−
Nickel in pearlitic steel	−	+	−	+++	?	+++	−
Nickel in Austenitic steel	−	+	−	+++	?	+++	−
Aluminium	?	?	?	?	?	?	−
Tungsten	+	+	+	+	?	+++	−
Vanadium	+	+	+	=	+	++	−
Cobalt	+	+	+	−	=	++	=
Molybdenum	+	+	+	−	?	++	−
Copper	+	+	+	=	?	+	=
Sulphur	?	?	?	−	?	?	++
Phosphorus	+	+	+	+	?	?	+

+ Increase, − Decrease, ? Not known, = Insignificant change

10.3.8 Heat Treatment

This improves micro constituents and as such machinability.

10.3.9 Tool Geometry

Its various aspects are dealt with in the chapter on tool life.

10.3.10 Cutting Fluid

Use of cutting fluid takes away heat generated during cutting operation and as such improves machinability.

10.3.11 Machine Tool, Tool and Work Factors

The rigidity of the system improves machinability.

10.4 MACHINABILITY INDEX

It is defined as a ratio of cutting speed of metal investigated for 20 minutes of tool life to cutting speed of a standard steel for 20 minutes of tool life.

$$\text{Machinability index \%} = \frac{\text{Cutting speed of metal investigated for 20 minutes tool life}}{\text{Cutting speed of standard steel for 20 minutes tool life}}$$

A free cutting steel, which is machined relatively easily, and the machinability index of which is arbitrary and fixed as 100%, is considered a standard tool. This steel has carbon content of 0.13 maximum, manganese of 0.06 to 1.1 and sulphur 0.08 to 0.03%.

The use of machinability index is made in selecting tool material, coolant, and in deciding other cutting parameters such as feed, depth of cut, speed, etc.

REVIEW QUESTIONS

1. Define the term machinability. How is it evaluated?
2. What are the four practical criteria of machinability?
3. What are the various factors which affect machinability?
4. What is a machinability index? What is its use?
5. Write a short note on machinability index.

CHAPTER 11

Economics of Metal Machining

11.1 COST ANALYSIS AND ECONOMICS

The goal of manufacturing is to produce items at lowest cost and maximum production rate. This can only be obtained by setting a suitable combination of machining parameters, leading to a most economical procedure of working.

In a machining operation, there are three elements of costs of production. These are:

C_m (Machine Costs): These include interest, depreciation, maintenance, etc. of the machine tool; wages, electricity, building rental, supervision and other administrative costs; but tool costs are excluded.

C_t (Tool Costs): These include purchase, grinding and cutting edge change costs.

C_f (Fixed Costs): These include "unproductive" or idle costs such as job change (handing), tool change (between operations), etc.

These are related as follows:

$$C_p = C_m + C_t + C_f \qquad (11.1)$$

where C_p is the total cost of production per piece.

Now, if C_m is the machine cost per minute in rupees;
C_t is the cost per cutting edge in rupees;
T_i is the idle time per piece in minutes;
T_m is the cutting time per piece in minutes, and
N is the number of pieces produced per cutting edge

Equation (11.1) can be re-written as

$$C_p = T_m + \frac{C_t}{N} + C_m \cdot T_i \tag{11.2}$$

where   Cost of machining per piece $= C_m \times T_m$ (11.3)

$$\text{Cost of tool per piece} = \frac{C_t}{N} \tag{11.4}$$

and $C_f = C_m \times T_i$ (11.5)

The non-productive costs are established by methods analysis and time study, while the tool costs are a function of the tool life selected and the cost of changing the tool. If a solid or brazed type tool is used, the cost of grinding must be included in the tool cost. The machining costs are dependent on the cutting speed and tool life. The overhead costs are generally calculated as a percentage of the direct labour cost. It is not possible to have maximum metal removeable as well as minimum cost, since the number of pieces produced between two grinds or tool changes are reduced as the metal removal rate is increased.

The general expression for the unit cost per piece on an operation, and equations for calculating the cutting speed and tool life for minimum cost or maximum production rate, are developed as given below. Before proceeding for this, the time C_p can further be split up as follows:

$$C_p = R_1(\text{loading and unloading time}) + R_2(\text{machine setting + gauging time})$$
$$+ R_3(\text{machining time}) + R_4(\text{tool changing time})$$

where R_1, R_2, R_3 and R_4 are different hourly rates in rupees.

The next step is to develop an equation for cost of machining in speed and feed rate values.

Let $V =$ velocity in m/min
  $D =$ diameter in m
  $f =$ feed rate in mm/rev
  $L =$ length of job in m.

Then $VT^n = C_t$

or $T = \left(\dfrac{C_t}{V}\right)^{1/n}$

Then

$C_p =$ Machining cost + Cost of tool regrind + Tool changing time cost + Idle time

or $C_p = C_n \times T_m + C_{tgr} \times T_{gr} + C_{tc} \times T_c + C_f \times T_i$

where C_n is the machining cost
  C_{tgr} is the grinding cost per cutting edge in rupees;
  T_{gr} is the grinding time per piece in minutes;
  C_{tc} is the changing time per cutting edge in rupees; and
  T_c is the changing time per piece in minutes.

Total time of machining or producing that job is given by

$$T_{\text{total}} = T_m + T_{gr} + T_c + T_i$$

where

$$T_m = \frac{\pi DL}{1000 fV}$$

$$\therefore \quad C_p = C_m \times \frac{\pi DL}{1000 fV} \cdot \frac{1}{\left(\frac{C_i}{N}\right)^n} + \text{Idle cost}$$

Substituting
$$k = \frac{\pi DL}{1000 f}$$

Then
$$C_p = \frac{kC_m}{V} + C_{tgr}\frac{k}{V}\frac{V^{1/n}}{C^{1/n}} + C_{tc}\frac{k}{V}\frac{V^{1/n}}{C^{1/n}} + \text{Idle cost}$$

$$= C_m\frac{k}{V} + C_{tgr}\frac{k}{C^{1/n}}V^{\frac{1}{n}-1} + C_{tc}\frac{k}{C^{1/n}}V^{\frac{1}{n}-1} + \text{Idle cost}$$

From this equation it is evident that C_p is a function of the velocity of cutting. Hence, for optimum cost of production we need to find the optimum cutting speed. Therefore, the optimum C_p can be found from

$$\frac{\partial C_p}{\partial v} = 0$$

$$\frac{d(C_p)}{dv} = k \cdot \left[C_m V^{-2}(-1) + C_{tc}\frac{\left(\frac{1}{n}-1\right)V^{\left(\frac{1}{n}-2\right)}}{C^{1/n}} + \frac{C_{tgr}\left(\frac{1}{n}-1\right)V^{\left(\frac{1}{n}-2\right)}}{C^{1/n}} + 0 \right]$$

or
$$= k\left[C_m V^2 + \frac{\left(\frac{1}{n}-1\right)}{C^{1/n}} \cdot V^{\left(\frac{1}{n}-2\right)} C_{tc} + \frac{\left(\frac{1}{n}-1\right)V^{\left(\frac{1}{n}-2\right)}}{C^{1/n}} C_{tgr} \right]$$

$$C_{tm} \times V^{-2} = \frac{\left(\frac{1}{n}-1\right)}{C^{1/n}}\left[V^{\left(\frac{1}{n}-2\right)} C_{tc} + C_{tgr} V^{\left(\frac{1}{n}-2\right)} \right]$$

$$C_{tm} \times C^{1/n} = V^{1/n}\left(\frac{1}{n}-1\right)(C_{tc} + C_{tgr})$$

$$C^{1/n} = V^{1/n}\left(\frac{1}{n}-1\right)\left(\frac{C_{tc} + C_{tgr}}{C_{tm}}\right)$$

$$V\left[\left(\frac{1}{n}-1\right)\left(\frac{C_{tc} + C_{tgr}}{C_{tm}}\right)\right]^n = C$$

Thus, the term in square bracket is $\boxed{T_{opt} = \left(\frac{1}{n}-1\right)\left(\frac{C_{tc} + C_{tgr}}{C_{tm}}\right)}$

where n is Taylor's toll life index which is

$\quad n = 0.125 \quad$ (for HSS tools)
$\quad \quad = 0.25 \quad$ (for Carbide tip tools)
$\quad \quad = 0.5 \quad$ (for Caramic tools)

If optimum tool life T_{opt} is calculated from n and cost data it is possible to obtain optimum cutting speed V_{opt} from Taylor's tool life equation:

$$V_{opt} T_{opt}^n = C$$

$\therefore$
$$V_{opt} = \frac{C}{T_{opt}^n}$$

The various time and cost elements identified in economics of machining are as under:

Loading and unloading time: This can be reduced by the use of properly designed fixtures, presetting the workpiece on pallets, automatic loading, quick acting chucks—especially demand quick loading facilities for full utilization of their advantages and for minimization of the effect of high investment.

Machine handling time: This can be reduced by such means as improved location stops, power feed shifts, multiple tool holders, etc. or may be practically eliminated by numerical tape control or by automated machine design.

Tool changing time (T_c): We find from the equation of cost of machining that the economy of machining system depends upon the above factor and it should be as minimum as possible. The machine is idle during the tool changing time. To study the tool changing time in greater depth, it could be broken down as follows:

$$T_c = T_w + T_k + T_h \qquad (11.6)$$

where T_w is the time taken for replacing the tool or indexing the insert
T_k is the time taken for resetting the job dimensions
T_h is the time taken for changing the tool holder or replacing the parts.

Tool Cost (C_t): In order to calculate the cost of the cutting edge for one tool life for a clamp tool holder, we have to consider the cost of insert as well as the cost of spare parts and the tool holder per cutting edge. For a brazed tool, the total cost of tool, the number of regrinds permissible and the cost of each regrinding will have to be considered.

To calculate the cost of one cutting edge (C_t) in a clamp tool holder system, the following equation can be used:

$$C_t = \frac{\text{Insert price}}{\text{Number of usable cutting edges}} + \frac{\text{Tool holder price}}{\text{Tool holder life in number of cutting edges}}$$

$$+ \frac{\text{Price of spare parts}}{\text{Life of spare parts in terms of number of cutting edges}}$$

$$+ \frac{\text{Chip breaker price}}{\text{Life of chip breaker in terms of number of cutting edges}}$$

i.e.,
$$C_t = \frac{C_P}{N_{tp}} + \frac{C_{th}}{N_{th}} + \frac{C_s}{N_{ts}} + \frac{C_{cb}}{N_{tcb}}$$

Sometimes, all the cutting edges of an insert cannot be used due to breakages. Specially in negative rake inserts, if one cutting edge breaks, the cutting edge immediately below may also get damaged.

As such, the total usable number of cutting edges in one insert can be calculated from

$$N_{tp} = n_s \cdot S$$

when n_s is the number of cutting edges per insert
S is the insert utilisation index.

In case of brazed tools, the following equation can be used:

$$C_t = C_{bt} - C_s + n_{rg} \times \frac{C_{rg}}{n_{rg} + 1}$$

where C_{bt} is the cost of brazed tool
C_s is the cost of scrap after tool has been used
n_{rg} is the number of regrindings per tool
C_{rg} is the cost of one regrinding.

Machine Cost (C_m):   Machine cost can be calculated as follows:

$$C_m = \frac{C_l + C_{mh} + C_o}{60}$$

where C_l is the labour cost in rupees/hour
C_{mh} is the machine hour cost in rupees/hour
C_o is the overhead cost of the plant in rupees/hour.

Again, C_{mh} can be calculated as follows:

$$C_{mh} = \frac{C_{mc} + C_i + C_e + C_r}{T}$$

where C_{mc} is the machine cost in rupees/year depending upon the method of depreciation
C_i is the cost of interest in rupees/year
C_e is the cost of energy for airconditioner and other machines in rupees/year.
C_r is the rent in rupees/year or cost of place
T is the average number of hours of machine utilisation per year (Generally 0.7 × number of working hours).

From the above discussions it will be easier to find out the optimum tool life (t_{opt}) by using the equation $\left(\frac{1}{n} - 1\right) t_c$, that governs the optimal solution for the machining and maximum productivity.

A production manager has to decide about the combinations of speeds and feeds so as to achieve t_{opt}. We can analyse the importance of tool change time that governs the productivity.

In view of the optimality of cost of machining and the times available with the production manager he has to further decide, in the light of the work material, the machine tool to be used and the tool material inventory in hand, the most economical proposition to use a brazed carbide tool, an indexable insert system, a HSS tool or any other tool available in stock. It is therefore necessary for the production manager to decide to use one of them. Let us try to analyse each one of them.

11.2 INDEXABLE INSERT SYSTEM

Let us assume that a tool holder costs ₹ 80.00 and lasts for 400 cutting edges, after which it may be written off. The cost per cutting edge comes out to be ₹ 0.20. However, the assumed value of 400 hours is too low, and the tool holder may last much longer than this. This price per insert can be taken as ₹ 8.00 and the price per cutting edge is, therefore, ₹ 1.

With a machine cost of ₹ 25 per hour the tool cost, if the tool change time is assumed to be equal to 1 minute, comes out as ₹ 0.42. As there is no grinding cost, the total cost per cutting edge in rupees will be

$$C_t = 1.20 + 0 + 0.42 = 1.62$$

11.3 BRAZED CARBIDE TOOL

It is necessary to analyse the economy of these tools for the sake of comparison. This can be done by taking up an example given below.

Suppose that the cost per cutting edge consists of the following:

Purchase cost: Brazed tool costs about ₹ 23.50 and the tool could be ground about 10 times; thus cutting is done 11 times and this gives a cost of ₹ 2.14 per cutting edge.

Grinding cost: If the tool has not worn excessively and tool grinding shop is well organised, then regrind of blazed tool involves the following operations:

  (i) Grinding of steel shank at two sides;
  (ii) Rough and finish grinding of two sides of the carbide tip;
  (iii) Nose radius forming;
  (iv) Lapping with diamond wheel and edge honing.

It may take about 13 minutes on a machine, the hourly operation cost of which is about ₹ 13. This gives a grinding cost of ₹ 2.36 for 10 regrinds spread over 11 cutting edges, but this may vary from ₹ 2.36 to ₹ 7.00 per grind.

Tool changing cost: With a reasonable resetting time of 5 minutes for a brazed tool of an hourly rate of machine of ₹ 25, the cost per cutting edge works out to be ₹ 2.08.

Hence total cost per brazed tool is ₹ (2.14 + 2.36 + 2.08) = ₹ 6.58.

11.4 VARIATION OF COSTS WITH SPEED FOR DIFFERENT TOOLS

Figure 11.1 shows the cost of a piece in rupees against cutting speed for HSS, carbide-brazed and carbide-indexable tools. The variation of the three costs, the fixed costs C_f, the machine costs C_m and the tool costs C_t is independently shown. If we study the C_p (production cost) equation, i.e.

$$C_p = C_m + C_t + C_f$$

we find that if a HSS or carbide-brazed tool is being replaced by carbide-indexable insert tools, the production cost per piece is reduced only to the extent of the reduction of C_t (the tool costs) or T_c (tool changing cost), as C_m and C_f will remain unaltered.

This is a case when we are using the carbide tools at a much lower rate than their rated value. Hence, if we want to use them economically, we must use them faster, for which a separate graph is to be plotted.

A cost comparison is also shown for ceramic tools on a separate cost-speed graph in Fig. 11.2. That gives an idea of how the costs change as speed changes.

If we change a machine, then the machine cost will change. However, fixed costs may remain the same, as shown in Fig. 11.3.

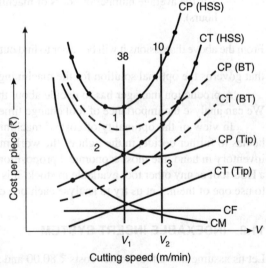

Fig. 11.1 Cost comparison for HSS, carbide-brazed and indexable tools.

Thus, the economics of machining has been tried to be explained here by comparing different tools.

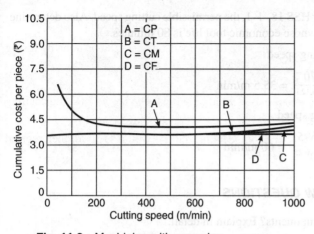

Fig. 11.2 Machining with ceramics—Cost variation with speed.

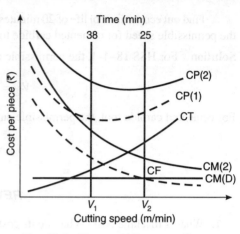

Fig. 11.3 How machine cost affects production costs.

EXAMPLE 11.1 For certain machining conditions the tool cost of operating the machine is ₹ 25 per hour and the total cost of a tool change is ₹ 15. If the depth and the rate of cutting are constant, cutting velocity employed is 36.5 m/min, and $n = 0.74$, find the economic cutting speed.

Solution
$$\frac{25}{60} \times T \times \frac{1}{15} = \frac{0.86}{0.14}$$

where $T = 221$ min (economic tool life).

Since $V_{60} = 36.5$ m/min for the cut employed,

$$V_{221} \times 221^{0.14} = 36.5 \times 60^{0.14}$$

$$V_{221} = 36.5 \left(\frac{60}{221}\right)^{0.14}$$

EXAMPLE 11.2 The following observations are made in an orthogonal cutting operation, on a centre lathe used in manufacturing of a HL 36 component shaft, used in switching off mechanism in high power control system of power transmission:

Material cut SAE 1050 (approx. 0.50% plain carbon) 201 KB, hot rolled.

Dimension of cut: $d = 2.5$ mm
$f = 0.25$ mm/rev.

Cutting fluid is soluble oil of $\frac{1}{15}$ ratio.

Tool material	Tool life equation
HSS 18–4–1 type	$VT^{0.125} = 170$
HSS 18–4–1, 5% cobalt	$VT^{0.135} = 186$
Cast non-ferrous alloy	$VT^{0.15} = 220$
Cemented (sintered) carbide type	$VT^{0.25} = 595$

Find out economic tool life of 20 minutes, for HSS 18–4–1, the permissible cutting speed. Also determine the permissible speed for cemented carbide tool, whose economic tool life is 80 minutes.

Solution   For HSS 18–4–1, the permissible cutting speed

$$V_{20} = \frac{170}{20^{0.125}} = 35.5 \text{ m/min}$$

For cemented carbide tool, the permissible cutting speed

$$V_{80} = \frac{595}{80^{0.25}} = 60 \text{ m/min}$$

REVIEW QUESTIONS

1. What is machine cost? What are its cost components? Explain in detail.
2. Compare the economics of indexable and brazed carbide tools. Give your recommendations for their uses with machine-work continuation.
3. Where is the function of time element in production cost? What are the various time elements and how can time be reduced? Explain.
4. How would we proceed to calculate tool cost per cutting edge in a clamp tool holder system? Explain.

Chapter 12

Metal Cutting and Metal Working Fluids

12.1 DEFINITION OF CUTTING FLUIDS

Cutting fluids, sometimes loosely referred to as lubricants or coolants, are liquids and gases applied to the tool and workpiece to assist and improve machining operations.

12.2 FUNCTIONS OF CUTTING FLUIDS

Cutting fluids are used for the following purposes or functions:

To act as coolant: It cools the cutting tool and workpiece. The heat produced is carried away by the fluid by supplying adequate quantity of cutting fluid. This makes more accurate production and measurement possible.

To act as lubricant: It lubricates the cutting tool and thus reduces the coefficient of friction between the chip and tool. This increases tool life.

To improve surface finish: The use of a cutting fluid results in better surface finish.

To act as cleaning agent: It washes away the chips from the tool, and keeps the cutting region free.

To act as chip breaker: It decreases adhesion between chip and tool and acts as a chip breaker and does not allow to form built-up edge.

To act as protector: It protects the newly generated surfaces from oxidation and corrosion.

To avoid expansion of workpiece: It prevents the expansion of workpieces by removing the heat from workpieces.

To improve machinability: It improves machinability and reduces machining forces.

To improve cutting speed: It allows maximum possible cutting speed to be used, thus reducing time and cost of production.

12.3 ADVANTAGES OF CUTTING FLUID APPLICATIONS

The use of cutting fluids results in the following advantages:

Increased tool life: The application of cutting fluid reduces tool wear rate, because of cooling and lubricating functions, and as such increases tool life.

Reduced tool cost: Longer tool life means less frequent recharging, resetting and regrinding of the tool; thus reduced overall tool cost and regrinding lost is achieved.

Increased production rate: Higher cutting speeds and feeds could be given because of higher heat dissipation rate. Hence increased production rate can be achieved.

Reduced power consumption: Reduction in overall power cost, since power requirement for machining reduces due to reduction in frictional forces.

Better surface finish: Surface finish of the work improves. A worn-out cutting tool produces irregular surfaces. Since use of cutting fluid avoids/reduces wearing out of tool, hence better surface finish is achieved.

Improved quality: Dimensional accuracy of the job is increased due to its protecting function.

Reduced overhead: Due to reduced tool cost and power consumption and increased production rate, the overhead of the plant gets reduced and cost per unit of production comes out less.

No change in physical properties: It does not allow the change in physical properties of the tool or work, because the temperature rise is not much.

12.4 CHARACTERISTICS OF GOOD CUTTING FLUID

A cutting fluid should have the following characteristics to fulfill its functions successfully:

Lubricating qualities: This quality prevents chip from touching and adhering to the tool face. Thus frictional force decreases and built-up edge formation is also stopped.

Cooling qualities: It should remove more heat quickly and cool down the tool and work. This will reduce tool wear and increase tool life and surface finish.

Rust resistance: The cutting fluid should prevent rusting or corrosion of the work or machine. It should not cause stain.

Low viscosity: It should have low viscosity so that chip and dirt can settle quickly.

Non-toxic: It should be non-toxic and should not be injurious to the skin.

Stability: It should have long life. It should not get spoiled quickly both in use and in storage.

Resistance to rancidity: It should not become rancid equally before and after in use.

Transparent: It should be transparent so that the operator can see the work while machining.

Nonflammable: It should have high flash point.

Smoke-free: It should not smoke or foam easily.

Small molecular size: It should have small molecular size so that diffusion and penetration to the chip tool interface can be easily achieved.

Odour-free: It should be free from undesirable odours.

Stable and non-volatile: It should be stable and should not volatise easily.

12.5 THEORY OF CUTTING FLUID

The main functions of the metal working fluids are cooling, lubricating and flushing away the chips.

By flooding over a tool, chip and the workpiece, a cutting fluid can remove heat and thus reduce the temperature in cutting zone. This reduction in temperature results in decrease in tool wear and an increase in tool life, because with the reduction in temperature, the rate of diffusion of constituents reduces and resistance to abrasion increases. The cooling action also results in small improvement in surface finish, increase in chip curl, and reduction in built-up edge formation.

Merchant has developed a theory to explain the penetration of cutting fluid. He assumed, based on photo-micrographic evidence that minute capillaries exist at the tool–chip interface, as shown in Fig. 12.1. The cutting fluid penetrates into the capillaries and reacts chemically with the clean surface of the chip under high pressure and temperature, and thus produces a soft film of a chemical compound between the chip and tool. This should prevent metal-to-metal contact, and reduce the friction and the formation of built-up edge.

The theory of cutting fluid, as suggested by Merchant, is valid only at low speeds of machining and never at higher speeds of the machining. At higher speeds, the cutting fluid acts only as a coolant and reduces the tendency of chip welding to the tool face.

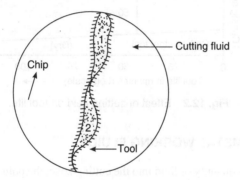

Fig. 12.1 Tool–Chip interface.

12.6 ACTION OF CUTTING FLUIDS AS LUBRICANT

It can be seen that, under the conditions of high temperature and pressure, and with the geometry and directions of moment involved in the metal-cutting action, full-film hydrodynamic lubrication cannot be expected and thus metal-to-metal contact occurs. In normal frictional circumstances, resistance to motion under these conditions arises from a combination of shearing at the contacting asperities and viscous shearing of any lubricating fluid which may be present. This condition is known as *boundary lubrication*. A lubricant boundary-layer reduces the area of intimate metallic contact between the two surfaces, and from general experience in frictional experiments, it is known that fatty acids are effective boundary lubricants: they can reduce friction even when present as non-molecular layers. The melting points of the fatty acids, or of the metallic soaps which they form by reaction with the metal, may be significant. No organic compound can be expected to lubricate satisfactorily at temperatures in excess of 200°C, and most fail at 100–150°C. Lubricants which contain sulphur, phosphorous or chlorine EP additives can afford greater surface protection during sliding, by the formation of a solid lubricant layer formed by chemical reaction between the lubricant and the metal surfaces. This remains effective at temperatures up to the melting point of the metallic sulphides, chlorides, etc.

12.7 CUTTING FLUIDS AND TOOL LIFE

The application of cutting fluid during the machining operation results in increase in tool life. Carbon steel tools that have less heat resistance have maximum increase in tool life (nearly 50%). The increase in tool life is nearly 25% in case of high-speed steel tools. The effect of cutting fluid on carbide tool is not known very well. Carbide tools may show slight improvement in tool life, but may also be damaged by thermal chipping. Figure 12.2 shows the effect of cutting fluid on tool life for different tool materials.

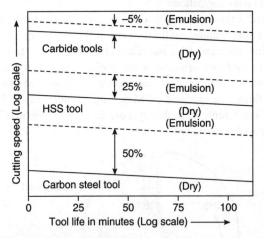

Fig. 12.2 Effect of cutting fluid on tool life.

12.8 APPLICATION OF METAL WORKING FLUIDS

It is only by forcing a sufficient quantity of fluid into the cutting zone, the point of contact between cutting tool and workpiece, that effective operation can be assured. In single-point turning, for example, tool life can be shown to improve as the quantity of cutting fluid is increased up to 15 litres/min. However, the proper delivery of an ample flow directed through suitable nozzles to the point of cutting is often neglected by machine operation sometimes, it must be admitted, because of the lack of adequate provision for splash guards.

On self-contained machines the cutting area drains down over various parts of the machine, often to a collecting pan, and thence to a sump or tank as shown in Fig. 12.3, which may be either separate or form part

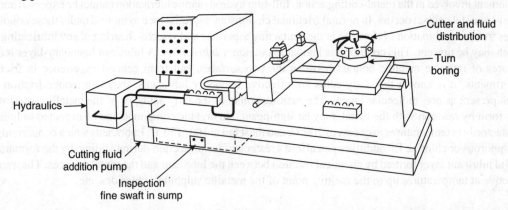

Fig. 12.3 Automatic turret-type lathe. (Cutting fluid tank indicated with dotted lines.)

of the machine base. The volume of the tank must be sufficient to allow time for cooling and for the setting of fine swarf an may be from 25 to 250 litres or more, depending on the type of machine. There is usually a coarse strainer on top of the collecting pan to prevent the larger swarf from entering the tank, and a strainer at the pump suction, but many machines have no other filtering devices. Important exceptions are grinding, honing, lapping and deep hole boring machines, where high quality of work depends upon removing the finer swarf and abrasive particles. Occasions can arise when the inclusion of filtration equipment on other machines can avoid the gross contamination and overloading of the coolant with metallic particles and help keep the fluid clean. Figure 12.4 shows the typical method of application of cutting fluids.

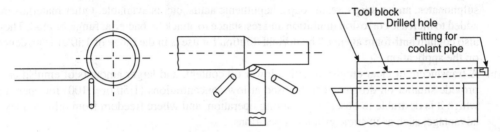

Fig. 12.4 Method of application of cutting fluids.

12.9 TYPE OF CUTTING FLUIDS

Cutting fluids sometimes are referred as coolant, cutting oil or lubricant. These can be in the form of either solid, liquid gas or mist.

12.9.1 Solid Cutting Fluids

Solids may either be included in the work material itself or be applied on the chip tool interface with some liquid (in which it remains in suspension), mainly to facilitate machining by reducing friction. They have little ability to remove heat. For example the solids may be graphite, molybdenum, disulphide etc. Stick waxes and bar soaps are also used as lubricants, specially in tapping and sawing operation. In grinding these are used as paste. Lead and copper are applied as coating on steel wire in drawing process as lubricant.

12.9.2 Liquid Cutting Fluids

These fluids are:

Plain water

Water as a metal working fluid was first used by Taylor. It shows improvement in tool life and mainly works as a coolant. It has inherent disadvantage of causing rusting of work and machine components.

Water with rust inhibitor

The addition of rust inhibitor in water suppresses the rusting affinity of water with ferrous metals; however, its inherent defect can not be completely eliminated. Hence its use is not recommended for important machining operations on costly machines.

Water emulsion and soluble oils

An emulsion is a suspension of oil droplets in water made by blending the oil with emulsifying agents and other materials. The addition of animal or vegetable fats, oils, or other esters produces super-fatted emulsions of greater lubricating value.

Water-miscible fluids form mixtures ranging from emulsions to solutions. These fluids are characterised to possess high specific heat, high thermal conductivity, and high heat vaporisation. These are understood to be the most effective cooling and lubricating fluids in metal working industry today.

Soluble oils, sometimes called 'coolants', 'suds' or 'mystics' in workshop terminology, are emulsions—suspensions of oil droplets in water maintained by the presence of emulsifying agents. The most common soluble-oil cutting fluids have droplets large enough to reflect almost all incident light (0.005–0.02 mm), and therefore appear opaque or milky. Usually a range of soluble oils is available as follows:

(a) A general-purpose emulsified oil simply based on a mineral oil with an emulsifier of petroleum sulphonates, amine soaps, resin soaps, haphthenic acids, etc. is available. Other materials may be added to increase corrosion inhibition and resistance to attack by bacteria, fungi, or mud. They may also contain anti-foam agents. The normal dilution for use is in the range of 1:l0 to 1:50, depending on the application.

(b) A translucent or clear emulsion with a lower oil content, and larger amounts of emulsifier which provide smaller oil droplets, can be used at low concentrations (1:50 or 1:100) for operations in which it is desirable to observe the cutting operation, and where freedom from oiliness may be an advantage, such as in some grinding operations.

Advantages and field of application of soluble oils: Soluble oil emulsions usually produce less fumes and possess more cooling power than do straight oils. Soluble oils undoubtedly play a leading part in grinding, turning, milling, drilling, sawing, and other operations, where cooling is primary requirement.

Limitations of soluble oils: Although one of the most important functions of a cutting fluid is to limit the temperature of the tool and work, soluble oils cannot in every instance meet all related conditions better than straight cutting oils. In many operations, the use of the most effective lubricant possible is often of paramount consideration, i.e. when it is not only necessary to reduce the temperature of the tool and stock and so minimise distortion, but also to facilitate the formation of the chip and lubricate the moving part of the machine. High-speed machining operations require quick heat dissipation and soluble oils are used to advantage, provided limiting factors do not prevent such use. Limiting factors might be the tendency of soluble oils to rust or corrode the parts of the machine, or to mix with oils required to lubricate machine bearings.

Qualities desired in soluble oils: These are as follows:
  (i) It should form a stable and permanent emulsion readily at ordinary temperatures with a variety of natural waters.
  (ii) It should not deteriorate in storage.
  (iii) It should not block the moving parts of the machine.
  (iv) It should not foam easily.
  (v) It should have wetting properties for the metal being machined.
  (vi) It must be harmless to the operator.

Troubles in use of soluble oils: When trouble is encountered with the use of soluble oils, it is usually traceable to one of the three following causes:
  (i) The hardness of the water
  (ii) The acidity in the system, or
  (iii) The inversion of the emulsion.

The latter is caused by a gradual loss of water through evaporation from the emulsion, thereby producing a mixture so rich in oil that the emulsion changes from the usual oil-in-water phase to the undesirable water-in-oil type.

Inverted emulsion. The inverted emulsion is skimy and may have a disagreeable odour. Nevertheless, inversion rarely occurs when the emulsion contains ten parts or more of water. Consequently, the percentage of water in soluble oil emulsions should be checked occasionally. This is usually done by breaking up a sample of the emulsion with concentrated hydrochloride or sulphuric acid in a graduated cylinder and calculating from the separated oil the amount of water necessary to make up for losses.

Most suitable oil emulsify readily with hard water. When it is necessary to soften the water, it may be done by adding tri-sodium phosphate, using not more than 1.5% by weight of this alkali. As a rule, less than 1% is sufficient, except for extremely hard water. Other alkalis such as soda-ash, borax, etc. may also be used in quantities less than 1% by weight. The water should always be softened before adding the oil.

Stable emulsions depend upon the alkalinity of the mixture. If marked acidity occurs, the oil separates from the water. If the separation of the oil from the emulsion is not visible, it can be checked for acidity with blue litmus paper. The cause of acidity may be found in a previous processing of the metal, such as, for example, pickling.

Straight cutting oils

Straight cutting oils include mineral oils of any desired viscosity and straight fatty oils. Mineral oils are suitable for light duty cutting, particularly on non-ferrous metals, where both lubrication and cooling may be necessary but neither requirement's severe. They are used to some extent for light duty cutting on automatics, where an emulsion might cause trouble by replacing the lubricating oil on machine parts, thereby necessitating a shutdown for cleaning. Straight minerals are not satisfactory for severe requirements. But when they are used, best results will be obtained from the relatively low viscosity grades which have good penetrating and cooling properties. At the present time, the greatest use of mineral oils in metal cutting is in blending them with cutting oil bases in the necessary proportions to suit specific requirements.

Conditions occurring in the machining of many metals demand a cutting fluid with both lubricating and cooling qualities to offset excessive pressures and temperatures. The various oils previously mentioned are used, with the selection depending upon the severity of the machining operations.

The viscosity of an oil is important when used as a cutting fluid. In general, the heavier the service, the heavier should be the viscosity; but consideration must be given to the fact that lighter oils are more readily circulated and will carry away heat factor. For best effectiveness, cutting oil should have a viscosity range of 100 to 200 Saybolt seconds at 38°C.

Gummy cutting oils hold small chips and other abrasives that gradually work into the bearings and slides of machine-tools, causing excessive wear and preventing accurate work.

Straight fatty oils, such as lard oil, were formerly used almost entirely for severe operations. However, the development and subsequent improvements in sulphurised cutting oils as well as high cost of animal fatty oils, has relegated fatty oils to a subordinate position, particularly for machining the tougher steels and alloys. When cutting oils containing sulphur objectionably tarnish the material being machined, lard and other fatty oils may be used or parts can be treated after machining to remove tarnish.

The primary objection to the use of straight animal fatty oils is their tendency to become rancid and when this happens the odour is objectionable. Such oils permit bacteria to breed and cause skin troubles among the operators.

Compounded or mineral lard oils

Blends of minerals and fatty oils are used for economic reasons, since for most machining operations where straight lard oils are suitable their cost is prohibitive. Therefore, under operating conditions requiring a cutting fluid of greater lubricating value than mineral oil, mineral lard oils are used. The fatty oil varies from 10 to 40%, the amount being increased with increase in hardness of the stock being machined or where deep cutting is essential. Lard oil, either as number one or prime lard, is the most commonly used animal fatty oil, in addition, many synthetic fats are employed. The blends using synthetic fatty oil are still referred to as mineral lard oils; they have the desired cutting characteristics of blends of mineral and animal fatty oils.

Mineral lard oils are used quite extensively in automatic screw machines, where a better surface finish is required in some operations than could be obtained by the use of a straight mineral oil. Their adaptability to tool cooling and lubrication in metal cutting is explained by the fact that they possess a certain degree of oiliness along with relatively low viscosity.

The percentage of lard oil in blended cutting fluids is a matter for determination under actual operating conditions. In general, it is desirable to use a cutting oil with the lowest percentage of lard oil that will do the work.

Oils with additives

In general, sulphurised oils may be classified into the three following groups:
- (a) Animal or vegetable oil with cooked in sulphur.
- (b) Sulphurised mineral oils consist of straight mineral oil in which the active sulphur content is usually not more than 3%. This oil is used as prepared, but may be blended further with a light mineral oil for some work.
- (c) Sulphurised base oils include a fatty oil to which 8 to 20% active sulphur is added, the mixture serving as a base, commonly called "sulphur base". This base is then diluted to approximate cutting strength and viscosity by blending with 5 to 20 parts of straight mineral oil of a viscosity such as 110 Saybolt seconds at 30°C. The resulting blend is quite transparent and has a low viscosity.

Sulphurised oils may be processed further with chlorine. Such sulphurised chlorinated cutting fluids are usually lighter-bodied than the sulphurised type, and are reported to retain higher quantities of active sulphur.

Oil refined from high sulphur crude is produced with a natural sulphur content, which is ineffective without added quantities of active sulphur. The natural sulphur is inert and of little value in cutting oils.

Sulphurised oils in the second and third groups mentioned above are used extensively for machining straight carbon or alloyed steels, stainless steel, high-nickel alloys, and copper. Ordinarily the soft stringy steels require an oil of high sulphur content, while the steels which are harder and more brittle require less sulphur. In general, broaching, threading and tapping require high sulphur oils, while drilling and reaming, shaping and turning, milling and hobbling require low sulphur oils.

Colloidal graphite, a chemically neural carbon, is used as an additive to oils and emulsions. It has been found to be very effective for increasing tool life, reducing cutting forces and improving surface finish in continuous operations on steel. Under certain conditions, the cooling action of the cutting fluid may be a question of volume and temperature rather than relative cooling ability.

In other words, a lard or mineral oil, or a mixture of these two, or a sulphurised oil may provide the requisite properties if supplied in sufficient volume and properly cooled and clarified.

Good grades of cutting oils have penetrating and adhesive properties, thereby promoting their value as cutting fluids. The penetrating quality is important because it enables the oil to enter infinitesimal figures in the outer surface of the chip, thus aiding its formation.

12.10 SYNTHETIC OR CHEMICAL CUTTING FLUIDS

These are mixtures of a number of chemical components dissolved in water. All of them are used as coolants, but some of them also act as lubricants. Basically they are water modifiers and are recommended particularly for machining operations in which the main function of the cutting fluid is to act as coolant.

12.10.1 Types of Synthetic Cutting Fluids

They are classified in three groups as follows:

Pure coolant: They are made up mostly of water softness and rust inhibitors and have little lubricating properties.

Coolant and lubricants: In addition to pure coolants, they also have qualities of lubrication.

Lubricating coolants: Coolant of such type are water softeners, rust inhibitors, wetting agents and also have chemical lubricating properties. Chlorine, sulphur, phosphorus, etc. are examples of this class.

12.10.2 Commonly Used Chemical Constituent Fluids

These are:
   (i) Phosphates and borates for water softening.
   (ii) Amines and nitrites for rust prevention.
   (iii) Soap and wetting agents for lubrication and reduced surface tension.
   (iv) Glycols as a mixing agent.
   (v) Germicides for controlling bacterial growth.

12.10.3 Advantages of Synthetic Fluids

Coolant life: In general the user can depend upon a longer fluid life with synthetics, particularly when compared with conventional soluble fluids. This not only manifests machine downtime for coolant changeover and reducing disposal problems.

Easing of odour problems: Again, this is in comparison with soluble oils. However, confusion often arises because of the occasional tendency of synthetics to pick up tramp oil from lubrication systems, eventually yielding odour problems from bacteria. Regular skimming of the coolant will generally maintain the advantage.

Working conditions: There should be improvements in visibility of work, cleanliness of components, machines and work area, and removal of oily smells.

Safety: Operating conditions should be made safer by: removal of misting often prevalent in large machine shops running on neat cutting oils; removal of possible fire risks with neat oils; and reduction of operator contact with mineral oil, which has a stigma related to cancer association, which is however not really so relevant now-a-days since most suppliers are utilising solvent-refined mineral oils for cutting oils.

Environmental problems: The latest types of synthetic fluids, after allowing any solids to settle out and skimming off tramp oil, are generally more acceptable to sewage authorities than conventional products.

Performance improvements: Depending on application and fluid, quite dramatic improvements in machining performance and tool life have been achieved with synthetics.

Coolant drag-out: The nature of the fluid means that less of it is lost in use, not only yielding obvious economy, but also indirectly leading to cleaner swarf.

12.10.4 Disadvantages of Synthetic Fluids

Machine lubrication: Since synthetics, by definition, do not contain mineral oil, they cannot provide the residual lubrication for which conventional cutting oils are sometimes relied on. However, relying on a coating of oil from the machining fluid often covers up inadequate machine lubrication procedures. Although it might be said that this characteristic is a bonus from the use of mineral oil products, a review of the machine manufacturer's recommendations will generally suggest routine lubrication procedures that will give more than adequate cover when synthetics are used. The introduction of semi-synthetics was partly aimed at this problem but, as with so many attempts to give 'the best of both worlds', these products are certainly not the ultimate answer. Whilst they do have merit in some applications, many of the problems with conventional soluble oils are reintroduced.

Removal of slideway lubricants: Many synthetics, particularly some of the early types, do have a tendency to emulsify or purge lubricating oil from slideways, etc. because of their general 'wetting-out, characteristics.' However, the more recent developments do not emulsify mineral oil, and the problems have been eased. In addition, improvement can also be achieved by the use of the more adhesive sideway lubricants, modifying coolant flow direction, and institution of the regular lubrication producers referred to previously.

Component protection: Products containing mineral oil obviously tend to leave a film of oil on the component after machining, which can provide useful inter-operational protection. The newer synthetics do, however, incorporate good rust prevention qualities. On the other hand, there is an increasing trend towards the use of dewatering rust preventives for inter-operational corrosion protection in the metal working industry, which obviously eases the requirement anyway.

Operator skin problems: Synthetics inherently incorporate a degree of detergency, and consequently can tend to remove the natural oils from the skin in the same way that washing-up liquids do. This can of course lead to skin problems of varying degrees. However, insistence on the use of the proper barrier creams will generally remove this problem.

Paint and seal attack: It is certainly true that standard machine tool paint and seal materials can be susceptible to attack by modem synthetic fluids. However, this problem is of decreasing relevance, since many machine tool manufacturers are now constructing their machines with a view to running with synthetics, and hence are utilising paints and seal materials which are immune to attack. This does not apply to all manufacturers. However, the changing of seals and repainting of machines with the right materials is not a major job, if it is found to be necessary, and can be a small price to pay for benefits elsewhere.

12.11   GASEOUS CUTTING FLUIDS

Air (still or compressed), carbon dioxide, and argon are used as gaseous type cutting fluids. Air is mostly used in grinding or machining process where the main function of fluid is cooling only. It is found effective if it is used at subzero cooled state. Carbon dioxide is used in low machinable and high strength thermal resistant alloys because of its high cost.

12.12   EXTREME PRESSURE (EP) LUBRICATION MECHANISM IN METAL CUTTING

Boundary lubrication conditions are achieved by adding some fluids, known as *extreme pressure* (EP) *additives*, which react chemically with the metal to form compounds on the metal surface. These compounds may either form layered structures which are easily showed in sliding, thus reducing friction, or these may inhibit the welding which would occur with bare metal surface in contact.

12.12.1   Advantages of EP

They result in:
    (a) Improved surface finish
    (b) Reduction in cutting force
    (c) Reduction in tool wear.

12.12.2 Conditions of Use of EP

These are:
- (a) To be effective the EP used must be in sufficient quantity.
- (b) The reactive species in the additive must be in proper form to become available at metal surface.
- (c) Temperature must be high enough to promote surface compound formation.
- (d) Sliding speed should be low so that surface reaction takes place.

12.12.3 Examples of EP

These are chlorinated paraffins, elemental sulphur or sulphurised fats.

12.13 CRITERIA OF SELECTION OF CUTTING FLUIDS

Table 12.1 shows the typical uses of cuttng fluids.

Table 12.1 Typical uses of cutting fluids

Operation	Type of fluid
Turning titanium	Synthetic coolant
For light cuts on aluminium alloys	Kerosene base oils
Copper alloys	Dry
Ferrous metals	Synthetic coolant
Steel or stainless steel	EP soluble oil
Finishing aluminium alloys	Mineral oil with friction additives
Heavy cut on aluminium alloys	Soluble oils
Copper alloys	Straight mineral oil
Steel	Dry

The following criteria are used in selection of cutting fluids:

Cost: The overall operation of the cutting fluids must be economic and cheap.

Viscosity: If viscosity is too low (mineral oils), the oil may be too volatile, while if it is too high the consumption will be higher due to its adhesiveness; their life is also less, so proper choice is to be made keeping this view.

Life: Life of coolant should be reasonably high.

Loss during operation: It should be minimum.

Operator's acceptance: It should be acceptable to the operators.

Type of finish: The cutting fluid must produce good finish.

Chemical action: Fluids with higher carbon chains give low cutting forces. Fluids with odd number of carbon atoms give less force while those with even number give high force.

12.13.1 Disadvantages of Using Cutting Fluids

Some of the side-effects of using cutting fluids are as follows:

Effect on health: In the extreme cases some fluids which are good for cutting process may produce toxic vapours which are harmful to health. An example of this class is an inorganic compound called tetrachloride.

Effect on work material: The work material may also get strained by certain fluids. For example, titanium alloy get cracks when chlorinated cutting fluids are used and rust forms on ferrous alloys when water is used.

Effect on machine tools: Fluids sometimes rust machine tool components and slideways, etc. In some cases the cutting fluid affect copper base bearing of the machine tools.

Effect on cutting fluids: In some cases the cutting fluids may get disintegrated, affecting the working conditions.

Water-miscible fluids are the most commonly used, probably for some 80% of all machining operations because they give a degree of combined cooling and lubrication suitable for the majority of metal-cutting operations carried out at higher speeds and lower pressures—general turning, milling, grinding, etc. In addition, they are more economical than neat oils because dilution with water brings the cost down, and the working conditions can be better, with cooler, cleaner pans, reduction of oil missing and fuming, and a reduction in fire hazard.

Synthetic water-miscible fluids have certain claimed advantages over emulsion-type water-miscible fluids. They have detergent properties and give a higher degree of cleanliness, resulting in clean machine tools and coolant troughs. Also, they leave only a very light residual film on the workpiece. They are easy to mix, and have improved stability and freedom from bacterial growth, which leads to a long working life. They also have the possibility of providing better lubricating properties. But while they have some admirable properties, in practice they behave differently from the usual emulsions and considerable difficulties can arise if this is not recognised.

Natural evaporation of the water content leads to solutions becoming too strong, for example and the detergent action will wash oil and grease from vital bearing surfaces. Many types of paint are attacked and thus high quantity epoxy or polyurethane types are needed to resist synthetic fluids. Material used for seals may also be attacked, leading to further bearing problems. The inclusion of some oil in these cutting fluids, helps to making them semi-synthetic fluids, avoid some of the difficulties with the lubrication of machine parts and the evaporation and dilution-control problems are also not so severe.

Neat oils have advantages over water-miscible fluids when both good lubrication and some cooling are required to provide an economic tool life and a good surface finish, especially when using high-speed steel tools at moderate speeds and feeds. They also have advantages whenever the combination of a slower cutting speed and a requirement for low surface roughness demands good lubrication. In operations such as broaching, especially with difficult work materials, the properties of the cutting fluid may be vital to the success of the operation. On complex machines such as multi-spindle and some single-spindle automatic lathes, where exposed slides and other parts are continually bathed in the cutting fluid, neat oils are traditionally used to provide lubrication.

Considering these general points, it can be said that the main criteria for cutting fluid selection for a particular application are type of machining operation, workpiece material and the machine conditions of speed, feed and depth of cut. The general effects of combination of machining operation and workpiece material is often conveniently expressed in chart form with the parameters ranged, roughly, in order of difficulty in machining, as in Table 12.2. There are some operations for which oils are specially formulated, for example, honing which demands a thinners paraffin-base oil, and some workpiece materials will also demand special consideration. Magnesium and its alloys, for example, should never be machined or ground with water based fluids because of possible fire hazards.

The effect of cutting conditions must also be considered. Here a major influence is the cutting speed. In general, the lubricating properties of the cutting fluid are of most importance at lower speeds, and the cooling properties at higher speeds. Thus, neat cutting oils cannot normally be expected to perform adequately at cutting speeds above 380 and below 30 m/min. An exception to this rule might be found in some grinding operations where the maintenance of wheel form—the reduction of wheel wear is of prime importance, as in form or thread grinding where neat oils are commonly used.

Table 12.2 Selection of metal working fluids for general workshop applications

Machining operation	Workpiece material			
	Free machining and low carbon steel	Medium carbon steel	High carbon and alloy steel	Stainless and heat resistant alloys
Grinding	Clear type soluble oil, semi-synthetic or chemical grinding fluid			
Turning	General purpose soluble oil, semi-synthetic or synthetic fluid	Extreme pressure soluble oil, semi-synthetic or synthetic fluid		
Milling	General purpose or fatty soluble oil, semi-synthetic or synthetic fluid	Extreme pressure soluble oil, semi-synthetic or synthetic fluid	Extreme pressure soluble oil, semi-synthetic or synthetic fluid (neat cutting oils may be necessary)	
Drilling	Fatty or extreme pressure soluble oil, semi-synthetic or synthetic fluid			
Gear shaping	Extreme pressure soluble oil, semi-synthetic or synthetic fluid	Neat cutting oils preferable		
Hobbing	Extreme pressure soluble oil, semi-synthetic or synthetic fluid (neat cutting oils may be preferable)	Neat cutting oils preferable		
Broaching	Extreme pressure soluble oil, semi-synthetic or synthetic fluid (neat cutting oils may be preferable)			
Tapping				
Thread or Form grinding	Extreme pressure soluble oil, semi-synthetic or synthetic fluid (neat cutting oils may be preferable)			

Table 12.3 Cutting fluid recommendations for six groups of metal on normal speeds, tool life and finish

Severity	Type of operation	Groups of metal based on Machinability Rating Classification					
		Ferrous metal				Non-ferrous metal	
		(1) 70% and more	(2) 50–60%	(3) 40–50%	(4) Below 40%	(5) Over 100%	(6) Below 100%
Greatest							
1	Broaching internal	Em SMO*	SMO*Em	SMO Em	SMO Em	MO Em	SMO ML
2	Broaching surface	Em SMO	EM SMO	SMO Em	SMO Em	MO Em	SMO ML
3	Tapping plain	SMO	SMO	SMO	SMO	Em Dry	SMO ML
3	Threading pipe	SMO	SMO ML	SMO	SMO		SMO +
3	Threading plain	SMO	SMO	SMO	SMO	Em SMO	SMO
4	Gear shaving	SMOL	SMOL	SMOL	SMOL		
4	Remaining plain	M1 SMO	ML SMO	ML SMO	ML SMO	ML MO Em	ML MOSMO
4	Gear cutting	SMO ML Em	SMO	SMO	SMO ML		SMO ML
5	Drilling deep	Em ML	SMO Em	SMO	SMO	MO ML Em	SMO ML
6	Milling plain	Em ML SMO	Em	Em	SMO	Em MO Dry	SMO Em
6	Milling multiple cutter	ML	SMO	SMO	SMO ML	Em MO Dry	SMO Em
7	Boring multiple head	SMO Em	SMO Em	SMO Em	SMO ML	K Dry ML	SMO Em
7	Multiple spindle automatic screw	SMO ML Em	SMO ML Em	SMO ML Em	SMO Em	Em Dry ML	SMO
8	High speed light feed automatic screw machines	SMO ML Em	SMO ML Em	SMO ML Em	SMO ML Em	Em Dry ML	SMO
9	Drilling	Em	Em	Em SMO	Em SMO	Em Dry	Em
9	Planing	Em	SMO Em	Em SMO	Em SMO	Em Dry	Em
9	Turning, single-point tool, form tools	Em SMOL ML	Em SMO ML	Em SMO ML	Em SMO ML	Em Dry ML	SMO SMO ML
Greatest							
10	Sawing— circular, hawk	SMO Em LM	SMO Em LM	SMO Em LM	SMO Em LM	Dry MO Mm	SMO Em ML
Least	Gripping						
	(i) Plain	Em	Em	Em	Em	Em	Em
	(ii) From (Thread, etc.)	SMO	SMO	SMO	SMO	MO SMO	SMO

O_k = Kerosene, L = Lard oil, MO = Mineral oil, ML = Mineral lard oil, Dry = No cutting fluid, SMO = Sulphurised oils, Em = Insoluble or emulsified oils and compounds.

*Oils containing both sulphur and chlorine, when carefully manufactured and sponsored, may be used where sulphurised oils are indicated.

+Preferred recommendations set in bold face type.

++In threading copper, palm oil is frequently used.

It has been reported by several observers that emulsions are usually unsatisfactory on some precision machine tools such as Fellows gears, shapers and Gleason gear generators.

A point which must not be overlooked in the application of EP cutting fluids is that the cutting oil should possess sufficient additive treatment and activity for the duty required and to prevent catering of the tool face and rough surface finish of the workpiece, but not so much as to lead to excessive wear through unnecessary elimination of any protective built-up edge, and overactive chemical reaction with the cutting tool material in the case of ferrous tools.

As shown in Table 12.3, there may often be situations where any one of various types of cutting fluids could be used for a given application; the interests of rationalisation in any one workshop, it may be deemed convenient and economic to choose one fluid for general purpose use, even if it does not give optimum for every operation.

While all the various characteristics of cutting fluids are of considerable interest in the workshop, the performance of the fluid in terms of the effect on tool life remains of prime importance. However, this criterion is difficult to specify in precise terms because of the complexities of the many interacting factors in metal-cutting operations.

Synthetic fluids, for example, give excellent performance in grinding operations, where generally the requirement for lubricating properties is less than in many metal working operations. Nevertheless, for optimum performance, the choice between a synthetic fluid, a semi-synthetic fluid and a soluble oil emulsion with EP additives may depend on the severity of the grinding operations. Some suppliers may not now recommend synthetic fluids for operations oilier than grinding, because semi-synthetic fluids are available with equivalent or better 'performance' with the added benefits of lubrication for machine slides and so on, and with easier maintenance of the correct concentration in use.

Opinion may be divided on the levels of performance achieved with synthetic and semi-synthetic fluids. Some suppliers would say that for general purpose machining, turning, milling, and drilling, the performance of synthetic fluids will equal or perhaps surpass that of general purpose oils, but not that of superior EP soluble oils, while semi-synthetics can be better than EP solubles. Other suppliers may claim that their synthetic fluids can outperform high-performance semi-synthetic fluids. Commercial formulations obviously differ and it is difficult to discuss general claims when test results are not published.

It will often be claimed that synthetic or semi-synthetic fluids are suitable for most of the applications for which neat oils are used, except for the most demanding operations such as broaching, deep-hole drilling and boring. Again, the absence of reported workshop data or results from tests or field trials can make objective judgment impossible, and some people doubt the sweeping claims made for water-mixed fluids. The exact phrases used in brochures may be indicative of the true situation and descriptive phrases like "in general, can equal the performance of a neat oil", 'for some operations better or equal to neat oils' or "will provide an acceptable tool life", should be considered carefully.

REVIEW QUESTIONS

1. What are the functions of a cutting fluid? Describe in brief.
2. How does the use of cutting fluids improve
   (i) Toole life
   (ii) Productivity
   (iii) Quality
   (iv) Surface finish.
3. What are the characteristics of a good cutting fluid? Explain.
4. How are the cutting fluids used and maintained for improving the characteristics listed in Exercise 2?
5. What are the different types of cutting fluids? Explain in detail.

6. What are the advantages and disadvantages in using soluble oils? Explain.
7. Differentiate between straight cutting oils and oils with additives.
8. What do you understand by the term "synthetic chemical cutting fluids"? Explain their advantages and disadvantages.
9. What are the criteria of selection of cutting fluids? Explain in detail.
10. What is cutting fluid?
11. Mention any four important requirements of cutting fluids.
12. How does the cutting fluid reach the tool-chip interface?
13. Why does the cutting fluid act only as a coolant at higher cutting speeds?
14. What are the main groups of cutting fluids?
15. Under what circumstances would you recommend the use of the cutting fluids listed as follows:
    (i) Extreme pressure emulsion
    (ii) Chemical coolant
    (iii) Liquid cardon dioxide
    (iv) Compressed air
    (v) Straight mineral oil
    (vi) Sulphurised fatty mineral oil blend.
16. Write a brief note on suitability of water as cutting fluid.
17. What are the different ways of applying cutting fluids?
18. How does cutting fluid improve tool life?

PART II
MACHINE TOOLS

Chapter 13

Introduction to Machine Tools

13.1 MACHINE TOOL—DEFINITION

The machine tool is a device in which energy is utilised in deformation of material for shaping, sizing or processing a product by removing the excess material in the form of chips. The form of surfaces produced and finished depend upon the shape of the cutting tool and the relative path of motion between the cutter and the workpiece.

13.1.1 Machine Tool Design Requirements

The following requirements must be met by the designer:

Functional requirements: The designer must consider the basic functional requirements—mechanical and climate environments, size, weight, reliability, service life, energy consumption and appearance.

Motor mountings: Another point of importance is motor mountings. They are often the most troublesome cause of vibration. Proper selection of motors, drives and structural design should be made.

Lifting attachment: It is necessary to provide holes for crane hooks for use when the complete machine is moved. The locations of support bolts and tic down bolts in stationary machine beds are of great importance.

Cost, weight and drive selection: When designing a machine tool, it is of importance to first establish diagrams of costs, weights, mechanical drives, machine elements, and electrical and hydraulic components.

Design of beds and columns: Large beds and columns must be designed with a minimum number of the large openings that are usually necessary to mount transmission boxes or other mechanical drives. The required openings should have ribbings on the inside to tie them to vertical and horizontal support ribs.

Lubrication systems: Provisions must be made for centralised lubrication of bed ways and slides.

Facility for chip removal: Facilities for chip removal must be considered from the start, since they will affect the overall machine design.

Optimisation: The design of machine is a compromise between physical constraints necessary for its proper functioning and the configuration aspects that identify its purpose, i.e. its operational capability. Thus the design engineer is continuously involved in the complex problems of optimisation of the weight-to-cost ratio, as well as the static and dynamic responses, within the physical boundaries prescribed by the function of the machine components, and the forces acting and the cost factors, and it is important that the design engineer becomes fully cognizant of the alternatives in the manufacture of such parts.

Safety: The machine tool design is such that its operation is fool proof from the safety point of view.

Higher productivity: It must be economical in use and must give higher cutting speeds, better chip disposal, etc.

Maintenance: As far as possible it should be maintenance-free.

Vibration: It should be vibration-free.

13.2 PRESENT TREND OF DESIGN OPTIMISATION

The present trend in machine design favours plan smooth lines, sharp square edges and clean joints. Well rounded contours and over-emphasised curves are often rejected. Frequently machine elements designed as weldments look elegant. Steel, welded into a homogeneous structure, stress-relieved and shot-blasted, makes a fine-looking machine part at reasonable cost. Holes may be located close to the edge of weldments without providing extra materials to make up for variations in fabricating techniques.

13.3 ELEMENTS OF MACHINE TOOLS

Every machine tool is basically composed of four elements:
  (i) Structure (It includes bed, column or frame)
  (ii) Slides and slideways
  (iii) Spindle
  (iv) Drive regulators.

13.4 CLASSIFICATION OF METAL CUTTING MACHINE TOOLS

The various machine tools, structure and design analysis have been discussed a lot in the prior note to this. The machine tools can be classified, depending on the nature of the work that should be carried out and cutting tools used therein. It can be divided in nine groups. These groups are given below:
  (i) Lathes;
  (ii) Drilling and boring machines;
  (iii) Grinding and polishing machines;
  (iv) Machine tools assembled with units;
  (v) Gear and thread-cutting machine;
  (vi) Milling machines;
  (vii) Planning and vertical shaping machines;

(viii) Cutting-off machines;
(ix) Other machine tools.

Each of these machine tools can further be sub-grouped, as shown below in case of lathes:
(a) Single-spindle automatic lathes;
(b) Multiple-spindle automatic lathes;
(c) Turret lathes;
(d) Drilling and cutting-off machines;
(e) Turning-and-boring lathes;
(f) Thread-cutting and face lathes;
(g) Multi-tooled lathes;
(h) Specialised engine lathes for profiled work; and so on.

By degree of specialisation: Metal-cutting machine tools can be divided into various groups as follows:
(a) Depending on the
    (i) *Universal.* different operations on parts of various names; e.g. turning, milling.
    (ii) *Specialised.* for machine parts of one kind (e.g. shafts).
(b) By weight: Ordinary to 10 tonnes.
(c) By accuracy:
    (i) Of normal precision;
    (ii) Of high precision.

Machine tools of high precision in turn are divided into:
(a) Machines of high precision;
(b) Machines of especially high precision;
(c) Special master-machines (e.g. these manufacturing load screws.);
(d) Automatics and semi-automatic
    (i) Copying machine tools;
    (ii) Programme-controlled machine tools.

13.5 MACHINE TOOL INDUSTRY—PROGRESS IN INDIA

During the last two decades, extensive research and development activity in the field of machine tool design has been carried out in Indian Institutes of Technology, Jadavpur University, University of Roorkee, Banaras Hindu University, etc. The pioneers in the field of manufacture of machine tools—Mysore Kirloskar, HMT, Praga Machine Tools, etc.—have attained so much competence that Indian-made machine tools can be compared to their counterparts from more industrialised countries, viz. England, Germany and many others.

A machine tool research laboratory called CMTL at Bangalore exclusively trains specialists in the field of machine tool design, and HMT is capable of producing NC machines. It is trusted that the day is not far enough when we shall be marketing NC and CNC machines to cater the needs of local as well as foreign markets.

An active design and development function is of utmost importance to link the technology of the laboratory to manufacturing functions. However, it appears that Indian researchers have not been able to develop confidence about their ability in the local industrialists' minds, due to which the latter still rely on foreign technical knowhow. It is difficult to comment about the fault as to where it lies.

To move with speed towards technological sophistication and socio-economic development in a developing country like ours, a coordinated and motivated effort towards self-reliance on the part of industrialists, government and academicians is of paramount importance, and it essentially depends on the concerned people and their approach towards this problem.

The future of machine tool design in the USA, Russia, and other developed countries is difficult to predict as they have already produced programme-controlled NC/CNC machine tools; however, the scope of machine tool manufacture in India is still wide open. Manufacture of defence goods requires more and more precision and interchangeability of parts, which can be met by such sophisticated machines. Therefore the demand for such machines is increasing in our country.

13.6 COMMON FEATURES OF MACHINE TOOLS

The main common features of all types of machine tools are as follows:

Bed or Body: It is usually made of C.I. Its main function is to carry and support all other parts of the machine and is bolted down on the floor. It is machined at places requiring accurate location of other components of the machine. In case of machine tools like lathe, planer, boring and grinding machines, it is called *bed* whereas in shaper or milling machine it is called *body*.

Main spindle: It is usually a rotating member carrying the workpiece in case of lathe and vertical boring mill, while in milling, drilling, horizontal boring and grinding machine the spindle carries the tool. Machines like shaper and planer have no spindle because of the reciprocating motion of the tool or a workpiece.

Power drives: All machine tools are driven by electric motors either directly or through the medium of belt and pulleys gears or chain drives. Power from the motor flows to spindle, tool and the machine table through various intermediate feed and speed change gears. Many of the production machines carry separate motors for imparting different motions to their various parts. The motors may be AC/DC-driven and are of three-phase type. AC motors up to 10 HP are mostly squirrel cage induction motors.

Mechanism for spindle speed variation: It includes either the set of gears, belt-pulley arrangement or other some typical mechanism.

Table: It is used for holding and supporting the work.

Accessories: They include tool holding or supporting devices. Tool attachments or standard jigs and fixtures are also some times included in accessories.

13.7 SELECTION OF MACHINE TOOLS

The primary considerations are:

(i) Metal removal rate
(ii) Dimensional accuracy
(iii) Surface finish
(iv) Operational cost
(v) Initial cost
(vi) Load or capacity utilisation
(vii) Maintenance aspects
(viii) Lubrication system
(ix) Ease of operation
(x) Vibration problems and rigidity
(xi) Power consumption
(xii) Tool consumption
(xiii) Economy and productivity.

13.8 SYSTEMS FOR CONTROL OF MACHINE TOOLS

Three types of systems are normally used for putting various controls on a machine into operation. These are:

Manual control systems: These utilise the services of the machine operator who handles various controls whenever needed. This is simple and mostly used, but suffers from the limitations of humans.

Automatic or semiautomatic systems: This type of control is achieved by using preset stops, templates, cams and other seven similar devices with the machine.

NC control systems: This employs punched tape cards or magnetic tape on which all necessary information or command is stored. This may be of either automatic or semi-automatic type.

REVIEW QUESTIONS

1. What is a machine tool?
2. Name some common machine tools.
3. Describe common features of a machine tool.
4. What factors are considered in selection of a machine tool?
5. Explain the systems of control for machine tools.
6. How are metal cutting machine tools classified?
7. Discuss the design requirements of a machine tool.

CHAPTER 14

Design of Machine Tool Beds

14.1 INTRODUCTION

The bed and other supporting elements of a machine tool form the backbone of the system. There are many points for consideration while designing these elements, viz. the material of these elements, their physical characteristics, shape, method of construction, dynamic performance of the material, etc.

These components have to perform many functions such as guiding a true required motion and transmitting the loads of its components and the dynamic loads to the foundation without getting deformed in doing so. The surface must be wear-resistant too, so as to maintain accuracy of motion even after long use.

Therefore, two materials for a machine tool bed have to be selected within the guidelines as mentioned above. Generally, guides are integral to beds in machine tools; sometimes, separate guides are also fabricated in modem machine tools.

Generally, C.I. is one of the most common machine tool bed materials. This is because of the fact that C.I. absorbs vibration and works as a good damper; furthermore, the presence of graphite flakes reduces friction and works as a lubricant.

But the fact that intricate shapes of casting as required in machine tool beds are liable to have defects and cracks have led to the development of other materials for machine tool beds and guides. Steel is one of these materials, due to the increased reliability of welded steel structures. This steel is free from inclusions, blow holes, porosity, thermal cracks and other defects. The presence of these defects in a C.I. machine tool bed weakens the structural strength of the machine tool.

Weldments and castings need proper stress relieving in order to ensure stability and alignment. This can be achieved either by seasoning or through heat treatment at temperatures ranging from 500 to 600°C.

14.2 FACTORS IN DESIGN OF BEDS

The choice of steel as a material of machine tool beds is dependent on many factors when compared to C.I. These factors are mostly the properties of the material and the behaviour of the fabricated structure when in use. Some of these factors can be listed as follows:

  (i) Strength
  (ii) Volume to weight ratio for the same strength
  (iii) Deflection and resonant frequencies
  (iv) Damping
  (v) Vibration and chatter

14.2.1 Strength

Consideration of strength is an important factor, but its measure as regards the machine tool structure is deflection and stress concentration at various points in such a structure. In machine design practice, stiffness rather than the load carrying capacity is the criterion of design. Thus if a section is designed for stiffness it will be stressed below the allowable stress. Furthermore it suggests that deflection in any part of the structure must not exceed the maximum allowable one and in actual working these limits of deflection are seldom reached. Therefore, for the same strength, only about half as much steel as C.I. is required, since steel has the value of Young's modulus of 2×10^6 kg/cm^2 while C.I. has 1×10^6 kg/cm^2. Table 14.1 shows the strength of steel when compared to C.I. for same size.

Table 14.1 Ratio of steel/cast iron

Type of section	(C.I.: Steel) bending moment	(C.I.: Steel) sustained torque
Circular tube, 80 mm inner, 100 mm outer	2 : 1	2 : 1
Box section, 75 mm × 100 mm, 10 mm thick	2.31 : 1	1.948 : 1
I section, 100 mm × 100 mm, 10 mm thick	3.1 : 1	1.72 : 1

Note: It will be noted that in case of box section, I section steel proves more than two times stronger than cast iron when bending moment is applied. In case of torsion steel, it is less than two times strong compared to cast iron, due to warping of ductile steel.

14.2.2 Volume to Weight Ratio

The volume required for steel and C.I. is inversely proportional to the modulus of elasticity. Thus the volume of steel structure will generally be about one half than that of C.I. Thus changing the design from C.I. to steel shall result in weight reduction. This can be further compared from the equation of weight.

$$W = \rho \frac{Pl}{\sigma_{max}}$$

where P = tensile load in kg;
 l = length of test specimen;
 σ_{max} = maximum allowable stress of the material in kgm/cm^2;
 ρ = specific weight of the material in kg/cm^2, i.e., Strain; E = Elastic modulus

Thus we see from the above equation that for a unit volume:

$$W \propto \frac{1}{\sigma_{max}/P} \propto \frac{P}{eE}$$

Thus $\dfrac{W_{Steel}}{W_{Cast\ iron}} = \dfrac{E_{C.I.}}{E_{Steel}} = \dfrac{1 \times 10^6}{2 \times 10^6} = 0.5$ i.e. 50%

Thus, as steel has a greater value of weight for a given volume than that of C.I. Use of steel saves material as shown in Table 14.2.

Table 14.2[*] Relative weight of Steel : C.I. as percentage

Type of section	Percent relative weight steel: C.I.
1 cm (square)	50%
1.193 cm (circle)	55.9%
1.58 / 1 cm (tube)	48.2%
1.97 / 0.06 cm (tube)	36%

[*]Table 14.2 shows practical values which differ from theoretical value of 50% for sections different from square section.

Design criteria for machine tool structures

1. **Strength considerations:** Consider a machine tool bed with two side walls as a simply-supported beam loaded with a concentrated load P at the centre.

   Stress σ is given by
   $$\sigma = \frac{Mz}{I}$$

   But
   $$M = \frac{Pl}{4}, \; z = \frac{h}{2} \text{ and } I = \frac{bh^3}{12}$$

   Therefore,
   $$\sigma = \frac{3Pl}{2bh^2}$$

   where b is the width, and
   h is the depth of rectangular cross-section.

   Now, Volume, $V_\sigma = bhl$

   Substituting $b = \frac{3}{2}\frac{Pl}{\sigma_{max}}$ (using maximum value of σ), then we have

   $$V_\sigma = \frac{3}{2}\frac{P}{\sigma_{max}}\left(\frac{l^2}{h}\right)$$

   Thus, for given material, if volume changes, then we have

   $$V_\sigma \propto \left(\frac{l^2}{h}\right)$$

   which is called strength criterion.

2. **Stiffness criterion:** For the beam considered, the deflection at centre is

   $$\delta = \frac{Pl^3}{48EI} = \frac{Pl^3}{48Ebh^3/12}$$

   $\Rightarrow$
   $$b = \frac{Pl^3}{48E\delta h^3/12}$$

   Now,
   $$V_\delta = bhl = \left(\frac{Pl^3}{4E\delta h^3}\right)h \cdot l$$

   $$= \frac{P}{4E\delta}\left(\frac{l^2}{h}\right)^2$$

   Thus,
   $$V_\delta \propto \left(\frac{l^2}{h}\right)^2$$

   Equating V_σ and V_δ for the two criteria

   $$\frac{3}{2}\cdot\frac{P}{\sigma}\left(\frac{l^2}{h}\right) = \frac{P}{4E\delta}\left(\frac{l^2}{h}\right)^2$$

   Thus,
   $$\frac{l^2}{h} = \frac{6E\delta}{\sigma}$$

Figure 14.2 shows the graphs of l^2/h vs. volume V_σ as well V_δ for cast iron and steel. The former is a straight line relation and the latter is a parabolic curve. The point of intersection A yields volume equal to 40 cm³ for steel. The point of intersection B for C.I. yields volume of 250 cm³. Thus, the material saving due to steel is 250/40, i.e. 6.25 times cast iron beams used. Figure 14.1 indicates a ratio of 3.5 reflecting saving of material of steel is cast iron for a rectangular section. The high value of 6.25 in this example is due to the use of 2 beams giving a 'box section effect' resisting bending. Other important results from Fig. 14.2 for the data assumed are:

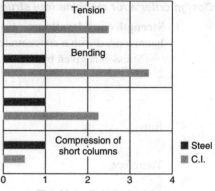

Fig. 14.1 Relative weights.

$\dfrac{l^2}{h}$ for steel = 300 cm corresponding to point A and volume

= 40 cm³

$\dfrac{l^2}{h}$ for C.I. = 500 cm and volume = 250 cm³

Referring to Fig. 14.2, it will be noted that OABG represents a curve of optimal design covering various materials from steel to cast iron including materials like semi steels, Meehanite, etc.

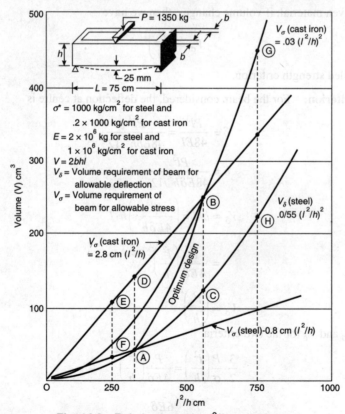

Fig. 14.2 Relation between l^2/h and volume.

14.2.3 Vibration Response

Cast iron beds usually have vibration response at lower frequencies under dynamic forces. But the steel structures on the other hand respond to higher frequencies under same dynamic conditions, thus resulting in smaller deflection. Therefore, forced vibrations, as they do occur in machine tools in operating conditions, may have resonance at low frequencies in case of C.I. beds as compared to steel beds under similar conditions of work.

Displacement amplitudes of structures are related to their resonant response frequencies. Cast iron structures resonate at lower frequencies compared to welded structures and hence have large amplitudes, assuming equal inputs. In an example, $f/f_0 = 1.506$. Reading up to the diagonal line and across to the scale, it is seen that the amplitude of the steel beam illustrated in Fig. 14.3 is 0.44 times that of an iron beam. Small amplitudes are better.

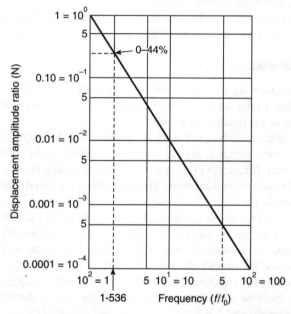

Fig. 14.3 Variation of displacement to amplitude with respect to frequency.

Figure 14.4 shows lathe beds made of cast iron and steel.

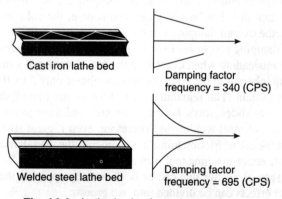

Fig. 14.4 Lathe beds of cast iron and steel.

14.2.4 Damping

The damping in the respective structures is indicated by the rate of decay of the vibration amplitude with time in response to a transient input. This shows that weldments, in addition to providing greater dynamic rigidity, can be effectively damp vibrations. The advantage of welded steel bed is that it has a high damping factor, so vibrations are less likely to effect work finishes. Weldments also provide greater dynamic rigidity, which is important for high accuracy.

In the past, the superior material damping characteristics of cast iron as compared to steel have been held to be of major design significance. Cast iron does have two to four times higher damping capacity than steel, but this higher damping capacity of cast iron is of relatively minor significance in the performance of most machine structures. In reality, slip or joint damping at mating surfaces, and slip damping at welded joints, are of much greater significance than hysteresis or internal damping.

The influence of shrinking forces in weldments increases the damping characteristics of steel structures. The added flat surfaces between adjacent walls maintain close contact and act as friction dampers.

14.2.5 Vibration and Chatter

Chatter and self-excited vibrations are more important in the machine tool structure, but in the theory of self-excited chatter the frequency ratio is useless, since chatter occurs at or very near the natural frequency of one of the modes of vibration of the machine structure.

To further explain the phenomenon of self-excited chatter, it may be helpful to think in terms of a bell struck by a mallet. The bell always responds at its characteristic frequency, the natural frequency, regardless of how many times per unit time (frequency) it is struck. The metal cutting process can be compared with a continuous striking of a mallet upon the bell structure. Therefore, the self-excited chatter frequency is governed by the structured natural frequency.

In the case of chatter caused by forced vibrations, the chatter frequency is determined by the frequency of an alternating force (unbalance, gears, pumps, etc.) acting on the structure. When the frequency of the alternating force occurs below the structure's natural frequency, the chatter amplitude is determined by the stiffness of the structure and the magnitude of the alternating force. For disturbing frequencies above the structure's natural frequency, the effective mass of the structure and the magnitude of alternating force determine the amplitude. In the case of frequency coincidence, the overall damping in the structure is the limiting factor on the chatter amplitude.

Quoting again from the study by Loxham et al., "It should be noted that damping studies were made on machine tools, basically one structure or one large component in which the rigidity of the machine is not a function of the slides, ways, and bolted joints, then the damping in the component becomes much more important. For example, a radial drill has basically one component, the column. Therefore, if damping is increased in this component, the overall damping is increased proportionately".

"A comparison of the damping properties existing in machine tool components made of cast iron and steel weldments can be very misleading when evaluating the tool damping in a machine tool. Investigations have shown that the damping inherent in the components contributes only 5 to 10% of the overall damping existing in the machine tool system. The remaining 90 to 95% of the overall damping is obtained at the junctions of components, such as slides, ways, bolted joints, etc., and these joints are common to a machine tool, whether it is cast or made of steel weldments. Therefore, even if great strides are made in increasing component damping through the use of friction joints in weldments, the overall increases in damping through the use of friction joints of the entire machine tool is of a much smaller degree."

There are numerous types of chatter that can be dealt with by applying the theory of the elementary vibratory system. The chatter effects can be divided into two groups:

Group 1 chatter occurs when the tool is working on a surface that has already undergone a machining operation. In these circumstances, there may be interaction between the movement of the tool point and the chatter marks (produced in the preceding operation) already existing on the surface of the work. Such interaction may lead to dynamic instability. This chatter is known as *regenerative chatter*.

Table 14.3 shows the damping factors for steel and cast iron machine spindles. Steel and cast iron machine spindles with different Young's modulus, have lowest damping factors: however, mounted spindle has highest. Material or internal damping is negligible in comparison to joint damping.

The chatter effect of Group 2 can arise when the work surface has not been previously machined. This group comprises of three types of chatter arising from the following physical sources — the friction conditions existing between the work and the wear marks on the tool; the dynamic modification of rake and clearance angles; and the tendency for the cutting force to vary sometimes with cutting speed.

Table 14.3 Damping factors for steel and C.I. machine spindles

Spindle	Young's modulus E	Damping factor
Steel	2×10^6 kg/cm^2	0.0001
	1.8×10^6 kg/cm^2	0.0002
Cast iron	1.4×10^6 kg/cm^2	0.0006
	1.4×10^6 kg/cm^2	0.001
Mounted		0.03

Equivalent section technique

The Equivalent Section concept involves a direct strength-static stiffness relationship that ignores vibrations, but which permits simple design conversion from existing functioning castings to fabricated steel. Though it is a useful technique, caution must be exercised in its application. Often the designer may allow himself to be limited by the configuration of the casting, instead of designing an altogether different and functionally more efficient product.

Every machine part has to maintain sufficient rigidity and strength in service. The elongation in tension e is given by:

$$e = \frac{Pl}{AE}$$

where P is the load in kg
l is the length of section in cm
A is the cross-sectional area in cm^2, and
E is the modulus of elasticity in kg/cm^2.

Cast iron is denoted by the subscript c and welded steel by the subscript s. For equal elongation e under a given load P for a given length l

$$e = \frac{Pl}{A_c E_c} = \frac{Pl}{A_s E_s}$$

$$A_s = A_c \left(\frac{E_c}{E_s} \right)$$

14.3 CALCULATION FOR DESIGN OF BEDS

Calculation for beds dimension is not sufficiently investigated, however it is usually fulfilled by one of the following schemes:

   (i) Calculation on the basis of assuming a bar;
   (ii) Bed calculation on the basis of a frame;
   (iii) Bed calculation on the basis of a space frame.

In the last two cases the compilation of a scheme depends on a designer's intuition. In general, the calculation of beds is of secondary importance. The main attention is paid to experiments. Model pieces of beds are thoroughly tested.

Admissible twisting stresses of beds range from 80 for C.I. and 150 to 200 kg/cm² for steel. To ensure proper bending, resistance, favourable ratio of b and h, etc. should be provided.

To ensure proper twisting resistance, ribs of beds are made at the angle of 45° to their axis, as the main stresses act in this direction. But such location of ribs is not always possible, due to the fact that chip falls on beds. In multi-tooled machines, the ribs are situated at the angle of 90° in order to support heavy loads.

14.3.1 Material of Beds

The following points are important:

1. Cast iron, alloyed C.I. and welded steel are used for beds.
2. Wall thickness of beds made of cast iron varies from 10 to 50 mm.
3. Wall thickness of beds made of light machines varies from 10 to 15 mm.
4. Wall thickness of beds for heavy machines is up to 50 mm.
5. Welded beds have wall thickness from 6 to 20 mm, whereas the bearing parts have 10 mm and more, and the ribs are of 6 to 8 mm and more.
6. Maximum rigidity is the main requirement for beds.

$$\text{Rigidity } K = \frac{Z}{\delta}$$

where $\delta \text{ (deflection)} = \dfrac{Wl^3}{4EI}$ and

$$Z = \frac{1}{12} bh$$

Therefore, $K = \dfrac{1}{3} EV \cdot \dfrac{h^2}{l^4}$ (where Volume = $b \times h \times l = V$)

or $K \propto EV$

This formula is good for designing. It shows that

1. The higher the rigidity, the greater is the volume of metal required for a bed;
2. The higher the bed, the better;
3. Short beds are preferable;
4. Steel beds of the same volume have higher rigidity. Steel beds are widely used now.

The complicated technology of welding is the main difficulty when a steel bed is made. However, a large number of devices are required here. It can be justified only in the event of batch production.

14.3.2 Shape of Beds

Cast iron possesses better vibration-proof characteristics than steel (at the same weight). Beds should meet the following requirements:

1. Geometrically exact travels of machine tool units. It is achieved by proper machining of slide ways.
2. Cavities provided in beds to accommodate electric hydraulic equipment, etc.
3. Convenient beds, transportation. Openings for crowbars of eyes (in three points) should be provided. Sometimes both of them are absent (the shape of a bed being convenient for transportation).

Also, dimension and weight should meet the transport technical requirements. Because of this, beds are sometimes made collapsible. Junctions of such parts are of great importance.

There are modern trends in bed design to make beds in the form of closed frames. It increases rigidity, e.g. in gear hobbling machines.

Figure 14.5 shows the position of ribs in welded beds.

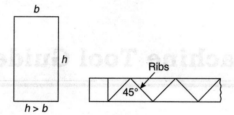

Fig. 14.5 Position of ribs in welded beds.

REVIEW QUESTIONS

1. Discuss the factors to be taken into account in selecting the material for machine tool beds/structures.
2. Discuss the types of sections used in design of a machine tool bed.
3. What is the equivalent Section Technique of design of machine tool bed?
4. What are the modern trends in bed design?
5. Why is a box section considered to be the best for beds and columns?
6. Discuss the advantages of C.I. over steel.
7. Among various cross sections, box arrangement for machine tool structure offers highest torsional stiffness and strength. However, opening and apertures have to be provided in box for housing bearing, chip flow, etc. and those have adverse effect. How can the strength and stiffness of box with apertures be then improved?
8. Explain the graphical approach to optimal design of a lathe bed.

CHAPTER 15

Design of Machine Tool Guides and Ways

15.1 INTRODUCTION

Straight line motions are guided on almost all kinds of machine tools. The guides not only produce accuracy of motion, but also withstand the heavy loads appearing when the machine tool is under operation.

The main requirements of a guideway can be listed as follows:

1. It should provide straight line motion under load as well as no load conditions.
2. It should not accumulate chips.
3. It should give minimum friction and good wear resistance.
4. It should have minimum deflection under cutting loads.

Guideways are mainly divided into three groups:

(a) Sliding
(b) Rolling, and
(c) Combination of sliding and rolling contacts in guide systems.

Guides are generally made of cast irons but can be made of steels, plastics, or reinforced concrete. But to cast a complicated profile is easier than to weld it; not only this, C.I. guides are preferable to any other material, as these possess better vibration-proof characteristics than steel for the same weight. On the contrary, plastic guides have been used for heavy machine tools after extensive research carried out in the erstwhile U.S.S.R., as well as the U.K. and the U.S.A., by machine tool research centres.

These guides show less wear and lower friction coefficient, better distribution of pressure, easy fabrication, etc. But they also have low strength and wear resistance, lower thermal conductivity, higher thermal deformation, etc.

Machine tool guides have to withstand the forces developed due to unfavourable conditions for the following reasons:

(i) Bad working surfaces
(ii) Uneven cutting forces causing uneven pressure and turning moments
(iii) Wear and abrasion due to entrapped chips between guide surfaces
(iv) Fluctuation of forces due to inhomogeneous work material
(v) Improper bearings used for carrying thrust force, which leads to excessive pressure on guides (circular).

15.2 WORKING SURFACE OF GUIDES

15.2.1 Flat Guideways

Figure 15.1(a) shows the plan view of guides. The ratio of the length of the movable part to the width should not be less than 1.5. Such guides should be well-lubricated. Chip disposal is not good.

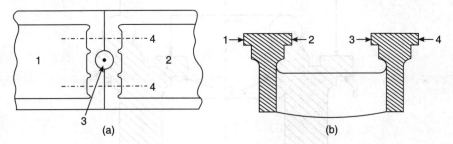

Fig. 15.1 Flat guideways.

Uses: These guides are used for heavy machines. Chip disposal is also not very good. Sometimes these guideways are used in medium and light machine tools too.

15.2.2 Inverted V-shaped Guideways

Inverted V-shaped guideways as shown in Fig. 15.2 are used in light power machine tools. Chip does not stick in such ways. Insufficient lubrication is not a serious problem in these guideways.

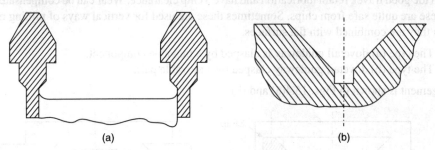

Fig. 15.2 Inverted V-shaped guideways.

These guideways have the following prominent features:

(i) They provide good longitudinal and transverse travel.
(ii) Chips are not accumulated on these ways.

(iii) Only requisite lubricant is retained.
(iv) Inverted ways are mainly used in planing and grinding machines, which must be kept clear from chips.

15.2.3 Combined Sliding Guideways (Flat and Inverted V)

The combined guideways are used for lathes, as these are suitable for mounting tailstock and carriage. The use of various elements of the combined guideways is shown in Fig. 15.3.

(i) Ways No. 1 serve for mounting the headstock and the tailstock. The wear is not considerable.
(ii) Ways No. 2 are used for carriage; the wear rate is quite high, but it is adjustable. Besides, it is uneven too.
(iii) Ways No. 3 are used to prevent the overturning of movable parts.
(iv) Flat surfaces are used for mounting a rack for longitudinal feed and for fixing the tailstock, steady rests, etc.

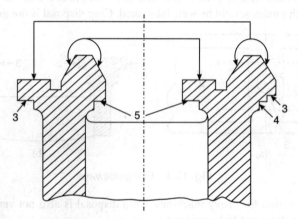

Fig. 15.3 Combined sliding guideways.

15.2.4 Dovetail Guideways

Such guides are used in lightweight machine tools and for movable parts such as lathe supports. Dovetail guides provide good travel, retain lubrication and have a chip clearance. Wear can be compensated to a certain extent. These are quite safe from chips. Sometimes these are used for vertical ways of milling machines and sometimes they are combined with flat surfaces.

(i) The type (a) dovetail guideway is clasped by a movable component.
(ii) The type (b) dovetail guideway clasped by a movable part.

The arrangement is shown in Figs. 15.4(a) and (b).

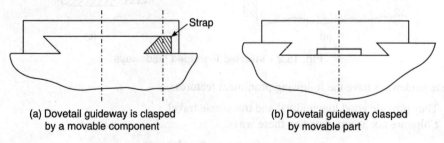

(a) Dovetail guideway is clasped by a movable component
(b) Dovetail guideway clasped by movable part

Fig. 15.4 Dovetail guideways.

15.2.5 Cylindrical or Round Guideways

This type of guideway is shown in Fig. 15.5. These ways are quite convenient for multi-tool machines and automates, of which the metal removal rate is quite high. Lubrication supplied by a movable part is sufficiently good. The wear is compensated for by the balls placed in an alternate layer; chips do not penetrate into the ball and cause no harm as balls are quite hard.

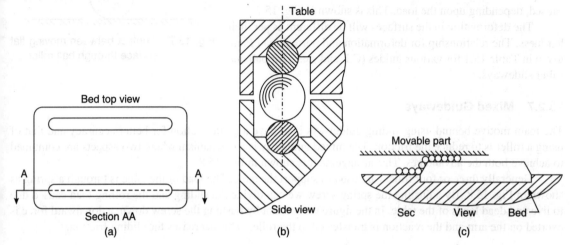

Fig. 15.5 Cylindrical or round guideways.

These ways are also used in presses, broaching and honning machines.

15.2.6 Roller Antifriction Guideways

In costlier machine tools, balls and rollers are used in the guideways of machine tools. These are also known as *antifriction guideways* (see Fig. 15.6). Such ways are being used as they offer the following advantages:

   (i) Load bearing capacity is higher.
   (ii) Lubrication requirement is moderate.
   (iii) Less wear but no compensation.
   (iv) Low coefficient of friction in comparison to conventional sliding guideways.
   (v) Chips can not penetrate and no disturbance to the motion to harm the guideway is created.
   (vi) Very accurate movements are obtained.
   (vii) Absence of stick-slip phenomenon.

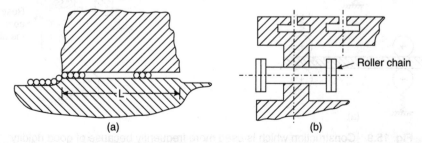

Fig. 15.6 Roller antifriction guideways.

The use of ball and roller ways require a smooth and hard guiding surface, thus increasing the cost of the machine tool.

The maximum travel in such guideways is limited to 2/3rds of the contact length of the rollers chain, as shown in Fig. 15.7.

When the moving surface comes in contact to the fixed surface through a ball and roller, a deformation in both the flat surfaces is caused, depending upon the load. This is shown in Fig. 15.7.

The deformation in the surfaces will be different as per their hardness. The relationship for deformation and normal load P are given in Table 15.1 for various guides (C.I. and steel) for ball and roller slideways.

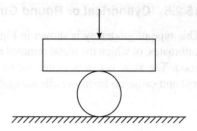

Fig. 15.7 Contact between moving flat surface through ball roller.

15.2.7 Mixed Guideways

The main motive behind using sliding guideways is to have a guide motion for better accuracy and that of using a roller is to reduce the friction. The mixed guideway is a mechanism where two aspects are combined to achieve both the advantages. This arrangement is shown in Fig. 15.8.

Generally three or four rollers are used to support the table. The load of the table is through a spring as shown in Fig. 15.8. By adjusting the spring screw we can divide the sliding and the rolling load caused due to live and dead loads of the table. In the figure we see that by pushing the screw down, a downward force is exerted on the arm and the reaction is transferred to the roller. This decreases the sliding friction.

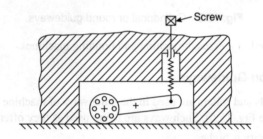

Fig. 15.8 Mixed guideways.

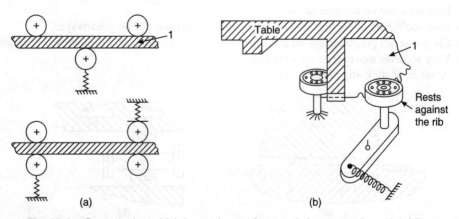

Fig. 15.9 Construction which is used more frequently because of good rigidity.

Functions of rib may also be fulfilled by other parts of the machine.

15.3 GUIDE AND SLIDEWAY MATERIAL

As the guided and the guiding surfaces are in direct contact, hence wearing of these surfaces is certain. If the mating parts are of the same material or have similar physical properties, then the wear of the upper part shall be more. The wear is caused due to several factors such as:

  (i) Surface condition of the mating parts or guideways.
 (ii) Deposition of dirt on the slideways.
(iii) Physical and mechanical properties of the mating parts.
(iv) Pressure exerted.

Usually C.I. and semi-steels with Ni and Cr are the most common guideway materials and possess sufficient strength, good damping capacity and requisite wear resistance. Surface treatment is generally given to guides.

Plastic guides are also used sometimes. Their merits and demerits have already been described in the beginning of this chapter.

Table 15.1 shows the deflection for different guideway materials.

Table 15.1 Deflection for different materials guideway

Guide and slideway material	Deflection (micron)	Formula for deflection
Steel guide with ball slideway	δ	$1.48 \sqrt[3]{\dfrac{p}{r}}$; r = Ball or roller radii in cm and p = Intensity of normal load in kg/cm^2
Steel guide with roller slideway	δ	$0.0028\, p\, \log e \left[8 \times 10^6 \left(\dfrac{r}{p} \right) \right]$
C.I. guides with ball slideways	δ	$1.9 \sqrt[3]{\dfrac{p}{r}}$
C.I. guides with roller slideways	δ	$0.0028\, p\, \log e \left(4.8 \times 10^6 \dfrac{r}{p} \right)$

15.4 PRESSURE ON SLIDING WAYS

15.4.1 Pressure Calculation

A designer has to consider for calculation the strength of all machine elements and the stresses coming up due to load under dynamic conditions. These components should be designed in view of the safe allowable stress on the machine element and its strength. In the previous chapters of this part of the book, it has been explained that the cutting forces are the governing factors for the design of a machine tool; but the surface pressure on the sliding guideways due to cutting forces decides the various parameters of the way.

In Fig. 15.10, it is evident that the guideways pressure is directly influenced by the cutting force system, i.e. F_y and F_z. The values of the pressure, i.e. P_z and P_z' can be calculated as follows:

$$P_z = F_z \times \left(\dfrac{l+d}{2l} \right) + F_y \cdot \dfrac{h}{l}$$

where d is the diameter of the job. Therefore

$$P'_z = F_z \times \left(\frac{l+d}{2l}\right) - F_y \cdot \frac{h}{l}$$

The magnitude of these pressures depend upon the instantaneous position of the cutting tool. As the cutting tool advances from the tailstock end towards the headstock, the centre point of the pressure on guideway also changes. Not only this, as the turning diameter of the job changes the point of action of the mean specific pressure also changes as shown in Fig. 15.11.

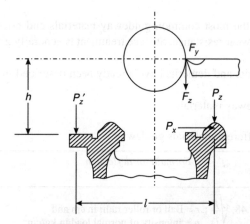

Fig. 15.10 Stresses on guideways due to cutting forces.

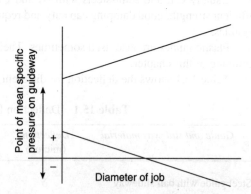

Fig. 15.11 Mean specific pressure on guideways with change in diameter of job.

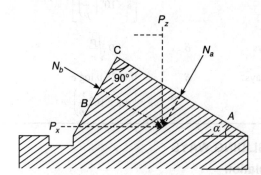

Fig. 15.12 Point of action of mean pressure on guideways.

Thus, it appears that the cutting process poses a dynamic force system on the sliding forces of the guideway, such that the pressure point as well as the magnitude of the force changes with turned diameter in case of a lathe.

If it assumed that the two forces A and B of the guideway make an angle of 90° at O (see Fig. 15.12), and that the surface A is inclined at an angle α to the horizontal. Then we have

$$N_a = P_z \cos \alpha - P_x \sin \alpha \tag{15.1}$$

and
$$N_b = P_z \sin \alpha - P_x \cos \alpha \tag{15.2}$$

If we substitute the value of P_z and P_x from Eqs. (18.1) and (18.2), we get the following relationship:

$$N_a = F_z \left(\frac{l+d}{2l}\right) \cos \alpha - F_y \left(\sin \alpha - \frac{h}{l} \cos \alpha\right) \qquad (15.3)$$

and

$$N_b = F_z \left(\frac{l+d}{2l}\right) \sin \alpha - F_y \left(\cos \alpha + \frac{h}{l} \sin \alpha\right) \qquad (15.4)$$

The angle α of guide face A is also a factor affecting the normal pressure N_a. If we analyse, we see that α should be limited to a value such that N_a does not become negative. Therefore

$$F_z \left(\frac{l+d}{2l}\right) \cos \alpha > F_y \left(\sin \alpha - \frac{h}{l} \cos \alpha\right)$$

and if it is not so, then N_a is negative, the implication of which is that the saddle may start lifting and leaving the guideway. In order to prevent the lifting of saddle even when there is no job on the chuck, we must have

$$\frac{F_z}{2} \cos \alpha > F_y \left(\sin \alpha - \frac{h}{l} \cos \alpha\right)$$

and to meet this condition α comes out to be 60° or less.

The example of a lathe guide has been considered in the present analysis; a combination of a flat and inverted 'V' guideway is taken as shown in Fig. 15.12. Usually in the case of the machine tools, we get non-uniform distribution of pressure; however, it can be assumed that pressure distribution is constant across the width of the guide and is a linear function of the length.

For non-uniform distribution of pressure, it is necessary that we must determine the maximum pressure for the calculation of the wear in guideways.

In Fig. 15.13(a), (b), (c) and (d), we consider four possibilities of the non-uniform distribution of pressure on the guideway. The area hatched by vertical pressure lines indicates the total force acting on a guide face of length l, and the centre of such a pressure is indicated by p_{ave} acting at a distance from the centre of the length, representing the centre of gravity of the pressure diagram.

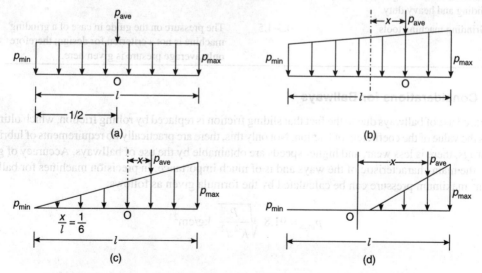

Fig. 15.13 Pressure distribution across guideways according to their profile.

From the fundamentals, the total force of any one of the cases can be obtained as follows:

$$\text{Force} = \frac{p_{max} + p_{min}}{2} \cdot lx \text{ where } x \text{ is the width of guide face under steady condition.}$$

Therefore, we get

$$p_{max} = p_{ave}\left(1 + \frac{6x}{l}\right) \text{ from Fig. 15.13(c).}$$

p_{ave} acts in between $0 < x < \dfrac{l}{6}$ for practically almost all machine tool guides.

The mean specific pressure can also be determined by the following procedure depending upon the guideway profile.

In all these cases the deformations are not accounted for, as specific pressures are insignificant. Such an assumption appears quite satisfactory and it is found that $p_{max} = 2.30$ kg/cm².

The higher the accuracy of a machine, the less is p_{max}. In machine tools of especially high precision, p_{max} sometimes is expressed in g/cm². Table 15.2 gives values of p_{max} for various machine tools.

Table 15.2 Permissible values of p_{max} for various types of machine tools

Type of machine tool	Value of p_{max} in kg/cm²	Remark
Lathe milling machine tools	25–30	This value of p_{max} is recommended for low velocity of sliding. For steel guides, p_{max} is 25% higher than this value.
Automates and special purpose machine tools	15–20	
Planers, heavy shapers, piano milling machine tools	10–12	These values of p_{max} are for very low velocity of sliding and for heavy duty machine tools.
Machine tools of high velocity of sliding and heavy duty	2.5–5	
Grinding machine tools	0.5–1.5	The pressure on the guide in case of a grinding machine is not a criterion for design; therefore only average presure is given here.

15.4.2 Considerations for Ballways

The increased use of ballways due to the fact that sliding friction is replaced by rolling friction, which ultimately decreases the value of the coefficient of friction. Not only this, there are practically no requirements of lubrication of guideways, there is less wear, and higher speeds are obtainable by the use of ballways. Accuracy of guided motion is the main characteristic of the ways and is of much importance. In precision machines for ballways, the kg/cm² maximum pressure can be calculated by the formula given as follows:

$$p_{max} = 91.8 \sqrt[3]{\frac{P}{K^2 d^2}} \text{ kg/cm}^2$$

where P is the load acting on each ball in kg
        d is the ball diameter in mm
        K is the coefficient accounting for elastic characteristics of materials being in contact given by

$$K = \left(\frac{1-\mu_1^2}{E_1} + \frac{1-\mu_2^2}{E_2} \right) \text{mm}^2/\text{kg} \qquad (15.5)$$

μ_1 and μ_2 are the ratios of Poisson's coefficient of the material
E_1 and E_2 are the moduli of elasticity. For material of ball and guide, kg/mm^2

For roller guideways

$$p_{max} = 0.8 \sqrt{\frac{q}{K \cdot d}} \text{ kg/mm}^2, \text{ for steel } p_{max} = 30{,}000\text{--}35{,}000 \text{ kg/mm}^2$$

where K and d are the same as above.
        q is the load in kg per mm width of the roller.

The obtained stress p_{max} is correlated with the yield limit.

15.4.3 Roller and Ball Guideways

As discussed previously in this chapter, the advantages of a combined guideway are that we get sliding and rolling frictions, and a normal load which acts normal to sliding and rolling contact surfaces is divided. Therefore total force is given as follows:

$$\text{Force} = \mu_1 N_1 + (\mu_2 \cdot N_2)\frac{2}{D}$$

where N_1 is the normal load acting on sliding contact;
        N_2 is the normal load acting on rolling contact;
        μ_1 and μ_2 are the sliding and rolling coefficients of friction;
        D is the diameter of the roller.

Thus, given the magnitude of this force, the average pressure on the guideway can be found out or the overall coefficient of friction can be found out.

$$\mu = \frac{F}{N} = \frac{\mu_1 N_1 + \mu_2 N_2 \left(\dfrac{2}{D}\right)}{N}$$

$$= \mu_1 \frac{N_1}{N} + \frac{\mu_2 N_2}{N} \cdot \frac{2}{D} = k \cdot \mu_1 + (1-k)\mu_2 \cdot \frac{1}{R} \qquad (15.6)$$

15.5 LUBRICATION AND PROTECTION OF GUIDEWAYS

15.5.1 Lubrication Theory

Considerable research has been done in this area in former USSR at the Machine Tool Institute Moscow, by Siebel, Mansurov and others by S.K. Basu and N.K. Mehta in India. The present discussion is based on the author's presentation of this topic.

The existence of an oil film, which completely separates the two bearing surfaces, helps in reducing the friction between the guiding surfaces. But in order to reduce the kinetic friction to a minimum, we must have hydrodynamic lubrication between the guiding surfaces. The force generated in the lubricating film supports all external loads. These external forces are of static and dynamic nature. The static loads are due to the weight of the moving and stationary parts directly above the film, e.g. table, workpiece and other accessories. The dynamic loads are due to cutting force. Practically, film lubrication occurs in planing, grinding, vertical boring turning and milling machines when the cutting motion is in the direction of table feed.

When the requirements for lubrication are met, we get what is called *thick film lubrication*. It implies that the surfaces are separated and the frictional forces present are only due to the internal resistance of the lubricant. This film lubrication is also called *hydrodynamic lubrication* or *perfect lubrication*.

The thin-film lubrication is obtained when some of the requirements are not wholly met. Although thin film lubrication exists widely, its exact nature has not been explained satisfactorily. But it is certain that the true character of the surfaces needs to be more thoroughly investigated by using thin film rather than thick film, as this may be a more practical case. This film lubrication is called *imperfect lubrication* or *boundary lubrication*.

In a straight flat guide the lubrication may be considered to be hydrodynamic, assuming that table and bed are separated by a film of considerable thickness. A film is formed in between the surfaces. This gives a pressure distribution along the film length as shown in Fig. 15.14.

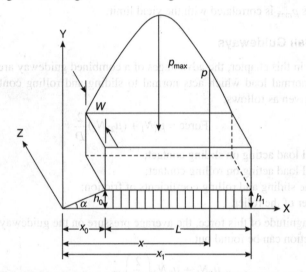

Fig. 15.14 Lubricant film and pressure distribution across it.

Here we develop the theory, assuming:

(i) The lubricant obeys Newton's law of viscous flow.
(ii) The lubricant is assumed to be incompressible.
(iii) The viscosity is assumed to be constant throughout the film.
(iv) The pressure is assumed to be constant on any line perpendicular to the direction of sliding. Thus from Fig. 15.15, we can analyse the force system, where an element of lubricant is considered with unit thickness in the z direction; thus we can determine the forces and formulate them to obtain an equation as follows:

$$p\,dy - \left(p + \frac{dp}{dx}dx\right)dy - s$$

$$dx + \left(S_s + \frac{\theta_{ss}}{\delta y}dy\right)dx = 0$$

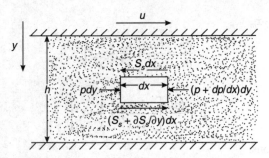

Fig. 15.15 Element of lubricant for analysis of force system acting over lubricating film.

This gives

$$\frac{dp}{dx} = \frac{\delta S_s}{\delta y}$$

where p is the oil film pressure, and
S_s is the shearing stress of the fluid given by

$$S_s = \frac{F}{A} = \mu \frac{du}{dy} \text{; } F \text{ is the force required to move the plate and } A \text{ is the area of the surface.}$$

Now, $\qquad S_s = \mu \frac{du}{dy} \quad$ or $\quad \frac{dS_s}{dy} = \mu \frac{d^2 u}{dy^2}$

After integrating this equation twice and putting the boundary conditions, we get

$$\frac{d}{dx}\left(\frac{h^3}{\mu}\frac{dp}{dx}\right) = 6u\frac{dh}{dx} \quad \text{or} \quad \frac{d}{dx}\left[\frac{dp}{dx}\right] = \frac{6\mu u}{h^3}\frac{dh}{dx} \tag{15.7}$$

This is the classic Reynolds equation for one-dimensional flow, neglecting side leakage in the z direction. In general, a situation is applicable when side leakage is not neglected, and this gives the following equation, called Newton's pressure distribution:

$$\frac{\partial}{\partial x}\left(\frac{h^3}{\mu}\frac{\partial p}{\partial x}\right) + \frac{\partial}{\partial z}\left(\frac{h^3}{\mu}\frac{\partial p}{\partial z}\right) - 6u\frac{dh}{dx} = 0 \tag{15.8}$$

where h is the thickness of the oil film at the point x where pressure is p
μ is the viscosity of oil
u is the velocity of sliding.

For the present analysis, side leakage is neglected $\frac{\partial p}{\partial z} = 0$ and Eq. (15.7) is utilized.

Since h_m is the thickness of the oil film at a point where p is any general value of pressure follows a parabolic distribution along x-direction and h_m is the thickness of oil film at x_m a point along x-axis where p is maximum.

Integrating Eq. (15.7)

$$\frac{\partial p}{\partial x} = 6\mu u\left(\frac{h_m}{h^3} - \frac{1}{h^2}\right)$$

Since $h = \alpha \cdot x$, $h_m = \alpha \cdot x_m$

Therefore,
$$\frac{\partial p}{\partial x} = \frac{6\mu u}{\alpha^2}\left[\frac{x_m}{x^3} - \frac{1}{x^2}\right] \qquad (15.9)$$

Integrating, we have
$$p = \frac{6\mu u}{\alpha^2}\left[-\frac{1}{2}\frac{x_m}{x^2} + \frac{1}{x} + C\right]$$

Putting the boundary conditions:
$p = 0$ when $x = x_0$ and $p = 0$ when $x = x_1$

We get
$$C = -\frac{1}{x_0 + x_1}$$

and
$$x_m = \frac{2x_1 x_0}{x_0 + x_1}$$

Thus,
$$p = \left(\frac{6u\mu}{\alpha^2}\right)\left[\frac{(x_1 - x)(x - x_0)}{(x_0 + x_1)x^2}\right] \qquad (15.10)$$

We write $p = A \cdot U$ where $A = \left(\frac{6u\mu}{\alpha^2}\right)$ independent of x coordinate and U depends on x, i.e.

$U = \frac{(x_1 - x)(x - x_0)}{(x_0 + x_1)x^2}$ for the above expression.

Therefore,
$$\frac{dp}{dx} = A\frac{dU}{dx} \qquad (15.11)$$

Comparing Eqs. (15.9) with (15.10) we can write:
$$\frac{dU}{dx} = \left(\frac{x_m}{x^3} - \frac{1}{x^2}\right) = \left\{\frac{(x_1 - x)(x - x_0)}{(x_0 + x_1)x^2}\right\}$$

Now, we substitute: $\frac{h_0}{x_0} = \alpha$, $m = \frac{x_0}{L}$, $M = \frac{x^2(1-E)E}{(2m+1)(m+E)^2}$; and $E = \frac{x - x_0}{L}$ in Eq. (15.10)

Thus,
$$p = \frac{6\mu u L}{h^2}M$$

For total force P, integrating p w.r.t. x between limits x_0 and x_1 for unit width of oil film.

$\therefore$
$$P = \frac{6\mu u}{h_0^2} \cdot \int_{x_0}^{x_1} M = \frac{6\mu u L^2}{h_0^2} \cdot K$$

where $K = m^2\left(\log_e \frac{m+1}{m} - \frac{2}{2m+1}\right)$.

Practical formula: For a finite breadth of guides, say W and assuming parabolic pressure distribution, Shiebel has shown that the total vertical force P can be equated to:

DESIGN OF MACHINE TOOL GUIDES AND WAYS

$$P = \frac{5\mu u}{h_{min}^2}\left[\frac{L^2 W}{1+(L/W)^2}\right] K = \frac{5K\mu u}{h_{min}^2}\left[\frac{L^2 W}{1+(L/W)^2}\right]$$

where K is a function of m and

$$K = m^2\left[\log_e \frac{m+1}{m} - \frac{1}{\left(m+\frac{1}{2}\right)}\right] \text{ and } m = \frac{x_0}{L}$$

If a graph is plotted as shown in Fig. 15.16, a parabolic curve which exhibits that the value K is maximum when very near $m = 0.7$ and $K = 0.27$ is obtained.

Hence, it is found that at a practical level, in most of plain guides of planing, shaping and grinding machine tools, $L \gg W$

Therefore,

$$\frac{1}{1+\left(\frac{L}{W}\right)^2} = \frac{1}{\left(\frac{L}{W}\right)^2}$$

$\Rightarrow$

$$\frac{L^2 W}{1+\left(\frac{L}{W}\right)^2} = \frac{L^2 W}{\left(\frac{L}{W}\right)^2} = W^3$$

Therefore, the load P is obtained by the equation

$$P = \frac{\mu u}{h_{min}^2} \cdot W^3 \cdot 5K \text{ or } P = \frac{0.133 \mu u W^3}{h_0^2} \quad (\text{Since } K = 0.0267)$$

for the above value.

Also, the value of h_0 can be found from the equation

$$h_0 = \Delta e_1 + \Delta e_2 + \Delta e_3$$

where $\Delta e_1, \Delta e_2, \Delta e_3$ are errors in geometric forms due to the errors of roughness, waviness and lack of parallelism. For the purpose of calculation

$h_{min} = 0.01\text{--}0.02$ mm   (in normal)

$\phantom{h_{min}} = 0.06\text{--}0.01$ mm   (in heavy duty machine tools)

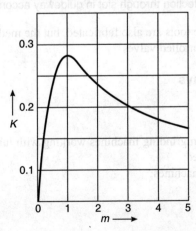

Fig. 15.16 Variation in values of K with respect to variation in m.

For longer life of a machine tool, it is better to have hydrodynamic lubrication, as this may balance a position of the load coming on the table, thus reducing wear on the guideway surfaces and increasing machine tool life.

In order to get hydrodynamic lubrication at the guideway surfaces, it is necessary to provide oil pockets by scraping the bed of the machine tools. These oil pockets help to fulfill the conditions to get hydrodynamic lubrication.

Other than this, for increasing the life of machine tool guideways, we protect these by providing guards, pleated cloth and felt packings as shown in Figs. 15.17 and 15.18.

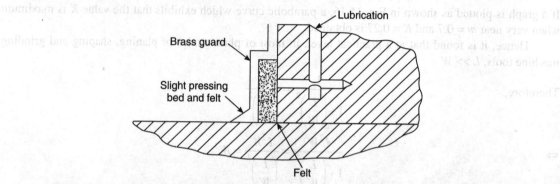

Fig. 15.17 Overload protection through brass guard, pleated cloth and felt packing.

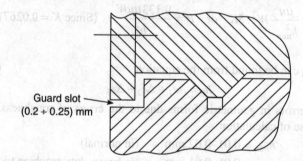

Fig. 15.18 Overload protection through slot in guideway accommodating film lubricant.

Plastic guideways for machine tools are also fabricated, but the method of lubrication, of such ways is further controlled by electrically controlled valves.

15.5.2 Protection of Guideways

The methods of protection are:
1. Guards;
2. Guards with air gas (e.g. in grinding machines working with lubrication). Such slots are effective only if lubrication is applied;
3. Pleated cloth in grinding machines;
4. Felt or fine-felt packings.

REVIEW QUESTIONS

1. What are the main requirements of a machine tool guideway?
2. What are the materials for construction of guideways?
3. How are guideways lubricates?
4. Discuss the methods of protection of guideways.
5. Discuss the types of guideways.
6. Discuss the factor tool guideways.
7. How is wear compensation achieved in case of V-guideways?
8. Describe the constructional features and important characteristics of the machine tool guideways.

Chapter 16

Design of Feed Power Mechanism and Screw

16.1 TRANSLATORY MOTION MECHANISMS

A kinematic pair of a translatory power motion system may consist of many types, but the main purpose of each is to provide exact movements to the main organs of a machine tool unit. Some of these pairs are discussed in the following sections.

16.1.1 Rack Gear and Rack-toothed Sector

This is used in feed motions and control mechanisms. This pair is rarely used in the main reciprocating motions of planning, broaching machines, etc.

16.1.2 Worm and Worm Rack

It is mostly used in the main drive of machine tools reciprocating motion.
    The length of travel is 4 metres and more.

16.1.3 Nut and Screw

The pair is widely used. It has a trapezoidal profile with a top angle of 30°. Such a profile provides easy closing of a split nut and the machining is simpler than that of a rectangular profile. Rectangular profile of the thread is used for the screws of high accuracy and the nut is unsplit as the run-out does not influence the accuracy of a nut travel. To improve the accuracy of a nut travel, corrective devices may be employed.
    The various kinematic pairs as described above are used depending upon the type of machine tools. However, these pairs can also be selected on the basis of accuracy of the pair and intended use as grouped in Table 16.1.

Table 16.1 Selection of kinematic pairs based on accuracy and intended use

Machine tool type	Accuracy
High precision machines	± 0.001 mm
Precision machines	± 0.004 mm to ± 0.008 mm
General workshop machines	± 0.015 mm to 0.025 mm
Machines not requiring high accuracy (saws, etc.)	± 0.045 mm and above

A screw transmitting power must be designed properly so that it may not fail (or may not remain accurate for transmitting motion) due to propelling fork, bending, twisting or due to combined stresses. The approximate intensity of pressure on the contact surface of the kinematic pair names from 30 kg/cm² to 80 kg/cm². However, in some cases it may exceed 120 kg/cm².

It is therefore necessary to decide upon the material of screw and the complementary part of the pair. Generally, screw is made of steel having hardness of 50–60 RC, and if it is a lathe the half nut is made up of brass or bronze. When high accuracy is required, power screws are made of high carbon steel.

V threads have the highest frictional resisting force and square threads have the minimum. Therefore, V threads are used for fastening purposes and square threads for power transmission purposes. Consequently it is always preferable to use square threads for lead screw of a lathe or for any other such use. The various kinds of threads used in power screws are shown in Fig. 16.1.

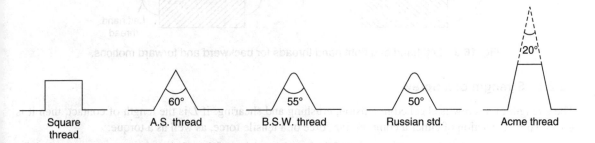

Fig. 16.1 Different shapes of threads of power screws.

16.2 DESIGN OF A SCREW

Take the case of a lead screw. It can be assumed as a beam loaded axially by a load equal to the propelling force. For a square thread lead screw as shown in Fig. 16.2, mean diameter d_m is determined as follows:

$$d_m = 8\sqrt{\frac{Q}{\lambda - p}}$$

where Q is the propelling force of the screw;
λ is the length of contact;
p is the specific pressure on the surface of contact (p varies from 30 to 80 kg/cm²).

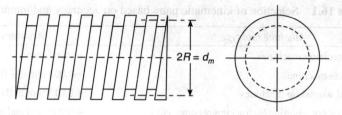

Fig. 16.2 Screw and nut motion.

The left hand and right hand threads for backward and forward motions are shown in Fig. 16.3.

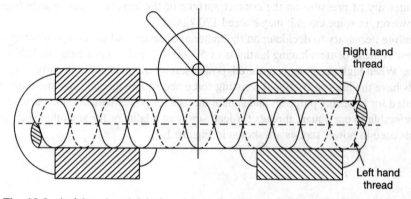

Fig. 16.3 Left hand and right hand threads for backward and forward motions.

16.2.1 Strength of a Screw

The strength of a screw is checked in tension, crushing and shearing. If L is the length of contact, then it is governed by the action of either a compressive force or a tensile force, as well as a torque.

1. **Direct stress:** As a starting point of design, the number of threads in contact are assumed 3–3.5.

$$n = \frac{L}{p_m} = 3 \text{ to } 3.5$$

where L is the length of contact
 p_m is the pitch of threads
 n is the number of threads in contact.

Then direct stress

$$\sigma_{max} = \sqrt{\sigma^2 + 4\tau^2}$$
$$= \sigma_{yield}$$

Again, $\dfrac{Q}{A}$ = Direct stress, where $A = \dfrac{\pi}{4} d^2$.

Therefore,
$$\sigma_{max} \leq \frac{Q}{\dfrac{\pi}{4} d^2}$$

while checking for direct stress.

2. **Bearing/Crushing stress (*p*):** The specific pressure on the surface of contact

$$p = \frac{Q}{\pi d_m \times n \times w}$$

where Q = applied load
d_m = mean diameter
w = depth of the thread.

3. **Shearing stress (τ):** It is defined as

$$\tau = \frac{16M}{\pi d_c^3}$$

where M is torque
d_c is diameter (core).

16.2.2 Rigidity Checking

The deflection caused in the power screw will cause error in pitch. It will be produced due to axial load Q and also due to torque M. Each will be calculated for supplying power.

1. Deflection Δ_1 due to Q:

$$\text{Strain }(e) = \frac{\Delta_1}{p} = \frac{Q}{AE}$$

$\Rightarrow$

$$\Delta_1 = \frac{Qp}{AE}$$

where A is the area and
E is Young's modulus.

2. Deflection Δ_2 due to Torsion: From the geometry of Fig. 16.4, since the angle of twist α' in 2π results in deflection Δ_2 in pitch p. Hence,

$$\frac{\Delta_2}{p} = \frac{\alpha'}{2\pi}$$

Now

$$\alpha' = \frac{Mp}{G} \quad \text{(from the theory of Torsion)}$$

where G is the modulus rigidity,
J is the polar moment of inertia,
α' is the angle of twist in radians, and
p is pitch.

We now introduce efficiency η in the equation

$$\frac{\text{Work output}}{\text{Effective input}}(\eta) = \frac{Qp}{2\pi M}$$

$\Rightarrow$

$$\eta = \frac{Qp}{2\pi M}$$

Therefore,

$$M = \frac{Qp}{22\pi\eta}$$

Also, $$\Delta_2 = \frac{Qp^3}{4\pi^2 \eta JG}$$

Now, $$G = \frac{4}{10}E \text{ for steel}$$

$$J = \frac{\pi}{32}d^4 = \frac{Ad^2}{8}$$

Hence, deflection Δ_2 due to Torque $M = \dfrac{Qp}{EA} \cdot \dfrac{p^2}{2\eta d^2} = \Delta_1 \cdot \dfrac{p^2}{2\eta d^2}$ (approximately)

Therefore, total pitch error $(\Delta) = \Delta_1 + \Delta_2 = \Delta_1 \left[1 + \dfrac{p^2}{2\eta d^2}\right]$

The major pitch error Δ_1, i.e. $\dfrac{Qp}{EA}$ is much greater than Δ_2. Except in multistart screw, the effect of Δ_2 to change value of p can be neglected because $\left(\dfrac{p}{d}\right)^2$ is quite small.

Figure 16.4 shows the calculation of error caused by torque.
For ordinary lead screws, admissible angle of twist adopted is 3° along the length of a screw.

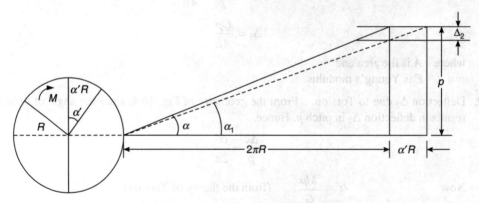

Fig. 16.4 Calculation of error due to torque.

Euler's Buckling formula of screws

Buckling in the case of a lead screw is caused due to axial thrust and can be calculated by Euler's formula when the distance between unsupported ends of the screw (l) is more as per the condition: $200 \geq \dfrac{l}{k} \geq 115$ for long columns (and beyond $\dfrac{l}{k} > 200$ the column is useless).

We can use

$$\sigma_{max} = \left[\frac{\pi^2 nE}{(l/k)^2}\right]$$

where k is the radius of gyration,
   $n = 1$ for hinged ends,
   $ = 2.25$ for fixed ends,
   $ = 1.6$ for both ends pinned or guided and partly restrained.

Long screws should be checked for longitudinal rigidity. Sometimes, the screw is supplied with steady rest barrels. In lathes, they are placed on both sides of an apron. Radial supports are used in the form of sliding bearings. Thrust supports are in the form of antifriction bearings. For high-speed screws, radial supports are made in the form of antifriction bearings. It is true for feed shafts too. High-speed lead screws have right hand and left hand threads. They serve for quick travels of heavy units of the machine tools.

The efficiency of transmission can be obtained using the formula:

$$\eta = \frac{\tan \alpha}{\tan (\phi + \alpha)}$$

where α is lead angle of screw.

This formula can be applied to ACME thread by putting

$$\phi = \tan^{-1}\left(\frac{\mu}{\cos \beta}\right)$$

where μ is coefficient of friction, and β is the semi thread profile angle of the ACME thread which is used in machine tools.

16.3 KINEMATIC PAIR—LINK MECHANISMS

A crank mechanism is a mechanism used for either oscillating or to and fro motion of a slider. It may be a quick return motion or a simple motion. It is used in shaping and planing, and sometimes a link mechanism is also used in control systems (Fig. 16.5).

Y_1 and Y_2 are not of the same measure if the guiding lever is made of C.I., and $Y_1 = Y_2$ if it is made of steel. The calculations are carried out for two supports at ends under the action of compressive and bending forces.

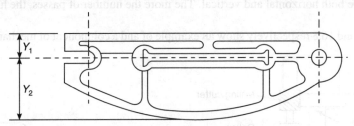

Fig. 16.5   Kinematic link pair for quick return mechanism.

16.4 KINEMATIC PAIR—CAM FOLLOWER AND LINK

Such pairs are used in automatic and ordinary control systems. Some forms of cams are as shown in Fig. 16.6.

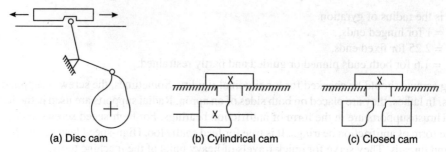

(a) Disc cam   (b) Cylindrical cam   (c) Closed cam

Fig. 16.6 Cam, follower and link mechanisms.

Design consists of the following:
1. Design of a cam profile;
2. Checking of cam mechanism (Pressure angles should not exceed permissible values);
3. Calculation for contact stresses.

In order to increase the distance of travels, leverage systems are used together with cams.

16.5 HINGED LEVERAGE SYSTEMS

Pantographs are used in copying machines. These are of high precision, and are mainly applied in machines of tool production.

Copying devices are divided into:
1. Mechanical
2. Electronic devices
3. Hydraulic devices
4. Electric–hydraulic devices.

The essence of these mechanisms consists of the following: A workpiece profile is completed by the face cam according to the drawing. The accuracy is not very high due to mechanical system, but nowadays volumetric copying is being employed on a large scale. The volumetric surface is divided into several plain profiles, and the machining is carried out by passes.

Passes may be both horizontal and vertical. The more the number of passes, the higher is the accuracy of a workpiece.

Figures 16.7 and 16.8 respectively show an example of and a component of mechanical copying device.

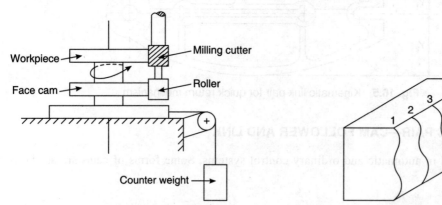

Fig. 16.7 Example of mechanical copying device.   **Fig. 16.8** Component of mechanical copying device.

Mechanical devices can not be successfully used to solve this problem, and hence electric or hydroelectric devices should be used.

16.6 MECHANISMS FOR INTERMITTENT MOTIONS

Intermittent motions are used where chain is being used as feed motion. There are two types of mechanisms:
  (i) Ratchet and pawl mechanisms.
  (ii) Maltese cross mechanisms.

Ratchet and pawl mechanisms are used in machine tools requiring rectilinear motions and other motions. Maltese cross is used in automatic cycle machine motions.

The intermittent motion comprises periods of fixation and motion. Ratchet and pawl mechanisms are an exception to this rule, as there is no fixation due to the self-stopping of the screw pair.

There are two systems of fixation:
  (i) Single fixation
  (ii) Double fixation.

Single fixation arrangements: The maximum accuracy of the single fixation system is of the order of 0.2 mm, and this system provides two points of support (see Fig. 16.9).

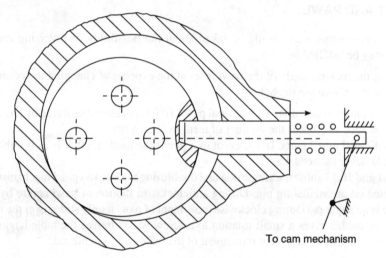

Fig. 16.9 Single fixation.

Double fixation arrangements: The maximum accuracy of this system is 0.1 mm. The operation for fixation or motion is usually effected by springs, and the removal of a stop and of a finger is carried out with cams A and C (see Fig. 16.10). The stop may be cylindrical or prismatic in shape. The cylindrical stop should be provided with a conical head, while the prismatic stop should have one or two chamfers in trapezium form.

The error in turning without a fixture is up to 1°, and in case of heavy machine tools it may be higher.

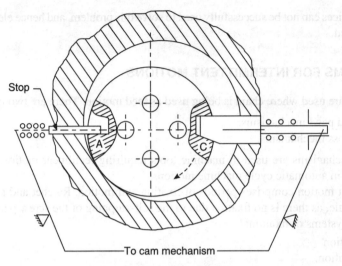

Fig. 16.10 Double fixation.

16.7 RATCHET AND PAWL

In designing such mechanisms, care should be taken about the possibility of reversing and controlling of travels. The latter may be fulfilled by

(a) Adjusting the rocking angle of the pawl lever at the expense of changing the eccentricity; and
(b) Placing plates above the ratchet.

The control or motions may be also done with several pawls (if it is necessary to divide by less than half a part).

Usually $Z_{max} = 150$, where Z is the number of teeth in the ratchet.

The angle is divided into 2 parts. In the event of 3 pawls the tooth number is as if trebled. The number of pawls on one axle never exceeds 4.

Figures 16.11 and 16.12 show the single dog and double dog ratchet and pawl mechanisms respectively. The pawl is mounted on an oscillating pin. During anticlockwise motion of pawl centre by α it rotates the ratchet wheel there is an angle α_0. During clockwise movement of pawl it only slides over the ratchet the tooth.

The rotation of ratchet gives a small rotation to a screw which moves the table having a nut through screw by a small amount. The intermittent movement of the table is thus achieved.

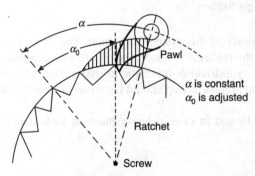

Fig. 16.11 Single dog ratchet and pawl mechanism.

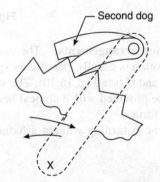

Fig. 16.12 Double dog ratchet and pawl mechanism for intermitted motion.

Figure 16.12 shows a system with 2 dogs to increase the reliability. The pawl may also be actuated by a hydraulic drive. Its merit is due to the scope of working with a rigid stop.

Ratchets are designed on the basis of crushing strength along the surface of contact, as well as bending of teeth and a pawl.

16.8 SAFETY DEVICES ON METAL CUTTING MACHINES

1. Block mechanisms are designed for preventing breakage, e.g. to prevent simultaneous engagement of the lead screw and the feed shaft.
2. Safety device against overloads, e.g. friction coupling as a safety machanism, will slip at a high torque; shear pins will be cut off, etc.
3. Safety devices for limiting deformations of the machine tool parts, e.g. if the traverse of the turning and boring mill is deflected to the amount exceeding the permissible sag, the motor of the machine automatically stops.
4. Temperature safety devices, e.g. if the heating of bearings exceeds the maximum limit, the motor stops.

Adjustable and limit stops belong to safety devices too. Adjustable stops serve the purpose of receiving the appropriate size of a part in a batch production. Limit stops serve as safety devices only, e.g. the limitation of the upper position of the traverse on the turning-and-boring mill is done either by this limit stop or by the coupling which is automatically disengaged if the upper position is reached.

16.9 SPINDLES AND SHAFTS

Shafts and spindles are designed to be run at high rotational velocity so that rigidity and accuracy are maintained, and no resonance is produced at any speed, and the life of the system is also enhanced.

Spindles and supports are made of steel containing carbon in the range of 0.4% to 0.45%. Cast iron is also used for spindles if and when stress and circumferential velocities are low, as in the case of facing, boring and turning mills, and other machines. Usually spindles and shafts are heat-treated (hardened or normalised), but sometimes nitriding is done whenever necessary to increase the life of spindle necks.

The diameter of a spindle is determined to provide good rigidity, while the other conditions of design are only to ensure strength.

Let us assume a spindle as a simply supported beam with a uniform load w all along its span of length l as shown in Figs. 16.13 and 16.14; then the admissible deflection y_{adm} varies from 0.0001 to $0.0005l$.

To provide for splines and keys, the diameter should be increased by approximately 10% as against the calculated value.

$$y = \frac{wl^3}{48EI} = y_{adm}$$

$$I = \frac{wl^3}{48E \cdot y_{adm}}$$

$$d = \sqrt[4]{\frac{64\, wl^3}{48E \cdot \pi \cdot y_{adm}}}$$

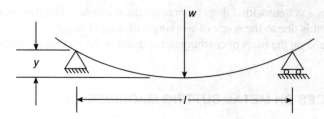

Fig. 16.13 Deflection of simply supported beam.

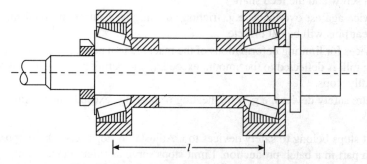

Fig. 16.14 Spindle supported on bearings.

When calculating the diameter of the spindle, all the forces acting on it should be taken into account. Besides, the permissible value of wedge angles of the supports and gears should be calculated. This angle is not important if the shaft is mounted on the self-setting bearings (spherical or ball bearings). Wedge angle β should be determined in case of the application of either sliding supports or supports with conical and cylindrical rollers.

The gear angle β may be neglected only if the gear is situated in the middle of the shaft; otherwise it should be determined.

This calculation may be a check, but sometimes it turns out to be the starting point for determining the shaft diameter.

16.10 EDGE EFFECT

The supporting edge of the spindle shaft influences the pressure distribution (see Fig. 16.15).

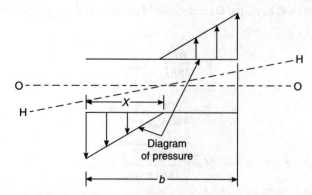

Fig. 16.15 Pressure distribution for spindle supported by edge.

The lower the value of X, the worse is the resistance of materials.

Such diagrams may be drawn for gears too (see Fig. 16.16). Cases I to III are applicable for designing the shafts. These diagrams are the first approach to the reality. Case IV demonstrates the presence of edge effect.

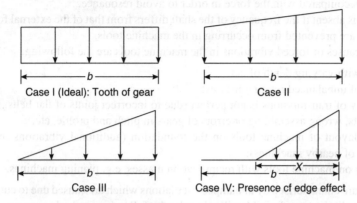

Fig. 16.16 Pressure distribution in case of edge support for gears.

The ratio of the maximum pressure to the mean pressure is the diagram coefficient K on the support, and b is the spindle support width.

$$\theta_{adm} = \frac{PK}{b^2 \cdot 10^4} \text{ radian} \quad \text{for the gear}$$

$$= \frac{PK}{10^4 \cdot P\left(\dfrac{250}{5000}\right)} \text{ radian}$$

where $P(250/5000)$ is the force under which the bearing will work for 5000 hours at 250 revolutions per minute.

Rake angle is determined by horizontal and vertical planes.

Spindles for drilling machines are designed according to the following equation:

$$\theta = \frac{ML}{GJ_0}$$

where Slenderness ratio of 20–25 is adopted in design times of lead screws $L/d = (20\text{–}25)$ assumed for lead screws

θ is the spindle deflection;

M is the bending movement;

G is the modulus of rigidity;

J_0 is the polar moment of inertia;

θ_{adm} is 3° on the whole length of the lead screw. This is permissible maximum angle of twist

16.11 WEAR

Wear is calculated on the basis of crushing strength of the material. Only 75% of the participating volume of the material in the work is taken into consideration in case of splined connections for wear calculation.

16.12 VIBRATION OF SPINDLES AND SHAFTS

Usually these calculations comprise the determination of vibration frequency for a shaft or a spindle; then the obtained frequency is compared with the force in order to avoid resonance.

The resonance is absent if the frequency of the shaft differs from that of the external force by 25 to 30%. Generally vibrations are prevented from occurring in the machine tools.

The principal causes of forced vibrations in the machine tools are the following:

  (i) Working with varying depth of cut;
  (ii) Presence of unbalanced rotating masses;
  (iii) The quality of transmissions is not perfect (due to incorrect joints of flat belts, incorrect stress of wedge belts, wrong assembling or errors of gears in pitch and profile, etc.);
  (iv) Incorrect layout of machine tools on the foundation (additional vibrations may appear due to vibrations of nearby machines);
  (v) Vibrations on machine tools with reciprocation masses, e.g. planing machines.

Machine tools are characterised by undamped oscillations which are caused due to cutting phenomenon. It is possible to eliminate undesirable vibrations by the following methods:

  (i) Attaching a flywheel is generally done in case of a milling machines and slotting machines. The weight of the flywheel varies from 80 to 100 kg for milling and from 200 to 300 kg for slotting machines; other than this, shaping machine has geared flywheel.
  (ii) Using the machines carefully and getting static and dynamic balance of rotating parts, as shown in Figs. 16.17, 16.18 and 16.19.
  (iii) Correct transmission through belts.
  (iv) In case of a gear drive, the gear wheels having the largest number of teeth should not be used last. Helical gear wheels should be used.
  (v) Proper foundation for any machine tool is a must to avoid vibration.

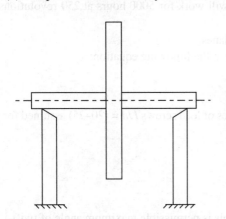

Fig. 16.17 Balancing of grinding wheel.

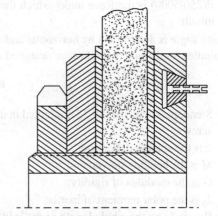

Fig. 16.18 Static balance gives good results only for parts in the form of discs. It is good for broad parts.

The part shown in Fig. 16.19 is statically balanced but it is under the action of two dynamic forces M_d in rotation. The part should be balanced by the dynamic force due to centrifugal action.

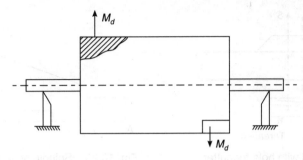

Fig. 16.19 Part in static balance and dynamic imbalance.

16.13 SPINDLE NOSE

Spindle noses are standardised for various machine tools. Threaded ends are applied in light and medium lathes, while flanged ends are used in heavy duty lathes. Turret and automatic lathes have special construction with a collet. Lathe spindles have screwed or flanged noses as shown in Fig. 16.20. Milling and drilling machines are provided with spindles which have tapered holes, using either a Morse taper or a taper shown in Fig. 16.21.

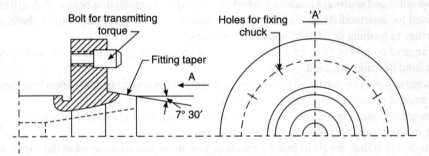

Fig. 16.20 Lathe spindles with screwed or flanged noses.

Let us consider the spindle nose for a drilling machine as shown in Fig. 16.21. The functions of the various components of the spindle nose (1, 2, 3 and 4) are described in detail below.

(a) In the simplest construction, elements 1 and 2 are omitted, while in some there are elements 1 (or 2); 3 and 4, and the nose looks as it is shown in Fig. 16.21, where a Morse taper is provided for holding the tools.

If element 1 in the nose is omitted, the torque is transmitted by friction only; slot 3 is for removal of tools.

Element 1 provides for a face key transmitting the torque (in heavy machine tools) and slot 2 serves for work while up-feeding. In this event the tool is provided with a hole for a cutter (see Fig. 16.22).

The spindle nose for grinding machines with medium power consumption is shown in Fig. 16.23.

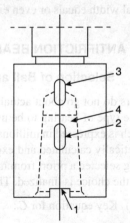

Fig. 16.21 Details of Morse taper in spindle of drilling machine.

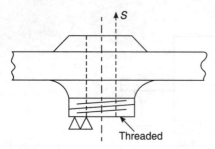

Fig. 16.22 Tool with hole for cutter.

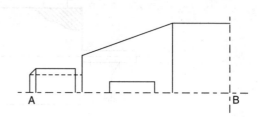

Fig. 16.23 Spindle nose for grinding machines with medium power consumption.

The threaded end A may be replaced by a threaded hole.

For light machines a capping is inserted between the grinding wheel and the spindle (for grinding wheels of small diameters).

16.14 SUPPORTS OF SPINDLES AND SHAFTS

Supports of spindles and shafts are made in the form of sliding and antifriction bearings. Antifriction bearings are usually used for intermediate shafts. Slow-speed shafts of control and other similar shafts are supported by plain bearings (a bushing or a bored hole in the housing).

If the range of control is 10 to 15 and more, and the speed of n_{max} is equal to 700 to 800 rpm and more, then amplification bearings are used.

If the range of control is narrow (2 to 5) and n_{max} is high, either sliding or antifriction bearings may be used.

If the range of control is wide, and n_{max} is small, sliding bearings are used.

If the range of control is narrow, and n_{max} is small, sliding bearings should be applied.

Sliding bearings are less prone to vibration than others.

Modern design trends are (i) to make a machine tool more vibration-proof at the expense of supports; and (ii) simplifying and reducing costs of spindle supports.

In order to obtain the equal vibration-proofing of sliding and antifriction bearings, it is necessary that their total width equals or even exceeds the width of sliding support. This is an empirical rule.

16.15 ANTIFRICTION BEARINGS

16.15.1 Selection of Ball and Roller Bearings

Designers do not go about actually 'designing' ball/roller bearings. They follow a systematic procedure, to select the type of bearing to be used to fulfill radial and axial load requirement to achieve a certain bearing life which is expressed in millions of revolutions. Apart from bearing life the dynamic load carrying capacity is theoretically calculated and expressed as C_t and compared with the actual load carrying capacity C_A of a bearining selected a priori from manufactures catalogue. If the selected bearing has a capacity C_A greater the C_t, then the choice is finalized. The various steps are given below:

  (i) Key equation for C_t:

$$C_t = P \cdot L_m^{1/b}$$

where, P is equivalent bearing load, Newtons

L_m is designed bearing life in millions of revolutions

and $b = 3$ for ball bearing and 3.34 for roller bearings.

Thus, to calculate C_t, P and L_m should be first determined.

(ii) **Calculation of P:** The equivalent bearing load P is given by equation:

$$P = xF_r + yF_a$$

where x and y are weightages to the radial load F_r and axial load F_a respectively. x and y are available from 'SXF'. Manufacturer's Catalogues or Design Data Books (e.g Mahadevan).

(iii) **Calculation of L_m:** The formula for L_m is

$$L_m = \frac{L_H \times 60 \times N}{10^6}$$

where L_H is bearing life in hours desired by the designer; N is revolutions per minute of the shaft on which the bearing is mounted. Using the values of P and L_m, the theoretical dynamics capacity C_t can be calculated. If the value of C_t is less than the actual capacity C_A of the bearing selected from the catalogue the selection is acceptable.

Illustrative example: To select a deep groove ball bearing for the following loading and RPM:

$$F_r = 6000 \text{ N}, F_a = 3500 \text{ N}, \text{RPM} = 1440 \text{ N}$$

Desired life = 550 hours.

Solution: Referring to standard tables:

$$x = 0.56, y = 1.25$$

∴ Equivalent load $P = xF_r + yF_a = 0.56(6000) + 1.25(3500) = 7742$ N

$$L_m = \frac{L_H \times 60 \times N}{10^6} = \frac{550 \times 60 \times 1440}{10^6} = 47.52$$

∴ From key equation:

$$C_t = PL_M^{1/b} = 7742(47.52)^{1/3} = 28064 \text{ N}$$

From Manufacturer's Catalogue, the standard ball bearing SKF = 6208 has capacity $C_A = 30700$ N. Thus the bearing SKF 6208 will more than meet the loading and life requirement.

16.15.2 Mounting of Antifriction Bearings

The following fits are applied for mounting the antifriction bearings into their housings:

(i) Snug fit
(ii) Drive fit
(iii) Wringing fit
(iv) Push fit.

In case the ring is rotating, the fit is quite tight; if it is stationary, the fit is loose.

In case the inner ring is placed on the shaft and the outer ring in the housing, then the fit in the housing is loose; but on the shaft it is tight. In this case, the outer ring must have the possibility of slight offset in axial direction.

Preliminary tightness is absent in some bearings. It should be provided by the unit construction. The preliminary tightness is not calculated, but it is determined in the course of mounting.

Axial tightness is carried out by two methods:

(i) Maximum preliminary tightness is limited beforehand; measuring distance sleeves are applied as shown in Fig. 16.24.
(ii) Preliminary tightness depends on the worker's skill during assembly of the antifriction roller bearing when manual assembly is done (see Fig. 16.24).

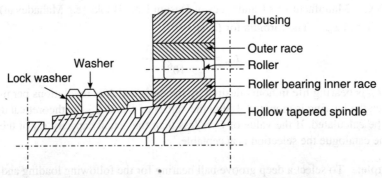

Fig. 16.24 Sectional view of mounted antifriction bearing.

Admissible run-out of sleeve faces should meet high requirement (see Fig. 16.25) of accuracy in case of deep groove ball bearings.

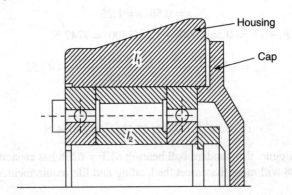

Fig. 16.25 Bearing assembly with two bearings in use.

Flange construction (see Fig. 16.26) is applied for providing displacement of the outer ring in respect of the inner one. The application of sealing eliminates oil leakage (board on nitro lacquer).

The shortcoming of this construction consists of the possible uneven tightness on the outer ring, due to which the edge effect may appear.

Figure 16.27 illustrates the construction made by the second method.

The thread with shortened pitch is applied in this construction (and so the regulation is more exact). The thread plug should be locked. Various methods of locking may be applied here (see Fig. 16.28).

DESIGN OF FEED POWER MECHANISM AND SCREW

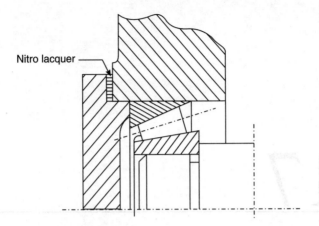

Fig. 16.26 Flanged coupling sealed by nitro lacquer for tapered roller bearing.

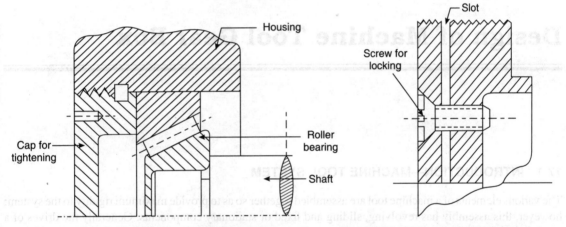

Fig. 16.27 Threaded construction for tightening—used to avoid uneven flange tightness.

Fig. 16.28 Method of locking thread plug used for tightening.

The stretching screw may be replaced by the stay screw. Sometimes the locking is fulfilled by means of straps. But the shortcoming of these constructions lies in the presence of a thread in the housing, leading to difficult production. In order to eliminate this shortcoming, steel threaded thimbles are inserted between the bearing and the housing.

REVIEW QUESTIONS

1. How will you design a lead screw of a lathe?
2. What are the mechanisms for intermittent motions?
3. What are the safety devices provided on metal cutting machine tools?
4. What is edge effect?
5. What are the safety precautions observed for the vibrations of spindle and shaft of a machine tool?
6. How are antifriction bearings mounted?

Chapter 17

Design of Machine Tool Gear Box

17.1 INTRODUCTION—MACHINE TOOL SYSTEM

The various elements of a machine tool are assembled together so as to provide maximum rigidity to the system; however, this assembly has revolving, sliding and fixed or stationary components. Generally the drives of a machine tool are covered and hidden, but operated by controls which are accessible to the operator. Various elements of a machine tool are made integral or fabricated and assembled together to make a system quite homogeneous in appearance and operation.

17.1.1 Drives and Regulation of Motion on Metal-cutting Machines

Metal-cutting machines receive working motions (speed and feed) from electric motors, which usually have constant revolutions per minute. In order to fulfil different operations, it is necessary to find out various numbers of spindle revolutions as well as different values of feed. For these purposes, speed and feed boxes which work by either stepped or unstepped principle of regulation are used.

17.1.2 Various Motions of Machine Tool System

To drive the various components of the machine tool system, we can have four methods:

  (i) Mechanical (ii) Electrical
  (iii) Hydraulic (iv) Pneumatic.

The choice of a particular method of drive will depend upon many factors such as cost, operating speeds and feeds, power to weight ratio, rigidity, reliability, maintenance costs, intended use, sophistication, and control.

DESIGN OF MACHINE TOOL GEAR BOX

Generally, today the user of machine tools demands better quality, improved performance, and higher operating speeds, and this has led to a design system quite complex in nature—consisting of any one of the drive methods stated above—of the machine tool system.

The field of machine dynamics, particularly that of the load bearing components of a system such as base, bed, table, saddle, columns and spindle support, are important for technological gains, as these components are made up of iron castings, but their design as steel weldments offers functional and economic advantages. The functional advantage is the possibility of using higher speeds and feeds, and the economic advantage is the low power to weight ratio and thus lower cost and ease of handling.

There are two types of motion in a machine tool systems—(i) the main motions, viz. cutting speed and feed, and (ii) the subsidiary motions such as fixing of workpiece, tool setting, machine control, etc.

Sometimes the primary motion is only in one axis as in the case of a broaching machine, but more often such motions are required in two or more than two planes. In such situations, we prefer to have an individual drive rather than a common drive. The line shaft drive is most obsolete, as there is not much control of speed in such a system of speed regulation.

The hydraulic or pneumatic speed regulation devices are used where an infinite number of speeds, within a range of maximum and minimum speed, are required; however, stepless regulation can also be achieved by a DC motor with a resistance control or mechanical drives using pressure variations, but for a very limited range.

Hydraulic speed regulation is good for straight line motions, e.g. broaching, grinding, milling and shaping machines.

17.2 FUNDAMENTALS OF MECHANICAL REGULATION

The ideal regulation of speed is stepless drive; but it has certain limitations, due to which it cannot be incorporated in all types of machine tools. Some idea regarding this fact has already been given in previous sections. The cost of a gearbox increases as the number of speeds increases, and this limitation hampers many users; therefore such machine tools are not commercially prospective.

In view of the above fact, it is customary to design (for a set of speeds) a gearbox which has a optimum numbers of speeds, with minimum speed loss. Alternative approaches to this problem have posed various solutions, out of which we have to pickup the best one.

If stepless regulation is not available, then the increment of speeds from a minimum level can be arranged in an AP, GP or HP series; it has been proved that the GP (Geometric Progression) gives minimum speed loss, and the GP series has other advantages too.

As we know that

$$\frac{v_a}{d_a} = \frac{\pi n}{1000} = k = \tan \phi$$

which reveals that to get v_a/n_i, a constant for the same diameter, we need to change n such that u_{a1}/n_1 is equal to $\pi d_1/1000$.

If we have a GP series such as

$$n_1, n_2, n_3, \ldots, n_{k-1}, n_k$$

then this can be written as

$$n, n\phi, n\phi^2, \ldots, n\phi^{k-1}, \ldots, n\phi^k$$

If you multiply this series by ϕ, you will have

$$n\phi, n\phi^2, n\phi^3, \ldots, n\phi^k, n\phi^{k+1}$$

Thus, we get another orientation of the speeds, just shifting the first speed to the next higher one and so on, without affecting the structural change in case of a geometric series with a common rates ϕ which is not possible for AP or HP series. Preferred numbers can be used in this series, which is not possible in other cases.

At present, we shall consider only rotary motions and other types of motions later on.

Let the RPM of the lathe spindle be

$$n_1; n_2; n_3; \ldots ; n_{k-1}; n_k; n_{k+1}; \ldots$$

What is the law governing these numbers?

For this purpose, let us examine the radial diagram. Let us turn a shaft which has a diameter d_A. According to the theory of metal cutting, we know that the cutting speed is given as under.

$$V_{an} = \frac{\pi d_{an}}{1000} \text{ m/min}$$

or

$$V_A = \frac{\pi n}{1000} d_A$$

If n is constant, then this relationship will be a straight line which passes through the origin (see Fig. 17.1).

$$\frac{V_A}{d_A} = \frac{\pi n}{1000} = K = \tan \phi$$

In our example, it is necessary to have the speed corresponding to the point A. But we must work either with the speed V_A or with the speed V_B (see Fig. 17.1). It is necessary to work with the speed V_B because it is near the speed V_N

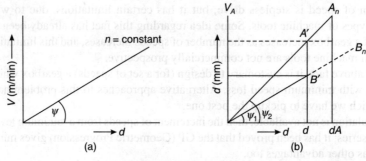

Fig. 17.1 (a) Relation between diameter d and speed V and (b) loss of cutting speed with change of diameter.

Let ΔV be the relative loss of the cutting speed, where

$$\Delta V = \frac{V_A - V_B}{V_A}$$

The maximum of this value will be with the diameter d_A:

$$\text{Max } \Delta V = \frac{V'_A - V'_B}{\Delta_A}$$

If we take this value as constant (as this is profitable for the exploitation of machine tools), then we shall get

$$\text{Max } \Delta V = \text{constant} = \frac{V'_A - V'_B}{V'_A} = 1 - \frac{V'_B}{V'_A}$$

$$= 1 - \frac{\pi d'_A \; n_{k-1}}{\pi d'_A \; n_k} \frac{1000}{1000} = 1 - \frac{n_k - 1}{n_k}$$

So the ratio $\dfrac{n_k - 1}{n_k}$ must be constant for V to be constant.

Then
$$n_k = n_{k-1} \; \phi$$

Let
$$\frac{n_{k-1}}{n_k} = \frac{n_{k-1}}{n_{k-1}\phi} = \frac{1}{\phi}$$

So, if max ΔV is constant, then the segment A′B′ must be constant too, for all of the values of rpm of the machine tool as shown in Fig. 17.1. Then the full radial diagram will be as shown in Fig. 17.2. This radial diagram is useful for determining the number of revolutions if d and V are known.

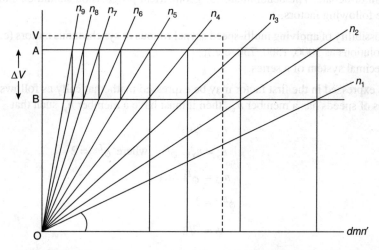

Fig. 17.2 Ray diagram or speed spectrum for geometric series.

It is the law of the geometrical progression with the common ratio ϕ.

$$n_k = n_{k-1} \; \phi \quad \text{where } \phi \text{ is a constant}$$

Therefore we have the number of speeds of a virtual gearbox as follows:

$$n_1$$
$$n_2 = n_1 \cdot \phi$$
$$n_3 = n_2 \cdot \phi = n_1 \phi^2$$
$$n_4 = n_1 \phi^3$$
$$n_x = n_1 \phi^{x-1}$$

Let us call the ratio n_{max}/n_{min} as the range of regulation denoted by R.

$$\frac{n_{max}}{n_{min}} = R = \frac{n_x}{n_1}$$

or
$$n_x = n_1 \phi^{x-1}$$

so
$$R = \phi^{x-1}$$

or
$$\phi = \sqrt[x-1]{R}$$

Even if the values n_{max} and n_{min} are the same, the number of speeds will be more if the constant ϕ is smaller. When choosing the denominator of the geometrical progression, we take into consideration the following two factors:

1. Desirability of having a higher number of speeds, as then it is necessary to have smaller values of ϕ;
2. Aspiration to have a compact structure of the speed box, as then it is necessary to have higher values of ϕ. This gives us a lesser number of speeds.

The values of ϕ are standardised, and the most common ones are 1.06, 1.12, 1.25, 1.41, 1.58, 1.87, and 2.

17.3 DEVELOPMENT OF SERIES OF NUMBERS

What requirements should the standard values of the constant of geometrical progression satisfy? This is an important point of debate. The denominator of geometrical progression should be chosen by taking into consideration the following factors:

(a) The possibility of applying multi-speed asynchronous or synchronous motors (e.g. with the number of revolutions as 3000, 1500, 750, etc.);
(b) The decimal system of a series.

The condition as expressed in the first factor may be expressed mathematically as follows:
If the series of speeds has a member n_{x3}, then it must have a member n_y, such that

$$n_y = 2n_{x3}$$

or
$$n_y = n_{x3} \cdot \phi^{E_1} \quad \text{where } \phi^{E_1} = 2$$

∴
$$n_{x3} = \phi^{E_1} \cdot n_{x3}$$

or
$$\phi^{E_1} = 2 \tag{17.1}$$

$$\phi = \sqrt[E_1]{2}$$

The condition of the second factor may be expressed mathematically as follows:
If the series has member n_x', then it should have a number such that

$$n_y' = 10n_x'$$

$$n_y' = \phi^{E_2} \cdot n_x'$$

$$10n_y' = \phi^{E_2} \cdot n_x'$$

$$\phi^{E_2} = 10$$

$$\phi = \sqrt[E_2]{10} \tag{17.2}$$

Let us take logarithms of expressions (17.1) and (17.2), then we have

$$E_1 \log \phi = \log 2 = 0.3 \tag{17.3}$$
$$E_1 \log \phi = \log 10 = 1.0 \tag{17.4}$$

Dividing Eq. (17.3) by Eq. (17.4), we get

$$\frac{E_1}{E_2} = \frac{3}{10}$$

Therefore, according to common understanding, if $E_2 = 40, 20, 10$ or 5, E_1 would be 12, 6, 3 or 1.5. Hence, the standard values of ϕ from Eq. (17.2) would be

$$\phi_{40} = \sqrt[40]{10} = \sqrt[12]{2} = 1.06$$

$$\phi_{20} = \sqrt[20]{10} = \sqrt[6]{2} = 1.12$$

$$\phi_{10} = \sqrt[10]{10} = \sqrt[3]{2} = 1.26$$

$$\phi_5 = \sqrt[5]{10} = \sqrt[1.5]{2} = 1.58$$

However, the series of two numbers for the values of ϕ was found to be insufficient; therefore, this series is to be supplemented with the following values of the series of number 2:

$$\phi = \sqrt{2} = 1.41$$

$$\phi = \sqrt[4]{10} = 1.78$$

The advantages of standardisation are the following:

(i) The decimal system of series is enough to erect a series from 10 to 100. All other numbers of this series may be obtained by multiplication or division of 10, 100, etc., and it is convenient for calculations.
(ii) If we put down every second member of the series with $\phi_{40} = 1.06$, then we get a series with $\phi_{20} = 1.12$; if we repeat this after every third member, the series with ϕ_{10} will be obtained.
(iii) The revolutions per minute of a synchronous motor can be fitted into this series very well. As mentioned earlier, R = Range of regulation = n_{max}/n_{min}, and if the number of speeds in the gearbox of a machine tool is denoted by k, then

$$R = \frac{n_k}{n_1} = \phi^{k-1}$$

We have

$$\phi = \sqrt[k-1]{\frac{n_k}{n_1}} \quad \text{or} \quad \phi = \left(\frac{n_k}{n_1}\right)^{\frac{1}{k-1}}$$

$$\log \phi = \frac{1}{k-1} \log \frac{n_k}{n_1}$$

or

$$k = 1 + \frac{\log \frac{n_k}{n_1}}{\log \phi}$$

or

$$k = 1 + \frac{\log R}{\log \phi}$$

Let us assume that $k = 2^{E1} \cdot 3^{E2}$, as this is one of the requirements of design of a speed regulation system. The value of k could be any number, as E_1 and E_2 are whole numbers starting from zero.

Thus, $k = 2, 3, 4, 6, 8, 9, 12, 16, 18, 24, 27, 36$, out of which the most widely used numbers are 3, 4, 6, 8, 9, 12, 16, 11, 24 and 36. Therefore, these are the steps of a stepped regulation or mechanical regulation system.

EXAMPLE 17.1 Design a gearbox having 6 speeds, i.e. 3 × 2 and 3 shafts indicating kinematic relationship.

Solution It is a common practice to limit the transmission ratio of gears in a gearbox in order to avoid excessively large diameters of the driven gears. The commonly used maximum and minimum values of transmission ratio are 2 : 1 and 1 : 4 for spur gears and 2.5 : 1 and 1 : 4 for helical gears.

When consecutive gear trains of transmission are engaged, the total transmission ratio of the drive is equal to the product of the transmission ratios of the simple trains that make up the drive.

In this example of transmission of speed at 6 steps, we require 10 gears as shown in Fig. 17.3. We can proceed for analytical investigation of the kinematic chain of gears as follows:

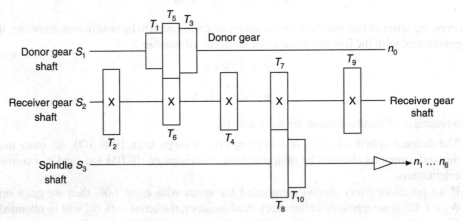

Fig. 17.3 Design of gearbox—kinematic layout diagram.

n_1 $\qquad n_1 = \dfrac{T_1 \, T_7}{T_2 \, T_8} n_0$

$n_2 = n_1 \phi$ $\qquad n_2 = \dfrac{T_3 \, T_7}{T_4 \, T_8} n_0$

$n_3 = n_2 \phi = n_1 \phi^2$ $\qquad n_3 = \dfrac{T_5 \, T_7}{T_6 \, T_8} n_0$

$n_4 = n_3 \phi = n_1 \phi^3$ $\qquad n_4 = \dfrac{T_1 \, T_9}{T_2 \, T_{10}} n_0$

$n_5 = n_4 \phi = n_1 \phi^4$ $\qquad n_5 = \dfrac{T_3 \, T_9}{T_1 \, T_{10}} n_0$

$n_6 = n_5 \phi = n_1 \phi^5$ $\qquad n_6 = \dfrac{T_5 \, T_9}{T_6 \, T_{10}} n_0$

Let us assume $\dfrac{T_1}{T_2} = e_1$, $\dfrac{T_3}{T_4} = e_2$, $\dfrac{T_5}{T_6} = e_3$, $\dfrac{T_7}{T_8} = e_4$, and $\dfrac{T_9}{T_{10}} = e_5$.

Therefore $n_1 = e_1 \cdot e_4 n_0$ $\qquad\qquad n_3 = e_3 \cdot e_4 n_0$
$\qquad\qquad\; n_2 = e_2 \cdot e_4 n_0$ $\qquad\qquad n_6 = e_3 \cdot e_5 n_0$
$\qquad\qquad\; n_4 = e_1 \cdot 3 \, e_5 n_0$

and
$$\frac{n_2}{n_1} = \frac{n_1 \phi}{n_1} = \phi = \frac{e_2 \cdot e_4 n_0}{e_1 \cdot e_4 n_0} = \frac{e_2}{e_1}$$

Hence $e_2 = e_1 \cdot \phi$

Similarly, we can obtain
$$e_3 = e_1 \cdot \phi^2$$
$$\frac{n_4}{n_1} = \phi^3 = \frac{e_1 \cdot e_5 \cdot n_0}{e_1 \cdot e_4 \cdot n_0} = \frac{e_5}{e_4}$$
$$e_5 = e_4 \cdot \phi^3$$

The link between the number of revolutions and the denominator ϕ is called the *main set*.

The first gearing set is that which has the power of denominator equal to the number of independent changes of speed in the main set.

$$n_1 = \frac{T_1 \, T_7}{T_2 \, T_8} n_0 = e_1 \cdot e_4 n_0 \qquad n_2 = \frac{T_1 \, T_9}{T_2 \, T_{10}} n_0 = e_1 \cdot e_5 n_0$$

$$n_3 = \frac{T_3 \, T_7}{T_4 \, T_8} n_0 = e_2 \cdot e_4 n_0 \qquad n_4 = \frac{T_3 \, T_9}{T_4 \, T_{10}} n_0 = e_2 \cdot e_5 n_0$$

$$n_5 = \frac{T_5 \, T_7}{T_6 \, T_8} n_0 = e_3 \cdot e_4 n_0 \qquad n_6 = \frac{T_5 \, T_9}{T_6 \, T_{10}} n_0 = e_3 \cdot e_5 n_0$$

It is evident from the above results that
$$\frac{n_2}{n_1} = \phi = \frac{e_5}{e_4}$$

$\therefore \qquad e_5 = e_4 \phi$

In this case, the double block of the gears is the main group, and the treble block of gears is the first gearing group.

The above analysis is called the method of investigating kinematic chains.

In cases where the number of sets is more than 2, such analysis becomes difficult because of bulky calculations.

EXAMPLE 17.2 Show graphically, the method of investigation of speed regulation by drawing ray diagrams.

Solution In general, we know that
$$n_x = n_1 \phi^{x-1}$$
$$\log n_x = \log n_1 + (x-1) \log \phi$$

Therefore, the above formula represents a straight line equation, hence it is possible to draw line diagrams which indicate speeds on the shafts of gearbox and are called ray diagrams. These ray diagrams are shown in Fig. 17.4.

These diagrams represent the speeds at output as well as intermediate shafts of gearbox and are developed from the kinematic arrangement of the drive. Shafts are shown by vertical equidistant and parallel lines. The speeds are plotted vertically on a logarithmic scale with log ϕ as a unit. Transmission engaged at definite speeds of the driving and driven shafts are shown on the diagrams by rays connecting the points on the shaft lines representing these speeds. Obviously, for transmission ratio of 1, the ray is horizontal. It is inclined up for transmission ratio greater than unity and inclined down for transmission ratio less than unity.

These diagrams indicate the distributive connections between input and output points and are of two types:
(i) Wide (open) diagrams 17.4(a) and (b) and (ii) Narrow (crossed) diagrams 17.4(c) and (d).

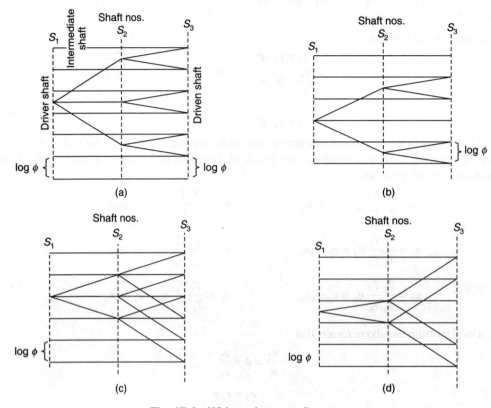

Fig. 17.4 Wide and narrow diagrams.

In open diagram the paths do not cross each other and in crossed diagram, the paths cross each other.
Figure 17.5 shows the possible open structure diagram for spindle to have nine speeds for a three-shaft gearbox.

These structural diagrams depict the range ratio R_n of transmission groups but give no information about transmission ratios (i). In order to determine the transmission ratios of all transmissions and the rpm values of speed-box shafts, it is necessary to plot the speed chart. The lines joining the points of adjacent shafts in a speed chart depict the transmission ratios only.

While plotting the speed graph, it is desirable to have the minimum transmission ratio, i.e. maximum speed reduction in the last transmission group. In this case, the remaining shafts of the speed box run at a relatively, higher speed and are, therefore, loaded to a lesser extent. Keeping in mind the restrictions,

$$i_{max} \leq 2 \quad \text{and} \quad i_{min} \geq \frac{1}{4}$$

the maximum number of intervals between the ends of lines joining the points on two adjacent shafts is also restricted; the limiting values are given in Table 17.1.

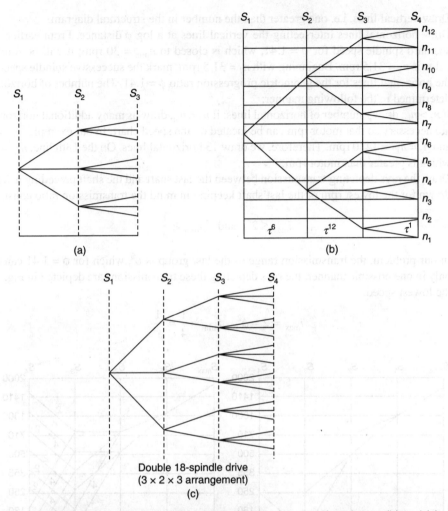

Fig. 17.5 Part (a) shows arrangement for 9 speed drive (1 × 3 × 3) and parts (b) and (c) show the arrangement for 12 and 18 speeds drive.

Table 17.1 Limiting values of transmission intervals for different ϕ

Types of transmission	Limiting number of intervals for various values of ϕ					
	1.06	1.12	1.26	1.41	1.58	1.78	2.0
Speed reduction, $i < 1$	24	12	6	4	3	2	2
Speed increase, $i > 1$	12	6	3	2	1	1	1

EXAMPLE 17.3 Plot a ray diagram for a 12-speed gearbox, which is a structure of 12 speeds, i.e. $12 = 2 \times 3 \times 2$ and $n_{min} = 30$, $n_{max} = 1500$ rpm, $Z = 12$, $\phi = 1.41$ and motor rpm = 1440.

Solution Let us elaborate the procedure of plotting the speed graph for the data given in Example 17.3 as an example for drawing Fig. 17.6.

(i) Draw vertical lines, i.e. one greater than the number in the structural diagram.
(ii) Draw horizontal lines intersecting the vertical lines at a log ϕ distance. From earlier discussion, standard spindle speed for $\phi = 1.41$, which is closed to $n_{min} = 30$ rpm; it will be found that this value is $n_1 = 31.5$ rpm. Beginning with $n_1 = 31.5$ rpm, mark the successive spindle speed values on the horizontal lines for the geometric progression ratio $\phi = 1.41$. The number of horizontal lines is determined in the following manner:

If $n_z > n_m$, draw z number of horizontal lines; if $n_z < n_m$, draw as many additional horizontal lines as are necessary so that motor rpm can be located on the speed chart. In our example, $n_m = 1440$ rpm and $n_z = n_{12} = 1410$ rpm. Therefore, we draw 13 horizontal lines. On the 13th line, $n_{13} = 2000$ rpm, which is greater than motor rpm.

(iii) Draw the rays depicting transmission between the last shaft and the shaft preceding it. The rays are drawn for the lowest rpm of the last shaft keeping in mind the transmission ratio constraints of

$$i_{max} \leq 2 \quad \text{and} \quad i_{min} \geq \frac{1}{4}$$

In our problem, the transmission range of the last group is ϕ^6, which for $\phi = 1.41$ can be divided only in one possible manner, the rays depicting these transmissions are depicted in Fig. 17.6(a) for the lowest speed.

$$i_{max} = 2 = \phi^2, \quad i_{min} = \frac{1}{4} = \frac{1}{\phi^4}$$

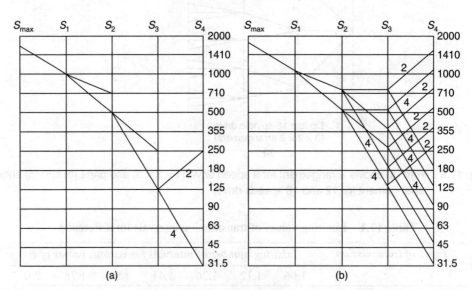

Fig. 17.6 (a) Rays diagram and (b) speed chart.

(iv) The transmissions between shafts (S_3) and (S_4) may be arranged in two different ways as shown in Fig. 17.6(a) and (b).
(v) If the rpm of the motor does not coincide with the standard spindle speeds, the motor rpm of shaft S_1 (mo) is reduced to a standard spindle speed on S_1 shaft by selecting an appropriate transmission ratio. While selecting the transmission ratio between shafts and S_{mo} and S_1 and S_2, it should be kept

in mind that the maximum speed reduction between shaft does not violate the condition $i_m \geq 1.4$. The rays are now drawn in Fig. 17.6(a) and the ray diagram is completed.

(vi) Having plotted the ray diagram, the rays of the remaining transmissions are also drawn keeping in mind the characteristic of each group. The completed speed chart is shown in Fig. 17.6(b).

EXAMPLE 17.4 Design a gearbox of a turret lathe having six speeds ranging from 25 to 800 rpm driving through a motor having 1500 rpm and using the value of $\phi = 2$.

Solution

1. To draw ray diagram draw four vertical lines representing motor shaft 1 (S_1), donor shaft 2 (S_2), intermediate shaft 3 (S_3) and spindle shaft 4 (S_4), as shown in Fig. 17.7(a).
2. Mark log values of speed on shaft 4, i.e. representing spindle shaft.
3. Mark log value of motor speed on shaft 1 as 1500.
4. As the value of ϕ is 2 and $i_{max} \leq 2 = \phi$ hence choose the type of a ray diagram. In this problem the ray diagram is going to be asymmetric due to provision of $i_{min} \leq 1/4$. As there are only 6 speeds, hence from shaft 1 to shaft 2 and from shaft 2 to shaft 3, we can go down to four ϕ values and toward up two ϕ values.
5. Select the points on a log ϕ diagram [Fig. 17.7(b)].

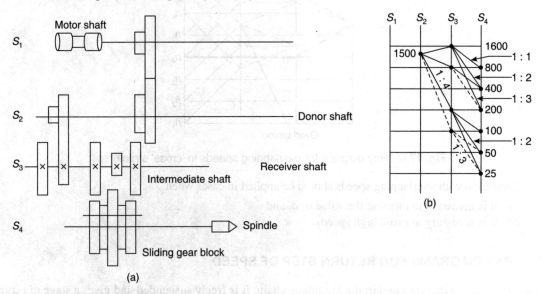

Fig. 17.7 Ray diagram and kinematic layout for gear train of the gearbox.

In this diagram, there are two ray diagrams, one in full lines and one in dotted lines.

(i) The size or dimension of a particular shaft is proportional to torque. If the torque is less, the diameter of the shaft also becomes less. To realise lower quantity of torque, it is necessary that the speed of the shaft should be high, if the power is constant. Therefore, it is a good proposition to have as high speed as possible on the intermediate shaft, so that the gearbox does not become voluminous.

(ii) The gear ratio should preferably be less than 1; though in many cases, it may be obligatory to use gear ratio more than 1 or equal to 1.

17.4 RAY DIAGRAM FOR OVERLAPPING SPEEDS

Overlapping speeds are obtained by reducing the power by one or more degrees in one of the gearing groups. Let us take a mechanism of $12 = 2 \cdot 3 \cdot 2$ speeds, the main group of which is the travel block of gears.

The ray diagram for overlapping speeds in 'cross structure' is shown in Fig. 17.8. Let us decrease the power of the first gearing group by 1 unit.

Thus, the overlapping of two stage speeds is obtained, and we have

$$\phi = \sqrt[4]{4} = 1.41 \quad \text{and} \quad \phi = \sqrt[4]{6} = 1.56$$

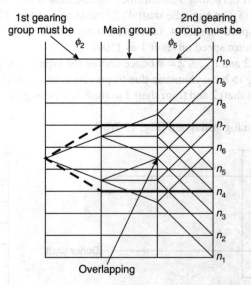

Fig. 17.8 Ray diagram for overlapping speeds in 'cross' structure.

Regulation with overlapping speeds should be applied in cases when

(i) It is necessary to increase the value of ϕ, and
(ii) It is necessary to avoid high speeds.

17.5 RAY DIAGRAMS FOR RETURN STEP OF SPEED

A barrel of gears is incorporated in the kinematic chain. It is freely suspended and gives a stage of return speeds.

Let $12 = 3 \cdot 2 \cdot 2$ be the kinematic chain with a stage of return. The structural diagram for this mechanism can be drawn up by anyone of the methods shown in Figs. 17.9 and 17.10. The overlappings are also obtained here.

The disadvantage in case of the arrangement shown in Fig. 17.10(a) is that a return step is not available. This arrangement is quite often used. The disadvantage in case of the kinematic arrangement in Fig. 17.10(b) is that visual control is difficult, but here a stage of return is observed, which is an advantage.

DESIGN OF MACHINE TOOL GEAR BOX **341**

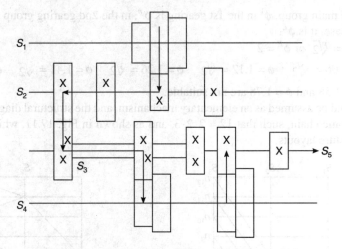

Fig. 17.9 Kinematic layout for versatile gear train.

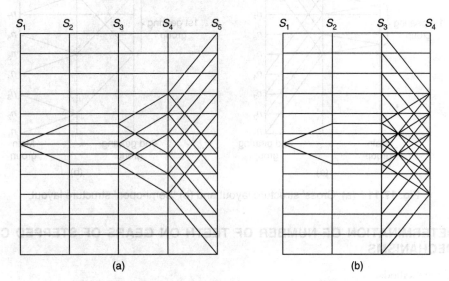

Fig. 17.10 Ray diagrams for kinematic layout of Fig. 17.9.

17.6 KINEMATIC ARRANGEMENT FOR TWO OR MORE SPEEDS AT INPUT SHAFT

While choosing a multispeed input, it is necessary to consider the following:
  (a) Economic aspect
  (b) Smaller size and simpler control.

In this case, the value of ϕ is limited to 2. That is

$$\phi_m = 2$$

The motor should be considered either the main group or as one of the groups.

If treated as the main group, ϕ^1 in the 1st gearing is ϕ^P, in the 2nd gearing group is ϕ^{pq}, and so on. In the general case, it is ϕ^E.

It means that $\phi = \sqrt[E]{2}$ or $\phi^E = 2$.

Hence $\phi = 1.06 = \sqrt[12]{2}$ $\phi = 1.12 = \sqrt[6]{2}$ $\phi = 1.26 = \sqrt[3]{2}$ $\phi = 1.41 = \sqrt{2}$ $\phi = 2$

The values $\phi = 1.58$ and $\phi = 1.78$ are not suitable.

The motor should be assumed as an elementary mechanism, and the structural diagram should be drawn for a 12-speed kinematic chain, such that $12 = 2 \cdot 2 \cdot 3$, and as shown in Fig. 17.11, which shows the 'cross' and 'semi-open' structure layouts.

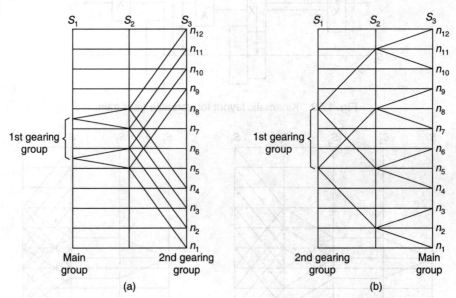

Fig. 17.11 (a) 'Cross' structure layout and (b) 'Semi-open' structure layout.

17.7 DETERMINATION OF NUMBER OF TEETH ON GEARS OF STEPPED CONTROL MECHANISMS

There are three methods:
  (i) Method of least common multiple;
  (ii) Method of difference;
  (iii) Constructive method.

17.7.1 Method of Least Common Multiple

Let us assume that all gears have the same module. From Fig. 17.12, we have
$$2A = m(T_1 + T_2) = m(T_3 + T_4) = mT_5$$
where T_0 is the total number of teeth.
$$T_0 = T_1 + T_2 = T_3 + T_4$$
$$e_1 = \frac{T_1}{T_2} \quad e_2 = \frac{T_3}{T_4}$$

e_1 and e_2 are thus known to us.

If T_0 is equal to a certain number, we can find $T_0 = T_1 + T_2$.
Now, $e_1 T_2 = T_1$; therefore $T_0 = e_1 T_2 + T_2 = T_2(e_1 + 1)$.

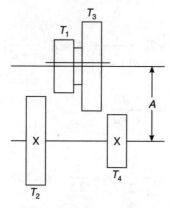

Fig. 17.12 Designing layout with least common multiple.

$$T_2 = \frac{T_0}{e_1 + 1}$$

Thus, in the same way we can find

$$T_0 = T_1 + T_2 \quad \frac{T_1}{T_2} = e_1 \quad T_2 = \frac{T_1}{e_1}$$

$$T_0 = T_1 + \frac{T_1}{e_1} = T_1\left(1 + \frac{1}{e_1}\right)$$

$$\therefore \quad T_1 = \frac{T_0}{1 + \frac{1}{e_1}} = \frac{T_0 e_1}{1 + e_1}$$

By analogy, we can also find

$$T_3 = \frac{T_0 e_2}{1 + e_2} \qquad T_4 = \frac{T_0 e_3}{1 + e_3}$$

Let
$$e_1 = \frac{a}{b} \qquad e_2 = \frac{c}{d}$$

where a, b, c and d are prime numbers. Then

$$T_1 = T_0 \frac{a}{b\left(1 + \frac{a}{b}\right)} \cdot K = T_0 \times \frac{a}{(a + b)} \cdot K$$

where T_0 is the least common multiple of the values ab and cd;
 K is the set for obtaining a minimum number of teeth of the smallest gear.

$$T_2 = T_0 \frac{1}{1+\frac{a}{b}} \cdot K = T_0 \times \frac{b}{a+b} \cdot K$$

$$T_3 = T_0 \frac{c}{d\left(1+\frac{c}{d}\right)} \cdot K = T_0 \left(\frac{c}{c+d}\right) \cdot K$$

$$T_4 = T_0 \frac{d}{c+d} \cdot K$$

EXAMPLE 17.5 Let

$$e_1 = \frac{1}{2} \quad e_2 = \frac{1}{2.5} \quad \phi = 1.26$$

$\therefore \quad e_2 = \frac{2}{5}$

then $\quad a = 1 \quad b = 2 \quad c = 2 \quad d = 5$

$\therefore \quad \begin{cases} a+b = 3 \\ c+d = 7 \end{cases}$ LCM $T_0 = 2$

$\therefore \quad T_1 = 21 \cdot \frac{1}{3} \cdot K = 7K \qquad T_2 = 21 \cdot \frac{2}{3} \cdot K = 14K$

and $\quad T_3 = 21 \cdot \frac{2}{7} \cdot K = 6K \qquad T_4 = 21 \cdot \frac{5}{7} \cdot K = 15K$

Let us take $K = 6$. Then $T_1 = 42$, $T_2 = 84$, $T_3 = 36$, and $T_4 = 90$.

If the modules of a pair of gears are different, then we have

$$2A_0 = m(T_1 + T_2) = m(T_3 + T_4)$$

If it is necessary to calculate by one module, say m_1, then

$$T_3 + T_4 = \frac{m_1}{m_2}(T_1 + T_2) = \frac{m_1}{m_2} T_0$$

EXAMPLE 17.6 Let

$$e_1 = \frac{1}{2} \quad e_2 = \frac{1}{2.5} \quad m_1 = 2 \text{ mm} \quad m_2 = 3 \text{ mm}$$

$$a = 1 \quad b = 2 \quad c = 2 \quad d = 5 \quad a+b = 3 \quad c+d = 7$$

$$T_1 = \frac{a}{a+b} \cdot T_0 K$$

$$T_2 = \frac{b}{a+b} \cdot T_0 K$$

$$T_3 = \frac{c}{c+d} \frac{m_1}{m_2} \cdot T_0 K$$

$$T_4 = \frac{d}{c+d} \frac{m_1}{m_2} \cdot T_0 K$$

where T_0 = LCM of values $a + b$, $c + d$ and m_1/m_2.
In this example, the LCM is 3, 7, and 2/3 $T_0 = 21$.

$$T_1 = \frac{1}{3} \cdot 21 \cdot K = 7K \qquad T_2 = \frac{2}{3} \cdot 21 \cdot K = 14K$$

$$T_3 = \frac{2}{7} \cdot \frac{2}{3} \cdot 21 \cdot K = 4K \qquad T_4 = \frac{5}{7} \cdot \frac{2}{3} \cdot 21 \cdot K = 10K$$

Let us take $K = 5$. Then

$$T_1 = 35 \qquad T_2 = 70 \qquad T_3 = 20 \qquad T_4 = 50$$

17.7.2 Method of Difference

This method has the disadvantage over the first method that it does not take into account the situation in which it is necessary that

$$T_2 = T_1 + \Delta T \qquad \text{where } T_{\min} \geq 5 \text{ teeth}$$

$$T_5 = T_4 + \Delta T$$

Since

$$T_0 = T_1 + T_4 = T_2 + T_5$$

and

$$e_1 = \frac{T_1}{T_4} \qquad e_2 = \frac{T_2}{T_5}$$

Also

$$e_2 = \frac{T_1 + \Delta T}{T_4 - \Delta T}$$

since

$$T_4 = \frac{T_1}{e_1}$$

$$e_2 = \frac{T_1 + \Delta T}{\frac{T_1}{e_1} - \Delta T} = \frac{e_1(T_1 + \Delta T)}{T_1 - e_1 \cdot \Delta T}$$

or

$$T_1 e_2 - e_1 e_2 \, \Delta T = e_1 T_1 + e_1 \cdot \Delta T$$

Therefore,

$$T_1 = \frac{e_1 \cdot \Delta T (1 + e_2)}{e_2 - e_1}$$

In the same way we can determine T_2, T_3, etc.

The kinematic layout with gear details is shown in Fig. 17.13. The corresponding ray diagram is shown in Fig. 17.17.

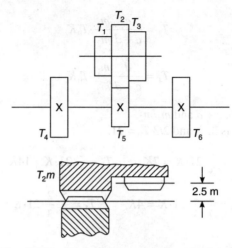

Fig. 17.13 Kinematic layout with gear details.

EXAMPLE 17.7

$$e_2 = \frac{1}{2} \quad e_1 = \frac{2}{5} \quad \Delta T = 5$$

Therefore,
$$T_1 = \frac{5 \cdot \frac{2}{5}\left(1 + \frac{1}{2}\right)}{\frac{1}{2} - \frac{2}{5}} = \frac{3 \cdot 10}{5 - 4} = 30$$

$$T_2 = T_1 + \Delta T = 35 \quad T_5 = 70$$

$$T_4 = \frac{T_1}{e_1} = \frac{30 \cdot 5}{2} = 75$$

17.7.3 Constructive Method

The shortcomings of the previous two methods are that the diameters of shafts, keyways, etc. are not taken into account. The constructive method consists of determining teeth numbers, while the above factors are also taken into consideration.

EXAMPLE 17.8 Design a gearbox of a machine tool (turret) having 9 spindle speeds ranging from 90 to 1800 rpm. The gearbox should be a compact one. Also

(a) Represent the speeds graphically.
(b) Draw the structural diagram.
(c) Show the layout of the gearbox.
(d) Find out the numbers of teeth on various gears.

$$n_1 = 90 \qquad n_2 = 130$$
$$n_3 = 190 \qquad n_4 = 280$$
$$n_5 = 400 \qquad n_6 = 585$$
$$n_7 = 850 \qquad n_8 = 1240$$
$$n_9 = 1800$$

Solution
$$\phi = \sqrt[9-1]{\frac{1800}{90}} = \sqrt[8]{20} = 1.455$$

The ray diagram, gearbox layout and structural diagram for Example 17.8 are shown in Figs. 17.14, 17.15 and 17.16 respectively.

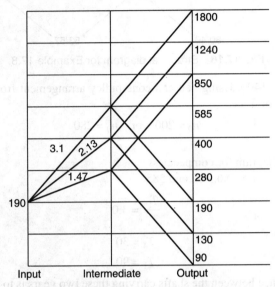

Fig. 17.14 Ray diagram for Example 17.8.

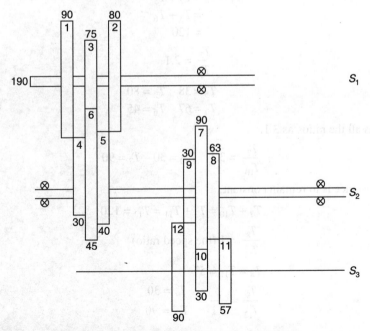

Fig. 17.15 Gearbox layout for Example 17.8.

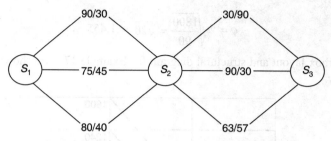

Fig. 17.16 Structural diagram for Example 17.8.

Let the motor rpm be 1400. Using belt and cone pulley arrangement from motor to the spindle and taking the reduction to be 7 or 5, i.e.

$$n_0 = 200 \quad \text{or} \quad n_0 = 280$$

So we take $n_0 = 190$

Now choose narrow diagram for compactness.

For speeds n_4, n_5 and n_6, i.e. 286, 402 and 585

$$\frac{T_1}{T_4} = 3.0$$

Let

$$T_4 = 30$$

∴

$$T_1 = 90$$

Since the centre distance between the shafts carrying these two gears is to remain constant,

$$T_1 + T_4 = T_2 + T_5$$
$$= T_3 + T_6$$
$$= 120$$

But

$$\frac{T_2}{T_5} = 2.1$$

$$T_5 = 38 \quad T_2 = 80$$

Similarly

$$T_3 = 67 \quad T_6 = 45$$

Now 2nd stage has all the ratios as 3.1.

$$\frac{T_7}{T_{10}} = 3.1 \quad T_{10} = 30 \quad T_7 = 90$$

Since the centre distance is to remain constant,

$$T_7 + T_{10} = T_3 + T_{11} = T_{12} = 120$$

$$\frac{T_8}{T_{11}} = 1 \text{ (the speed ratio)}$$

or

$$T_8 = T_{11} = 57$$

Since

$$\frac{T_9}{T_{12}} = \frac{1}{3.1} \quad \begin{array}{c} T_9 = 30 \\ T_{12} = 90 \end{array}$$

DESIGN OF MACHINE TOOL GEAR BOX 349

EXAMPLE 17.9 (a) A manufacturing concern takes up the demand of supplying turret lathes to its customers having 9 speeds powered by a 8 kW motor. The speed range is from 90 to 1500 rpm. Design a suitable gearbox giving all details.

(b) After a few years, the customers demand that the working speed range may be increased to 2500 rpm, as lower speeds are rarely used. Suggest a workable alteration in the gears so as to meet this demand without changing the structure of the gearbox.

Solution (a) Gearbox for turret lathe:
   Data—
   1. Spindle speeds: 9
   2. Range: 90 to 1500
   3. Capacity: 8 kW
   4. Preferred number: 1.06, 1.12, 1.26, 1.41, 1.58, 1.78, 2.

The nearest preferred number in the list of 1.41, out of the given listed numbers. The various speeds in descending order shall be:

$n_1 = 1500$ $\quad\quad n_2 = 1062$
$n_3 = 755$ $\quad\quad n_4 = 535$
$n_5 = 380$ $\quad\quad n_6 = 270$
$n_7 = 191$ $\quad\quad n_8 = 135$
$n_9 = 90$

According to the considerations of slip and strength, etc., the speed ratio allowable at the pulleys are 5 to 7.
So the primary speed is between $1440/7 = 206$ and $1440/5 = 288$.
Assuming motor rating to be 444 V at 1440 rpm and
Hence we take 270 as the primary speed. For economy and compactness, narrow ray diagram will be preferred.
The ray diagram is shown in Fig. 17.17.

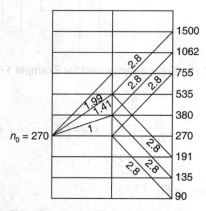

Fig. 17.17 Ray diagram for Fig. 17.13 and Example 17.9(a).

Let the minimum number of teeth on the gear be 20. Then

$$\frac{T_1}{T_2} = 1.99 \quad T_2 = \boxed{20} \quad T_1 = \boxed{40}$$

Now, as the centre distance remains constant,

$$T_3 + T_4 = T_1 + T_2 = 60$$

$$\frac{T_3}{T_4} = 1.41$$

$$T_4 = \frac{60}{2.41} = 24.9 \approx \boxed{25}$$

$$T_3 = \boxed{35}$$

$$T_5 + T_6 = T_3 + T_4 = T_1 + T_2 = 60$$

$$T_5 = T_6 = 30$$

The gear arrangement is shown in Fig. 17.18.

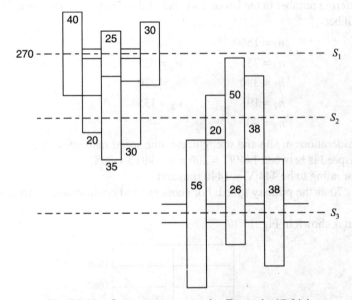

Fig. 17.18 Gear arrangement for Example 17.9(a).

Now for speeds 1500, 1062, 755

$$\frac{T_7}{T_{10}} = 2.8, \quad T_{10} = \boxed{20}$$

Therefore

$$T_7 = \boxed{56}$$

The structural diagram is shown in Fig. 17.19.

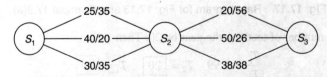

Fig. 17.19 Structural diagram for Example 17.9(a).

(b) Now, if after a few years the company wants to rearrange the speed to have maximum speed of 2500 rpm approximately, the ray diagram shifts towards the high value of speeds.

Then, using the same preferred number

$$1500 \times 1.41 = 2118 \tag{i}$$
$$2118 \times 1.41 = 2980 \tag{ii}$$

There is no restriction on minimum speeds as they are rarely used.

Let us assume that the minimum number of gear teeth is 20.

$$\frac{T_1}{T_2} = 3.95$$

$$T_2 = \boxed{20}$$

$$T_1 = \boxed{79}$$

$$T_1 + T_2 = 99 = T_3 + T_4$$
$$= T_5 + T_6$$

$$\frac{T_3}{T_4} = 2.8$$

$$T_3 = \boxed{73}$$

$$T_4 = \boxed{26}$$

$$\frac{T_5}{T_6} = 1.98$$

$$T_5 = \boxed{66}$$

$$T_6 = \boxed{33}$$

The ray diagram is shown in Fig. 17.20.

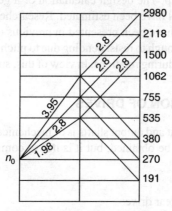

Fig. 17.20 Ray diagram for Example 17.9(b).

For second stage

$$\frac{T_7}{T_{10}} = 2.8 \quad T_{10} = \boxed{20} \quad T_7 = \boxed{56}$$

$$\frac{T_8}{T_{11}} = 1 \quad T_8 = T_{11} = \boxed{38}$$

$$\frac{T_9}{T_{12}} = 2.8 \quad T_{12} = \boxed{20} \quad T_8 = \boxed{56}$$

The structural diagram is shown in Fig. 17.21.

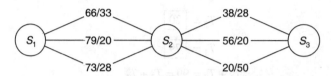

Fig. 17.21 Structural diagram for Example 17.9(b).

$$T_7 + T_{10} = T_8 + T_{17} = T_9 + T_{12} = 76$$

$$\frac{T_8}{T_{11}} = 2.8 \quad T_{11} = \boxed{20} \quad T_8 = \boxed{56}$$

$$\frac{T_9}{T_{12}} = 1 \quad T_9 = T_{12} = \boxed{38}$$

17.8 PRACTICAL ASPECTS IN THE DESIGN OF DRIVES

The theoretical calculations of the speeds and the number of teeth have been duly provided in the preceding section; however, the actual speeds available on the machine tool generally vary by $\pm\phi n$ in such a way that ϕn is equal to zero. Not only this, it is limited to $\pm 10\%$ (0–1). If deviation exceeds this limit, re-design should be done.

After all speeds, the number of teeth are determined, the number of gears and layout of gearbox, location of input, output shaft, etc. are taken up. The design calculation of a gear wheel is taken up after the forces acting on the spindle of a machine tool have been estimated. Researchers have given empirical relationships for estimating these forces, and these have been presented in previous chapters.

The gear tooth must be strong enough to resist bending due to pitch line pressure, and also hard enough to resist the surface wear that takes place during rotations. In view of this, suitable gear material should be chosen.

17.9 MECHANICAL REGULATION OF DRIVES

If the kinematic chain between the input and output shafts is of mechanical type, we will get stepped regulation; however, stepless regulation can also be obtained, but it is not so commonly used. The methods of stepped regulation are as follows:

(a) Belt and cone pulley drive;
(b) Belt and pulley with back gear drive;
(c) Gearbox drives.

For stepless regulation there are various methods, as mentioned previously and as will be discussed in Chapter 18.

17.9.1 Belt and Cone Pulley Drive

The belt and cone pulley drive (see Fig. 17.22) is supposed to be the simplest and oldest mechanism for transmitting power from one shaft to another. It has certain inherent disadvantages such as large size, small range of speeds, and that the torque transmitted by the driven shaft is proportional to the speed at which it is rotating. However, it has the advantage of being simple in design and cheap. Thus it is still found useful for various purposes.

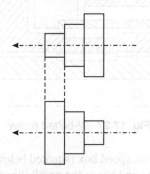

Fig. 17.22 Belt and pulley drive.

There are many types of belts as follows:
(i) Flat belt;
(ii) Wedge belt;
(iii) Round belt (for low power).

If a speed box is used there, the belt drive may be situated either in the place where the motion is transmitted to the speed box or in the place of transmitting the motion from the speed box to the spindle (see Figs. 17.23 and 17.24). Two methods of transmission from speed box to spindle are shown in Figs. 17.25 and 17.26.

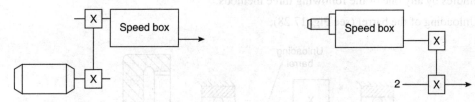

Fig. 17.23 Transmission from motor to gearbox. **Fig. 17.24** Transmission from gearbox to spindle.

Wedge-belt drives are usually employed in the first variation and flat-belt drives in the second one.

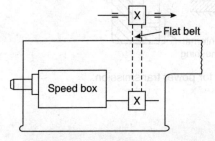

Fig. 17.25 Transmission from speed box to spindle—Method 1.

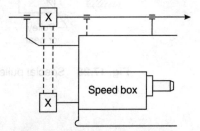

Fig. 17.26 Transmission from speed box to spindle—Method 2.

A V-belted pulley is shown in Fig. 17.27.

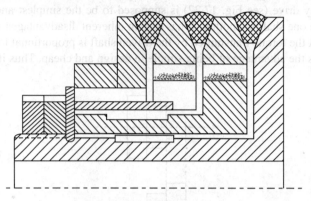

Fig. 17.27 V-belted pulley.

The second variation is applied if the speed box (situated below) and the spindle are separate units. In this case the machine tool is more steady, and hence the possibility of vibrations is reduced. This scheme is used for machines of higher precision. It necessitates the application of flat belts (as it is difficult to place wedge belts here).

There are three methods of regulating the stress of a belt in such drives. They are:

(i) By moving a motor (in advance of swivel);
(ii) Belt-tightening pulleys;
(iii) Special construction of a regulated pulley (one of the two in Fig. 17.28).

More than 3 slots are usually not provided.

The bending of shafts can be eliminated by making the belt transmit the torque alone. This is done first of all for spindles by any one of the following three methods:

(i) Unloading of the barrel (see Fig. 17.28);

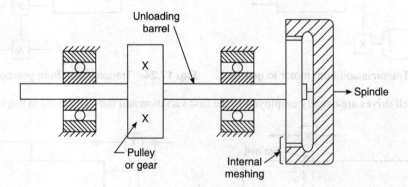

Fig. 17.28 Special pulley or gear for power transmission.

(ii) The spindle does not participate in transmitting the force (see Fig. 17.29); and

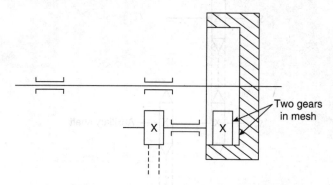

Fig. 17.29 Use of internal gear of small diameter.

(iii) Unloading of the first shaft of the speed box.

In designing belt and especially chain transmissions, great attention should be paid for their location. Generally, they can be located in two ways:

(i) Vertical location (see Fig. 17.30).
(ii) Horizontal location (see Fig. 17.31).

The horizontal location of transmission is better as the gravity force g changes the character of belt stress, and the character of oscillations are more favourable than in the vertical locations. It is clearly revealed in chain transmissions with large distance between centres. But horizontally located belts require larger dimensions of the chain or belt; therefore angular location of belts is preferable as shown in Fig. 17.32.

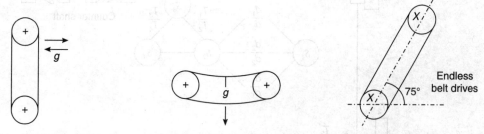

Fig. 17.30 Vertical pulley belt transmission.

Fig. 17.31 Horizontal pulley belt transmission.

Fig. 17.32 Inclined pulley belt transmission.

To improve the transmission system, one more shaft is added as shown in Fig. 17.29, and by this the oscillations of one chain are absorbed by the other.

17.9.2 Belt Pulley Drive with Back Gear

A typical example of such a mechanism is shown in Fig. 17.33, with the structural diagram in Fig. 17.34 and the ray diagram in Fig. 17.35.

356 TEXTBOOK OF PRODUCTION ENGINEERING

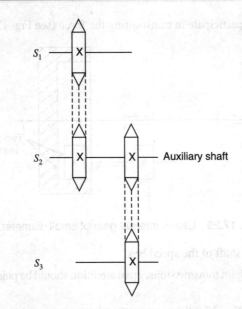

Fig. 17.33 Gear drive with stepped pulley.

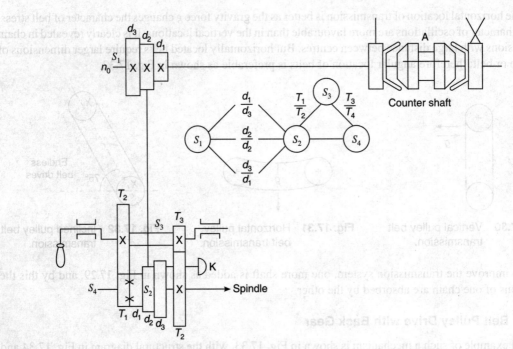

Fig. 17.34 Structural diagram for gear drive with stepped pulley.

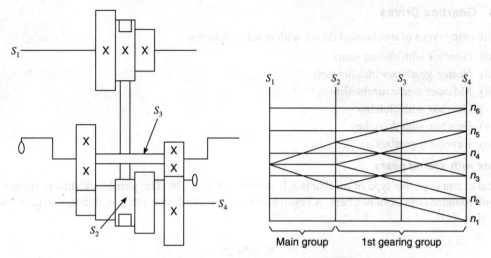

Fig. 17.35 Ray diagram for gear drive with stepped pulley.

This type of system has the following advantages in comparison to plain belt and pulley drive mechanisms:

Advantages
  (i) Smoothness of spindle rotation
  (ii) Absence of vibration
  (iii) Safety in use
  (iv) Compactness and greater number of speeds.

Disadvantages
  (i) Cannot be automatised.
  (ii) Distribution of power is directly proportional to the diameter of the stepped pulley.

If T is the tension and H is the power on the spindle, and in particular, $Hs_1, Hs_2 \ldots$ is power on spindles S_1, S_2, etc.

$$Hs_1 = TV_1 \frac{1}{60.102} \text{ kW} \qquad \text{where } V_1 = \frac{\pi d_1 n_0}{1000} \text{ mpm}$$

$$Hs_2 = TV_2 \frac{1}{60.102} \text{ kW} \qquad \text{where } V_2 = \frac{\pi d_2 n_0}{1000} \text{ mpm}$$

$$Hs_3 = TV_3 \frac{1}{60.102} \text{ kW} \qquad \text{where } V_3 = \frac{\pi d_3 n_0}{1000} \text{ mpm}$$

$$Hs_1 = T \frac{\pi d_1 n_0}{60.102 \cdot 1000} \text{ kW}$$

$$Hs_2 = T \frac{\pi d_2 n_0}{60.102 \cdot 1000}$$

$$Hs_3 = T \frac{\pi d_3 n_0}{60 \cdot 102 \cdot 1000}$$

It means that
$$H_1 : Hs_2 : Hs_3 = d_1 : d_2 : d_3$$
$$Hs_1 > Hs_2 > Hs_3.$$

17.9.3 Gearbox Drives

There are many types of mechanical drives with gears as follows:
  (i) Gearbox with sliding gears
  (ii) Norton gearbox with idler gear
  (ii) Meander's gear mechanism
  (iv) Gearbox with clutches
  (v) Gearbox with drive key
  (vi) Reversing gearbox.

Gearbox with sliding gears

A typical example of this type of mechanism is shown in Fig. 17.36. The gear block shift in splines on the shaft with rounded cogs as shown here. A keyed block of gears mesh alternately with different gears mounted on the other shaft.

$$n_1 = n_0 \frac{T_1}{T_2}$$

$$n_2 = n_0 \frac{T_3}{T_4}$$

The gear wheels Z_2 and Z_4 are either made of one gear blank or are permanently connected with each other. There are 2, 3 and seldom 4 gear wheels. These are keyed together to form a sliding gear.

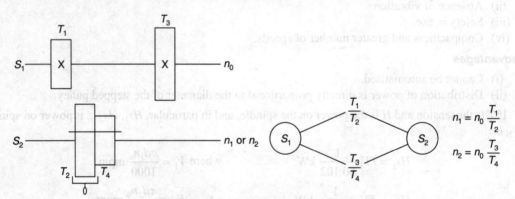

Fig. 17.36 Mechanism with gears.

Advantages

  (i) Much higher values of power can be transmitted;
  (ii) Such drives give constant power except for frictional losses;
  (iii) High efficiency;
  (iv) Provide a large range of speeds;
  (v) Can be automated.

Disadvantages

 (i) Difficulty in operation;
 (ii) It is possible to shift only when the machine is stopped.

In order to make the shiftings easier, cogs are made with roundings, as shown in Fig. 17.36.

Norton gearbox mechanism

This mechanism is shown in Fig. 17.37(a) and (b), and is generally used for low speeds, i.e. for feed boxes. The various speeds that can be obtained are shown in the same figure.

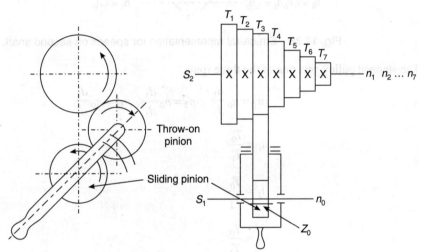

(a) Kinematic layout of Norton gearbox.

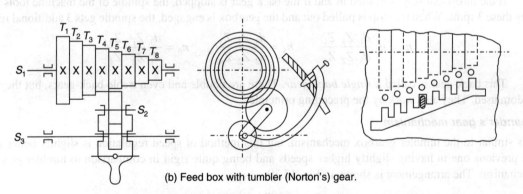

(b) Feed box with tumbler (Norton's) gear.

Fig. 17.37 Norton gearbox mechanism.

The gear wheel is fixed on the shaft by means of the sliding key (see Fig. 17.38). It may be clutched (through the idler) with any one of the gear wheels which are fixed on the shaft.

The advantage of this mechanism is compactness, while the shortcoming is low rigidity which leads to failure to transmit large power. It is generally used for various feed boxes.

The driving shaft gets either clockwise or counter-clockwise reversing, depending on the position of the double cone-clutch (on the right or on the left side).

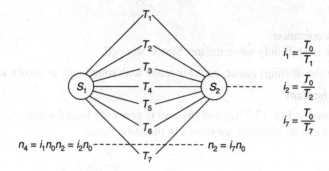

Fig. 17.38 Structural representation for speeds on second shaft.

The stepped pulley gets one of the three rpms.

$$n_1 = n_0 \frac{d_1}{d_2} \quad n_2 = n_0 \frac{d_2}{d_3} \quad n_3 = n_0 \frac{d_3}{d_1}$$

$$i_1 = \frac{T_0}{T_1}$$

$$i_2 = \frac{T_0}{T_2}$$

$$i_3 = \frac{T_0}{T_3}$$

If the throw-out stop is pushed in and if the back gear is stopped, the spindle of the machine tools gets only these 3 rpms. When the stop is pulled out and the gearbox is engaged, the spindle gets 3 additional rpms.

$$n_4 = n_0 \frac{d_1}{d_3} \frac{Z_1}{Z_2} \frac{Z_1}{Z_4} \quad n_5 = \frac{d_3}{d_2} \frac{Z_1}{Z_2} \frac{Z_3}{Z_4} \quad n_6 = \frac{d_3}{d_1} \frac{Z_1}{Z_2} \cdot \frac{Z_3}{Z_4}$$

This back-gear is called a *single back-gear*. There are double and even treble back-gears, but they are seldom used. They also work by the preceding principle.

Meander's gear mechanism

It is similar to the tumbler gearbox mechanism, but this method of speed regulation is slightly better than the previous one in having slightly higher speeds and being quite rigid in comparison to tumbler gearbox mechanism. The arrangement is shown in Fig. 17.39.

$$n_1 = \left(\frac{T_1}{T_2}\right)^2 n_0$$

$$n_2 = \left(\frac{T_1}{T_2}\right)^1 n_0$$

$$n_3 = \left(\frac{T_1}{T_2}\right)^0 n_0$$

$$n_4 = \left(\frac{T_1}{T_2}\right)^{-1} n_0$$

$$n_5 = \left(\frac{T_1}{T_2}\right)^{-2} n_0$$

$$n_6 = \left(\frac{T_1}{T_2}\right)^{-3} n_0$$

$$n_7 = \left(\frac{T_1}{T_2}\right)^{-4} n_0$$

$$n_8 = \left(\frac{T_1}{T_2}\right)^{-5} n_0$$

It uses geometrical progression with the denominator T_1/T_2.

The advantages, the field of usage and the shortcomings are the same as in Norton's mechanism.

Mechanisms with the chain of gears are used in various feed boxes, speed boxes and in other machine-tool mechanisms.

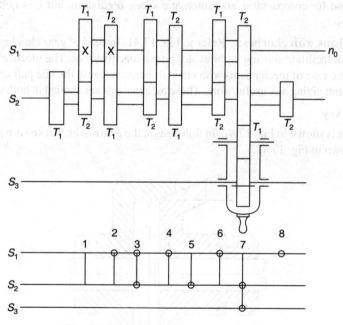

Fig. 17.39 Meander's gear mechanism.

Gearbox drive with clutches

This arrangement is used (see Fig. 17.40) where speed is regulated by the use of mechanical or friction clutches. Furthermore, the use of more than one clutch can increase the range of speed regulation to a much higher value. A clutch is engaged or disengaged by moving the key.

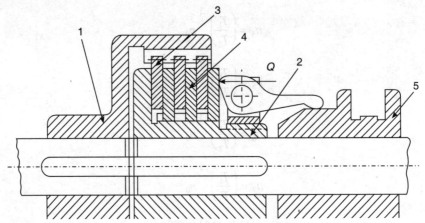

1: Flywheel, 2: Driven member, 3: Friction rings, 4: Friction rings, 5: Actuating member, Q: Lever

Fig. 17.40 Multiple-disc friction clutch.

$$n_1 = \frac{T_1}{T_2} n_0 \,;\, n_2 = \frac{T_3}{T_4} n_0$$

This type is good for compactness and automatic speed regulation, but it is quite expensive and the speed loss is quite high.

Elementary mechanisms with clutches: Refer to Fig. 17.41. Instead of gear clutches, it is possible to use friction clutches which facilitate shifting without stopping a machine tool. The mechanical efficiency in this case is lower than in the case of mechanisms with chain of gears, because the idle pair of the gear consumes a part of the power without giving any useful work. These mechanisms are applied in both speed and feed boxes.

Gearbox with drive key

A typical arrangement is shown in Fig. 17.42. In this design, the gears to be marked are placed in a key which slides in or out as shown in Fig. 17.42.

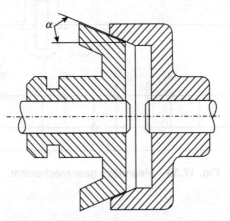

Fig. 17.41 Cone-type friction clutch.

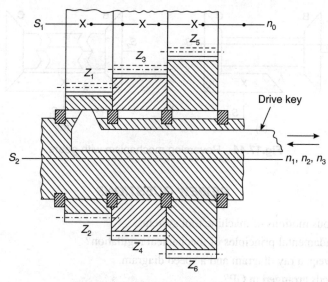

Fig. 17.42 Gearbox with drive key.

$$n_1 = n_0 \frac{T_1}{T_2} \qquad n_2 = n_0 \frac{T_3}{T_4} \qquad n_3 = n_0 \frac{T_5}{T_6}$$

These mechanisms are used only in various feed boxes.

The advantages and shortcomings are the same as in Norton's and Meander's mechanisms.

Speed reversing mechanism

Sometimes the direction of speed is to be reversed without wasting much time; for this purpose, an arrangement is used as shown in Figs. 17.43 and 17.44.

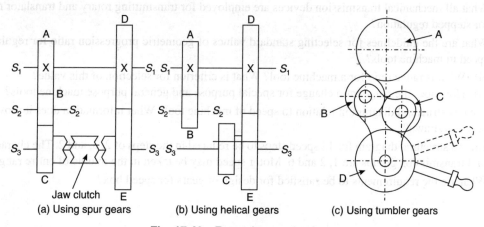

(a) Using spur gears (b) Using helical gears (c) Using tumbler gears

Fig. 17.43 Reversing mechanism.

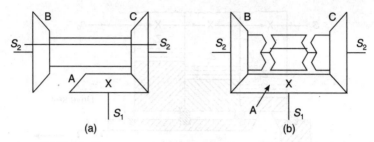

Fig. 17.44 Reversing mechanism with gears.

REVIEW QUESTIONS

1. Discuss the various motions of machine tool systems.
2. What are the fundamental principles of mechanical regulation?
3. Distinguish between a ray diagram and a speed diagram.
4. Why are the speeds arranged in GP?
5. Describe the principle of stepped speed regulation as applied to machine tools.
6. Discuss the methods of stepped regulation.
7. Discuss the advantages and limitations of gearbox drive.
8. Write short notes on
    (i) Norton gearbox mechanism.
    (ii) Meander's gear mechanism.
9. How the speed boxes for machine tool can be classified according to (i) general layout and (ii) method of changing speed?
10. How the speed steps required machine tool are determined?
11. What all mechanical transmission devices are employed for transmitting rotary and translator motions for stepped regulation?
12. What are the guidelines for selecting standard values of geometric progression ratio for regulation of speed in machine tools?
13. (i) What is range ratio for a machine tool? What is criterion for selection of this value?
    (ii) How does the range ratio change for special purpose and general purpose machine tools?
14. What is structural diagram in relation to speed of machine tool? What information does it provide and how is it drawn?
15. Draw a structural diagram for 12 speed steps to be reduced in the steps of 2, 3, and 2. The characteristic of 3 transmission groups are 1, 2 and 6. Motor speed may be taken in the middle of entire range?
16. What are the requirements to be satisfied for design of gears for speed box?

CHAPTER 18

Stepless Regulation of Speeds

18.1 INTRODUCTION

Power has a definite role to play in the pattern of our national development. At the same time, transmission of power is equally important. Stepless transmission is still more important for the design of modern machine tools. Stepless regulation has certain advantages over stepped regulation, such as:

(i) There is no loss of speed when change of speed takes place.
(ii) Speed can be adjusted without stopping the machine tools.
(iii) Time loss is minimum.
(iv) The process of regulation can be completely automatised.

On the other hard, it has certain limitations such as:

(i) It is a costly system.
(ii) Maintenance is also costly.

Stepless regulation can be obtained by any one of the methods as discussed in the following section.

18.1.1 Classification

There are three types of system:

(i) Systems with direct current (rectifiers or Leonard systems), which have a range of regulation of speed from 80 to 2000 rpm. This is also known as *electrical system*.
(ii) Hydraulic systems, which have a range of regulation of speed from 200 to 300 rpm.

(iii) Mechanical variators have a range of regulation of speed from 3 to 6 rpm (sometimes from 10 to 15 rpm). Variators should be situated in the place of greatest velocity where the torque is small. There are two types of variators for their application on machine tools:
(a) Where the chain consists of variators connected in series;
(b) Where the chain consists of a variator and a stepped mechanism.

18.1.2 Reversing of Motion in Machine Tools

The reversing of motion involves the following aspects:

(i) Electrical motor reversing is frequently used. It has the advantage of having a simplified construction of the machine, which leads to quick reversing. On the other hand, it is impossible to reverse only a part of the kinematic chain.
(ii) Hydraulic drive reversing is quick, reversing up to 1200 reverses per minute, and in this case rotor mass does not participate in reversing. Therefore losses are less than that of in electric motor reversing.
(iii) Mechanical reversing mechanism should meet the following requirements:
(a) It should transmit the power required.
(b) Forces of inertia, due to reversing, accelerate the wear of parts of the mechanism. The mechanism should be designed so that premature wear is eliminated.
(c) It should not require much power.

18.1.3 Mechanical Regulation

There are two methods of stepless regulation:
(i) Speed regulation with flexible coupling; and
(ii) Speed regulation with direct metal-to-metal contact.

Speed regulation with flexible coupling

This involves a belt drive with flat belt and two variable speed belt cones. The arrangement is shown in Fig. 18.1, wherein two frustums or cones are used to regulate speed without any step.

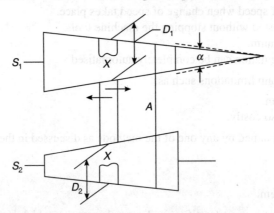

Fig. 18.1 Frustum or cones as pulley variators with flat belt.

The speed ratio obtained in this case is

$$i = \frac{D_1}{D_2}$$

It is necessary to satisfy the condition

$$\tan \alpha < \mu$$

where, μ is the coefficient of friction.

If this condition is not fulfilled, the belt will slip.

Usually $\alpha < 5.6°$; hence the range of the transmission ratio of this variator is considerably limited.

In practice, such belt cones can be about 1.5 m long.

Belt drive with adjustable pulleys: In order to increase the cross-stiffness of the belt, the wooden cleats are fastened to the belt with copper bevels (see Fig. 18.2). The wooden cleats are made with the bevels (chamfering) corresponding to those of the variable speed belt cones. Sometimes wedge belts are also used. The regulation range of these variators may reach 9. Usually the power does not exceed 5 kW.

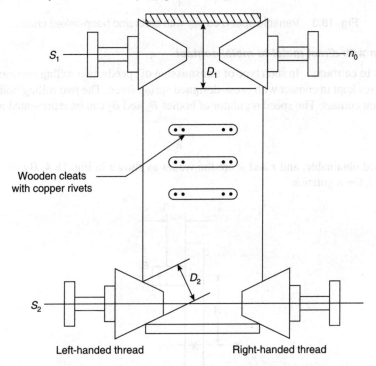

Fig. 18.2 Cone pulleys and flat belt used as variator with wooden cleats.

Chain variators: In this case the variator works by the principle of engagement. The scheme of transmission of rotary motion is the same as in the previous case. Variable speed belt cones are made with the open type slots and the hoop-linked chain is used instead of the belt in this case (see Fig. 18.3).

The cones are displaced to half of the tooth pitch. The plates are wider than the chain.

These variators may transmit power up to 75 kW. The range of control is 10.

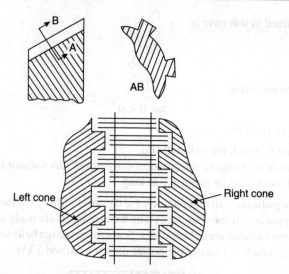

Fig. 18.3 Variable speed cones with slots and hoop-linked chain.

Speed regulation with direct metal-to-metal contact

Two rolling discs in contact: In such type of transmission of speeds, two rolling friction discs rotate about their supporting axes kept in contact with press-designed spring force. The two rolling bodies may roll under line contact or point contact. The speed regulator of bodies B_1 and B_2 can be represented as

$$i = \frac{r}{R}$$

where i is the speed obtainable, and r and R are the values as shown in Fig. 18.4. By changing the value of R, we can change i, the regulation.

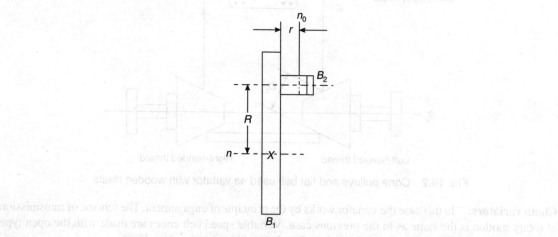

Fig. 18.4 Two rolling friction discs in contact acting as variator.

Ball and closed cup type variator: The ball variators require a very high degree of manufacturing accuracy, but the results are promising and have good scope of use. The range of regulation is up to 10 and can take up power up to 35 kW. The arrangement is shown in Fig. 18.5.

STEPLESS REGULATION OF SPEEDS **369**

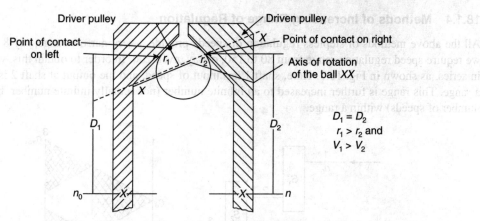

Fig. 18.5 Rolling ball and closed cup type speed variator.

Referring to Fig. 18.5, XX is axis of rotation of ball. r_1 and r_2 are lengths of perpendicular from points of left and right contact of cup surfaces upon XX.

Thus speed reduction $= i = \dfrac{V_2}{V_1} = \dfrac{r_2}{r_1}$

If r is greater than r_1, $V_2 > V_1$

Ball and open cup type variator: An arrangement similar to the previous one is shown in Fig. 18.6. We change the speed by changing the point of contact of intermediate rollers. The same range is obtained as in the case of ball and cup type variators as explained here.

$$n = n_0 \dfrac{D_1}{D_2}$$

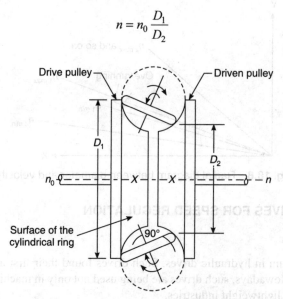

Fig. 18.6 Rolling ball-like cylindrical ring and open cup type variator.

18.1.4 Methods of Increasing Range of Regulation

All the above methods of stepless regulation of speeds provide a very narrow range, say 8 to 10, whereas we require speed regulation range from 80 to 150, or say, even 200. In order to obtain this, we use variators in series, as shown in Fig. 18.7. Here, shaft 1 has input of speed, and the output at shaft 2 is in the form of a range. This range is further increased to an infinite number (not actually infinite number, but a very large number of speeds) within a range.

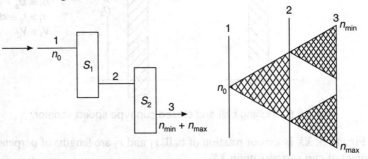

Fig. 18.7 Schematic arrangement of variators in series.

An example of this arrangement of increasing the range of regulation is the drive-in lathe having a pulley belt drive with a back gear. This has been discussed in the previous chapter with a ray diagram.

A graph between diameter and velocity is shown in Fig. 18.8, which reveals the increase of range of regulation to $n_{2\,min}$ from the range $n_{1\,min}$ for a particular diameter to be turned. The radial diagram for the variator shows a stepped mechanism.

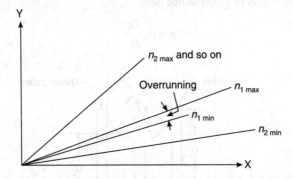

Fig. 18.8 Radial diagram between diameter and velocity.

18.2 HYDRAULIC DRIVES FOR SPEED REGULATION

18.2.1 Introduction

Fluid is the working medium in hydraulic drives. Such drives found their first application in grinding and broaching machine tools. Nowadays, such drives are being used not only in machine tool industry but also in aircraft, artillery and many lightweight industries.

All hydraulic drives contain two parts—a pump and a motor with fluid as the link between them. The pump driven by a motor converts electrical energy into the kinetic energy (K.E.) of fluid, and in turn converts this K.E. into the mechanical energy used up in a certain operation.

There are two types of hydraulic drives:

(i) Drives using K.E. of the fluid directly, viz. turbines.
(ii) Drives using static force, i.e. volumetric change.

Machine tools use the latter case of hydraulic power, which employs the principle of fluid displacement under pressure. This phenomenon is shown in Fig. 18.9.

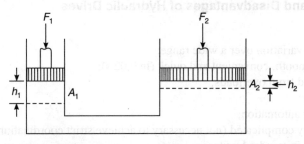

Fig. 18.9 Schematic diagram exhibiting principle of fluid displacement.

Let us suppose that the pump leakage is zero, and the fluid is incompressible. Then we have

$$Q = A_1 h_1$$
$$E = A_2 h_2$$
$$h_1 A_1 = h_2 A_2$$
$$\frac{h_1}{h_2} = \frac{A_2}{A_1} \quad \text{(equation of travel)}$$
$$F_1 = pA_1$$
$$F_2 = pA_2$$

Therefore,
$$\frac{F_2}{F_1} = \frac{A_2}{A_1} \quad \text{(equation of force)}$$

where Q is the volume of liquid moved
A_1 and A_2 are the cross-sectional areas
h_1 and h_2 are the displacements as shown
F_1 is the force at the pump end
F_2 is the force at the motor end
p is the intensity of pressure in kg/cm^2.

If Q is expressed in L/min, the power of the hydraulic drive
$$= p \cdot A_1 \cdot h_1 = Q \cdot p$$
$$\text{Horse power} = \frac{p \times Q}{(60) \cdot (75) \cdot (100)}$$

In the hydraulic drive the work done E is calculated as:
$$E = F_1 h_1 = p \cdot A_1 \cdot h_1 = Q \times p$$

or
$$H = \frac{p \cdot Q}{450} \quad \text{(in H.P.)}$$

This power H is called *indicated power*, and the motor power is expressed as under:

$$\text{H.P. of motor} = \frac{H.\text{indicated power}}{\eta}$$

where η is efficiency.

18.2.2 Advantages and Disadvantages of Hydraulic Drives

The advantages are:

  (i) Stepless speed variation over a wide range;
 (ii) Reversing is smooth, convenient and quick (in 0.02 s);
(iii) Easily protected from overloads;
 (iv) Self-lubricating;
  (v) Convenient for automation;
 (vi) System not very complicated (not necessary to achieve strict coordination);
(vii) Can be applied in standard units.

The disadvantages are:

  (i) Lack of uniform characteristics;
 (ii) Leakage of fluid;
(iii) Change in fluid property with temperature;
 (iv) Visual control over hydraulic drive not possible.

18.2.3 Requirement of Fluids Applied in Hydraulic Systems

Working fluids are under the influence of variable temperatures, pressures and speeds; therefore they should meet the following requirements:

   (i) These fluids should not vapourise at the operating temperatures;
  (ii) They must not contain, absorb or reduce air;
 (iii) They must not foam;
  (iv) They must not cause corrosion;
   (v) They must lubricate very well;
  (vi) They should have chemical stability;
 (vii) They should have adequate viscosity;
(viii) They should meet fire prevention requirements.

18.2.4 Mineral Oils as Fluids for Hydraulic Systems

Refined mineral oils fulfil these requirements quite satisfactorily. This is evident from Table 18.1.

Table 18.1 Mineral oils

Kind of oil	°E 50	Ignition temp.	Thickening temp. (°C)
Spindle oil (industrial "15")	1.8–2.2	160°	−25°
Spindle oil (industrial "20")	2.8–3.2	170°	−20°
Turbine oil	2.9–3.2	180°	−15°

If the temperature is reduced the viscosity increases, resulting in higher losses while in use.

18.2.5 Properties of Hydraulic Oils

(i) For such oils, the compression of the fluid is neglected up to 150 kg/cm² pressure, but above this value it is taken into consideration.

(ii) The velocity with which the elastic waves move in the fluid are approximately equal to the velocity of sound.

Density

Density is expressed as mass per unit volume and has $\dfrac{kg}{m/s^2 \cdot m^3}$ or $\dfrac{kg \cdot s^2}{m^4}$ as its unit.

Viscosity

There is a relative viscosity and an absolute viscosity.

Absolute viscosity, written as η, is a stress arising between two adjacent layers of fluid which are moving with different velocities. In technical units

$$\eta = \frac{kg \cdot s}{m^2}$$

The unit of η is $\dfrac{dyne \cdot s}{m^2}$, written as poise. 0.01 poise is equal to 1 centipoise.

Kinematic viscosity (v): The unit of this is

$$v = \frac{\mu}{\delta} = \frac{\frac{kg \cdot s}{m^2}}{\frac{kg \cdot s^2}{m^4}} = \frac{m^2}{s}$$

μ is the stress (measured in kg) arising as a result of motion of two layers of the fluid at a distance of 1 metre, on an area of 1 m² with a velocity of 1 m/s.

In physical units
$$v = \frac{cm^2}{s} = \text{stoke}$$

0.01 stoke = 1 centistokes

Thermal conductivity

This is given by $\lambda.t. \ c = a + b.t$ kcal/cm.°C. For spindle oils, $a = 3 \times 10^{-4}$ and $b = 1.25 \times 10^{-2}$.

Heat capacity

For mineral oils
$$C_p = 0.4 \frac{cal}{kg \cdot °C}$$

$$H_c = C_{p \ max} \quad \text{(in kcal/kg/°C-kg)}$$

Oil circulation

Oil circulation system may be of two types:
(a) Open type
(b) Closed type.

Open circulation for a drive system is good because of setting and cooling conditions, but it is characterised by large dimensions, and air also has access to the system.

18.2.6 Efficiency in Stages

This is given by the following derivation:

Unit of Q = litre/min

Unit of pressure = P kg/cm^2

Therefore
$$H = \frac{P \text{ kg}.L \text{ cm}}{\min. \eta_{tot}} = \frac{P \cdot Q}{450 \cdot \eta_{tot}} \text{ H.P.}$$

$$\eta_{tot} = \eta_{vol} \, \eta_{mech} \, \eta_{hydr}$$

$$\eta_{tot} = \frac{Q - Q_{leakage}}{Q}$$

18.2.7 Cylinders for Hydraulic Drive Pumps

The requirement of a cylinder for a hydraulic drive mechanism is necessary, as it converts the kinetic energy of the moving fluid into the mechanical energy required for operating the piston.

These cylinders can be classified into two main categories.

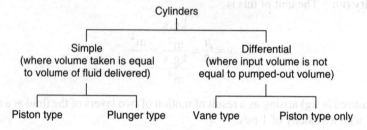

Simple piston type cylinder

This type of piston cylinder is shown in Fig. 18.10, which is generally used for reciprocating movements of tables. If the length of table is L and that of the piston is H, then

$$L = H + 2l$$

where l is shown in Fig. 18.10.

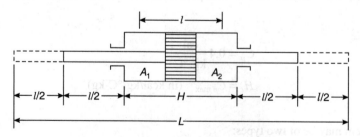

Fig. 18.10 Piston type hydraulic cylinder.

$$A_1 = A_2 \quad V_1 = \frac{Q}{A_1} = \frac{Q}{A_2} = V_2$$

Plunger type cylinder

The schematic arrangement of this type of pump cylinder is shown in Fig. 18.11. The table movement is limited to a value equal to the plunger stroke length given by the equation as given below. This type of cylinder is capable of dealing with higher fluid pressures and thus more power is available at the table, and the motion at ends are quite smooth; but it has short stroke lengths.

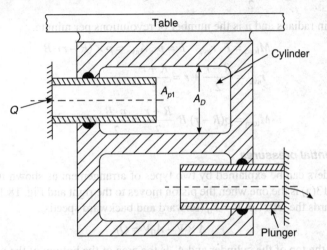

Fig. 18.11 Plunger type cylinder.

$$L = \frac{Q}{A_D - A} = \frac{Q}{A_{p1}}$$

$$F = P \cdot A_{p1}$$

Vane type cylinder

This type of cylinder is quite compact in construction and therefore suitable for rotary motions or reciprocating movements such as in the case of broaches, planers, etc. The schematic arrangement is shown in Fig. 18.12. For a particular type of design of such a cylinder, the fluid pressure is responsible for the output torque as shown in the equation.

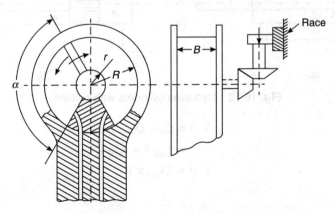

Fig. 18.12 Vane type cylinder.

q is the discharge of oil per movement of cylinder vane and is given by

$$q = A_{sect}. B \quad A_{sect.} = \frac{\alpha}{2}(R^2 - r^2) \quad q = \frac{\alpha}{2}(R^2 - r^2) \cdot B$$

$$Q = \frac{\alpha \cdot n\beta}{2}(R^2 - r^2)$$

where α is expressed in radians and n is the number of revolutions per minute.

$$M_{torq} = F \cdot r_{ave} \quad F = p \cdot A_{ave} = p \times (R - r) \cdot B$$

$$r_{ave} = \frac{R-r}{2} + r = \frac{R+r}{2}$$

$$M_{torq} = p(R-r) B \frac{R+r}{2} = \frac{p \cdot B}{2}(R^2 - r^2)$$

Cylinder with differential pressures

The differential cylinders can be explained by two types of arrangement as shown in Fig. 18.13(a) and (b). The circuit in Fig. 18.13(a) is the one when the piston moves to the right and Fig. 18.13(b) is the circuit when the piston moves towards the left. Let $V = V_0$ (forward and backward speeds).

$$A \times V - A_0 V = Q_P$$

where A is the area of the top of the cylinder and A_0 is the area of the bottom of the cylinder

$$V(A - A_0) = Q_P$$

or

$$A - A_0 = \frac{Q_P}{V}$$

where Q_P is the volume of oil, or

$$A_{rod} = \frac{Q_P}{V}$$

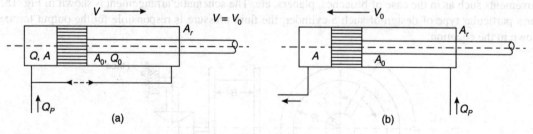

Fig. 18.13 Hydraulic cylinders with pistons.

But if $V = V_0$

$$V \cdot A = Q_P + Q_0$$
$$= A_{rod} V + A_{rod} V$$
$$V \cdot A = 2 A_{rod} \times V$$

Therefore,

$$A_{rod} = \frac{A}{2}$$

18.2.8 Piston and Piston Rod Seals

Loss of hydraulic power is unavoidable due to many reasons viz. leakage pass through joints, and friction in the piping. The latter is unavoidable; however, leakage could be reduced to zero at higher pressures some troubles may arise, whereas high pressures are not required in machine tools operation and control.

In order to prevent leakage, proper seals must be designed for pistons and piston rods. There are three types of seals as given below:

Seals for pistons and piston rods

The methods of sealing are as follows:

1. Cup seals
2. Piston ring seals
3. Groove seals.

Cup seal

The various components of a cup type of seal is shown in Fig. 18.14 in position with the cylinder walls, so as to seal the leakage. The various components of the complete assembly are given below. Various materials for seals which are generally used are:

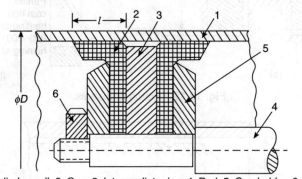

1. Cylinder well, 2. Cup, 3. Intermediate ring, 4. Rod, 5. Cup-holder, 6. Nut

Fig. 18.14 Cross-sectional view of piston rod seal.

The frictional force that is offered by such a seal can be calculated as follows:

$$F_r = \mu F = \pi D l p \times \mu$$

where p is in kg/cm^2
 l and D are in cm
 $\mu = 0.006$ to 0.01.

Such seals are suitable to be used for pressures up to 20 kg/cm^2 safely.

There are other types of seals, as shown in Figs. 18.15 to 18.18. Rubber seals have the disadvantage of being worn out quickly and are to be changed quite often to ensure no leakage. Piston ring seals are also used for higher pressure up to 100 kg/cm^2, as shown in Fig. 18.17.

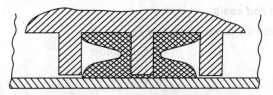

Fig. 18.15 Bell-shaped seals.

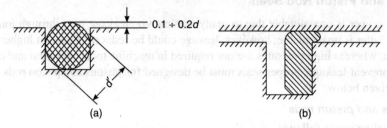

Fig. 18.16 Rubber ring seals.

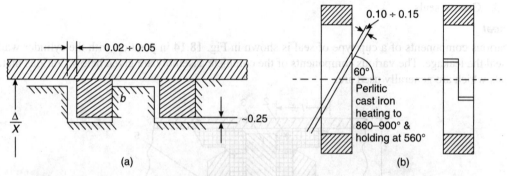

Fig. 18.17 Piston ring seals.

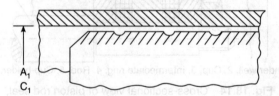

Fig. 18.18 Groove type seals.

They are applied for both low and high pressures

$$F = \pi Db(ZP_r + p)\mu$$

where z = Number of rings

P_r is the elastic pressure of ring on the wall

p is the specific pressure in kg/cm^2 and has a range of 0.6–0.9 kg/cm^2

b is contact length

It is assumed that the pressure is applied to only one side of the ring.

$$\mu = 0.12 \text{ to } 0.15$$

Specially designed piston rod seals

The components of such a seal are given below (see Fig. 18.19):

1. Front top (box)
2. Cup holder
3. Cups
4. Press flange
5. Coupling bolt
6. Rod

The force that this seal can withstand can be calculated as follows:
$$F = \pi dlp$$
where d and l are in cm and $p = 0.4$ to 1.3 kg/cm^2.

This type of seal is shown in Fig. 18.19 in position. Force calculations are also shown. The force that a seal can withstand depends on the resisting length, l, which can be increased or decreased as per requirement.

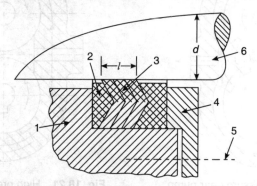

Fig. 18.19   Cross-sectional view of special type of seal.

18.2.9 Pumps for Hydraulic Drives of Machine Tools

Geared, reciprocating and vane pumps are mainly used in hydraulic drives. Pumps may have both regulated and constant capacity.

Gear pumps

Here the pump capacity is given by
$$Q = 2 F \cdot b \cdot Z \cdot n \text{ mm}^3/\text{min}$$

where F is the tooth space in mm^2
b is the tooth length in mm
Z is the number of teeth
n is the number of revolutions per minute.

$$F = \frac{h \cdot t}{2} = \frac{(d_{out} - d)t}{2}$$

where d_{out} is the outside diameter of gears
d is the pitch circle diameter
t is the circular pitch
h is the tooth height.

$$Q = (d_{out} - d) \cdot t \cdot Z \cdot b \cdot n \cdot \frac{1}{10^6} \frac{1}{\text{min}}$$
$$= \frac{\pi d(d_{out} - d) \cdot b \cdot n}{10^6} \cdot \frac{1}{\text{min}}$$

Pumps of low pressure: ≤ 10 kg/cm^2
Pumps of medium pressure: $10-25$ kg/cm^2
Pumps of high pressure: $25-100$ kg/cm^2
Pumps of high pressure have balanced construction.

Low pressure and high pressure gear pumps are shown in Figs. 18.20 and 18.21 respectively.

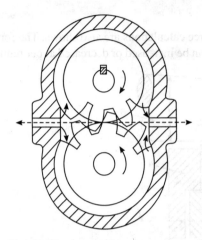

Fig. 18.20 Low pressure gear pump.   **Fig. 18.21** High pressure gear pump.

Characteristics of gear pumps: These are as follows:
1. Q varies from 5 to 125 mm³ per min.
2. P_{max} is up to 100 kg/cm²
3. H.P. ranges from 1.3 to 6.5 kW
4. $\eta_{tot} = \eta_{vol}\, \eta_{mech}\, \eta_{hydr} = 0.6$–$0.7$ $\eta_{vol} = 0.8$–0.92

Pulsation of flow is up to 14%.

Gear pumps may be used for machine tools working at medium and high speeds, and with small cutting forces. They are applied for grinding, tool grinding, and honing machines, as well as for the travel of grinding-machine tables, and as a drive for subsidiary mechanisms too.

Rotary vane pumps

Here, one revolution of the shaft involves two suctions and two deliveries of a fluid volume between two vanes, the rotor, and the stator.

Therefore, due to this principle of work, such a pump is called a *double-acting pump* (see Fig. 18.22).

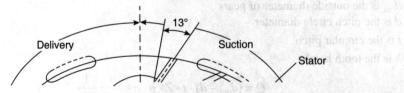

Fig. 18.22 Rotary vane pump.

The double-acting pump has relieved construction, and P_{max} may reach 65 kg/cm². Vanes are pressed to the stator due to the pressure from the compression chamber and the centrifugal force. The stator ring is made of steel, and the vanes are also made of steel. The slots of the rotary disc are broached.

$$Q = \frac{2B \cdot n}{10^6}\left[\pi(r_{maj}^2 - r_{min}^2) - \frac{S \cdot Z \cdot (r_{maj} - r_{min})}{\cos \gamma}\right] \text{ per minute}$$

where B is the width of vanes in mm;
n is the number of revolutions of the shaft per minute;
r_{maj} is the major semi-axis;
r_{min} is the minor semi-axis;
S is the vane thickness;
Z is the number of vanes;
γ is the take angle of vane.

Characteristics of rotary vane pumps: These are as follows:
1. Q is between 5 and 100 mm³ per min.
2. p_{max} is 65 kg/cm².
3. η_{vol} is between 0.85 and 0.92.
4. η_{tot} is between 0.72 and 0.76.

These pumps are more expensive when compared with gear pumps.

Radial piston pumps
A radial piston pump is shown in Fig. 18.23.

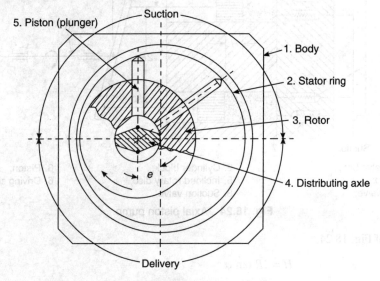

Fig. 18.23 Radial piston pumps.

Characteristics of radial piston pumps: These are as follows:

Productive capacity:
$$Q = \frac{\pi d^2}{4} \cdot 21 \cdot Z \cdot n \cdot \frac{1}{10^6} \text{ per min}$$

$$Q = 50\text{–}400 \text{ per min}$$

$$\eta_{max} = 75\text{–}200 \text{ kg/cm}^2$$

Pulsation of flow is 2%.

$$\left. \begin{array}{l} \eta_{vol} = 0.92\text{–}0.95 \\ \eta_{vol} = 0.89 \end{array} \right\} \text{ for } P_{max} = 75 \text{ kg/cm}^2$$

$$\eta_{vol} = 0.92\text{–}0.95 \brace \eta_{vol} = 0.89} \quad \text{for } P_{max} = 75 \text{ kg/cm}^2$$

Reversing time is 0.5 s.
Power ranges from 20–100 kW.
These pumps are more expensive than the rotary vane pumps.

The moment of inertia of these pumps is considerably less than that of electric motors. It allows change in the number of revolutions, as well as reversing, very quickly.

The capacity of a pump is regulated by varying its eccentricity e.

Axial piston pumps

In this pump the axes of pistons are parallel to the rotor axis. Therefore we shall call these *axial piston pumps* (see Fig. 18.24).

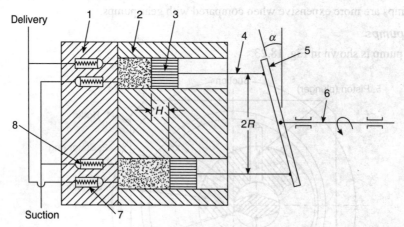

1. Control valve box; 2. Cylinder block; 3. Piston;
4. Piston rod; 5. Inclined rotary disc; 6. Driving shaft;
7. Delivery valve; 8. Suction valve.

Fig. 18.24 Axial piston pump.

From Geometry of Fig. 18.24,

$$H = 2R \tan \alpha$$

Pump capacity:

$$Q = \frac{\pi d^2}{4} \cdot Z \cdot n \cdot (2R \cdot \tan \alpha) \frac{1}{10^6} \text{ (in L/min)}$$

The characteristics of axial piston pumps are as follows:

1. Q is between 4 and 1200 L/min.
2. P_{max} is up to 200 kg/cm^2.
3. N is between 1 and 300 kW.
4. η_{vol} is between 0.82 and 0.84.
5. The speed (n) ranges of hydraulic drives are from 30 to 3600 rev/min.

18.2.10 Throttle Valves for Hydraulic Drives

A fluid forced out in a piping by a pump runs at a pressure higher than what is required to be used in the cylinder. Furthermore, quite often we need to convert a pressure head into the velocity head, so as to vary the volume of fluid delivered to the cylinder.

A throttle valve must fulfill the following requirements for its successful use:
  (i) It must produce minimum contamination at low rate of discharge.
  (ii) It must produce a minimum change in discharge rate when pressure differential is changed.

The output of a fluid through an orifice can be expressed from the standard equation in hydraulics, i.e.

$$Q = \mu \times A \times Y = \left(\mu A \cdot \sqrt{\frac{2g}{\rho}}\right) p^\alpha$$

where μ is the coefficient of friction for mineral oils $\mu = 0.73$;
   A is the cross-sectional area of the throttle slot;
   g is 9.81 m/s^2;
   p is the pressure drop of the throttle;
   ρ is the specific weight of oil;
   Y is 0.009 kg/cm^3.

In the general case

$$\text{Putting} \left(\mu A \sqrt{\frac{2g}{\rho}}\right) = C$$

$$Q = C \cdot p^\alpha$$

where $\alpha = 0.5–1$.

Thus, discharge rate can be calculated from the preceding equations. For those oils which are liable to contamination, take $\alpha = 0.5$.

Slotted throttle

This throttle answers both requirements well and therefore is widely used in practice; for such a case, $\alpha = 0.06$.

The various types of throttle valves that are used in speed regulation of hydraulic drives are:

  (a) Throttle valve with eccentric slot;
  (b) Diaphragm throttle valve;
  (c) Slotted throttle valve.

These are shown in Fig. 18.25(a), (b) and (c).

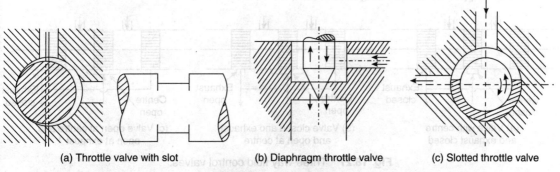

(a) Throttle valve with slot   (b) Diaphragm throttle valve   (c) Slotted throttle valve

Fig. 18.25 Throttle valves.

18.2.11 Fluid Control Valves

Fluid control valves are required to change the direction of flow. Such valves are of two types, classified as per the method of operation, as:

(i) Axial valves
(ii) Rotary valves.

An axial flow control valve is shown in Fig. 18.26(a) and (b). It consists of three components: (1) the delivery pipe, (2) the exhaust pipe, and (3) the plunger. As per the position of the plunger, the cylinder port is either open or close, and the fluid either comes out of the cylinder or enters the cylinder as the case may be. This is well indicated in Fig. 18.26(a) and (b). The port opening is from 0.15 mm to 0.04 mm, and the maximum velocity of the fluid in the pipe is limited to 3 m/s to 4.5 m/s.

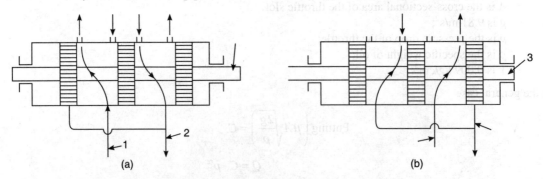

Fig. 18.26 Control valves for oils and fluids.

Two-way valves are applied in cases where it is necessary to obtain either a rectilinear reciprocating motion or a rotary motion of a certain machine tool unit, and three-way valves are used where either a reversing reciprocating motion, or a rotary motion with its stop in the central position of the valve plunger, is required.

Types of three-way valve

There are three types of three-way valves, which are used according to system requirements. These are:

(a) Valve with centre and exhaust closed in its neutral position. It is blocked in this portion.
(b) Valve open at centre and closed at exhaust.
(c) Valve open at centre and open at exhaust.

The three types are shown in Fig. 18.27.

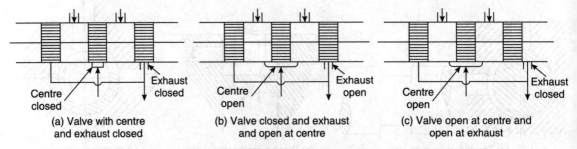

Fig. 18.27 Three-way fluid control valves.

Methods of operating valves

A valve can be operated either manually, electrically or mechanically; a particular method is suited most to a certain situation, while it may not suit the others. For example, a thermostat (for temperature control) which is set to operate at a certain temperature will cut off the supply of F_{12} when the temperature is reached. Similarly, a solenoid valve is used for reversing the motion of a table which closes or opens the port of the valve. These valves are shown in Fig. 18.28, explaining the principle of working for different situations involving machine tool drives.

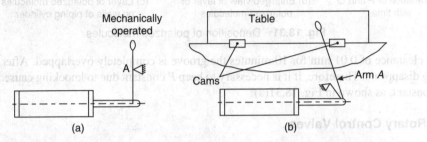

Fig. 18.28 Mechanisms of operation of mechanically operated value.

Arm A oscillates between 1° and 30° to either open or close the valve port, as shown in Fig. 18.29.

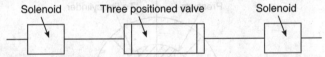

Fig. 18.29 Details of mechanism as shown in Fig. 18.28(a) and (b).

Figure 18.30 shows the working model based on the principle of operation of the mechanically operated value shown in Figs. 18.28 and 18.29.

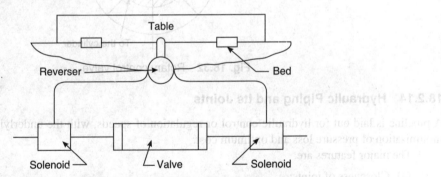

Fig. 18.30 Working model based on principle shown in Figs. 18.28 and 18.29.

18.2.12 Resistance to Flow due to Obstruction

Obstruction is likely to be caused due to deposition of contaminants with the elapse of time and use of piping, irrespective of whether piping runs full or partially full. Leakage in course of time through valves is reduced and the force required for operating valves increases.

A graph plot exhibits the variation of P and Q with time, as plotted in Fig. 18.31(a). A layer of polarized molecules gets deposited on the walls of the piping cylinder, as shown in Fig. 18.31(c).

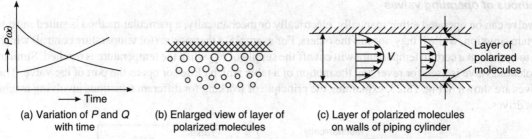

(a) Variation of P and Q with time   (b) Enlarged view of layer of polarized molecules   (c) Layer of polarized molecules on walls of piping cylinder

Fig. 18.31 Deposition of polarized molecules.

With clearance of 0.01 mm for 10 minutes the groove is completely overlapped. After knocking, this overlapping disappears. Therefore, if it is necessary to keep P constant due to knocking caused by vibrations, P is kept constant, as shown in Fig. 18.31(a).

18.2.13 Rotary Control Valves

A rotary valve is shown in Fig. 18.32. It has ports at 90° at 45° from the axes (vertical and horizontal). Therefore, turning the knob through 45° changes the direction of flow as shown in Fig. 18.32.

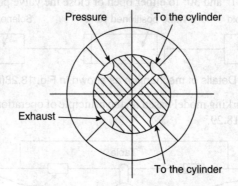

Fig. 18.32 Rotary control valve.

18.2.14 Hydraulic Piping and Its Joints

A pipeline is laid out for hydraulic control or regulation of speeds, with the underlying principle being the minimisation of pressure loss and optimum cost.

The major features are:

(a) Closeness of joints;
(b) Minimum length;
(c) Minimum number of bends.

The types of joints for copper and steel piping are shown in Fig. 18.33(a) and (b), with various components used therein.

Flanges are used for higher pressure pipes having more fluid in them, viz. pipes having 50 mm or more diameter. Flanged joints as shown in Fig. 18.34 are used.

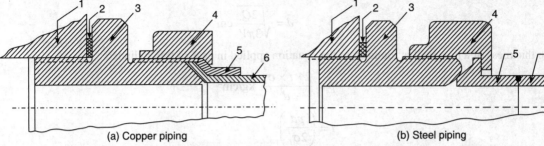

(a) Copper piping
1. Wall of body, 2. Red copper seal, 3. Connection hose, 4. Nut, 5. Nipple, 6. Red copper pipe

(b) Steel piping
1. Wall of body, 2. Red copper seal, 3. Connection hose, 4. Nut, 5. Nipple (steel), 6. Steel pipe

Fig. 18.33 Laying of piping and joints.

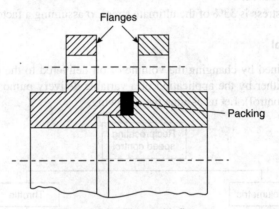

Fig. 18.34 Pipes with flanges.

Pipelines should be tested for a maximum of 1.25 kg/cm² pressure.

18.2.15 Cross-section Calculation

If the discharge of oil is Q L/min and its velocity V m/s is known, then the cross-section of a pipe can be determined by the formula

$$a = \frac{Q}{V}\frac{L \cdot s}{\min . m} = \frac{1000 \text{ cm}^2 \cdot s}{60 \text{ s} \cdot 100 \text{ cm}}\frac{Q}{V} = \frac{Q}{6V}\text{cm}^2$$

The recommended velocities in pipelines are:

V_{suction} : 1.5 to 2 m/s

V_{delivery} : 3 to 5 m/s.

$$a = \frac{\pi d^2}{4}$$

or

$$\frac{Q}{6V} = \frac{\pi d^2}{4}$$

or

$$\frac{4Q}{6V\pi} = d^2$$

or
$$d = \sqrt{\frac{2Q}{3\pi V}} \text{ cm}$$

The thickness of a pipe wall is found from the equation applied in case of thin cylinders:
$$p = \frac{2 \times t \times \sigma_t}{d} \text{ kg/cm}^2$$

∴
$$t = \left(\frac{pd}{2\sigma_t}\right)$$

where t is the tube thickness
  σ is the ultimate stress
  d is the diameter of the tube
  i.e. σ_t or induced stress is 33% of the ultimate stress. σ assuming a factor of safety of 3

18.2.16 Speed Control

Speed control can be obtained by changing the volume of oil delivered to the cylinder. The change of oil volume can be achieved either by the application of a variable delivery pump (volumetric control), or by throttling the oil (throttle control). Let us assume the following:

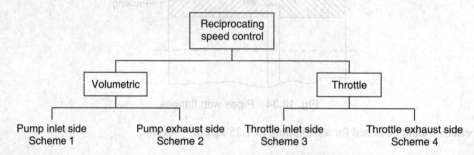

We shall compare these four methods by the following factors: efficiency, cost, stiffness, and vibration-proofing. The values of the leakages are the following:

Leakage through pump: $\delta_p = 5\text{–}6 \text{ cm}^3/\text{min per kg/cm}^2$
Leakage through valve: $\delta_v = 0.2\text{–}0.1 \text{ cm}^3/\text{min per kg/cm}^2$
Leakage through cylinder: $\delta_c = 0.2\text{–}0.1 \text{ cm}^3/\text{min per kg/cm}^2$

Speed loss due to loads

V_w is the operating speed with load.
V_u is the operating speed without load.

$$V_w = V_u - \Delta V = \frac{Q_w}{A} - \frac{\Delta Q}{A}$$

Let the speed loss be denoted by δ.
Therefore, speed loss can be computed as follows:

$$\delta = \frac{V_u - V_w}{V_u} = \frac{\Delta V}{V_u} = \frac{\Delta Q}{A \cdot V_u} = \frac{\Delta \cdot p}{A \cdot V_u} = \frac{\Delta \cdot F}{A^2 \cdot V_u}$$

where   F is the resistance offered by the system in kg
        A is the area in cm^2
        Δ is the change in Q per kg/cm^2.

Thus, from the equation

$$\delta = \frac{\Delta F}{A^2 \cdot V_u} \qquad \left(\text{we see that } \delta \propto \Delta F \text{ and } \propto \frac{1}{V_u}\right)$$

The relationship is shown in Fig. 18.35(a) and (b) for speed loss with respect to resistance and velocity respectively, and the conclusions is therefore that speed loss increases as load or resistance increases, while speed loss decreases as speed without load increases.

$$\delta = \frac{\Delta \cdot F}{A^2 \cdot V_u}$$

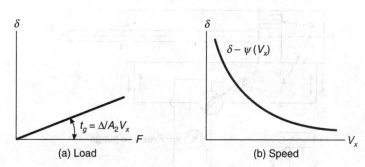

Fig. 18.35 Relationship of speed loss with respect to load and speed.

Hence, in the volumetric control by scheme 1 (variable delivery pump on the inlet side) the speed loss of the system is directly proportional to the load and inversely proportional to the operating speed without load.

Let us consider the following case.

Let $F = 1000$ kg
       $\delta = 5$ cm^3/min per kg/cm^2
       $A = 100$ cm^2
       $V_{u1} = 2$ cm/min

$$\delta_1 = \frac{5 \cdot 1000}{10000 \cdot 2} \cdot 100 = 25\%$$

and $V_{u2} = 50$ cm/min

$$\delta_2 = \frac{5 \cdot 1000}{10000 \cdot 50} \cdot 100 = 1\%$$

Scheme 1. *Volumetric control with variable delivery pump on inlet side*

1. The oil in full volume is supplied to the cylinder. The velocity of the piston is regulated by varying the pump delivery.
2. Safety valve 2 works only during overload of a hydraulic system.
3. Pressure in the pump exceeds that in the cylinder by the amount of losses, resistance in pipes and in hydraulic mechanisms.

Figure 18.36 shows the schematic for this scheme.
Let us suppose that there is no leakage.

Then
$$\Delta Q = 0$$

$$V = \frac{Q_p}{A}$$

Q_p is regulated by eccentricity e.

$$Q_p = q\psi n, \text{ where } \psi = \frac{e}{e_{max}} \quad \text{(varies from 0 to 1)}$$

$$V = \frac{q\psi n}{A} \quad \text{(ranges from 0 to } V_{max})$$

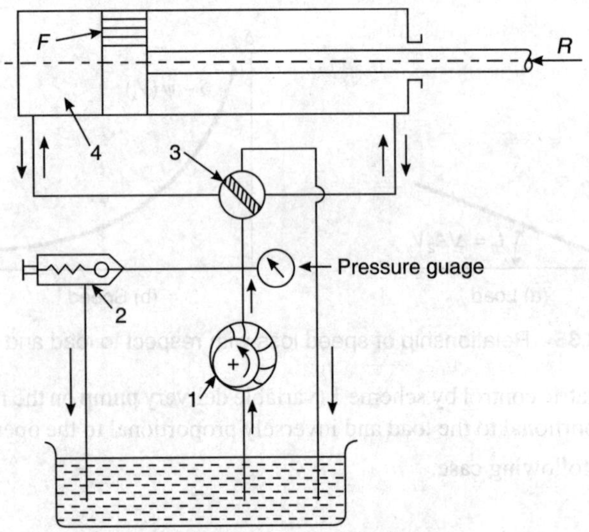

1. Variable delivery pump, 2. Safety valve,
3. By-pass valve, 4. Working cylinder

Fig. 18.36 Schematic 1 of pump, valve, cylinder and oil supply system.

In practice, leakage can not be avoided. Let us find operating speeds with load and without it.

$$V_w = \frac{Q_p - \Delta Q}{A} \qquad V_u = \frac{Q_p}{A}$$

$\Delta Q_p = \delta_p \cdot p$ – leakage in the pump

$\Delta Q_v = \delta_v \cdot p$ – leakage through valve

$\Delta Q_c = \delta_b \cdot p$ – leakage through cylinder

$$\Delta Q = \sum_{l=1}^{l=k} \Delta Q_i = p(\delta_p + \delta_v + \cdots + \delta_i)$$

The value of p is maximum; hence Q is also maximum.

Efficiency

$$N = \frac{p \cdot Q_p}{450} \text{ H.P.}$$

$$N = \frac{p \cdot V_w \cdot F}{450} = \frac{R \cdot V_w}{450} = \frac{R \cdot V_u}{450} = \frac{R \cdot \Delta V}{450}$$

$$\frac{RV_u}{450} = \left(\frac{1 - \Delta V}{V_u}\right) = \frac{RV_u}{450}(1 - \delta)$$

Pump capacity is directly proportional to the load.

It is necessary to point out that such a system has no losses in oil throttling; therefore, its efficiency is rather high.

The field of application involves machine tools of high productive capacity which work with wide control range (planing and broaching machines).

Scheme 2. *Volumetric control with variable delivery pump on exhaust side*

Here $\quad Q = 0$

Stroke speed $\quad V_i = \dfrac{Q_p}{A_2}$

Cutting stroke speed $\quad V_c = \dfrac{Q_{p_2}}{A_1}$

If $\Delta Q = 0$, $\quad V_i = \dfrac{Q_{p_2}}{A_1}$ and $V_c = \dfrac{Q_{p_2} - \Delta Q}{A_1}$

The piston balance provision is given by

$$P_p A_1 = P_b \cdot A_2 + R + T$$

$$P_b l = \frac{P_p A_1 - R - T}{A_2}$$

where $\quad P_p$ is the pump pressure;

P_b is the back pressure;

Figure 18.37 shows the schematic for this system.

Here, F_1 and F_2 represent the resistance offered by the system using backward and forward strokes, while P_1 and P_2 represent the pressures during the idle and working strokes.

The pump capacity is spent on the motion of the table and in running off the excess oil. Part of the oil is returned to the circuit.

The efficiency is less than in Scheme 1, while the cost is higher and the vibration-proofing is better.

Scheme 3. *Throttling control with throttle on inlet side*

The delivery of Pump 1 is, to some extent higher than that required for $V_{\max}$. The surplus oil runs off through Valve 2. Pressure between the pump and throttle p_v is constant, both with load and without. This pressure is adjusted by means of Valve 2. Pressure p_f behind the throttle varies, depending on load R.

As in this scheme, the pump works with constant pressure and the oil leakage does not influence the piston speed motion when the load is changed.

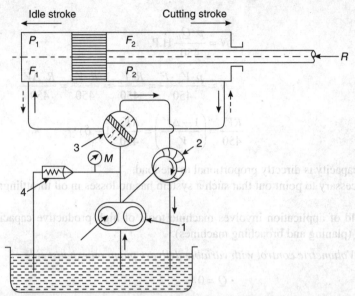

1. Constant delivery pump, 2. Variable delivery pump, 3. By-pass valve, 4. Safety valve

Fig. 18.37 Schematic 2 of pump, valve, cylinder and oil supply system.

But at the same time, the speed depends on the load, since p_c depends on R and the quantity of oil passing through the throttle depends on $\Delta p = p_v - p_c$.

As the load is increased, the speed is reduced and vice versa. This fact is insignificant with small loads. That is why these schemes are widely applied for grinding machines.

Due to this fact, that oil constantly runs off under pressure, and the efficiency of this system is not high. Figures 18.38 and 18.39 show the schematic of this system.

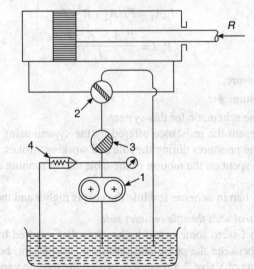

1. Constant delivery pump, 2. By-pass valve,
3. Throttle, 4. Safety valve

Fig. 18.38 Schematic 3 of pump, valve, cylinder and oil supply system.

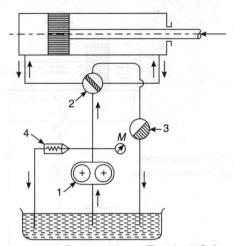

1. Constant delivery pump, 2. By-pass valve, 3. Throttle, 4. Safety valve

Fig. 18.39 Schematic 4 of pump, valve, cylinder and oil supply system.

Scheme 4. *Throttling control with throttle on exhaust side*

Pressure on the right of the cylinder depends on load R. It means that the volume of exhaust oil passed through the throttle depends on R too. The load being increased, the pressure in the right part of the cylinder drops. This is followed by less discharge of oil through the throttle and decreased speed. If the load is reduced, then the speed increases.

This scheme, in comparison with the previous ones, gives the highest smoothness of motion. Therefore it is practised on a large scale.

18.2.17 Design of the Components of Grinding Machine Hydraulic Drive

The following are the various components of a hydraulic drive for grinding machines:

1. Gear pump
2. Safety valve
3. Throttle
4. Piston valve
5. Cylinder
6. Control levers
7. Cylinder and cylinder rod diameter
8. Oil discharge (backward and forward strokes)
9. Piping diameters
10. Pump pressure
11. Motor HP.

Figure 18.40 shows the arrangement of pumps, valves, piston and cylinder, while Fig. 18.41 gives a full working view of the machine with its various components.

For the design of this drive system, if it is given that

$$V_{\text{forward}} = 15 \text{ m/min}$$
$$V_{\text{return}} = 1.2\, V_{\text{forward}}$$
$$F = 10 \text{ kg}$$

and if the cylinder has one rod. Then we have net force on the piston, denoted by P, from

$$P = F + T + P_i$$

where F is the force due to oil pressure;
T is the force due to friction;
P_i is the force of inertia, i.e. weight of table x.

394 TEXTBOOK OF PRODUCTION ENGINEERING

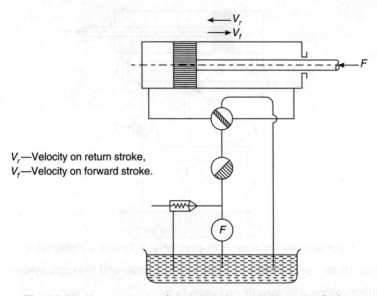

V_r—Velocity on return stroke,
V_f—Velocity on forward stroke.

Fig. 18.40 Arrangement of pumps, valves, piston and cylinder.

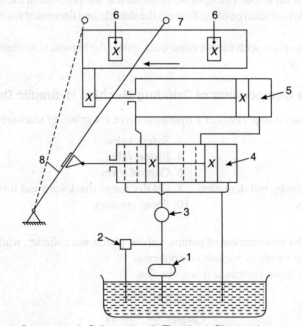

1. Gear pump, 2. Safety valve, 3. Throttle, 4. Piston valve,
5. Single-acting cylinder, 6. Stops, 7. Work table, 8. Control lever

Fig. 18.41 Working view of grinding machine hydraulic drive.

Given that: $F = 10$ kg, $T = \pi dl\mu$

If $l = 6$, $d = 1$ and $\mu = 0.01$ then

Force of inertia on the piston = $\pi \times 6 \times 1 \times 10 \times 0.01 = 1.8$ kg ≈ 2 kg
Weight of the table = 500 kg
Coefficient at friction between table and guides is taken as 0.15 for rest and 0.08 in motion.

$$T_{rest} = 500 \times 0.15 = 75 \text{ kg}$$
$$T_{motion} = 500 \times 0.08 = 40 \text{ kg}$$

Let t be 0.25.
For speed from 0 m/min to 15 m/min

Therefore, acceleration $\alpha = \dfrac{V}{t} = \dfrac{15}{60 \times 0.25} = 1 \text{ m/s}^2$

$$P_i = \dfrac{500}{981} \times \alpha = \dfrac{500}{981} \times 1 = 0.5 \text{ kg}$$

Therefore, total force on the cylinder pivoting = $10 + 40 + 2 + 24 = 76$ kg
Return = $40 + 2 + 24 = 66$ kg
Therefore, total force experienced by the cylinder

$$P_{working} = 10 + 40 + 2 + 24 = 76 \text{ kg}$$
$$P_{return} = 40 + 2 + 24 = 66 \text{ kg}$$

Pressure on the pump $P = \dfrac{(151)4}{\pi 6^2} = 5.34 \approx 6 \text{ kg/cm}^2$

Figure 18.42 shows the distribution of pressure during the pump stroke of a grinding machine hydraulic drive.

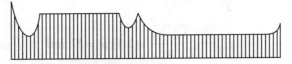

Fig. 18.42 Pressure distribution during pump stroke.

Let us assume the pressure drop in the throttle to be

$$P_{th} = 5 \text{ kg/cm}^2$$

Determination of cylinder and rod diameters

We have

$$F_w = \dfrac{\pi D^2}{4}; \quad F_\tau = \dfrac{\pi (D^2 - d^2)}{4}$$

$$\dfrac{V_w}{V_t} = \dfrac{F_w}{F_\tau} = \dfrac{D^2 - d^2}{D^2} = \dfrac{1}{1.2} \qquad \text{(As } V_{return} \text{ equals 1.2 times forward)}$$

$$1.2 D^2 - 1.2 d^2 = D^2$$
$$0.2 D^2 = 1.2 d^2$$
$$D^2 = 6d^2$$

For grinding machines $d = 16 - 25$ mm

Let us take $d = 25$ mm

Then $6d^2 = 6 \times 625 = 3750$ mm²

Therefore $D = \sqrt{3750} \approx 60$ mm $= 6$ cm

Determination of oil discharge in cylinders

$$Q_w = V_w \frac{\pi D^2}{4} = 1500 \frac{\pi \cdot 36}{4} = 42400 \text{ cm}^3/\text{min} = 42.4 \text{ L/min}$$

$$Q_r = V_t \frac{\pi(D^2 - d^2)}{4} = 1.2 \times 1500 \times \frac{\pi \times 29.75}{4} = 42.4 \text{ L/min}$$

Pump delivery

$$Q_r = \frac{Q_w}{\eta_{\text{tot}}} = \frac{42.4}{0.9} = 47 \text{ L/min}$$

Pipeline cross sections

$$Q_p = fV$$

Let us take $V_{\text{suct}} = 1.5$ m/s and
$V_{\text{del}} = 3$ m/s

Now

$$A_{\text{suct}} = \frac{Q_p}{V_{\text{suct}}} = \frac{47.1000}{60.150} = 5.2 \text{ cm}^2$$

$$\frac{\pi d_{\text{s.p.}}^2}{4} = 5.2 \text{ cm}^2$$

$$f_{\text{suct}} = \sqrt{\frac{5.2 \times 4}{\pi}} = 2.7 \text{ cm} = 27 \text{ mm}$$

$$A_{\text{del.}} = \frac{Q_p}{V_{\text{suct}}} = \frac{47.1000}{60.300} = 2.6 \text{ cm}^2$$

$$d_{\text{d.p.}} = \sqrt{\frac{2.6.4}{\pi}} = 1.8 \text{ cm} = 18 \text{ mm}$$

Pressure drop before the throttle, $p = 3$ kg/cm².
Pressure which should be developed by the pump

$$P_p = 6 + 5 + 3 = 14 \text{ kg/cm}^2$$

Motor power for driving the pump

$$\text{H.P.} = \frac{R_p \times Q_p}{450 \times \eta_{\text{system}}} = \frac{14 \times 47}{450 \times 0.85} = 1.7 \text{ hp} = 1.25 \text{ kW}$$

According to the oil discharge, an appropriate hydraulic equipment is chosen.

18.2.18 Hydraulic Drive for Rotary Motion

This consists of
1. Safety valve
2. Return valve.

The system works by closed cycle. The reservoir serves only the purpose of additional feeding.

$$Q_p = f(e_p)$$

If $e_p = 0$, $Q_p = 0$.

Reversing is carried out by the stator displacement from e_p to e_p.

If the hydraulic drive per revolution discharges the oil volume q, then the number of its revolutions will be

$$n_{motor} = \frac{Q_p}{q} \text{ rev/min}$$

The number of revolutions are regulated by both the numerator and the denominator, as n_{max}/n_{min} is the range of control R.

In hydraulic drives, R is up to 30.

The regulation of the speed is carried out by changing the value of the eccentricity e (as it regulates the quantity of oil discharge).

18.2.19 Advantages and Disadvantages of Hydraulic Drive

After having considered the design and operation considerations, we must know the limitations and benefits of such drives.

Figure 18.43 shows the schematic of a hydraulic drive for rotary motion.

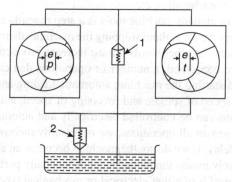

Fig. 18.43 Schematic of hydraulic drive for rotary motion.

Figure 18.44 shows the working diagram of a rotary motion pump.

Advantages
  (i) This method of speed regulation ensures stepless control.
  (ii) Inertial forces are minimum.
  (iii) Reversing of speed is smooth.

Limitations
  (i) Such drives do not maintain constant speed control characteristics.
  (ii) Efficiency of such drives is very low.
  (iii) This method of speed regulation is costly.

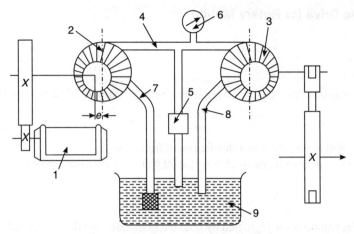

1. Electric motor, 2. Wing pump, 3. Hydraulic motor, 4. Hydraulic line, 5. Safety valve,
6. Pressure gauge, 7. Suction line, 8. Discharge pipe, 9. Oil receiver

Fig. 18.44 Working diagram of rotary motion pump.

18.3 ELECTRICAL AND ELECTRONIC REGULATION OF SPEEDS

18.3.1 Introduction

Inputs from electrical technology are essential to understand the working of various kinds of electric motors and devices. In addition, it is also essential to link concepts from applied mechanics to develop mathematical models explaining speed-torque characteristics, etc.

Use of stepless regulation in modern machine tools is a step towards automation, as by incorporating this we can change the speeds as we desire without stopping the machine; therefore it must reduce the working time needed to develop the product—in other words it must increase productivity.

So far as machine tools are concerned, a number of operations that can be mechanised on machines of various type are: stopping and starting the machine; automatic loading and unloading; and instantaneous stopping of the motor. Variable speed of spindle and reversing of speed, automatic gauging of dimensions, and many more similar operations can be controlled electrically and automatically. Whenever we use any electrical system for controlling any or all operations, we necessarily incorporate a self-operating switch in order to minimise its operating delay. If we do so, the machine becomes an automatic machine.

Therefore, automation merely means the replacement of manually performed operations by automatic switches or mechanisms; the control is of either electrical or mechanical type, as the case may require.

Electrical control switches and the complete system to be incorporated in a machine tool are not only expensive but complex too; hence, for an automatic machine to be an economical proposition, the individual or company using it must be able to operate it continuously at maximum output; this is a necessary condition.

18.3.2 Power Required by a Machine Tool

The phenomenon of metal removal is the process of plastic deformation for shaping the job to the required dimensions. This deformation consumes the energy available from the motor. The resultant force R, acting as a tool, can be resolved to the following three components:

F_x—a horizontal force perpendicular to the direction of feed (feed force).

F_y—a force perpendicular to the feed force or thrust force.

F_z—a force operating in a vertical plane and perpendicular to both f_x and f_y.

This force F_z is a function of velocity feed and depth of cut, and can be determined as
$$F_z = C_p \cdot d^9$$
Thus, knowing the cutting force, we can show the moment due to this to be
$$M_{\text{cutting}} = F_z \times \frac{D}{2}$$
where D is the diameter of the job in a turning operation. Similarly for other operations, these forces can be determined theoretically. Cutting force being the major power consumer, we can estimate the power requirement for any machine tool provided we know the cutting force, as follows:
$$\text{HP}_{\text{cutting}} = \frac{F_z \times V_c}{7.5 \times 60}$$
or
$$\text{HP}_{\text{cutting}} = \frac{F_z \times V_c}{7.5 \times 102} \text{ kW}$$
where V_c is the cutting speed and is given by
$$V_c = \frac{\pi DN}{1000} \text{ m/min}$$
Therefore, the power of the motor is equal to $\dfrac{F_z \cdot V_c}{60 \times 102 \times \eta}$ (in kW).

where η is the overall efficiency of the motor.

The dynamic forces such as cutting force, feed force, accelerating force, and frictional forces, and the static forces such as inertia force, static frictional forces, etc., create turning moments in the opposite direction to that of the rotation of the motor. Let these be denoted by dynamic moment (M_d) and static moment (M_{sta}) respectively.
$$M \geq M_d \cdot M_{sta}$$
If ω is the angular speed and J is the moment of inertia, then we have:

Moment = Polar moment of inertia × Angular acceleration
$$M_d = J \cdot \frac{d\omega}{dt}$$
and
$$J = \frac{W}{g} k^2$$
where k is the radius of gyration, W is the revolving weight and $\dfrac{d\omega}{dt}$ is the angular acceleration.

Let
$$M = M_d + M_{sta}$$
then
$$M_d = M - M_{sta}$$
$$J \cdot \frac{d\omega}{dt} = M - M_{sta}$$
$$\int_0^{t_s} dt = \frac{J}{(M - M_{sta})} \int_0^{\omega} d\omega$$

Thus, from the above integration we can find out the starting time t_s to attain angular speed ω which is given by the equation below.

$$\because \quad \omega = 2\pi N$$

$$t_s = \frac{J \cdot 2\pi \cdot N}{60(M - M_{sta})}$$

This is the time taken to achieve a speed of N rpm.

18.3.3 Selection of Motor for Speed Regulation

A very large variety of electric motors are available, as rapid progress has been made in the field of electric motor manufacturing. The best machine is one which has proper coordination between the drives and controls on one hand and the prime over on the other.

18.3.4 Classification of Drives

Various drives in a machine tool can be classified as follows:
- (a) Main drives
- (b) Traverse drives
- (c) Feed drives
- (d) Auxiliary drives.

Main drives: Such drives are responsible for metal removal, cutting and forming operations. They consume the maximum energy. Speed regulation is most desirable and can be obtained mechanically, hydraulically or electricating.

Traverse drives: The positioning of the tool, moving along the work, and stopping and reversing of the tool or the table can be performed by hand, but an automatic power drive is quite preferable when a rapid and constant tool feed or table movement is necessary. In case of such drives, sometimes individual drives are used for automatic machine tools. Motors having characteristics of constant speed, high starting torque and provision for reversing are used.

Feed drives: This type of drive is almost the same as the traverse drive, but sometimes an independent tool feed drive is used, e.g. in boring and drilling machines. The feed drives must be of constant torque drive, but it requires a wide range of speed variation.

Auxiliary drives: For auxiliary drives or the accessories of a machine tool used for positioning, circulating, cooling or lubricating, and for chucks, generally torque motors are used. There torque motors are generally DC squirrel cage motors or heavily compounded DC motors.

18.3.5 Characteristics of Electric Motors

Modern trends in machine tool design favour the use of individual drives for the various moving components of machine tools. This allows the operator greater flexibility over the machine tool operations; not only this, the motor and control drive equipment can be tailormade for the drive. But before we use the motor for such drives, we must look into the requisites for this type of use of a motor. Some of the characteristics of a motor that must be fulfilled for such a use are:
- (i) Starting characteristics
- (ii) Running characteristics

(iii) Braking of motors
(iv) Other miscellaneous activities.

A motor employed in any system has to overcome all the inertia forces of the machinery before it can do some useful work. Therefore, for a motor to be used for a drive, the inertia forces on the motor, together with the load that will be effective when the machine is in use, and accelerations to achieve the required speed within the specified time for the design of the motor, must be considered. The motor must not draw heavy starting current.

The running losses due to eddy current should be minimum and the power factor should be such that losses do not increase. The speed and torque characteristics of the motor should be the same as that of the load. When load is applied, speed changes in series motors; DC shunt or induction motors have constant speeds, while induction motors with several pole groupings can be used for one or more definite speeds independent of load; DC compound wound motors with field control or a wounded-rotor. Induction motors with rheostatic speed control can be used for speed regulation of drives, depending upon their suitability for a particular system.

For the selection of a motor suitable for a drive, we must know the power need for the drive. This can be determined by the equation

$$P_{kw} = \frac{F \cdot V}{102 \times 60 \times \eta}$$

where V is the velocity of the components;
η is the efficiency of the motor;
P_{kw} is the power in kilowatts;
F are frictional forces.

The turning moment or the torque of the motor can be written as:

$$T = 975 \frac{P_{kw}}{\eta} = 975 \frac{F \cdot V}{102 \times 60 \times \eta}$$

where η is the efficiency of the motor, and the starting torque

$$T_{start} = T \times \frac{\mu_0}{\mu}$$

where μ_0 is the friction between the moving parts when the velocity is zero and μ is coefficient of friction at speed is more

$$\frac{\mu_0}{\mu} \leq 2$$

The speed of a induction electric motor can be empirically expressed by the following relation:

$$N = \frac{60 \, f(1-s)}{2p}$$

where f is the frequency of the AC current;
p is the number of poles in the coil;
s is the slip.

The slip can be found from

$$s = \frac{N_s - N}{N_s}$$

where N_s is the synchronous speed of the motor, N is rotor speed.

If we examine the speed equation for N, we find that
$$N \propto f$$
$$N \propto \frac{1}{P}$$
and
$$N \propto (1-s)$$

Generally, frequency of the AC supply is constant; hence there are only two factors on which the speed of the motor depends, i.e. the number of poles and the slip. Slip is often kept fixed and hence it is only the number of poles that control the speed of the motor. Therefore, if $f = 50$ cps and $p = 2$.

$$n_s = \frac{60 \times 50}{2} = 1{,}500 \text{ rpm}$$

if $p = 1$, then

$$n_s = \frac{60 \times 50}{1} = 3{,}000 \text{ rpm}$$

Thus, we see that changing number of poles p is one of the methods of speed regulation in AC electrical motors. However, better control of speed can be achieved with DC motors and particularly with compound wound machines; therefore, these machines are increasingly used in machine tools for speed regulation work.

Whether we use a DC motor or an AC motor, a control equipment is to be used. This will render good service to the user. For the analysis of this, we have to treat the motor, the load and the control system as an integrated unit. The control equipment is to be chosen so as to match the desired controller requirements in view of the motor load and the steady and transient characteristics of the machine. For example, the acceleration cycle imposed by the operator will largely depend upon the inertia and torque characteristics of the load and voltage regulation of the power system. Deceleration is a similar phenomenon known as *braking*. Therefore the control system for such a drive must be evaluated from the following three points of view:

(i) The effect of the changes initiated by the operator through the control system upon either the load, the torque or the speed of the machine;
(ii) The effect of these changes on the motor and its control system, as one knows the nature and duration of the transients;
(iii) The effect of these changes, such as change in the current drawn, upon the power supply system.

Figure 18.45 shows the motor–machine load drive system.

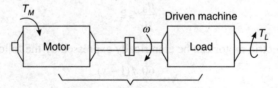

Fig. 18.45 Motor-machine load drive system.

The driven machine generally does not rotate at the same speed as that of the motor; but for the convenience of calculations it may be taken for granted that the motor load system has the same angular velocity. However, in the cases where the load is driven through a considerable gear reduction, the inertia of the load is negligible compared to that of the motor. The converse is true when the speed of the load greatly exceeds that of the motor. Here this distribution is not of much interest, as we are concerned with the behaviour of the entire system.

Let there be a system representing a motor and a load. The load is put in the form of torque T_L where T_M is the motor torque;

T_L is the load torque;
ω is the angular velocity;
J is the polar moment of inertia.

We can write:
$$J = \frac{W}{g} k^2$$

where W is the weight of rotating masses;
k is the radius of gyration.

Also, the fundamental torque equation for any motor load system is

$$T_M(\omega) = T_L(\omega) + J \frac{d\omega}{dt}$$

$$T_M = T_{(static)} + T_{(dynamic)}$$

As an example of this, when in a metal removal operation the steady state is reached, then

$$T_{L(static)} = \text{Moment of torque clue to cutting tool}$$

Therefore, $\quad \frac{d\omega}{dt} = 0$

Therefore, $\quad T_M = T_L$

It will be noticed that ω_g and T_g are the break-even values of angular velocity and torque respectively (Fig. 18.46).

In general the torque $T_M > T_L$, and the acceleration is positive and increasing. If brakes are applied where the torque is of frictional character and oppose the rotation of the motor, such torques are termed 'passive' torques, and where $T_M < T_L$, acceleration is negative and velocity decreases (e.g. hoist, elevators, and railroad locomotives), loads have the ability to drive the motor under equilibrium conditions, and such torques are called 'active' torques.

The speed-torque characteristics are shown in Fig. 18.46 and equilibrium occurs at that velocity for which the two torques are equal in magnitude.

The torque equation, $T_M = T_L + J \cdot d\omega/dt$ is termed the equation of motion, the solution of which will describe the behaviour of the speed regulation system. The values of T_L and T_M are represented in Fig. 18.46 as a function of speed ω.

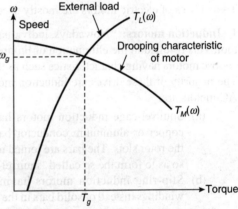

Fig. 18.46 Torque speed characteristics for Motor and external load.

A rotational system (refer Fig. 18.53) contains inertia and load torque components. If (ω_L/ω_M) is designated by γ, it will be seen that the inertia torque and load torque T_L is made up of the components of torque in three stages as under:

Motor power $\rightarrow \left[T_M \omega_M = \gamma J \left(\frac{d\omega_M}{dt} \right) \omega_M + \gamma^2 T_1 \omega_L + \gamma^3 T_2 \omega_L \right] \rightarrow$ From equating power flow

where ω_L is angular velocity of load and ω_M is angular velocity of motor. Based on Newton's 2nd law applied to a rotary system.

Thus, we can find out the T_M for a system of linkages transmitting power from the motor to a power consuming unit of a machine tool.

The selection of a motor depends on the requirement, i.e. the rating of the motor, which in turn depends upon: (i) heating, (ii) overload. Also, T_{max} is 2 to 3 times T_{rated}.

But an electric motor is to be selected by the load at a steady state

$$T_M = T_L + T_f \quad \left(\text{as } \frac{d\omega}{dt} = 0\right)$$

If T_L does not include frictional load, then T_f is the frictional load torque.

$$T_M = T_L + T_f = T_{max}$$

The load is the metal cutting load; hence the power consumed is given by

$$\text{HP}_{cutting} = \frac{T_L \times N}{71620 \times \eta}$$

where N = speed of rotation

and

$$\eta = \frac{\text{HP}_{cutting}}{\text{HP}_{rated\ motor}} = \text{mechanical efficiency}$$

Thus, we can determine the rated HP of the motor.

Electric motors for machine tools

Two classes of electric motors are mostly used for machine tools—Induction motors and DC motors.

1. Induction motors: Nowadays, individual electric motors are used for separate drives within a machine tool. This resulted in the elimination of linking mechanisms, simplification of design and the opportunity to choose motors having characteristics such as the speed-torque characteristics suited for specific applications. The majority of these drives are induction motors, since they are the simplest and least costly. Two typical AC motors are:

(a) Squirrel-cage induction motors having solid copper or aluminium conductor bars through the rotor slots. The bars are joined to end rings so as to form the so called 'squirrel-cage'.

(b) Slip-ring induction motors having 3-phase windings instead of solid bars in the rotor slots.

The stator is of same type for both types of motors. The squirrel-cage motor with its solid conductor bars is cheaper as well as more robust than the wound-type slip-ring motor.

Induction motors can run for short periods at torques greater than their full-load rating. The output torque may be as much as 250% of full load. A typical torque-speed curve for a single squirrel-cage motor is shown in Fig. 18.47.

The wound-rotor slip-ring motors can have the position of maximum torque displaced during operation by suitable control device, as shown in Fig. 18.48. If the motor is to start against full load the starting device is so arranged that the motor gives 1½–2½ times full-load torque as shown by dotted lines in Fig. 18.48. This provides excess

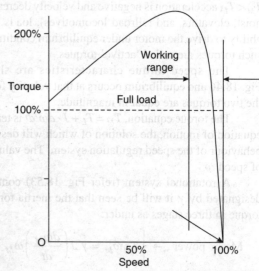

Fig. 18.47 Typical torque-speed curve for squirrel-cage motor.

torque for acceleration, and as the motor speed increases, external resistances are reduced to zero until the motor runs like a squirrel-cage machine.

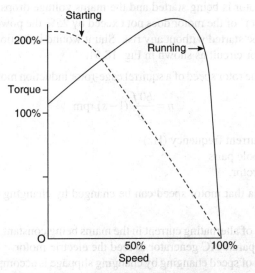

Fig. 18.48 Torque–speed curve for slip-ring motor.

A modification in the rotor design such as having a double-cage will result in a different torque–speed curve. A double squirrel-cage rotor, Fig. 18.49(a), provides a high starting torque and good running characteristics as shown in Fig. 18.49(b).

Within the working range of a motor (see Figs. 18.47 and 18.48), the torque–speed curve approximates to a straight line, hence in this range, torque is proportional to slip, i.e. deviation from synchronous speed. For example, if a motor of 1500 rev/min synchronous speed runs at 1450 rev/min under full-load conditions it will rotate at 1400 rev/min under 200% of full-load torque and so, the starting torque $M_s < M_{r2}$ is sufficient. Motors which are started up under load should develop higher starting torques.

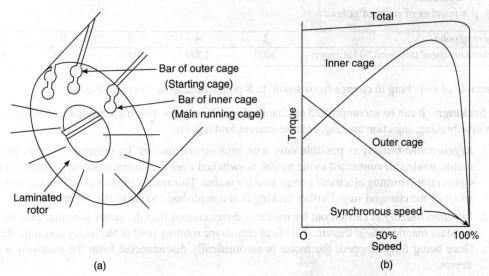

Fig. 18.49 (a) Double squirrel cage rotor (b) torque–speed characteristic of double squirrel cage motor.

Starting of induction motors: When a motor with a squirrel-cage rotor is started, the intensity of the starting current exceeds the rated value by 4 to 8 times. This rise in the current causes the voltage in the mains to drop. When a high-power motor is being started and the mains voltage drops, its starting torque also drops. In cases when the 'rated power' of the motor does not exceed by 25% the power supplied by the transformer. Only then such a motor can be started without any risk. Shunt wound induction motors are started by using a rheostat connected to the rotor circuit as shown in Fig. 18.51.

Changing motor speed: The rotor speed of a squirrel cage-rotor induction motor is obtained from the formula

$$n = \frac{60f}{p}(1-s) \text{ rpm}$$

where f is the alternating current frequency (Hz)
 p is the number of pole pairs
and s is the slippage of rotor.

It is evident from the formula that motor speed can be changed by changing current frequency, slippage or the number of pole pairs.

1. With the frequency of alternating current in the mains being constant, the first method can be applied only if there is a separate AC generator to feed the electric motor.
2. The second method of speed changing by changing slippage is accomplished by introducing effective resistance into the rotor circuit, which can be done only with wound-rotor induction motors.

Speed variation: Change of speed may be obtained by changing the number of pole pairs as in the case of pole change motor.

Pole change motors are usually squirrel-case induction motors. By switching arranged to change the number of poles, several fixed speeds can be obtained; the general relationship is without considering slip

$$n = \frac{60f}{p}$$

where n = rev/min
 f = cycles/s
 p = number of pairs of poles.

Number of poles	2	4	6	8	12
Synchronous speed (rev/min) 50 Hz supply	3000	1500	1000	750	500

The principle of switching to change from 4-pole to 8-pole windings is shown in Fig. 18.50.

Motor braking: It can be accomplished mechanically or electrically. Electrical methods of braking include regenerative braking, injection braking, reverse-current braking, etc.

1. *Regenerative braking* is possible only with multi-speed motors. Its principle consists in that the motor, while still connected to the mains, is switched over to a lower speed step, starts working as a generator returning electrical energy into the mains. This results in the motor slowing down to the speed of the changed step. Further braking is accomplished mechanically or otherwise.

2. *Injection braking* is carried out by injecting direct current into the stator winding, thus forming a constant magnetic field therein. This field retards the rotating field of the motor and stops the latter. Once being fully stopped, the motor is automatically disconnected from the mains by a special device.

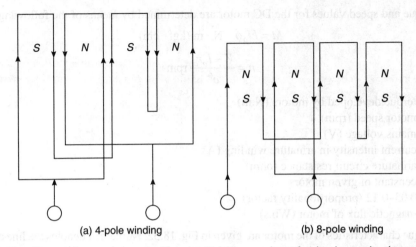

(a) 4-pole winding   (b) 8-pole winding

Fig. 18.50 Change in supply connections to alter the number of poles in a pole-change motor.

3. *Reverse-current braking* is carried out by interchanging two phases in the stator winding. Here, the direction of the rotating magnetic field is reversed, which retards the inertially rotating rotor. When the braking ends, the motor is automatically disconnected from the mains. The drawback of this method is that it is attended with load peaks and produces impacts in machine transmissions. However, this method is widely used in machine tools due to its simplicity and reliability.

Motor reversal is accomplished by changing any two external terminals (phases) of the motor.

2. Direct current motors: Direct current motors with shunted excitation (shunt-wound motors) are extensively used in heavy machine-tool drives. They are connected according to the circuit diagram shown in Fig. 18.51. The armature winding A is connected to the mains through starting rheostat Rh_1, exciting (shunt) winding SW through rheostat Rh_2 used for speed variation.

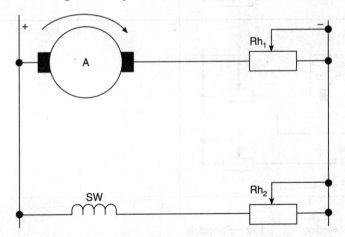

Fig. 18.51 Circuit diagram of shunt-wound motor connection.

The torque and speed values for the DC motor are determined by means of the following formulae:

$$M = kI_a\phi \quad \text{N} \cdot \text{m} \; (\text{kgf} \cdot \text{cm})$$

$$n = \frac{V - I_a r_a}{c\phi} \; \text{rpm}$$

where M = torque developed by motors (N.m)
  n = motor speed (rpm)
  V = mains voltage (V)
  I_a = current intensity in armature winding (A)
  r_a = armature circuit resistance (ohm)
  c = constant of given motors
  k = 0.05–0.12 (proportionality factor)
  ϕ = magnetic flux of motor (Wb.s)

The speed-torque characteristics of the motor are given in Fig. 18.52. Numeral 1 denotes the line corresponding to the rated speed-torque characteristic.

The relatively small value of armature winding resistance determines a sufficiently rigid rated characteristic of the shunt-wound motor which is expressed graphically by the modest slope of line 1.

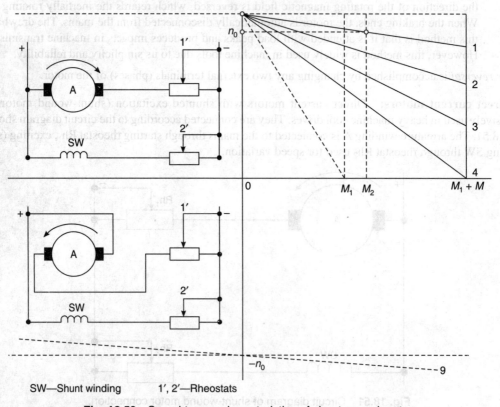

SW—Shunt winding 1', 2'—Rheostats

Fig. 18.52 Speed-torque characteristics of shunt-wound motor.

With the motor in operation, the resistance of rheostat 1 can be increased; this will result in an increase of the total armature circuit resistance and the slope of the characteristic's line. In this way, a number of modified rheostat-produced characteristics 2, 3, 4 are obtained. Power losses in the exciting circuit depend on the motor power and are within the limits of 1–8%; the lower the power, the higher the losses.

The rated value of armature current is determined as the difference between the rated value of motor current and the value of exciting current. However, the exciting current value in the shunt-wound motor is small, so, it is often not taken into account in design.

Shunt-wound motors can endure short-term operation under overload with permissible overload factor $h = 2$–2.5. This value is limited by the appearance of considerable brush sparking.

Shown in Fig. 18.52 by dotted line 9 is the speed–torque characteristic of motor with reversed armature polarity, in which the direction of motor rotation is reversed.

Starting the shunt-wound motors: It is carried out only with the aid of a starting rheostat. At start-up, rheostat V is connected to the circuit with all its steps and the motor begins speeding up in accordance with characteristic 4. The resistance of the rheostat steps is determined in such a way that on being started up the motor should develop predetermined torque M_1 (usually $M_1 \approx 2M_r$). As the motor accelerates, its torque drops and when it reaches the predetermined value M_2 ($M_2 \approx 1.1 M_r$), one rheostat step is disconnected. The motor then changes over to run in accordance with characteristic 3. Gradually, the steps of the rheostat are disconnected until the motor runs according to its rated speed-torque characteristic.

In machine tools, this operation is accomplished automatically.

Changing the speed of DC motors: It can be effected by changing the armature circuit resistance, magnetic flux, and the input voltage.

The first method is rarely used because it involves energy losses. The second method is by changing the magnetic flux which is the most commonly used method. The magnetic flux value is changed by rheostat 2' (Fig. 18.52). The rheostat resistance being increased, the exciting current and magnetic flux are reduced which results in an increase in idling motor speed and slope of motor speed-torque characteristics, represented by a number of straight lines (2, 3, 4). These lines are not parallel to the rated characteristic 1 of the motor and have the greater inclination, the lower values of magnetic fluxes they correspond to. The number of these characteristics depends on the number of steps on rheostat 2', where the number of rheostat steps is large, motor speed changing becomes practically stepless.

The third method of speed changing is by changing the input voltage which involves the use of special circuitry and is employed in generator-motor systems.

The braking of DC motors is carried out by the methods similar to those used for braking induction motors. Regenerative braking is accomplished by the means of a shunt circuit rheostat, whose operation causes the armature speed to drop to a minimum. Here, the motor begins operating as a generator returning electric energy into the mains. The motor is brought to a full stop by being disconnected from the mains. Dynamic braking, the most common method of braking, consists in the armature being disconnected from the mains, while the exciting current is on, and closed through the ballast resistor (or rheostat). Reverse-current braking is done by changing over the direction of current in the armature circuit.

Figure 18.53 shows the motor power transmission mechanism and flow of power from motor to load.

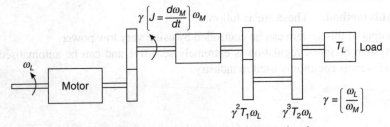

Fig 18.53 Motor power transmission mechanism.

From Fig. 18.54, we see that the AC motor connected to the mains (3φ) runs continuously almost at a constant speed; by changing the resistance of the rheostat, we change the load on the exciter motor which excites the fields of the generator. This changes the output of the generator, consequently changing the input to the DC motor driving machine tool organ.

Therefore the speed of a DC motor (N) can be seen to be as follows:

$$N = \frac{E_G - I_a \cdot R_a}{A \cdot \phi}$$

where
- E_G is the output of the generator G
- I_a is the armature current
- R_a is the armature resistance
- ϕ is the magnetic flux produced and passing through the motor
- A is the constant of the motor.

Thus, the speed can be controlled by controlling the armature resistance or the voltage drop in the armature circuit ϕ and the power circuit voltage E, which is done in the Ward Leonard system of speed control.

Speed regulation in this system is carried out, for a given current I_a with a constant flux ϕ of the DC motor, by changing the excitation E_G. Thus the motor torque T_m is given by

$$T_m = A \phi I_a$$

Thus we see that

$$HP_{motor} = 2\pi N T_m$$

But

$$N = \frac{E_G - I_a \cdot R_a}{A \cdot \phi}$$

or

$$N \propto E_G$$

Hence, $HP_{motor} \propto E_G$.

Thus the motor output varies as the voltage E_G across the generator changes.

Furthermore, we see that N varies with $(1/\phi)$ and therefore speed can also be increased by decreasing motor flux ϕ and keeping generator voltage E constant.

Regulation of speed by Ward Leonard system: Refer to Fig. 18.54

This system is also popularly known as a *motor generator system* of speed regulation, as this uses three machines, i.e. an induction motor, a DC generator and a DC motor. An exciter is connected to excite the field of the DC generator G and the DC motor M_2.

It is first necessary to convert the AC supply to DC by means of the motor generator set. The output voltage of the DC generator is supplied to the armature of a separately excited DC motor, and by varying the generator current–voltage output the motor speeds are regulated. An exciter set is usually coupled directly to the motor generator set. However, a metal rectifier for this purpose, i.e. for an exciter can be used for small sets and it is a common practice.

Advantages of this method: These are as follows:

(i) The output of this system can be controlled by using very low power.
(ii) This system of speed regulation is extremely sensitive and can be automatised. Therefore, this method is most commonly used in industry.

(iii) This is possible to control accurately the speed, torque, acceleration and position by this method of speed regulation. This is achieved by automatic regulating circuits, or electronic or magnetic amplifiers controlling the shunt field of the main generator.
(iv) Ward Leonard system cannot be ignored where accuracy and quick response are the overriding requirements, e.g. in proficiency machines.

A schematic diagram of the Ward Leonard system is shown in Fig. 18.54.

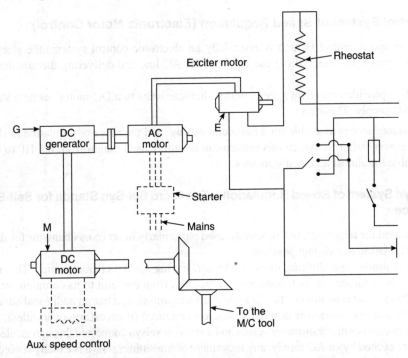

Fig. 18.54 Ward Leonard system.

The characteristic curve (see Fig. 18.55) of the Ward Leonard system shows the variation of power and torque with speed.

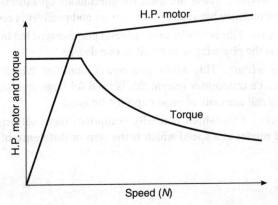

Fig. 18.55 Characteristic curve of Ward Leonard system of speed control.

18.3.6 Speed Regulation by Amplidyne

This is a quick response Direct Current generator requiring a very small amount of field power for its excitation—in contrast to the ordinary excitation in Ward Leonard system. The use of this device assures automatic speed regulation of the machine. The small field power of the amplidyne can easily be supplied from valves; thus a combination of amplidyne and electronic amplifier makes it possible to obtain the extremely accurate control, particularly for automatic regulation of speed.

18.3.7 Emotrol System of Speed Regulation (Electronic Motor Control)

In this system of speed control (which is essentially an electronic control system) the electronic gas tube diode works as a power rectifier, taking current from the AC line and delivering the armature and the field power as DC.

This system provides excellent speed control characteristics to a DC motor, using a static system fed by a standard AC supply. These are:

1. Constant speed is available for a particular setting, and the range of regulation is 10 : 1 to 100 : 1.
2. The power rating for which this system can be safely used ranges from 0.5 HP to 6.0 HP. This is suitable for almost all electric motors.

18.3.8 Selsyn System of Speed Regulation—The Word Sel-Syn Stands for Self-Synchronous Device

This is a device used for remote control of speeds, used particularly in servo mechanisms for the transmission of electrical information to a distant position.

A selsyn is similar to a slip ring induction motor, having special characteristics. This is so made that the angular motion or torque can be transmitted electrically, from one unit to its complementary unit placed at a distance. This system is analogous to a flexible shaft transmission; it has an additional advantage of being infinitely flexible and may workover as much distance as required (if necessary, over miles).

This system consists of a transmitter selsyn and a receiver selsyn connected through an electrical tie. Both of the selsyns are excited by an AC supply, any movement of transmitter selsyn is exactly copied by the rotor of the receiver selsyn. As per the analogy, this transmission is a flexible shaft transmission; hence there is a phase lag between the receiver and the transmitter selsyn. This angle increases with an increase in transmitted torque.

There are many types of selsyn available, such as

1. *General purpose selsyn:* These are used for continuous operations or rotations, where a great accuracy of correspondence between the transmitter and receiver is not essential.
2. *Precision use selsyn:* This is similar to the general purpose type but in this case the internal friction is minimum so that the phase lag is as small as one degree.
3. *Power-amplifying selsyn:* This works as a power amplifier, which combines the function of a torque amplifier and a transmitter selsyn; this is used for large and heavy machines which need to be operated with small amounts of power applied by hand.
4. *Differential selsyn:* This which normally comprises three selsyns. This records the angular displacement and rotates at a speed which is the sum or difference of the speeds of the other two selsyns.

In Fig. 18.56 the use of selsyn to control a planer feed is shown.

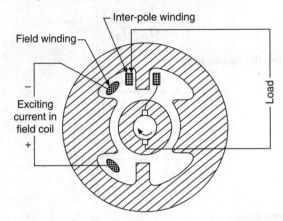

Fig. 18.56 Use of selsyn for control of speed of planer.

18.3.9 Braking

Braking is employed to effect tool change quicker, thus reducing non-productive time and reducing the cost of the component, as in the case of traverse or carriage drives. Those drives which have the largest stored energies will be required to show the greatest reduction in stopping time with the application of brakes. On the contrary, a drive having a constant load torque and negligible inertia may stop itself in a quite short time of working.

The braking systems in which the motor supplies the braking torque are known as *dynamic braking systems*. These are different from the magnetic or mechanical brakes in which a friction torque is produced.

18.3.10 Analysis of Braking

For studying the dynamics of braking, we must find out the various torques acting on the motor, load, etc. during the braking period. These are:

(i) The motor torque, if it is arranged to provide braking, will be denoted as $-T_M$.

(ii) The torque due to frictional brakes is denoted by $+T_B$. This is assumed as constant.

(iii) The load torque consists of active load (e.g. when cutting is taking place) and inactive load denoted by T_{LA} and T_{LI} respectively. These work as passive and active portions of the load torques and are denoted by $T_L = T_{LI} \pm T_{LA}$.

(iv) The inertia torque is denoted by $J\dfrac{d\omega}{dt}$.

We consider one example where a motor is to be stopped by braking. If there is motor torque, T_L and T_B are independent of speed and therefore constant, and $T_M = 0$.

Since Load Torque T_L and braking torque decelerate the motor:

We can write:
$$(T_L + T_B)\, dt = -J\int_{\omega_0}^{\omega} d\omega$$

or
$$(T_L + T_B)t = -J(\omega - \omega_0)$$

where ω_0 is the initial velocity and ω the final velocity.

$$t = \frac{-J(\omega - \omega_0)}{(T_L + T_B)}$$

If t_z is the time required to stop the motor, then

$$t_z = \frac{-J(0 - \omega_0)}{(T_L + T_B)}$$

$$\frac{J\omega_0}{(T_L + T_B)} = \frac{2\pi N_0 \left(\dfrac{w}{g} \cdot k^2\right)}{T_L \left(1 + \dfrac{T_B}{T_L}\right)}$$

where w is the weight inertia mass
 k is the radius of gyration
 N_0 is the initial velocity of motor in rpm.

From the above equation, we see that

$$t_z \propto \frac{1}{(T_L + T_B)}$$

Also, from the above equation it can be proved that

$$\omega = \omega_0 - \frac{T_L + T_B}{J} \cdot t$$

Therefore, the total number of revolutions required to stop a machine may also be calculated:

$$n = \frac{1}{2\pi} \int_0^{t_s} \omega \cdot dt$$

$$= \frac{1}{2\pi} \int_0^{t_s} \left(\omega_0 - \frac{T_L + T_B}{J} t\right) dt$$

$$= \frac{1}{4\pi} \frac{J\omega_0^2}{(T_L + T_B)}$$

The variation of time of stopping with the ratio of $\dfrac{T_B}{T_L}$ has been shown in Fig. 18.58.

The kinetic energy of the motor and the load is dissipated by the brake in a way that is given by the equation

$$2\pi n_s (T_B + T_L) = \frac{1}{2} J \omega_0^2$$

Figure 18.57 shows the principle of working of a braking system for a machine tool.

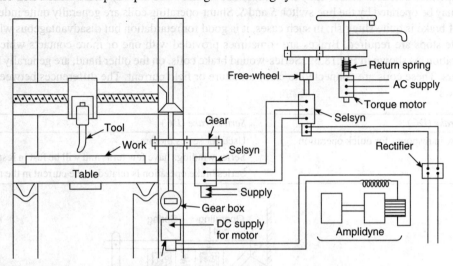

Fig. 18.57 Principle of working of braking system for machine tool.

So far we have seen that a running motor is to be stopped for the purpose of loading and unloading a job. Sometimes, braking is necessary in case of power failures, as the motor shaft is to be held stationary when the motor is to be stopped. For this task, magnetic brakes are used. The principle of working of such a braking system is shown in Fig. 18.58.

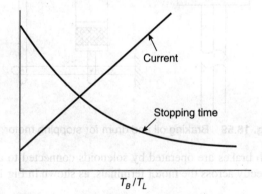

Fig. 18.58 Current and stopping time characteristics of braking system.

By pressing a button the brakes can be operated. The brake operating mechanism is an AC or DC solenoid, a torque motor or a pneumatic cylinder. The movement of the brake is as small as possible and generally kept as a few hundredths of a multimetre. The heat dissipating properties are of paramount importance in case of the brake material. The control circuits must be so arranged that the brake permits the shaft rotation whenever the power is applied to the shaft motor, and is set in operation when power supply from the motor is removed. Thus, brake control should be intimately associated with the supply line.

Secondly, the circuit should be designed so as to avoid hazardous conditions of brake failing. Thirdly, the brake should start working softly.

DC-operated solenoid brakes may have (a) shunt-wound coils and (b) series-wound coils. The shunt brake coil may be operated by the line switch *S* and *S*. Shunt-operating coils are generally quite inductive and operation of brake is quite sluggish; in such cases, it is good for retardation but disadvantageous where rapid and accurate stops are required. Brakes are sometimes provided with one or more contacts which operate during operation, as shown in Fig.18.59. Series-wound brake coils, on the other hand, are generally used with series motors. These coils are connected to carry armature or field current. The differences between the two are as follows:

Shunt brake (DC)	Series brake (DC)
Too slow in response for quick operation	Used for series motors.
	Series windings have few turns and will be fast in response.
	Series brake operation is related to the current in the motor.

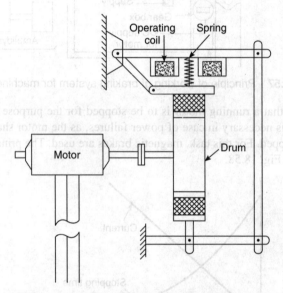

Fig. 18.59 Braking oil and drum for stopping motor.

AC Operated Brakes: Such brakes are operated by solenoids connected to an AC line operated on full voltage, and are connected directly across the motor terminals, as shown in Fig 18.60.

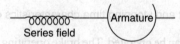

Fig. 18.60 Connector between AC motor and solenoid for stopping motor.

In such systems the peak solenoid current comes at the same time as the peak motor current, which is objectionable. Not only this, such brakes also produce noise due to AC solenoid coils.

In order to overcome these discrepancies, DC solenoid brake is used with an AC motor.

18.3.11 Starting and Stopping of Motors

The simplest speed control is attained by starting and stopping the machine tool or the motor. There are various methods of speed control; here we will confine ourselves to only this type of speed regulation.

In machine tools the drive starts under the 'almost no load' condition, and load is applied only when full speed of the motor is reached. Contrary to this, there are drives in which the load is only inertia load and once the machine is at full speed the torque requirements are quite low. Thus it suggests that the control requirements of a particular drive may be quite varied; therefore a speed control mechanism must be designed keeping in mind these surrounding requirements.

The following are the two methods of controlling the starting, accelerating and stopping torques of the motor:

(i) Fall voltage starting;
(ii) Reduced voltage starting.

Full voltage starting involves the application of full line voltage to the motor terminals at no speed. The motor speed regulation is then determined by the load, torque and voltage regulation of the supply.

Reduced voltage starting involves the application of a reduced voltage to the motor, or other means of reducing motor torque.

The drive motor used in a machine tool may be DC or AC, but generally AC motors are much more common in use as compared to DC motors; therefore, we shall discuss the starting and stopping of only AC motors.

The second method is most commonly used for starting motors. Resistance may be connected in series with the motor windings to reduce the current and voltage. A circuit is shown in Fig. 18.61. This method uses the working principle of push button control of 3-phase motor. Solenoid A is active when the start button is pushed in, by putting the motor in line, by keeping the circuit of the solenoid closed at point S. The coil circuit is interrupted by pushing the stop button in, which causes the solenoid to disconnect the motor and stop it.

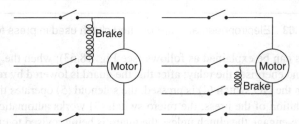

Fig. 18.61 Series resistance across the motor for reducing current and voltage.

By the addition of a resistance in the motor circuit of a DC motor at the time of starting and then gradually cutting as the motor speeds up.

18.3.12 Clutch Control

Nowadays, clutch control is used in machine tools. The clutch may be of electromagnetic type (see Fig. 18.62). Resistance may be connected in series with the motor winding to reduce the current and voltage. Circuit in Fig. 18.62 shows push button method of control of a 3 phase motor. Solenoid A is active when the start button is pushed by putting the motor in line by closing the circuit of solenoid closed at point S. The coil circuit is interrupted by pushing that stop button in which causes solenoid to disconnect the motor and stop it or electropneumatic type (see Fig. 18.63). In press tools an electropneumatic air-operated clutch is used to obtain only a single stroke. A guard, which must be closed when engaging the clutch and disengaged after one revolution, is used. The guard after this is lifted and lowered again before the next operation can be carried out.

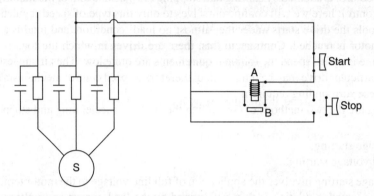

Fig. 18.62 Electromagnetic type clutch for starting and stopping motors.

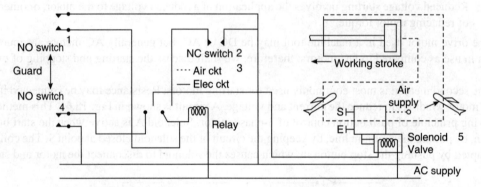

Fig. 18.63 Electropneumatic air-operated clutch used in press tools.

The operation of this can be explained as follows (see Fig. 18.63): when the guard is in top position, the NO switch (1) works and energises the relay; after this the guard is lowered by relay works with the help of the NC switch (2); when the NO switch (4) is pressed, the solenoid (5) operates the air valve and engages the clutch. After one revolution of the press, the micro switch (3) works automatically and disengages the clutch. It is impossible to re-engage the clutch unless the guard is being revised to close the NO switch (1).

The clutch is similar in construction to the ordinary auto-friction clutch, and therefore the constructional details are not given here.

Interlocking of motors is generally used in various systems; e.g. in milling machines it is often used to protect the machine against overload by providing interlocking in motors for feed drives and main drives. The feed motor is interlocked with spindle motor, so that the feed motor can not be started unless the spindle motor is started. This can be done with the electromagnetic clutch.

18.3.13 Relays

Relays are used in machine tools to protect the motor from overload burning. However, the use of such devices is not only limited to this; there are definite time delay relays, current-operated relays (as in our case), and back emf type relays in small DC motors.

REVIEW QUESTIONS

1. What are the various types of stepless regulation system? Explain in brief.
2. What are the stepless regulation methods utilising mechanical systems?
3. What are the advantages of stepless regulation over stepped regulation?
4. How is the reversing of spindle in machine tools achieved? Discuss.
5. Discuss the methods of universal range of regulation.
6. What are the advantages and limitations of hydraulic drives? What are the requirements for using hydraulic drive systems?
7. What are the various components of a hydraulic drive element in a drive system?
8. How would you prefer to select a hydraulic drive for linear or rotary motions?
9. How would you select a motor for speedless regulation? Discuss in brief.
10. How is selection of motors used in speedless regulation achieved? Explain in detail the magnitude of torque and the number of revolutions required to stop a motor.

Chapter 19

Machine Tool Vibrations

19.1 INTRODUCTION

Damping is essentially a major factor of consideration in the design of machine tools, particularly in the case of metal cutting machines, where a substantial portion of energy is transmitted through natural vibratory movements of the machine elements. As natural damping of the various elements of the machine tools is very low, a small portion of this energy is sufficient to produce large amplitudes of vibration. Thus the oscillations produced due to energy supplied are known as *metal cutting vibrations of machine tools*.

Metal cutting vibrations give rise to severe production problems, e.g.

(i) Cutting tool vibrations are set up and a rough quality of surface is generated.
(ii) These vibrations are harmful to tool life; however in some cases, purposely introduced tool vibrations increase the tool life.
(iii) The lower frequency vibrations are harmful to the machine and deteriorate its bearings. Not only this, vibrations may be transmitted through the structure of the foundation or building to other machines. For example, a milling machine working satisfactorily may induce vibrations in a nearby grinder, which may therefore produce defective parts.

19.2 TYPES OF MACHINE TOOL VIBRATION

In machine tools, vibrations are of two types:

(i) Forced vibration
(ii) Self-excited vibration.

Forced vibrations are caused due to a oscillating force away from the cutting zone e.g. due to oscillating nearby machines, unbalanced shafts gears, motors, etc. The machine vibrates at the forced frequency and not at its natural frequency.

The regenerative vibrations are a combination of forced and self-excited vibrations. The frequency of such vibrations is neither constant nor even proportional to operating speed. These vibrations are typical in turning operation, where the defect generated in one revolution is not identical to that of the preceding or succeeding revolution.

Self-excited vibrations also called *chatter*. These occur at or very near the natural frequency of the machine. The presence of this type of vibration is the indicative of instability of the process or the system as a whole.

The problem of machine tool vibration is not so simple, as these categories of vibration can be distinguished easily. But these vibrations are very often mixed together.

The machine tool which vibrates is a very complex, rigid and under-damped assembly of many elements such as spindle, moving components, and fixtures. Therefore damping is one of the methods to improve the performance of a machine tool.

The factor of rigidity or stiffness, defined as the *displacement to force ratio*, is a very common idea for designers. Many people, however, still consider damping only as a means to improve bad machine tools, so much so that many customers consider a machine tool, in which damping is provided, unfavourable.

We may agree to the idea that stiffness is a very important factor, even the only factor to be considered in the matter of static behaviour of a machine. The machining accuracy is a direct function of the static stiffness. But with respect to chatter as well as surface roughness, and accuracy in dynamic loading conditions, damping is as important as static stiffness is for working condition.

Moreover, we must consider that the machine tool accuracy is measured up to 1 micrometer, and we know by experience that nothing is stiff anymore at such small dimensions. A machine tool is not only a structure but an assembly of elements, either bolted together or moving with respect to each other, i.e. the spindle mounting, the carriage, and the fixtures.

As proof that we do not understimate the importance of increasing first the static performances, we give in this chapter an example of the improvement in dynamic response obtained only by reconsidering the bed of a lathe.

Secondly, we want to mention an important feature brought up by Tlusty—the reorientation of the direction of the vibration modes with respect to the cutting force. A new design of tailstock doubles the performance and the chatter depth. This is a matter of position coupling of the modes.

Thus, we will find in general that under any given set of conditions the machine will have an infinite number of modes of vibration, each one having its own natural frequency. Not only this, the modes are often coupled such that a vibratory motion in one direction produces a movement in another. Furthermore, the mechanically uncoupled modes can be coupled by the cutting process. Therefore the problem is very complex and can not be analysed so easily. Practically the problem is overlooked by the users of machine tools. However, this can not continue for long, as it is very important from the point of view of machine tool design and also has strong academic value.

19.2.1 Forced Vibration

The simplest type of forced vibration in a machine tool is that of the spring and mass system as shown in Fig. 19.1.

The equation of motion of such a system can be written as

$$M\frac{d^2 x}{dt^2} + Kx = F(t)$$

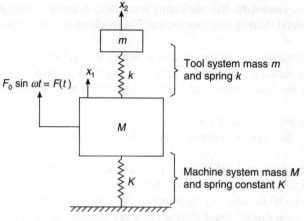

Fig. 19.1 Force vibrations.

Somehow, this class of vibrations can be eliminated by packings in the foundation, if vibrations are forced from outside the machine itself. Thus, vibration within the machine can be eliminated by improving the dynamic balance characteristic and concentricity of the moving parts. Large machine tools render great difficulties in overcoming the forced vibrations outside the machine. This can be done by using spring-supported reinforced concrete foundation, or by employing active servo type suspension elements.

If ω_n is the natural frequency of vibration of the machine elements, where ω is the forcing frequency, then

$$\frac{\omega}{\omega_n} = \mu_1$$

In order to prevent resonance or deflection from becoming infinite, we need to have

$$\omega < \omega_n$$

The foundation of the machine tool is generally separated from the ground by a layer of rubber, cork, felt, glass wool and even sand.

Generally a case of metal cutting vibrations is one of self-excited vibrations and is dealt in Section 19.3 onwards.

The equations of motion can be written as

$$M\ddot{x}_1 + Kx_1 + k(x_1 - x_2) = F_0 \sin \omega t$$

and

$$m\ddot{x}_2 + k(x_2 - x_1) = 0$$

These can be solved for x_2 and x_1.

19.2.2 Self-excited Vibration

Regenerated vibrations in practical machining operations are also a category of self-excited vibrations. Self-excited vibrations in machining operations are divided into the following two groups, as per the conditions that prevail and the vibrations that occur:

(a) Non-regenerative
(b) Regenerative.

Non-regenerative vibrations or chatter: This primary chatter occurs in cutting, when there exist no interaction between the vibratory motion of the system and the undulatory surface produced during the cut.

earlier to cut. This type of chatter vibration occurs during machining, due to the fluctuation in horizontal cutting force that lags slightly behind the horizontal vibration of the work. Due to this lag, a certain amount of energy per cycle of vibration is available for repetitively exciting the vibration of the work piece.

Regenerative vibrations or chatter: This class of vibration occurs during machining, when the vibratory motion of the system is in effect of the undulatory surfaces produced during the preceding revolution. This effect of surface undulations on the vibration system is called *feedback*. It is important to note that in all practical machining operations, regenerative chatter is likely to be present.

Therefore it can be summed up that *chatter* implies oscillations between the tool and the work that are sufficient in magnitude to cause a perceptible irregularity in the tool marks on the finished surface. Its frequency appears to be determined by the frequency of pulsation of the cutting pressure, and the natural vibration frequencies of the work, tool and machine. Its severity undoubtedly is determined by the periodic vibrations in the cutting force and vibrational frequencies, such as those of the work, the tool and the tool support.

While it is known that the presence of chatter makes it impossible to obtain a high quality of machined surface, its effect on the cutting speed is uncertain. Taylor states that the presence of chatter necessitated a reduction in cutting speeds from 10 to 15% when high-speed steel tools were used. Other experiments have indicated that a slight amount of chatter may be advantageous in roughing cuts, as it breaks the chip without appreciably affecting cutting speed. Chatter is definitely objectionable when sintered carbide tools are used, as they will chip. Severe chatter tends to cause excessive wear on feed screws, bearings, etc. of the machine tool and to loosen all fastenings.

19.3 CAUSES OF CHATTER

Chatter is created by several variables, of which the following probably are the most important:

    (i) Material to be cut;
   (ii) Chip proportions as affected by
        (a) depth of cut;
        (b) feed;
        (c) tool contour;
  (iii) Cutting speed;
  (iv) Stiffness of work;
   (v) Stiffness of tool;
  (vi) Stiffness of work support;
 (vii) Stiffness of tool support;
(viii) Vibration caused and multiplied by nature and design of machine tool, such as the gear ratios and tool forms used.
  (ix) Setting of tool with respect to the work.

Material to be cut: Chatter can seldom be controlled by a change in the material to be cut, the choice of which usually is determined by considerations other than ease of machining. Soft materials having low strength in shear appear to have less tendency towards chatter than those having high strength. Chatter is possible when cutting soft cast iron, but the conditions tending to set up vibrations must be much more severe than when cutting a medium steel.

Chip proportions: Chip proportions, whether affected by a change in the relation between depth of cut and feed or by a change in tool contour, have a marked effect on the tendency towards chatter. It has been noted that the thinner a chip is in proportion to its width, the greater will be the tendency to chatter when all cuts are run at the cutting speeds, giving a constant tool life. For this reason, thick chips should be cut rather than wide, thin ones, whenever possible. While the relation involved is much more complicated than a simple change in chip proportion, this viewpoint may be useful in planning tool designs to avoid chatter.

Let chip width be denoted by L, and the chip thickness by T. Little tendency towards chatter exists with a machine tool in ordinarily good condition, and with reasonably stiff work, when L/T is less than 15. It may be possible to work without chatter under favourable conditions, when cutting steel, with an L/T ratio of 20 to 25. In the range of L/T from 25 to 50, some chatter will usually be present, depending on the stiffness of the work, the machine and the cutting speed. With values of L/T greater than 50, chatter will be serious unless machine and work are particularly rigid or the cutting speed is very low.

When cutting cast iron, chatter seems to begin when L/T equals approximately 50. A tool with a large nose radius or a large side cutting edge angle is prone to chatter, as L/T is large and the chip is fairly uniform in thickness. If chatter occurs, either of these or both should be reduced. By providing a curved cutting edge, the chip is made much thicker at the larger diameter of the work than at the nose end of the tool. The variable thickness tends to reduce vibrations.

Cutting speed: Chatter undoubtedly is affected by cutting speed; a decrease in speed tends to reduce its severity, where it does not absolutely suppress it. Several experimenters report that under some circumstances an increase in cutting speed will reduce chatter. No way of determining the change in speed required to eliminate it is known, except to make an actual trial under shop conditions.

Stiffness of work: Chatter is affected by the stiffness of the work. Work supported on centres may be quite free from chatter at the beginning and the end of the cut, but have considerable chatter midway, between the ends, with the intensity increasing as the distance from the point of support increases. A steady rest or follows rest will reduce or suppress chatter, particularly in long, slim work on centres. The use of opposed cutting tools, taking cuts approximately equal in area, will tend to prevent chatter. Under certain conditions such as in thread chasing with light cuts and low speeds, so-called *spring tools* are also useful.

At times, a small cross-section of metal between the point of cut and the point of drive of the work may reduce the torsional stiffness and cause a tendency towards chatter, unless the size of the chip is reduced.

Stiffness of tool: To avoid chatter caused by the vibration of the cutting tool, the tool should always be clamped well in the tool holder, with as small an overhang beyond the point of support as possible. In general, the overhang of the tool beyond a solid support should not be more than one and a half times the height of the shank of the tool holder. A relation between the size of the cut, the size of the tool shank, and the tool point overhang, etc. shall be discussed later. While these data are recommended for carbide-tipped tools, they will serve as good guides for high-speed steel and cast non-ferrous tools. The tools should be clamped in a rigid, flat-base tool holder.

Stiffness of work support: The tendency towards chatter may be aggravated by the failure property of the support of the work. Lathe work on centres should have the centres as tight as possible. Chucked work should be tightly clamped. Planer and shaper work should be supported for the full length of cut to prevent deformation of the work by the cutting pressure.

Stiffness of machine tool: Little can be done with a machine tool which tends to cause chatter, because of an inherent lack of stiffness, except to make certain that all bearings, particularly the tool slides and main spindle bearings, are kept in adjustment with no more freedom than is necessary for proper operation of the machine.

Other causes of chatter: Chatter may be affected by the gear ratios in the machine tool, by the accuracy with which the gears are cut and the natural periodic vibrations of the driving mechanisms of the machine, the work, and the tool supporting structure. Chatter so caused can be reduced only by the use of a viscous gear lubricant. Such a lubricant may be of value in a machine with badly worn out driving gears and bearings.

Tool setting: Chatter in turned work sometimes may be avoided by setting the nose of the tool above the work centre, so that the flank at the end of the tool tends to ride the finished surface at the work. While this appears to give a steadying effect which may avoid chatter, it is not recommended as a regular method of operation.

19.3.1 Mathematical Analysis of Chatter Vibrations

Furthermore, the phenomenon of chatter can be explained by the formation of discontinuous chips and presence of built-up edge during cutting. Due to this fact, a periodic fluctuation in cutting force occurs. Frequency of this fluctuation in force corresponds to the frequency of discontinuity in the chip; or if BUE exists, it correspond to the frequency of shedding fragments. In a complex system like that of a machine tool work system, it is quite possible that this frequency of force fluctuation may coincide with some natural frequency of the tool or its supports. If this situation arises, resonant forced vibrations in the tool are set up.

Such vibrations of tool are indicated in Fig. 19.2. The severity of these vibrations are of much importance when resonance occurs. That will happen when the frequency ratio of the fluctuating force to that of the natural frequency of the tool equals unity, as shown in Fig. 19.3.

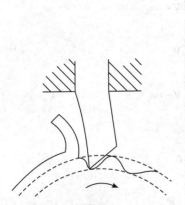

Fig. 19.2 Cutting tool and work.

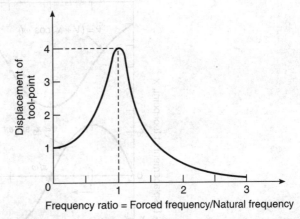

Fig. 19.3 Variation of amplitude of tool vibration under chatter.

Occurrence of self-excited vibration is subject to no such restriction, it depends primarily upon cutting speed and cutting force. Cutting force and velocity curve has a negative slope as speed goes down. It so happens that velocity coinsides with the simple harmonic velocity of the cutting tool at its natural frequency.

Let the amplitude of the tool point be x_0, and mean cutting speed be V.

Thus the displacement equation for the tool is

$$x = x_0 \sin \omega t$$

$$\text{Velocity of tool point} = \frac{dx}{dt} = \omega \cdot x_0 \cdot \text{constant}$$

Velocity of workpiece relative to tool varies between $V + \omega x_0$ and $V - \omega x_0$.

Forces acting on tool at a particular value of relative cutting velocity may be obtained from the force-speed curve (see Fig. 19.4).

When the tool is moving downwards through mid-position, relative velocity is $V - \omega x_0$, and force acting on tool is P_1 (See Fig. 19.5).

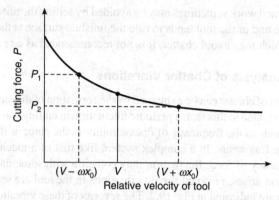

Fig. 19.4 Variation in relative cutting speed of tool with force acting on tool.

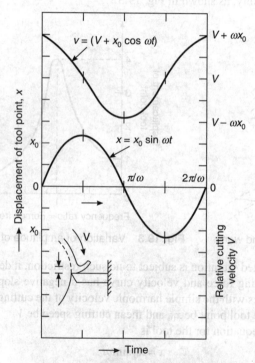

Fig. 19.5 Variation in relative cutting speed.

When the tool is moving upwards through mid-position, relative velocity is $V + \omega x_0$, force is P_2 and P_1 is greater than P_2, so that the greater force acts in the direction of motion of tool.

Therefore, a net amount of work is done on vibratory system in a cycle, and if this energy is greater than the amount of energy lost from system per cycle due to damping, then the amplitude of velocity will tends to increase, and the vibration is said to be *self-excited*. The vibration is called so because the motion of the vibratory system itself controls the varying force which, in turn, maintains the motion. The velocity can be regarded as *a free* vibration subject to negative damping. A limit to the amplitude of vibration is set when vibrational velocity of tool point becomes slightly greater than surface velocity of workpiece.

When this occurs, relative cutting speed becomes negative over a part of each cycle and frictional force between tool required from driving motor of the machine. This produces a powerful positive damping effect on vibration, which prevents any appreciable increase in its amplitude beyond the value.

$$x_0 = \frac{V}{2\pi f}$$

where V is the cutting speed in m/s;
f is the natural frequency of tool in cycles/s.

A similar vibration can take place between sliding surfaces which are elastically supported. Due to the fact that the friction between the surfaces varies with the sliding speed in the same way that the cutting force varies with the cutting speed for a tool, e.g. in the cases of violin strings, squeals of stoves, rusty door hinges, and chalks on blackboard.

As shown in Fig. 19.6(a), wear on clearance face causes additional periodic force to act on the tool, which produces a further stimulus to increase vibrations. Increase in tool wear increases the tendency of chatter vibrations to occur, or to increase the severity of these vibrations when they already exist. Figure 19.6(a) shows the effect of progressive tool wear on amplitude of vibrations of a flexible cutting tool. Hence in practice, wear should be limited.

Limit amplitude of vibration, in case of maximum vibrational velocity, just exceeds the surface speed of workpiece. This is also shown in Fig. 19.6(a).

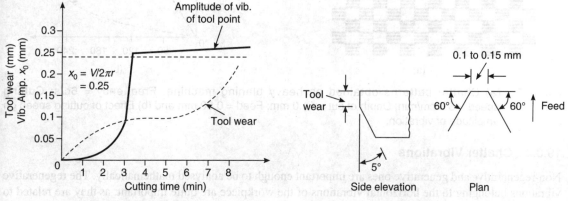

(a) Cutting time (6 min). Effect of tool clearance wear on amplitude of vibration

(b) Cutting speed 167.5 m/min. Depth of cut 0.12 mm freq. of vib. 1800 c/s Mat: low carbon steel tool: Tungsten carbided tool overhang: 75 mm.

Fig. 19.6 Chatter occurring as a result of flank wear and tool overhang.

The important principle established by this result is that the greatest amplitude of vibration which can be attained varies in direct proportion to cutting speed.

Fluctuating force due to clearance wear, which acts to increase negative damping, is directly proportional in magnitude to the active length of the cutting edge, which commonly occurs with round-nosed tools or form tools.

Tool vibration appears to be independent of depth of cut for a given cutting speed. At the smallest values of depth of cut, the vibration may be sufficient to cause the tool to be out of contact with work material during part of each cycle.

Figure 19.6(b) shows the orderly patterns produced on work surface by chatter vibration. Patterns are caused because the tool, during cutting, overlaps with small amounts of undulation formed during its previous passage. During cutting under these surface conditions, the cutting tool is subject to a periodic disturbing force, the frequency of which is almost exactly the same as the natural frequency of the tool. The characteristic effect of this additional disturbing force on the tool in a one-oscillation displacement of the tool lags 90° in phase behind the disturbing force.

Surface finish is affected by cutting speed because surface wave length is proportional to it. As clearance wear of tool may be considered constant at any given speed at low cutting speed, the trailing edge of wear flat will remove more metal, resulting in a superior finish. Indeed, at low cutting speeds there may be no trace of undulation, even though tool is vibrating [see Fig. 19.7(a) and (b)].

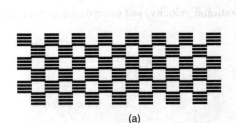
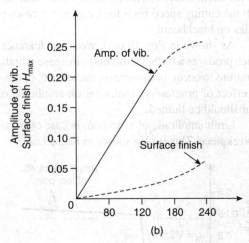

Fig. 19.7 (a) Chatter patterns obtained on heavy planing machine; Frequency = 6c/s; Cutting speed = 5.5 cm/min; Depth of cut = 11.0 mm; Feed = 0.50 mm and (b) Effect of cutting speed on amplitude of vibration.

19.3.2 Chatter Vibrations

Non-regenerative and generative ones are important enough to be analysed mathematically. The regenerative vibrations belonging to the horizontal vibrations of the workpiece are quite important, as they are related to the frequency of chatter. This can be analytically derived by looking into Fig. 19.8. Let C be the centre of a workpiece before cutting; when cut is initiated the centre C is displaced to O due to a thrust applied to the tool, denoted by a.

It is assumed that x is positive when measured in the direction away from the cutting edge. In regenerative chatter, the instantaneous force acting on the work is not only a function of the present state of workpiece motion of a preceding revolution. For simplicity, it is assumed that the system is a single-degree-freedom vibration; then we have the chattering equation as

$$m\ddot{x} + c\dot{x} + R(x + a) = f\{d + \lambda x(t + t_0) - x\}$$

where λ is the overlap factor.

Thus λ defines the portion of the previous cut which overlaps the present cut. The symbol t_0 represents the time lag corresponding to the phase difference between the undulatory work surface produced in the preceding revolution and the present motion, and is given by θ_ω, ω being the frequency of chatter vibration.

C is the centre of work before cutting;
O is the centre of work during or after application of steady force;
a is the displacement of centre;
x is the vibrational displacement;
m is the equivalent mass of the system in vibration;
R is spring constant like stiffness factor;
t_0 is the time lag;
t is the time of cycle of vibration during cut;
d is the undeformed chip thickness;
F_C is the horizontal thrust force for a chip thickness d;
c is the damping coefficient for chip thickness d;
F denotes function of $\{...\}$.

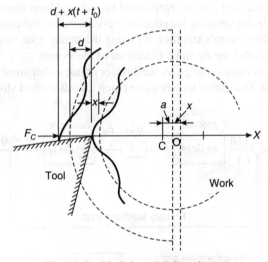

Fig. 19.8 Model explaining chatter.

Thus in case of regenerative chatter, t would be

$$m\ddot{x} + c\dot{x} + K(x+a) = F_C\{d + \lambda x_{(t+t_0-t')} - x_{(t-t')}\}$$

where t' is the time lag of the cutting force.

Assuming F_C is proportional to d, the undeformed chip thickness

then
$$F\{d + \lambda x_{(t+t_0-t')} - x_{(t-t')}\} = K\{d + \lambda x_{(t+t_0-t')} - x_{(t-t')}\}$$

where $d + \lambda x_{(t+t_0-t')} - x_{(t-t')} \geq 0$.

K is a coefficient (stiffness factor) depending upon the various conditions such as mechanical properties of work, tool geometry, width of cut, etc. Then in steady state

$$F_C(d) = Kd$$

Substituting
$$Ka = Kd$$

we get, for steady state:
$$\ddot{x} + 2h\dot{x} + \omega_n^2 x + \omega_n^2 x(h) - \lambda \omega_n^2 x(H) = 0$$

where $\omega_n^2 = \dfrac{K}{m}$

$x(h) = x(t - t')$
$x(H) = x(t - t_0)$
$H = t' - t_0.$

Thus the regenerative vibration can be treated on the basis of this equation and solved for x, f, etc.

19.4 CLOSED LOOP REPRESENTATION OF THE METAL CUTTING PROCESS

The present section is a very useful and fertile area for research. For a thorough understanding the reader should first have an initial exposure to Laplace transforms.

The process of metal machining can be represented by a closed loop model. The advantage of such a representation is that it helps in determining the structural dynamics and the consideration for metal cutting phenomenon, and in finding the system's stability. By using the steady state behaviour of the metal cutting process, it is also possible to predict the dynamic behaviour of the system.

Two feedback paths are indicated—a negative feedback or primary vibrational path and a positive feedback or regenerative vibrational path. The chatter loop or closed loop for self-excited vibration is shown in Fig. 19.9.

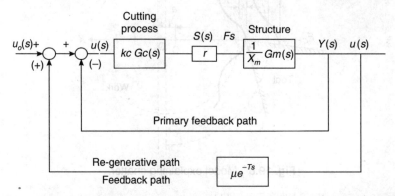

$u(s)$—Actual depth of cut; $u_o(s)$—Reference input signal (set uncut depth); kc—Spring constant; $Gc(s)$ and $Gm(s)$—Forward path transfer functions; X_m—Displacement; $Y(s)$, $u(s)$—Output signal; μe^{-Ts}—Feedback transfer function; $S(s)$—Combined forward function; r—Multiplying factor for system frequency response.

Fig. 19.9 Closed loop system of metal cutting process.

$$\dfrac{u(s)}{u_o(s)} = \dfrac{1}{1 + (1 - \mu e^{-Ts})\dfrac{r}{km} Gm(s)}$$

The transfer function relating to $u_o(s)$ and $u(s)$ can be obtained directly from the characteristic equation of the close-loop system, which is

$$1 + (1 - \mu e^{-Ts}) \dfrac{r}{km} Gm(s) = 0$$

It is known that a linear system closed loop is stable only when

(i) All the roots of its characteristic equation have –ve real parts.

(ii) All the roots have +ve real parts.
(iii) If the real part of any root is zero and rest are –ve, then the stability is in transition.

It is the requirement for system stability to get into the mechanism of the machine tool vibration, which explains the reasons for unusual high stability at low speed and the phase between the force and displacement of the tool.

In such studies, generally it is assumed that (see Fig. 19.10).

(i) The oscillation of the system in the direction of speed in negligible.
(ii) The movement of work under dynamic loading is negligible and can be represented as follows:

if F is the input force (harmonic);
M_2 is the tool (dynamometer) mass;
K_2 is the spring constant for the tool spring;
M_1 is the machine system mass;
K_1 is the spring constant for the machine;
ω_d is the forcing frequency;
ω_n is the natural frequency of the system;

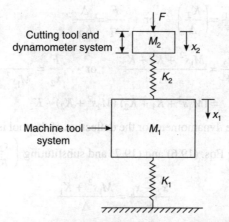

Fig. 19.10 Spring-mass system for metal cutting process using cutting tool dynamometer.

then we have
$$M_2 \ddot{x}_2 + K_2(x_2 - x_1) = F(t) \tag{19.1}$$
and
$$M_1 \ddot{x}_1 - K_2(x_2 - x_1) + K_1 x_1 = 0 \tag{19.2}$$

By Laplace transformation to change from time domain to S plane or frequency doman:
$$[M_2 s^2 + K_2]x_2(s) - K_2 x_1(s) = F(s) \tag{19.3}$$
$$-K_2 x_2(s) + (M_1 s^2 + K_1 + K_2) x_1(s) = 0 \tag{19.4}$$

Solving Eq. (19.3) and (19.4)

Multiply Eq. (19.3) by K_2 and Eq. (19.4) by $(M_2 s^2 + K_2)$ we have a set of simultaneous equations as belwo:
$$K_2[M_2 s^2 + K_2]x_2(s) - K_2^2 x_1(s) = K_2 F(s)$$
and
$$-K_2[M_2 s^2 + K_2]x_2(s) + (M_2 s^2 + K_2)\{M_1 s^2 + K_1 + K_2\} x_1(s) = 0$$

On adding the above equations we get:
$$[(M_2 s^2 + K_2)\{M_1 s^2 + K_1 + K_2\} - K_2^2] x_1(s) = K_2 F(s)$$

Designating the terms in the square bracket as Δ, we have a simple expression

$$(\Delta)x_1(s) = K_2 F \qquad (19.5)$$

From Eq. (19.5)

$$\frac{x_1}{F} = \frac{K_2}{\Delta} \qquad (19.6)$$

Substituting the value of $x_1 = \dfrac{K_2 F}{\Delta}$ in Eq. (19.4) we have,

$$-K_2 x_2 + (M_1 s^2 + K_1 + K_2) \cdot \frac{K_2 F}{\Delta} = 0$$

On simplifying the above,

$$\frac{x_2}{F} = \frac{M_1 s^2 + K_1 + K_2}{\Delta} \qquad (19.7)$$

$$\frac{x_1}{F} = \left(\frac{K_2}{\Delta}\right) \quad \text{or} \quad \frac{F}{x_1} = \frac{\Delta}{K_2}$$

and $\qquad \dfrac{x_2}{F} = \dfrac{M_1 s^2 + K_1 + K_2}{\Delta} \quad \text{or} \quad \dfrac{F}{x_2} = \dfrac{\Delta}{M_1 s^2 + K_1 + K_2}$

and $\qquad = (M_1 s^2 + K_1 + K_2)(M_2 s^2 + K_2) - K_2$

The strain measured by the dynamometer or the deflection of the tool is $x_2 - x_1$, that is, the compression of the spring K_2. Hence utilizing Eqs. (19.6) and (19.7) and substituting $\left(\dfrac{x_2}{F}\right)$ and $\left(\dfrac{x_1}{F}\right)$, we have,

$$\frac{x_2 - x_1}{F} = \frac{M_1 s^2 + K_1}{\Delta}$$

and $\qquad \omega_n = \sqrt{\dfrac{K_1}{M_1}}$

and $\qquad \omega_d = \sqrt{\dfrac{K_2}{M_2}}$

For more information, the reader is advised to study Armarego and Brown[*] due to tool-dynamometer system.

19.5 VIBRATION ELIMINATION

Sometimes it's observed that a machine tool may give rise to vibrations, leading to the production unacceptable machined items. It may so happen that the machine tool may be vibrating at a frequency very close to or at the natural frequency of the machine itself. If it is not so, the resonance shall not occur.

---

[*]Armarego, E.J. and G. Brown, "Machining of Metals" Printice-Hall, 1985.

The stiffness of the machine tool as a whole is quite important in order to avoid dangerous events or bad surface finish. Thus, stiffness and damping are of utmost importance for obtaining good dynamic characteristics of the machine tool and its components. Therefore it is important to ensure that while designing a new machine tool, unwanted vibrations should be eliminated as far as possible.

The stability is primarily affected by the maximum value of the real part of the response-forcing function equation. Due to these maximum values, the dynamic stiffness is lower than the static stiffness. This can be improved upon by damping out the vibrations of the machine tool. This is only possible if we know the frequency and mode of vibration of the machine tool system. Hence the dynamic performance evaluation of machine tools or any other mechanical machinery is of utmost importance, particularly from the design point of view, as the demands for higher productivity, higher speeds, feed and accuracy have increased. Till recently it was not possible to measure and predict accurately the dynamic performance of mechanical machinery. However, the development of new measurement tools have opened up the possibilities of analysing the dynamic behaviour of the machine tools.

REVIEW QUESTIONS

1. What are the causes of vibrations in metal cutting machine tools?
2. What are the harmful effects of vibrations in machine tools? Explain.
3. What do you understand by *self-excited vibration*? How are these vibrations set up in machine tools? Explain.
4. From first principles, derive the equation of motion of forced vibration. Indicate the forcing functional the displacement due to the same.
5. Describe the Waltham analysis of chatter vibration.
6. How would you eliminate vibration in a machine? Explain.

Chapter 20

Mechanization and Automation

20.1 MACHINE TOOLS FOR QUANTITY PRODUCTION

Automatic and semi-automatic machine tools are used for large-quantity production, where labour is often the greatest element of cost in the total cost of a product. The machine operator becomes virtually a material feeder, in order to reduce the labour cost to a minimum. It is assumed that the operator has background knowledge and experience of standard machine tools such as centre lathes, milling, shaping, and drilling machines, and cylindrical and surface grinding machines, which in their standard form are not ideal production machines. These automatic and semi-automatic machines are suitable for either short runs or long runs, and often the operator has sufficient skill to set up the machine before commencing production.

20.2 SEMI-AUTOMATIC MULTI-TOOL CENTRE LATHES

In this type of lathe, the work is mounted between centres (on mandrel if bored). Several tools are used, so that the entire turning is completed in one cycle. The traversing of the tools across the workpiece is carried out automatically, and the operator merely loads and unloads the work between the centres. Therefore, depending upon the cycle time, an operator might feed more than one machine.

The turning tools are carried in a series of tool boxes at the front and back of the machine, with the tools being carried on cam-operated slides which traverse on dovetail and round slideways. The rear tools are usually arranged for cross-traversing, and carry out the plunge-turning operations such as undercutting, chamfering or forming. The front tools then traverse longitudinally and carry out the plain turning of diameters. One feed can be chosen for turning and another for fast return.

The general configuration of the tools is as shown in Fig. 20.1.

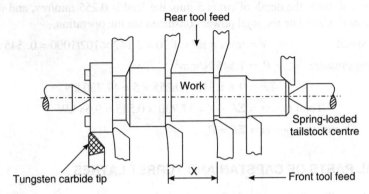

Fig. 20.1 Multi-tool centre lathe tool set-up.

The total machining time per component is the time to turn the longest shoulder length. If uneven tool wear takes place on the tools, then inaccuracies will quickly develop in the work. This can be avoided to some extent by using high speed steel tools for smaller diameters, and tungsten carbide tools for the larger diameters, allowing higher cutting speeds amd hence increasing the production rate.

The work may involve casting or forging a bar which has already been centred and machined to length by facing. A spring-loaded tailstock is used for quick release and loading of the workpiece. Consideration must be given to the method of driving the component between centres from the headstock. This must be fast, simple and positive; the method used for centre lathes is not suitable here. Advantage must be taken of any holes, slots or lugs in the workpiece to obtain a positive drive from the headstock. In such a case the sequence of operations would be arranged so that the driving medium is machined prior to the multi-turning operation. Also, a special driving plate would be required to be designed in the jig and tool drawing office for each batch of components to be turned. In the case of bar work, the component can be driven from the outside diameter using a quick release type of chuck, which simply drives the bar between centres from two jaws. Alternatively, if the component is to be turned over its full length in one operation, a commercial driver of the Kosta type can be used. This has driving pins which press against the end-face of the bar, and it may be necessary to fit a pressure gauge to the tailstock, in order to ensure that the pressure between the driving pins and the tailstock centre is not excessive. Long, slender work will need steadying during machining, and the use of these will increase the set-up time and hence increase the fixed costs.

Consideration must be given to the power requirement for a multi-turning operation. This in many cases will be very high and it would be wise to check that the motor is not being overloaded.

The power W required for a single tool is given by

$$W = P_{sc} f d V$$

where P_{sc} is the specific cutting pressure (N/mm^2)
 f is the tool feed (in mm/rev)
 d is the depth of cut (in mm)
 V is the cutting speed (in m/s).

Power for several tools in a multi-tool set up is

$$W_T = \Sigma(P_{sc} f d) \times V$$

EXAMPLE 20.1 A 102 mm diameter mild steel bar is being turned at 1.70 rev/s on a multi-tool centre lathe having nine tools. For all tools the depth of cut is 5 mm, the feed is 0.255 mm/rev, and the specific cutting pressure is 1.545 N/mm². Calculate the total power consumed for the operation.

Solution Cutting speed $V = \text{rev/s} \times \pi d = 1.70 \times 3.142 \times 102/1000 = 0.545$ m/s

Specific cutting pressure $P_s = 1.545$ N/mm²

$$\Sigma P = 9 \times 1.545 \times 0.255 \times 5 = 17.730 \text{ kN}$$

∴ Total power $= \Sigma P \times V = 17.730 \times 0.545 = 9.68$ kW

Cumulative power $= \Sigma P = P_{cp}$

20.3 PRINCIPAL PARTS OF CAPSTAN AND TURRET LATHES

A turret lathe has essentially the same parts as the engine lathe, except the turret and its complex mechanism, incorporated for making it suitable for mass production work. Figures 20.2 and 20.3 illustrate the different parts of capstan and turret lathes.

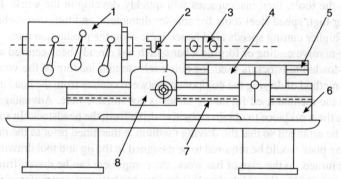

1. Headstock, 2. Cross-side toolpost, 3. Hexagonal turret, 4. Saddle for auxiliary slide,
5. Auxiliary slide, 6. Lathe bed, 7. Feed rod, 8. Saddle for cross-slide

Fig. 20.2 Capstan lathe parts.

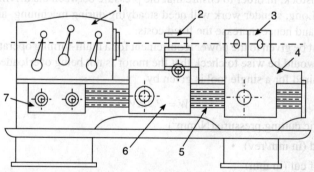

1. Headstock, 2. Cross-slide toolpost, 3. Hexagonal turret, 4. Turret saddle,
5. Feed rod, 6. Saddle for cross-slide, 7. Feed gearbox

Fig. 20.3 Turret lathe parts.

Bed: The bed is a long box-like casting provided with accurate guideways, upon which are mounted the carriage and turret saddle. The bed is designed to ensure strength, rigidity and permanancy of alignment under heavy duty services. Like in engine lathes, precision surface finishing methods must be applied to keep it resistant to wear during the service period.

Headstock: The headstock is a large casting located at the left hand end of the bed.

The headstock of a capstan or turret lathe may be one of the following types:

1. Step cone pulley driven headstock;
2. Direct electric motor driven headstock;
3. All geared headstock;
4. Preoptive or preselective headstock.

Step cone pulley driven headstock. This is the simplest type of headstock and is fitted with small capstan lathes, where the lathe is engaged in machining small and almost constant-diameter workpieces. Only three or four steps of the pulley can cater to the needs of the machine. The machine requires special countershaft unlike that of an engine lathe, where starting, stopping and reversing of the machine spindle can be effected by simply pressing a foot pedal.

Difference between Capstan and Turret Lathes

Capstan lathe	Turret lathe
In capstan lathe, the turret carries out its to and fro motion on an additional slide which is clamped on the bed of the lathe and can be adjusted along the bed.	In turret lathes, the hexagonal turret is mounted on a saddle which slides directly on the bed in the same way as a lathe saddle.

Except this distinguishing characteristic, in all other respects these two types of semi-automatic machines are fundamentally the same. Turret lathes are said to be of *saddle type* and capstan lathes are said to be of *ram type*. As the turret of a capstan lathe slides over a slide mounted on bed, its length of travel is restricted, although its position may be adjusted along the bed. But in turret lathes, the saddle carrying the turret slides over the lathe bed and can operate over its full length. For this reason, larger machines developed as turret lathes and capstan lathes are most common and useful for small and medium-size work. Figures 20.4 and 20.5 show a capstan lathe and a turret lathe respectively.

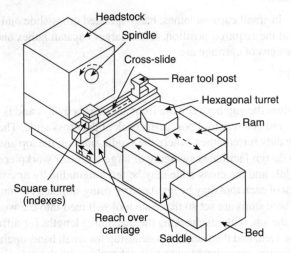

Fig. 20.4 Capstan lathe.

438 TEXTBOOK OF PRODUCTION ENGINEERING

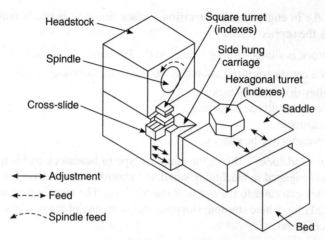

Fig. 20.5 Turret lathe.

Motor driven headstock. In this type of headstock, the spindle of the machine and the armature shaft of the motor are one and the same. Any speed variation or reversal is effected by simply controlling the motor. Three or four speeds are available, and the machine is suitable for small-diameter workpieces rotated at high speeds.

All geared headstock. On the larger lathes, the headstocks are geared and different mechanisms are employed for speed changing by actuating levers. The speed changing may be effected without stopping the machine.

Preoptive or preselective headstock. It is an all-geared headstock, with provisions for rapid stopping, starting and speed changing for different operations by simply pushing a button or pulling a lever. For different operations and for turning different diameters, the speed of the spindle must change. The required speed for the next operation is selected beforehand, and the speed changing lever is placed at the selected position. After the first operation is complete, a button or a lever is simply actuated, and the spindle starts rotating at the selected speed required for the second operation without stopping the machine. This novel mechanism is effected by the friction clutches.

Cross-slide and saddle: In small capstan lathes, hand operated cross-slide and saddle are used which are clamped on the lathe bed at the required position. In the larger capstan lathes and heavy duty turret lathes, usually the two types of designs of carriage are:

1. Conventional type
2. Side hung type.

The conventional type bridges the gap between the front and rear bedways and is equipped with four station type tool posts at the front, and one rear tool post at the back of the cross-slide. The side hung type carriage is generally fitted with heavy duty turret lathes, where the saddle rides on the top and bottom guideways on the front of the lathe bed. The design facilitates swinging of large-diameter workpieces without being interfered by the cross-slide. The saddle and the cross-slide may be fed longitudinally or crosswise, by hand or power. The longitudinal movement of each tool may be regulated by using stopbars or shafts set against the stop fitted on the bed and carriage. These stops are set so that each tool will feed into the work to the desired length for the purpose of duplicating the job, without checking the machining lengths for different operations each time. These stops first trip out the feed, and then serve as a deadstop for small hand-operated movements of the tool to complete the cut. The stopbars are indexed so as to synchronize with the indexing of the tool. The tools are mounted on the tool post, adjusting the correct heights by using rocking or pacing pieces.

Turret saddle and auxiliary slide: In a capstan lathe, the turret saddle bridges the gap between two bedways, and the top face is accurately machined to provide the bearing surface for the auxiliary slide. The saddle may be adjusted on lathe bedways and clamped at the desired position. The hexagonal turret is mounted on the auxiliary slide. In turret lathe, the turret is directly mounted on the top of the saddle, and any movement of the turret is effected by the movement of the saddle. The movement of the turret may be effected by hand or power. The turret is a hexagonal-shaped tool holder intended for holding six or more tools. Each face of the turret is accurately machined. Through the centre of each face, accurately bored holes are provided for accommodating shanks of different tool holders. The centre line of each hole coincides with the axis of the lathe when aligned with the headstock spindle. It addition to these holes, there are four tapped holes on each face of the turret for securing different tool holding attachments. At the centre of the turret on the top of it, there is a clamping lever which locks the turret on the saddle. Six stop bars are mounted on the saddle, which restrict the movement of each tool mounted on each face of the turret to be fed up to a predetermined amount for duplicating workpiece. After one operation is completed, as the turret is brought back from the spindle nose, the turret indexes automatically by a mechanism incorporated on the bed in the turret saddle, so that the tool mounted on the next face is aligned with the work.

20.4 AUTOMATION MECHANISMS ON CAPSTAN AND TURRET LATHES

The carriage, cross-slide, turret slide, and the saddle holding the turret may be fed into the work by hand or power. Separate feed rods transmit power to the carriage and turret saddle for this purpose. In addition to the apron mechanism and complicated geared headstock, such lathes are incorporated by a novel device by virtue of which the turret indexes are automatically brought away from the spindle nose after the first operation is completed. This mechanism is explained in the following section.

20.4.1 Indexing Mechanism

A simple line sketch of the mechanism is shown in Fig. 20.6. The figure illustrates an inverted plan of the turret assembly. Turret 1 is mounted on spindle (5), which rests on a bearing on the turret saddle (not shown in the sketch). The index plate (2), the bevel gear (3) and an indexing ratchet (4) are keyed to the spindle (5). The plunger (14) fitted within the housing and mounted on the saddle locks the index plate by spring (15) pressure, and prevents any rotary movement of the turret as the tool feeds into the work. A pin (13) fitted on the plunger (14) projects out of the housing. An actuating cam (10) and the indexing pawl (7) are attched to the lathe bed (9) at the desired position. Both the cam and the pawl are spring-loaded. As the turret reaches the backward position, the actuating cam (10) lifts the plunger (14) out of the groove in the index plate due to the riding of the pin (13) on the bevelled surface of the cam (10), and thus unlocks the index plate (2). The spring loaded pawl (7), which by this time engages with a groove of the ratchet plate (4), causes the ratchet to rotate as the turret head moves backward. When the index plate or the turret rotates through one-sixth of the circle of revolution, the pin (13) and the plunger (14) drop out of the cam (10), and the plunger locks the index plate at the next groove. The turret is thus indexed by one-sixth of the revolution and again locked into the new position automatically. The turret holding the next tool is now fed forward, and the pawl is released from the ratchet plate by the spring pressure.

The synchronized movement of the stop rods with the indexing of the turret can also be understood from Fig. 20.6. The bevel pinion (6) meshes with the bevel gear (3) mounted on the turret spindle. The extension of the pinion shaft carries a plate holding six adjustable stop rods (8). As the turret rotates through one sixth of the circle of revolution, the bevel gear (3) causes the plate to rotate. The ratio of the teeth between the pinion and the gear are so chosen that when the tool mounted on the face of the turret is indexed to bring it to the cutting position, the particular stop rod for controlling the longitudinal travel of the tool is aligned with the stop (12). The setting of the stop rods (8) for limiting the feed of each operation may be adjusted, by unscrewing the lock nuts and rotating the stop rods on the plate. Thus, six stop rods may be adjusted for controlling the longitudinal travel of tools mounted on the faces of the turret.

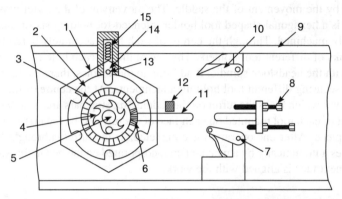

1. Hexagonal turret, 2. Index plate, 3. Bevelled gear, 4. Indexing ratchet, 5. Turret spindle,
6. Bevelled pinion, 7. Indexing pawl, 8. Screw stop rods, 9. Lathe bed, 10. Plunger-actuating cam,
11. Pinion shaft, 12. Stop, 13. Plunger pin, 14. Plunger, 15. Plunger spring.

Fig. 20.6 Turret indexing mechanism.

20.4.2 Bar Feeding Mechanism

The capstan and turret lathes require some mechanism for bar feeding. The long bars protrude out of the headstock spindle to be fed through the spindle upto the bar stop, after the first piece is completed and the collet chuck is opened. In simple cases, the bar may be pushed by hand. But this process unnecessarily increases the total operational time, because the spindle and the long bar must come to a dead stop before any adjustment can be made. Thus in each case, unnecessarily long time is taken in stopping, setting, and starting the machine. Various types of bar feeding mechanism have therefore been designed, which push the bar forward immediately after the collet releases the work without stopping the machine, enabling the setting time to be reduced to the minimum. Figure 20.6 illustrates a simple bar feeding mechanism. The bar (6) is passed through the bar chuck (3), the spindle of the machine and then through the collet chuck. The bar chuck (3) rotates in the sliding bracket body (2) which is mounted on a long slide bar. The bar chuck (3) grips the bar centrally using two set screws (5), and rotates with the bar in the sliding bracket body (2). One end of the chain (8) is connected to the pin (9) fitted on the sliding bracket (10), and the other end supports a weight (4), with the chain running over two fixed pulleys (7) and (11) mounted on the slide bar. The weight (4) constantly exerts end-thrust on the bar chuck, while it revolves on the collet chuck released. Thus, bar feeding may be accomplished without stopping the machine.

For gripping a very irregular-shaped work, soft jaws which grip the contoured profile efficiently are used. Tapered components are also suitably held by soft jaws.

20.4.3 Bar Holding Mechanism

Type of collet chuck: The collet chucks are used for gripping bars introduced through the headstock spindle of a capstan or turret lathe. This is one of the most common methods of holding work. They are much more suitable than a self-centring chuck in mass production work, due to its quickness in action and accurate setting. The chucks may be operated by hand or by power. Different sizes of spring collets, having bores of either square, hexagonal, round or any other shape, are fitted in the chuck body for holding different sizes of bar having different sections. Collets grip the work by the spring action of its split jaws. The collets are classified by the methods used to close the jaws on the work.

Push out type collet. To grip the work, the tapered portion of the spring collet is pushed into the mating taper of the chuck. There is a tendency of the bar to be pushed slightly outward, when the collet is pushed into the chuck body for gripping. If the bar is fed against a stop bar fitted on the turret head, this slight outward movement of the bar ensures accurate setting of the length of machining. Figure 20.7 illustrates a push type collet in use.

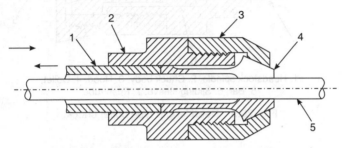

1. Push tube, 2. Headstock spindle, 3. Hood, 4. Collet, 5. Bar

Fig. 20.7 Push out type collet chuck.

Draw-in type collet. To grip the work, the tapered portion of the spring collet is pulled back into the mating taper of the chuck, which causes the split end of the collet to close in and grip the bar. The machining length of the bar in this type of chuck cannot be accurately set, as the collet while closing will draw the bar slightly towards the spindle. Figure 20.8 illustrates a draw-in type collet chuck.

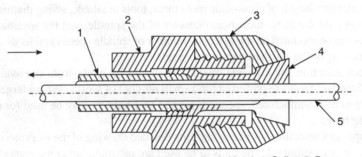

1. Draw tube, 2. Headstock spindle, 3. Hood, 4. Collet, 5. Bar

Fig. 20.8 Draw-in type collet chuck.

Dead length type collet. For accurate positioning of the bar, both the push out and draw-in type collet present some error due to the movement of the bar along with the collet while gripping. This difficulty is removed by using a sationary collet on the bar. A sliding sleeve closes upon the tapered collet which is prevented from any end-movement by the tapered collet which is prevented from any end-movement by the shoulder stop. Figure 20.9 illustrates a dead length type collet chuck.

Fixture. A fixture may be described as a special chuck built for the purpose or holding, locating and machining a large number of identical pieces, which cannot be easily held otherwise. Fixtures also serve the purpose of accurately locating the machining surface.

442 TEXTBOOK OF PRODUCTION ENGINEERING

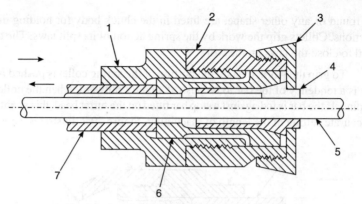

1. Headstock spindle, 2. Chuck body, 3. Hood, 4. Collet,
5. Bar, 6. Sliding sleeve, 7. Push tube

Fig. 20.9 Dead length type collet chuck.

20.5 TOOLING LAYOUT FOR CAPSTAN AND TURRET LATHES

Proper planning for systematic operations is carried out in advance before setting the work on a capstan or turret lathe. The following procedures should be adopted to plan and execute a work:

1. To organize for effective planning and control, an up-to-date capacity chart is essential The chart supplied by the manufacturers contains all working details of the machine, such as the maximum and minimum diameter of the work that can be mounted, maximum length of stroke of the turret and saddle, maximum length of cross-slide movement, tools available, swing diameter over the carriage, bore diameter on the turret face, bore diameter of the spindle, and the maximum size of the collet chuck that can be mounted on the machine, number of spindle speeds and feeds available, power, etc.
2. Proper drawing of the finished part is also needed.
3. Proper tool selection for different operations should be made from the available tools and tool holders. Standard tools are preferred for a small number of works. Where large number of identical pieces are to be manufactured, special tools and tool-holders may be used for reducing setting and machining time.
4. After proper tool selection has been made, the finished drawing of the workpiece is superimposed on the capacity chart supplied. The tools to be used are mounted out at the respective positions on the turret face, and on the cross-slide tool posts, in a pre-finalised feed sequence. The length of travel of tools for each turret face is now calculated from the chart and position of stops decided. Difficulty in setting and operation, if any, is chalked out on paper.
5. Proper spindle speeds, feeds and depth of the cut are determined for each operation.
6. Finally, the work tool machine system is set as per the plan chart.

EXAMPLE 20.2 Describe hexagonal bolt production on a capstan/turret lathe.

Solution Complete planning in case of hexagonal bolt production is given as follows:
1. The capacity chart of the machine is made available.
2. The drawing of the finished hexagonal bolt is taken into consideration.
3. Tools and equipment such as bar stop, roller steady turning tool holder, roller steady bar-ending tool holder, self-opening die head, chamfering tool, parting tool, etc. are acquired.

MECHANIZATION AND AUTOMATION **443**

4. The sketch of the work and tools are superimposed on the capacity chart to decide the length of travel of the tool and the position of stops. Figure 20.10(a) and (b) shows a typical tooling layout.
5. Proper speeds and feeds for each operation are next determined.
6. Various operations are performed in the following order:

Setting of the bar stops: The bar stop is set at a distance of 70 mm from the collet face by using a slip gauge. An extra length of 100 mm is taken. An allowance of 10 mm for parting off and clearance of the collet face is done, so that the parting-off tool may penetrate deep into the work without any interference [see Fig. 20.10(c)].

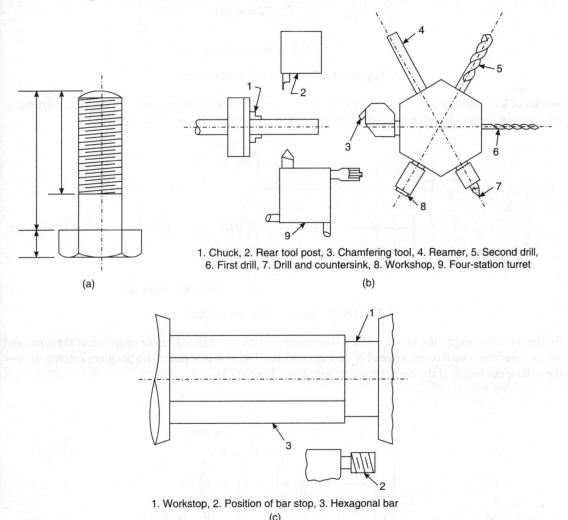

1. Chuck, 2. Rear tool post, 3. Chamfering tool, 4. Reamer, 5. Second drill, 6. First drill, 7. Drill and countersink, 8. Workshop, 9. Four-station turret

(a) (b)

1. Workstop, 2. Position of bar stop, 3. Hexagonal bar
(c)

Fig. 20.10 (a) Hexagonal bolt; (b) Typical capstan and turret lathe tooling layout; (c) Setting bar stop.

Setting of the roller steady box turning tool: The roller steady box turning tool is set on the next turret face for turning a diameter of 16 mm. The stop for turning the tool is set at 20 mm from the collet face by a slip guage. The rollers are set slightly approximately 1.5 mm behind the cutting edges (see Fig. 20.11).

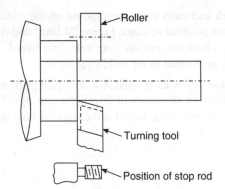

Fig. 20.11 Setting of box turning tool.

Setting of bar ending tool: The bar ending tool is set on the next turret face and is brought into operation after turning the bar. The stop is adjusted in the position by using a slip gauge (see Fig. 20.12).

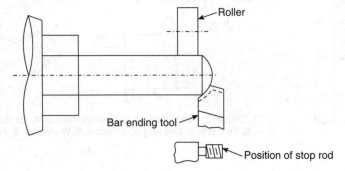

Fig. 20.12 Setting of bar ending tool.

Setting of self-opening die head: The self-opening die head is mounted on the next face of the turret and the dies are fitted into it to cut a thread of 16 mm diameter. The stop is adjusted in a position, keeping in view the pulling out length of the die to be self-released (see Fig. 20.13).

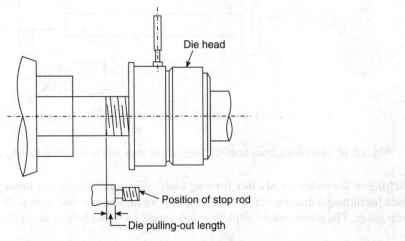

Fig. 20.13 Setting of self-opening die head.

Setting of chamfering tool: The chamfering tool is mounted on the four-station turret mounted on the cross-slide. The extreme longitudinal position of the saddle is adjusted by a stop. So also is the cross-feed movement of the cross-slide (see Fig. 20.14).

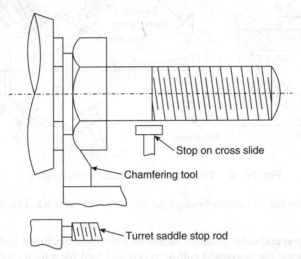

Fig. 20.14 Setting of chamfering tool.

Setting of parting-off tool: The parting-off tool is set on the rear tool post on the cross-slide, and the longitudinal position of the parting-off tool is adjusted by the stop set at a distance of 6 mm from the turret face (see Fig. 20.15).

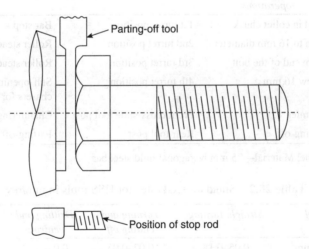

Fig. 20.15 Setting of parting-off tool.

Tooling schedule chart: A tooling schedule chart for each workpiece is of great importance in a capstan or turret lathe work for ready reference. A tooling schedule chart for the foregoing job operations is given in the following discussion.

See Fig. 20.16 for tooling layout of a bush bearing with table.

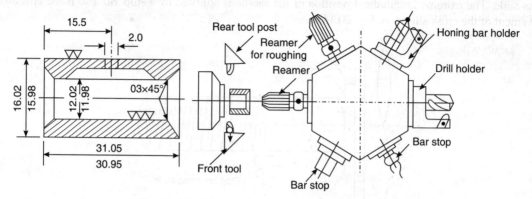

Fig. 20.16 Tooling layout for plain bush bearing.

Feed: It is the distance that the tool moves through per revolution of the work. This is expressed in millimetres per revolution.

Depth of cut: It is the perpendicular distance measured between a machined and an unmachined surface. Tables 20.1 and 20.2 indicate the suggested cutting speed and feed for a high speed steel tool for different operations.

Table 20.1 Tooling schedule chart

Serial no. of operations	Description of operations	Tool positions for mounting	Tool type
1.	Hold in collet chuck	1st turret position	Bar stop
2.	Turn to 16 mm diameter	2nd turret position	Roller steady box-turning tool
3.	Form end of the bolt	3rd turret position	Roller steady bar-ending tool
4.	Screw 16 mm	4th turret position	Self-opening die head with chasers for 16 mm
5.	Chamfer	Front cross-slide tool post	Chamfering tool
6.	Parting-off	Rear tool post	Parting-off tool

Machine—Capstan lathe; Material—75 mm hexagonal mild steel bar

Table 20.2 Standard feed rates for HSS tools in mm/rev

Material	Straight turning	Turning and cutting off	Drilling and centring	Boring
Steel 50 kg/mm^2	0.05–0.18	0.02–0.05	0.04–0.14	0.10–0.30
Steel 70 kg/mm^2	0.06–0.19	0.015–0.04	0.03–0.10	0.08–0.18
Free cutting steel	0.06–0.19	0.02–0.05	0.04–0.12	0.10–0.30
Stainless steel	0.03–0.08	0.005–0.03	0.03–0.08	0.06–0.15
Brass	0.10–0.22	0.02–0.10	0.08–0.25	0.15–0.35
Aluminium	0.10–0.22	0.02–0.08	0.03–0.18	0.18–0.45

20.6 SINGLE-SPINDLE AUTOMATIC LATHES

Automatic lathe is the advanced version of the capstan lathe. The other main types of single-spindle automatic lathe, in addition to the turret type, is the sliding head type and the multi-spindle automatic type.

Turret type automatics are sometimes known as *automatic screwing machines*, because of the facility with which small screwed parts can be produced using such machines in high quantities. Some can accept bar up to 50 mm diameter in a collet chuck, and the general configuration is that of a capstan lathe, except that a third overhead tool slide is provided. This is used for parting off if the other two cross-slides are used for other tools. Also, the other main difference is that a six-station round-tool turret is provided, which indexes about a horizontal instead of vertical axis.

The general layout of a turret automatic lathe is shown in Fig. 20.17. A simple plan view showing the principle of operation of the main features is shown at Fig. 20.18. It is not the intention in this book to analyse machine tool design. However, some knowledge of the principle of operation of an automatic lathe is required before one can understand how it can be controlled and tooled up for mass production.

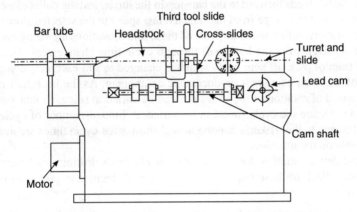

Fig. 20.17 General layout of turret automatic lathe.

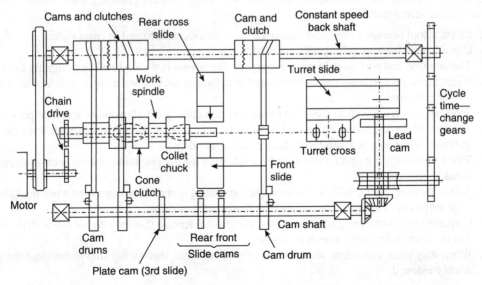

Fig. 20.18 Plan view showing principle of operation of turret automatic lathe.

Specially manufactured plate cams are used to operate the tool and the turret slides. This latter cam is called the *lead cam*. They all originate as circular discs, and are machined to the required profile. Each tool slide carries one tool and the turret carries six tools, only one turret station being in operation at once. Therefore the slide cams have a single lobe, and the lead cam will have as many lobes as many turret stations in use. The radial throw (lift) of each cam lobe is equivalent to the required length of travel of the tool for which it is designed. The feed of each tool is also controlled by the rate of lift of the cam lobe. The turret indexes automatically for one (or two if required) station, by means of a Geneva plate mechanism, at the end of each reverse stroke of the turret slide.

Figure 20.18 shows the cam shaft at the front of the machine. At the back of the machine is the backshaft which rotates at a constant speed (usually 2 rev/s or 4 rev/s). The backshaft carries dog clutches in three positions which are operated through drum cams, with these in turn being actuated through levers from cam drums carried on the front cam shaft. By setting trip dogs in the correct angular positions on the cam drums, the dog clutches on the back shaft can be made to operate when required in the cycle of operations. These three clutches when operated will respectively cause (a) the turret to index one (or two) stations; (b) the collet chuck to open, hence the bar feeds forward to the bar stop in the turret, and the collet closes again gripping the bar; and (c) the spindle speed to change from the selected fast speed to the selected slow speed, or vice versa. The backshaft rotates one revolution while any one of these idle operations are being carried out.

The cam shaft is driven from the backshaft through cycle time change gears (see Fig. 20.18); this is usually a compound train of pick-off spur gears which is changed to suit each new component being tooled up. One revolution of the front cam shaft produces one component. As the backshaft rotates at a constant speed, therefore the speed of rotation of the cam shaft can be varied to suit each new component, by means of the cycle time gears. Hence the cycle time can be varied. A limited number of cycle time change gears are provided, and reference to the maker's handbook will show what cycle times are available at any of the spindle speeds available on the machine.

Many innovative devices such as slot sawing attachments, cross-drilling attachments, etc., are available to increase productivity. Each of these requires a cam to operate them, and provision is made on the front cam shaft for this purpose.

General principles for design of cams

Cams rotate continuously with uniform velocity. The following rules are generally used during designing a cam for automatic screw machines:

1. Circular tool arrangements, best suited for the workpieces, are used in cross slides.
2. Considerably high spindle speeds are used.
3. The quickest and best method of arranging the operations is decided before designing the cams.
4. When the same operation can be performed both by a turret tool and a cross slide tool then the later will be preferred.
5. Work approach allowance, make-up clearances and cutting-off allowances are to be provided for.
6. The feed of parting tool is decreased at the end of cut when the piece is separated from the bar.
7. Sufficient clearance is given for tools to pass one another.
8. When a thread is cut up to a shoulder by a die, then for the provision of clearance a grooved neck is made near the shoulder.
9. A number of drills or a drill, withdrawn and introduced several times, are used when the hole length to be drilled is more than four times the drill diameter.
10. Care must be taken during selecting tools and tool holders, so that they do not fall with the machine bed or turret slide when turret is revolved.
11. When only three operations are to be done by turret tools, then swing stop is used and the turret is doubly indexed.

12. During threading operation, the nearest spindle speed most suitable is adopted (by special arrangement) to attain the quality of thread required.
13. Standard tools are used as far as possible.
14. While designing for special tools having intricate movements, springs are avoided as much as possible, and positive actions are used.

Designing the cam actually means the determination of the number of spindle revolutions required for each operation and idle movement, with overlaps whenever and wherever possible for such operations and idle movements that can be performed simultaneously.

This reduces the actual time of completing a job. Hence, efforts are always to be made for overlapping operations, so that time to complete a job may be reduced to minimum. Lobes are developed on cam surface to feed the tools on work. The radial height or throw of these lobes will represent the total travel of slides or tools, and the amount and shape of cam surface will govern the rate of feed of tool. The rise of lobes is followed by drops during which the returning of tools occur idly. Operations performed by cross-slide tools are plotted on cross-slide cams.

Use of tables

For designing the cam, data are compiled from standard tables showing speeds, feeds, hundredths of cam surface needed for feeding stock, turret indexing, and clearance between turret tool and cross-slide tools.

Dividing cam surface in hundredths

The purpose of dividing the cam surface in hundredths is to locate readily and easily the operations and movements on the drawing of cam. This simplifies the problem of drawing cam.

Layout of cams

All the operations for processing a job are sequentially arranged and are to be performed either by the turret, front slide or rear slide cams. The following steps are followed for cam layout:

(i) Prepare the sequence of operations;
(ii) Calculate throw of cams for each operation;
(iii) Calculate required number of job revolutions;
(iv) Calculate clearances;
(v) Calculate total revolution and cycle time;
(vi) Calculate cycle time for change of gears;
(vii) Convert revolutions into hundredths;
(viii) Prepare cam design sheet;
(ix) Design cams in terms of sizes, configurations, rises and drops;
(x) Draw cam profiles.

EXAMPLE 20.3 Consider a problem of machining a pin as shown in Fig. 20.19. (Bhattacharya and Sen, 1973).

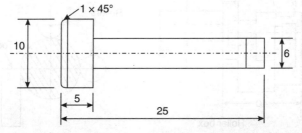

Fig. 20.19 Reduction of 10 mm bar to circular-headed pin.

List of operations

The list of operations is:

1. Feed bar to stop (on turret)
2. Clear and index
3. Roller box turn (on turret)
4. Clear and index
5. Form square on front slide
6. Part of on back slide
7. Clear.

Procedure of calculations

Calculating spindle rpm: Let the cutting speed be 20 m/min for roller box turning and 25 m/min for forming. Taking the minimum valid for all operations, we have

$$N = 312\left(\frac{V}{D}\right) = 312\left(\frac{20}{10}\right) = 624 \text{ rpm}$$

The nearest available rpm in the machine is 600. Hence, this value is chosen for calculations.

Calculating the throw of cams: This is done as follows:

Turning by "Box Tool". The box tool leaves an amount of length unmachined because of the design of the box, which requires subsequent machining as shown in Fig. 20.20.

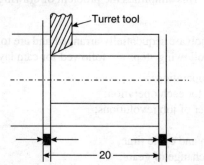

Fig. 20.20 Turning by box tool for 20 mm length.

The dimensions of a roller box are shown in Fig. 20.21(a).

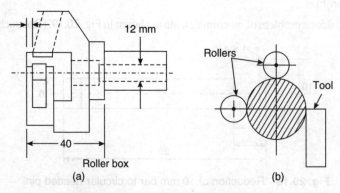

Fig. 20.21 Roller box configuration.

Hence, length to be turned = 20 mm
Length left over for square operation = 1 mm
Approach = 1 mm
Hence, total throw = 20 – 1 + 1 = 20 mm.
The dispositions of roller supports and cutting tool are indicated in Fig. 20.21(b).

Form square. Figure 20.22 shows the squaring of the neck from the front slide tool.
Approach (to take care of cam tolerance, clearance, etc.) = 0.5 mm
Total throw = 2.5 mm.

Parting-off and chamfering. Both of these operations are performed by a single tool as shown in Figs. 20.22 and 20.23.

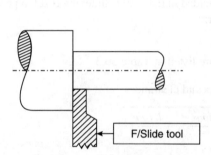
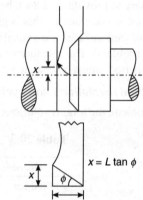

Fig. 20.22 Squaring of neck.   **Fig. 20.23** Parting-off and chamfering.

The total depth required to be parted = $\frac{10}{5}$ = 5 mm

Approach = 0.5 mm
Further advance part centre = 0.1 mm
Length to be parted off as additional throw = 2.5 tan 15° = 0.66
Total throw = 5.76 mm
The angle ϕ depends on bar size, D and material of workpieces.
For D = 10 mm, L = 2.5 mm, ϕ =15° for steel, iron, etc. and 20° for brass, copper, etc.

Calculation of required job revolutions

(i) Roller box turning:
Travel = 20 mm
Feed = 0.2 mm/rev
Required revolutions = 20/0.2 = 100
(ii) Forming a square from front slide:
Travel = 2.5 mm
Feed = 0.05 mm/rev
Required revolutions = 50
(iii) Parting-off and chamfering:
Travel = 5.76 mm

Feed = 0.06 mm/rev
Required revolutions = 5.76/0.06 = 96

(iv) Index/bar feeding, etc.

These are operated by one revolution of the auxiliary shaft which rotates at a constant speed of 120 rpm.
Time of index/bar feed = 1/2 s.
During this period the job rotates by: $1/2 \times (600/60) = 5$ rev.
Allowing a little extra to clear trips, etc., the number revolutions accepted for indexing and clearing or bar feed, would be 6.

Calculation of clearances

It is often necessary to provide clearance between the events of tool withdrawal of travel and tool approach of the slides or vice versa. Usually, this is given in hundredths of cam profile. Between the box tool and the form tool, 6/100 of cam surfaces is assumed for clearance, and between part-off and feed bar-to-stop, 4/100 is provided (in this problem). These values of clearance are provided in the tables. Indexing is often performed during this clear period unless a separate time is provided for.

Calculation of total revolutions and cycle time

In the present problem, the total revolutions and clearances are listed in Table 20.3.

Table 20.3 Total revolutions and clearances

Operation	Revolutions	1/100
Feed bar	6	
Index	6	
R/B turn	100	
Index and clear	—	6
Form tool (50)	(60)+	
Part-off tool	96	
Clear	—	4
Total	208	10

+Overlapped with parting.

If cam surface is divided into 100 divisions, 90 divisions correspond to 208 revolutions. Hence, total revolutions required will be given by

$$\text{Total revolutions} = \frac{208 \times 100}{100 - 10} = 208 \times \frac{100}{90} = 231$$

$$\text{Cycle time} = \frac{231}{600} = 0.385$$

$$\text{Cam shaft rpm} = \frac{600}{231} \approx 2.6$$

Correction of cycle time for coordinating cam shaft rpm with respect to auxiliary shaft rpm

From Fig. 20.24, an equation for kinematic balance can be established for compatible cam shaft rpm for a given cycle time with respect to auxiliary shaft rotating at 120 rpm.

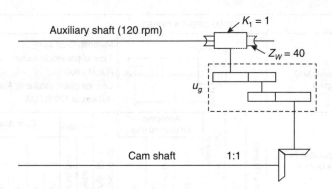

Fig. 20.24 Calculating change-gear ratio u_g.

From kinematic balance

$$120 \frac{1}{40} u_g = \frac{600}{231} = \frac{200}{77}$$

From which

$$u_g = \frac{200}{77} \cdot \frac{1}{3} = 0.866$$

If the gear having 77 teeth is not available, but a gear having 78 teeth is available, then the change-gear ratio is changed to (20/78) × (30/90). This changes the total of revolutions to 234. This excess is distributed by adding one revolution to roller box turning and two revolutions to the parting-off process.

Conversion into hundredths

In the present example, the total of 234 revolutions represents 100 divisions of the cam surface. Hence, each revolution equals 0.425 hundredth. All revolutions are systematically multiplied by this factor to get the respective hundredths, making marginal adjustments wherever necessary, in order that the total comes to 100. This is shown in Table 20.4.

Table 20.4 Converting revolutions into hundredths

Operation	Revolutions	Hundredth
1. Feed bar to stop	6	3
2. Index	6	3
3. Roller box turn	101	42
4. Index and clear	6	
5. Form square	50+	21+
6. Part-off and chamfer	98	42
7. Clear	4	
Total	—	100

+Overlapped.

Preparation of cam design sheet

A cam design sheet has to be prepared as shown in Fig. 20.25, in which all these steps of calculations are systematically entered.

454 TEXTBOOK OF PRODUCTION ENGINEERING

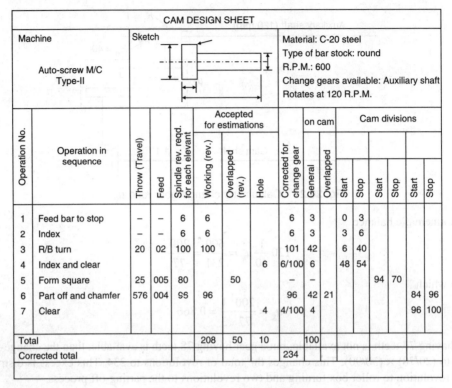

Fig. 20.25 Cam design sheet.

Drawing of cams

Step 1—Sizes and configurations: Prior to drawing of cams, the configuration of the cam system must be known in order that the radii of the cam profiles are known both at the beginning and at the end for each event.

Usually, when the follower roller is on the minimum radius of the cam, the maximum admit is obtained (see Fig. 20.26). When the roller is on the maximum radius of the cam profile, minimum admit is obtained. Thus

$$(R_{max} - R_{min})_{cam} = \text{Turret travel} = A_{max} - A_{min}$$

Suppose for a particular size of automatic screw cutting machine, the following data are known or given:

Cam size: $R_{max} = 60$ mm

               $R_{min} = 25$ mm

               $R_{shaft} = 10$ mm

Follower: Roller radius, $r = 15$ mm

               Distance of pivot from centre of rotation of cam = 120 mm

               Lever lengths $a = b = 80$ mm

Admit: $A_{max} = 75$ mm

               $A_{min} = 40$ mm

               Maximum travel = 35 mm.

MECHANIZATION AND AUTOMATION 455

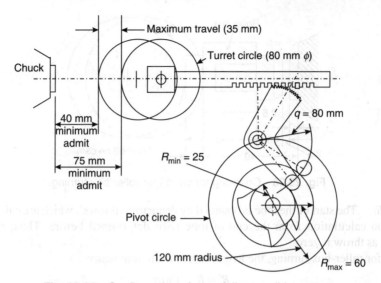

Fig. 20.26 Configuration of cam–follower–slide system.

Step 2—Lobe radii: At this stage, the lobe radii are determined for each event, from which the 'cut-off' of the blank size is determined. There are the following two turret events in the example under consideration:
(i) Feed bar stopping
(ii) Roller-box turning
from which start-of-lobe radii are calculated.

Feed bar stopping. The cut-off on the blank is calculated with respect to the distance moved by the turret, in order to attain the objective of the event with respect to the configuration shown in Fig. 20.27.

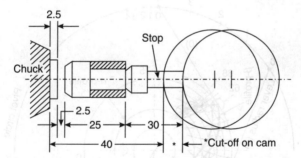

Fig. 20.27 Checking for cut-off at feed bar to stop.

Thus R for 'feed-bar' lobe = R_{max} − 'cut-off'
$$= 60 - (2.5 + 2.5 + 25 + 30 - 40)$$
$$= 40 \text{ mm}$$

The roller must rise to this lobe radius of 40 mm and have a dwell until the bar feed is completed.

Roller-box turning. By considering the advance of the roller-box turning tool as shown in Fig. 20.28, the calculated value of cut-off for this case is found to be $40 + 5 + 2.5 + 2.5 - 40 = 10$ mm.

Hence, R for roller-box turning is $60 - 10 = 50$ mm.

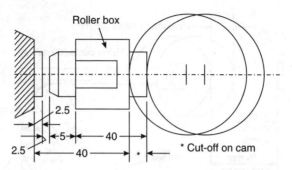

Fig. 20.28 Checking for cut-off for roller-box turning.

Start-of-lobe radii. The start of the lobe is obtained by deducting 'throws', which are calculated beforehand during revolution calculation from the end-of-lobe radii determined before. Thus, for feed bar stop, $R' = R = 40$ mm, as throw is zero.

Similarly, for roller-box turning, the length turned is 20 mm; hence

$$R' = R - \text{'throw'}$$
$$= 50 - 20 = 30 \text{ mm}$$

This value of R' should be greater than the minimum radius of the cam.

Step 3—Method of drawing cams: The cams are drawn by using the technique of 'reversed rotation', and assuming uniform velocity, i.e. equal rises in equal intervals of cam rotation.

As an illustrative example for the correct procedure for profiling cams, consider that a rise of 20 mm from minimum radius has taken place in a travel from the 45/100 position to the 62/100 position, followed by a drop to the minimum radius at the 68/100 position. The construction is indicated in Fig. 20.29.

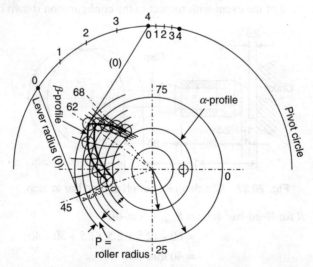

Fig. 20.29 Constructing cam profile for given rise and drop.

The steps of construction are given in the following lines:

(i) The cam blank diameter is drawn along with the lower lobe radii, which is R minus the throw.
(ii) The pivot circle is drawn.

(iii) Concentric circles to include roller radii are drawn.
(iv) The throw is now divided into a few equal parts.
(v) Accordingly, the corresponding positions of the fulcrum on the pivot circle are determined and divided into the same number of equal parts.
(vi) Concentric circles are drawn through radial divisions, with the cam centre as the origin.
(vii) Arcs are drawn from the pivot circles with the lever radius a as the radius of arc.
(viii) At the intersecting points roller centres are situated. If the points are joined, theoretical cam profile (β-profile) is obtained.
(ix) At the tangent points of the rollers, the actual cam profile is drawn.
(x) The same procedure is followed for drawing both rise and fall of the cam profile.

By following the above procedure, the turret and cross-slide cams for the present example can be drawn. Only the outlines are shown in Fig. 20.30.

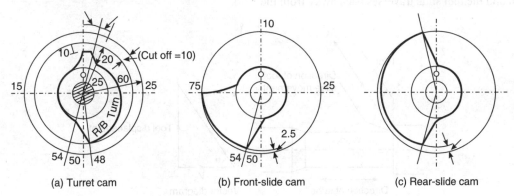

Fig. 20.30 Outlines of turret and cross-slide cams.

20.7 HYDRAULIC COPYING SYSTEMS

Copying systems are used extensively on centre lathes. Taper turning attachment is an example of a simple mechanical straight line copying device which is available on most standard lathes. Special production copying lathes are now easily available and being used to produce the most complex forms including threads on components product in quantity, the use of expensive form tools, with a limitation on the length of their cutting edge, is eliminated. It is claimed that copy turning lathes are preferable to multi-tool lathes because of greater accuracy, less work deflection and less power required at the spindle.

Many ingenious hydraulic systems have been designed for lathe tracer-controlled copying units, all having the advantage of having very little contact pressure between stylus and template. These hydraulic units are basically servo mechanisms, i.e. a control system which magnifies a relatively small input force or signal to a larger output force or signal for operating the mechanism. This output signal from the servo mechanism must be continually and automatically modified to suit variations in the input signal; thus we then have an automatic control system. A simple pilot valve controlling the stroke of a hydraulic cylinder is not a servo mechanism. The hand (small input force) opens the pilot valve, allowing oil to one end of the cylinder giving movement of a ram. However, the position of the ram cannot be controlled and monitored precisely by means of the pilot valve once the ram is moving. A servo mechanism can however give this precise degree of control. Figure 20.31(b) shows in simple form the principle of operation of a hydraulic servo copying system for a lathe.

Modern copying systems designed on this principle do not have a linkage system between the piston and spool as shown, and the valve and hydraulic system are more complex.

In Fig. 20.31(b), as the lathe saddle traverses along the bed, the stylus will follow the template edge always being kept in contact by spring pressure on the LH end of the valve spool. If the stylus, in following the template form, moves to the right, then the spool will move to the right. This will allow the oil to the left of the piston to exhaust, and will allow oil in to the right of the piston, hence moving it to the left. Therefore, the tool slide and tool will move to the left reproducing the template shape upon the workpiece. A stylus movement to the left produces the opposite effect. It is essential that there is immediate tool response in answer to stylus movement, in order to give accurate copying of the template. Also, the hydraulic force acting upon the piston must be great enough to overcome the radial cutting force on the tool.

The tool slide may be set at an angle to the work axis in order to allow square shoulders to be reproduced on the work as shown at Fig. 20.31(a). To produce the square shoulder shown, the saddle traverses from right to left and the tool slide traverses back away from the work.

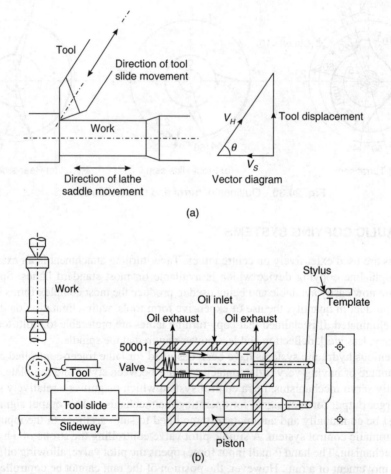

Fig. 20.31 (a) Tool slide set at angle to work axis and (b) Hydraulic copying attachment.

In the vector diagram shown in Fig. 20.31.

V_S is the saddle velocity (feed);

V_H is the tool slide velocity (feed);

θ is the tool slide angle.

If the ratio $V_S/V_H = 1/2$ and $\theta = 30°$, the tool point will be displaced along a straight line perpendicular to the work axis.

The responsiveness and accuracy of the copying system depends upon the accuracy with which the spool valve is manufactured. The spool shoulder lengths and port openings must be of precise length, and the resistance to oil flow must be identical at any spool position.

20.8 ELECTRIC COPYING SYSTEM

These are used extensively but require a high degree of skilled maintenance for the circuits. Mostly they use magnetic clutches with a 'make' and 'break' action to traverse the machine slides in the direction required. Electric systems are used for copying on lathes, but we will consider here their use on profile or die sinking milling machines. These are used in tool rooms for the production of metal dies and moulds required for thermoplastic moulding, pressure die casting, sheet metal drawing, drop forging, etc. An electronically coupled tracer is traversed in lines along the template (or three-dimensional model), causing the end-milling cutter of appropriate shape to mill the profile out of a metal die block. Increasing increments of feed are applied by the operator until the profile is completed to the correct depth.

The well-known Keller system uses this principle of die sinking, although the manner in which it is applied to their various types of machines is different (see Fig. 20.32).

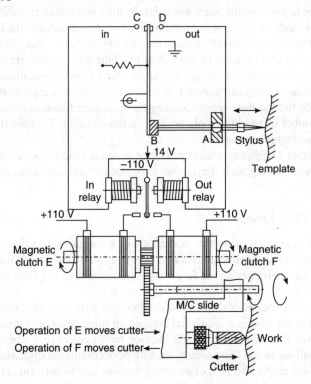

Fig. 20.32 Electrical copying system.

The figure illustrates in simplified form the principle of operation of an electric copying system for a die-sinking milling machine.

The tracing stylus is held centrally in a universal ball joint A and is free to move in any direction. The back end of the stylus is centred in a cup, which is an integral part of the switch lever B. When the stylus is fully extended the contacts at C will be closed, thus completing the IN circuit and operating the IN relay. As soon as the stylus moves inwards by a very small amount, the contacts at D will close, thus completing the OUT circuit and operating the OUT relay. Hence the magnetic clutch F will be operated, which moves the machine slide and cutter head along its slideway in a direction to correspond with the stylus movement.

As the stylus is traversed, say in a vertical plane, up and down the template, it will move in and out to correspond with the profile. Either clutch E or clutch F will be energized, causing the cutter to move in the appropriate direction. The 14-volt circuit has an earthed negative and since switch lever B is also earthed, closing of either contact C or D will cause a relay to be energized, hence operating a magnetic clutch. Three-dimensional work will obviously require a more complex system than two-dimensional work.

In the Keller system the action of the stylus is so light and sensitive that a force of approximately 1 N only is required to operate the controls. A movement of less than 0.025 mm in the stylus movement is sufficient to close contact C or D and cause a change in the direction of the cutter travel. The model or template can be made from any suitable material, and such is the sensitivity of the system that the grain pattern on a wooden model will be reproduced upon the metal copy.

20.9 TRANSFER MACHINES

20.9.1 Transfer Machines and Automated Flow Lines

A production line may be non-automatic, semi-automatic or fully automatic. In semi-automatic production the loading and unloading of blanks, inspection and handling are not done automatically. In an automatic production line the operator loads the work and usually checks the job, but all the handling devices are automatic.

A transfer device is a combination of individual machine tools which are arranged sequentially and integrated with interlocked controls. Although not applicable in every operation, this type of mechanism solves production problems in many cases. Application of transfer devices leads to increased productivity; the number of machine tools and the floor space requirement have been found to reduce by 33% to 50%. There is a reduction in the number of machine tool operator on this account. Transfer devices also lead to better quality and reduced manufacturing costs.

The decision whether to employ a transfer device is made strictly on economic considerations. The initial cost of such devices is very high. Employment of transfer devices also requires high-quality blanks and skilled personnel.

20.9.2 Automated Flow Lines

An automated transfer line consists of several machines which are linked together and transfer parts between the stations and represents a transfer line automation. The transfer of the work parts occur automatically and the work stations carry out their specialized functions automationally. The flow lines can be symbolized as shown in Fig. 20.33(a).

A raw work part enters one end of the line and the processing steps are performed sequentially as the part moves from one station to the next in a sequential manner. In some cases, it is possible to incorporate buffer storage zone into the flow lines, either at a single location or between every workstation. It is also possible to include inspection stations in the line to automatically perform intermediate checks on the quality of the work part. Manual stations might also be located along the flow line to perform certain operations which are difficult or uneconomical to automate.

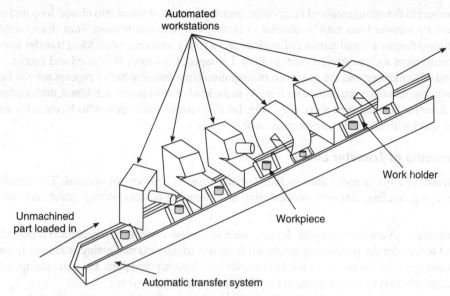

Fig. 20.33(a) Automated transfer line.

The objectives of flow line automation are the following:
   (i) To reduce labour cost
   (ii) To increase production rates
   (iii) To reduce work-in-progress
   (iv) To minimise distance moved between operations
   (v) To achieve specialisation of operation
   (vi) To achieve integration of operations.

20.9.3 Classification of Transfer Lines

Transfer lines fall into five categories:
   (i) Depending on the mode by which blanks are fed into the line and finished parts are ejected piece by piece as a continuous flow out of the line;
   (ii) Depending on the rate of production, transfer lines are classified as *single-flow* (progressive action) lines and *multiple-flow* (parallel progressive) lines;
   (iii) As to the type of machine tools employed, there are transfer lines with special machine tools designed and built for a given line, with unit-built machine, with semi- and fully-automatic general-purpose machines, and with modernised (automated) universal machines;
   (iv) Transfer lines are classified according to the type of intermachine transfer system:
      (a) The *pass-thought type*, in which the workprice passes through the clamping zone (this type being used for machining housing-type parts on automatic unit-built machines);
      (b) The *overhead type*, where the work piece is conveyed horizontally in a longitudinal direction and vertically in a transverse direction;
      (c) The *side-loading (frotal) type* with longitudinal and transverse conveying movements;
      (d) The *combined-transfer* type;
      (e) The *rotary conveying type*, used in rotary transfer lines.

462 TEXTBOOK OF PRODUCTION ENGINEERING

With respect to the arrangement of equipment, transfer lines are divided into closed-loop and open-loop types. Closed-loop transfer lines may be circular or rectangular. Circular transfer lines (for example, rotary transfer machines) feature a small number of stations and a rotary indexing table. Most transfer lines have an open-loop arrangement such as straight-line (in-line), L-shaped, U-shaped, W-shaped and zigzag.

The kind of workpiece and the sequence of operations in a manufacturing process are the key factors which determine the required type of transfer line. As to the kind of workpieces machined, the transfer lines are classified as those for housing-type parts, for shafts, for disc-shaped parts (gear wheels, etc.), for antifriction bearing races, and for small parts (screws, pins, rollers, etc.).

20.9.4 Elements of Transfer Lines

Along with machine tools, transfer lines include transfer systems and control systems. The transfer system contains conveying, loading, turnover, reorientation and clamping devices, storage units and chip disposal conveyers.

Conveying devices: Various conveying devices, such as diverse conveyers, mechanical hands and chutes and tubes are used to transfer the parts being machined from one to the next machining station in transfer lines. Intermittent conveyers, for instance, are used to transfer both housing type parts and parts clamped in pallets. Chain, band, and other types of conveyers are used as intermittent conveyers.

An intermittent conveyer called *transfer bar* (incorporated as disappearing finger), is shown in Fig. 20.33(b), to move workpieces (2) along the line. This transfer bar is imparted periodic reciprocating motion.

A transfer bar incorporated as rotating finger provides higher accuracy of workpiece movement and positioning on machining stations. The transfer bar with fingers 1 is imparted a periodic reciprocating motion and a rotating motion about the bar axis. The workpieces being machined are moved only during the forward stroke of the bar.

A transfer bar mechanism like the walking beam is shown schematically in Fig. 20.33(b). To convey the workpieces, bar (3) is imparted consecutive reciprocation motions in the horizontal and vertical planes.

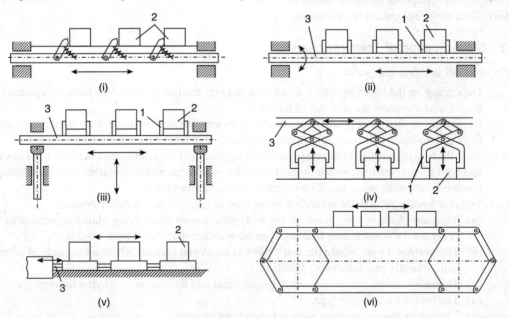

Fig. 20.33(b) Conveying devices.

An overhand transfer bar with grips as a complicated version of the walking beam type, in which workpieces (2) are transferred by means of grips 1. The grips are carried by bar (3), which moves above the workpieces.

An intermittent conveyer of the pushing type transfers workpieces (2) by means of piston rod (3) of a hydraulic or pneumatic cylinder. During the forward stroke, the piston rod pushes all the workpieces simultaneously pallets located on the conveyer.

Chain conveyers are used in many transfer lines where continuous motion of the workpieces being machined is required. This type of conveyer is hardly ever used for intermittent motion, because it cannot provide accurate work transference for location and clamping at the machining stations.

Workpieces locating and clamping devices, namely stationary fixtures and pallets, are used in transient lines. Stationary fixtures are intended for one machining operation on a given machine tool or station. The functions accomplished by this type of fixtures are: tentative orientation of the workpiece, its final clamping and release, and guidance of the cutting tools.

During machining, the workpieces are automatically loaded into stationary fixtures. All the operations involved in loading the workpieces into the fixture from the conveyer and unloading it from the fixture into the conveyer (such as gripping of the workpiece by the loading device, placing it on datum surfaces in the fixture, releasing it, etc.) are accomplished by relatively simple movements of the loading device. Stationary fixtures are mainly used in transfer lines for parts which are immovable during machining (cylinder heads, engine block, etc.).

Individual travelling fixtures or pallets are used in many transfer lines. They serve to clamp complex shape parts which have no surfaces suitable for automatic location in machining and transfer. Such pallets with workpieces are transfer-oriented: they are conveniently located and fixed at the machining stations or machine tools. The workpieces are placed into the pallets and clamped therein by hand, and released and unloaded after machining. Sometimes this is done by special automatic devices installed at the entry of the transfer line.

Storage units: To reduce time losses resulting from the need for re-setting, adjustment, and maintenance of some machines, transfer lines are split into sections, each section being capable of operating independently. In addition to this, workpiece stockpiles are introduced at workpieces from preceding operations to store them and feed them into the assembly line. There are two types of storage units: (i) the progress-through type and the (ii) blind-alley type. In the first type, the workpieces travel though a storage unit when the transfer line functions normally, i.e. for a single workpiece to be ejected from the storage unit, all the workpieces in the unit must be moved. The second type of storage unit is arranged in such a way that the flow of the workpieces, transferred from one line section to the next one, bypasses the storage unit during normal operation. The storage unit starts functioning only when the preceding section of the transfer line is shut down.

Transfer line control systems: To ensure sequential action of component machine tools and mechanisms built into automatic transfer lines. The automatic control systems used include the following:

(a) A system for controlling all the movements and sequence of operation of all the main and auxiliary units in the transfer line;
(b) An interlocking system ensuring trouble-free operation of all the machine tools, mechanisms, and cutting tools;
(c) A system for adjustment of the machines and cutting tools;
(d) An inspection system for checking the dimension of the workpieces being machined; and
(e) A signalling system facilitating transfer line servicing.

The automatic control systems employ electrical, hydraulic, and pneumatic interconnecting devices, the inter-connecting devices are, in turn, classified as external, internal, intermediate, and auxiliary interlinkages.

External control interlinkages serve to co-ordinate operation of independent transfer line sections. Intermediate control interlinkages ensure the co-ordinate operation of separate machine tools and mechanisms within each section of the transfer line. Internal control interlinkages comprise control devices and circuits, which provide sequential operation of machines in machine tools of the transfer line. Auxiliary control interlinkages co-ordinate operation of separate transfer line units with that of other control systems. External and auxiliary interlinkages are mainly electrical, while intermediate ones are combinations of electromechanical, electrohydraulic, and electropneumatic devices. Internal interlinkages employ various mechanical, electrical, pneumatic and hydraulic devices, and their combinations.

Centralized, decentralized, and combined control systems are used in automatic transfer lines, depending on the component equipment, dimensions of the line, and its work cycle time.

Chip disposal: The following methods of chip disposal are used in transfer lines: a mechanical method by means of scraper, brush, and screw conveyers; a gravity method, where chip is conveyed to an inclined chute and slides down into a special chip collector; removal of the chip by means of liquid or compressed-air jet; and by means of electromagnets.

Coolant supply of transfer lines: This is accomplished by several methods which are the following:
   (a) Centralized supply from the main coolant system the plant.
   (b) Supply from a special coolant system built for given transfer line (used where centralized supply is not available).
   (c) Supply from coolant tanks attached to the component machine tools, where machining requires the use of coolant.

Types of transfer devices
There are two configurations in which work flow can take place:
   (i) Rotary type
   (ii) Inline type.

The rotary transfer device moves the workpieces in a circular path, which usually has a maximum diameter of about 3 metres. The workpieces are placed individually in holding fixtures which are attached to the table of the machine. In general, they are not used to move very heavy pieces when the total number of operations required is rather low.

Most of the rotary transfer devices are dial indexing machines, but some other rotary mechanism, generally more rudimentary, can also be considered for rotary transfer.

Dial indexing machines are circular tables on which the workpieces are positioned. The table rotates and carries the pieces from station to station in such a way that work can be performed simultaneously, on all the pieces that are on the table, as per the following sequence (see Fig. 20.34):

1. Perform screw driving.
2. Inspect for presence of first part.
3. Eject assembly and load next component.
4. Feed and escape second part.
5. Inspect for presence of second part.
6. Drive.

The dial indexing machine is an ideal choice for a transfer device, when the total number of production operations required is rather low and not too complicated, due to the fact that the workstations can be mounted in a central column or around the periphery of the table.

If the number of workstation increases, it becomes necessary to increase the diameter of the table to provide room to mount the extra number of stations. A point is reached where table mass and tooling become

MECHANIZATION AND AUTOMATION

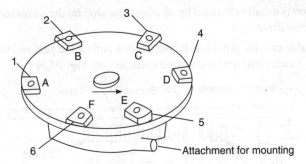

Fig. 20.34 Six-station dial indexing table.

difficult to accelerate and decelerate, and extensive floor space is consumed. In this case the rotary table is not a good choice, and the possibility of using another type of transfer device should be considered.

Multiple-station dial indexing tables are built by commercial manufacturers in standard sizes, and are available in almost any number of station from 2 to 24 or more indexes per revolution. The most common are 6, 8, 12, and 24 stations. Rotation can made in a multitude of diameters and designs with electric, pneumatic or hydraulic drive units.

The cost of the rotary table usually increases with the degree of accuracy required. The accuracy depends on the type of drive used in conjunction with the method of locking and locating the dial before performing the work on the pieces.

According to the type of drive, the rotary indexing machines can be divided into:

(i) Rotary table
(ii) Indexing table
(iii) Dial table.

Rotary table: This is the least expensive of the rotary devices, with prices starting at about ₹ 10,000. It is a non-precision rotary device but one that is adjustable.

The rotary table is driven by a pneumatic or hydraulic cylinder, which actuates a rack and gear mechanism to convert linear motion to rotary motion. As the gear rotates, the ratchet pawl is driven into teeth of the ratchet gear attached to the dial table, causing the dial to index (see Fig. 20.35). Retraction of the cylinder causes the pawl to override the ratchet gear.

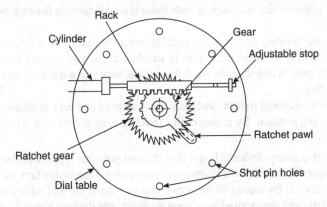

Fig. 20.35 Top view of indexing table.

Accuracy of location is usually obtained by an adjustable stop on the travel rod or the piston. This type of drive is called the *ratchet drive*.

Indexing table: This table uses the pin drive. It consists of a table drive gear and slides up down on a spline to engage and disengage a rack powered by indexing cylinder (see Fig. 20.36).

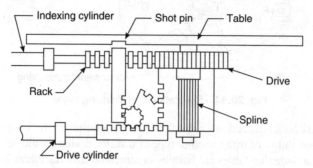

Fig. 20.36 Front view of indexing table with driving mechanism.

Prior to indexing, the engagement cylinder raises the gear engaged with the rack, and it then disengages the shot pin. The indexing cylinder advances the rack, causing the table drive gear to rotate the table to the next index position. Then the engagement cylinder disengages the gear from the rack and raises the shot pin into the table. The indexing cylinder resets the mechanism on the pin into the table. The indexing cylinder retracts to reset the mechanism on the next indexing operation. This type of drive gives high accuracy of operation, making the index table a more accurate transfer device compared to the rotary table.

The index table is more expensive than the rotary table, with prices starting at about ₹ 20,000. The table is adjustable but more difficult to use than the rotary table.

One of the problems with these two types of rotary device is a tendency to over-index or have free travel during the index cycle due to the ratchet action.

Another problem is the hammering action of the drive cylinder, although this condition can be corrected by using a cushioning device at the end of the stroke. Such a drive should have a value to regulate the escapement flow.

The rotary table gives an accuracy of 0.125 mm; the index table gives better accuracy depending on the specific type of drive, reaching in some cases an accuracy close to ± 0.25 mm.

Floating position. To improve the accuracy in both these types of table, a floating position of the holding fixtures could be used.

In order to obtain the floating position, the holding fixture is built with a floating side in which holes for securing pins are drilled. A set of securing pins is attached to the tool in such a way that they reach the holding fixture before the tool. When the pins reach the fixture, they will enter the holes drilled for them, thus securing the holding fixture.

When the tool is at the holding fixture (and piece), it is secured in a very accurate position. Figure 20.37 shows one station in floating position. By means of this technique, an accuracy of ± 0.25 mm can be obtained in either type of table.

Dial table: Generally, this rotary device is larger than the two previous types; it is also more expensive. It is driven by a motor that operates a Geneva or Ferguson drive motion. Due to this fact, the dial table is the most accurate rotary transfer device. By means of these two types of drive the dial table can give accuracy, high speed, high torque capacity, and any desired acceleration, dwell, and declaration characteristic. However, the drives can not be adjusted; they are built for a given number of indexes per revolution, which is practically impossible to change.

Other rotary transfer devices: As mentioned previously, there are other rotary transfer devices besides the dial indexing machines. Two other types of rotary transfer devices are shown in Figs. 20.37 and 20.38. Two nests are built equidistant from a pivot point with an included angle of 120°.

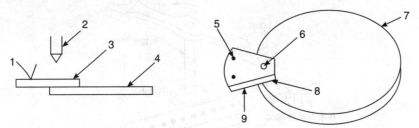

1. Securing pin, 2. Tool for operations, 3. Fixture, 4. Dial table, 5. Holes for securing, 6. Workpiece, 7. Dial table, 8. Fixture, 9. Floating plate

Fig. 20.37 Pneumatic rotary transfer device.

A pneumatic or hydraulic cylinder drives a rack- and gear-mechanism that changes the linear motion of the drive unit into an oscillating motion. Parts are alternately loaded and unloaded at two positions, while work is performed on the part at third position. With this device, the cycle time is reduced to the time required for indexing the fixture, plus either the part unloading/loading time or the work-performing time, whichever is longer.

If the rack and gear drive is replaced with an unidirectional ratchet drive or something similar, the oscillating motion can be changed to a rotary motion in one direction. Its additional nests are added at either 120, 90, 72, 60 or 45 degrees apart; the transfer machine works as a rudimentary 3-, 4-, 5-, 6-, or 8-station dial indexing machine. In this case, the cycle time is equal to the time required to index plus the ongoing operation time at any one station.

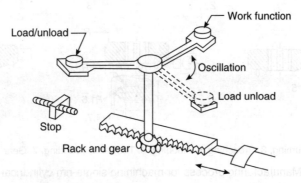

Fig. 20.38 Mechanical work transfer device.

The two previous transfer devices are rudimentary mechanisms, useful in cases where the required accuracy is not high and investment in more expensive transfer devices is not justified.

Figure 20.39 shows the mechanism of a work transfer device.

Rotary transfer lines: Rotary transfer lines consist of rotary machine tools linked by a conveyer and a single drive. This type of transfer line offers a high productive output; it can be easily changed over from job to job, and can be used in both batch and mass production. Rotary transfer lines operate on the continuous action principle. The machining and transference of workpieces are carried out in parallel, with the cycle phases either coinciding or being shifted in time in respect to each other.

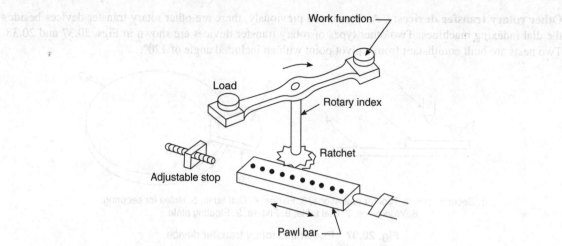

Fig. 20.39 Mechanism of work transfer device.

Figure 20.40 shows the manufacturing process for machining single-rim cylindrical gear wheels.

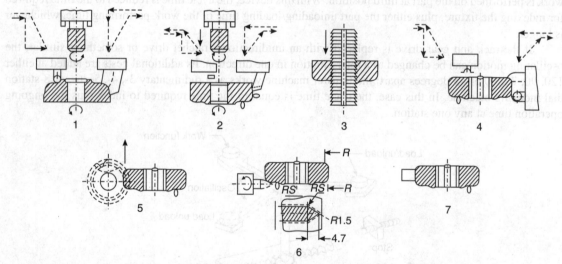

1, 2, 4. Turning, 3. Broaching, 5. Hobbing, 6. Tooth-end rounding, 7. Gear shaving

Fig. 20.40 Manufacturing process for machining single-rim cylindrical gear wheel.

Figure 20.41 shows the diagram of a rotary transfer line, in which machining and conveying the workpieces are done simultaneously. The workpiece being machined and the cutting tools are imparted the necessary movement while being rotated about the central axis in each component rotary machine. The workpiece is transferred from machine (2) to (4) through conveying device (3), which is continuously rotating.

A number of rotary machine tools making up a rotary transfer line is shown in the figure. The necessary rotating movement is imparted from an electric motor to the cutting tool, workpiece, machining station, and conveying device, through appropriate gear transmission, worm gearing, etc. Translator movements of the cutting tools and workpiece are effected by corresponding cams or by a hydraulic system.

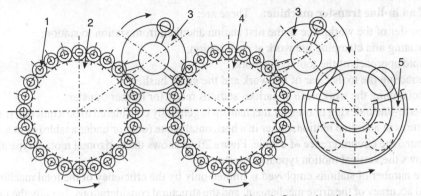

1. Work spindles, 2. Drilling rotary machine, 3. Conveying device,
4. Reaming rotary machine, 5. Heat-treatment rotary machine

Fig. 20.41 Diagram of rotary transfer line.

Figure 20.42 shows a rotary machine in a rotary transfer line.

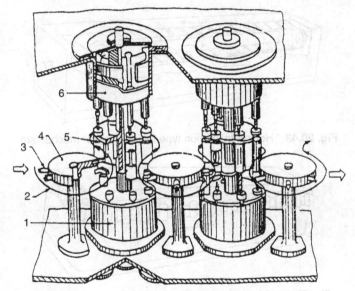

1. Rotary machine, 2. Trajectory of workpiece passage through
machines and conveying devices, 3. Grips of conveying device,
4. Conveying device, 5. Tool block, 6. Cam for vertical movement of tools

Fig. 20.42 Rotary machine in rotary transfer line.

In-line transfer devices

The first multiple-station transfer devices were of the rotary type. However, the demand for more stations led to the development of transfer mechanisms with many stations stretched out on a straight line. These are *in-line transfer devices*. These devices have two main types of mechanism:

(a) Pallet type transfer mechanism
(b) Plain type transfer mechanism (also known as non-pallet).

Functions of an in-line transfer machine: These are:
  (i) Transfer of the workpiece to the first station and then from station to station;
  (ii) Locating and clamping the work at each station;
  (iii) Rapid approach of the tool to the work;
  (iv) Retracting the tools clear of the work and the guide bushes;
  (v) Unclamping the workpiece at various stations ready for further transfer;
  (vi) This in-line indexing or transfer mechanism is generally considered to be a unit that transports work fixtures in an inline motion, either in a horizontal plane (over or under a table) or in a vertical plane around the circumference of a table. Figure 20.43 shows the horizontal motion type and Fig. 20.44 shows the vertical motion type mechanism.
  (vii) The number of stations employed is limited only by the efficiency of the total machine, the power and accuracy of the drive mechanism, and the structural considerations. Usually the in-line transfer devices are chosen when the number of stations exceeds 24.
  (viii) The production cycle time for all types of indexing units is equal to the sum of the time required for the indexing mechanism to advance one station and the time for the slowest work station to complete the cycle. Since indexing time is considered unproductive time, it should be minimised as much as possible.

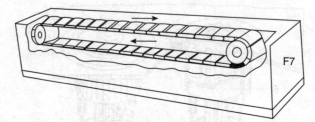

Fig. 20.43 Horizontal motion type in-line transfer mechanism.

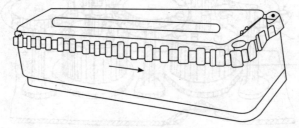

Fig. 20.44 Vertical motion type in-line transfer mechanism.

Pallet type transfer machine: In this type of transfer machine, the parts are transferred from station to station while being clamped in holding fixtures. The holding fixtures are called *pallets* and hence the name *pallet type transfer device.*

After the operations are completed, the pallets are returned by conveyor to the loading station, where the finished part is unloaded and an unfinished piece is loaded on the empty pallet.

The pallet type of transfer machine is more accurate than the plain type, due to the fact that pallets can be built with very close tolerances; and as the part once clamped will not be removed from one pallet until it is finished, the accuracy of positioning will be very good along the whole process. The two-line indexing machines are shown in Fig. 20.45. At the start of the transfer, the transfer rails are against a fixed stop A.

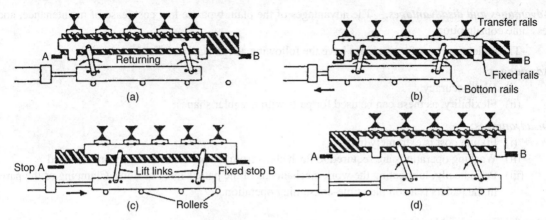

Fig. 20.45 Two-line indexing machine.

The cylinder then advances, moving the bottom rails, transfer rails and pallets to the right until fixed stop B is contacted. As the cylinder continues its forward stroke, the bottom rails continue to the right, causing the lift levers to shift position and lower the transfer rails from the pallets. The cylinder then retracts, moving the transfer mechanism to the left against the fixed stop A. The cylinder is retracted, moving the bottom rails to the left and causing the transfer rails to rise vertically under the pallets as the lift links shift their positions.

Plain type transfer machine: In this type of transfer machine, the parts move in an unclamped condition from station to station. At the machining station, the fixed or disappearing type dowel locators and the hydraulically actuated fixture clamp as well as hold the parts. In other words, the fixtures are fixed and only the workpiece moves throughout the station.

In Fig. 20.46, a fixture arrangement for one such workstation is shown. The plain type transfer machines are used, when the parts are held in identical positions in each machining station, and when it is not necessary to change over from one part to another frequently. Also, they are used for parts with rather regular shape.

This type of machine is less expensive than the pallet type, but it gives less accuracy.

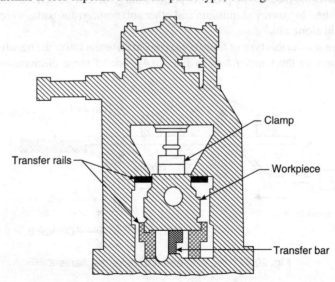

Fig. 20.46 Fixture arrangement for plain type transfer machine.

Advantages and disadvantages. The advantages of the plain type are low cost, ease of maintenance, and less time consumption.

The pallet type transfer machines have the following advantages and disadvantages:

Advantages:
  (i) High accuracy.
  (ii) Flexibility, as these can be used for parts with irregular shapes.

Disadvantages:
  (i) Fixture cost is extremely high.
  (ii) Washing operations are required to flush chips from the pallets before they are reloaded.
  (iii) Periodically, lubricating the work mechanisms of the pallets must be done. Clamping of the parts in the pallets is a complex time-consuming operation.

20.9.5 Transfer Mechanisms

There are various types of transfer mechanism used to move parts between stations. These mechanisms can be grouped into two types: those used to provide linear travel for in-line machines, and those used to provide rotary motion or dial indexing machines.

Linear or in-line transfer mechanisms

The three typical mechanisms are as follows:
  (i) Walking beam transfer bar system
  (ii) Pawl chain drive conveyer system
  (iii) Rotating bar mechanism.

Walking beam transfer bar system: With the walking beam transfer mechanism, the work parts are lifted up from their workstation location by transfer bar and are moved on position ahead to the next station. The transfer bar then lowers the parts into nests which position them more accurately for processing. This type of transfer device is illustrated in Fig. 20.47.

This type of transfer bar mechanism provides positive movement of parts from station to station without sliding. It is generally used to convey aluminium and other soft non-ferrous parts, which would be subjected to too much wear if slid along rail.

Two cylinders are used in this type of transfer bar; one to raise and lower the transfer bar and workpieces, and the other to reciprocate the transfer bar. In Fig. 20.47, one of these mechanisms can be seen. Here,

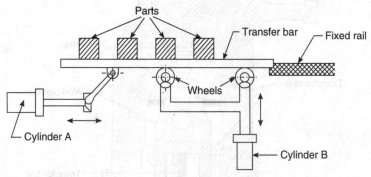

Fig. 20.47 Walking beam transfer mechanism.

cylinder B lifts the transfer bar and the workpieces; then cylinder A retracts, pulling the transfer bar which slides on the wheels attached to end of the cylinder B. Then cylinder B goes down, lowering the transfer bar and leaving the workpieces in an advanced station. Cylinder A advances, returning the transfer bar to its original position.

There are many other types of in-line transfer devices, which generally are combinations of two or more of the transfer mechanisms. Some are rudimentary and cheap, which sometimes determines their selection, and some are necessary for certain situations.

Pawl type transfer mechanism: This is a simple and inexpensive type of transfer mechanism, in which the parts are slid from one machining station to the next. A single transfer bar, either round or rectangular in cross-section, has a series of pivoted fingers mounted upon it. These fingers are either spring-loaded or weighted, so as to latch against the rear surfaces or the parts (see Fig. 20.48).

Thus, a forward stroke of the bar transfers the parts. The fingers pivot upward and slide along the top of the parts during the return stroke. The main disadvantage of this system is that transfer rates are limited, in order to avoid excessive skidding of the parts when the bar is stopped at the end of the forward stroke. To help solve this problem, some finger designs include an angle in the back, so as to retain the forward ends of the piece parts. Such designs have met with varying degrees of success. The pawl type provides an adequate method if the parts have good sliding surfaces.

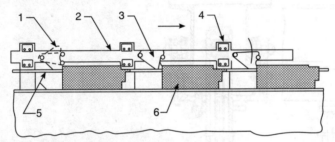

1. Pivoted finger, 2. Transfer bar, 3. Finger, 4. Bar support, 5. Guide rails, 6. Piece part

Fig. 20.48 Pawl type transfer mechanism.

Rotating bar mechanism: This transfer bar is used where parts are to be trapped. The bar has fingers attached to it and are rotated by a separate hydraulic cylinder that reciprocates the bar (see Fig. 20.49). Each part is trapped to completely avoid the part overshooting the machining station.

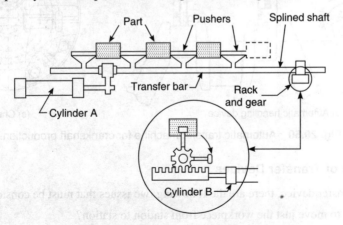

Fig. 20.49 Rotating bar mechanism.

Cylinder A advances to move the transfer bar to the right, and the pushers connected to the bar advance the parts by one station. At this point, Cylinder B advances, causing the rack to drive and rotate the transfer bar through 90°. The pusher are rotated clear of the parts, and Cylinder A retracts to move the transfer bar back to its home position. The transfer is completed with Cylinder B rotating the transfer bar by 90° and inserting the pusher in between the parts again.

The automatic transfer machine for crankshaft production is shown in Fig. 20.50.

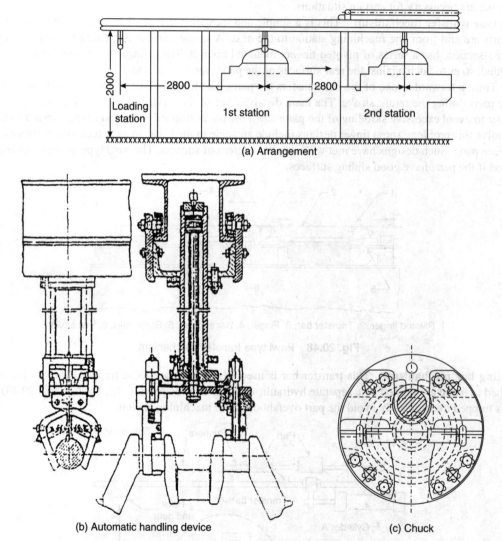

(a) Arrangement

(b) Automatic handling device   (c) Chuck

Fig. 20.50   Automatic transfer machine for crankshaft production.

20.9.6 Selection of Transfer Devices

While selecting a transfer device, there are the following two issues that must be considered:
  (i) Is it better to move just the workpiece from station to station?
  (ii) Is it better to move the fixture and the workpiece from station to station?

In some complicated assembly operations, both principles are employed. In order to make a decision about the specific type of transfer device to be used, the following factors should be considered:

1. Number of operations required;
2. Physical size and weight of parts as well as the total assembly;
3. Weekly, monthly or yearly volume of parts and hourly production rate desired;
4. Permissible transfer time as part of total operating time;
5. Accuracy of transfer required to assure proper location and placement of parts or other station functions;
6. Inertial load involved for acceleration and deceleration of transfer mechanism;
7. Forces to be applied at various workstations;
8. Need for variable speed drives to meet varying production requirements and assembly speeds;
9. Flexibility of transfer device timing to permit changing of operating cycle.

20.9.7 Methods of Work Transfer

The transfer mechanism of the automated flow line must not only move the partially completed workpiece or assemblies between adjacent stations, but it must also orient and locate the parts in the current position for processing at each station.

The transfer methods can be classified into following three categories:

(i) Continuous transfer;
(ii) Synchronous or intermittent transfer;
(iii) Asynchronous or power-and-free transfer.

They are categorized by the type of motion that is imparted to the workpiece by the transfer mechanism. These transfer mechanisms are used for both processing and stable operations. In case of assembly operations the mechanisms are used for transferring the partially completed workpieces to the stations. The mechanism used for feeding the new workpieces is not the transfer mechanism but is integral to the workstations.

The type of transport system to be used depends on:

(i) Operation to be performed;
(ii) Manual stations present or not;
(iii) Number of work stations online;
(iv) Production rate required;
(v) Weight and size of work parts.

Continuous transfer mechanism

In this type of system the work parts are moved continuously, which requires that the work heads should also move. This becomes sometimes difficult as in the case of machining operation, because of the inertia problem due to size and weight. But for example, in case of manual beverage bottling operation (where the operator can move with a moving flow line) the system is used efficiently. Continuous transfer mechanisms are in fact easy to design and fabricate, and high production can also be achieved.

Synchronous or intermittent transfer mechanism

As the name suggests, in this method the workpieces are transported with an intermittent or discontinuous motion. The workstations are fixed in position, and the parts are moved between the stations and registered at the proper location for processing all the work parts transported at the same time, which leads to the term *synchronous transfer system*. The system is used in machining operations, press working operations and mechanized assembly.

Asynchronous transfer or power-and-free system

This system allows each work part to move to the next station when processing at the current station has been completed. Each part moves independently with respect to the other parts. Hence, some parts are processed on the line at the same time that the others are transported between stations.

Figure 20.51 shows the tooling layout of an automatic transfer machine for cast iron sealing members.

Disadvantage: Cycle times are generally shorter than that of the other types.

Advantages: These are as follows:
  (i) Greater flexibility than in the other two systems;
  (ii) Larger work parts can be handled;
  (iii) No problem due to cycle time variation.

Figure 20.52 shows the machine tool arrangement in an automatic transfer machine for a cast iron sealing member.

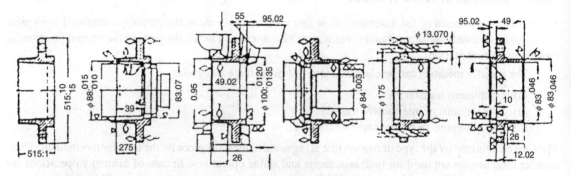

Fig. 20.51 Tooling layout of automatic transfer machine for cast iron sealing member.

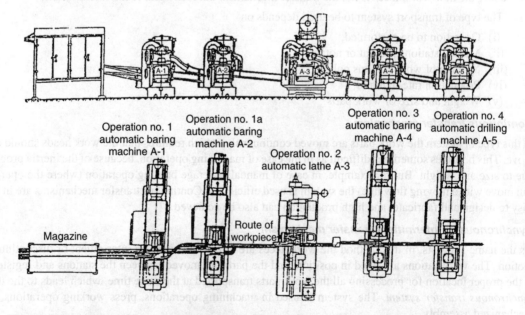

Fig. 20.52 Machine tool arrangement in automatic transfer machine for cast iron sealing member.

20.9.8 Arrangement of Transfer Lines

Transfer lines containing automatic composite-built machines are used for machining housing-type parts. Unit-built machines are very popular because they allow the use of up to 70% of normalised units. Figure 20.53 shows the diagram of a typical unit-built transfer machine. Here, the workpieces being machined are transferred through all the machining stations without being removed for that purpose from the conveyer. The workpieces are located and clamped at each machining station by means of stationary fixtures.

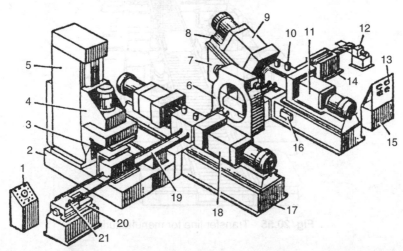

1. Control console, 6. Workpiece turnover drum, 12. Chip conveyer drive unit, 13. Attached hydraulic unit, 15. Hydraulic power station, 16. Pump for automatic lubrication, 19. Turntable, 20. Transfer bar; standard units, 4. Separate-feed unit head, 5. Upright, 8. Power slide, 9, 11 and 18. Self-contained unit heads, 10. Hydraulic cylinder to clamp workpieces, 17. Base plate; specially designed units, 2, 14 and 21. Stands, 8. Fixtures, 7. Inclined base plates

Fig. 20.53 Composite automatic transfer line.

Figure 20.54 shows an automatic transfer line for machining single-rim cylindrical gear wheels.

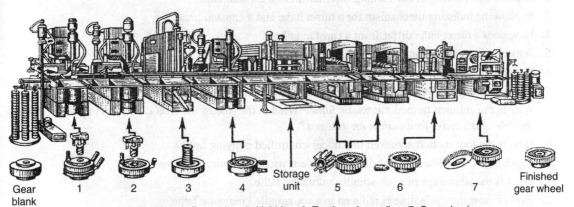

1, 2, 4. Turning, 3. Broaching, 5. Hobbing, 6. Tooth end rounding, 7. Gear shaving

Fig. 20.54 Automatic transfer line for machining single-rim cylindrical gear wheels.

Figure 20.55 shows a typical transfer line for manufacturing.

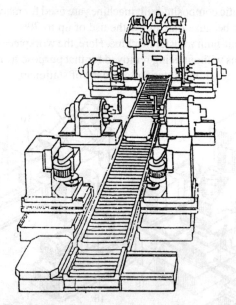

Fig. 20.55 Transfer line for manufacturing.

REVIEW QUESTIONS

1. How does capstan lathe differ from engine lathe in the following aspects:
   (i) Work holding;
   (ii) Tool holding;
   (iii) Number of tool stations?
2. Explain the working of bar feeding mechanism of a capstan lathe.
3. Explain the indexing mechanism for a turret lathe and a capstan lathe.
4. How does a turret lathe differ from a capstan lathe?
5. Why is the capstan lathe called the *ram type turret lathe*?
6. How are the tool motions achieved in a centre lathe, a turret lathe, and an automat?
7. Why is the capstan lathe also called *semi-automatic lathe*?
8. Why is the automat termed *automatic lathe*? What is the role of the *lead cam* in an *automat*? What is the role of the cross-feed cam in an automat?
9. How is the tool motion achieved in a tracer-controlled copying lathe?
10. Explain how the Wheatstone bridge principle is used in a hydraulic lathe.
11. What is the advantage of a six-spindle automatic lathe?
12. Explain how the cycle time is reduced in a six-spindle kinematic lathe.
13. In a capstan lathe the length of the stroke is controlled by ———.
14. The tool setter adjusts the length of the stroke by adjusting ———.
15. Draw the kinematic layout of an automatic lathe.
16. Sketch the tooling layout for a capstan lathe for mass production of hexagonal bolts.

CHAPTER 21

Numerical and Computer Numerical Controlled Machines

21.1 INTRODUCTION

Numerical control, also referred to as NC, is a form of automation that is applied to conventional (general purpose) machine tools such as lathes, milling machines, drilling machines and presses. These are the types of machine used in average machine shops for production of different parts in relatively small quantities. Compared to the automatic systems of production used in industry for mass production, the NC machines are more flexible to accommodate changes in design of components, simpler to operate and less expensive.

More than 50% of machine tools today are engaged in production of small lots and piecework jobs. At present, production in small lots is carried out either on capstan and turret lathes, single- or multi-spindle automatic machines or by attachment of accessories such as tracer control unit to the existing machine tools. The mechanisms used on these machines require cams, templates, stops, etc. which require longer set-up time for every new lot or job. A need, therefore, was felt for machine tools which would automatically produce parts rapidly and economically in small lots. This has been successfully met by the development of numerically controlled machine tools.

It will be noticed from the above description that the NC machine is a classic union of the automatic machining system and the conventional machining process. Unlike automatic machines, it has the added flexibility to accommodate changes in operations and elimination of errors due to the human element, which crops up in conventional machining processes. Thus accuracy, repeatability and consistency in operations are the strong points of these machines.

Since the introduction of numerical control in the 1950s, there have been dramatic advances in digital computer technology. The physical size and cost of computers were reduced drastically; as a result, it was logical to use the capabilities of digital computers in numerical control. The use of computer technology initially involved large mainframe computers in the 1960s, followed by minicomputers in the 1970s and microcomputers in the 1980s. Today, numerical control implies Computer Numerical Control (CNC).

21.2 HISTORY OF NC

The first development work in the area of NC is attributed to John Parsons and his associate Frank Stulen at Parsons Corporation in Michigan (USA). Parsons was a a sub-contractor to the United States Air Force (USAF). He was toying with the idea of utilizing digital computers, which were just then becoming popular, to reduce the complexity of computation. Milling of complex curvature is a highly skilled job. He proposed the concept of coordinate position data contained on punched cards to define and machine the surface contours of airfoil shapes. He named his system the *Cardamatic Milling Machines*, since the numerical data was stored on punched cards.

The idea was presented to the USAF by Parsons in 1948. The USAF accepted his proposal, and a contract was awarded to him to develop such a machine. Then the project was sub-contracted to the Servomechanism Laboratory of Massachusetts Institute of Technology in 1951. The Servomechanism Laboratory finally demonstrated a working milling machine in 1952.

The first NC machine was developed by retrofitting a Cincinnati Hydro-Tel Vertical-Spindle Contour Milling Machine (24 in. × 60 in. conventional tracer mill). In retrofitting, all motion elements were removed and replaced by three variable-speed hydraulic transmissions, and connected to three lead screws of the table. A feedback control system was incorporated and the resolution of the machine was 0.0005 in. The controller of the machine was a combination of analogue and digital computers. It consisted of 292 vacuum tubes. It occupied a floor space area greater than that occupied by the machine tool itself.

A patent was awarded to MIT for the machine tool system in 1962, entitled *numerical control servo system*. The Bendix Corporation produced the first commercial production base NC unit in 1954 after purchasing the patent rights from MIT. The first controller with transistor technology was introduced in 1960. There was a tremendous revolution in the field of NC when Integrated Circuits (ICs) came into being in 1967. The use of ICs allows a 90% reduction in the number of components and 80% reduction in wiring connections.

A brief period-wise listing of the historical development of NC is given in Table 21.1.

Table 21.1 Historical development of NC

Period	Developments
1940s	First concept of using coordinate position data, contained on punched cards to define and machine the complex contours of airfoil shapes, developed by John Parsons, a contractor for the U.S. Air Force.
1948	Idea presented to U.S. Air Force.
1949 (June)	Initial Air Force contract awarded to Parsons.
1949 (July)	Sub-contract awarded by Parsons to Servomechanism Laboratory at Massachusetts Institute of Technology.
1951	Name *Numerical Control* adopted.
1952 (March)	First NC machine developed and experimental machine put in operation.
1952 (August)	Patent for the machine tool system filed by MIT.
1962 (December)	Patent entitled *Numerical Control Servo System* awarded to MIT.
1952 (May)	Patent also filed by John Parsons and Frank Stulen, entitled *Motor Controlled Apparatus for Positioning Machine Tool*.
1958 (January)	Patent issued to John Parsons.
1954	First commercial production-based NC unit produced by Bendix Corporation after purchasing patent rights from MIT.

(Contd.)

NUMERICAL AND COMPUTER NUMERICAL CONTROLLED MACHINES

Table 21.1 Historical development of NC (*Contd.*)

Period	Developments
1959	MIT announces *Automatic Programmed Tools* (APT) programming language.
1960	First controller with transistor technology introduced.
1960	*Direct Numerical Control* (*DNC*) eliminated paper tape punch programs and allowed programmers to send files directly from computer to machine tool controllers.
1967	Use of *Integrated Circuits* (ICs) in NC reduced 90% components and 80% wiring connections.
1968	*First Machining Centre* by Kearney and Trecker (machine tool builders) marketed.
1970s	CNC machine tools developed.
1980s	Graphics-based computer applications developed.
1990s	Price drop in CNC technology.
1997	PC Windows/NT-based Open Modular Architecture Control (OMAC) systems introduced to replace "firmware" controllers.

Major changes in manufacturing technology can generally be traced back to historical causes. Figure 21.1 shows the four main lines of development that led to the first NC tool.

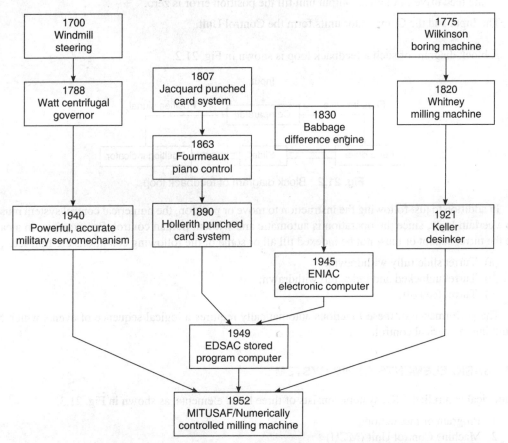

Fig. 21.1 Historical development of numerical control.

21.3 WORKING PRINCIPLE OF NC MACHINE

A typical NC machine receives information in digital form as coded numbers—indicating dimensions of the parts to be produced—punched in cards or paper tapes, or electronically recorded on magnetic tape, in a coded language understandable to the machine. Upon receipt of such information the various machine motions are controlled accordingly. Many manufacturing operations are successfully executed using numerical control. These include punch presses, riveting machines, spot welding machines, pipe bending machines, electronic assembly machines and electric wiring machines.

Normally there are two basic modes in which mathematical processing can be made: the mode of proportions and the mode of numbers, leading to graphical solutions or numerical solutions. Simulation of these techniques led to analogue computing and digital computing. We are presently interested in digital processes.

The main units in numerical control are:

- Input—where the required movement or position is fed in as a signal.
- Feed drive—which moves the machine tool slide to the required position.
- Position indicator on the machine tool slide—which indicates the actual position in the form of a signal.
- Comparator—which compares the input signal with the actual position signal and continues to actuate the feed drive through an output unit till the position error is zero.

The Input and the Comparator units form the Control Unit.

A block diagram of such a feedback loop is shown in Fig. 21.2.

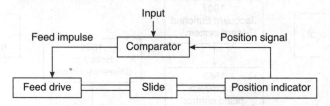

Fig. 21.2 Block diagram of feedback loop.

In addition to just following the instruction to move or position, the numerical control system must also have a certain logic, since the operation is automatic and without human control. For example, in a capstan lathe the turret cannot or must not be indexed till all or some of the following conditions are fulfilled:

(a) Turret slide fully withdrawn;
(b) Turret unlocked and index pin withdrawn;
(c) Turret feed off.

The performance of these functions automatically requires a logical sequence of events which has to be built into numerical control.

21.4 BASIC ELEMENTS OF NC SYSTEM

A numerical controlled (NC) system consists of three basic elements, as shown in Fig. 21.3:

1. Program of instructions
2. Machine Control Unit (MCU)
3. Processing equipment of machine tool.

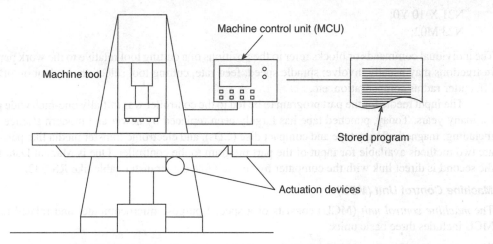

Fig. 21.3 Basic units of numerically controlled machine.

Program of instructions

The program of instructions consists of the detailed step-by-step commands that direct the actions of the processing equipment. In machine tool applications, the program of instructions is called a *part program*, and the person who prepares the part program is called a *part programmer*. A part program can be defined as *a series of commands which direct the cutter motion and support systems of the machine tool*.

A part program is a detailed plan of manufacturing instruction required for machining the work part as per the drawing. One complete instruction set is known as an *NC block*, and usually written in a single line. The format for the part program is standardized by ISO, which is followed by various controller manufacturers with little variation. The standard format is

N-G-X-Y-Z-A-B-C-F-S-T-M

where   N Code is the program sequence number
         G Code is the set of preparatory commands or functions
         XYZ and ABC Codes are the dimension words or co-ordinate data
         F Code is the feed command or function
         S Code is the speed command or spindle function
         T Code is the tool selection or tool call
         M Code is the set of miscellaneous functions
         Code is the End of Block (EOB).

A sample program is given below:

N01 G90 G 80;
N03 G00 T 12 M06;
N05 G00 X 0 Y0 Z0. 1 F10 S2500 N13;
N07 G01 Z-0.5;
N09 G02 X-10 I0 J0 F20;
N13 X0 Y0;
N17 X10 Y0;
N19 X0 Y-10;

N21 X-10 Y0;
N23 M02;

The individual commands or blocks refer to the positions of a cutting tool relative to the work part. Additional instructions may usually involve: spindle speed, feed rate, cutting tool selection, coolant on/off, spindle on/off, cutter radius compensation, etc.

The input media for the part program to be fed to the controller was initially one-inch wide punched tape for many years. Today, punched tape has largely been replaced by newer and modern storage technologies including: magnetic floppy tape and compact disc (CDs), and electronic transfer media like pen drives. There are two methods available for input of the part program to the controller. One is Manual Data Input (MDI); the second is direct link with the computer by any suitable data transfer cable like RS-232.

Machine Control Unit (MCU)

The *machine control unit* (MCU) consists of a special purpose microcomputer and related hardware. The MCU includes three basic units:

Memory unit: The function of the memory unit is to store and receive the instructions related to the process. It communicates with other units and devices of MCU to supply and receive the instructions when needed.

Data Processing Unit (DPU): The Data Processing Unit reads the part program, decodes it, processes the information and passes it to the CLU.

Control Loop Unit: The Control Loop Unit converts the information into control signals and drives the servomechanism, receives feedback (position and velocity), and instructs the DPU to read new instructions.

The MCU also includes control system software, calculation algorithms, and translation software to convert the NC part program into MCU-usable format. Today, virtually all new MCUs are based on computer technology; hence the term NC is replaced by CNC.

Every NC machine tool, unlike the conventional one, is fitted with a machine control unit (MCU). The MCU may be housed in a separately situated cabinet-like body, or may be mounted on the machine itself, or it may be like a pendant which could swing around (see Figures 21.4(a) and 21.4(b)). It is with this control unit that the path to be followed by the cutting tool, spindle speeds, feed rate, tool changes, and several other functions of the machine tool are automatically controlled.

MCU organization: The circuitry for the above-mentioned requirements of a controller is hardwired on PCBs (printed circuit boards). Accordingly, the MCU may contain several such hardwired cards for sending signals to each of the actuation systems. A typical MCU may be organized as shown in Fig. 21.4.

The MCU accepts the part program fed from its keyboard, or a tape reader or a floppy drive, or a cassette drive, or even via a direct link with a computer. It decodes and converts the coded part program into binary electrical signals. [Holes and no-holes combinations are converted into bytes (groups of eight binary signals) of pulse and no-pulse.]

The part program data in binary is now transferred to a buffer storage. The purpose of this storage is to permit faster transfer of data to other active areas of the MCU. Without buffer storage, the MCU must wait for the tape reader to read a block of information. If this read time happens to be too long, the machine motion pauses, thereby resulting in marks on the part being machined. In non-buffered MCUs, therefore, the tape reading speed is of great importance, although the length of data blocks should also not be overlooked. For proper machine operations, internal data processing in the MCU must be properly synchronized. It has to follow a logical sequence and has to perform many logical tasks.

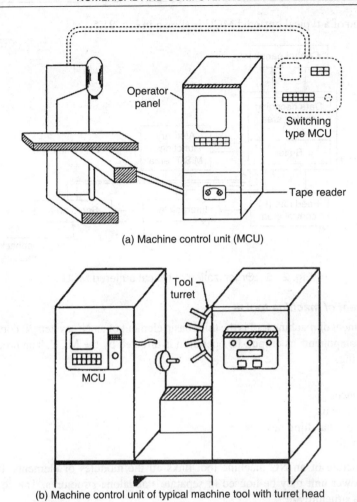

(a) Machine control unit (MCU)

(b) Machine control unit of typical machine tool with turret head

Fig. 21.4 Machine-mounted MCU on turning centre.

Processing checks: The circuitry for this logical control is built from the basic AND, OR and NOT circuits. Diodes, transistors and capacitors are therefore extensively used in the control circuits. By the use of integrated circuits (IC chips), entire sets of circuitry can be contained on solid state modules that offer compactness and a high degree of reliability. By the large-scale integration (LSI) technique, several thousands of circuits are built on small (typically 10 mm × 20 mm) chips. A major part of this data processing takes place in the data decoding and control area.

The auxiliary function area is machine-related, e.g. spindle and coolant on/off, speed range of spindle, program stop, rewinding of tape, and tool changes. The output from this control area to the machine tool can be of BCD, decimal or varying voltage, or a combination of these output types.

The organization of a typical buffered MCU is shown in Fig. 21.5.

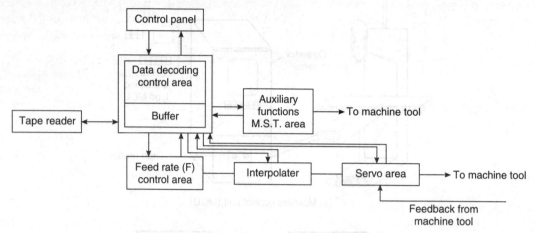

Fig. 21.5 Organization of typical buffered MCU.

Processing equipment of machine tool

The processing equipment of machine tool is the third basic element of an NC system. It converts the input raw material into useful components as output. Its operation is directed by the MCU. The processing equipment consists of the following:

(i) Structure
(ii) Driving devices
(iii) Actuation systems
(iv) Tool and work handling devices
(v) Controller unit.

Structure: The structure of an NC machine tool links all the modules or elements. In some cases, the controller unit and power unit may be housed in separate standalone structures. The considerations taken during the design of a structure are:

(a) Productivity and precise performance for continuous production;
(b) Thermal distortions caused by hot chips and removal of chips;
(c) Multi-directional forces during machining.

NC machines tend to be heavy structures of alloyed cast iron or fabricated steel construction, leading to better swarf disposal and easy access to the workpiece and tooling.

NC drives: The drives for NC machine tools are usually either hydraulic, DC or stepping motors. The selection of type of drive is determined by:

(a) Power requirements of machine tool
(b) Power sources available
(c) Desired dynamic characteristics.

Hydraulic drives. Hydraulic systems can range in size up to hundreds of HP; they operate with efficiencies that are close to that of DC motors and are used in relatively high-powered NC systems.

Hydraulic systems develop much higher maximum angular accelerations than those of electric motors of the same peak power. They have small time constants, which results in smooth operation of the machine tool slides. They are, however, more expensive and cause some inconvenience by working with oil lines rather than with wires.

DC motors. DC motors are commonly used for stepless control of speed. The DC motor, since its voltage and speed can be varied smoothly and easily regulated, is the most commonly used drive for NC machine tool systems. However, the main supply is usually a 3-phase AC type; but none of the various types of AC motor have such suitable and continuous control possibilities. The rectifiers are used today are quite simple devices.

Stepping motors. The stepping motor is an incremental digital control device which translates an input pulse sequence into a proportional angular movement, rotating one angular increment, i.e. step for each input pulse. The shaft position is determined by the number of pulses, and its velocity is determined by the pulse frequency. The shaft speed in steps per second is equal to the incoming frequency in pulses per second (pps). The stepper motor is able to operate in open-loop control systems, which are the simplest forms of incremental servo-systems.

Actuation systems: Corresponding to each of the control functions of an NC machine tool, an actuation system acts on receiving a suitable command signal from the MCU. Though electromechanical, hydraulic and pneumatic types of actuation system are available, the electromechanical type is most common on NC machines. Whatever may be the type used, an efficient, stiff and responsive actuation system, which is virtually free from backlash, influences favourably the accuracy of its NC machine tool. The elements of actuating mechanisms for the slides, i.e. for positional control are as follows:

1. Stepper motor/servomotor;
2. Ball screw and nut with support bearings; harmonic drives;
3. Feedback devices on closed loop systems;
4. Linear bearings/linear motion systems wherever the sliding motion of guideways is desired to be changed to rolling friction.

The group of above-mentioned elements used to control the position of a machine slide is also known as a *servo*. Servo control is basically approached in two ways—either as an open loop servo system or as a closed loop servo system.

Control loops: As described later in Section 21.10, NC machine tools are classified on the basis of open loop or closed loop control systems. The actuation system is that part of the mechanism of an NC system which sets the machine in action, and the control loops are the medium of transmitting information for such an action. A few important characteristics of these loops are described here; for detailed description, see Section 21.10.

Open loop servo systems. In open loop system, generally the stepping motors are employed as the driving component to provide machine-slide motion. This motor responds in incremental steps for 10–20 V (each impulse) of the digital input signal; its rotor moves precisely by one step. Usually a step angle of 1.8° is very common. Thus, one step of stepper motor rotates the screw by 1/200th of a revolution. This corresponds to a linear movement of 0.02 mm for a screw with a lead of 4 mm, if the stepping motor is directly coupled with the screw.

The open loop application is generally restricted to smaller machines, because of the limited power output availability with the stepping motors (the maximum being normally around 5 kW and a torque of 200 Nm). Further, the number of pulses per second restrict the speed of the drive. A typical maximum for stepping motors is 800 pulses per second.

Closed loop servo systems. A block diagram of a closed loop positioning control system is shown in Fig. 21.26. In this case, the control system makes use of a position transducer, referred to as a *feedback device*. This sensing device measures the actual position of the slide. The difference between the required position and the actual position is detected by the comparator circuit, and the necessary action is taken within the servo to minimize this difference/error. Such a control system can typically have capabilities of up to 0.0001 mm resolution and speeds of up to 20 m/min.

Recirculating ball screws. These are widely used for both open and closed loop systems to convert the rotary motion of the motor and the screw into the linear motion of the slide. The balls keep circulating between the nut and the screw, providing rolling friction and thereby a low coefficient of friction (see Fig. 21.6). The efficiency of these screws is rarely below 90%. It is very high as compared to that of the sliding friction type lead screws, which is rarely above 25%. An additional advantage of the ball screws is that these can be preloaded. This is helpful in reducing backlash in the two-directional motion of the slide which is attached to the screw.

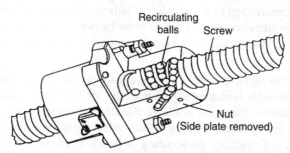

Fig. 21.6 Recirculating ball screw.

The other elements of the actuation system such as feedback devices and transducers are devices routinely described in standard textbooks. Hence it is needless to take up the description of these elements over here.

Tool and work holding devices: The machining centre is a machine tool capable of performing several different machining operations on a work part in one set-up under program control. The machining centre is capable of milling, drilling, reaming, taping, boring, facing and other similar operations. In addition, the features that typically characterize the NC machining centre include the following:

Automatic tool changing capability. A variety of machining operations means that a variety of tools are required. The tools are contained in the machine in a tool drum. When a tool needs to be changed, the tool drum rotates to the proper position and an automatic tool changing mechanism, operating under program control, exchanges the tool in the spindle and the tool in the drum.

Automatic work part positioning. Most machining centres have the capability to rotate the job relative to the spindle, thereby permitting the cutting tool to access four surfaces of the part.

Pallet shuttle. Another feature is that the machining centre has two (or more) separate pallets that can be presented to the cutting tool. While machining is being performed with one pallet in position in front of the tool, the other pallet is in a safe location away from the spindle. In this way the operator can unload the finished part from the prior cycle and fixturing the raw work part for the next cycle while machining is being performed on the current workpiece. Some of the shuttles are shown in Fig. 21.7.

21.5 COORDINATE SYSTEM IN NC MACHINE TOOLS

The sequence of tool transfer operations is shown in Fig. 21.7.

The coordinate system for designating the axes is the conventional 'right hand coordinate system' as shown in Fig. 21.8(b). The labelling of the axes is carried out using a right hand coordinate system where the fingers of the right hand are aligned with the positive X-axis and are then rotated (through the smaller angle) toward the positive Y-axis; then the thumb of the right hand points in the direction of the positive Z-axis. Otherwise, the orientation is a 'left hand coordinate system'.

NUMERICAL AND COMPUTER NUMERICAL CONTROLLED MACHINES 489

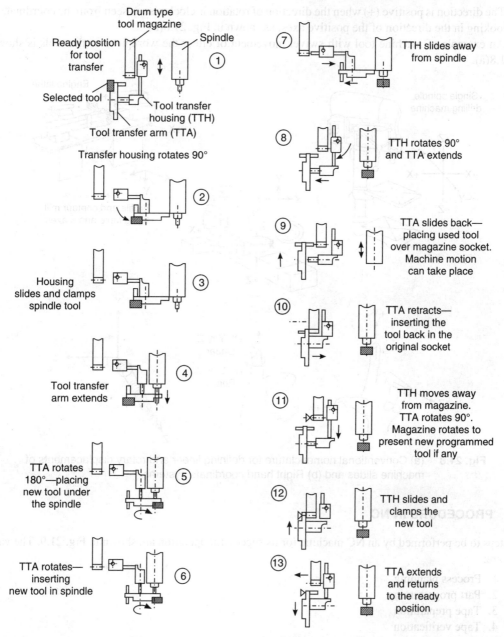

Fig. 21.7 Sequence of tool transfer operations.

The right hand coordinate system is also known as the 'clockwise rotating coordinate system'. The reason for this is the sequence of the axis definitions: If the X-axis is rotating in the direction of the Y-axis, the movement is the same as if a screw is being turned in the Z-direction as shown in Fig. 21.8(b).

Some machining operations require programming angles of rotation about one or several coordinate axes. The rotations about the coordinate axes are identified by the angles of rotation A, B and C as shown in Fig. 21.8(b).

The direction is positive (+) when the direction of rotation is clockwise as seen from the coordinate zero point looking in the direction of the positive axes, as shown in Fig. 21.8(b).

An example of machine tool with swivel movement of either the workpiece or the tools is shown in Fig. 21.8(a).

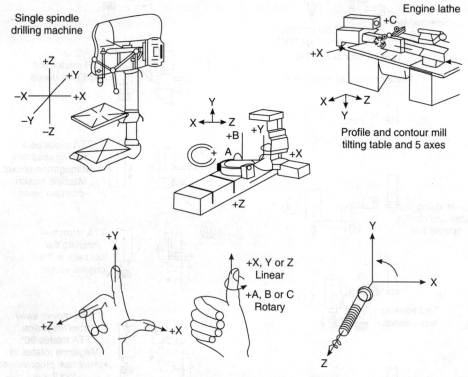

Fig. 21.8 (a) Conventional nomenclature for defining linear and rotary displacements of machine slides and (b) Right hand coordinate systems.

21.6 PROCEDURE IN NC

The steps to be performed by an NC machine for its successful operation are shown in Fig. 21.9. The various steps are:

1. Process planning
2. Part programming
3. Tape preparation
4. Tape verification
5. Production.

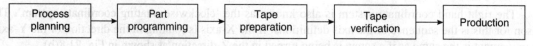

Fig. 21.9 Block diagram representing NC procedure.

NUMERICAL AND COMPUTER NUMERICAL CONTROLLED MACHINES 491

Process planning: Process planning is concerned with preparation of route sheet. The route sheet describes the sequence of operations to be performed on the work part. It also lists the machines through which the part must be routed to complete the sequence of operations.

Part programming: It is an important software element in the NC manufacturing system. It looks like a computer printout containing a number of lines/statements/instructions (called *NC blocks*). It is therefore the plan proposed for machining the part. It is written keeping in view the various standard words, codes and symbols. It is dependent on the machine tool hardware and the MCU. Some NC blocks written in the word address format punched on a paper tape are shown below:

N030 G81 Z-85

N040 G17

N050 G83 × 200 Y 200 I-100 J0

Tape preparation: In older NC systems, a tape is prepared according to the part program written. In manual part programming, a tape is prepared on a typewriter-like device equipped with tape punching capabilities. In modern NC systems, the part program prepared by computer can be directly fed to the controller of the machine, which converts into machine-understandable language the mechanical actions of slides and tools.

Tape verification: After the punched tape has been prepared, the tape is verified for checking the accuracy. During the verification process, the computer plots the various tool movements or table movements on paper. If any error is found, a new tape is prepared and again verified till errors are eliminated. Sometimes foam or plastic material is used for this tryout.

Production: Production is the final and useful step in NC procedure. It involves different operations on different machines, so that the raw material gets converted to the finished component. The tools and fixtures are set up. The operator loads the work part on the table and specifies the part program to be run to the machine control unit. Then the machine performs the necessary operations according to the part program. When the part is completed, the operator unloads it from the machine table and loads the next work part for machining.

21.7 STRUCTURE OF NC PROGRAM

An NC program consists of blocks of information which correspond to a particular step in the process operation. The working instructions contained in a block are divided into 'words' consisting of a combination of identifying letters (address) and a series of digits with or without a sign.

The meaning and order of presentation of such words are given in the programming instructions of the particular NC control system. Normally the presentation is given by a block of words preceded by an address character. A block is of the form

N001 G54 X 0000750 Y 0010000 Z 0015000 F 00600 0009000 S.. T..M03

where N is the sequence number;
       G are preparatory functions;
       XYZ are primary motion dimensions;
       F are feed functions;
       S are spindle speed functions;
       T are tool functions;
       M are miscellaneous functions.

The number of digits and the implied decimal point positions are specified in the input format of the particular control system.

The sequence of operations are as follows:

(a) Read an input block from the punched paper tape.
(b) Decode it and send the words of information, distance to be moved, feed rate, spindle speed etc., to the appropriate part of the machine control unit.
(c) Different parts of the hardware obey the following instructions using built-in logic to control the sequence: Select current tool (T), set spindle speed (S), switch coolant on (M), and then move axes in synchronism (interpolation) to make a straight line or circular arc cut for specified distance (X, Y, Z).
(d) On completion of moves, dwell while the next input block is read in and so on till M00 (Program Stop).

The process of tape preparation can be either manual or computer-aided, depending upon whether the shape or contour is simple or complex. In manual part programming, coordinate information is extracted from the component drawing, and written out on a process layout sheet with miscellaneous information, to regulate the operation of the machine tool. The required information is then punched on the control tape; each block or line of information is numbered in sequence for reference and search purposes.

To enable set functions, e.g. start, stop, etc. to be readily programmed, the coordinate information is supplemented by code numbers. The control medium commonly used is one-inch wide paper tape, on which information is punched in eight tracks in binary code. Each character is represented by a combination of holes and number of holes in one row across the tape. As the tape is reached, a pulse is created for each punched hole, these pulses then being transmitted through the control system and amplified to actuate the machine tool motions. A number of alphanumeric codes are in use, of which the two most widely used are the Electronics Institute of America (EIA) and the American Standard Code for Information Interchange (ASCII) which is the reverse of the International Standards Code (ISO) (see Fig. 21.10).

The EIA tape format, which has now become obsolete, is a seven-bit odd parity code with the parity bit in the fifth track. ASCII is a seven-bit even parity code with the parity bit in the eight-track. Both are eight-track tapes with identical physical dimensions.

21.8 TAPE FORMAT

The compiling of NC data into suitable blocks of information for the machine control systems follow these standard formats:

1. Fixed sequential format
2. Word address format
3. Tab sequential format.

In the fixed sequential format, the instructions in the blocks are always recorded in the same sequence. All instructions are given in every block, including those instructions which remain unchanged from the preceding block.

In the word address format, this being the most widely used format, each word is preceded and identified by a character called an *address*. The system enables instructions, which remain unchanged from the preceding block, to be omitted from succeeding blocks.

The tab sequential format requires the instructions to be punched in the same sequence, and each word is preceded by a tab character. If instructions remain unchanged in succeeding blocks, the instructions need not be repeated (see Fig. 21.11).

NUMERICAL AND COMPUTER NUMERICAL CONTROLLED MACHINES 493

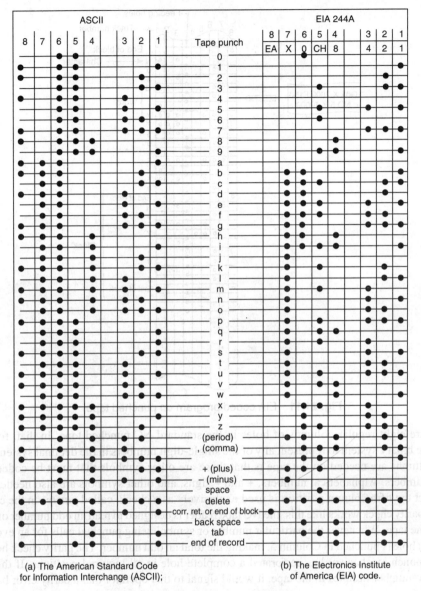

(a) The American Standard Code for Information Interchange (ASCII);
(b) The Electronics Institute of America (EIA) code.

Fig. 21.10 Types of tape code.

Tape coding: The coding of the tape is provided by either the presence or absence of the hole in various positions, because these are two possible conditions for each position—either the presence or the absence of a hole. This coding system is called the *binary code*. It uses the base 2 number system, which can represent any number in the familiar base 10 or decimal system. In the binary system there are only two numbers—0 and 1. The meaning of successive digits in the binary system is based on the number 2 raised to successive powers.

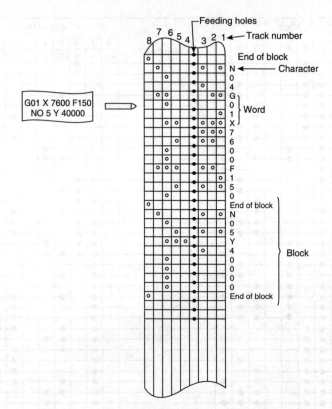

Fig. 21.11 EIA coded program on punched tape.

There are eight regular columns of holes in the standard NC punched tape. In that, four digits are required in the binary system to represent any of the single-digit numbers in the decimal system. The reason why eight columns are needed on the tape is that these are other symbols that must be coded on the tape besides the numbers. Alphabetical numbers, + and – signs, and other symbols are also needed in NC tape coding. In fact the fifth column position is used exclusively as a check, called *parity*, on the correctness of the tape. The parity check works like this: the NC tape reader is designed to read an odd number of holes across the width of the tape. Whenever the particular number or symbol being punched calls for an even number of holes, an extra hole is punched in Column 5, making the total an odd number. The parity check helps to assure that the tape punch mechanism has perforated a complete hole in all required positions. If the tape reader counts an even number of holes in the tape, it would signal to that operator that a parity error has occurred.

Magnetic tape: Certain types of NC machine tools use digital magnetic tape as a memory device which stores the program of the part. The information may be recorded on the tape explicitly or in a coded form.

Usually the magnetic tape has seven channels. Recording information in a coded form means that the magnetic pulses on the tape correspond precisely to the holes on the first 7 tracks. In the standard perforated tape, since when using the EIA code the 8th track is punched only in the End of Block (EOB) character (which is noted as a single hole in the 8th track), only the EOB character should be described differently on a 7 channel magnetic tape.

NUMERICAL AND COMPUTER NUMERICAL CONTROLLED MACHINES 495

The input media for MCU is shown in Fig. 21.12.

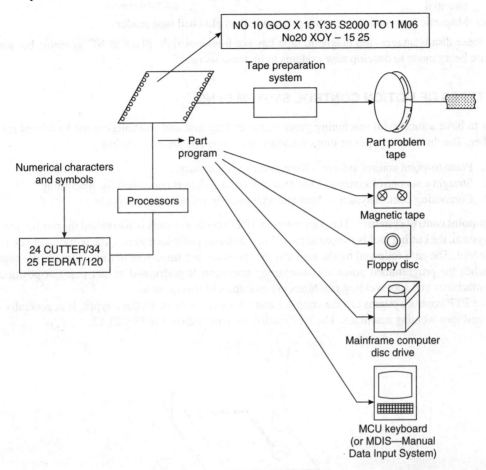

Fig. 21.12 Input media for MCU.

Usually the information on a magnetic tape is not written in a coded form. That is, the pulses used for advancing the arcs of motion are produced directly from the tape, where each recorded pulse is an instruction for moving one basic length unit, e.g. 0.001 inch or 0.01 mm. Each axis of motion has a separate channel on the magnetic tape. By increasing the density of pulses in a certain channel, the feed along the appropriate axes is increased. By changing the reading speed, the cutting feed can be varied.

The magnetic tape is inserted into a reader which reads the tape at a constant speed, and the tape is advanced with the actual working procedure. The cutting time being longer, the length of the tape will increase.

Compared with the perforated tape method, the Data Process Unit (DPU) of the NC Unit can be elevated, as the need for decoding circuits and interpolators is avoided. Therefore such systems are called *non-interpolating* or *off-line interpolator* systems, and are less expensive.

On the other hand, a magnetic tape-controlled system has the following disadvantages:

(a) The magnetic tape has to be produced by a computer. Therefore, it does not seem to be economical for use on a point-to-point system.

(b) Errors in programming can be corrected only with difficulty and the use of a computer seems to be essential.
(c) Magnetic tape readers are more expensive than punched tape readers.

Due to these disadvantages, the magnetic tape has not found its right place in NC systems; but presently, efforts are being made to develop new systems to increase its use.

21.9 TYPES OF MOTION CONTROL SYSTEM IN NC

In order to have a successful machining process, the cutting tool and workpiece must be moved relative to each other. The three basic types of motion control system in NC are given below:

1. Point-to-point control system—No contouring capability.
2. Straight-cut control system—One-axis motion at a time is controlled for machining.
3. Contouring control system—Multiple axes are controlled simultaneously.

Point-to-point control system: The point-to-point (PTP) control system is also called the *positioning system*. In this system, the cutting tool is moved at predefined point-to-point locations in the working envelope of the machine tool. The path followed by the tool and the speed are not important in this system. Once the cutting tool reaches the programmed point, the machining operation is performed at that point or position. CNC drilling machines and spot welding machines are examples of this system.

The PTP control systems are the simplest and least expensive in all three types. It is generally used in drilling and spot welding machines. The PTP control system is shown in Fig. 21.13.

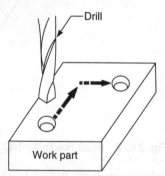

Fig. 21.13 Point-to-point (positioning) control in NC.

Straight-cut control system: The straight-cut control system is also known as the *linear control system*. In this system, the cutting tool is moved along any one axis at a time. The path followed by the cutting tool is important in this system. With this type of control system, it is not possible to control movement in more than one axial direction at a time. Therefore, angular cuts are not possible in this system. The straight-cut control system is shown in Fig. 21.14(a) and (b). An example of this system is the NC lathe machine.

Contouring control system: The contouring control system is also known as the *continuous-path NC system*. In this system, the cutting tool is capable of performing motion in more than one axis simultaneously. The path of the cutting tool is continuously controlled to generate the desired geometry of the work part. This system is capable of generating straight or plane surface at all orientations, angular surfaces, two-dimensional curves, and three-dimensional contours, as well as conical shapes in the work parts. The continuous or contouring control system is shown in Fig. 21.15(a) and (b). An example of this system is the three-axis milling machine.

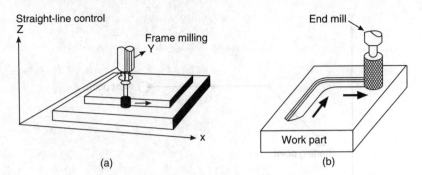

Fig. 21.14 Straight cut control in NC.

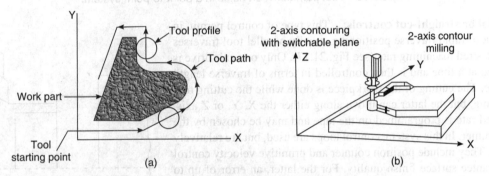

Fig. 21.15 Continuous path (contouring) control in NC. (Note that the cutting tool path must be offset from the part-out line by a distance equal to its radius.)

21.10 CRITERIA FOR CLASSIFICATION OF NUMERICAL CONTROLLED SYSTEMS

The classification of NC machine control tool systems can be done in four ways:

(a) According to type of machine control: point-to-point/positioning, straight-cut and contouring (or continuous path);
(b) According to structure of control circuits: analogue or digital;
(c) According to programming: incremental or absolute;
(d) According to type of control loops: open-loop or closed-loop.

According to type of machine control

Point-to-point control: Point-to-Point (PTP) is also called the *positioning system*. This type of control permits positioning of the tools along programmed points at rapid traverse rate, with the cutting tool not in engagement. The path of the cutting tool and its federate while travelling from one point to next are of no significance. Therefore, the system would require only the position counter for controlling the final position of the tool upon reaching the programmed point. The path from the starting point to the final position is not controlled (see Fig. 21.16).

The simplest example of point-to-point NC machine tool is a drilling machine. In a drilling machine the workpiece is moved along the axes of motion till the centre of the hole to be drilled is exactly beneath the drill. Then the drill is automatically moved towards the workpiece (with spindle speed and feed rate which can be controlled), and the drill moves out in a rapid traverse speed. The workpiece moves to a new point and the above sequence of actions is repeated.

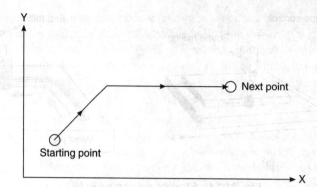

Fig. 21.16 Tool path between holes in a point-to-point system.

Paraxial or straight-cut controls: This type of control permit, in addition to rapid traverse positioning, axially parallel tool traverses at the desired machining rate (see Fig. 21.17). Only one axis drive is operated at a time and is then controlled in terms of traverse length and rate. The cutting of the workpiece is done while the cutting tool is moving, but the latter can move along either the X, Y, or Z axis. The feed rate is programmed on the tape and may be chosen by the programmer. In this system, control loops are used, but are relatively simple. They include position counter and primitive velocity control to guarantee surface finish quality. For the latter, an error of up to ±5% from the programmed feed is still permitted. An interpolator is not required for the straight-cut system, since no simultaneous operation of the axes is required. Straight-cut controls are used on simple milling machines and lathes.

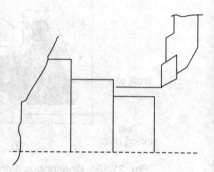

Fig. 21.17 Straight-cut system.

Contouring controls: This type of control permits positioning at rapid traverse rate, axially parallel feeds, and feed traverses to any arbitrary point on the workpiece, e.g. along straight and circular paths as shown in Fig. 21.18. All axes of motion might move simultaneously, each at a different velocity. When a nonlinear path is required, the axial velocity changes even within the segment. For example, cutting a circular contour requires a sine rate of change in one axis, while the velocity of the other axis is changed at a cosine rate.

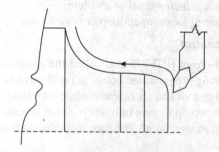

Fig. 21.18 Contouring (continuous path) system.

In contouring machines, the position of the cutting tool at the end of each segment, together with the ratio between the axial velocities, determines the desired contour of the part, and at the same time the resultant feed also affects the surface finish. Each axis of motion is equipped with a separate position loop and counter. Dimensional information is given on the tape separately for each axis, and is fed through the Data Process Unit (DPU) to the appropriate position counter. The programmed federate, however, is that of the contour and has to be processed by the DPU, in order to provide proper velocity commands for each axis. This is done by means of an interpolator. The function of the interpolator is to obtain intermediate points lying between those taken from the drawing.

Interpolation: The method by which contouring systems move from one programmed point to another is called *interpolation*. This is a feature that merges the individual axis commands to a predefined tool path. Basically, there are three types of interpolation: linear, circular and parabolic. Most CNC controls provide both linear and circular interpolation. A few controls use parabolic interpolation.

Linear interpolation. In linear interpolation (straight-line interpolation) as illustrated in Fig. 21.19(a), the interpolator within the control unit calculates a chain of points along a straight (linear) connection line between two tool positions. With this method of programming, any straight-line path, can be traced which will include all taper, cuts. During tool movement from point-to-point, the axis movements are continuously corrected, so that the tool does not deviate from these points by more than the permitted tolerance. With linear interpolation, the circle or arc must be broken down into a number of straight-line segments. When the segment becomes smaller, the circle or arc will be smoother as shown in Fig. 21.19(b). Linear interpolation control requires that the ends of each of these segments be provided.

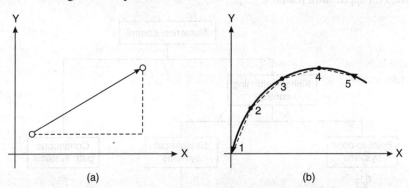

Fig. 21.19 (a) Linear interpolation and (b) Continuous path approximation.

Circular interpolation. In circular interpolation, the CNC system calculates a chain of points along the desired circular path between two tool positions. These points are then used to correct axis movements in such a way that the tool does not deviate from the precise circular path beyond the prescribed tolerance. Circular interpolation is illustrated in Fig. 21.20.

The programming of circles and arcs has been greatly simplified by the development of circular interpolation. Arcs up to 90° may be generated, and these may be further joined together to form a half, three-quarter or full circle as desired. Circular interpolation is limited to one plane at a time. Thus a circle can be traced in the X–Y plane, the X–Z plane or the Y–Z plane, but not in a combination of planes. A contouring control can also be used as a straight-cut control, and a straight-cut control as a point-to-point control, but not vice versa.

Parabolic interpolation. Parabolic interpolation (see Fig. 21.21) is used to provide approximations of free-form curves using higher-order equations. They require considerable computational power and are not as common as linear and circular interpolations.

500 TEXTBOOK OF PRODUCTION ENGINEERING

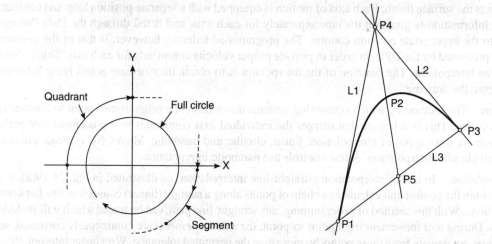

Fig. 21.20 Circular interpolation.

Fig. 21.21 Parabolic interpolation.

NC manufacturing system based on position control system: Figure 21.22 gives the NC systems classification based on application features.

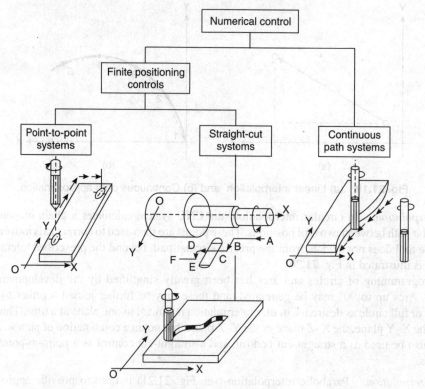

Fig. 21.22 Classification of NC systems.

With a point-to-point system, the machine performs machining operations in specific positions and does not affect the workpiece while moving from one point to the next. An example is that of a drilling machine where coordinates are positioned by the system.

Straight-cut or straight-line system is an extension of the point-to-point control system, with the provision of straight-cut milling capability. This is obtained by providing movement at controlled feed rate in one axial direction at a time. Instances of this are used in face milling, pocket milling, picture frame milling, stepped turning on lathe, etc.

The continuous-path NC system calls for continuous, simultaneous and coordinated, or in other words, interrelated motions of the tool and workpiece along different coordinate axes. This enables profiles, contours and curved surfaces to be machined. This means that the position of several slides must be controlled on the machine tool, which means that their relative positions and velocities must be established at every point and continuously throughout the entire operation.

According to structure of control circuits

The NC system according to structure of control circuit may be divided into:

(a) Analog system
(b) Digital system

In an analog control system the quantities may be varied continuously, while in digital systems the quantities are varied discretely such as in case of the presence or absence of a quantity. The shortest cycle of presence and absence is the resolution of the digital system and establishes its accuracy. In digital systems of machine tools, each such cycle of information provides a voltage-pulse where each pulse represents a basic length unit which determines the system resolution. Therefore, theoretically the digital system is of a finite accuracy, while analog information can accept any value. The accuracy achieved in the analog system depends on the accuracy of the various components used to construct the electronic circuits and on the precision of sensors which are used.

Both digital and analog controls are used in NC systems of machine tools. Deciding the control type may sometimes be a little difficult, as any digital control contains an analog component and vice versa. The input to NC systems is always digital, as the dimensions that are taken from the drawing are given in numbers, which is a digital form. On the other hand, the output of a NC system is always analog, as the slides of the machine tool are moving in a continuous and smooth form. Therefore, each one of the system types contains an analog and digital information unit. However, the type of control, digital or analog, is referred to by the type of information appearing at control loop inputs. Whenever a sequence of pulses is applied, the control is digital, and if the input is continuously variable, the control is analog.

According to programming

Increment system: An incremental system (see Fig. 21.23) is one in which the reference point to the next instruction is the end-point of the preceding operation. Each dimensional data is applied to the system as a distance increment, measured from the preceding point at which the axis of motion was present.

Incremental dimensions are distances between adjacent points (see Fig. 21.23b). These distances are converted into incremental coordinates, by accepting the last dimension point as the coordinate origin for the new point. This may be compared to a small coordinate system, i.e. shifted consequently from point-to-point (P1, P2, through P9) as shown in Fig. 21.23(a).

Absolute system: An absolute system (see Fig. 21.24) is one in which all moving commands are referred to one reference point, which is the origin, and will be called the *zero point*. All position commands are given as absolute distances from that zero point.

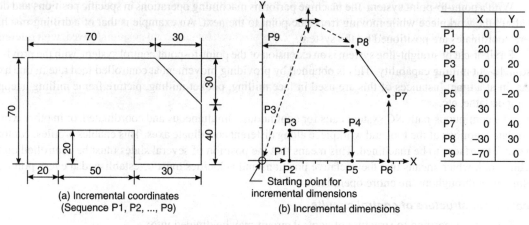

Fig. 21.23 Incremental system.

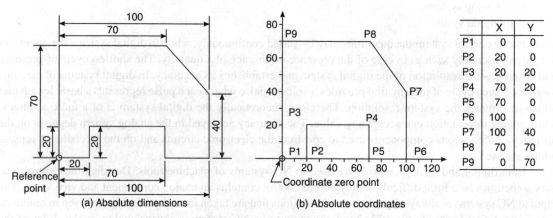

Fig. 21.24 Absolute system.

According to type of control loops

Open-loop control system: Every control process, including NC systems, may be designed as an open or a closed-loop control. The term *open-loop control* means that since the loop is open, there is no feedback and the action of the controller has no reference to the result it produces (see Fig. 21.25).

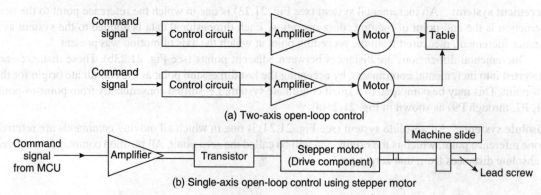

Fig. 21.25 Open-loop machine slide control system.

When a command signal is given to the drive unit to perform the motion through a certain distance, the slide moves for the distance desired; this is irrespective of whether the moving member arrived at the desired position. This is an open-loop NC system, where there is no check on the actual position arrived, with reference to the position desired and as directed by the system.

Closed-loop control system: Figure 21.26 shows a closed-loop control for a single axis of motion. Closed-loop control systems measure the actual position and velocity of the axis, and compare them with the desired references. The difference between the actual and the desired values is the error. The control is designed in such a way as to eliminate the error or to reduce it to a minimum.

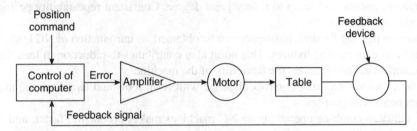

Fig. 21.26 Closed-loop system for one axis.

In NC systems, both the input to the control loop and feedback signals may be a sequence of pulses. The digital comparator correlates the two sequences and gives, by means of a digital-to-analog converter (DAC), a signal representing the position error of the system, which is used to drive the de-servomotor. The feedback device, which is an incremental encoder, supplies a pulsating output. The incremental encoder consists of a rotating disc divided into segments, which are alternately opaque and transparent. A photocell and a lamp are placed on both sides of the disc. When the disc rotates, each change in light intensity falling on the photocell provides an output pulse. The rate of pulses per minute provided by the encoder is proportional to the revolutions per minute of the leadscrew. Such a control can typically have capabilities of upto 0.0001 mm resolution and speeds upto 10 m/min.

Extreme care must be taken during the design of a closed-loop control system. By increasing the magnitude of the feedback signal (more pulses per revolution of the leadscrew), the loop is made more sensitive.

The command signal is constantly compared with the feedback position signal. The error which is the difference between the command reference and the actual position is fed through an amplifier to the actuating drive system. When the required position is reached, the error will become zero and the drive will stop. Built-up tolerance and wear in a machine tool can be reason of high inaccuracy in an open system, but are nearly put aside in the closed loop.

For contouring NC systems, closed-loop drives are to be preferred for obvious reasons. Straight-line contouring can actually be carried out with open-loop systems, but in general a closed-loop NC will be more effective, more accurate and more suitable for nearly all operations.

21.11 ADVANTAGES AND LIMITATIONS OF NC MACHINES

Higher productivity in small-batch production, reduced costs, higher repetitive accuracy, and comparatively less skill required in the operation of NC machines are among the principal advantages of the application of NC machines. The introduction of NC, CNC, DNC and CAM is tending to restructure the metal working industry in a big way.

The following are the specific advantages in the usage of NC machines:

(i) The machine can switch over to different jobs, as set-up times are very low. Tool setting is done on the presenting devices and off the machine. Cutting speeds, feeds, depths of cut, dimensions, etc. are stored in the form of tape.

(ii) Increased effective utilization of machine is achieved, as all non-cutting movements of the slide are performed at a rapid feed rate of 5 to 10 m/min, and dimensional measurements are eliminated.

(iii) Greater quality control is there, since the dimensions of a component are produced by automatic movement of the slides, and hence the mid-tolerance can be achieved for all the dimensions.

(iv) Eliminates rework and scrap to a very great degree. Consistent repeatability performance of the machine ensures this.

(v) Savings in jigs and fixtures, cost/space can be obtained, as introduction of NC leads to elimination of jigs and complicated fixtures. This point also contributes to reduction in lead time to produce the component and increases the flexibility of the machine.

(vi) Machines with contouring controls dispense with form tools and templates required to produce components with profiles.

(vii) The ability to combine operations on NC machines makes production faster, and eliminates the waiting time of components in between machines and stage inspection.

(viii) Change in design can be easily incorporated, as it means only the change of tape.

(ix) In-process inventory gets reduced due to reduction in lead time and faster set-up time.

(x) Reduced floor space, number of men, and handling results in better management control over production.

(xi) Development work gets faster with the use of NC machines.

(xii) Cost accounting and production control becomes very precise.

(xiii) Dependence on skilled operators can be dispensed with.

(xiv) Optimum utilization of HP of the machine.

(xv) Cutting technology is stored in taped form.

There are a number of problems, inherent in conventional NC, which have motivated tool builders to seek improvements in the basic NC system. Among the difficulties encountered in using conventional NC machines are the following:

NC training requirements: NC manufacturing requires training of personnel for software as well as hardware. Part programmers are trained to write instructions in desired languages for the machines on the shopfloor. They also need to be acquired along with the manufacturing process. Similarly, machine operators have to be prepared for the new NC culture. These factors are important for the successful adoption and growth of NC technology.

NC manufacturing establishment and running costs: The cost of NC manufacturing set-ups could be several times more than that of their conventional counterparts. As NC is a complex and sophisticated technology, it also requires higher investments for maintenance in terms of wages of highly skilled personnel and expensive spares. The automatic operations of NC machines itself implies relatively higher running costs. Moreover, the requirements of airconditioned environment for operating NC technology adds further to the running cost.

Part programming mistakes: In preparing a punched tape, part programming mistakes are common. The mistakes can be of either syntax or numerical form, and it is not uncommon for three or more passes to be required before the NC tape is found correct. Another related problem in part programming is to achieve the optimum sequence of processing steps. This is mainly a problem in manual part programming.

Punched tape: Paper tape is especially fragile, and susceptibility to wear and tear causes it to be an unreliable NC component for repeated use on the shopfloor; more durable tape materials such as Mylar and aluminium foil are utilised to help overcome this difficulty; however, these materials are relatively expensive.

21.12 COMPUTERS AND NC MACHINES

It is in this context that the dedicated minicomputers came to the market as Computer Numerical Controlled (CNC) systems. The inherent feature of memory in the computer numerical control system made the new technology software-oriented. This system is flexible, unlike the rigid hardwired circuit NC systems used hitherto.

The appearance of the microprocessor on the scene has brought a new range of CNC systems which are highly cost-effective. These, along with the semiconductor memory devices, have reduced the size of the system as compared to the core memories used in the earlier minicomputers. The distinguishing feature of the minicomputers—the relative capability for faster processing—is gradually disappearing, as microcomputers are steadily entering the arena.

Computers have been associated with NC right from its inception, and over the years the interaction of computers with NC has given rise to the concepts of Adaptive Control (AC), Computer Numerical Control (CNC), and Direct Numerical Control (DNC), and the trend now is towards integrated manufacturing systems. A few other field terms that have been developed in this direction are computer-aided design (CAD) and computer-aided manufacturing (CAM).

Computer numerical controlled (CNC) systems

In the initial stages the computer was associated as an offline data processing aid in the computer programming of parts. However, from 1970 onwards, computers have joined NC manufacturing both as offline support and online data processing controls. Powerful programming languages such as APT III (Automatically Programmed Tools) developed have made NC programming easy for complex jobs, and the popularitry of APT and APT-derived processes demonstrated the acceptance of computers in NC manufacturing during the 1960s itself.

However, the later computers' small size but high power in processing ability occupied the place of NC systems as hardware, giving rise to CNC systems, and the phenomenon of the computer as a central unit controlling an array of NC/CNC machines came to be known as Direct Numerical Control (DNC). Thus the computer established itself as the main tool for the future in the integrated manufacturing system.

NC systems used on machine tools can be broadly classified as:

(a) Hard-wired or conventional NC systems
(b) Computer Numerical Controlled (CNC) or soft-wired NC systems

In the hard-wired system the entire data input and data handling sequence including control functions are determined only by the fixed circuit interconnections of decision elements and storage devices. Hence the hard-wired NC becomes a right system functionally, where any additions or changes in feature need a corresponding addition or change in system circuitry.

Conversely, in the Computer Numerical Controlled system, a dedicated and stored computer program is used to perform all the basic NC functions as per the control program stored in the memory of the computer, called the *Executive Program*. Thus, the machine control data comes directly from the computer memory and not from the continuously read tape as in the case of the hard-wired NC system. Also, the executive program resident in the memory of the computer makes such a system flexible in that a new executive can be input (through paper tape), and the fact that the control logic for the machine tools is generated by a software program rather than by wired logic circuits. Figure 21.27 is the schematic of a typical modern CNC system.

506 TEXTBOOK OF PRODUCTION ENGINEERING

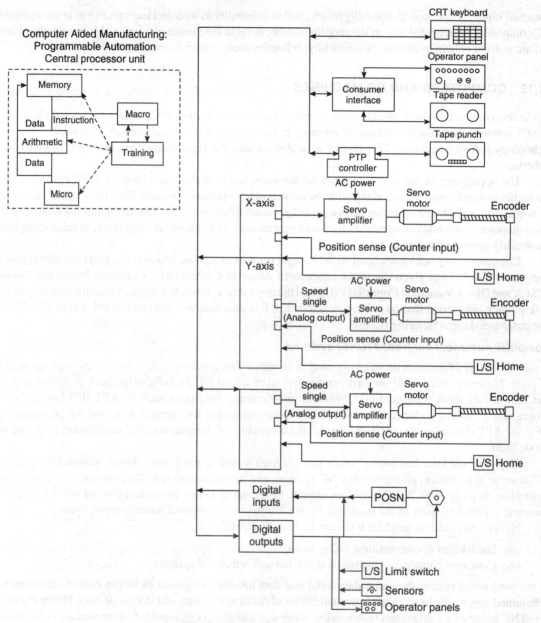

Fig. 21.27 CNC system layout.

Because of the memory availability, users can also store part program in the computer. The capability to edit these part programs is a real advantage to the part programmer right at the machine station. By using the ability of the computer, additional features such as tape punching and user-oriented sub-routine program are possible on the CNC. Diagnostic programs available on the CNC make the troubleshooting function easier. Flexibility, increased capability and reliability, together with reduction in cost over the years, make CNC today the most popular system.

Due to advancement in microelectronics, the current generation of CNC systems use VLSI and microprocessor-based computers. Quite a few of the microprocessor-based CNC systems currently use semiconductor memory elements. These have made CNC machine tools available to the small and general engineering shops.

Another distinct advantage of the CNC system is that it lends itself easily for adaptation to the DNC set-up or to the integrated manufacturing system. This is so because of the ability of the minicomputer to communicate with a larger central computer of the DNC system.

Direct numerical controlled (DNC) systems

The association of computer with NC right from its inception stage is quite noteworthy. The inclusion of a minicomputer within the NC system came much later than the involvement of digital computers with NC manufacturing.

As early as 1967, the Japanese manufacturer FANUC developed a computer-operated machining group control system called *System K*. Under this group control system, seven NC lathes were controlled using the part program, in their loading, batch quantity regulation and delivery schedules. As more and more computers came to be used in such a way, the term Direct Numerical Control became popular. By 1974 there were 70 such DNC installations in Japan. In the USA. too, DNC was catching up fast, pioneered by NC machine tool system builders and computer manufacturers. However, DNC installation, being expensive and requiring highly skilled software knowledge, could not be made easily except in large corporations. Figure 21.28 shows the typical block diagram of a DNC installation.

The main computer (A) is the one used for part programming and for other data processing operations. With this is interlinked the process control computer (B), which is assigned to control a group of NC systems and machine tools (C and D) through data transmission links (I_1, I_2, I_3, etc.). The process control computer (B) with its own data files can store active programmes and supply to the NC systems as required. New programmes can be introduced through the tape reader (E) or provided by the main plant computer (A).

Two distinct approaches were noticed in the DNC field. The first, in which the process control computer feeds the program and functions such as character check, decoding, buffer store, internal interpolations, comparators, etc., were retained in the NC system. In other words, the conventional NC system, excepting the tape reader input, remained there as part of DNC.

The second possibility is the transfer of tasks normally performed by the NC system, such as decoding, interpolation/comparisons, etc. direct to the process control computer. This calls for a powerful computer. This in the early stages accounted for savings in the cost of hardware at individual NC systems. However, the question of reliability and flexibility of production in the event of a major computer breakdown remained a big question. This to a great extent has been overcome later on by the introduction of CNC systems, where the hardware cost also came down substantially.

The ongoing development of NC/CNC/DNC points to the future towards the integrated manufacturing system wherein design, planning, scheduling and manufacturing will all be integrated in the automated factory. Industrial robots will play an important role in this link-up.

The major challenge of the near future in the field seems to be in the area of software technology. The need to harness the full potential of the powerful computers that are being built now would make the highest demand on the expertise of the software engineers. Another aspect would be the adoption of more and more digital controls, even on small and simple machine tools and other process applications. This would be feasible because of cost reduction in the electronic elements used.

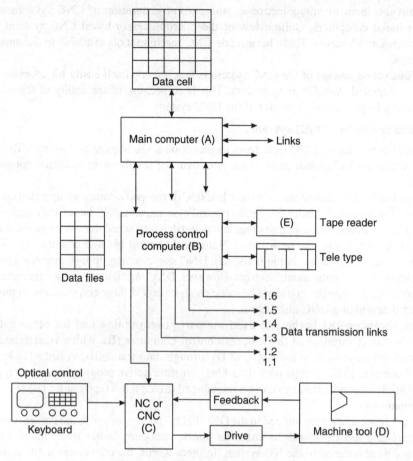

Fig. 21.28 Block diagram of DNC.

Already, many leading industrial houses have started use of NC/CNC in their manufacturing plants. The recently installed equipment are showing good results in production. Some of the leading machine tool builders have started manufacturing CNC machine tools and a few of these have been installed in the metal working plants. However, fear, mental blocks and resistance to change will need to be eliminated before NC can be made popular in India. Signs of growth of this technology are already evident. The import of indigenous NC machine tools in the past 4 to 5 years have accounted for ₹ 75–100 lacs in purchase value. With the growth of electronics in this country, NC holds great promise for the future of manufacturing in India.

21.13 COMPUTER AIDED PART PROGRAMMING (ON CNC LATHE MACHINE HAVING FANUC CONTROL)

Miscellaneous and preparatory functions and codes on CNC machines

Miscellaneous functions (M codes): These are instructions describing miscellaneous functions like calling the tool, spindle rotation, coolant on, etc. Table 21.2 shows the codes and their corresponding functions.

Table 21.2 Miscellaneous functions (M codes)

M Codes	Function
M00	Program Stop
M01	Optional Stop
M02	Program End
M03	Spindle Forward
M04	Spindle Reverse
M05	Spindle Stop
M06	Tool Change
M08	Coolant On
M09	Coolant Off
M10	Vice Open
M11	Vice Close
M12	Spindle Forward and Coolant On
M13	Spindle Reverse and Coolant On
M14	Drill Extend
M25	Drill Extend
M26	Drill Retract
M30	Program Rest and Reward
M38	Door Open
M39	Door Close
M62	Output 1 On
M63	Output 2 On
M64	Output 1 Off
M65	Output 1 Off
M66	Wait for Input 1 On
M67	Wait for Input 2 On
M76	Wait for Input 1 On
M77	Wait for Input 2 On
M98	Subprogram Call
M99	Subprogram Exit

Preparatory function (G codes): A2-digit number following address G determines the meaning of the command of the block concerned (see Table 21.3). The G codes are divided into the following two types:

Type	Meaning
One-shot G codes	The G code is effective only at the block in which it was specified.
Modal G codes	The G code is effective until another G code in some group is commanded.

Note on G *codes.* These are:

(a) Maximum spindle speed setting (G50) is valid when the constant surface speed control option is provided.
(b) The G codes marked * are set when the power is turned on.

(c) The G codes in group 00 are not modal. They are effective only in the block in which they are specified.
(d) A number of G codes can be specified in a block even if they do not belong to some group. When a number of G codes of some group are specified, the G code specified last is effective.
(e) All the G codes may not apply to each machine.

Table 21.3 Preparatory function of G codes

G Code	Group	Function
G00*	1	Positioning (Rapid traverse)
G01	1	Linear Interpolation (Feed)
G02	1	Circular Interpolation (CW)
G03	1	Circular Interpolation (CCW)
G03	0	Dwell
G20	6	Inch data input
G21	6	Metric data input
G28	9	Reference point return
G32	1	Thread cutting
G40*	7	Tool nose radius compensation cancel
G41	7	Tool nose radius compensation left
G42	7	Tool nose radius compensation right
G50	0	Work coordination, Change maximum spindle speed setting
G70	4	Finishing cycle
G71	4	Stock removal in turning
G72	0	Stock removal in facing
G73	0	Pattern repeating
G74	0	Peck drilling in Z-axis
G75	0	Grooving in X-axis
G76	0	Thread cutting cycle
G90	1	Cutting cycle A
G92	1	Thread cutting cycle
G94	1	Cutting cycle B
G96	2	Constant surface speed control
G97*	2	Constant surface speed control cancel
G98	11	Feed per minute
G99*	11	Feed per revolution

The functions of the post-processor are:
1. Reading CL data.
2. Converting cutter location points to machine tool co-ordinates.
3. Checking if side movements exceed the limits of travel of the particular machine.

4. Checking that specified speeds and feeds do not exceed the capabilities of the machine tool.
5. Correcting any parameter which is beyond limits in a machine tool.
6. Assigning suitable G and M codes.
7. Calculating tool axis orientations in multi-axis machining.
8. Preparing suitable output operation sheets.
9. Providing list of errors and warnings if necessary.
10. Calculating machining time and length of tape.

Computer aided part programming languages: Several part programming languages are available today, and among them the APT (Automatically Programmed Tools) system is the most widely used and comprehensive. It was first developed at the Servomechanisms Laboratory of MIT in 1955. The language was further developed by MIT and later by IIR 1 and CAM–I. The present version of APR is APT IV. Some of the other popular languages are listed in Table 21.4.

Table 21.4 List of part programming languages

Program	Developer	Features
APT	MIT/II TRI/CAM-I	P, C
ADAPT	IBM	P
AUTOSPOT	IBM	P
CINTURN	Cincinnati Milacron	T
COMPACT–II	MDSI	P, 2 1/2C
EXAPT–I	TH, Aachen	P
EXAPT–II	TH, Aachen	T
EXAPT–II	TH, Aachen	3C
GENTURN	General Electric	T
MULTURN	Metal Institute, The Netherlands	T
NEL 2PL	Ferranti	P
NEL 2C	Ferranti	T
NEL 2CL	Ferranti	2C
PROMPT	Weber	C
NELAPT	NEL	C
SPLIT	Sundstrand	C
UNIAPT	United Computing Co/MCAUTO	P,5C
FATPT	FANUC	P,C
UCC–APT	University Computing Company	T,C

P—Point-to-point control
C—Contour control
T—Turning

Geometric statements in APT. The APT language permits the definition of a variety of geometric entities, with the added capability that each of the allowable types can be defined in several ways. For instance, a point in space can be defined simply by listing the X, Y and Z coordinates, or as the intersection of two lines, or still as one of the two possible points of intersection of a line and a circle. The geometric types allowed, together with the APT vocabulary work used to indicate each, are given in Table 21.5. Figure 20.29 shows various steps involved in the processing of information in APT.

Table 21.5 Geometric types and vocabulary use

Geometric type	Vocabulary word
Point	POINT
Line	LINE
Plane	PLANE
Circle	CIRCLE
Cylinder	CYLINDER
Ellipse	ELLIPSE
Hyperbola	HYPERBOLA
Cone	CONE
General Conic	GCONIC
Loft Conic	LCONIC
Vector	VECTOR
Matrix	MATRIX
Sphere	SPHERE
Quadric	QUADRIC
Tabulated Cylinder	TABCYL
Polyconic Surface	POLCON
Ruled Surface	RLDSRF

An example of these parameters is

FEDRATA

which is a scalar or a floating point number. This statement indicates the desired feed rate that should be used for best finish, chip removal, etc. The actual feed rate will be as close to the desired rate as possible without exceeding the dynamic limitations of the machine tool.

ON
FLOOD
COOLANT/MIST
TAPKUL
OFF

The first four modifiers turn the coolant on in a variety of ways; the last modifier specifies that the coolant is to be turned off.

Four of the post-processor words with modifiers affect other parts of the processor, in addition to the post-processor. They are MACHINE, INDEX, COPY and TRACUT.

The APT processor will allow any combination of vocabulary modifier words, floating point numbers, arithmetic expressions, and scalar symbols as minor modifiers, except where an alphanumeric string of characters or other restrictions are specified.

21.14 PART PROGRAMMING IN APT EXAMPLE

Figure 21.29 illustrates a part for which a program for contour machining is to be written. The part program manuscript written to produce this part is also given in the figure.

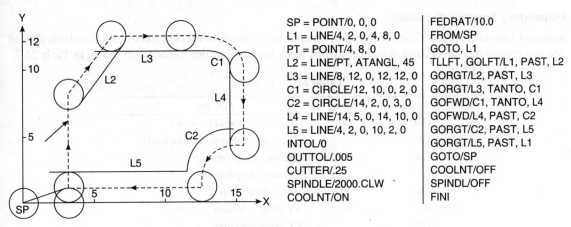

Fig. 21.29 Illustration for part programming.

21.15 MISCELLANEOUS AND PREPARATORY FUNCTONS AND THEIR CODES FOR CNC MILLING

Miscellaneous function (M codes)

When a 2-digit figure is specified following the address M, a 3-digit BCD code signal and a strobe signal are transmitted. These signals are used for ON/OFF control of a machine function such as tool change, spindle rotation change, and coolant ON/OFF (see Table 21.6). One M code is specified in one block. The selection of M codes for functions varies with the machine tool builder.

Table 21.6 M codes

M code	Function	M code	Function
M00	Program Stop	M62	Output 1 On
M01	Optional Stop	M63	Output 2 On
M02	Program End	M64	Output 1 Off
M03	Spindle Forward	M65	Output 1 Off
M04	Spindle Reverse	M66	Output 2 Off
M05	Spindle Stop	M67	Wait Input 1 On
M06	Tool Change	M70	X Mirror On
M08	Coolant On	M71	Y Mirror On
M09	Coolant On	M76	Wait for Input 1 Off
M10	Vice Open	M77	Wait for Input 2 Off
M11	Vice Close	M80	X Mirror Off
M13	Coolant, Spindle Forward	M81	Y Mirror Off
M14	Coolant, Spindle Reverse	M98	Subprogram Call
M30	Porgram Stop and Rewind	M99	Subprogram Exit

Preparatory function (G codes)

A number following address G declaring the meaning of the command of the concerned block G codes is divided into two types—one-shot G and modal G—which were described earlier. G codes are shown in Table 21.7.

Table 21.7   G codes

G code	Group	Function
G00*	01	Positioning (Rapid traverse)
G01*		Linear interpolation (cutting feed)
G02*		Circular interpolation
G03		Circular interpolation
G04	00	Dwell, Exact stop
G17*	02	XY plane selection
G18		ZX plane selection
G19		YZ plane selection
G20	06	Input in inch
G21		Input in mm
G28	00	Return to reference point
G40*	07	Cutter compensation cancel
G41		Cutter compensation left
G42		Cutter compensation right
G43	08	Tool length compensation + direction
G44		Tool length compensation – direction
G49*		Tool length compensation cancel
G73	09	Peck drilling cycle
G74		Counter tapping cycle
G76		Fine boring
G80*		Canned cycle cancel
G81		Drilling cycle, spot boring
G82		Drilling cycle, counter boring
G83		Peck drilling cycle
G84		Tapping cycle
G85		Boring cycle
G86		Boring cycle
G87		Back boring cycle
G88		Boring cycle
G89		Boring cycle
G90*	03	Absolute command
G91		Incremental command
G92	00	Programming of absolute zero point
G94	05	Feed per minute
G95		Feed per rotation
G98*	10	Return to initial point in canned cycle
G99		Return to R point in canned cycle

Illustrative examples on NC programming

ILLUSTRATION 21.1 Write and create NC programs for the new components as per Fig. 21.30 given below.

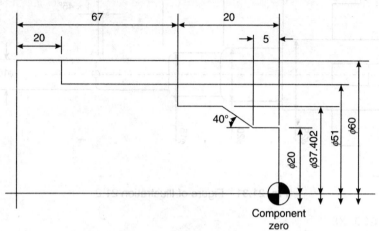

Fig. 21.30 Figure of Illustration 21.1.

Stock: 65 mm dia. Dimensions in mm
The above example is to be faced and then turned.

```
    N10  G00 X62 Z0
    N20  G01 X0 F0.1
    N30  G00 X62.0 Z1.0
    N40  G71 U1.5 R0.5
    N50  G71 P60 Q130 U1.0 W0.15 F0.1
    N60  G01 X20.0 F0.4
    N70  Z-5.0 F0.1
    N80  X37.402 W-10.369
    N90  Z-20.0
    N100 X51.0
    N110 Z-63.75
    N120 X60.0
    N130 Z-83.75
    N140 G70 P60 Q130
```

ILLUSTRATION 21.2 Write and create NC programs for the new components as per Fig. 21.31 given below.

```
     [BILLET X25 Z65.................BILLET SIZE
    N10   G99 G21 G40 G97 S2000 M13..............Set cutting conditions
    N20   M06 T0101...............Tool call
    N30   G00 X 26 Z2
    N40   G01 Z0 F0.3................Face and retract
    N50   X-1 F0.1
    N60   G00 X25 Z1
    N70   G71 U1.5 R0.5..................Set parameters for
    N80   G71 P90 Q190 U1 W0.1 F0.125...............Canned cycle
```

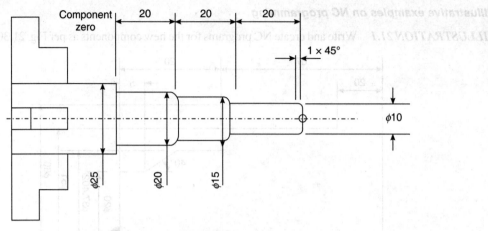

Fig. 21.31 Figure of Illustration 21.2.

```
N90     G00 X8
N100    G01 Z0 F0.1
N110    X10 Z-1 F0.05
N120    Z-20 F0.1
N130    X14
N140    X16 Z-21 F0.05
N150    Z-40 F0.1
N160    X18
N170    X20 Z-41 F0.05
N180    Z-60 F0.1
N190    X25
N200    G70 P90 Q190..................Finish profile
N210    G28 U0 W0..................Reference point return
N220    M30..................Program reset and rewind
```

ILLUSTRATION 21.3 Write and create NC programs for the new components as per Fig. 21.32 given below.

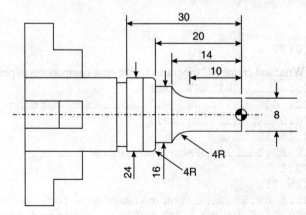

Fig. 21.32 Figure of Illustration 21.3.

```
[BILLET X25 Z35................BILLET SIZE
G99 G21 G41 G97 S2500 M13................Set cutting conditions
M06 T0101................Tool Call
G00 X26 Z2
G01 Z0 F0.3................Face and retract
  X-1 F0.1
G00 X25 Z1
G71 U1.5 R0.5 Set parameters for canned
G71 P1 Q2 U1 W0.1 F0.125  Cycle
N1 G00 X8
G01 Z0 F0.1
G02 X16 Z-14 R4.0
G01 Z-20
G03 X24 Z-24 R4.0
G01 Z-30
N2 X25
G28 U0 W0
M06 T0202 S3000
G00 X25 Z1
G70 P1 Q2
G28 U0 W0
M06 T0303 S2000
G00 X26 Z-30
G01 X-2 F0.075................Part off
G00 X26................Rapid retract
G28 U0 W0................Home position
M30................Program reset and rewind
```

ILLUSTRATION 21.4 Write and create NC programs for the new components as per Fig. 21.33 given below.

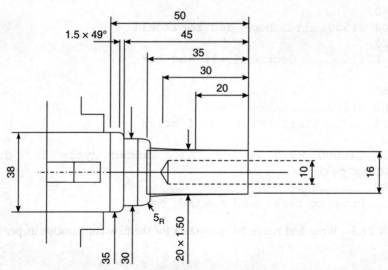

Fig. 21.33 Figure of Illustration 21.4.

```
[Billet X38 Z55............Billet size
G99 G21 G40 G50 S3000............Set cutting conditions, Max RPM
G96 S150 M13............Constant surface speed, start
M06 T0101      Spindle, Coolant On, Tool call.
G00 X39 Z2............Rapid to start position
G01 X-1 F0.125............Face and retract
G00 X38 Z1
G71 U1.5 R0.5 Set parameters for canned
G71 P1 Q2 W0.1 F0.125............Cycle
N1 G00 X16
G01 Z0 F0.1
X20 Z-20
 Z-35
G03 X30 Z-40 R5.0
G01 Z-45
X32
X35 Z-46.5
Z-50
N2 X38
G28 U0 W0............Home position and tool call
M06 T0202
G00 X43 26
G01 G42 X38 Z1 F0.3............Positioning, and applying cutter compensation
G70 P1 Q2 ............Finishing cycle
G28 G40 U0 W0............Cancel cutter compensation and home
G97 M06 T0303 S900............Tool call and change speed to RPM
G00 X20.25 Z0............Rapid to thread start position
G76 P030060 Q100 R0.05
G76 X-18.76 Z-33 P0920 Q250 F1.5............Threading cycle
G28 U0 W0
M06 T0404 S1500............Home and tool call
G00 X0 Z2
G01 Z06 F0.1............Centre drill and home
G00 Z2
G28 U0 W0
M06 T0505 S1200............Tool call
G00 X0 Z2............Start position of drill
R74 R1.0
G74  Z-30 Q10000 R0.0 F0.125............Canned cycle for drilled hole
   including peck
G28 U0 W0............Home
M30............Program reset and rewind.
```

ILLUSTRATION 21.5 Write and create NC programs for the new components as per Fig. 21.34 given below.

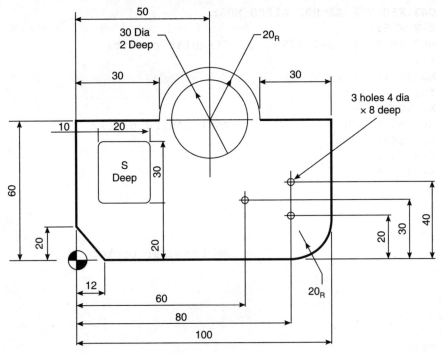

Fig. 21.34 Figure of Illustration 21.5.

```
[BILLET X110 Y90 Z10;
[TOOLDEF T1 D8.0 Z0;            Used in offline software only
[TOOLDEF T2 D6.0 Z0;
[TOOLDEF T3 D4.0 Z0;
00003;                          Program number
G91 G21 G28 X0 Y0 Z0            Home position
G49 G40 M06 T01;                Tool call
G90 G43 G00 X-10 Y-8 Z5 H01;    Tool positioning
G01 Z-4 S1000 M03 F100;         Blocks
G41 X0 H09;
Y60 F80;
X30;
G17 G02 X70 Y60 I20 J0;
G01 X100;
Y20;                            Outside profile
G02 X80 Y0 R20;
G01 X12;
X0 Y20;
X-8;
G00 Z5 M05;
G91 G28 X0 Y0 Z0;               Home position
G49 G40 M06 T02;                and tool change
```

```
G90 G43 X50 Y62 Z2 H02 A1200 M03;
G01 Z-2 F35;
G91 G03 X0 Y0 I0 J-2 F75        Circular pocket;
G01 Y5;
G03 X0 Y0 I0 J-7;
G01 Y5;
G03 X0 Y0 I0 J-12;
G90 G01 Z5 F1000;               Positioning blocks
G00 X27 Y47;
G01 Z-5 F35;
G91 X-14 F75;
Y-5;
X-14;
Y-5;
X-14;
Y-5;                            Rectangular pocket
X-14;
Y-5
X14;
Y-5;
X14;
G90 Y47;
X13;
Y23;
X27;
G00 Z5 M05;
G91 G28 X0 Y0 Z0;_____Home position
G49 G40 M06 T03;_____Toolchange
G90 G43 X60 Y30 Z10 H03 S1000 M03;__Posi. 1st hole
G99 G83 X60 Y30 Z-8 Q4 R2 F100;_____Hole cycle
X80 Y40;                        Positioning blocks
Y20;                            for remaining holes
G00 G80 Z20 M05;
G91 G28 X0 Y0 Z0;_____Home position
M30;
```

REVIEW QUESTIONS

1. Explain the essential difference between point-to-point and continuous path types of numerical controlled machine tools.
2. What is the function of the planner or process engineer in part programming?
3. List some of the important advantages of numerical controlled machining with automated tool changing capablities.
4. What is numerical control? List the basic components of a numerical controlled system.
5. Give a brief description of how numerical control works.
6. What are the different G codes and M codes used for numerical controlled machining?

OBJECTIVE TYPE QUESTIONS

Tick the most correct response:

1. The instruction on the tape of the NC machine is prepared in
   - (a) Numeric form
   - (b) Coded form
   - (c) Binary coded decimal form.

2. The numerical control machine tool is operated by
   - (a) Feedback system
   - (b) Numerical controls
   - (c) Series of coded instructions.

3. Numerical control can be applied to
   - (a) Lathe
   - (b) Drilling machine
   - (c) Milling machine
   - (d) All of the above.

4. The numerical controlled machines are controlled by a tape whose width is
   - (a) 20 mm
   - (b) 30 mm
   - (c) 50 mm.

5. In an NC machine, programmed instructions are stored on
   - (a) Punched tape
   - (b) Head box
   - (c) Graphic terminal.

6. Which of the following names is associated with NC machines?
   - (a) Machining centre
   - (b) Head box
   - (c) Graphic terminal.

7. Based on the control system feature, NC machines can be classified as
   - (a) Point-to-point system
   - (b) Straight line system and continuous path system
   - (c) Point-to-point system, straight line system and continuous path system.

8. The machine tool in which the point-to-point numerical controlled system is applied is the
   - (a) Drilling machine
   - (b) Grinding machine
   - (c) Milling machine.

ANSWERS

Objective Type Question

1. (c) 2. (c) 3. (d) 4. (a) 5. (a) 6. (a) 7. (c) 8. (a)

CHAPTER 22

Gear Cutting, Broaching and Thread Cutting

22.1 INTRODUCTION

Gears are used for transmission of power in automobiles, industrial motion and machinery, machine tools, engines, etc. With increasing tindustrialization, the gears are now required to render free operation to ensure high load bearing capacity at a constant velocity ratio. Wear and fatigue strength of the gear tooth are the criteria determining durability and reliability. These depend to a great extent on the production technique and the material of the gear.

22.2 VARIOUS METHODS OF GEAR PRODUCTION

The various methods are as follows:

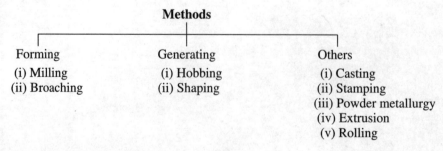

22.3 VARIOUS KINDS OF GEARS

Figure 22.1 shows the configuration and meshing of various kinds of gears which are in common use.

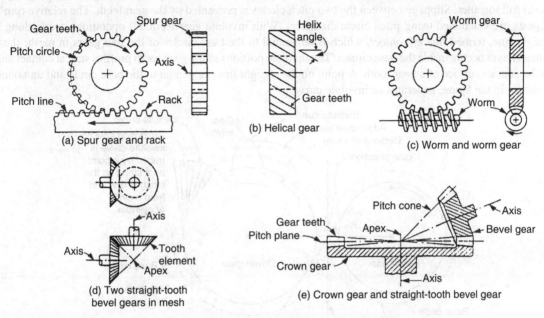

Fig. 22.1 Sketches illustrating important kinds of gears.

Table 22.1 shows the standard tooth proportions. These proportions are functions of module of a gear tooth.

Table 22.1 Standard tooth proportions

Gear terms	Proportions
Addendum	1.00 m
Dedendum	1.25 m
Tooth thickness	1.570 m
Tooth space	1.570 m
Working depth	2 m
Total depth	2.25 m
Clearance	0.25 m
Pitch diameter	Z m
Outside diameter	(Z + 2) m
Root diameter	(Z + 2.5) m
Fillet radius	0.4 m

where, m = module of gear tooth and
Z = Number of gear teeth

Module m is a measure of size of gear tooth
For automotive applications m = 4 to 5
For watches it is very small, m need not be a round number it could be in fraction also like, 5y, 4.3 mm.

22.4 INVOLUTE GEAR TOOTH FUNDAMENTALS

As shown in Fig. 22.2, the pitch circles (which are the two theoretical imaginary circles of the two gears in mesh) roll together. Slippage between the two pitch circles is prevented by the gear teeth. The relative rpm's of gears are calculated using pitch circle diameters. With involute gear teeth, all operations occur along a straight line, termed *line of contact*, which is tangential to the base circles of the two gears in mesh; thus, contact never occurs inside the base circles. The involute portions of the gear-tooth profiles start at contact and on the contact surface of a gear tooth. A point on the straight line, tangential to the base circle and unwound as shown in the figure, generates an involute curve.

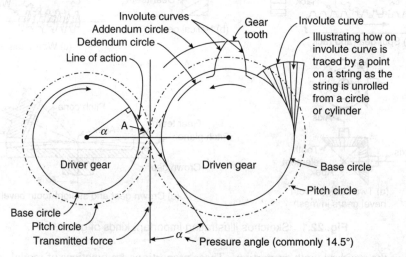

Fig. 22.2 Fundamentals of involute gears.

The diameter of the pitch circle of a gear is known as its *pitch diameter*. The gear tooth size, normally expressed as diametral pitch DP, is obtained as follows:

Diametral pitch (DP) = Number of teeth on the gear/pitch diameter in inches

This is an important relationship in gearing.

22.5 GEAR TEETH MANUFACTURING

The basic methods for machining gear teeth include the forming–cutting and generating methods. The *forming–cutting method* is usually used on a horizontal milling machine with the gear blank indexed between cuts with a dividing head. Each tooth space is cut with a form cutter making the gear teeth change slightly as the number of teeth on the gear is changed; theoretically different forms of cutters should be used for cutting gears with different numbers of teeth, In practice a set of cutters, including 8 and sometimes 15 cutters, is used to cut all numbers of teeth from 12 up to rack. A set of form cutters is arranged so that the gear teeth will be cut as correctly as possible. The forming–cutting method is used when only small quantities of gears are cut and less accuracy is needed, because this method is relatively slow. Gear milling is shown in Fig. 22.3.

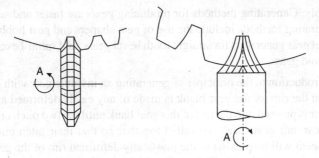

Fig. 22.3 Gear milling.

The set of 8 cutters are:

Cutter No. 1	23 teeth and more
Cutter No. 2	55 teeth to 134 teeth
Cutter No. 3	35 teeth to 134 teeth
Cutter No. 4	26 teeth to 134 teeth
Cutter No. 5	21 teeth to 134 teeth
Cutter No. 6	17 teeth to 134 teeth
Cutter No. 7	14 teeth to 134 teeth
Cutter No. 8	12 teeth to 134 teeth

The following additional cutters are included in another set:

Set of 15 cutters:

Cutter No. 1½	80 teeth to 134 teeth
Cutter No. 2½	42 teeth to 54 teeth
Cutter No. 3½	35 teeth to 34 teeth
Cutter No. 4½	23 teeth to 25 teeth
Cutter No. 5½	20 teeth to 19 teeth
Cutter No. 6½	19 teeth to 15 teeth
Cutter No. 7½	13 teeth to 12 teeth

The areas of application are:

1. All types of gears, i.e. spur, helical, worm, and in special circumstances, bevel gears, can be produced on milling machines.
2. Gear cutting can be carried out on the conventional type of milling machine, which is normally available in a modern workshop.
3. This method can be employed for both roughing and finishing operations.
4. It is economical and suitable for one-off and small batches.

The limitations are:

1. Milling is not a production process.
2. The pitch accuracy is very much dependent upon the dividing head. The tooth form is not accurate.
3. Internal teeth can not be cut.

Generating methods for producing gear teeth make use of certain relative motions between the work gear and the cutter during the machining. Generating methods produce theoretically correct gear tooth profiles with one cutter, regardless of the number of teeth desired on the gear. A cutter for a given diametral pitch may be used to cut gears with any desired number of teeth. All teeth will have theoretically correct profiles, and all gears will mesh interchangeably. Generating methods for producing gears are faster and suited for the production of large quantities. Generating methods include the use of gear shapers and gear hobbers for spur and helical gears, straight tooth bevel gear generators for straight tooth bevel gears, and spiral-bevel and hypoid generators for spiral-bevel and hypoid gears.

Gear shaping: An introduction to the principle of generating an involute gear with gear shaper is provided in Fig. 22.4. Assume that the rim of the gear blank is made of any easily deformed and mouldable material. An ordinary involute gear is pressed into the rim of this gear bank until the two pitch circles are just in contact. At the same time the gear and gear blank are rolled together so that their pitch circles roll on each other without slippage. Gear teeth will be pressed in the plastically deformed rim of the gear blank, and they will be theoretically correct involute gear teeth. The gear teeth plastically formed in the gear blank rim will mesh correctly with the teeth of any other involute gear of the same diametral pitch and pressure angle, regardless of the number of teeth of the gears, and can be rolled using this cold working process. This is sometimes done proactively. It is included here to explain the principle of generation.

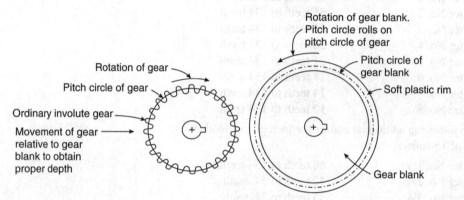

Fig. 22.4 Theoretical example to introduce generating action used by gear shapers. (Ordinary involute gear producing theoretically correct involute gear teeth in easily deformed rim of gear blank.)

In gear shaping illustration practices as shown in Fig. 22.5, the shaping tool gear, which was used above—in the two methods, are the same. The cutter and the gear blank are connected kinetically, so that they will roll together as the cutter reciprocates for the cutting. First the cutter must cut to the desired depth. The cutter and gear blank then rotate slowly together, as the gear teeth are cut in the gear blank. Slightly more than one complete revolution of the work gears is required. Often, two separated cuts are made completely around a work gear. Cutters to be used in roughing and finishing processes for producing helical gears should have helical teeth, and as a cutter reciprocates, a helical guide which is installed on the machine causes the cutter to twist during its cutting strokes, so that the cutter teeth follow helical paths.

Gear hobbing: Theoretically, the gear hobbing process is similar to that used for gear shaping, and it is illustrated in Fig. 22.6. Assume, for this purpose, that the gear blank has a rim made of a material which can be readily, plastically and drastically deformed. As shown, an ordinary involute rack is pressed into the soft plastic gear blank. As this is done, the rack is moved lengthwise and the gear blank is rotated, so that it rolls with the rack without slipping. At the proper depth, the pitch circle of the gear blank just contacts and rolls on the pitch line of the rack. Theoretically correct involute gear tooth profiles are formed in the plastically deformed rim of the gear blank. The number of teeth obtained depends upon the pitch diameter of the work gear. The pitch or size of the gear teeth produced on the work gear will be the same as that of the rack used.

GEAR CUTTING, BROACHING AND THREAD CUTTING

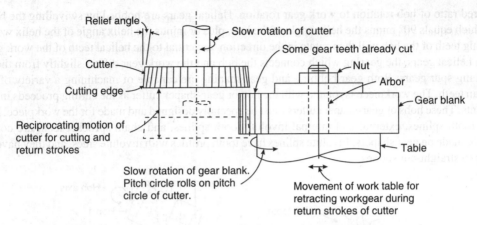

Fig. 22.5 Schematic sketch showing operation of gear shaper.

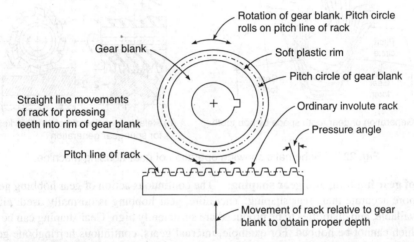

Fig. 22.6 Theoretical example to introduce generating action used by gear hobbing machines. (Ordinary involute rack produces theoretically correct involute gear teeth in easily deformed rim of gear blank.)

The above method is not practical for commercial use, but is presented merely as a theoretical introduction to gear hobbing. In gear hobbing, the rack becomes a cutter, and the soft plastic rim is not needed on the gear blank. The cutter, known as a *hob*, basically accomplishes the same result as the rack used above. The generating action of the rack is diagrammatically represented in Fig. 22.7, that shows the successive position of the gear relative to the gear blank developed involute profile.

A hob has a helical thread and thus resembles a worm. Since the sides of the rack teeth are straight lines, hob threads are relatively easy to produce and inspect. Gashes parallel to the axis are cut across the thread of a hob at convenient intervals to provide cutting teeth. These cutting teeth are relieved intervals behind their cutting edges for clearance, like in a forming milling cutter. Rotation of the hob in effect causes the theoretical rack to move along a straight-line path, as shown. Thus a rotating hob represents a continuously moving rack as it cuts, like in a milling cutter. When hobbing a spur gear as shown, the hob on its spindle must be swivelled to an angle which equals 90° minus the helix angle of the hob thread; this is done so that the hob cutting teeth rotates in the spaces of the gear teeth being cut. The hob spindle and work-gear arbor are connected to give the desired ratio of hob rotation to work gear rotation. Helical gears are hobbed by swivelling the hob to an angle which equals 90° minus the helix angle of the hob, plus or minus the helix angle of the helix work gear; the cutting teeth of the hob will then rotate in the direction tangential to the helical teeth of the work gear. For hobbing helical gears, the gearing which connects the hob and the work gear differs slightly from those used for hobbing spur gears. Both gear hobbing and gear shaping are capable of machining a variety of profiles other than teeth. The workpiece is rotated with the hob or gear-shaper cutter as the cutting proceeds in suitable increments. These hobs or gear-shaper cutters must be specially designed and made for the workpiece. External straight-tooth splines, external and internal involute-tooth splines, and chain sprocket teeth are commonly hobbed or made on gear shapers. Involute splines have tooth profiles with involute sides, and they have proven superior to straight-cut splines.

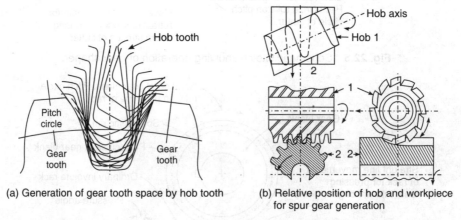

(a) Generation of gear tooth space by hob tooth

(b) Relative position of hob and workpiece for spur gear generation

Fig. 22.7 Schematic showing operation of gear hobbing machine.

Comparison of gear hobbing and gear shaping: The continuous action of gear hobbing generally makes it faster and more accurate than gear shaping. Therefore, gear hobbing is normally used; either process is suitable and available, and quantities to be produced are sufficiently high. Gear shaping can be used for some workpieces which cannot be hobbed. For example, internal gears, continuous herringbone gears, and parts with gears positioned close to another gear or a shoulder must be gear-shaped; internal splines in blind holes must be machined with a gear shaper.

Machining of bevel gears by generation process: The key to understanding the generation and cutting process of bevel gears is as follows:
- A circular face plate termed *cradle* represents an imaginary crown gear which meshes with the gear blank to be cut and slowly rolls within.
- The cradle has the cutter mounted on it, as shown in Fig. 22.8.
- The cutting tool blade reciprocates and removes metal from the gear blank to produce the gear tooth. Figure 22.8 shows the generation of straight-tooth bevel gear, whereas Fig. 22.9 shows the generation of curved-tooth bevel gear by the Gleason process; a single tooth of the face mill represents a single tooth of an imaginary crown gear which meshes with the workpiece gear blank.

Gear finishing: Gear-cutting machines are not capable of finishing gears to the close tolerances demanded after gear cutting; it is rare that a gear is entirely free of all errors in index, tooth profile, eccentricity, surface smoothness, and helix angle. The purpose of gear finishing is to eliminate or substantially reduce these possible errors, so that gears run quietly and smoothly when transmitting large amounts of power at high rotational speeds. After gear cutting, only a few thousandths of an inch of excess material should remain on the gear teeth

GEAR CUTTING, BROACHING AND THREAD CUTTING

for finishing. Gear-finishing operations include gear shaving and gear burnishing for gears which are not to be hardened, and gear shaving, gear lapping, or gear grinding for gears which are to be hardened. A commonly used sequence for gear finishing include gear shaving, surface hardening, and gear lapping.

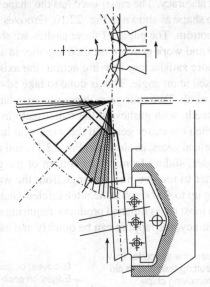

Fig. 22.8 Cutting straight-tooth bevel gear in reciprocating tool bevel gear generator.

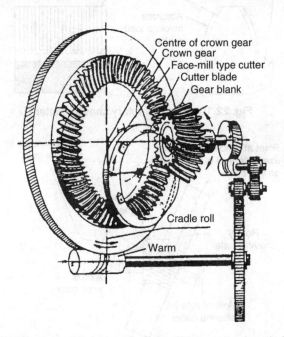

Fig. 22.9 Cutting curved-tooth bevel gears with face-mill type cutter in spiral bevel gear generator.

Gear shaving: The most economical means for correcting gear-cutting errors is gear shaving, which may be accomplished by two or more slightly different methods. The method which is briefly described here is known as *rotary shaving*, and it is applicable for shaving super and helical gears, either external or internal. Rotary shaving gives gears an overall high accuracy. The cutter used has the shape of a spur gear which meshes with the work gear. The cutters, made to shape as shown in Fig. 22.10, Grooves are cut on the sides of each cutter gear tooth, extending from top to bottom. The edges of these gashes are sharp and so they are able to shave the teeth of a work gear. The cutter and work gear are meshed together in a gear-shaving machine, as shown in Fig. 22.10. In order to obtain more satisfactory shaving action, the axis of the cutter and the axis of the work gear are not parallel but crossed at an angle. This is done to take advantage of the sliding which takes place between two gear teeth in the mesh. Two spur gears in parallel axes have a small amount of unavoidable sliding from top to bottom of the teeth, with gashes on the cutter teeth as shown in Fig. 22.10, practically as shaving would be carried out, when the cutter gear and the work gear have parallel axes. However, if the cutter, the work gear, or both, are helical gears, as shown in Fig. 22.11, and their axes are set at an angle equal to the difference of their helix angles, sliding from top to bottom of the gear teeth occur diagonally. This allows the cutting edges of the cutter to remove fine shavings from the work rotated entirely by the cutter. Rotational speeds are high, reaching up to 400 sfpm at the pitch circles. Gears which have been rotary-shaved and hardened by heat treating are in most cases finished products, requiring no further finishing. If it is desired to lap these gears for increased accuracy, the lapping can be quickly and easily done.

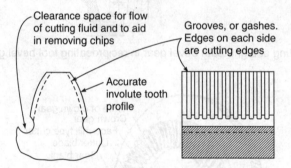

Fig. 22.10 Tooth of rotary shaving cutter.

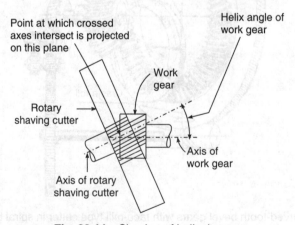

Fig. 22.11 Shaving of helical gear.

22.6 BROACHING OPERATION

Broaching is done with a multiple-point cutting tool called *broach*. The important cutting and grinding elements of a broach are shown in Fig. 22.12. The roughing and finishing teeth and the first sizing teeth are designed to cut and thus remove a certain amount of material as the broach moves along a straight-line path for the cutting stroke. Each succeeding roughing and finishing tooth is made to rise slightly higher than the tooth which precedes it. Thus, in effect, a broach with its cutting teeth becomes progressively larger in section, parallel to the direction of its travel for the roughing and finishing teeth. The increase in tooth size is uniform, and it determines the amount of material to be removed by each tooth. The roughing teeth remove the majority of the material, and each roughing tooth may remove a depth of cut up to about 0.025 mm. The depth of cut for the finishing teeth is less (up to about 0.015 mm).

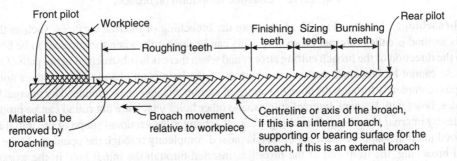

Fig. 22.12 Important features of simple broach (Front and rear pilots are not needed for external broaches).

During broaching, the broach moves past the workpiece only once. Thus, each roughing and finishing tooth takes only one cut for each workpiece. All of the sizing teeth are made to be exactly the same size. Only the first sizing tooth will cut. The sizing teeth behind the first one help to ensure the production of dimensions. A broach may be resharpened until the last sizing tooth becomes very small.

Burnishing teeth are sometimes used for producing smoother and denser surfaces. Burnishing teeth do not cut because they are well-rounded and extremely smooth. Burnishing is a cold-working operation in which the burnishing teeth compress the metal at the surface and cause a light amount of plastic flow. Cutting-tooth marks are removed by burnishing. During burnishing, a workpiece surface may change dimensionally by about 0.005 mm.

Rake and clearance angles suitable for the material to be broached are provided on each tooth. Sizing teeth are provided with a small straight land, 0.04–0.8 mm in width, immediately behind their cutting edges. This land is parallel to the direction of broach travel. Clearance is provided behind the land. The amount of material removed per tooth will be about 0.015 mm. At least two and preferably three teeth should be in contact with a workpiece during broaching. The distance between adjacent broach teeth is expressed as the *pitch*. The pitch of the broach teeth should provide for sufficient space for the chip. The broach teeth may be nicked at appropriate places to help break up the chips. Broaching is done at relatively low cutting speeds, which seldom exceed 10 metres per minute. These low cutting speeds help in creating less wear, and as a result, keep tooth wear at a minimum.

Figure 22.13 shows the enlarged tooth form on a broach.

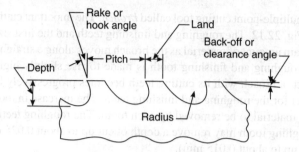

Fig. 22.13 Enlarged tooth form on broach.

Internal broaching: Refer to Fig. 22.14 showing the broaching of internal surfaces, such as the sides of holes. This method is called *internal broaching*. This can be done only when the surfaces to be broached are parallel to the direction of the broach cutting stroke, and when there is no obstruction in the path of the broach. A blind hole cannot be broached, because the bottom of the hole is an obstruction and does not allow the breach to pass completely through the hole. Openings of various shapes can be produced by internal broaching. Square holes, holes with keyway, internal splines, and other holes which are not round can be produced from drilled holes by internal broaching. Broaching gradually changes the sectional shape of the initial round hole to the desired internal shape. A sectional broach is passed completely through the opening. At the beginning of internal broaching, the front end of the broach is inserted through the initial hole in the workpiece. The front pilot of an internal broach serves to locate the broach properly in the workpiece opening at the start of broaching. Some internal broaches are pulled or pushed through an opening in the workpiece at the start of broaching. Some internal broaches are designed to fit the pull head of a broaching machine. Internal broaches, if they are not too slender, may be pushed through the workpiece opening. When an internal broaching is pulled or pushed through an opening in the workpiece, it is ordinarily free to follow the initial opening. A rear pilot serves to keep the internal broaching straight during the operation until the last tooth has passed through the opening. After its cutting stroke, a broach must not be returned to its starting position by forcing it backwards through the workpieces. This would cause excessive wear on the broaching teeth.

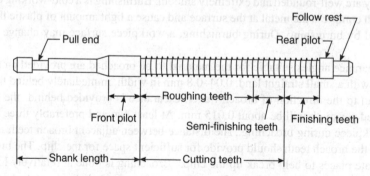

Fig. 22.14 Internal pull broach elements.

Horizontal broaching machines: Nearly all horizontal machines are of the pull type. They may be used for either internal or external broaching, although internal work is more common. A horizontal broaching machine shown in Fig. 22.15 consists of a bed or a base of a little more than twice the length of the broaching stroke, a broach pilot, and the drive mechanism for pulling the broach.

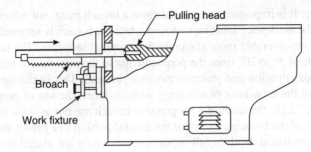

Fig. 22.15 Horizontal pull broaching machine.

Horizontal broaching machines are used primarily for broaching keyways, splines, slots, round holes, and other internal shapes or contours. They have the disadvantage of taking more floor space than vertical machines.

Surface broaching: The broaching of external workpiece surfaces, known as *surface broaching*, was developed after internal broaching and has greatly broadened the field of broaching (refer to Fig. 22.16). In surface broaching, a broach is securely mounted on the sliding member of a surface broaching machine. In operation, this sliding member moves in a straight line for a distance sufficient to carry the broach completely past the workpiece surface. The sliding member is well supported by bearing surfaces to present deflection under the enforcers of broaching. A suitable fixture is provided to locate the workpiece in relation to the broach. The workpiece surfaces to be broached are located so that the teeth of the broach can properly contact

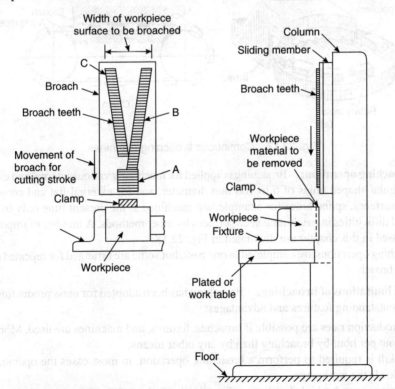

Fig. 22.16 Schematic illustrating external broaching with vertical surface broaching machine (A progressive type of broach is used).

them on the cutting stroke. It is important to note that since a broach must not rub on the workpiece surface during its return stroke, the workpiece must be removed when the broach is returned to its starting position. Surface broaches may be considerably more elaborate than internal broaches. The teeth of a broach perform oblique cutting at an angle of 5° to 20° from the perpendicular to the direction of broach travel. This helps a broach to cut with reduced vibration and process smoother surfaces. This is similar in principle to those of helical milling cutters. For the broaching of relatively wide surfaces, the use of progressive broach may be helpful. As shown in Fig. 22.16, the teeth of a progressive broach are not as wide as the workpiece surface to be broached; Section A of the broach cuts first at the central portion to a predetermined depth. Sections B and C cut from this predetermined depth to full depth, and since they are placed at an angle as shown, they will also broach the entire width of the workpiece surface, because broach sections such as B and C cut at the start at greater depths. Cutting at greater depths allows cutting edges to get underneath hard, rough surfaces that scale more quickly. A progressive broach has a disadvantage in that it must be long. The surface being broached in Fig. 22.16 is flat and often the surface to be broached may have special contours. Larger surface broaches are commonly made in two or more sections, which are assembled in a holder.

Continuous broaching operation: Refer to Fig. 22.17(a) and (b). In these broaching operations the workpieces are loaded into fixtures which carry them past one or more stationary broaches. Several fixtures may be fastened at intervals to an 'endlessly' moving chain. As a workpiece is moved through a horizontal tunnel supporting the broaches, broaching is completed. Continuous broaching with rotary table is shown in Fig. 22.17(a), while Fig. 22.17(b) shows an arrangement with endless moving chain.

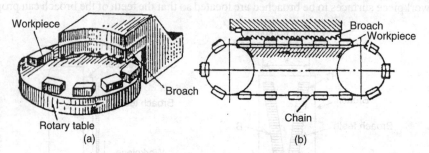

Fig. 22.17 Continuous broaching machines.

Summary of broaching operation: Broaching is applied for machining various internal and external surfaces, for round or irregular shaped holes of 6 to 100 mm diameter, and for external flat and contoured surfaces. Certain types of surfaces, spline holes for example, are machined at the present time only by broaching due to the exceptional difficulties in machining such surfaces by other methods. A number of important broaching operations discussed in this chapter are illustrated in Fig. 22.18.

Most broaching operations are completed in one pass, but some are arranged for repeated cuts to simplify the design of the broach.

Advantages and limitations of broaching: Broaching has been adopted for mass production work because of the following outstanding features and advantages:
1. High production rates are possible if broaches, fixtures, and machines are used. More pieces can be turned out per hour by broaching than by any other means.
2. Little skill is required to perform a broaching operation; in most cases the operator merely loads and unloads the workpiece.
3. High accuracy and a high level of surface finish of about 0.8 microns can be easily obtained in broaching.

GEAR CUTTING, BROACHING AND THREAD CUTTING

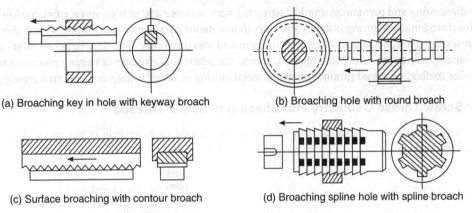

(a) Broaching key in hole with keyway broach

(b) Broaching hole with round broach

(c) Surface broaching with contour broach

(d) Broaching spline hole with spline broach

Fig. 22.18 Important broaching operations.

4. Both roughing and finishing cuts are completed in one pass of the tool.
5. The process can be used for either internal or external surface finishing.
6. Any form that can be reproduced on broaching can be machined.
7. Cutting fluids can be readily applied where it is most effective, because a broach tends to draw the fluid into the cut.

Certain reasons, however, limit the application of the broaching process. They are:

1. There are various difficulties in design and production, as well as high tool cost. A broach usually does only one job and is expensive to make and sharpen.
2. There are limitations in terms of the size of the workpiece.
3. The surfaces to be broached should not have an obstruction.
4. The stock removal rates are low.
5. There are difficulties in the design of broaching fixtures.

22.7 PRODUCTION OF SCREW-THREADS

Threads are of great importance in engineering. They are used as fasteners, to transmit power or motion and for adjustment in machine elements. The subject of thread manufacturing has attained a great importance because of the rising demand for high precision fastening devices and power transmission devices. At present, threads are manufactured by the following processes:

1. Casting
2. Thread chasing with single chaser and multiple chaser die head
3. Thread rolling
4. Thread milling
5. Thread grinding

Out of these thread production processes, thread chasing and thread rolling will be discussed in the present chapter in detail.

Casting

The accuracy and finish of thread made by casting will depend upon the casting process used. Threads made by sand casting are rough and are not used much, except occasionally in non-precision machinery. Threads made by die-casting and permanent mould casting are very accurate and of high surface finish. However, die-casting for thread manufacturing is applicable only to low melting point non-ferrous metals and, therefore, are not fit for repeated use, because of their low strength and less durability. The drawbacks of sand casting can be overcome by using shell-moulding method. Due to the inherent drawbacks of casting processes mentioned above, other method of thread production using metal cutting or metal rolling is much preferred.

22.7.1   Screw Thread Geometry and Chasing of Screw Threads

A screw is produced when a helical groove with a vee shaped cross-section is cut away or formed in a cylindrical rod or blank.

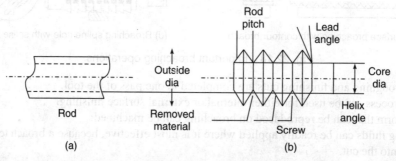

Fig. 22.19   (a) Bar stock and (b) Geometry of screw thread.

The present discussion is confined to the cutting of screw threads in a rod to generate a male thread, as shown in Fig. 22.19. Threads in a nut are formed in an existing hole and the scheme of metal removal can be understood once the scheme for a male thread is understood. The various schemes of metal removal to produce a thread are discussed below:

1. **Plunge cutting:** The scheme of cutting comprises removing metal as shown in Fig. 22.20. Successive cuts as shown remove metal in three passes. This can be achieved on a lathe machine.

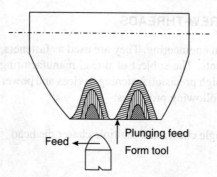

Fig. 22.20   Plunge feeding.

2. **Cutting by swiveling compound slide:** In this method, successive cuts are taken as shown in Fig. 22.21. Only the side cutting edge removes metal while the end cutting edge burnishes the right flank.

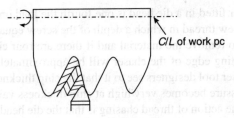

Fig. 22.21 Cutting with side cutting edge.

3. **Metal removal using a thread chaser:** To increase tool life, a single tool with a number of cutting teeth is used. Figure 22.22 indicates the progressive metal removal with a single thread chaser, usually called a threading comb in workshop terminology. Figure 22.23 shows the action of chaser teeth 1, 2, 3 and 4. It will be seen that the cutting load is shared by each of the four teeth. The shaded zones show the metal removal by each of the four teeth. This is made possible due to the provision of a chamfer angle on a chaser as shown in Figs. 22.22 and 22.23. Usually, the first two to three teeth do the cutting and additional teeth are provided which do not have a chamfer and they only do the sizing and burnishing of the thread grooves work piece. In the figure, the burnishing teeth are not shown.

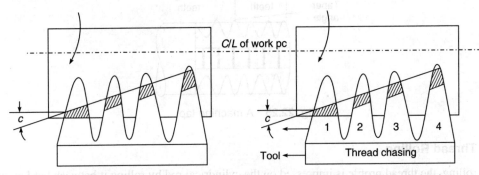

Fig. 22.22 Scheme of metal removal in thread chasing.

Fig. 22.23 Thread chasing.

4. **Die heads:** The die heads employ four thread chasers fitted inside a circular body either tangentially or radially as shown in Fig. 22.24(a), (b) and (c). Each chaser is displaced axially with respect to one another by a distance of one fourth pitch of the screw to share the cutting load.

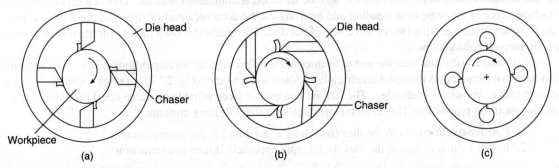

Fig. 22.24 (a) Radial chaser, (b) Tangential chaser and (c) Circular form chaser.

The cutting load on chaser fitted in a die head is one fourth the load on a single isolated chaser. Let us take an example of cutting a screw thread in which a depth of the screw equals 0.5 mm to be cut. If there are three threads along the chaser to remove the material and if there are four chasers on the die head, the chip thickness to be cut by each cutting edge of the chaser will be approximately $0.5/(3 \times 4) = 0.041$ mm chip load per cutting edge of the chaser tool designers see to it that the chip thickness does not fall below 0.02 mm because the specific cutting pressure becomes very high at chip thickness values below 0.02 mm.

An interesting aspect of the action of thread chasing is that the die head is automatically pulled into the cut as the die head acts like a hardend nut as the workpiece starts rotating, due to the screw and nut action. On the completion of the cut, the chasers are withdrawn and travel back as the direction of rotation of the workpiece is reversed.

22.7.2 Internal Screw Threads

A tap which is used for cutting internal thread in a pre-drilled hole is called threading tap. A tap is just like a screw on which gashes are provided to accommodate the chips. It is also provided with a taper angle on its front end so as to obtain a gradual cutting action by each of its teeth. Figure 22.25 shows various geometrical features of a tap.

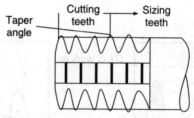

Fig. 22.25 A machine tap.

22.7.3 Thread Rolling

In thread rolling, the thread profile is impressed on the cylindrical rod by rolling it between hard metal dies. Usually, flat dies, as shown in Fig. 22.26 are used or circular dies as shown in Fig. 22.27 are used for higher production.

In the reciprocating flat die machine, one die is stationary and the other reciprocating. The part to be threaded is rolled between the dies, as the moving die reciprocating in reference to the stationery die. The stroke of the reciprocating die will depend upon the diameter of the tread being produced since during one stroke, the blank makes one complete revolution and the threaded is completely formed. This is a highly versatile machine, since at the same time treading and knurling can be done on a part of right and the left hand tread can be rolled, by assembling two or three sets of flat dies. This method is used mainly for the manufacturing of commercial bolts and nuts.

In cylindrical die machine, the part to be threaded is rolled between rotaing cylindrical dies. The machine can have two circular dies located diametrically opposite to each other (Fig. 22.27) or three round dies located diametrically opposite to each other. The flat die machine is more suitable for large sized precision threads and for short run production. This machine operates with the following motions:

1. Positive rotation of both the dies (in a two die machine) in the same direction.
2. Radial motion of one of the dies for its rapid approach, infeed and retraction.

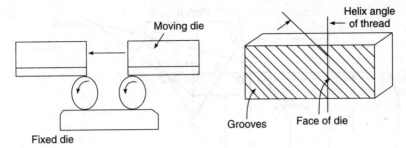

Fig. 22.26 Thread rolling with flat dies.

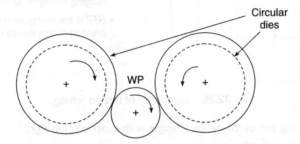

Fig. 22.27 Thread rolling with circular dies.

This method has the main application of rolling the thread on taps.

Advantages of thread rolling
1. Because the rolls are made of 1.5% carbon alloy steel containing vanadium and chromium (12%), their life is more.
2. The grain flow lines in rolled threads make them stronger than machine cut threads.
3. The process is fast and production rate is more.
4. Accuracy and finish is better.

Limitations
1. Initial blank size needs to be calculated with accuracy.
2. Internal threads cannot be rolled.

Determination of the blank size in thread rolling

It is very essential to determine the initial blank diameter in thread rolling if the final specification of the screw, particularly its major diameter, root diameter, angle of the vee and the material to be rolled are specified in advance. A particular geometrical analysis becomes necessary to arrive at the Initial blank dimensions before rolling. Figure 22.28 and the analysis presented thereafter is often used to arrive at the accurate blank dimensions.

Let B_d be rolling diameter of the blank.

M_d be major diameter of the thread.

R_d be root diameter of the thread.

C_2 be the diameter of centroid of triangle PQS.

C_1 be the diameter of centroid of triangle STZ.

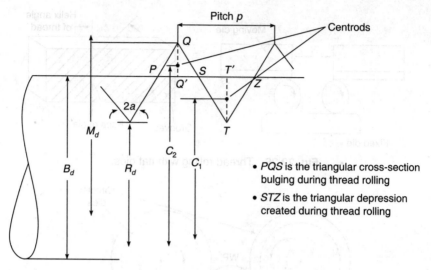

Fig. 22.28 Geometry of thread rolling.

Therefore, volume of the ring presented by the triangle with cross section PQS

= Area of $\triangle PQS$ × circumference

$$= \frac{1}{4} PS(M_d - B_d) \times \pi C_2$$

Similarly, the volume of the ring presented by the cross-section STZ

$$= \frac{1}{4} SZ(B_d - R_d) \times \pi C_1$$

$$= (M_d - B_d) \tan \alpha = (B_d - R_d) \tan \alpha$$

$C_2 = B_d + \frac{1}{3} \cdot QQ'$, where QQ' is the height of thread above rod blank

$$= B_d + \frac{1}{3}(M_d - B_d)$$

$$= \frac{1}{3} \cdot B_d + \frac{1}{3} M_d$$

$C_1 = B_d - \frac{1}{3} \cdot TT'$, where TT' is the depth of thread below initial rod blank

$$= B_d - \frac{1}{3}(B_d - R_d)$$

$$= \frac{1}{3} \cdot B_d + \frac{1}{3} R_d$$

Then we can write the following equation:

$$PS(M_d - B_d)\, \pi C_2 = SZ(B_d - R_d)\, \pi C_1$$

Now putting $PS = (M_d - B_d) \tan \alpha$ and $SZ = (B_d - R_d) \tan \alpha$, we get the equation below:

$$(M_d - B_d) \tan \alpha \cdot (M_d - B_d) \cdot \frac{1}{3} \cdot (2B_d + M_d) = (B_d - R_d) \tan \alpha (B_d - R_d) \cdot \frac{1}{3} \cdot (2B_d + R_d)$$

or
$$(M_d - B_d)^2 \cdot (2B_d + M_d) = (B_d - R_d)^2 \cdot (2B_d + R_d)$$

or
$$\boxed{M_d^3 - R_d^3 - 3M_d \cdot R_d^2 + 3B_d^2 \cdot R_d = 0}$$

The above equation is of great importance to calculate the blank diameter, if the root diameter and the major diameter are given. The example below illustrates the application of the equation.

EXAMPLE 22.1 Calculate the blank diameter for cold rolling 12 mm diameter 60°V threads with 1 mm pitch on a mild steel blank.

Solution

Depth of thread = $\frac{1}{2}$ cot 30° = 0.866

Major diameter = 12 mm
Root diameter = 12 – 1.732 = 10.268 mm

On substituting the values in the above equation, the blank diameter is calculated as below:
$$12^3 - 10.268^3 - 3 \times 12 \cdot B_d^2 + 3 \times 10.268 \times B_d^2 = 0$$
$$5.196 \, B_d^2 = 647$$

or
$$B_d = 11.4 \text{ mm}$$

REVIEW QUESTIONS

1. On what basis would you classify gear cutting methods?
2. Describe the principle of gear generation.
3. Sketch a hob and label its various elements.
4. How are the right hand or left hand characteristics of a hob determined?
5. How is a gear made by the process of shaping?
6. Distinguish between hobbing, shaping and milling.
7. What are the various methods of finishing gears?
8. When is the gear grinding method preferred?
9. How does gear shaping differ from honing?
10. What kinds of gears are made by broaching?
11. Define hobbing time.
12. Calculate the time taken to hob a spur gear of 72 teeth with a single threaded hob running at 75 rpm and having a feed of 2.5 mm per blank revolution. The width of the face of the gear is 35 mm.
13. Describe the cutting action of a broach.
14. How are broaching machines classified?
15. Describe the main considerations while heat-treating a broach.
16. Find out the power consumed in broaching a keyway in a cast steel pulley. The dimensions of the keyway are 10 mm width, 5 mm depth and 30 mm length. If broaching is performed at 3 m/min and the feeds for rough and finish broaching are respectively 0.10 mm/tooth and 0.03 mm/tooth.

PART III
PRECISION MEASUREMENT AND MANUFACTURING

Part III

Precision Measurement and Manufacturing

CHAPTER 23

Metrology and Precision Measurement

23.1 LINEAR PRECISION MEASUREMENT

Since modern production is concerned with interchangeable products, a great precision of dimensional control is required in industry. For this to be achieved the importance of precision measuring instruments can be well understood. In this introductory section on precision measurement, we shall discuss the constructional details and procedure of measurement with two most basic measuring instruments, the Micrometer screw gauge and the Vernier calipers.

23.1.1 External Micrometer

The external micrometer is primarily used to measure external dimensions like diameter of shafts, thickness of parts, etc. to an accuracy of 0.01 mm. The essential parts of the instrument as shown in Fig. 23.1 are as follows:

1. **Frame:** The frame is made of steel, cast steel, malleable cast iron or light alloy.
2. **Hardened anvil:** The anvil protrudes from the frame for a distance of at least 3 mm or so.
3. **Screwed spindle:** The spindle does the actual measuring and possesses thread of 0.5 mm. pitch.
4. **Graduated sleeve or barrel:** It has datum horizontal facial line and has fixed graduations along its length as shown in Fig. 23.1.
5. **Thimble:** This is a tubular cover fastened with the spindle and moves with the spindle. The bevelled edge of the thimble is divided into 50 equal divisions, every fifth division being numbered.
6. **Ratchet or friction stop:** This is a small extension to the thimble. The ratchet slips when the pressure on the screw exceeds a certain amount. This produces uniform reading and prevents any damage or distortion of the instrument.
7. **Spindle clamp or clamp ring:** This is used to lock the instruments at any desired setting.

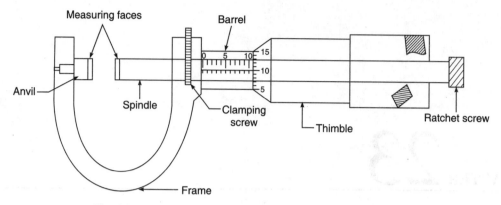

Fig. 23.1 Different parts of an external micrometer (metric).

Internal details of a micrometer

The barrel 2 and frame 1 (Fig. 23.2) constitute one piece. The thimble has an internal thread (as shown in Fig. 23.2) which guides the hardened measuring spindle 3 and forms an integral unit. As already stated, the thimble has 50 divisions along its circular scale .Other details, such as threaded collar 4 for the adjustment of the inside thread, anvil 5, clamp ring 6 and ratchet stop 7 are also shown in Fig. 23.2.

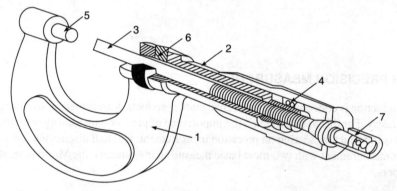

Fig. 23.2 Internal details of a micrometer.

Procedure for reading: The thread of the measuring spindle has a pitch of 0.5 mm which means that giving one turn to the thimble, results in an axial movement of 0.5 mm of the hardened measuring spindle along with the thimble. The bevelled part of the thimble is divided into 50 parts. If the thimble is moved by one graduation line, the spindle moves forward by 0.5/50 = 0.01 mm. This is called the least count of the instrument. It will be seen in Fig. 23.3 that even half graduations are shown on the barrel scale.

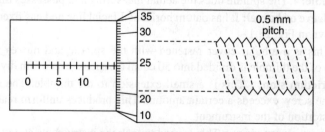

Fig. 23.3 Reading a micrometer.

The formula for using a micrometer is:

Resultant reading = Complete reading on the barrel scale + Least count × Spindle reading

which coincides with the horizontal line of the barrel scale. Thus, the final reading in the case of Fig. 23.3 is

$$13.55 + 0.01 \times 25 = 13.75 \text{ mm}$$

The external micrometers have a range of 25 mm and are available with the following ranges of measurement:

0 to 25, 25 to 50, 50 to 75, 75 to 100, 100 to 125 and so on.

23.1.2 Vernier Calipers

The underlying principle of the vernier is that when two scales or their divisions which are slightly different are used, the difference between them can be utilized to determine the accuracy of measurement. This principle was invented by the French Scientist Pierry Vernier and is called the 'Principle of differential measurement'. The outside Vernier Caliper is illustrated in Fig. 23.4. It comprises a beam or main scale which carries the fixed graduation, two measuring jaws, a vernier head having a vernier scale engraved on it, and an auxiliary head of a vernier clamp which is used for a specified dimension by a micrometer screw.

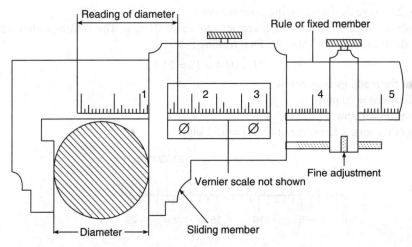

Fig. 23.4 Vernier calipers.

Principle of the vernier

As stated earlier, the vernier is a precision measuring instrument. The smallest dimension which can be measured by a vernier is called its least count. The least count of a vernier is the difference between one main scale division and one vernier scale division.

For example, if 20 divisions of the vernier scale coincide with 19 divisions of the main scale, then

$$20 \text{ VSD} = 19 \text{ MSD}$$

$$\Rightarrow \quad 1 \text{VSD} = \frac{19}{20} \text{ MSD}$$

Thus, Least Count, $\text{LC} = 1\text{MSD} - 1\text{VSD} = 1 - \dfrac{19}{20} = \dfrac{1}{20}$ MSD

Now, if 1MSD = 1 mm, then the least count

$$LC = \frac{1}{20} \times 1 \text{ mm} = 0.05 \text{ mm}$$

The procedure for reading a vernier, the formula used is:

The measurement = (Main Scale Reading on the left of the zero of the vernier scale)
+ (LC × Vernier reading exactly coinciding with any main scale division)

The use of the formula is illustrated with an example below:

Reading a Vernier scale: An enlarged diagram of the metric vernier scale is shown in Fig. 23.5. On the main scale, 1 cm is divided in 10 parts, each being 1 mm. This is again divided into two parts giving 0.5 mm. The vernier scale has 25 divisions which are numbered. The length of 25 divisions on the vernier is equal to the length of 24 divisions on the main scale. It is seen that the difference between each small division on the main scale and one division on the vernier is

$$\frac{1}{25} \times 1 \text{ mm} = 0.04 \text{ mm}$$

A study of Fig. 23.5 reveals the following information:

Main scale reading on the left of the vernier zero reads 21 mm. The vernier scale reading coinciding with a main scale division is 18. Thus, the final reading is

$$21 + 0.04 \times 18 = 21.72$$

The break-up of the readings is as below:
2 main divisions = 20 mm
1 sub-division = 1 mm
18 vernier divisions × Least count (0.04) = 0.72 mm

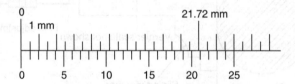

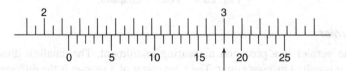

Fig. 23.5 Taking a vernier reading (note that 18th vernier division is coinciding with a main scale division).

23.2 LENGTH STANDARDS

The international prototype metre (see Fig. 23.6) is a line standard. The length is defined by two fine terminal lines. A form of line standard, now commonly used in machines, employs an optical measuring system. The causes of variation in the distance defined by the terminal lines of the line standards are:

(i) Temperature variation due to expansion on heating;
(ii) Curvature variation caused by changes in weight distribution, stress distribution and grain structure of material;
(iii) Long-term or secular changes.

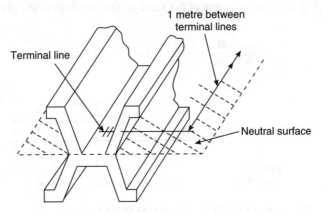

Fig. 23.6 International prototype metre of 1889.

To qualify for use as a working line standard, a material must have the following properties:

(i) It should be able to take a high surface polish and be free of rusting.
(ii) It can have a suitable coefficient of expansion.
(iii) It should be able to maintain its size over long periods of time.

Standard length is defined at standard temperature; the international standard temperature for this purpose is 20°C.

Pure nickel is very stable (Table 23.1), takes a suitable polish, does not rust and has a coefficient of expansion of 12.8×10^{-6} per degree celcius. Invar, an alloy of 58% nickel and 42% iron, has similar properties but a coefficient of expansion of 11.2×10^{-6} per degree celcius in the temperature region of 20°C, a figure very close to that for steels. Measurement of steel components by reference to such a line standard is therefore attractive; it is no longer necessary to work at exactly 20°C, provided the work and the standard are at the same temperature. Many line standards are now made from this alloy.

Table 23.1 Coefficients of linear expansion

Material	Approx. expansion per degree celcius units $\times 10^{-6}$
Aluminium	22–24
Brass	18–20
Bronze	16–18
Cast iron (grey)	9–10
Copper	16–17
Magnesium	28–30
Nickel	12–13
Steel	11–12

23.3 POINTS OF SUPPORT

Sir G.B. Airy was able to show that supporting a bar at two critical points gives a condition for which the ends of the bar remain horizontal and the slope remains zero.

With point O (see Fig. 23.7) as the origin of the x and y axes, the slope of the portion of the bar between A and B is given by

$$\frac{dy}{dx} = \frac{W}{EI} \times \frac{1}{6}[l(l^2 - 3a^2) - (l - x)^3] \tag{23.1}$$

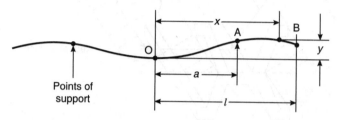

Fig. 23.7 Airy points of support.

$$\text{Deflection} = y = \frac{W}{EI} \times \frac{1}{24}[4la^3 - l^4 + 4lx(l^2 - 3a^2) + (l - x)^4] \tag{23.2}$$

where W is the weight per unit length;
 E is Young's modulus;
 I is the second moment of area.

For the ends to be horizontal, $\frac{dy}{dx} = 0$, and from Eq. (23.1)

$$l(l^2 - 3a^2) - (l - x)^3 = 0$$

putting $x = l$, since we are dealing with point B

$$l^2 = 3a^2 \quad \text{or} \quad \frac{a}{l} = \frac{l}{\sqrt{3}} = 0.577\, l$$

The positions so defined are known as the *Airy points of support*.

It can be shown that minimum curvature occurs when the deflections at O and B are equal, i.e., when y is 0 at B.

From Eq. (23.2), when $y = 0$

$$4la^3 - l^4 + 4lx(l^2 - 3a^2) + (l - x)^4 = 0$$

and when $x = l$

$$4la^3 - 12l^2a^2 + 3l^4 = 0$$

or

$$4\left(\frac{a}{l}\right)^3 - 12\left(\frac{a}{l}\right)^2 + 3 = 0 \tag{23.3}$$

Solving Eq. (23.3) gives

$$\frac{a}{l} = 0.554$$

The deflection, at its maximum value, is about twice as large for the Airy points of support $\frac{a}{l} = 0.557$ as that for the points $\frac{a}{l} = 0.554$.

In order to be useful as end standards, standard end lapped bars are supported at the airy points, with the result that the lapped end faces lie parallel when $a/l = 0.557$. On the other hand, ideally, line standard should be supported at $a/l = 0.554$, so that the deflection is minimal. In an experiment to determine the flatness of a surface plate, a straight edge is supported at $a/l = 0.554$ to minimize its deflection. $a/l = 0.557$ and $a/l = 0.554$ suggested theoretical points of support.

The international metre is defined in terms of the wavelength of monochromatic light. The International Committee of Weights and Measures recommended Krypton 86 as the radiation source: under standard conditions the metre equals 1650763.73 wavelengths. The metre had its origin in the International Prototype Metre of 1889, a line standard of the form shown in Fig. 23.6. Laser interferometry enables direct reading to be taken of the lengths of the end-standards.

23.4 INTERFEROMETRY

Light is a form of energy radiation having wave nature. Suitable sources can emit monochromatic rays (rays confined to a very narrow variation of wavelength λ), which provide a basis for interferometry.

As they leave a common source, rays are 'in phase'; but by making them travel paths of differing length before they re-combine, two such rays can be made to interfere. Figure 23.8 illustrates graphically the results of such re-combination

(i) When the path difference is $N\lambda$ (i.e., in phase), as in Fig. 23.8(a)
(ii) When the path difference is $(N + 1/2)\ \lambda$ (i.e., out of phase), as in Fig. 23.8(b)

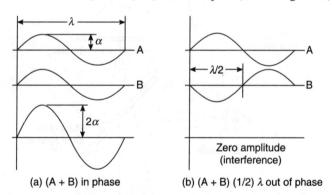

(a) (A + B) in phase   (b) (A + B) (1/2) λ out of phase

Fig. 23.8 Wave concept of interference.

The dark bands produced by interference provide a physical basis for precision measurement.

Optical flat: The simple practical method of using the interference effect is by means of an optical flat. This is a thick disc of either glass or quartz with parallel faces ground and polished flat to a very high order of accuracy. Usually only one face is of specified flatness; it may be coated to increase the light reflected from the surface. It then becomes a more efficient beam splitter. Refer to Fig. 23.9.

The gap between the slip planes gives rise to a path difference of $1/2\lambda$; as the air gap is crossed twice, interference bands will be seen where the width of air gap is $N\lambda + 1/4\lambda$, $N\lambda + 3/4\lambda$, $N\lambda + 5/4\lambda$, etc.

Two inferences can be drawn from the appearance of the bands:

(i) As they are similar to the contour lines on a map, any curvature or any irregularity in their successive pitches (p) indicates that the slip gauge surface is not flat.

(ii) From the pitch of the interference bands and the known wavelength of the light, angle α (in radians) can be determined as follows:

$$\alpha = \frac{\lambda}{2p} \qquad (23.4)$$

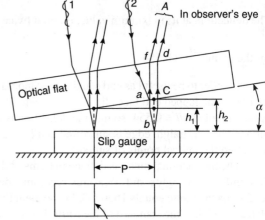

Fig. 23.9 Optical flat resting at very small angle α.

This principle is used for accurate measurement of angles.

Figure 23.10 shows the optical arrangement of an NPL-type interferometer for testing the flatness and parallelism of slip gauges. The slip is wrung to a flat-lapped rotatable plate. The comparison of the interference bands on the face of the slip and on the plate is used to monitor the parallelism of the slip gauge.

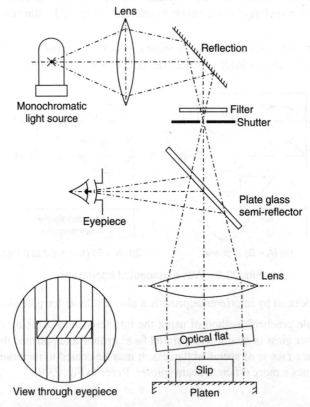

Fig. 23.10 NPL interferometer for testing flatness and parallelism of slip gauges.

23.5 CALIBRATION OF LENGTH STANDARDS

Figure 23.11 shows the principle of an NPL type length interferometer capable of measuring end-standards up to 100 mm to an accuracy of 0.025 μm or 0.000025 mm. Usually there are light sources providing monochromatic light of three different wavelengths with this system. Figure 23.11 shows how the interference bands are measured to obtain fractional increments of $½\lambda$, where $a/b = f$.

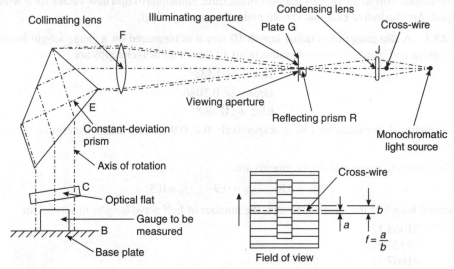

Fig. 23.11 Optical system of NPL gauge length interferometer (0–100 mm).

The geometry of contacting surfaces is shown in Fig. 23.12.

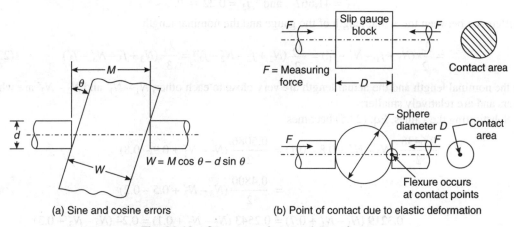

(a) Sine and cosine errors (b) Point of contact due to elastic deformation

Fig. 23.12 Geometry of contacting surfaces.

The height of the gauge must be some multiple of the wavelength used. This most probably incorporates a fractional element; so for the three different wavelengths, $\lambda_1, \lambda_2, \lambda_3$, belonging to three different colours (say)

$$h = \frac{1}{2}(N_1 + f_1)\lambda_1$$

$$= \frac{1}{2}(N_2 + f_2)\lambda_2$$

$$= \frac{1}{2}(N_3 + f_3)\lambda_3$$

The value of h should be known to a fair degree of accuracy, thus enabling the N-values to be estimated. These values must be rounded off to the nearest integers in the three equations to find new values for h which should be almost equal. An illustrative example should make this concept clear.

EXAMPLE 23.1 A slip gauge of nominal length 10 mm was measured on a gauge length interferometer using the red, green and blue light of a cadmium lamp, of which the wavelengths are

Red λ_1: 0.6438

Green λ_2: 0.5086

Blue λ_1: 0.4800

The observed fractional displacements were respectively 0.8, 0.9 and 0.5 units. Determine the error in the length of the gauge.

Solution The observed fractional displacements are

$$f_1 = 0.8 \qquad f_2 = 0.9 \qquad f_3 = 0.5$$

Since the nominal length of the gauge = 10 mm, the number of half wave lengths in 10 mm are

Red   31063.12
Green   39324.81
Blue   41667.22

∴ $N_1' = 31,063$ and $f_1' = 0.12 \simeq 0.1$

                    $N_2' = 39,324$ and $f_2' = 0.81 \simeq 0.8$

                    $N_3' = 41,667$ and $f_3' = 0.22 \simeq 0.2$

The difference between the actual length of the gauge and the nominal length

$$= \frac{\lambda_1}{2}(N_1 + f_1 - N_1' - f_1') = \frac{\lambda_2}{2}(N_2 + f_2 - N_2' - f_2') = \frac{\lambda_3}{3}(N_3 + f_3 - N_3' - f_3') \qquad (23.5)$$

Since the nominal length and the actual length are very close to each other, $N_1 - N_1'$ and $N_2 - N_2'$ are whole numbers and are relatively smaller.

Substituting the values, Eq. (23.5) becomes

$$\frac{0.6438}{2}(N_2 - N_1' + 0.8 - 0.1) = \frac{0.5086}{2}(N_2 - N_2' + 0.9 - 0.8)$$

$$= \frac{0.4800}{2}(N_3 - N_3' + 0.5 - 0.2)$$

or $0.3219 (N_1 - N_1' + 0.7) = 0.2543 (N_2 - N_2' + 0.1) = 0.24 (N_3 - N_3' + 0.3)$

By hit and trial, it will be seen that substituting

$$N_1 - N_1' = 1 \qquad N_2 - N_2' = 2 \qquad N_3 - N_3' = 2$$

makes the three errors nearly equal, i.e., 0.3219×1.7, 0.2543×2.1 and 0.24×2.3 or 0.55 μm, 0.54 μm and 0.55 μm respectively.

Thus the actual error can be taken as 0.55 µm.
This method is slow and is now superceded by laser interferometer.

23.6 SLIP GAUGES—BS 888 AND BS 4311

These are the working length standards of industry. BS 888 contains useful information on the care and use of slip gauges. There are also details of several useful accessories such as measuring jaws for internal as well as external work and holding devices. All this information results in wide application of slip gauges in industry.

An idea of the accuracy to which such gauge blocks are manufactured can be obtained by noting the permissible errors laid down in BS 888 for gauges up to and including 25 mm; the permissible errors are as shown in Table 23.2.

Table 23.2 Maximum permissible errors of slip gauges up to and including 25 mm (unit 0.00001 mm)

	Workshop grade	Inspection grade	Calibration grade	Reference grade
Length	+20–10	+20–10	±12	±5
Flatness	25	10	8	8
Parallelism	25	10	8	8

23.7 SOME SOURCES OF ERROR IN LINEAR MEASUREMENT

Accuracy in measurement depends upon the method and the cleanliness level as well as upon the equipment available. The reference laboratory of a firm is extremely accuracy-conscious. Where slip gauges are used as a basis of reference, errors arising from slip gauges exceed the error of the reference standard for a majority of the precision measurements made in the average inspection department.

In addition to temperature variations, errors in metrological measurement can arise from the following causes:

(i) Curvature at contacting surfaces;
(ii) Errors of alignment;
(iii) Parallax effects and vernier acuity.

Figure 23.12(b) illustrates the difference in the contact geometry between the flat anvils of a comparator, a slip gauge or a precision ball. Since there must be some contact force, there is also some deflection due to stress in each case. Due to elastic measuring force the reading obtained for a ball will contain a larger error, due to elastic deflections at the contact points, than the reading obtained for the slip gauge. In the case of a sensitive comparator, the geometrical conditions of contact which occurs for each of two comparative readings should be similar. The first object is achieved for the bench micrometer by the use of fiducial indicators, and for comparators by the spring load on the moving anvil.

Figure 23.12(a) illustrates errors of alignment with respect to work measured between the flat anvils of a bench micrometer. These errors are frequently called the sine and cosine errors.

Figure 23.13 illustrates the parallax error for the reading of scales. Errors from this source tend to fall as the magnification factor of the instrument rises. Reference to BS 887: Vernier Callipers and BS 870: External Micrometers will show that the dimensions indicated (such as t in Fig. 23.13) is minimized for reducing parallax effects. One way of overcoming the parallax effects is to project the scale and the index on to the same plane, as in most optical measuring equipment.

Figure 23.14 illustrates the well-known principle of double reticle lines and what is meant by *vernier acuity*. It is ergonomically convenient to take a reading between two parallel lines.

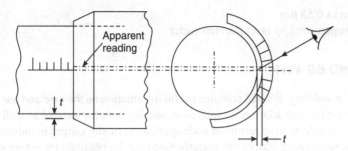
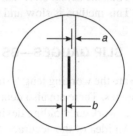

Fig. 23.13 Parallax error of micrometer reading.

Fig. 23.14 Double reticle lines.

23.8 ANGULAR MEASUREMENT

The basis of most angular measurements is the divided circle, as in the case of the scale of a vernier protractor. In its most sophisticated form, this circle is a silver-coated or glass disc upon which the division lines are etched, and the scale is read through an optical system of considerable magnifying power. Optical dividing heads operating on this principle are now available, with readings up to 1 second of arc and maximum cumulative error not exceeding 5 seconds of arc.

Table 23.3 illustrates the angle slip gauges.

Table 23.3 Angle slip gauges

Degree series	1° 3° 9° 27° 41° 90° (square)
Minute series	1′ 3′ 9′ 27′
Second series	3″ 9″ 27″

In addition to the divided circle, angle slip gauges are available; these are generally made to a tolerance of 2 seconds of arc. Since they may be wrung together by sum or by difference, a small number of gauges can give a large range of combinations.

Figure 23.15 shows how an angle of 13° 24′9″ can be built up from such gauges. The precision polygon (see Fig. 23.16) is a further piece of standard equipment in angular measurement; it is the three-dimensional equivalent of the divided circle.

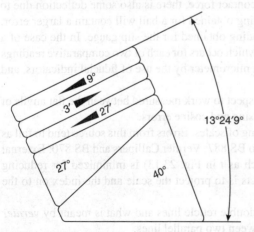
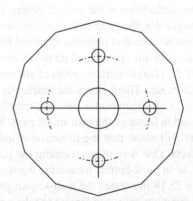

Fig. 23.15 Wringing of angle slip gauges.   **Fig. 23.16** Standard twelve-sided precision polygon.

There are two main sources of error in angular measurement:
  (i) Error of centering, e.g. of a divided scale;
  (ii) Error between the plane in which an angle is defined and the plane in which it is measured.

The sine curve of errors has been plotted in Fig. 23.17. Here

$$\text{Error in } \theta = \Delta = \frac{e}{R}\sin\theta$$

EXAMPLE 23.2 An accurately divided circular scale, of 100 mm diameter, rotates about a centre displaced 0.0025 mm from the centre of graduation on a line joining the 150° and 330° positions. Draw a graph showing the error of centering.

Solution We have

$$\frac{e}{R} = 0.00005 \text{ rad}$$
$$= 10 \text{ s of arc (very nearly)}$$

Figure 23.17 shows a graphical construction for the sine curve.

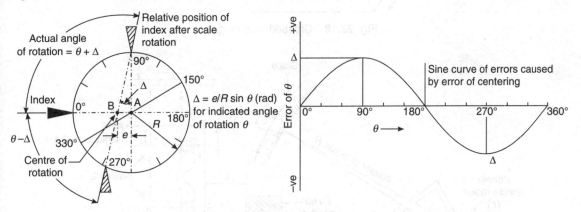

Fig. 23.17 Centering error in circular division.

23.9 MEASUREMENT OF SMALL LINEAR DISPLACEMENTS

A comparator is an instrument which magnifies small linear displacements in order to make them visible. In addition to high magnification, a comparator must have the following attributes:
  (a) It must be robust and have repeatability;
  (b) The magnification factor must be high;
  (c) It must operate from a uniform force exerted on the moving anvil.

There are several types of linear comparator now available; the main types are mechanical, optical, pneumatic and electrical. Examples of each should be studied in a metrology laboratory and their relative advantages and disadvantages assessed. The following discussion of some of the main operating principles is not intended to be exhaustive, but to draw attention to a few of the most important points.

Mechanical method: Levers are among the chief means of magnification. Crossed-strip hinges (see Fig. 23.18) are often employed in place of pivots in order to avoid "play". Figure 23.19 illustrates the operating principle of the SIGMA mechanical comparator. Unit A is displaced against a light spring by the movement of the measuring anvil. The knife edge of this unit causes B to rotate about the centre of the "crossed-strip" hinge, so as to rotate the long arms attached to it. These arms connect a metallic tape, which is partly round and secured by screws, to the spindle which carries the pointer. The pointer moves against a suitably divided fixed scale. A feature of the arrangement is the method of mounting the knife edge of A in such a way that dimensional can be adjusted by means of the screws E. This enables the desired magnification factor to be set.

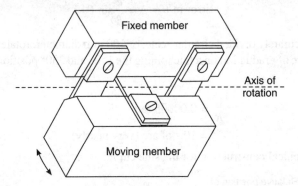

Fig. 23.18 Crossed-strip hinges.

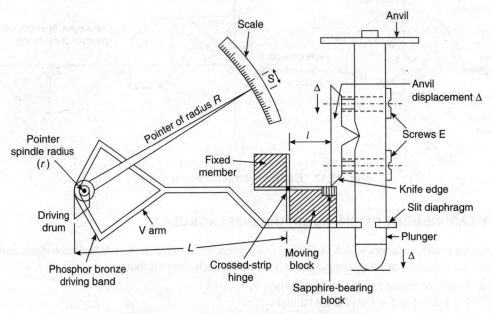

Fig. 23.19 Operating principle of mechanical comparator.

An eddy-current damping device is attached to the pointer spindle and makes the instrument "dead-beat". Figure 23.19 shows the various distances S, L, R, Δ, l and r. It will be seen that the magnification is given by the ratio of pointer displacement S to anvil displacement Δ.

$$\text{Magnification} = \frac{S}{\Delta} = \frac{L}{l} \times \frac{R}{r}$$

Optical method: The main feature of optical type comparators is that a beam of light, as a magnification lever, has no inertia and may be contained within a compact space by reflecting it between mirror surfaces.

Most of the comparators which use a beam of light are improved versions of the optical lever illustrated in Fig. 23.20, since the change of the angle on reflection is twice the change of the angle at which the incident ray enters; there is a multiplying factor of two each time the beam is reflected.

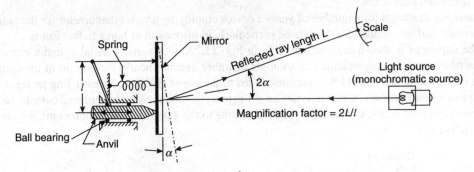

Fig. 23.20 Principle of optical lever.

The movement of the plunger causes tilting of the mirror, with a corresponding deflection of the parallel beam projected from the objective lens. Referring to Fig. 23.21, the reflected parallel beam is brought to focus in the same plane as the scale itself, and slightly to one side of it. On the same graticule as the scale is an index, and the reflected beam forms an image of the scale against this index. The position of the scale image relative of the index depends on the angle of the mirror and the movement of the plunger, which can therefore be calibrated in terms of scale divisions. The image of the scale and the index are viewed through a magnifying eyepiece, or in some cases through a projection system.

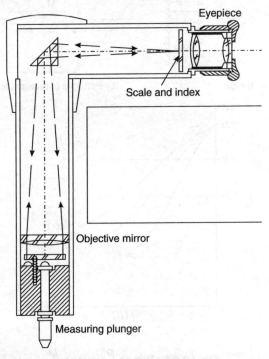

Fig. 23.21 Optical comparator.

560 TEXTBOOK OF PRODUCTION ENGINEERING

Pneumatic gauges: The general principle of a pneumatic comparator is to apply a jet or multiple jets of air to the surfaces being measured, with the orifices close to but not in actual contact with the surface. Variations in size affect the aperture of escape of the air, and the corresponding variation in back pressure is utilised to indicate the actual dimension.

Solex has marketed for a number of years a device employing a water manometer for the indication of back pressure, and this method has been used particularly in inspection of bores to fine limits.

The apparatus is shown diagrammatically in Fig. 23.22(a). A vertical metal cylinder contains water to a prescribed level; a tube, attaching to an upper chamber, reaches nearly the bottom of the cylinder, and air from the airline at about 140 kN/m pressure is fed to the upper chamber. A restricting jet leads from this chamber to a second cylinder, which also has an outlet to a glass manometer tube placed outside the cylinder. This second chamber is also connected by a flexible tube to the gauging plug or other unit, which itself has two restricting jets.

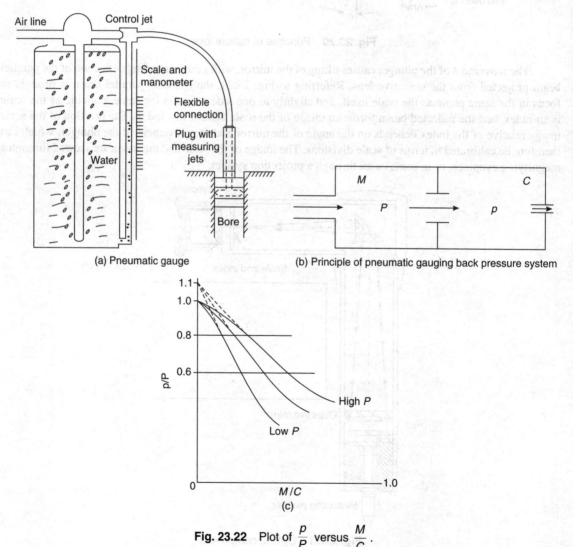

Fig. 23.22 Plot of $\dfrac{p}{P}$ versus $\dfrac{M}{C}$.

The principle of operation is as follows. The pressure in the first chamber behind the jet is equal to the head of water forced down the vertical tube, with excess air bubbling out from the bottom of the tube and escaping freely. The slots on the measuring plug are so proportioned, relative to the control jet, that partial closure of these jets causes various pressures in the second chamber, which lead to variation in the height of water in the manometer tube, which is measured with a calibrated scale set beside this tube. In other words, the pressure in the second chamber is regulated by the relative rates of escape of air through the first jet and the plug jets; if the plug jets are completely closed, the manometer level is depressed to the bottom of the tube.

The size of jets and plugs have to be related to the size of the bore and the limits to which it is required to gauge; it is for this reason that the method is more applicable to quantity production than to general gauging or metrology work. For external measurement of parallel surface, however, it is possible to use jet units which can be separated by varying amounts with slip gauges.

Basically, the "back pressure" system, as exemplified by the Solex gauge in Fig. 23.22(a), can be represented by two chambers connected by the control jet, with the measuring jet functioning as an escape point to the atmosphere in the second chamber. Thus, the constant pressure chamber in Fig. 23.22(a) is the volume above the dipper tube behind the control jet, while the second chamber is the volume between the control jet and the gauging or measuring jet.

The two pressures are P and p respectively, while the cross-sectional areas of the jets are the control jet C and the measuring jet M. The values are shown diagrammatically in Fig. 23.22(b).

If $\dfrac{p}{P}$ is plotted against $\dfrac{M}{C}$, a curve of the shape shown in Fig. 23.22(c) is obtained. The slope and position of the curve vary only slightly for variations of the applied pressure P, so that the same principles apply to both low- and high-pressure systems. A portion of the curve is straight for the value of $\dfrac{p}{P}$ between 0.6 and 0.8. Where non-linearity is compensated for by non-linear scale, the useful portion can be extended to the range between 0.5 and 0.9, and this is usually done with the Solex instrument. The same can be done with high-pressure systems by utilising a non-linear scale on the dial of the pressure gauge.

From the well-known equation

$$\frac{p}{P} = A - b\frac{M}{C} \quad \text{where } A \text{ and } b \text{ are constants}$$

A is the value of $\dfrac{p}{P}$ when $\dfrac{M}{C}$ is zero, i.e. when the measuring jest is completely closed, and this has proved to be almost constant at the value $A = 1.1$ for all practical values of P (15 to 1500). The value of b is the slope of the straight portion of the curve. By substituting the value of A and the maximum and minimum values of 0.6 and 0.8 for $\dfrac{p}{P}$ in the equation in turn

$$\frac{b}{C} M_{\max} = 1.1 - 0.6 = 0.5$$

$$\frac{b}{C} M_{\min} = 1.1 - 0.8 = 0.3$$

$$\therefore \quad \frac{M_{\max}}{M_{\min}} = \frac{5}{3}$$

and

$$\frac{M_{\max} - M_{\min}}{M_{\max} + M_{\min}} = \frac{1}{4}$$

$$\therefore \quad M_{av} = \frac{1}{2}(M_{max} + M_{min}) = M_{max} - M_{min}$$

From the above relations, $\frac{1}{2}(M_{max} + M_{min})$ is the average value of M_{av}; it follows that the range of variation of M available with a linear response is half the mean value of M. This means that if the measuring jets in the plug gauge of Fig. 23.22(a) have a diametral clearance of, say, 0.05 mm in the mean bore to be measured, the gauging system has a linear range of ± 0.0125 mm (0.025 mm total) for this value.

In the case of the plug gauge illustrated, the measuring jets' cross-sectional area is not the same as that of the orifice in the plug; this must be quite large in relation to the control jet. The jet area is the circumference of the orifice multiplied by the total clearance (the sum of both sides) between the ends of the orifices and the two imaginary cylindrical surfaces projected from the bore of the orifices at each end.

Still keeping in mind the linear range, where $\frac{p}{P}$ lies between 0.6 and 0.8, and suitably differentiating the initial equation, it can be shown that the pneumatic sensitivity is

$$\frac{dp}{dM} = -\frac{b}{C} \cdot P$$

when $M = M_{av}$, $\frac{p}{P} = 0.7$.

$$\therefore \quad 0.7 = 1.1 - b\frac{M_{av}}{C}$$

i.e.
$$\frac{b}{C} = \frac{0.4}{M_{av}}$$

Hence
$$\frac{dp}{dM} = 0.4 \frac{P}{M_{av}}$$

and that, on using a head of 500 mm of water and linear gauging range of 0.025 mm, the overall magnification is 4000. The calculations are: For $P = 500$ and $M_{max} - M_{min} = 0.025$, i.e. $M_{av} = 0.05$

$$\text{Overall magnification} = \frac{dp}{dM} = \frac{0.4 \times 500}{0.05} = 4000$$

When a high-pressure system is used, the water manometer is obviously unsuitable as a measuring device. The pressures may be 200 or 300 kN/m^2 and a pressure gauge, incorporating a bourdon tube or bellow system and an operating pointer over a scale, is the usual form of indication device. A bellow can be arranged within a chamber subject to the initial pressure P, although this pressure is assumed to be constant since the system compensates for slight variations as it indicates differences between P and p.

Electrical principles: There are several electrical principles which can be applied to the measurement of small displacements, e.g. strain gauges may be used. Electronic amplifiers can give magnifications of extremely high order. However, for stability and reliability, the most robust comparator of this type operates on variable inductance measured via a bridge network. Figure 23.23 shows the measuring head. The displacement of the laminated iron armature between the induction coils L_1 and L_2 puts the bridge circuit out of balance, causing the ammeter A to move and to indicate the magnitude of the displacement.

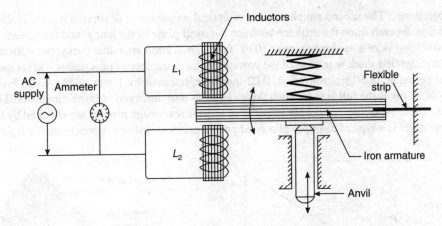

Fig. 23.23 Measuring head of electrical comparator.

23.10 OPTICAL MAGNIFICATION OF WORKPIECE

The comparators described above magnify small linear displacements in order to make them visible; an alternative to this is to magnify the workpiece so that direct measurement may be made to a high order of accuracy. There are alternative ways of achieving this.

Measuring microscope: Refer to Fig. 23.24. The objective lens is a magnifier which produces an image CD of the workpiece AB; the eyepiece is a further magnifier which makes the image CD, acting as a virtual object, form the virtual image EF. If the objective lens magnifies 6 times, and the eyepiece 10 times, the virtual image will be 60 times the full size.

In a toolmaker's microscope or measuring microscope, the work is mounted on a rectangular coordinate table having micrometer control of the displacements made by either slide. Cross lines in the focal plane of the eyepiece measurements can be made either by having a graduated circular table as part of the work stage, or by having a rotatable graticule line in the eyepiece. It is generally more convenient to measure very small work by means of a measuring microscope than by contact methods, e.g. the measurement of the smaller BA screw threads.

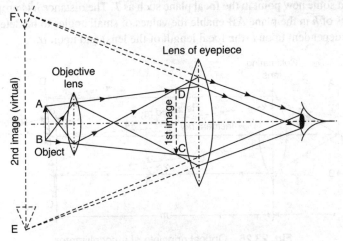

Fig. 23.24 Optical principle of microscope.

Optical projection: The second simple system of optical projection is illustrated in Fig. 23.25; the degree of magnification depends upon the distance between the focal plane of the lens P and the screen. Instruments which can handle work of a moderate size at 50 to 100 magnification are rather bulky; but optical projection, resulting in a magnified shadow outline of the workpiece, has a number of plus points. Direct measurements by scale can be made; at 50 magnification, 0.02 mm is represented by 1 mm on the scale. Profiles can be drawn at 25 or 50 times the full size for such things as press tool dies even of three-dimensional complexity, and this is the cheapest method of profile inspection. Most screw gauge profiles are checked by this method. Baty shadowmaster is a typical example of a good profile projector using a monochromatic light source.

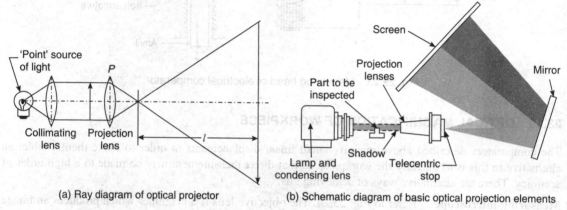

Fig. 23.25 Optical projection.

23.11 MEASUREMENT OF SMALL ANGULAR DISPLACEMENTS

The optical device for amplifying an angular displacement is the *autocollimator*. Figure 23.26 illustrates the optical principle upon which the instrument works. Suppose a ray of light is emitted from source S at the middle of the focal plane of the collimating lens. The lens will convert this into a parallel beam which is then reflected from some working surface such as CD. If CD is inclined at angle α, from the principle of the optical lever the reflected ray makes an angle of 2α with the incident ray. The reflected beam re-enters the collimating lens to be refocused at some new point in the focal plane such as T. The distance h is proportional to the angle, and the measurements of h in the plane AB enable the values of small angles to be determined. Note that the value of h is directly dependent upon l (the focal length of the lens) and upon α.

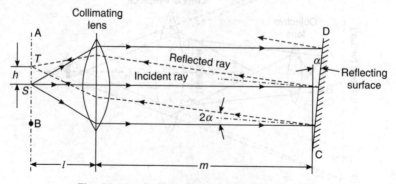

Fig. 23.26 Optical principle of autocollimator.

The earth's gravitation has a fixed direction for any small geographical area, and may be employed as a datum for measurement of angles. The plumb-bob has its counterpart in instruments based upon a pendulum. Figure 23.27 shows such an instrument, called the *Talyvel*, designed by Taylor and Hobson of the UK. At the end of the pendulum there is a soft iron portion a which displaces, under gravitational force, between inductance coils C_1 and C_2 depending on angle θ. A bridge circuit of the kind shown in Fig. 23.27 feeds a signal to the meter which measures displacement from a datum in either minutes (or seconds) of angle or millimetres per metre. The pendulum and indicating meter have a damping system; the instrument has a range of about $\pm 2°$ and can be read to increments of one second of arc.

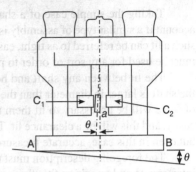

Fig. 23.27 Talyvel (developed by Taylor and Hobson).

The spirit level is a well-known alternative instrument for measuring small angular displacements relative to the horizontal datum—the level of a liquid at rest. Figure 23.28 shows the main features of this instrument. A 20-second level has a displacement of 2 mm for a tilt of 0.01 in 1000 and is representative of the precision class of this inexpensive and very useful piece of equipment. The principal use of both Talyvel and spirit level is testing straightness and flatness.

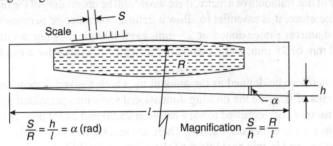

$$\frac{S}{R} = \frac{h}{l} = \alpha \text{ (rad)} \qquad \text{Magnification } \frac{S}{h} = \frac{R}{l}$$

Fig. 23.28 Precision spirit level.

23.12 LIMITS, FITS AND TOLERANCE

Limits and fits

In the design and manufacture of engineering products, a great deal of attention has to be paid to the assembly and fitting of various components. In the early days during the nineteenth century, the majority of such components were actually mated together, with their dimensions adjusted until the required type of fit was obtained. These methods called for craftsmanship of a high order, and very fine work was and is produced during the crafts era as well as today.

The present-day standards of quantity production, interchangeability, and continuous assembly of many complex components could not exist under such a system. Many of the exacting design requirements of modern machines can be fulfilled without reproduction with precision on a mass scale.

Modern mechanical production engineering is based on a system of limits and fits, which forms a schedule or specification to which manufacturers can adhere.

Throughout this chapter, frequent reference will be made to holes and shafts, and it should be inferred that these terms refer to the two main classes of production engineering components. A shaft is taken to refer to any 'male' component, generally circular in section. But this is not necessarily so, as there are components such as a plug, axle, spindle, locating spigot, etc., which are located in a hole, bore recess, or other form of 'female' location.

Taking the simple case of a shaft and hole fitting together, a number of factors have to be taken into account if a similar type of assembly is to be repeated on a number of different components. Speaking loosely, such a fit can be referred to as tight, easy, smooth, loose, and so on, but is obvious that more precise definitions must be used for any sort of order to prevail.

The fit between any shaft and hole must fall into one of two categories: interference and clearance. If the shaft is larger in diameter than the hole, they will assemble with an interference fit, and a certain amount of force will be necessary to fit them together. If the shaft is smaller than the hole, they will assemble more easily, and this will be a clearance fit. The limiting case is where the shaft and the hole have the same diameter, but even in this case, accurate measurement usually reveals some difference in diameter between them.

The foregoing description must not be confused with classes of fit, of which there are three—clearance, interference and transition fits; these will be defined and described later in this chapter.

Limits and tolerances

It is never possible to make anything exactly to a given size or dimension. However carefully it may be made, some method or measurement could be found which would detect an error from nominal size or variations in dimension. The most accurate method of measurement available would leave some uncertainty.

For example, a shaft made to a nominal size of 25 mm might appear to be exactly 25 mm when measured with caliper and rule, whereas a micrometer might detect either an error in size or variation in diameter. Even if measured to an accuracy of one millionth of a metre, there would still be uncertainty of the order of a fraction of this.

In engineering, therefore, it is essential to allow a definite tolerance or permissible variation on every critical dimension. A diameter dimensioned at 25 mm, even though it is not a fitting dimension, is not expected to be, say, 22 mm or 28 mm, and a commonly accepted tolerance for a machined diameter of this size is ± 0.25 mm.

A tolerance can therefore be defined as the amount by which a given dimension is permitted to vary. Limits, or more fully limits of size, are the limiting dimensional variations permitted by a given tolerance; the 25 mm shaft may, for instance, be permitted to vary between 25.00 and 24.99 mm. In this case, the tolerance is 0.01 mm and the limits are the dimensions stated. There are several ways in which such limits can be stated on drawing or specification, and in this example the following are most likely to be used:

$$\begin{cases} 25.00 \\ 24.99 \end{cases} \quad \text{or} \quad 25.00 \begin{matrix} +0 \\ -0.01 \end{matrix} \quad \text{or} \quad \begin{matrix} 25.00 \text{ (max)} \\ -0.01 \end{matrix}$$

Tolerances on dimension may be either unilateral or bilateral. A unilateral tolerance is one which lies wholly on one side of the basic dimension; a bilateral tolerance is one whose limits lie on either side of the basic dimension, but which are not necessarily equally displaced about it.

The above example represents a unilateral tolerance.

An example of bilateral tolerance would be:

$$25 \pm 0.01 \quad \text{or} \quad 25.00 \begin{matrix} +0.01 \\ -0.01 \end{matrix}$$

In some cases, the nominal size lies right outside the limiting dimensions. For example:

$$25.00 \begin{matrix} -0.05 \\ -0.10 \end{matrix}$$

This method of tolerancing is frequently used in preference to the more obvious one:

24.95	+0.05
or 25.00	
24.90	−0.01

since it contains the nominal size (25 mm).

Fits and allowances

We have seen how the dimension of a shaft or hole can be controlled with desired limits, at least by specification, and it should therefore be possible to arrange for any desired fit between the hole and the shaft.

A clearance fit could be obtained by making the lower limit on the hole equal to or larger than the upper limit on the shaft. Any hole and any shaft made to these tolerances would assemble with a clearance fit with certainty.

An interference fit could be obtained with equal certainty by making the lower limit on the shaft equal to or larger than the upper limit on the hole.

Between these two conditions lies a range of fits known as *transition fits*. These are obtained when the upper limit on the shaft is larger than the lower limit on the hole, and the lower limit on the shaft is smaller than the upper limit on the hole. It must be realized that transition fits exist only as a class; any actual hole and shaft must assemble with either clearance or interference fit. The three classes of fit are illustrated diagrammatically in Fig. 23.29(a). The conventional diagram of fits is shown in Fig. 23.29(b).

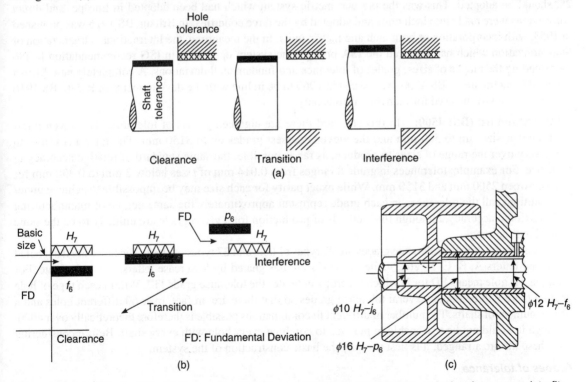

Fig. 23.29 (a) Types of fit, (b) Conventional diagram of fits, and (c) Belt drive unit showing appropriate fits.

The difference between the maximum shaft and the minimum hole is known as the *allowance*. In a clearance fit, this is the minimum tolerance and is positive. In interference fit, it is the maximum interference and is negative.

Figure 23.29(c) shows fits suitable for a pulley on a shaft: $\phi 10$ H_7–j_6 highlighting a transition fit (sliding fit); a bush press-fitted in housing highlighting an interference fit $\phi 16$ H_7–p_6; and the shaft having clearance fit with the inside diameter of the bush indicated as $\phi 12$ H_7–f_6 facilitating lubrication due to formation of oil film in the clearance.

Systems of limits and fits

In theory, a designer may choose any fit and tolerance which he thinks will suit his requirements, but common experience has made it possible and desirable to have certain standardized systems of limits and fits. The first British standard in this field was BS 164, issued in 1906. This was revised in 1924 and again in 1941, but never gained really wide acceptance.

The Limits and Fits Committee of the British Standards Institution had again commenced revision of this specification shortly after the war with a view to widening its usefulness and application. Much work had already been done when it become apparent that much wider standardization between Britain, the USA and Canada would be possible, due to discussions which had been taking place between the armed services of the three countries. Further co-operative discussions in the field of standardization among these countries have been known as the ABC Conference.

It was decided that the International Standards Association (ISA) system, detailed in ISA Bulletin 25, should be adopted. This was the pre-war metric system which had been adopted in Europe, and direct conversions were made into inch units and adopted by the three countries. In Britain, BS 1916 was published in 1953, with incorporation of both inch and metric units. In the meantime, the International Organization of Standardization which had replaced the ISA brought the system up-to-date in ISO recommendation R 286 by extending the ranges of sizes, grades of tolerance and fundamental deviations. A completely new British standard in metric units, BS 4500, was issued in 1969 to be in line with the developments in R 286. BS 1916 must therefore now be used for inch dimensions only.

British Standard (BS) 4500: In this standard there are eighteen grades of tolerance. The seven finest grades cover sizes up to 500 mm and the eleven coarsest grades up to 3150 mm. The tolerances in each grade vary over the range of sizes to produce, as far as possible, the same standard of relative accuracy in each size. For example, tolerances in grade 8 ranges from 0.014 mm of sizes below 3 mm to 0.330 mm for sizes between 2500 mm and 3150 mm. While exact parity for each size may be impossible to achieve under all conditions, all the tolerances in each grade represent approximately the same degree of manufacturing and functional accuracy, although the methods of production for a given grade are unlikely to be the same throughout the range.

Associated with the tolerance grades are 27 types of hole and 27 types of shaft. The holes are designated by capital letters A, B, C, H, etc., while the shafts are designated by lowercase letters, e.g. a, b, f, etc. For example, H_7 hole means FD for H which is zero; 7 indicates the tolerance grade IT7. Within each type of hole or shaft there will be one or several tolerance grades so that there are, in fact, over 60 different holes and a similar number of shafts. The number of different fit combinations possible is therefore theoretically over 4000, although it would not be reasonable in practice to combine every hole with every shaft. Before considering how these fits are arranged, it is best to study the basic construction of the system.

Grades of tolerance

These are built up from fundamental tolerances which are multiples of the fundamental tolerance units for a diameter (or size) D.

$$i = (0.45 \times \sqrt[3]{D} + 0.001 \times D) \times 0.001 \text{ mm}$$

or
$$i = (0.052 \times \sqrt[3]{D} + 0.001 \times D) \times 0.001 \quad \text{(in B.S. 1916 only)}$$

where D is measured in millimetres and inches respectively.

D is not always exactly equal to the size under consideration, but is the geometric mean of the range given in the tables of BS 4500, which includes the size. For example, the size 25 mm falls in the range of 24 to 30 and the corresponding value of D is $\sqrt{24 \times 30} = 26.8$.

The eighteen tolerance grades are designated IT1 to IT16. For the standard limits and fits specified, grades IT6 to IT16 only are used. The value of IT6 is $10 \times i$ and the higher values decrease from this in geometrical progression, based on the five-series preferred numbers in which the factor is the fifth root of 10. Thus

$$IT7 = IT6 \times 10^{1/5} = IT6 \times 10^{0.2}$$
$$IT8 = IT7 \times 10^{0.2} = IT6 \times 10^{0.4} \quad \text{and so on}$$

Hence $\quad IT11 = IT6 \times 10$

and $\quad IT16 = IT6 \times 100$

and so on.

Grades IT1 to IT5 are arranged in different steps which have been arbitrarily chosen; these values were intended for gauge tolerances in the international system, although they have not been adopted for this purpose in the British Gauge Standards.

EXAMPLE 23.3 Working from the basic principles, find suitable tolerances for (a) 82 mm IT6 and (b) 440 mm IT12. Compare the calculated values with the rounded values given in BS 4500.

Solution 82 mm is in the size range 80–120, while 440 mm is in the size range 400–500. From BS 4500 Appendix A, IT6 = 10 and the tolerances increase in accordance with the R5 series as the IT number increases. Hence

IT series	6	7	8	9	10	11	12	13
R5 series	1	1.6	2.5	4	6.4	10	16	25

and so on. Therefore, IT12 tolerances are 16 times the IT6 tolerances.

For 82 mm IT6 D for range $= \sqrt{80 \times 120} = 98$

$$i = 0.45 \sqrt[3]{98} + (0.001 \times 98) = 2.173 \text{ } \mu m$$

$$IT.6 = 10i = 21.73 \text{ } \mu m$$

For 440 mm IT12

$$D \text{ for range} = \sqrt{400 \times 500} = 447$$

$$i = 0.45 \sqrt{447} + (0.001 \times 447) = 3.888 \text{ } \mu m$$

$$IT12 = 16 \times IT6 = 160i$$

$$= 160 \times 3.888 = 622 \text{ } \mu m \quad \text{(BS 4500 gives 630 } \mu m\text{)}$$

ISO system of limits and fits: The International system of limits and fits is derived by the International Standards Organization in Geneva, Switzerland. This organization takes the best features of standards from UK, USA, Germany, Russia, etc. to arrive at an International standards. For limits and fits, it covers sizes generally upto 3150 mm. As shown in Fig. 23.30 there are 28 tolerance bands for each basic size: lower case letters are for shafts and upper case letters are for holes. The upper deviation for shafts *a* to *g* is below the zero lines as shown. The lower deviation for holes A to G is above the zero line. The shaft h is called the basic shaft and hole H is called basic hole. Both h and H have deviation zero.

Fits: Shaft a to g give clearance fits in general. j shafts tends to give transition fits. Shafts k to z_c give an interference fit because the tolerance bond falls above zero line. It will be noted after studying Fig. 23.30 that a very large number of fits is possible if any shafts is mated with any hole. Looking to this problem various standardizing holes have been limited and standardized the type of fits (such as the BS 4500) taken up in the next section. It may be noted by the reader that we have preferred to use those British Standards which are accepted by the ISO.

Deviation for holes designated A to H correspond exactly in value with those for shafts a to h, but are in opposite directions. A and a are the largest deviations, holes being positive and shafts negative, and the deviations for both H and h are zero. Thus the first five designations represent a clearance fit system. The remaining groups J to Z (holes) and j to z (shafts) do not correspond in their deviation in quite the same way; they are intended for use in interference and transition fits, although some are intended for other uses where fits are not involved at all. The diagram in Fig. 23.30 shows the tolerance zones for both holes and shafts specified in BS 1916. These will also apply to BS 4500. The rules for calculating FD values are rather complicated. BS 4500 Appendix A gives the rules. A typical example is f, $F = 5.5\ D^{0.41}$, where f is negative and F is positive. Full details are beyond the scope of this section.

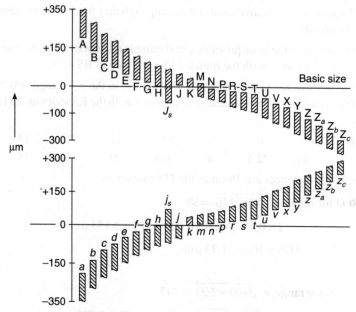

Fig. 23.30 Tolerance for holes and shafts as per ISO standard.

Selection of fits

The various types of fits could be obtained from BS 4500 from the association of two variables:

   (i) 18 tolerance grades;
  (ii) 27 fundamental deviations, i.e. from the zero line.

It is highly impractical to have such a large number of combinations. Keep in mind that

   (i) IT1 to IT5 are meant for precision standards;

(ii) IT6 to IT11 can be used for fits, gauges, etc; and
(iii) IT12 to IT16 are too coarse to be used for fits; and BS 4500 has recommended only a few select fits as shown in Fig. 23.31.

These are referred to as *selected fits* and may be either hole-based or shaft-based. Figure 23.31 shows these fits on a hole basis.

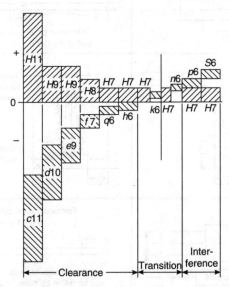

Fig. 23.31 BS 4500 selected fits on hole basis.

23.13 GEOMETRICAL TOLERANCES

It is quite easy to specify tolerance to a linear dimension. However, specifying tolerance or limits of variation permissible to cover geometrical features of a component such as circularity, squareness and symmetry is not that easy. BS 308 has developed a clearcut scheme to express such dimensions. First of all, taking the case of a tapered shaft the tolerance is to be specified, so that an acceptable shaft must have a top surface which should lie between two perfect and concentric cones as shown in Fig. 23.32 within a defined zone. Other geometrical tolerances in BS 308 are:

(i) Straightness
(ii) Squareness
(iii) Concentricity
(iv) Symmetry
(v) Position.

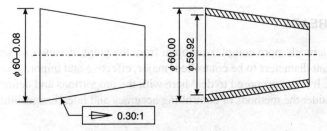

Fig. 23.32 Taper specifications and its interpretation.

Geometrical tolerances given before are shown in Fig. 23.33.

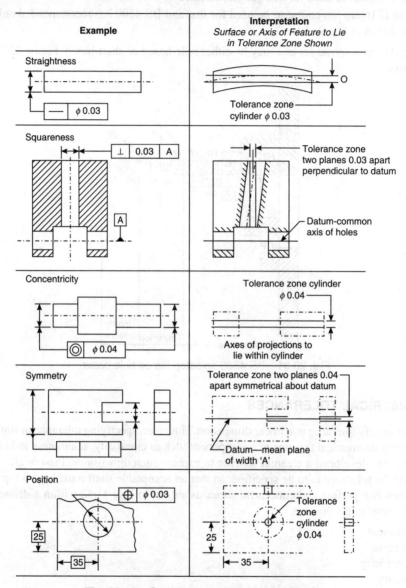

Fig. 23.33   Geometrical tolerances in BS 308.

23.14 SCREW THREADS

Screw threads present a much more complex problem from the point of view of tolerance than plain work. There are three important diameters to be controlled: major, effective and minor, as well as pitch, angle, and general form of thread. It is not proposed to deal here with the proportions and definitions of screw threads. It is appropriate to consider the methods of controlling accuracy and interchangeability.

ISO metric designation BS 3643 provides three main thread grouping—a COARSE thread series, a FINE thread series and a CONSTANT pitch series. Nominal series showing 1st, 2nd and 3rd choices and ranging from 1 mm to 300 mm are derived from a preferred size series. For average work coarse pitch series, each nominal size has its designated pitch (range 2.25 mm to 6 mm). The fine and constant pitch provides design conditions; the fine pitch series does give a tremendous range of ostensibly standard sizes, but it will be a very long time, if ever, before threading equipment and screwed parts of all these sizes are available from stock. Wherever possible, thread sizes should be selected from the coarse pitch series.

BS 3643 also provides for different qualities of screwed work as shown below:

Close fit	Designation 5H	(nut) 4th
Medium fit	Designation 6H	6g
Free fit	Designation 7H	8g

The numbers relate to the tolerance magnitude and the letters to the deviation from basic size.

The full designation M8 × 1.25 6H/6g stated on drawing would imply an ISO metric thread, nominal diameter 8 mm, pitch 1.25 mm, nut to 6H bolt to 6g to (i.e. medium fit.)

23.14.1 Tolerance for ISO Metric Threads

Figure 23.34 illustrates the tolerance zone for the close fit (5H/4h). The H/h symbols are for zero deviation and so the maximum material condition is controlled by the basic form. Notice also the following:

(a) The tolerance zones are greater at the root and the crest than along the flanks.
(b) A unilateral system is used, with bolt tolerance below the basic size and nut tolerance above the basic size.
(c) A clearance exists at the minor diameters.
(d) The upper limit of major diameter of the nut is not specified; it is controlled by tap dimension conforming to BS 949.
(e) Because the basic form is not symmetrical about the pitch line (3/8H above, 1/4H below), the relatively large-diameter tapping holes are breakages. Figure 23.35 shows the large amount of extra material which need to be removed to tap-symmetrical form. This is the principal difference between the ISO metric form and the earlier metric SI thread form.

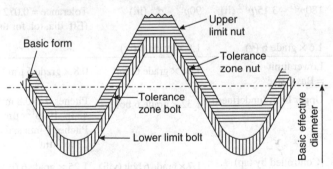

Fig. 23.34 Tolerance zones for close fit (5H/4h).

(f) From the tolerance zones shown in Fig. 23.34, it follows that three different types of errors may be allowed:
   (i) Error of flank angle;
   (ii) Error of pitch over the length of fitting;
   (iii) Error of effective diameter.

It is equally necessary to realize that no combination of these errors should cause the profile to lie outside the tolerance zone at any point along the fitting length of an assembly.

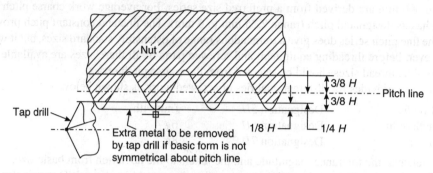

Fig. 23.35 Material to be removed to tap-unsymmetrical form of ISO metric thread.

23.14.2 Magnitudes of Tolerance and Deviation

As for plain work, tolerances and deviations are in proportion to the basic size to maintain constant quality for different work sizes. The rational basis of the system used is complicated by the fact that the error of the pitch becomes more serious with the flank angle error, and the range of the fitting length of screwed parts becomes more serious with the progressive pitch error. It is not possible to relate tolerances only to the work tolerance, as in plain work.

The tolerances and deviations given in BS 3643 are based on ISO specification 965/1: 1973(E) and are derived as shown in Table 23.4.

Table 23.4 Screw thread tolerances

	Tolerance grade	Major diametre in μm	Effective diameter in μm	Minor diameter (unit as shown)
Bolts	4	0.63 × grade 6	0–63 × grade 6	Upper limit: nom–1.2268p × FD mm (vii)
	6	$180p^{2/3} - 3.15/p^{1/2}$ (ii)	$90p^{04} - d^{01}$ (iii)	Tolerance = 0.072 p + (Eff. dia. tol. for the grade) mm (vi)
	8	1.6 × grade 6 (v)	1.6 × grade 6 (iv)	
Nuts	5	Lower limit = Basic size	1.06 × grade 6 bolt	0.8 × grade 6 μm
	6	Upper limit (undefined)	1.32 × grade 6 bolt	Pitches 0.2–0.8 mm $433p - 190p^{1.22}$ μm Pitches 1 mm and above $230p^{.7}$ μm
	7	(Controlled by tap)	1.7 × grade 6 bolt (viii)	1.25 × grade 6 (ix)

FD deviation g (for bolts); FD = 15 + 11 p μm (p = pitch)   (i)

Note: The above tolerances apply where the nominal length of engagement L_N is within the limits as shown below:

$$L_{N\,min} = 2.24\, pd^{0.2}\ \text{mm}$$
$$L_{N\,max} = 6.7\, pd^{0.2}\ \text{mm}$$

where d is the smallest diameter of the range of which the tolerances are common.

The value of d in the tolerance formula $90P^{0.4}d^{0.1}$ is the geometric mean of the diameter range. The calculated values have to be rounded according to the R40 series. However, the information given here is to show the fundamentally logical basis of a screw thread system of limits and fits, and reference to BS 3643 should always be made to obtain the actual dimensions. Figure 23.36 shows the three different grades of work by tolerance zones drawn to scale. The 'close' class should only be used where the highest quality is essential, because it is relatively expensive to produce the condition where quick and easy assembly is required and where threads may become dirty or damaged.

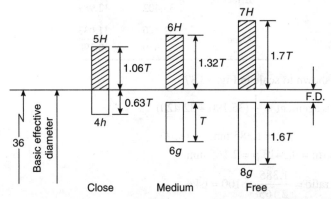

Fig. 23.36 Three classes of fit reprinted to scale.

EXAMPLE 23.4 Using Table 23.4 and Formulae (i) to (ix), find limits for the three important diameters of a M36 × 4 "free"class screw thread assembly. The diameter falls within the range 30–39; the length of engagement is within L_N limits. Find the least possible depth of engagement and express this as a percentage of the basic depth.

Solution The required assembly is 7H/8g free class.

Bolt dimensions

(i) FD = 15 + (11 × 4) = 59 μm (rounded value = 0.06 mm)

(ii) Mean diameter of range $(d) = \sqrt{30 \times 39} = 34.2$ mm

(iii) Effective diameter tolerance grade 6 = $90 \times 4^{0.4} \times 34.2^{0.1} = 223$ μm

Tolerances (grade 8)

(iv) Major diameter = $1.6 \left(180 \times 4^{2/3} - \dfrac{3.15}{\sqrt{4}} \right) = 724$ μm (rounded value = 0.75 mm)

(v) Effective diameter tolerance = 1.6 × 223 = 357 μm (rounded value = 0.355 mm)
(vi) Minor diameter tolerance = (0.072 × 4) + 0.355 = 0.643 mm
(vii) Upper limit of minor diameter of bolt = 36 − {1.2268 × 4) + 0.060} = 31.033 mm

Nut dimensions

Tolerances (grade 7)

(viii) Effective diameter tolerance = 1.7 × 223 = 379 μm (rounded value = 0.375 mm)
(ix) Minor diameter tolerance = 1.25 (230 × $4^{0.7}$) = 758 μm (rounded value = 0.750 mm)

The limits for M36 × 4–7H–8g thread assembly are shown below.

Diameter	Nut	Bolt
Major	36.000 +	$\dfrac{35.940}{35.190}$
Effective	$\dfrac{33.777}{33.402}$	$\dfrac{33.342}{32.987}$
Minor	$\dfrac{32.420}{31.670}$	$\dfrac{31.033}{30.390}$

These limits are shown to scale in Fig. 23.37.

Least depth of engagement = $\dfrac{1}{2}(35.190 - 32.420)$

= 1.385 mm

Depth of basic form = $0.5413\, p = 2.165$ mm

Minimum depth ratio = $\dfrac{1.385}{2.165} \times 100 = 64\%$

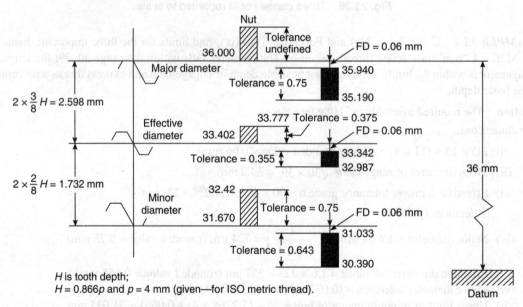

H is tooth depth;
$H = 0.866p$ and $p = 4$ mm (given—for ISO metric thread).

Fig. 23.37 Example 23.4 with tolerances shown.

23.15 LIMIT GAUGING

There are several methods available for the control of component dimensions in a system of limits and fits. Each component could, for example, be measured with an instrument giving a suitable accuracy, and this method is often adopted, particularly for closely limited work. The system of limit gauging has the advantage that it can be operated in many cases by quite unskilled persons. This system is called *Taylor's principle of gauging*.

Considering the simplest case as an example, the limit gauges for a hole consist of two cylindrical plugs which are made to scale for the limiting values of the hole dimension. The plug made to the lower limit of the hole is known as a Go gauge, and this will obviously enter any hole which is not smaller than the lower limit allowed. The plug made to the upper limit is known as a Not Go or No Go gauge, and will not enter any hole which is not larger than the upper limit allowed.

A pair of gauges used in this way will control the production of components strictly between the limits specified.

For the gauging of shafts, the ring or gap gauges can be used in a similar manner. A ring gauge consists of a piece of metal in which a hole of the required size is bored, and a gap gauge usually consists of a placate or frame with a parallel-faced gap of the required dimension.

Go and Not Go plugs are often arranged at either end of a common handle. Go and Not Go gaps are also usually included in the same frame. Typical plug and ring gauges are shown in Fig. 23.38.

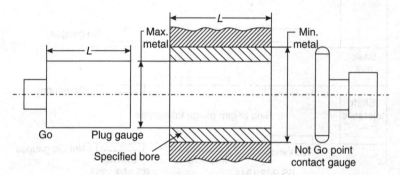

Fig. 23.38 Limit gauging to bore.

We have so far assumed that the Go and Not Go gauges have been made exactly to the upper and lower limits of the work tolerance, but it is obvious that the gauge maker must have some tolerance, however small, for the manufacture of the gauges. There is also the fact that, as soon as a gauge is put into service, it starts to wear. This necessitates a system of gauging which will ensure, as far as possible, that all work passed by the gauges will be within the specified tolerances, and also that the gauge tolerances do not encroach more than is necessary on the work tolerances.

23.15.1 Gauge Tolerances

Before the First World War, almost the only people to carry out and organize inspection of components by gauging were government departments who purchased armaments and other stocks from manufacturers. Since all these supplies were the subject of legal contracts, the government departments were under an obligation to accept every component which was strictly within the specifications and tolerances laid down. Hence, any limit gauges used had to ensure that no component which was within its tolerance, even though it might be right on the extreme limit, should be rejected. Since every gauge has to have a manufacturing tolerance, it was therefore decided that official gauges used should not encroach on the work tolerance in any way. Consequently, gauge tolerances were always outside the work tolerances, and such gauges were known as 'inspection' gauges. The manufacturer, on his part, used gauges during manufacture and possibly for final inspection, and his greatest concern was to ensure that the work he submitted to the official inspectors would pass their gauges. This led to the establishment of 'workshop' gauges, and the tolerances were so that they did encroach on the work tolerance.

This system of gauging held sway for many years up to 1953, and in almost every case the tolerance of the gauges was about 10% of the tolerance of the work they were gauging. The disposition of the tolerances of the workshop and inspection gauges, relative to work tolerances for holes and shafts, is shown in Fig. 23.39.

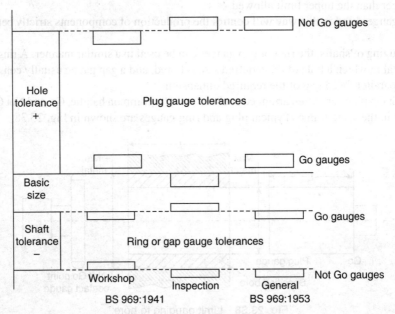

Fig. 23.39 Tolerance zones for workship, inspection and general gauges.

Plain gauges: Until 1941, there was no British standard for plain limit gauge tolerances, although the NPL had for many years specified its own tolerances for screw gauges. The workshop and inspection principle using a 10% gauge tolerance was however in almost universal use and in 1941 a British standard, BS 969, was published on these lines.

A few years after the Second World War, this standard was under review by BSI and it was found that while there was considerable merit in the old 'workshop' and 'inspection' gauge system, it had a number of serious defects when viewed in the light of modern precision engineering practices. One objection, which may be seen as more apparent than real, was that the use of inspection gauges permits the acceptance of work which is outside tolerance. It was felt that this may have influenced many designers to reduce their work tolerances so as to ensure that work would not exceed somewhat wider tolerance which they would otherwise have in mind. One could of course argue that full use of the workshop gauge system would safeguard this point, and in fact, some organizations have used workshop gauges for inspection as well as production. The main advantage of this was economic rather than technical, since a workshop Go gauge could be allowed to wear right through its workshop tolerance and into the inspection tolerance zone before being scrapped, whereas an inspection gauge had at most only its own tolerance zone for wear. The use of workshop gauges for both production and inspection necessitated careful selection and placing of these gauges since, if friction between the two functions were to be avoided, the gauge used for inspection had to be nearer the work limit than the gauge used in production.

The outcome of the revision of BS 960, which was published in 1953, was the specification of a new single type of gauge to be known as the *general* gauge. Again, the economics of gauging was seriously considered and it was felt that industry as a whole would gain if gauge manufactures stocked only one series of gauges instead of two for every standard size required. The general system only partly follows the principle of ensuring that all work is strictly within tolerance. The Go gauge tolerance lies within the work tolerance zone and maintains this principle on the maximum metal condition which is critical for assembly and interchangeability. A hole and a shaft will always assemble it they do not exceed their respective limits at the maximum metal condition. To ensure that the gauging system should not encroach unduly on the tolerance available for manufacture, it was decided to compromise with the Not Go gauge tolerance and place this outside the work tolerance zone. These gauge tolerance zones are also shown in relation to hole and shaft tolerances in Fig. 23.40.

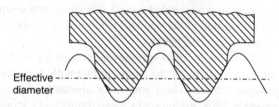

Fig. 23.40 Truncated form of effective Not Go gauge.

It will be seen that the Go gauge has wear allowance: this is specified where the value of tolerance is sufficient. No limit gauging system can be devised that would be perfect (unless of course gauges did not wear and could be made exactly to size zero) and this general system seems to be the best compromise. There can be very few component designers who are so sure of the exact value of the tolerance which can be allowed on a particular feature or component that they cannot permit the possibility of a 10% extension of tolerance due to the Not Go gauge at the minimum metal condition; there it cannot affect interchangeability. There are very few cases where this would be important; the designer must obviously specify slightly closer tolerances than he would otherwise do.

One point must always be borne in mind in gauging inspection and measurement—that a certain tolerance is specified. It does not mean that work or gauge will be at one or the other extreme. For example, under normal conditions of production, it is most unlikely that extremes of fit of a hole and a shaft will be found when any pair comes together on assembly. Statistically, it is much more likely that any assembly will be fairly close to the average fit from the tolerances. Similarly, gauge tolerances which lie outside the work tolerance will only occasionally lead to the acceptance of work which is outside tolerance: on average, the gauge is likely to lie outside work tolerance by only half its own tolerance or even less, since wear always takes place.

23.16 SCREW GAUGE

In parallel with limit gauge systems for plain work, screw threads were for many years gauged under the workshop and inspection systems of limit gauging. Long before there were any published standards on limit gauge tolerances, the NPL had issued in 1916 a publication, *Notes on Screw Gauges*, which included limits of tolerance for workshop and inspection Go and Not Go screw gauges, covering both plugs and rings. As with plain limit gauges, a wartime emergency specification, BS 919, was published in 1940 for screw gauges and the previous NPL tolerances were reproduced in this.

At the same time as the revision of plain limit gauge tolerances was undertaken by the BSI, screw gauge tolerances were also reviewed, and BS 919 was re-issued in 1952, specifying general gauges in place of the old workshop and inspection gauges. Exactly the same principles apply for general crew gauges as for plain limit gauges, that is to say, the Go gauge is placed inside the work tolerance and the Not Go gauge is placed outside it. Some deviation for the plain limit gauge system was necessary in the provision of tolerances for setting plugs for adjustable caliper or ring gauges and the provision of reference plug gauges and reference setting plugs. Go and Not Go plain plug gauges for minor diameters of threaded bolts and similar caliper gauges of major diameters of threaded shafts were also specified.

The old NPL tolerances on effective diameter of plug gauges, which were in the form of single figures which had to include the effects of pitch and angle errors, encouraged manufacturers to keep simple effective diameters near the lower limit to allow for possible pitch and angle errors. The 1940 edition of BS gave a simple effective diameter tolerance, which included a wear allowance and a separate figure for the effects of pitch and angle errors. The 1952 edition of the standard continued this policy, making the pitch and angle equivalents vary with the pitch and size of the gauge.

A further revision of BS 919: Part 1: 1960 specifies tolerances for unified threads only, and further separates tolerances on the pitch from those on the angle, in line with American practice. It also includes tolerances for adjustable ring gauges. Separate check gauges for adjustable and solid ring gauges are available.

Part 3 of BS 919 was issued in 1968, giving screw gauge limits and tolerances for ISO metric threads. This give tolerances for pitch (effective) diameter for Go and Not Go plug, ring and caliper gauges and also for check plugs for setting calipers and checking ring gauges. Tolerances on major and minor diameters were also given wherever appropriate. Flank angle tolerances were given for each pitch, as well as pitch tolerances for various lengths of thread. As in the case of the earlier parts of the standard, effective diameter tolerances on the gauges are related to the effective diameter tolerances on the work.

23.16.1 Gauging Principles

While discussing Taylor's theory of gauging in Section 23.15, it has already been shown that the Go gauge operates always at the maximum metal condition of the component for the safeguard of the assembly and the interchangeability of the mating components. It follows that, as far as possible, it should also be the geometrical equivalent of the mating part. It will not be the dimensional equivalent where clearances or interferences have to be ensured, since its one dimension will have to include these allowances. Thus, the Go gauge should be of full form.

Not Go gauges, on the other hand, should check only one dimension of the component or feature at a time, in order to search out any dimension which is outside the minimum metal limit. The very fact that a Go gauge does go in the component ensures that none of the dimensions involved in the component, including errors of form such as straightness, roundness, squareness, and so on, is within the maximum metal limit. However, the Not Go gauges were designed to check several dimensions simultaneously; one of these dimensions, which is within the minimum metal limit, would be sufficient to keep the gauge out, even though some or all of the other dimensions are outside the limit. Strictly speaking therefore, each dimension must have its own Not Go gauge, which must be so designed that it can check out any part of the dimension which may be outside the limit.

The simplest example to illustrate this is the plug gauge designed to check a cylindrical hole. The Go gauge should be of the same diameter (within its own tolerance) as the lower limit of the hole and should also be of the same length as the hole or the mating component, whichever is the least. The Not Go gauge should be of very restricted form, possibly a pin gauge, so that it can be used to search for any possible dimension surface and to check this at any position along the length of the hole and not only at the outer part. Similarly, the Go gauge for a shaft should be a full-length ring gauge, while the Not Go gauge should be a narrow-gap gauge.

This principle of limit gauging was the subject of a patent by William Taylor, of the firm of Taylor and Hobson, in 1905. It is not clear whether any royalties were paid for its use and the principle has been used widely since then, with very few people realizing that any patent was attached to it or, in fact, that it was a definite idea of one man. The term *Taylor's Principle* has been generally applied to it only since the Second World War.

As with most principles, it is seldom possible to keep to the ideal completely in practice and it would, in fact, be very inconvenient to do so in most applications. The usual Go plug gauge, so often, is not the full length of the component and, in the smaller sizes which form the greatest bulk of the work done, the Not Go end is a fully cylindrical ring gauge for shafts which are seldom used, except when absolutely foolproof interchangeability must be assured (for example, where one manufacturer's product has to be assembled with that of other makers, perhaps in several different parts of the world). While a gap gauge is ideal as a Not Go, it is imperfect as a Go. How then can accurate inspection be carried out with this equipment?

The answer is that it cannot—if the whole responsibility for quality is left to the final inspection of the work done. The whole production process quality must be under full quality control and those responsible for both production and inspection must always be aware of the standards of accuracy being achieved and the trends of error which are present. Taking the idea to its logical conclusion, no system of checking or measurement is one hundred per cent certain. Inspection, no matter how thorough, is not solely a matter of digital counting or assessment; it is a sampling procedure. It is true that a really full-form Go gauge is as near as one is likely to get to complete inspection at one end of the tolerance zone, but even here the certainty depends on 100% accuracy of the gauge which has almost certainly been checked by direct measurement. Direct measurement, using a micrometer on shaft, for example, involves taking simple measurements at a chosen but nevertheless finite number of places and deciding on the probable size of the piece. Experience tells us how likely or unlikely it is that we have missed a particularly high or low spot. From our knowledge of the production process. say turning or grinding, we probably know that can be discounted.

There are other traps for the unwary, such as the effects of lobing, A cylindrical part, either hole or shaft, can appear to be round to a high degree of accuracy when measured or gauged by a direct diameter-measuring device, such as a micrometer or gap gauge, but may be a long way from being round and may completely fail to assemble with a mating component which apparently ought to fit easily. Centreless grinding can produce this effect on a shaft, and a boring or internally ground component held in a three-jaw chuck will almost certainly produce a three-lobed effect to some degree. Similarly, ovality cannot be detected with a 120° three-point internal micrometer; these imperfect methods do, nevertheless, suffice to provide accurate practical controls in production when used intelligently. Screw threads are a case in point. Few manufacturers use more than a full-form Go gauge and a Not Go effective diameter gauge in runtime inspection, yet there are many things which could go wrong if not checked by these gauges. Major and minor diameters depend on blank diameters, drill sizes and so on, and the general form of thread depends on accurate threading tools. All these should be under control, and regular samples of the threads, should be inspected analytically and by projection.

A real need for a standard has to be apparent before work is started on it, and requests are made by interested bodies. The ranges on measuring equipment was started at the request of the Institution of Production Engineers of the UK, shortly before the Second World War. Some standards are based on existing NPL test schedules which have been used by the laboratory for certification; examples of this sort of development are the British standards for micrometers, squares, block levels, and several similar types of instruments.

The BSI is not a British government institution, although it receives some support from public funds, and government departments are frequently represented on its committees. Committee work is voluntary and, of course, unpaid. The institution is financed by subscriptions from industry, the British State and other interested bodies, as well as from the sale of its publications. Firms and individuals are eligible for membership and, in return for their subscriptions, receive certain privileges in obtaining specifications. Since the war the name of

BSI and its approval symbol, the kite mark, have become widely known among the general public, due to the extension of British Standards into many fields of products in domestic and general use.

A new organization, part of the Ministry of Technology of the UK, came into being in 1966. This is the British Calibration Service (BCS). Its purpose is to approve testing and calibrating laboratories in a number of fields of technology, including engineering metrology. Approved laboratories may issue BCS certificates which will have universal acceptance, for example, for gauges and measuring equipment. Much of this type of work has been done for some years by manufacturers under AID, and other official approval schemes by the BCS will unify these schemes and reduce still further the necessity for NPL certification of routine calibrations.

23.16.2 Screw-thread Gauging

Due to the complex geometry of a helical groove, a full assessment of the accuracy of a screw thread by direct measurement is a long procedure. During large-scale production the accuracy of screw threads is controlled by limit gauging.

If a tapped hole is considered, the gauges required are:

  (i) A full-form Go gauge, made to basic sizes;
  (ii) An effective-diameter Not Go gauge;
  (iii) A minor-diameter Not Go gauge.

The full-form Go gauge is a screw plug of length equal to the work length of engagement; it defines the maximum metal condition of the tolerance zone. The effective Not Go gauge has restricted contact with the workpiece as shown in Fig. 23.40. It defines the minimum metal condition at the pitch points of the flanks.

The full truncated form is expensive to manufacture and is used only for very coarse pitches where, due to the longer length of the thread flank, the angle error is of greater significance than for fine pitches. Ideally, the effective Not Go gauge should not be influenced by the errors of pitch or the angle of the thread gauged. It has just two to three truncated threads.

The minor-diameter Not Go gauge is a plain cylindrical plug, sometimes called a *core Not Go gauge* or *core plug*.

Reference to Taylor's principle of gauging will show that the following gauges are required for a complete check of an external thread:

  (i) A full-form Go ring gauge;
  (ii) An effective-diameter Not Go gauge;
  (iii) A major-diameter Not Go gauge;
  (iv) A minor-diameter Not Go gauge.

Screw ring gauges are expensive and slow in use; they have limited application for final inspection purposes, and for gauging screw threads on thin-walled components which get distorted easily under the contact forces exerted by a caliper-type gauge. Caliper-type gauges having 'edge' type anvils are more commonly used.

For effective inspection of diameters, a Not Go gauge must be of a caliper type, and have a thread form similar to that shown in Fig. 23.41 in order to conform to Taylor's principles.

The major diameter can be gauged by using a plain caliper gauge. made to the lower size, as the Not Go gauge; alternatively, blanks may be limit-gauged for size prior to threading, a method which generally provides sufficient control.

The common form of caliper gauge for external screw-thread work is shown in Fig. 23.41. For very large work, this type of gauge becomes unwieldy, and gauges which measure the effective diameter in terms of the radius of an arc are then substituted.

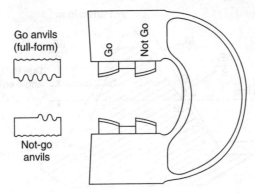

Fig. 23.41 Screw-thread caliper gauge.

Effectiveness of the gauging method

The gauging methods described restrict the work to the tolerance zone, because the magnitude of the zone is fixed by the major diameter, minor diameter and effective diameter limits. Gauging methods do not, however, enable the particular sources of error to be distinguished, and for this purpose, direct measurements are necessary and various formulae need to be used for arriving at the correct effective diameters.

Figures 23.42 and 23.43 illustrate the effect upon the fitting conditions of error of pitch and error of flank angle. Parts having such errors will assemble, provided there is sufficient difference between the simple effective diameters to absorb these errors.

Such errors give rise to the concept of virtual effective diameter. As defined in BS 2517, virtual effective diameter is the effective diameter of an imaginary thread of perfect form and pitch, having full depth of thread but clear at the crests and roots, which will just assemble with the actual thread over the prescribed length of engagement. This diameter exceeds the diameter by an amount related to the combined effects of errors of pitch and errors of flank angle.

The gauging value of a screw plug or ring will be influenced by errors of both pitch and flank angle; so, both must be very accurately measured: pitch directly on a special type of measuring machine, and angle by optical projection or measuring microscope.

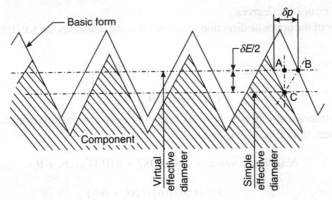

Fig. 23.42 Influence of pitch error on effective diameter required in mating part.

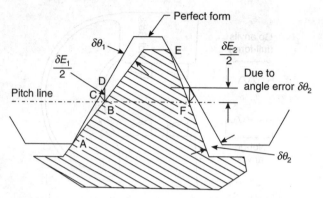

Fig. 23.43 Effect of angle error on effective diameter.

It follows directly from the geometry of Fig. 23.42 that for errors of pitch, the virtual change in effective diameter = $\delta p \cot \theta$, where δp is the maximum pitch error over the length of engagement and θ is the semi-angle of the vee. For the ISO metric system, the virtual change in effective diameter is $1.732\, \delta p$.

Figure 23.43 shows the effects arising from the angle errors $\delta\theta_1$ and $\delta\theta_2$, which occur because the depth of the thread is not symmetrical with respect to the pitch line; the effects will differ slightly according to whether contact is made at A or E.

Considering point A, $AB = 0.25p$ from the ISO metric standard form. Therefore

$$BC \approx AB \times \delta\theta_1 \qquad BD = 2BC$$

Hence $\qquad BD \approx (2 \times 0.25p \times \delta\theta_1)$ rad $= \dfrac{\delta E_1}{2}$ rad (i.e., error in radius)

It follows from the geometrical conditions of error $\delta\theta_1$ on both sides of the axis of the bolt that the resulting change of effective diameter would need to be

$$\delta E_1 = 2 \times 0.25 p\, (\delta\theta_1 + \delta\theta_1) \frac{\pi}{180}$$

$$= 0.0087 p\, (2\delta\theta_1) \qquad (23.4)$$

where the flank angle error is in degrees.

If the angle error of the opposite direction is considered, the contact point will be at E and the expression changes to

$$\delta E_2 = 2 \times 0.375 p\, (\delta\theta_2 + \delta\theta_2) \frac{\pi}{180}$$

$$= 0.0131 p\, (2\delta\theta_2) \qquad (23.5)$$

In general, the non-symmetrical aspect is ignored, and the positive or negative angle errors are regarded to have equal effect; hence, from Eqs. (23.4) and (23.5)

$$\delta E(\text{average value}) = \frac{1}{2}(0.0087 + 0.0131) p\, (\theta_1 + \theta_2)$$

$$= 0.0109 p\, (\delta\theta_1 + \delta\theta_2)$$

The virtual effective diameter could now be clearly defined as the sum of the simple effective diameters of the component plus the error dE (average) which equals $0.0109p\, (\theta_1 + \delta\theta_2)$. It is the effective diameter of the smallest nut of perfect form which will mesh with the bold (refer to Fig. 23.42). If the error calculated is excessive, the bolt will have to be rejected.

Effective diameter of screw thread

Best size measuring cylinder wire method: The latest method of determining effective diameter is based on geometry as shown in Fig. 23.44. D_w is the diameter measured by the anvils over the cylinder. It can be seen from the geometry of Fig. 23.44 that effective diameter E_s is given by the following equation:

$$E_s = D_w - 2(r + x)$$
$$= D_w - 2(r + r \sin \theta)$$
$$= D_w - 2r(1 + \sin \theta)$$

where θ is the semi-angle of the thread;
       r is the radius of the measuring cylinder wire.

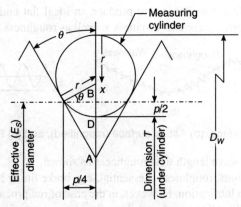

Fig. 23.44 Best size measuring cylinder wire method.

NPL method: In this method, the machine is set from an axial cylinder held between machine centres. Figure 23.45 shows the basic geometry of the NPL and the *P*-value method of effective diameter measurement. Referring to Fig. 23.45

$$\frac{P}{2} = AB - AD = \frac{1}{4} p \cot \theta - r(\operatorname{cosec} \theta - 1)$$

where $AB = \dfrac{1}{4} p \cot \theta$

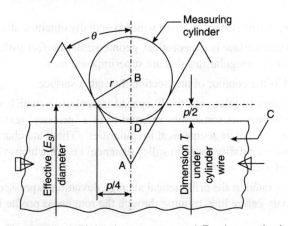

Fig. 23.45 NPL basic geometry and *P*-value method.

$$\therefore \quad P = \frac{1}{2} p \cot \theta - r (\operatorname{cosec} \theta - 1)$$

where $AD = r \operatorname{cosec} \theta - r$.

E_s = effective diameter = $T + p$, where T is the dimension under the axial cylinder. The NPL method has the merit that there is no need to have an accurate measuring wire of radius r as in the best size measuring method.

23.17 SURFACE FINISH AND ITS MEASUREMENT

It is not possible by any manufacturing process to produce an ideal flat and smooth surface as shown in Fig. 23.46(a), most of the real surfaces contain waviness as well as roughness as shown in Fig. 23.46(b).

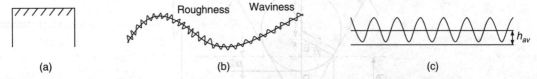

Fig. 23.46 (a) Ideally flat surface, (b) Actual surface (magnified), and (c) Roughness average value.

Waviness has much more wave length than roughness as shown.

In some practical applications, roughness is essential, e.g. brake lining inside of an engine cylinder to maintain oil film which assists in lubrication. However, in the case of reciprocating parts, a film of oil is to be maintained to prevent wear and tear. The roughness depends on:

1. Vibrations during cutting,
2. Type of machining process,
3. Rigidity of MTFW complex, i.e. Machine tool—Fixture—Workpiece complex.
4. Speed, feed, depth of cut,
5. Coolant used.

23.17.1 Surface Finish Terminology

Surface: The surface of a part is confined by the boundary which separates that part from another part, substance or space.

Actual surface: This refers to the surface of a part which is actually obtained after a manufacturing process.

Nominal surface: A nominal surface is a theoretical, geometrically perfect surface which does not exist in practice, but is an average of the irregularities that are superimposed on it.

Profile: Profile is defined as the contour of any section through a surface.

Roughness: Roughness refers to relatively finely spaced irregularities which might be produced by the action of a cutting tool. Roughness is sometimes referred to as a 'primary' texture. This is also known as 'Microgeometrical deviation' or 'Microgeometrical irregularities'. These are characterized by a low ratio of their pitch 's' to height 'h' of irregularity (peak to valley distance) i.e. s/h which is less than 50. Lower values of s/h are the irregularities of great height.

Roughness height: This is rated as the arithmetical average deviation expressed in microinches or micrometers normal to an imaginary centre line, running through the roughness profile Fig. 23.46(c).

Roughness width: Roughness width is the distance parallel to the normal surface between successive peaks or ridges that constitute the predominant pattern of the roughness.

Roughness width cut-off (Sampling length): This is the maximum width of surface irregularities that is included in the measurement of roughness height. This is always greater than roughness width and is rated in inches or centimetres.

Waviness: Waviness consists of those surface irregularities which are of greater spacing than roughness and it occurs in the form of waves. It may be caused by vibrations, machine or work deflections, warping, etc. it is also referred as 'secondary texture'. The wavelength is greater than about 1 mm. The pitch or spacing exceeds the sampling length chosen for measurement of roughness.

Flaws: Flaws are surface irregularities or imperfections which occur at infrequent intervals and at random intervals. Such imperfections are scratches, holes, cracks, pits, checks, porosity, etc. These may be observed directly or with the aid of a penetrating dye or other materials that make them visible for examination and evaluation.

Flaws may also be termed '**Macro-geometrical irregularities**'. These are random, not regularly repeated deviation from the theoretical surface. These are characterized by a large ratio of the length '*l*' over which the deviation extends to the amount of deviation $l/h > 1000$, e.g. out of roundness, taper, etc.

Lay: Lay is defined as the direction of the predominant surface pattern produced by tool marks. Symbols used to indicate the direction of lay are shown in Fig. 23.47.

The various types of 'Lay' are shown diagrammatically in Fig. 23.47. It should be noted that surface roughness is measured at 90° to the direction of lay.

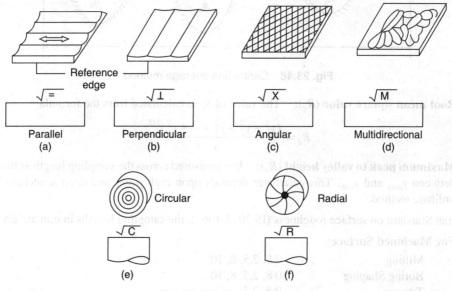

Fig. 23.47 Types of lay.

Traversing length: It is the length of the profile (measured in a direction parallel to the general direction of the profile) necessary for the evaluation of the surface roughness parameters. The traversing length includes from 3 to 10 roughness widths cut-off, which is also called sampling length, cut-off wavelength or metre (instrument) cut-off. It is apparent that the sampling length is also measured in a direction parallel to the general direction of the profile.

It is clear from above that any finished surface could contain both types of irregularities (waviness and surface roughness) superimposed on each other. When measuring surface roughness, the problem of separating waviness from it is usually encountered. However, conventionally, the surface roughness is defined within the area where waviness and deviations are eliminated.

23.17.2 Evaluation of Surface Roughness

There are several methods of expressing surface roughness. In the present section, three methods are explained:
1. **Centre line average (CLA) or (R_a):** It is the arithmetic average value of the heights from the mean line measured within the sampling length L ignoring the sign of the ordinate as shown in Fig. 23.48.

   Thus,
   $$R_a = \frac{(y_1 + y_2 + y_3 + \cdots + y_n)}{n}$$

   The mean line is chosen so that areas
   $$A_1 + A_3 + A_5 = A_2 + A_4 + A_6$$
   above and below the mean line as shown in Fig. 23.48.

   y_1, y_2, y_3, etc. are the values of the ordinates at the peak and trough points not shown in Fig. 23.48.

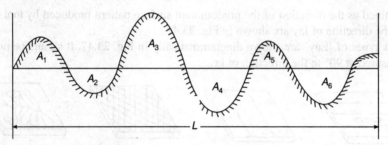

Fig. 23.48 Centre line average method.

2. **Root mean square value (R_q):** The value of R_q is calculated from the formula
$$R_q = \frac{(y_1^2 + y_2^2 + y_3^2 + \cdots + y_n^2)^{1/2}}{n}$$

3. **Maximum peak to valley height (R_y):** It is measured across the sampling length as the difference between y_{max} and y_{min}. This parameter depends upon high peaks and deep scratches. This is the ordinate method.

As per Indian Standard on surface roughness (IS 3073-1967), the sampling lengths in mm are given below:
1. **For Machined Surface:**

Milling	:	0.8, 2.5, 8, 10
Boring Shaping	:	0.8, 2.5, 8, 10
Turning	:	0.8, 2.5
Grinding	:	0.25, 0.8, 2.5
Planning	:	2.5, 8, 10, 25
Reaming	:	0.8, 2.5
Broaching	:	0.8
Diamond boring	:	0.25, 0.8

Diamond turning	:	0.25, 0.8
Honing	:	0.25, 0.8
Lapping	:	0.25, 0.8
Super-finishing	:	0.25, 0.8
Buffing	:	0.8, 2.5
Polishing	:	0.8, 2.5
Surface	:	0.8, 2.5, 8, 10
Spark machining	:	0.8

2. **For Non-Machined Surfaces:** Burnishing, Drawing, Extrusion, Moulding, Electro-polishing: 0.8, 2.5.

Preferred values for R_a (μm): 0.025, 0.05, 0.1, 0.2, 0.4, 0.8, 1.6, 3.2, 6.3, 12.5, and 25. There is a close relationship between surface roughness and tolerance. In general, tolerances must be greater than R_y (and waviness if any), unless the fit is a force fit and the surface roughness can be at least partially smoothed out in the fitting process. R_y can be taken approximately as 10 times R_a. The ratio of R_t (height of trough) and R_a ranges from 4 to 5. It has been established that the mean height of irregularities should not exceed 10 to 25% of the machining tolerance.

23.17.3 Representation of Surface Roughness

The I.S.O. has recommended a series of preferred roughness values and corresponding roughness grade numbers to be used when specifying surface roughness on drawings. These values and grade numbers are listed in Table 23.5.

Table 23.5 Roughness values

Roughness values, R_a (μm)	Roughness grade number	Roughness symbol
50	N 12	~
25	N 11	∇
12.5	N 10	
6.3	N 9	
3.2	N 8	∇ ∇
1.6	N 7	
0.8	N 6	
0.4	N 5	∇ ∇ ∇
0.2	N 4	
0.1	N 3	
0.05	N 2	∇ ∇ ∇ ∇
0.025	N 1	

23.17.4 Relationship of Surface Roughness to Manufacturing Process

Each manufacturing process ordinarily produces surfaces in a certain range. This variation is due partly to the obvious control over roughness, i.e., at the operator's disposal, such as the shape of the tool and speed of the cut, partly due to factors beyond the operator's control, variation in hardness of material or changes in grinding wheels and largely due to differences in shop practice of the individual plants. Table 23.6 gives the range of surface roughness found for the various production methods.

Table 23.6 Typical ranges of surfaces roughness, R_a (μm)

Process	Roughness CLA
Flame cutting	12.5–25
Snagging	6.3–25
Sawing	1.6–25
Planing/Shaping	1.6–12.5
Drilling	1.6–6.3
Chemical milling	1.6–6.3
EDM	1.6–6.3
Milling	0.8–6.3
Broaching	0.8–3.2
Reaming	0.8–3.2
EBM	0.8–6.3
LBM	0.8–6.3
ECM	0.2–3.2
Boring/Turning	0.4–6.3
Barrel finishing	0.2–0.8
Electrolytic grinding	0.2–0.6
Roller furnishing	0.2–0.8
Grinding	0.1–1.6
Honing	0.1–0.8
Electro-polishing	0.1–0.8
Polish	0.1–0.4
Lapping	0.05–0.4
Superfinishing	0.05–0.2
Sand casting	12.5–25
Hot rolling	12.5–25
Forging	3.2–12.5
Permanent mould casting	1.6–3.2
Investment casting	1.6–3.2
Extruding	0.8–3.2
Cold rolling, drawing	0.8–3.2
Die casting	0.8–1.6

The ranges given in Table 23.6 are typical of the processes listed. Higher or lower value may be obtained under special conditions. It is clear from the table that the values for surface roughness overlap for different manufacturing processes but it has its own surface pattern. Turning and shaping have parallel feed lines, and face milling produces curved or crossed lines on a surface. The large number of small cutting edges acting at random in grinding give a directional pattern of small scratches that vary in length and often overlap. Honing and lapping may produce multidirectional or crisscross pattern.

23.17.5 Measurement of Surface Roughness

Shop and inspection departments in industry employ two principal methods to evaluate surface roughness:

1. The surface is compared visually and by a feel with a standard.
2. Roughness is measured with indicating, recording or optical instruments.

Visual inspection: A visual examination is the simplest means of judging the finish on surfaces of any shape, size or material, but this method alone is unreliable because appearance is not a true indication of roughness. The reliability of this method may, however, be improved by comparing the work with graded standards which are the copies of master surfaces finished by shaping, grinding and milling, etc. and the roughness is graded in micrometres AA to agree with standard numbers. The machining method is indicated on the standards. The comparison of work and specimen of this kind by sight and feel is adequate for many requirements.

Instruments for inspection of surface roughness: According to the method employed for evaluating the surface roughness, instruments are classified as (a) Profilometers and (b) Profilographs. Profilometers are used for a quantitative evaluation of surface roughness by means of a single numerical rating. It may measure either the maximum, root mean square, or arithmetical mean height of the irregularities. A profilograph, on the other hand, reproduces the microirregularities in a magnified scale for subsequent analysis of the profilogram, so obtained, to determine the value of the microirregularities.

Most instruments, in general, used for the measurement of surface roughness are designed to respond to the irregularities of the surface through the agency of a stylus, which rests on the surface and is traversed across it. It is necessary to have a datum level to which the vertical movements of the stylus can be referred. The datum most commonly used is provided by a skid which has a relatively large radius of curvature in the direction of the traverse. This also rests on the surface but follows its general contour, riding over the crests of smaller irregularities without responding to them individually. These movements of the stylus normal to the surface, measured relative to a datum corresponding to the path followed by the skid, are recorded by an instrument.

Electrical stylus instruments

The Talysurf: The Talysurf is an electronic instrument developed by Taylor and Hobson Co. of UK and works on carrier modulation principle. A good understanding of the working of a Talysurf is possible only if a student understands certain basics of the electronics principles underlying the various forms of waves encountered in the working of the instrument. They are:

1. *Carrier wave:* It is a wave produced by an oscillator. It has high frequency and in case of Talysurf it is sinusoidal in shape.
2. *Modulated carrier wave:* This type of wave is produced when a signal wave howsoever small created by the displacement of the surface measuring probe rides over the carrier wave.
3. *Demodulated wave:* This wave is produced when during the act of extracting the information like surface undulations, the carrier wave is eliminated.
4. *Filtered wave:* This type of wave is produced when the demodulated wave is passed through an electronic filter of which the function is to remove unwanted frequency component, in the case of surface roughness measurement, it is the waviness. Thus, it is essential to understand the various stages which result in a final wave which is recorded by the sensitized paper recorder. Now, the working of the Talysurf can be discussed. The bridge circuit of a Talysurf is fed with a high frequency AC supplied by an oscillator. The oscillator supply is essential instead of a DC source or battery because in the latter case, the signal generated by the stylus movement will produce a signal voltage of very small magnitude insufficient to be measured. In actual practice, with Talysurf, any vertical movement of the stylus of the instrument (which is a fine diamond tip of 0.002 mm) is traversed across the surface of the work, creates a change in the air gap between the armature and E-shaped stamping, shown in Fig. 23.49. This changes the value of the inductances in the two arms of the wheatstone bridge. Due to this, a signal is created which creates a change in the nature of the wave fed by the oscillator to the bridge. Thus the final wave coming out of the bridge is a modulated carrier wave. In the next stage, this is demodulated so that only the peaks of the wave appear as

592 TEXTBOOK OF PRODUCTION ENGINEERING

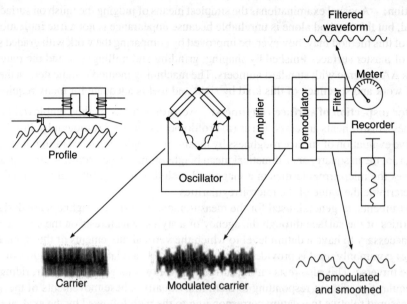

Fig. 23.49 System of working of Talysurf.

output. Now, the current is only a function of the vertical movement of the stylus. After a filer is used to filter out waviness, the pen recorder produces a permanent record of the surface roughness topology. This is because a tiny spark occurs due to the effect of the filter (which is usually a R–C circuit) and blackens the sensitized paper of the recorder.

The Talysurf provides three motorized speeds for measurement giving respectively × 20 and × 100 horizontal magnification and a speed suitable for average reading. A neutral position in which the pick-up can be traversed manually is also provided. In this case, the arm carrying the stylus forms an armature which pivots about the centre piece of E-shaped stamping as shown in Fig. 23.49. On two legs of (outer pole pieces) the E-shaped stamping, there are coils carrying an AC current. These two coils with other two resistances form a bridge circuit. As the armature is pivoted about the central leg, any movement of the stylus causes the air gap to vary and, thus, the amplitude of the original AC current flowing in the coils is modulated. The output of the bridge thus consists of modulation only as shown in Fig. 23.49. This is further demodulated so that the current now is directly proportional to the vertical displacement of the stylus only.

As stated earlier, the demodulated output interacts with pen recorder to produce a permanent record and in addition to a meter to give a numerical assessment directly as shown in Fig. 23.49.

Piezo-electric instrument: This instrument is shown in Fig. 23.50. This is a voltage generating instrument based on the Piezo-electric principle. In piezo-electric effect, whenever certain materials like quartz are subjected to mechanical compression, they produce a current of very high frequency. The stylus arm is shown in Fig. 23.50. Profilometer is connected to the piezo-electric crystal which is pivoted to the body of the instrument. The arm or the body of the instrument with a reference shoe (skid) is drawn across the surface; the stylus following the finer surface details. The stylus movements are transmitted through the stylus arm to the crystal. Due to this, mechanical vibrations are caused in the crystal. These vibrations are transformed into proportional voltages based on the piezo-electric principle. From the amplified voltage signals, either a meter reading or a graphical trace can be obtained. The radius of curvature of the skid ranges from 5 mm to 50 mm. For wavy surface, the skid length should be greater than twice the wavelength.

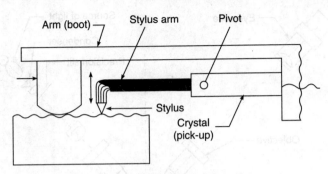

Fig. 23.50 Piezo-electric instrument.

Optical instruments

Profilograph: The principle of working of a tracer-type profilograph is shown in Fig. 23.51. The work to be tested is placed on the table in a fixture. The table can move to and fro because of the lead screw and the nut combination attached with the table. The combination is driven by a motor shown in Fig. 23.51. A tracer stylus is attached at the end of the lever which carries a mirror and is hinged at a point as shown. A beam of light is focused on the mirror and gets reflected on the drum rotated through a set of bevel gears. A sensitive film is wrapped over the drum so that light rays trace the profile of the work surface as the stylus moves over the work and at the same time, the drum rotates. As the stylus moves over the rough surface of the work piece test piece, there are vibrations in the vertical plane in the stylus due to undulations in the surface. Due to these vibrations, the mirror also changes its positions and the rays of light get deflected and a true form of the surface of the test piece is traced on the sensitive paper. In this instrument, the horizontal magnification can be changed by decreasing the drum diameter, or by changing the lead screw and pinion pitch or by changing the speed of the motor. For increasing the vertical magnification, the length of the lever should be increased or the distance of the drum from the mirror should be increased.

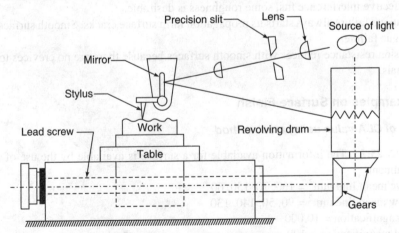

Fig. 23.51 Tracer-type profilograph.

Double microscope: This is an optical method. Figure 23.52 shows the set-up of a double microscope. From the light sources, a beam of light passes through the condenser and precision slit and is directed at an angle of 45 degrees to the test surface. The observing microscope having objective and eyepiece is also inclined at 45 degrees to the test surface. The eyepiece contains a scale by which the height of the irregularities are

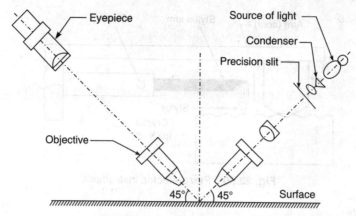

Fig 23.52 Double microscope.

measured through the micrometre. The surface roughness is evaluated by the light cross-section method which is based on the following principle. Light is focused on the surface through an optical system at 45 degrees. A band of this light is reflected and received through the eyepiece. The edge of this band reproduces the profile of these irregularities which means that it will show an optical cross-section. This is a magnified view. A double microscope working on the above principle is shown in Fig. 23.52. The field view of the eyepiece contains a scale called eyepiece microscope which is used to measure the height of the surface irregularities.

Effect of surface finish on the performance of machine elements

1. Wear resistence: Rough surfaces with more contact pressure are prone to wear rapidly. In the case of automobiles, it is desirable that for the first few thousand kilometres, it is run within a speed limit to permit a 'wear-in' after the smoothness is achieved between piston ring surface and cylinder wall and the lubrication is improved.
2. For effective interference fits, some roughness is desirable.
3. The fatigue failure always starts its propagation farm surface cracks. Smooth surfaces are less liable to fatigue failures.
4. Corrosion resistance reduces with smooth surfaces because there are no crevices to hold corrosive materials.

Numerical Examples on Surface Finish

Determination of CLA value using area method

ILLUSTRATION 23.1 The information available for a surface is available by the use of a surface finish measuring instrument.

Area above mean line mm^2 = 160, 70, 150, 60
Area below mean line mm^2 = 90, 50, 140, 130
Vertical magnification = 10,000
Horizontal magnification = 100
Traversing length = 1 mm

To calculate the value of suface roughness:

Here, total area = 160 + 70 + 150 + 60 + 90 + 50 + 140 + 130 = 850 mm^2

The CLA value can be obtained by:

1. Determining the true area by dividing the area (850 mm²) by the overall magnification (10,000 × 100) which yields = $\dfrac{850}{10,000 \times 100}$.

2. Obtaining the actual CLA in microns, multiplying (850 × 10⁻⁶) by 1000 and dividing it by the traversing length of 1 mm.

Thus, the CLA = $\dfrac{850 \times 1000}{1 \times 10,000 \times 100} = 0.85$ mm

ILLUSTRATION 23.2 Determination of CLA and RMS from given data:

The data derived from a Talysurf is as given in terms of 6 readings: 25, 20, 18, 24, 27, 31 microns.

Therefore, $\quad \text{CLA} = \dfrac{(25+20+18+24+27+31)}{6} = 24.16\ \mu m$

$$\text{RMS} = \sqrt{\dfrac{(25)^2 + (20)^2 + (18)^2 + (24)^2 + (27)^2 + (31)^2}{6}} = 602.5\ \mu m$$

23.18 TESTING OF MACHINE TOOLS

Machining processes for the production of different parts can be classified into two categories. One category is termed forming in which the shape produced is a direct replica of the tool shape. The second category is termed Generation. Majority of parts produced in a machine shop are through Generation. Generation is the result of relative motion between the tool and the workpiece. For example, a cylinder is the result of rotation of the workpiece between centres in a lathe combined with a linear motion of the tool held in the toolpost on the cross slide which is traversed parallel to the axis of the rotating workpiece. It is obvious that to generate a true cylinder, the movement of the tool must be absolutely parallel to the axis of the workpiece. The components produced by the machine tools is mostly by generation. As a result, the quality of surface produced is governed by the accuracy of the relative motions of the machine tool members concerned. It is important to know the capability of the machine tool by evaluating the accuracy of the various mechanisms that are directly responsible for generating the surface. To this end, a variety of tests which have been developed are described in this chapter.

23.18.1 Equipment Used for Testing

The accuracy of the machine tools should be higher than the accuracy of the components it will be producing. Taking an example, a Jig is produced on a Jig Borer. It is a machine tool having the maximum accuracy level and is kept in air-conditioned environment. Similarly, the quality of the measuring equipment used for machine tool testing should be in conformance with the quality expected from such testing. A few of the more commonly used equipment is detailed below.

Dial gauges: This is the most common instrument used for measuring the accuracy of the machine tool elements. A least count of 0.01 mm is generally sufficient for carrying out the tests. The plunger pressure should not be too low, which may interfere with some measurements involving the swing over, where the dial indicator may have to be in an inverted position. The dial indicator should preferably have a magnetic base.

Mandrels: Test mandrels used in machine tool testing are hardened and ground cylindrical bars. The roundness and straightness of the mandrel is important for measuring various accuracies. In case of long mandrels, careful consideration should be given to the deflection of the mandrel due to its self weight while testing. It may, therefore, be desirable to have hollow mandrels for large lengths. The diameter of the mandrel should be increased to improve its rigidity. The measuring length of the mandrel range from 100 mm to 500 mm for most cases.

Straight edges: Heavy and stress-relieved straight edges of the sufficient length made of steel or cast iron should be used. They should be ground to close tolerances with a squareness of the order of ± 0.01 mm in case of tri squares.

Bubble tubes or spirit levels: It is necessary to install the machine tool with the help of bubble tube or spirit level in the longitudinal direction and the lateral direction. For the spirit level, the sensitivity of the order of 0.01 mm per metre length is acceptable.

23.18.2 Test Procedures

Elaborately detailed tests for machine tools have been developed in the form of charts suitable for specific machine tools. However, some simple tests are discussed in this chapter as indicated below:

1. Testing the quality of the slideways and the locating surfaces.
2. Testing the accuracy of the main spindle and its alignment with respect to other parts of the machine tool.

Before the tests are conducted, it is important to ensure that the machine tool warms up in order to test the accuracy of the machine tool in the normal working conditions. For this, the machine tool should be run for a period (at least 30 minutes) which will bring all the bearing surfaces and spindles to the normal working temperature. The test when conducted under this condition will be a true indicator of the actual accuracy that can be ensured along its working range.

23.18.3 Tests for Acceptance

Slideways accuracy

To test the quality of the slideways, it is necessary to mount the dial indicator on a good benchmark surface. Then the plunger is moved along the longitudinal direction of the slideways which will provide an indication of the flatness error present on the surface of the slideways. The reading can be taken at least in 10 intervals along the length.

Spindle accuracy

These tests check the true running of the spindle and the centre located in the spindle with the other axes or slideways of the concerned machine tool.

Eccentricity check: The live centre may be loaded into the lathe spindle and a dial indicator is mounted as shown in Fig. 23.53. This test is required only for the machines where the workpiece is held between centres. The readings of the dial indicator are taken while rotating the spindle through a full rotation. This test enables the verification of avoidance of eccentricity error.

Eccentricity and inclination check of the spindle: Lack of true running of spindle is determined by eccentricity and inclination error. Taper shank of the test mandrel of about 300 mm length is mounted into the spindle as shown in Fig. 23.54. The plunger of the dial gauge rests on the surface of the mandrel. The spindle is revolved slowly and the readings of the dial gauge are noted. A deviation less than 0.01 mm is acceptable. The test is repeated with the dial gauge positioned near the spindle bore as well as other end of the test mandrel.

Test procedure for milling machines is shown in Fig. 23.55, while in case of the radial drilling machine, it is shown in Fig. 23.56. This test is carried out on all such machines, as have the main running spindle.

METROLOGY AND PRECISION MEASUREMENT 597

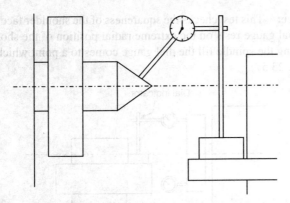

Fig. 23.53 Checking the true running of the centre in the lathe spindle.

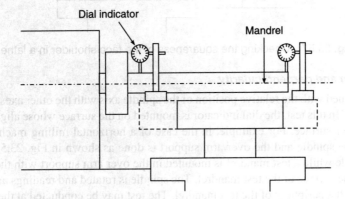

Fig. 23.54 Checking the true running of the spindle of a lathe.

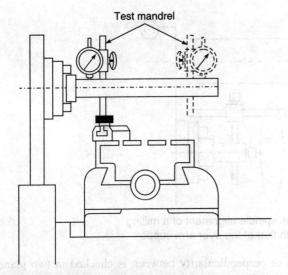

Fig. 23.55 Checking the true running of the spindle of a milling machine.

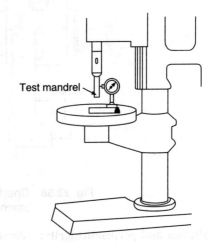

Fig. 23.56 Checking true running of the spindle of a radial drilling machine.

Perpedicularity of the face: This test checks the squareness of the shoulder face with respect to the spindle axis. The plunger of the dial gauge rests on the extreme radial position of the shoulder face and the reading taken. It is repeated, rotating the spindle till the dial gauge comes to a point which is diametrically opposite to the earlier position (Fig. 23.57).

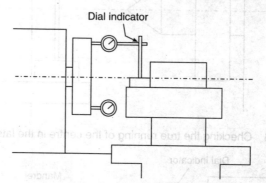

Fig. 23.57 Checking the squareness of the face shoulder in a lathe.

Tests for parallelism and perpendicularity

These tests are concerned with the relative position of the spindle axis with the other axes of the machine tool.
Spindle alignment: In this test, the dial indicator is mounted on the surface whose alignment is to be tested with respect to another surface. For example, in the case of a horizontal milling machine, a check for the alignment between the spindle and the over arm support is done as shown in Fig. 23.58. The dial gauge is mounted on the spindle while a test mandrel is mounted in the over arm support with the plunger of the dial indicator resting on the surface of the test mandrel. The spindle is rotated and readings are taken when it is at different positions on the periphery of the test mandrel. The test may be conducted at the two extreme points on the mandrel.

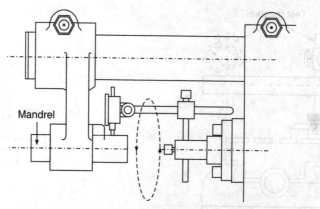

Fig. 23.58 Checking the spindle alignment of a milling machine with that of the over arm support.

Parallelism and perpendicularity: Parallelism or perpendicularity between is checked in two planes, horizontal and vertical. For this purpose, the test mandrel is mounted in the spindle, as shown in Fig. 23.59, in case of a lathe with the dial indicator mounted on the saddle/carriage. The plunger of the dial indicator will be contacting the mandrel surface as shown in Fig. 23.59. The saddle is moved for a specified distance and the dial reading noted. This test is repeated in the horizontal direction as well.

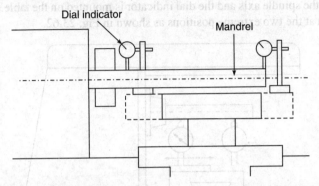

Fig. 23.59 Checking the parallelism between the spindle axis and the slideways in a lathe.

In the case of lathes, other tests that can be conducted pertaining to Tailstock, Leadscrew, etc.

1. Parallelism between the outside diameter of the tailstock sleeve (Fig. 23.60) and the slideways.
2. Parallelism between the tailstock sleeve taper in the slideways by fitting the test mandrel into the tailstock sleeve.

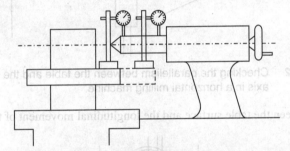

Fig. 23.60 Checking the parallelism of the tailstock sleeve.

3. Parallelism between the leadscrew axis and the slideways by mounting the dial indicator on the leadscrew itself.
4. Parallelism between the line of centres and the slideways (Fig. 23.61).

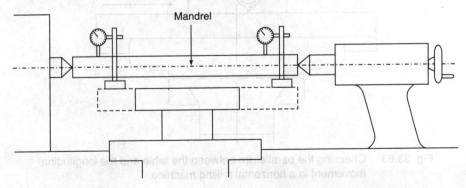

Fig. 23.61 Checking the parallelism of the line of centres in a lathe.

In the case of a milling machine, the tests are:

1. Parallelism between the table and the spindle axis (Fig. 23.62). A test mandrel with 300 mm length is mounted in the spindle axis and the dial indicator is mounted on the table. The reading of the dial gauge is taken at the two extreme positions as shown in Fig. 23.62.

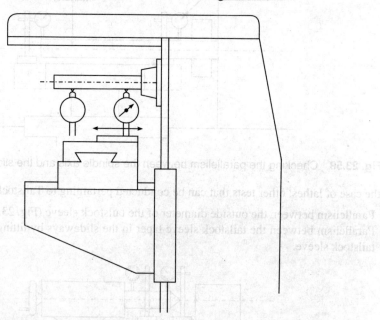

Fig. 23.62 Checking the parallelism between the table and the spindle axis in a horizontal milling machine.

2. Parallelism between the table surface and the longitudinal movement of the table (Fig. 23.63).

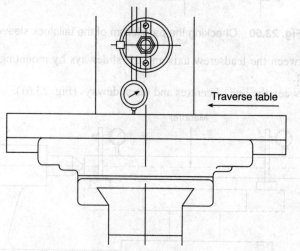

Fig. 33.63 Checking the parallelism between the table and the longitudinal movement in a horizontal milling machine.

3. Parallelism between the spindle axis and the transverse movement of the table. The test set-up is similar to the earlier test but this time the reading is taken with the table moved to the extreme positions in the transverse direction (Fig. 23.62).
4. Perpendicularity between the spindle and the vertical columnways by moving the knee. The dial indicator is fixed to the spindle and a Tri-square is fitted to the table as shown in Fig. 23.64. The dial indicator plunger will be touching the vertical side of the tri-square. The knee is then raised to take the reading at the two extreme ends of the vertical leg of the square.

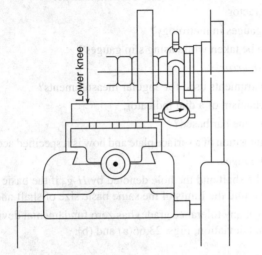

Fig. 23.64 Checking the perpendicularity between the spindle and the vertical movement in a horizontal milling machine.

In the case of radial drilling machines, the following tests can be conducted:

1. Parallelism between the drilling head slideways and the base plate.
2. Perpendicularity between the spindle axis and the table top (Fig. 23.65). The spindle is rotated by 360° and the reading of the dial indicator is taken while its plunger touches the table top.

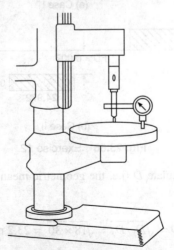

Fig. 23.65 Checking the perpendicularity between the spindle and the base plate of a radial drilling machine.

REVIEW QUESTIONS

1. Explain the difference between accuracy and precision.
2. Define the term *least count* applied in various measuring devices.
3. Explain the constructional features of the following instruments:
   (i) Micrometer
   (ii) Vernier callipers
   (iii) Dial indicator
   (iv) Combination set
   (v) Vernier bevel protractor.
4. What is the use of slip gauges in metrology?
5. What precautions are to be taken while using slip gauges?
6. Define the term *backlash error*.
7. What are the various instruments used for angular measurements?
8. Sketch the internal mechanism of a dial indicator.
9. On what principle is the sine bar based?
10. What is the material composition of a surface plate and how is it specified according to ISO specifications?
11. Explain the use of limit gauges.
12. Calculate the limits of the shaft and the hole denoted by $H_7 g_6$ if the basic size is 25 mm, $H, h = 0$, and $G, g = \pm 0.007$ mm. Also find the limits of the same basic size of shaft and hole denoted by $H_7 g_6$.

    Solution: The H hole (of any tolerance grade) has zero fundamental deviation, and the shaft of g type has negative fundamental deviation, Figs. 23.66(a) and (b).

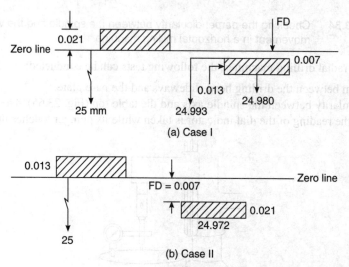

Fig. 23.66 Exercise 12.

To calculate i, we should calculate D (i.e. the geometric mean of the diameter steps, in this case 18 and 30), first.
Therefore

$$D = \sqrt{D_1 \cdot D_2} = \sqrt{18 \times 30} = 23.2 \text{ mm}$$

Fundamental tolerance unit:

$$i = 0.45 \, D^{1/3} + 0.001 \, D \text{ micron}$$
$$= 0.45(23.3)^{1/3} + 0.001 \times 23.2$$
$$= 1.3 \, \mu m$$
$$= 0.0013 \text{ mm}$$

(a) For Case I ($H_7 g_6$):
Fundamental tolerance of hole:

$$H_7 = 16i$$
$$= 16 \times 1.3 = 20.8 \, \mu m$$
$$= 0.021 \text{ mm}$$

Fundamental tolerance of shaft:

$$g_6 = 10i$$
$$= 10 \times 1.3$$
$$= 13 \, \mu m$$
$$= 0.013 \text{ mm}$$

In the problem the fundamental deviations for hole and shaft are given as follows:
For hole of h type: 0
For g type shaft: -0.007
For G type hole: $+0.007$

Limits of hole H_7:
    Lower deviation = 0
    Upper deviation = 0.021 mm
    Minimum size of hole = 25.00 mm
    Maximum size of hole = 25.021 mm

Limits of shaft g_6:
    Upper deviation = -0.007
    Lower deviation = $-(0.007 + 0.013) = -0.020$
    Maximum size = $25 - 0.007 = 24.993$ mm
    Minimum size = $25 - (0.007 + 0.013) = 24.980$ mm

(b) For Case II ($H_6 g_7$):

$$i = 1.3 \, \mu m$$
$$= 0.0013 \text{ mm}$$
$$IT \, 6 = 10i = 0.013 \text{ mm}$$

Limits of hole H_6:
    Lower deviation = 0
    Upper deviation = 0.013
    Minimum diameter of the hole = 25 mm
    Maximum diameter of the hole = 25.013 mm

Limits of shaft g_7:

$$\text{Fundamental deviation} = -0.007 \text{ mm}$$
$$\text{Fundamental tolerance} = IT7 = 16i = 16 \times 1.3 = 0.021 \text{ mm}$$
$$\text{Maximum diameter of the shaft} = 25 - 0.007 = 24.993 \text{ mm}$$
$$\text{Minimum diameter of the shaft} = 25 - (0.007 + 0.021)$$
$$= 24.972 \text{ mm}$$

13. The fit for a shaft and hole combination is designated in the following way:

$$30 \; H_6 \; d_5$$

The fundamental tolerance unit i is given by the following relationship:

$$i = 0.45 \; D^{1/3} + 0.001 \; D$$

The basic diameter, 30, lies between the diameter steps of 18 and 30.
The fundamental deviation for d type shaft is equal to

$$-16 \; D^{0.44}$$

Calculate:
1. Minimum and maximum diameter of the hole;
2. Minimum and maximum diameter of the shaft;
3. Tolerance;
4. Allowance;
5. Deviations.

Also, state the type of fit.

14. A slider block shown in Fig. 23.67 is moving in guides. Calculate the maximum and minimum guide width and block width dimensions as per the following information.

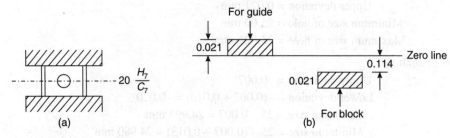

Fig. 23.67   Exercise 14.

The standard tolerance is given by

$$i = 0.45 \sqrt[3]{D} + 0.001 \; D$$

where D is the diameter (in mm) of the geometric mean step, and i, the multiplier for the grade 7, is 16. The fundamental deviation for fit C is

$$C = -(95 + 0.8 \; D)$$

The diameter step range is 18 to 30.

The fit is given by $20 \dfrac{H_7}{C_7}$.

Solution: The geometric mean diameter

$$D = \sqrt{D_1 \times D_2} = \sqrt{18 \times 30} = 23.2 \text{ mm}$$

The standard tolerance unit

$$i = 0.45(D)^{1/3} + 0.001D$$
$$= 0.45(23.2)^{1/2} + 0.001 \times 23.2$$
$$= 1.3 \text{ µm}$$
$$= 0.013 \text{ mm}$$

The tolerance for grade 7 (both the hole and shaft, i.e. the guide and the block, have the same tolerance grade 7)

$$= 16i$$
$$= 16 \times 1.3$$
$$= 20.8 \text{ µm}$$
$$= 0.021 \text{ mm}$$

The fundamental deviation for C block (or shaft)

$$= -(95 + 0.8D) \quad \text{(given in the problem)}$$
$$= -(95 + 0.8 \times 23.2)$$
$$= -(95 + 18.56$$
$$= -113.56 \text{ µm}$$
$$= -0.114 \text{ mm}$$

Now with the help of Fig. 18.17, we can calculate the minimum and maximum diameters easily. These are:
(i) The minimum diameter of the guide
$$= 20 \text{ mm}$$

The maximum diameter
$$= 20 + 0.021$$
$$= 20.021 \text{ mm}$$

(ii) The minimum diameter
$$= 20 - 0.114 - 0.021$$
$$= 19.865 \text{ mm}$$

15. A gear ring is fitted on a hub having $H_6 i_6$ fit. The bore of the gear ring is 50 mm diameter. Calculate
    (i) The gear bore size
    (ii) The hub dimensions
    (iii) The allowance
    (iv) The type of fit.

The fundamental deviations are:

$$H = 0 \quad j = -9 \text{ µm}$$

The multipliers for standard tolerance are 10 from IT 6.

The standard tolerance is given by

$$i = 0.45 \sqrt[3]{D} + 0.001D \quad \text{(in μm)}$$

where D is the diameter of the geometric mean of sizes ranging between 30 and 50 mm.

16. A gear as shown in Fig. 23.68 is keyed to the axle with the fit designated as 30 H_7/K_6. Calculate:
    (i) The hole and shaft dimensions
    (ii) The tolerances, mentioning also whether the tolerances are unilateral or bilateral
    (iii) The class of fit.

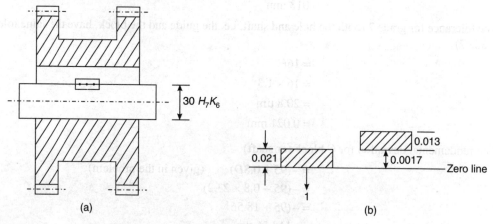

Fig. 23.68 Exercise 16.

The standard tolerance is given by

$$i = 0.45 \sqrt[3]{D} + 0.001D \text{ (in μm)}$$

where D is the geometric mean diameter, the diameter steps are 18 mm and 30 mm, and the multiplier for grades of tolerance 6 and 7 are 10 and 16 respectively.
The fundamental deviation for shaft K is

$$K = +0.6 \sqrt[3]{D}$$

Solution: The geometric mean of the diameter steps

$$D = \sqrt{D_1 \times D_2} = \sqrt{18 \times 30} = 23.2 \text{ mm}$$

The fundamental tolerance unit $i = 0.45 D^{1/3} + 0.001D$

$$= 0.45(23.2)^{1/3} + 0.001 \times 23.2$$
$$= 1.3 \text{ micron}$$
$$= 0.0013 \text{ mm}$$

The tolerance for hole of grade 6

$$= 16i$$
$$= 16 \times 0.0013 = 0.021 \text{ mm}$$

Tolerance for shaft of grade 6 = $10i$

$$= 10 \times 0.0013 = 0.013 \text{ mm}$$

The fundamental deviation for the H hole = 0
The fundamental deviation for K shaft
$$= + 0.6 D^{1/3}$$
$$= 0.6 \,(23.2)^{1/3}$$
$$= 0.6 \times 2.85$$
$$= 1.71 \, \mu m$$
$$= 0.0017 \text{ mm}$$

Therefore
(i) Minimum diameter of hole = 30 mm
Maximum diameter of the shaft = 30 + 0.021 = 30.021 mm
(ii) Maximum diameter of the shaft = 30 + 0.0017 = 30.0017 mm
Maximum diameter of the shaft = 30 + 0.0017 + 0.013 = 30.0147 mm
(iii) Tolerances:
Hole tolerance = + 0.021 mm
Unilateral shaft tolerance = 0.013 mm

17. A hole and shafting system is described in the following manner:
$$60 \frac{H_7}{m_6}$$

The standard tolerance is given by
$$i = 0.45 \sqrt[3]{D} + 0.001 D$$

where D is the diameter (in mm) of the geometric mean step.
The multiplier for grade 7 is 16; for grade 6 it is 10.
The fundamental deviation for fit is
$$FD = IT7 - IT6$$

The diameter lies between 50 and 80 mm.
Calculate:
(i) The diameter of the hole and the shaft
(ii) The tolerances
(iii) The allowances
(iv) Also, state the class of fit and sketch the fit.

18. A machined shaft is to rotate in a reamed hole at low speed. The IT8 grade has been selected for tolerance. The fit is
$$50 \frac{H_8}{C_8}$$

The fundamental deviations are:
$$h = 0$$
$$C = -(95 + 0.8 D)$$
where D is the diameter of the shaft.

The standard tolerance unit

$$i = 0.45\sqrt[3]{D} + 0.001D$$

The multiple of standard tolerance for grade IT8 is 25.
The diameter of the shift lies in the size range of 50–80.
Calculate:
  (i) The dimensions of hole and shaft
  (ii) The deviation.
Also, state the class of fit and sketch it.

19. An idler gear is to rotate over a 40 mm shaft. For this, the fit is supposed to be sliding fit. Calculate:
  (i) The base and shaft dimension
  (ii) The limit
  (iii) The allowance.

The following additional information are given:
  The fundamental deviations are

$$h = 0$$
$$C = -(95 + 0.8D)$$

The multipliers for standard tolerances are 25 for grade IT8 and 16 for IT7.
The standard tolerance is given by

$$i = 0.45\sqrt[3]{D} + 0.001D$$

where D is the diameter (in mm) of the geometric mean state.

20. Choose the type of fit:
  (i) Shaft rotating in bush
  (ii) Cam shaft rotating in bearing
  (iii) Bush fixed in housing
  (iv) Oil-lubricated pump shaft rotating in bearing
  (v) Dowel pin secured to base plate
  (vi) Gear with bush rotating on pin
  (vii) Valve seating
  (viii) Gear ring fit on hub
  (ix) Collar passed to shaft
  (x) Carriage-moving glideways.
21. What is the importance of surface finish?
22. Explain the various elements of surface roughness.
23. Explain the terms—primary texture and secondary texture.
24. What is meant by lay? What are the different types of lay and give the symbol for each type of lay?
25. Discuss the various methods of valuating surface roughness.
26. What are the units of surface roughness?
27. Discuss the various methods of measuring surface roughness.
28. What is the difference between Profilograph and Profilometer?

29. Discuss the effect of surface finish on functional properties of a surface.
30. How is the surface finish of a lapped shaft measured?
31. Explain the optical method of measuring surface finish.
32. Find
    (i) The maximum clearance or interference.
    (ii) The minimum clearance or interference.
    (iii) The maximum and minimum diameters of the hole and the shaft.

    Solution: For the following shaft and hole combination:
    (a) Hole: $20 + 0.1$
                $ - 0$
        Shaft: $20 + 0.5$
                $ + 0.1$
    (b) Hole: $20 + 0.1$
                $ - 0$
        Shaft: $20 + 0.2$
                $ - 0$
    (c) Hole: $20 + 0.1$
                $ - 0$
        Shaft: $ + 0.0$
    (d) Hole: $20 + 0.1$
                $ - 0$
        Shaft: $20 - 0.2$
                $ - 0.3$

33. Sketch the following type of fits and show the tolerances on them:
    (i) $H_7 C_5$
    (ii) $H_8 g_7$
    (iii) $H_7 h_6$
    (iv) $H_7 k_8$
    (v) $H_6 m_7$.

34. A hole and shaft pair designated by 50 $H_7 h_6$ are under production. The following other relevant information are given:
    The 50 mm diameter lies in the diameter step of 30 and 50. The standard tolerance unit

    $$z = i = 0.45 \sqrt[3]{D} + 0.001D$$

    where D (in mm) is the geometric mean of diameter steps. Also, IT6 = $i\,10i$, and above the IT6 grade the tolerance magnitude is multiplied by 10 at every fifth step. Calculate the dimensions of the gauges to inspect the products.

    Solution: Both the hole and the shaft are under inspection. The hole will be inspected with plug gauge and the shaft with ring gauge.
    We shall calculate first the tolerance and the standard deviation for both shaft and hole, then the tolerance for the Go and No Go gauges, and finally the dimensions of these gauges.

    $$D = \sqrt{30 \times 50} = 38.7 \text{ mm}$$

    The standard tolerance unit

    $$i = 0.45(D)^{1/3} + 0.001D$$

$$i = 0.45(38.7)^{1/3} + 0.001 \times 38.7$$
$$i = 0.45 \times 3.38 + 0.001 \times 38.7$$
$$= 1.56 \ \mu m = 0.00156 \ mm$$

Now, tolerance for the hole
$$IT7 = IT6 \times 10^{0.2}$$

(If it were not given, we could have taken IT7 = $16i$)
$$= 10i \times 10^{0.2}$$
$$= 10 \times 0.00156 \times 1.58$$
$$= 0.0247 \ mm$$
$$= \text{Work tolerance or tolerance for hole}$$

Tolerance for the shaft   IT – 6 = $10i$
$$= 10 \times 0.00156$$
$$= 0.0156 \ mm$$

The fundamental deviation for hole and shaft of h type is 0.

Design of plug gauge:

Draw the tolerance zones for gauge and No Go gauge for the general gauge as shown in Fig. 23.69(a). In this case $T_1 = 0.0247$
$$0.1 \times T_1 = 00247$$
$$a = \text{Wear allowance}$$
$$= 0$$

(since work tolerance is less than 0.0875 mm).

The maximum diameter of the No Go plug gauge
$$= 50 + 0.0247 + 0.002747$$
$$= 50.0274 \ mm$$

The minimum diameter
$$= 50.00 + 0.0247$$
$$= 50.0247 \ mm$$

For Go gauge:

Maximum diameter = 50 + 0.0247 = 50.0247 mm

Minimum diameter = 50.00 mm

Consider the ring gauge. The tolerance zones for h type shaft is shown in Fig. 23.69(b).

The minimum diameter of No gauge = $20 - a - 0.1 \times T_2$
$$= 50 - 0 - 0.0015$$
$$= 49.9985 \ mm$$

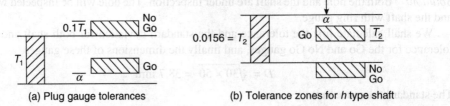

(a) Plug gauge tolerances   (b) Tolerance zones for h type shaft

Fig. 23.69   Exercise 34.

Maximum diameter of No gauge = 50.00 mm.
Dimension of No Go gauge:

Maximum diameter	= 50 – T
	= 50 – 0.0156
	= 49.9844 mm
Minimum diameter	= 50 – 0.015 – 0.1 × 0.0156
	= 50 – 0.0156 – 0.00156
	= 49.983 mm

35. A 20 mm m_6 shaft is to be checked by a Go No Go snap gauge. Assume 7% wear allowance and 10% gauge maker's tolerance (as percentages of the tolerance of the shaft). The fundamental deviation for m is obtained from IT7–IT6, where the multiples for grade IT7 is 16 and for grade IT6 is 10. Calculate the maximum and minimum sizes of the gauge.

36. Calculate the maximum and minimum dimensions of Go and No Go snap gauges to check g_9 shafts of basic size 25 mm. The fundamental deviation is given as follows:
Standard tolerance unit
$$i = 0.45D^{1/3} + 0.001D$$
where D is the geometrical mean diameter.
The shaft diameter lies between 18 mm and 25 mm steep. The gauge maker's tolerance is 1% of shaft tolerance.

Solution: The geometric mean diameter
$$= \sqrt{18 \times 30} = 23.2 \text{ mm}$$

The fundamental tolerance unit
$$i = 0.0013 \text{ mm}$$

The tolerance for g_6 shaft
$$= 10i$$
$$= 10 \times 0.0013$$
$$= 0.013 \text{ mm}$$

The fundamental deviation
$$= -0.007 \text{ mm}$$

Gauge tolerance
$$= 10\% \text{ of Shaft tolerance}$$
$$= 0.1 \times 0.013$$
$$= 0.0013 \text{ mm}$$

Now draw Fig. 23.70(a) and (b) to show the respective tolerances.
Design of snap gauge:
(i) Go gauge:
$$\text{Maximum size} = 25 - 0.007 - a$$
$$= 24.993$$

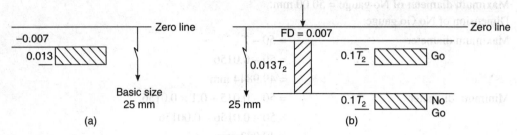

Fig. 23.70   Exercise 36.

$$\text{The minimum size} = 25 - 0.307 - a - 0.1T_2$$
$$= 25 - 0.007 - 0.1 \times 0.013$$
$$= 24.9917 \text{ mm}$$

(ii) No Go gauge:
$$\text{Maximum size} = 25 - T_2$$
$$= 25 - 0.013$$
$$= 24.987 \text{ mm}$$

37. Tolerance on a product is the difference between
    (i) The maximum and minimum permissible sizes of the product
    (ii) The actual size and the minimum size of the product
    (iii) The maximum permissible size and the actual size of the product.
38. Limits are
    (i) The two extreme permissible sizes
    (ii) The upper permissible size and the nominal size
    (iii) The lower permissible size and the nominal size
    (iv) The nominal size and the actual size.
39. On a product, the tolerance can be
    (i) Negative   (ii) Positive
    (iii) Zero   (iv) All of these.
40. The deviation in the dimension of the products is
    (i) The algebraic difference between the actual size and the nominal size
    (ii) The difference between the actual size and the nominal size
    (iii) The maximum permissible size and the nominal size
    (iv) The minimum permissible size and the nominal size.
41. The fundamental deviation of the product for basic holes with clearance fit has the lower deviation either
    (i) Positive   (ii) Zero
    (iii) Negative   (iv) All these.
42. Basic holes should be
    (i) Larger than the basic size   (ii) Smaller than the basic size
    (iii) Equal to the basic size.
43. Basic shafts are always
    (i) Bigger than the basic dimension
    (ii) Smaller than the basic dimension
    (iii) Equal to the basic dimension.

44. The basic shaft can be represented by (assuming basic size 15)
    (i) 15^{+0}_{-x}
    (ii) 15^{-x}_{+0}
    (iii) 15^{-x}_{-x}
    (iv) $15^{-x}_{-x_2}$.

45. The factor for getting a desired fit are
    (i) The magnitude of tolerance and the fundamental deviation of the mating parts
    (ii) The tolerance of the mating parts
    (iii) The allowance for the parts
    (iv) The amount of deviation.

46. $H_8 d_5$ represents
    (i) Interference
    (ii) Press fit
    (iii) Loose running fit.

47. $H_{11} C_{11}$ represents
    (i) Interference fit
    (ii) Clearance fit
    (iii) Positional fit
    (iv) Tight assembly.

48. $H_8 h_7$ represents
    (i) Normal location fit
    (ii) Interference fit
    (iii) Clearance fit
    (iv) None of these.

49. $50 \dfrac{H_7}{g_6}$ represents
    (i) Basic size of hole and shaft equal to 50
    (ii) $H_7 - H$ type hole with tolerance grade 7
    (iii) $g_6 - g$ type shaft with tolerance grade 6
    (iv) All of these
    (v) None of these.

50. Holes A to G are (with respect to basic size)
    (i) Undersized
    (ii) Of basic size
    (iii) Oversized
    (iv) None of these.

51. There are (as per BS 4500 standards)
    (i) 10 tolerance grades
    (ii) 18 tolerance grades
    (iii) 20 tolerance grades
    (iv) No tolerance grades.

52. For quality gauge (plug or gap gauge, etc.) the tolerance grades selected are
    (i) IT16 – IT12
    (ii) IT10 – IT5
    (iii) IT1 – IT4.

53. For automobile piston, fuel injection equipment, etc., the tolerance grades selected are
    (i) IT4 – IT5
    (ii) IT10 – IT12
    (iii) IT14 – IT16
    (iv) IT18.

54. The tolerance grades are in
    (i) Arithmetic series
    (ii) Geometric series
    (iii) Logarithmic series
    (iv) Hyperbolic series.

55. For limit gauging, the plug made to the upper limit is called the
    (i) Go gauge
    (ii) No Go gauge.

56. For limit gauging the plug made to the lower limit is called the
    (i) Go gauge
    (ii) No gauge
    (iii) No Go gauge.
57. For limit gauging the ring made to the lower limit is called the
    (i) Go gauge
    (ii) No gauge
    (iii) No Go gauge.
58. For limit gauging the ring made to the lower limit is called the
    (i) Go gauge
    (ii) No Go gauge
    (iii) Both of these.
59. During limit gauging for acceptance of a product
    (i) The Go plug gauge should enter the product (if it is hollow)
    (ii) The Go ring gauge should pass over the product (if it is solid)
    (iii) Both of these
    (iv) None of these.
60. During limit gauging for acceptance of a product the
    (i) No Go plug gauge should not enter the product
    (ii) No Go ring gauge should not pass over the solid
    (iii) Both of these
    (iv) None of these.
61. The sine bar should not be used to set up angles greater than
    (i) 30°
    (ii) 45°
    (iii) 60°
    (iv) 75°.
62. The sine bar measures angles
    (i) Without the use of any other instrument
    (ii) With the help of slip gauges
    (iii) With the help of slip gauges and dial gauges
    (iv) With the help of dial gauge only.
63. Comparators are instruments to measure
    (i) Angular dimensions
    (ii) Any dimension
    (iii) Linear dimension.
64. Comparators are instruments similar to
    (i) Steel rules
    (ii) Micrometer and vernier caliper
    (iii) Slide rule
    (iv) None of these.
65. A comparator needs the help of the
    (i) Slip gauge
    (ii) Angle gauge
    (iii) Both of these
    (iv) Autocollimator.
66. A comparator records the
    (i) Absolute dimension
    (ii) Difference between the standard value and the value of the element under measurement
    (iii) Symbolic value only.
67. Comparators are used to measure the
    (i) Length of a bar
    (ii) Diameter and thickness of a product
    (iii) All of these
    (iv) None of these.

68. By using a mechanical comparator the magnification attainable is
    (i) Not more than 500
    (ii) Not more than 1000
    (iii) Not more than 2000
    (iv) As high as 5000.
69. The mechanical comparator can be used to measure the
    (i) External diameter
    (ii) Internal diameter
    (iii) Angular dimension
    (iv) All of these.
70. Pneumatic comparators can be used to measure the
    (i) Internal diameter
    (ii) External diameter
    (iii) Surface roughness and ovality error
    (iv) All of these.
71. In pneumatic comparators, air
    (i) Above atmospheric pressure is used
    (ii) At atmospheric pressure is used
    (iii) At negative pressure is used.

Chapter 24

Jigs and Fixtures

24.1 DEFINITIONS

Jig: A jig may be defined as a device which holds and locates a workpiece, and guides and controls one or more cutting tools. The holding of the workpiece and the guiding of the tool are such that they are located in true position relative to each other. The jig comprises a plate, a structure, or a box made of metal or nonmetal, depending upon the applications, having provisions for holding the components in identical positions one after the other and then guiding the tool to the correct position on the workpiece, in accordance with the drawing, specification, operation, or layout.

Fixture: A fixture may be defined as a device which holds and locates a workpiece during an inspection or for a manufacturing operation. The fixture does not guide the tool. A fixture comprises standard or specifically designed work-holding devices, which are clamped on the machine table to hold the work in position. The tools are set in position on the work by using gauges or by manual adjustment.

Tool: A tool may be defined as a device used to remove material from a workpiece under controlled or stable conditions.

Tooling up component: The expression *tooling up a component* means to design and supply all tools like jigs, fixtures, cutting tools and gauges required for manufacturing and inspection.

24.2 DISTINCTION BETWEEN JIG AND FIXTURE

	Jig	Fixture
Function	It holds and locates the workpiece and guides the cutting tool.	It holds and positions the workpiece but does not guide the cutting tool.
Construction	These are made light in weight and clamping is not called for in most of the cases.	They are made heavy in construction and are bolted rigidly on the machine tools.
Handling	Easy and quicker.	Difficult in handling and removal.
Application	They are used in drilling, reaming or tapping operations.	They are used in milling, grinding, planing, welding and turning operations.

24.3 ADVANTAGES

The following are the advantages of using jigs and fixtures in mass production work:

Interchangeability: The application of jigs and fixtures ensures the interchangeability of parts, which facilitates the assembly operation.

Economy: It reduces the overall cost of machining by fully or partially automatising the process.

Rejection: It practically eliminates rejection. Reproducibility is ensured.

Elimination of operations: The application of jigs and fixtures eliminates the marking-out, measuring, and other setting operations before machining.

Increased machining accuracy: As the workpiece is automatically located and the tool is guided without involving manual operations, the machining accuracy obtained is of high degree.

Quality control: It ensures quality; as such, quality control cost is minimised. Even operations pertaining to inspection are minimised.

Fatigue reduction: It reduces the fatigue of the operators to a great extent.

Handling operations made easy: Handling operations becomes easy and simplified with the use of jigs and fixtures.

Reduced labour cost: It enables semi-skilled operators to perform the operations as the setting and guiding of the tool is mechanised. This reduces labour cost.

Increased production capability: Due to quick setting and locating of workpieces, manipulation and handling time gets reduced; as such, production capability of the machine gets enhanced.

Use of better or higher cutting parameters: Due to high rigidity of clamping, the range of speed, feed and depth of cut can be increased on the workpiece for matching operations.

Enhanced safety: Application of jigs and fixtures ensures enhanced in-built operation safety.

Elimination of adverse effect of chip formation: The use of jigs and fixtures minimises the adverse effects of chip formation on the workpiece.

24.4 PRINCIPLES OF DESIGN OF JIGS AND FIXTURES

The following design principles must be considered at the time of design of jigs and fixtures:

Principle of rigidity: Jigs and fixtures should be rigid enough to bear intermittent forces and absorb shocks.

Principle of adequate clearance: There should be plenty of clearance between the jig base and the work to take care of slight variations in the dimensions of the job. In case of mass manufacture, such a clearance is also necessary for chips to be accommodated, so as not to disturb the alignment of work. The clearance should be enough so as to pass through the opening between the work and the jig plate rather than choke the jig bush.

Principle of swarf clearance: In large jigs, adequate swarf clearance can be provided by designing the jig with card holes at points where swarf is likely to accumulate.

Principles of locating points and supports: *Locating* refers to the establishment of the desired relationship between the workpiece and the jig or fixture. The following points should be considered when designing the locating points:

   (a) Locating and supporting surfaces should, wherever possible, be removable. Generally, surfaces should be of hardened material.
   (b) Locating points should be clearly defined. The use of large flat machined surfaces should be avoided.
   (c) Locating or supporting pins should be fitted into through holes and not blind holes.
   (d) Points more than necessary should not be provided.
   (e) Locating points should be chosen as far apart as possible.
   (f) Most satisfactory locating points are those in the mutually perpendicular planes.
   (g) All locating points which require replacement due to wear and tear should be easily replaceable.
   (h) Design location points should be accessible to workers.
   (i) Location should be provided at a proper place, so that swarf does not disturb the alignment.
   (j) Sufficient degree of movement should be provided by the locator.
   (k) Adjustable locators should be provided for rough surfaces.
   (l) Locating pins should be tapered.
   (m) At least one datum surface should be established at the first opportunity.
   (n) Locating surfaces should be as small as possible and the location must be done from the machined surfaces.
   (o) Sharp corners in the locating surfaces should be avoided.

Principle of foolproofing: If possible, special arrangements should be made in the design of the jig so that it is impossible to insert the piece in any way but the correct one. Every effort should be made to make the jig or fixture foolproof.

Principle of clamping: The functions of clamping is that of applying and maintaining sufficient counteracting holding force to a workpiece to withstand all tooling forces. The following points should be noted for the designing of any clamp:

   (a) The clamping arrangement should be as simple as possible.
   (b) Quick-acting clamps should be used wherever possible.
   (c) The clamping arrangement should be easily removable from the work.
   (d) Fixed stops should be used against the cut which will take the direct thrust on the cutters.
   (e) Arrange all the clamps and adjustments on the side nearest to the operator.
   (f) Clamping should always be arranged directly above the points supporting the work.

(g) Fibre pads should be riveted to clamp faces where metallic contact with the work would cause damage.
(h) Compression spring should be used to lift the clamp away from the workpiece.
(i) Clamping devices, when subjected to vibrations or heavy pressure, must not get distorted.
(j) Thick sections should be chosen for bearing clamping forces.
(k) Placement of nut or hand wheel should be such as to control the amount of pressure to be exerted on the workpiece.
(l) The movement of the screw, lever or cam of the clamping device, whether of the rotary or reciprocating type, should be strictly limited to make the device quick-acting.

Principle of reduction of idle time: The main aim of the designer should be to reduce the idle time of the manufacturing job. The method of location and clamping should be such as to reduce the idle time.

Principle of safety: All the sharp edges should be removed from the jig or fixture unit. All precautions on safety must be built into the jig or fixture.

Principle of simplicity: As far as possible, standard components or parts should be used, and similar operations should be grouped. The design should be as simple as possible.

Principle of spring location: Springs should be used as far as possible to lift the clamps.

Principle of economy: Unnecessary parts should not be incorporated; in other words, the design should not be complicated. It should be economic to reproduce.

Principle of maintainability: As far as possible the jig or fixture should be maintenance-free.

Principle of loading and unloading: It should be simple, easy and quick.

Principle of accuracy: A jig or fixture designer has to determine what variations can be permitted in an operation.

Principle of cooling: Adequate arrangements must be made to provide cooling to the cutting edge.

Principle of jig base or jig feet: A jig which is not bolted to the machine must be provided with four jig feet fitted to a jig base. A base is therefore necessary, whether or not feet are to be provided.

24.5 DESIGN PROCEDURE

The following design procedure for jigs or fixtures is adopted:
(i) Study the job carefully.
(ii) Draw the outline of the job to be performed by the machine.
(iii) Decide the sequence of operations.
(iv) Identify the limiting dimensions and special features.
(v) Develop the location system.
(vi) Develop the clamping system.
(vii) Determine the position and type of tool guides.
(viii) Arrange the position of the feet or base.
(ix) Connect the locations, clamps, tool guides and feet together within a suitable framework.
(x) Carefully check all the possible danger points.
(xi) Draw a scale-size view of the component in red ink, dotted for the scale drawing of the jig.
(xii) Make the final drawing.

24.6 LOCATIONS

In the design of jigs and fixtures, the location of the component is a very important aspect as the correctness of the location influences the accuracy of the finished product. *Location of piece* means placing it in a definite position for further operations so that the effect produced on all the pieces will be the same.

24.6.1 Factors Affecting Location

These are as follows:
  (a) The condition of the workpiece surfaces;
  (b) The shape of the workpiece;
  (c) The required reference point from the surfaces to the workpiece;
  (d) The tolerance of the location itself relative to the reference surfaces.

24.6.2 Principles of Location

A workpiece in space is free to move in any direction and can have 12 such directions. It can move in either of two opposed directions along three mutually perpendicular axis clockwise or counterclockwise. Each direction of movement is considered as one degree of freedom. To locate a workpiece accurately, it must be confined to restrict it against movements in any of the 12 degrees of freedom (see Fig. 24.1), except those called for the operation.

Principle of least points and principle of extreme positions: According to the principle of least points, no more points than necessary should be used to secure location in any one plane. (More points can only be used for finished surfaces. In such a case, the extra support should not interfere with the location and should therefore be made adjustable.)

According to principle of extreme position, the locating points should be chosen as far apart as possible on any one surface. This principle would ensure minimum misalignment for a certain displacement of any point with respect to another, if the distance between the points is increased.

Six-point location principle or 3–2–1 principle of location: The 3–2–1 principle of location (see Fig. 24.2) states that to locate a workpiece fully, it has to be placed and held against three points in a base plane, two points in a vertical plane and one point in a plane square with the first two. It is important that the above planes be square with each other and the points spaced as far apart as possible.

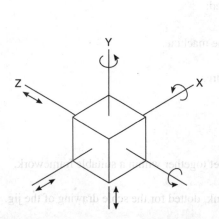

Fig. 24.1 Twelve degrees of freedom of object in space.

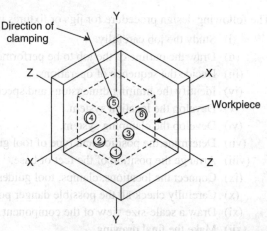

Fig. 24.2 Six-point location principle (3–2–1 principle).

24.6.3 Six-point Location of Rectangular Block

For the location of a rectangular job, three pins A, B and C are provided in the base of the fixed body of the jig at the horizontal level (refer to Fig. 24.3). This constrains one downward movement along axis Y–Y, two rotational movements along axis X–X and two rotational movements along axis Z–Z. Further, two pins D and E check one movement along axis X–X, two rotational movements along axis Z–Z and finally one more pin F at the back restricts the linear movement along axis Z–Z. Thus, all the essential 9 degrees of freedom which are to be restricted are constrained. The locating points for an uneven object can be determined by different arrangements, but the guiding principle remains the same. This is called the well-known 3-2-1 principle which states:

$$3 = \text{Rest buttons restrict 5 d.o.f}$$
$$2 = \text{Rest buttons restrict 3 d.o.f.}$$
$$1 = \text{Rest button restricts 1 d.o.f.}$$

Total d.o.f. restrained = 9 (by locations in 3 planes)

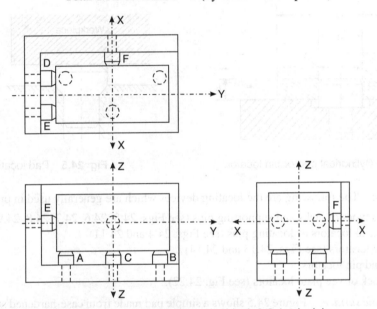

Fig. 24.3 Location of rectangular block 3-2-1 principle.

24.6.4 Types of Locators

The various types of locators are as follows:

  (i) Edge locators
 (ii) Pin locators
(iii) V-locators.

24.6.5 Locating Devices and Their Choice

When there is a choice of location systems and location points in a job, the most effective location system must be selected. Out of all, the cylinders are the most effective location devices and must be selected first, because a cylindrical locator is least difficult to produce. Furthermore, a single cylindrical locator will eliminate eight out of the twelve degrees of freedom. The ease of loading and unloading the workpiece is also a criterion for the choice of a locator. This is explained in Figures 24.12 and 24.15.

In Fig. 24.12, a clamping bolt holds the job and constrains all the 12 degrees of freedom, even the one of rotation. But the fact is that only a limited area of the job is workable. This limitation restricts the use of this type of locator.

The diameter of one cylindrical peg shown in Fig. 24.4, must be of height h, such that

$$h = \sqrt{2D\phi}$$

where h is the height or length of the plug;

D is the outer dimension of the workpiece (diameter in case of a round job); and

$\phi = d' - d$; with d', the diameter of the hole in the job and d is the diameter of the plug.

Figure 24.5 shows a pad locator which, with the required clamping force, will constrain all the six degrees of freedom but for the degree of rotation. A second locator may be installed for this purpose to achieve complete location.

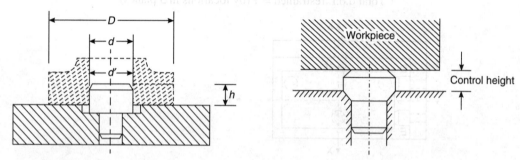

Fig. 24.4 Cylindrical peg or pin locator. **Fig. 24.5** Pad locator.

Locating devices: The following are the locating devices which are generally used in practice:
- (i) Various types of jacks and supporting pins (see Figs. 24.5, 24.6, 24.7, 24.8, 24.9, 24.10)
- (ii) Cylindrical locators or locating pins (see Figs. 24.4 and 24.11)
- (iii) Conical locators (see Figs. 24.13 and 24.14)
- (iv) Diamond pin locators
- (v) Vee block or vee plate locators (see Fig. 24.22).

Location from plane surface.   Figure 24.5 shows a simple pad made from case-hardened steel and located in the body of the jig or fixture. Point contact is obtained by using the locating surface of spherical radius shape. Several locators of this type would be used to produce a location system. When locating from an uneven surface, some of the locators will need to be made adjustable; adjustable support points would also be used when plenty of support is necessary and where, if the additional support be made fixed, it would produce redundant location. A very simple adjustable support is shown in Fig. 24.6. More elaborate arrangements are shown in Figures 24.7 and 24.8. In this system, the support point can be in a recess or similar difficult place for the operator to reach, because the adjustment point is some way from the support point. The support can be spring-loaded, so that it adjusts itself to the uneven surface, with the pin being locked in position as shown in Fig. 24.9. The clamping bolt for location post is shown in Fig. 24.12.

Location from a profile.   When, for example, a group of holes is to be drilled in a best position, relative to a cast profile, the component can be best positioned on the base of the jig or fixture by using a sight plate, as shown in Fig. 24.16.

At the early machining stages, a workpiece may be positioned from a profile by means of pins that form the location system as shown in Fig. 24.17. The arrangement shown in Fig. 24.18 produces a form of 'vee'

JIGS AND FIXTURES **623**

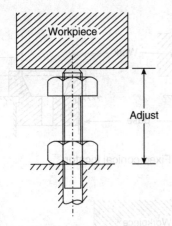

Fig. 24.6 Simple adjustable locator.

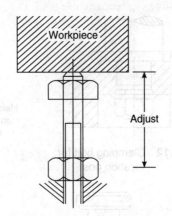

Fig. 24.7 Provision of recess.

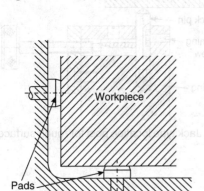

Fig. 24.8 Location of pads at 90°.

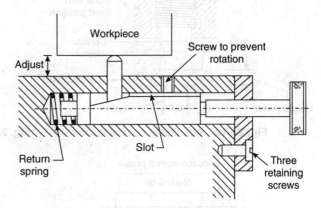

Fig. 24.9 Spring-loaded adjustable locator.

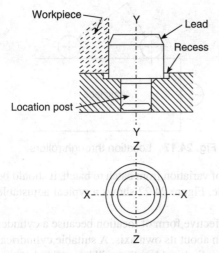

Fig. 24.10 Location post (location by plane surface).

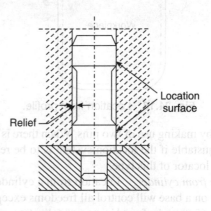

Fig. 24.11 Location post (location by cylindrical surface).

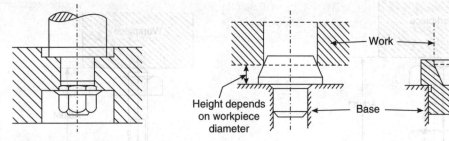

Fig. 24.12 Clamping bolt for location post.

Fig. 24.13 Fixed conical locator.

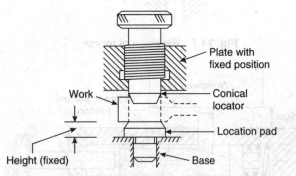

Fig. 24.14 Adjustable conical locator.

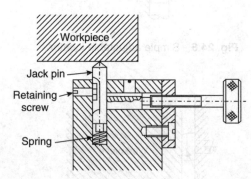

Fig. 24.15 Jack type location post for rough surfaces.

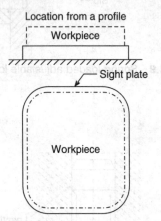

Fig. 24.16 Location from profile.

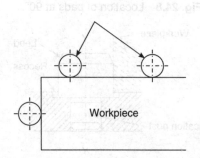

Fig. 24.17 Location through rollers.

location by making use of two pins. When there is a large amount of variation from batch to batch, it should be made adjustable if the location system is to be reasonably reliable. Figure 24.22 shows a typical adjustable vee type locator of this type.

Location from cylinder. Location from a cylinder is the most effective form of location because a cylinder mounted on a base will control all freedoms except that of rotation about its own axis. A suitable cylindrical location point can be found in many applications, partly because a cylindrical location will be used to locate it upon assembly and partly because cylindrical features will be present where pins and cylinders are involved.

Figure 24.19 shows a long locator, and it will be seen that the lead is very generous and that the post is relieved slightly, so that location only occurs at the extreme ends. This shows how the jamming problem can be eased by using a specially-shaped chamfer lead; the recess will prevent jamming from taking place.

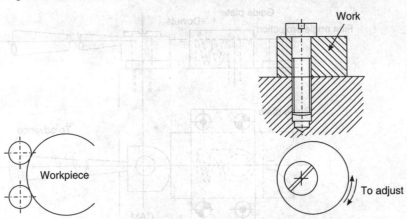

Fig. 24.18 Cylindrical work located by roller.

Fig. 24.19 Screw locations.

Figure 24.20 shows the error arising due to wrong location, while Fig. 24.21 shows a cylindrical locator in combination with a diamond-shaped locator which is smaller in size than the hole.

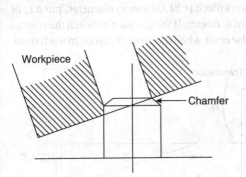

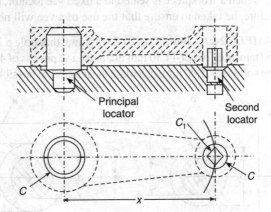

Fig. 24.20 Error due to wrong location.

Fig. 24.21 Cylindrical locator combined with diamond-shaped locator smaller in size than hole. The merit of the diamond pin locator is that it allows the designer to use wider open tolerance on 'x', i.e., as $\pm (C + C_1)$ instead of $\pm 2C$ which is lesser. Diamond pin locator permitting a clearance C_1 greater than C are widely used in industry.

Vee location. This type of location can be used for location from a cylinder or from a part of the cylindrical profile. The vee locator can be a fixed one as shown in Fig. 24.22, or a sliding one. A vee location system can consist of two fixed vees for an approximate location, or of one fixed and one sliding vee, when the effect of profile variation can be compensated for. A vee location can be used as the second locator in conjunction with a cylindrical locator.

Figure 24.22 shows an arrangement for the sliding vee location, with the cam-operated system, being rapid to operate. A vee location system that includes a sliding vee can be made to produce a small downward clamping force by inclining the sides of the vee as shown in Figs. 24.22 and 24.23.

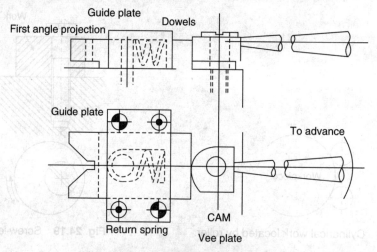

Fig. 24.22 Adjustable vee plate location.

When a workpiece is seated in a fixed vee locator, its position will depend upon its diameter; care must, therefore, be taken to ensure that the use of a vee will not cause errors due to component variation.

EXAMPLE 24.1 From Fig. 24.23, assume that hole a in the workpiece is 51.00 mm in diameter, pin a is of 50.60 mm, hole b is of 44.50 mm diameter and pin b is of 44.25 mm diameter. If the distance between the centres of the holes is 305.00 mm, calculate the maximum possible angular error when the piece is placed in the fixture.

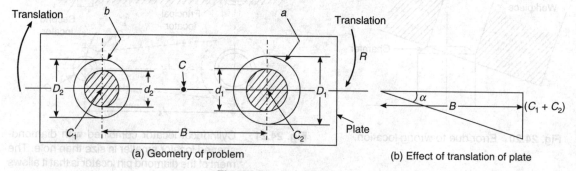

Fig. 24.23 Example 24.1.

Solution If the plate is translated about C_1 (treating the slight arc movement as linear movement) by clearance $(D_1 - d_1)$ on R.H.S. and $(D_2 - d_2)$ on left, the resulting angle error α is given by

$$\tan \alpha = \frac{(D_1 - d_1) + (D_2 - d_2)}{B}$$

Rearranging, the error is

$$\tan \alpha = \frac{(D_1 + D_2) - (d_1 + d_2)}{B}$$

where B is the distance between the pins;
    D_1 is the diameter of hole a;
    d_1 is the diameter of pin a;
    D_2 is the diameter of hole b;
    d_2 is the diameter of pin b;
    α is the error angle.
Now $D_1 = 51.00$ mm;
    $D_2 = 44.50$ mm;
    $d_1 = 50.60$ mm;
    $d_2 = 44.25$ mm;
    $B = 305.00$ mm.

$$\tan \alpha = \frac{(51 + 44.50) - (50.60 + 44.25)}{305.00} = 0.0021$$

$$\alpha = 0°7'$$

EXAMPLE 24.2 A rectangular plate 100 × 200 mm is in full contact with the base of a fixture. The hole in the workpiece has been machined to 38.10 mm diameter at the exact centre of the plate. Figure 24.24 shows this plate. The plug diameter is 37.95 mm. Calculate:

(a) The maximum height of the plug if the 100.00 mm side is lifted off the plug.
(b) The maximum height of the plug if the 200.00 mm side is lifted off the plug.
(c) Compare the two plug designs.
(d) Calculate the maximum diameter at the plug for the 200.00 mm lift which will permit the use of a 5.08 mm height plug.
(e) Compare the clearance in part (d) with part (b).

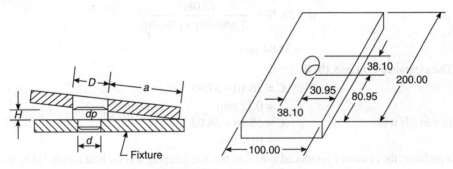

Fig. 24.24(a) Example 24.2.

Solution The maximum height of the plug should be

$$H = \sqrt{2(2a + D)(D - d)}$$

where H is the height or length of the plug in mm;
    a is the distance from the end-point to the edge of the hole in mm;
    D is the diameter of the hole in mm;
    d is the diameter of the plug in mm.

(a) If the 100 mm side is raised, the maximum height of the plug is given by

$$H = \sqrt{2(2a + D)(D - d)}$$

Since $D = 38.10$; $d = 37.95$; $a = 30.95$

$$H = \sqrt{2(2 \times 30.95 + 38.10)(38.10 - 37.95)}$$
$$= \sqrt{2 \times 100 \times 0.15}$$
$$= \sqrt{30} = 5.5 \text{ mm}$$

(b) If the 200 mm height is raised, the minimum height of the plug should be
$$H = \sqrt{2(2a + D)(D - d)}$$
Since $D = 38.10$, $d = 37.95$, and $a = 80.95$ mm, we have
$$H = \sqrt{2(2 \times 80.95 + 38.10)(38.10 - 37.95)}$$
$$= \sqrt{2 \times 200 \times 0.15}$$
$$= \sqrt{60} = 7.74 \text{ mm}$$

(c) The greater the distance a, the longer be the acceptable interference swing radius permitted.

(d) The maximum plug diameter is 5.08 mm; the height retained may be calculated from
$$H = \sqrt{2(2a + D)(D - d)}$$

Solving for d, we have
$$d = D - \frac{H^2}{2(2a + D)}$$

Since $D = 38.10$ mm, $a = 80.95$ mm and $H = 5.08$ mm, we have
$$d = 38.10 - \frac{(5.08)^2}{2[2(80.95) + (38.10)]}$$
$$= 38.04 \text{ mm}$$

(e) The clearance C in part (b) is
$$C = 38.10 - 37.95$$
$$= 0.15 \text{ mm}$$
In part (d) it is
$$C = 38.10 - 38.03$$
$$= 0.07 \text{ mm}$$

In this problem, the clearance is reduced to half, so that the side plug and the hole is reduced by as much. This means the centre of hole-to-machined surface dimensions can be located with twice the accuracy.

EXAMPLE 24.3 Using the data from Example 24.2
  (a) Calculate the plug height for the 203 mm workpiece length if a 3.20 mm chamfer is machined on the workpiece hole.
  (b) Compare the plug heights for Examples 24.2 and 24.3.
  (c) Assume a plug of the height obtained in Example 24.2(b). What new plug diameter may be used if a 3.20 mm chamfer is added to the workpiece hole?

Solution
  (a) The height of the plug will be
$$H = C + \sqrt{(2a + D)(D - d)}$$

where
    C is the height of the chamfer in mm;
    a is the distance from the point to the edge of the hole in mm;
    D is the diameter of the hole in mm;
    H is the height of the plug in mm;
    d is the diameter of the plug in mm.

Now $$H = C + \sqrt{(2a + D)(D - d)}$$

Since $C = 3.20$ mm, $a = 80.95$ mm, $D = 38.10$ mm, and $d = 37.95$ mm, we have

$$H = 3.20 + \sqrt{(2 \times 80.95 + 38.10)(38.10 - 37.95)}$$
$$= 3.20 + \sqrt{30}$$
$$= 8.70 \text{ mm}$$

(b) When a chamfer of 3.20 mm is added the comparison of plug height is
$$= 8.70 - 7.74$$
$$= 0.96 \text{ mm}$$

(c) The new plug diameter may be
$$d = D - \frac{(H - C)^2}{(2a + D)}$$

Since $c = 3.20$ mm, $D = 38.10$ mm, $a = 80.95$ mm and $H = 7.74$ mm, we have

$$d = 38.10 - \frac{(7.74 - 3.20)^2}{2 \times 80.95 + 38.10}$$
$$= 37.997 \approx 38.00 \text{ mm} \quad \{\text{From Example 24.2(b)}\}$$

Alignment groove. An alignment groove [Fig. 24.24(b)] may be machined into the plug to ensure right angle between plug axis and fixture top face. The depth of the groove, d_g, may be about 95% of the diameter of the body of the plug. That is
$$d_g = 0.95 d$$

The diameter of the pilot end of the plug is given by
$$d_p = \frac{2d^2}{D} - d$$

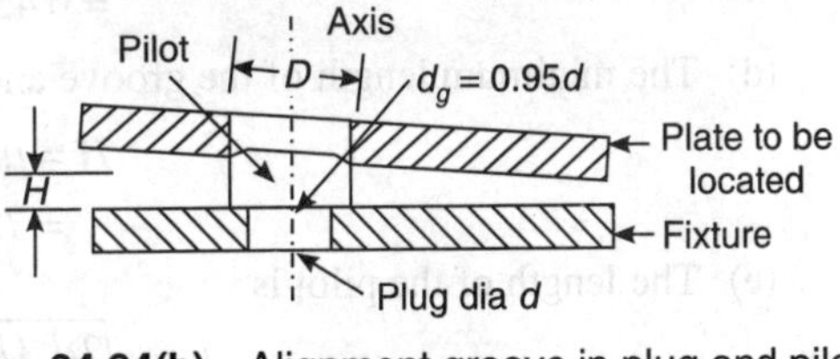

The clearance between the hole and the pilot is
$$C = d_p$$

Fig. 24.24(b) Alignment groove in plug and pilot.

The length of the pilot and the groove is a function of the coefficient of friction of the material in the plug and that in the workpiece. For steel-against-steel it usually is 0.20. The equation for the minimum length of the pilot and groove is given by
$$H = \mu D$$

This may be any desired length longer than H, the equation for the length of the pilot is

$$h = \sqrt{2d_p(D - d_p)}$$

where d is the diameter of the plugs;
d_p is the diameter of the pilot;
d_g is the diameter of the groove; where $d_g = 0.95d$
h is the pilot length;
H is the minimum length;
D is the diameter of the hole;
μ is the coeficient of friction;
C is the clearance.

EXAMPLE 24.4 Using the plug and hole diameters from Example 24.2, calculate:
  (a) The groove diameter in the plug;
  (b) The pilot diameter;
  (c) The clearance between the hole and the pilot;
  (d) The minimum length of the groove and the pilot;
  (e) The length of the pilot.

Solution We have the following data: $d = 37.95$ mm, $D = 38.10$ mm and $\mu = 0.20$ mm. Therefore
  (a) The groove diameter in the plug is

$$d_g = 0.95d = 0.95(37.95)$$
$$= 36.05 \text{ mm}$$

  (b) The pilot diameter is

$$d_p = \frac{2d^2}{D} - d = \frac{2(37.95)^2}{38.10} - 37.95 = 37.65 \text{ mm}$$

  (c) The clearance is

$$C = D - d_p = 38.10 - 37.65$$
$$= 0.45 \text{ mm}$$

  (d) The minimum length of the groove and the pilot is

$$H = \mu d = 0.20 \times 37.95$$
$$= 7.59 \text{ mm}$$

  (e) The length of the pilot is

$$h = \sqrt{2d_p(D - d_p)}$$
$$= \sqrt{2 \times (37.65)(38.10 - 37.65)}$$
$$= 5.82 \text{ mm}$$

EXAMPLE 24.5 Using the plug and hole diameters from Example 24.2, calculate
  (a) The distance from the flat to the centre line of the plug;
  (b) The distance from the flat to the opposite bearing surface;
  (c) The height of the plugs;
  (d) The additional error when an equivalent plug is used;

Solution Here $D = 38.10$ mm, $d = 37.95$ mm and $a = 80.95$ mm. Therefore

(a) The flat should be cut, so that the distance from the flat to the centre line of the plug is
$$F = 0.35 \times 37.95 = 13.28 \text{ mm}$$

(b) The distance from the flat to the opposite bearing surface will be
$$d_f = F + \frac{d}{2} = 13.28 + \frac{37.95}{2} = 32.25 \text{ mm}$$

(c) The height of the plug should be
$$H = 2.4(2a + 8.5d)(D - d)$$
$$= 2.4[2(80.95) + 8.5(37.95)](38.10 - 37.95)$$
$$= 174.41 \text{ mm}$$

(d) The additional error when an equilateral plug is used instead of a full round plug is
$$E = 0.20(D - d) = 0.20(38.10 - 37.95)$$
$$= 0.03 \text{ mm}$$

24.7 CLAMPING

24.7.1 Functions of Clamps

The function of a clamping device is that of applying and maintaining sufficient counteracting holding force to a workpiece to withstand all tooling forces.

The clamping arrangements should be as simple as possible, without sacrificing strength. The main expectation from clamps is that they should be convenient for the operator. Therefore, it is advisable to avoid complicated clamping arrangements; clamps should be quickly operated to reduce the idle time of the machine.

The following points are to be noted for the designing of any clamp:

(i) Clamping should always be arranged directly above the points supporting the work;
(ii) Fibre pads should be riveted to clamp faces where metallic contact with the work would cause damage;
(iii) Arrange all the clamps and adjustment on the side nearest to the operator;
(iv) Clamping arrangement should be easily removable from the work;
(v) Fixed stops against the cut which will take the direct thrust of the cutters should be provided.

24.7.2 Clamping Devices

Some sort of clamping device is essential for both jigs and fixtures. Clamping may be complex or simple, but it must fulfill the following design requirements:

(i) The clamping devices must hold the workpiece rigidly against all distributing forces;
(ii) It should also keep the workpiece firmly in contact with locating pins or locating surfaces;
(iii) The time required to loosen the clamp on the workpiece and lighten it again on the next piece should be minimum. Compression spring should be used to lift the clamp away from the workpiece;
(iv) The clamping devices, when subjected to vibrations or heavy pressure, must be positive and should not get distorted.
(v) The clamp, while holding the workpiece, should not damage it. Thick sections should be chosen for bearing clamping forces.
(vi) The placement of nut or hand wheel should be such as to control the amount of pressure to be exerted on the workpiece.

(vii) The movement of the screw, lever or cam of the clamping device, whether of the rotary or reciprocating type, should be strictly limited to make the device quick-acting.

24.7.3 Types of Clamps

There are many types of clamps used in production work; some of them are given below:

(i) Strap clamps or solid clamps;   (ii) Wedge clamps;
(iii) Latch type clamps;   (iv) Screw clamps and C-clamps;
(v) Equalising clamps;   (vi) Quick-acting nut type clamps;
(vii) Toggle clamps;   (viii) Quick-acting cam-operated clamps;
(ix) Bayonet type quick-acting clamps;   (x) Multiple clamps.

Solid clamps: Figure 24.25 shows a flat plate clamp; in this system the clamp is supported by a heel pin. This clamp is rotated about the stud axis to remove the workpiece and held against the rotational forces by the heel pin engaged in the plate.

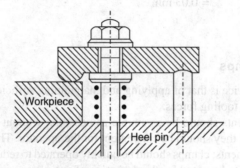

Fig. 24.25   Flat plate clamp.

Figure 24.26 also shows a clamp with heel pin, but in this arrangement the heel pin engages in a slot in the clamp plate to produce a better clamp location.

Large variations in workpiece height can be compensated for by having an adjustable heel pin, as shown in Fig. 24.27.

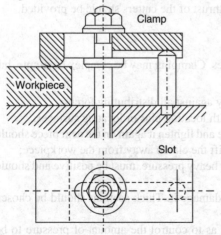
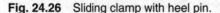
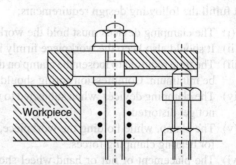

Fig. 24.26   Sliding clamp with heel pin.   **Fig. 24.27**   Solid or strap clamp with adjustable heel pin.

The clamp can be strapped as shown in Figs. 24.28 and 24.29 to suit the workpiece.

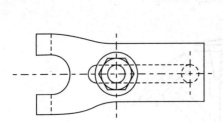

Fig. 24.28 Two-point strap clamp.

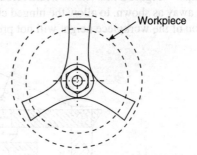

Fig. 24.29 Three-point strap clamp.

The clamps shown so far are all secured by hexagonal nut and spanner; this is acceptable when a high clamping force is necessary as and when milling, and when turning out a hand-operated nut it is better if it permits the required force to be applied. The hand nut should be large enough to allow the operator to grip it with comfort.

Wedge clamps: A typical wedge clamp is shown in Fig. 24.30. Smaller variations in workpiece height can be allowed for by using a pair of spherical washers, so that the clamp plate can be at an angle to the stud axis and still permit the nut to clamp it correctly (see Fig. 24.31). Figure 24.46 shows another form of wedge clamp.

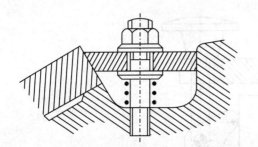

Fig. 24.30 Wedge clamp.

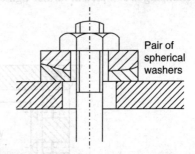

Fig. 24.31 Spherical washer for small adjustment.

Figure 24.32 shows the arrangement of a clamp plate with a slot for fly nut.

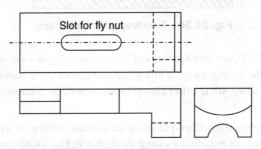

Fig. 24.32 View showing arrangement.

As an alternative to the slot shown in the previous examples, the plate can be shaped as shown in Fig. 24.25, so that it can be rotated to remove the workpiece.

Latch clamps: Figure 24.33 shows a latch clamp; to release the workpiece, the nut is slackened off and the bolt swung away as shown, to allow the hinged clamp to be turned away. The clamping pad is shaped so that any variation of the workpiece height will not prevent correct clamp seating.

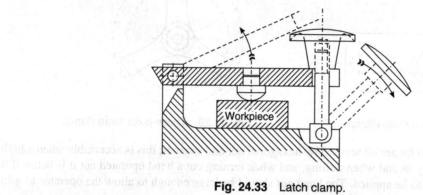

Fig. 24.33 Latch clamp.

The latch clamp can be adopted to produce two-way clamping as shown in Fig. 24.34. The swinging bolt is attached to a link member that is in turn attached to the body. When the clamping nut is lightened, it causes this link member to rotate about the fixed pin, so that a horizontal clamping force is applied to the workpiece. When this clamping is completed, further tightening will produce a vertical clamping force.

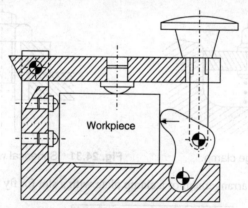

Fig. 24.34 Two-way latch type clamp.

Screw clamps: Figure 24.35 shows another form of direct clamping device, the screw clamp. The clamp is supported by a post, and can be swung away to allow the workpiece to be removed.

Figures 24.36 to 24.39 show some other clamp types as well as components and mechanisms used in clamping.

Equalising clamps: When clamping at two places on an uneven surface, or when clamping two workpieces whose heights are likely to vary, an equalising clamp as shown in Fig. 24.40 can be used. This system can be extended to enable several workpieces to be clamped at once.

Figure 24.41 shows two types of eccentric clamping system.

Direct clamping: The bottom clamp shown in Fig. 24.42 is a simple direct clamping device that can be swung away to allow the workpiece to be removed. When the clamping strap must be removed completely, one of the arrangements shown in Fig. 24.42 can be used.

JIGS AND FIXTURES **635**

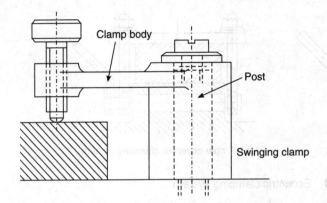

Fig. 24.35 Screw clamp—the bottom clamp.

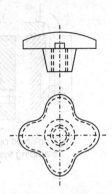

Fig. 24.36 Cast hand nut.

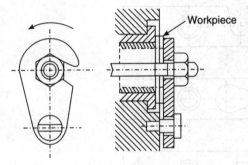

Fig. 24.37 Swing washer clamp.

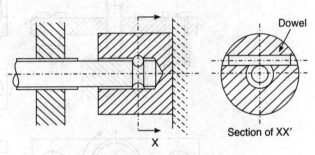

Fig. 24.38 Details of clamping while using dowel and swing washer.

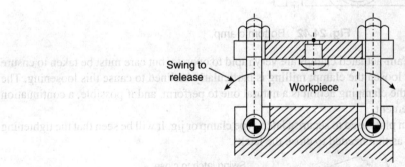

Fig. 24.39 Removable clamp with swing bolts.

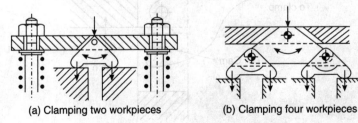

(a) Clamping two workpieces   (b) Clamping four workpieces

Fig. 24.40 Equalising clamps.

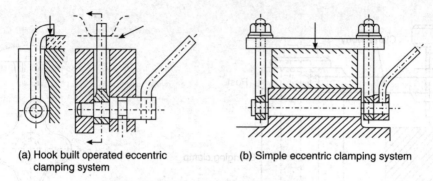

(a) Hook built operated eccentric clamping system

(b) Simple eccentric clamping system

Fig. 24.41 Eccentric clamping system.

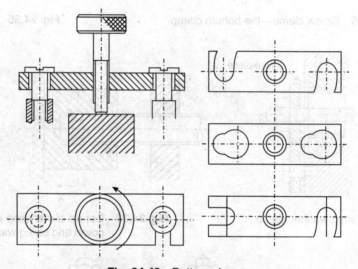

Fig. 24.42 Bottom clamp.

Cam-actuated clamping: Cam-actuated clamps are very rapid to operate, but care must be taken to ensure that the cutting action will not loosen the clamp; milling is particularly inclined to cause this loosening. The cam must be arranged so that the clamping action is a natural one to perform, and if possible, a continuation of the clamp-positioning movement.

Figure 24.43 shows a cam plate used to secure a latch-type clamp or jig. It will be seen that the tightening action is in the same direction as that needed to close the latch.

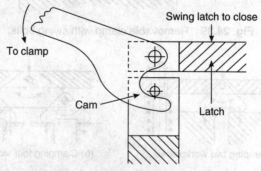

Fig. 24.43 Cam-actuated latch clamp.

Cams can be used to secure plate clamps. For example, the system shown in Fig. 24.43 is a cam-actuated version of the clamp shown in Fig. 24.33. Figure 24.44(b) shows another type—the common cam-actuated clamp.

Toggle clamps: Toggle clamps are very easy to open and give secure clamping action. Figure 24.44(a) shows the toggle part of a plunger-operated clamp; the linkage diagram shows the relative movements, and indicates that when the linkage is in the clamping position a large angular movement is necessary to unlock the clamp. Figure 24.44(b) shows another toggle clamping system, in which the clamping lever is quickly moved clear of the workpiece.

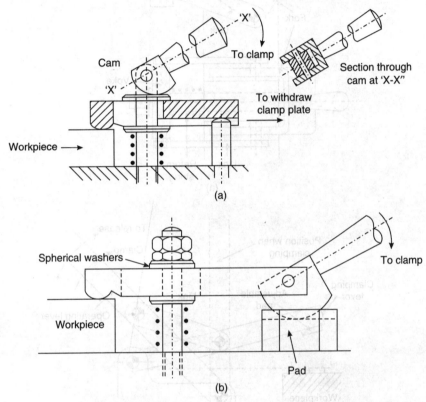

Fig. 24.44 (a) Toggle part of plunger-operated clamp and (b) Cam-actuated clamp using fastener and toggle cam.

Pneumatic clamping: Compressed air is used extensively for clamping, and for operating location and indexing devices. The advantages claimed for pneumatic clamping are that there is less wear on the clamps and associated parts, there is less tendency for damage to occur to the workpiece, that the operation is more rapid, and that the clamping pressure can be controlled more accurately. Figure 24.45 shows two different types of pneumatic toggle clamp.

The basic features of a pneumatic system are a control valve, a reducing valve to control the pressure, and a cylinder unit to operate the mechanical part of the arrangement. An air-flow regulator and pressure gauge may also be included in the system, and several cylinders may be operated by one system. The air can be tapped from an airline or supplied by a small compressor.

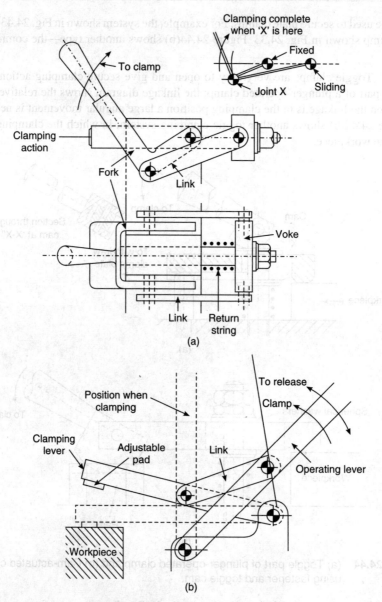

Fig. 24.45 Pneumatic toggle clamp.

Figure 24.46(a) shows an air-operated clamp. The push rod can be used to clamp the workpiece directly or to operate a rack and pinion, wedge or toggle clamp.

To safeguard against accidents following air-pressure failure, a non-return valve can be incorporated into the system, or as an alternative the clamping force can be produced by a powerful spring, and the air can be used to release the clamp as shown in Fig. 24.46(a). Care must be taken to ensure that the operator can not place his fingers between the clamp and the workpiece; the clamp workpiece clearances should therefore be made as small as possible, and if further safeguard is necessary the valves can be arranged so that the operator can operate both valves with his hands before clamping can take place.

Figure 24.46(b) shows a type of wedge clamp.

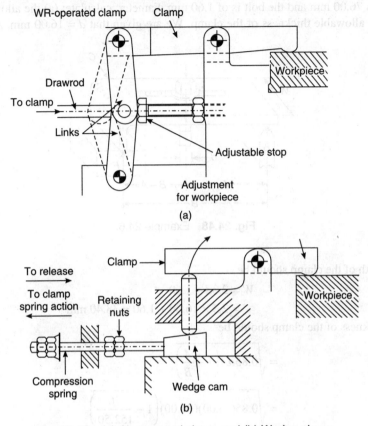

Fig. 24.46 (a) Air-operated clamp and (b) Wedge clamp.

Figure 24.47 shows the methods adopted for the introduction of compressed air between sliding surfaces.

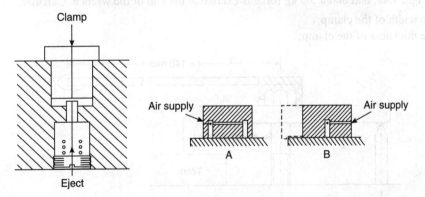

Fig. 24.47 Methods adopted for introduction of compressed air between sliding surface.

EXAMPLE 24.6 Assume the bolt in Fig. 24.48 to be halfway between the pivot point P and the workpiece. If the distance A is 76.00 mm and the bolt is of 1.60 mm diameter, calculate: (a) the allowable width of the clamp, and (b) the allowable thickness of the clamp. We are given that $d = 16.00$ mm, $A = 76.00$ mm and $B = 152.50$ mm.

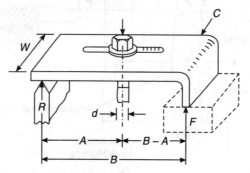

Fig. 24.48 Example 24.6.

Solution

(a) The width of the clamp should be

$$W = 2.3d + 1.60$$
$$= 2.3 \times 16.00 + 1.60 = 38.40 \text{ mm}$$

(b) The thickness of the clamp should be

$$= \sqrt{0.85dA\left(1 - \frac{A}{B}\right)}$$

$$= \sqrt{0.85(16.00)(76.00)\left(1 - \frac{76}{152.50}\right)}$$

$$= 22.77 \text{ mm}$$

EXAMPLE 24.7 Assume that a 140.00 mm long open-end wrench is used to tighten a bolt of 1.60 mm diameter in Fig. 24.49, and that a 5.5 kg force is exerted at the end of the wrench. Calculate

(a) the width of the clamp;
(b) the thickness of the clamp;

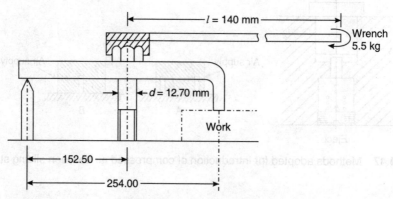

Fig. 24.49 Example 24.7.

(c) the load on the bolt;
(d) the moment on the stop;
(e) the working stress on the clamp;
(f) the safety factor if the ultimate stress of this material is 4570 kg/cm²;
(g) the maximum radial load which can be applied to this bolt.

Solution

(a) The width of the clamp should be
$$W = 2.3d + 1.60$$
$$= 2.3(12.70) + 1.60$$
$$= 30.81 \text{ mm}$$

Given that
$l = 140.00$ mm
$d = 12.70$ mm
$A = 152.50$ mm
$B = 254.00$ mm
$C = 12.70 + 1.60 = 14.30$ mm

(b) The thickness of the clamp should be
$$t = \sqrt{0.85dA\left(1 - \frac{A}{B}\right)} = \sqrt{0.85 \times 12.70 \times 152.50 \times \left(1 - \frac{152.50}{254.00}\right)}$$
$$= 25.00 \text{ mm}$$

(c) The load on the bolt, F, from torque T, assuming a factor of safety of 5 for design torque (770×5)
$$F = \frac{5T}{d} = \frac{5 \times 770}{12.70}$$
$$= 303.15 \text{ kg mm} \quad (\because \text{ Nominal torque } T = 140 \times 5.5 = 770 \text{ kg mm})$$

(d) The moment on the strap is
$$M = \frac{FA(B - A)}{B} = \frac{303.15 \times 152.50 \times (254.00 - 152.50)}{254.00}$$
$$= 18{,}473.95 \text{ kg mm}$$
or $= 18{,}47.41$ kg cm
or $= 18.47$ kg m

(e) The stress on the clamp is a function of the section modules of the strap.

Section modules (Sec. Mod.) $= \dfrac{(W - C)t^2}{6}$ here t is thickness of the strap

$$= \frac{(30.81 - 14.30)(25)^2}{6} = 1718.75 \text{ mm}$$

The stress on the clamp is
$$s = \frac{M}{\text{Sec. Mod.}} = \frac{18473.95}{1718.75}$$
$$= 10.85 \text{ kg/mm}^2 = 1085 \text{ kg/cm}^2$$

(f) The safety factor = $\dfrac{4570}{1085}$ = 4.2

(g) The maximum radial force which can be placed on this clamp is

$$d = 1.35\sqrt{F/S}$$

Solving for maximum force

$$F_{max} = \dfrac{Sd^2}{(1.35)^2}$$

Given that

d = 12.70 mm

S = 10.85 kg/m^2

Therefore, $F_{max} = \dfrac{10.85 \times (12.70)^2}{(1.35)^2}$ = 960.22 kg

EXAMPLE 24.8 A cam has a circle of diameter 75.00 mm and the total rise which takes place through 100°. The middle of the rise is to clamp the average block size. Design the cam.

Solution
1. The total rise is

$$D = 0.001 \times D/2 \times \text{angle of rotation}$$
$$= 0.001 \times 75/2 \times 100 = 3.75 \text{ mm}$$

2. Half of the size is to be above the base circle and half below the base circle, as shown in Fig. 24.50.
3. The 100° portion of the base circle in which the rise is to take place is divided into an equal number parts. The rise is divided into the same number of parts. This is shown in Fig. 24.50. In this case the base circle is divided into 10 parts (10° each) through 100°, and the rise is divided into 10 parts of 0.375 mm each, five divisions to the left of the mean position and five to the right.

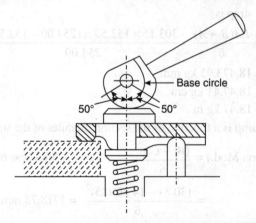

Fig. 24.50 Cam-operated spring-loaded clamp.

4. Starting at 0°, consective rise and degree divisions are intersected as shown in Fig. 24.50. These intersections are connected to give the cam surface.

Another method used to lay out a clamping cam is to determine the rise needed through 100°. Since the average clamping action takes place at the mid-point of the 100° portion (see Fig. 24.34), the radius r of the average clamping circle is

$$r = \frac{r_1 + r_2}{2}$$

where r is the radius of the average base circle;
r_1 is the radius at the beginning of the rise;
r_2 is the radius at the end of the rise.

Once the rise and the radius r of the average base circle are determined, the procedure of Example 24.8 are followed.

EXAMPLE 24.9 In Example 24.8, calculate r_1 and r_2.

Solution The total rise is given as

$$\Delta r = \text{Total rise}$$
$$= r_2 - r_1$$
$$= 3.75 \text{ mm}$$

The radius, $r = 37.50$ mm

The maximum and minimum radii are given by $r \pm \dfrac{\Delta r}{2}$.

Therefore the minimum radius is

$$r_1 = r - \frac{\Delta r}{2} = 37.50 - \frac{3.75}{2} = 35.63 \text{ mm}$$

and the maximum radius is

$$r_2 = r + \frac{\Delta r}{2} = 37.50 + \frac{3.75}{2} = 39.38 \text{ mm}$$

If it is possible to purchase cams which will serve the designer's purpose, then he should buy them. If it is really less expensive to make the cam, then it is useless to buy them commercially. Table 24.1 shows the various commercially available cams and their dimensions.

Table 24.1 Dimensions of commercial cams

A (in mm)	r (in mm)
3.97	29.37
5.16	37.30
7.94	56.36
19.32	72.23
11.11	76.99

EXAMPLE 24.10 Assume that a certain material is used to make the pins in a toggle, shown in Fig. 24.50, to an allowable shear stress of 420 kg/cm². If the toggle is to deliver a 180 kg force to the work, calculate the pin diameter.

Solution The pin diameter is

$$d = \sqrt{\frac{\mu F_0}{S}}$$

Given $\mu = 0.22$
$F_0 = 180$ kg
$S = 420$ kg/cm^2 = $\frac{420}{100}$ kg/mm^2

Therefore

$$d = \sqrt{\frac{0.22 \times 180 \times 100}{420}}$$

$$= 3.07 \text{ mm}$$

EXAMPLE 24.11 A toggle clamp (described in Example 24.5) is to deliver a 180 kg force to a workpiece in a fixture (see Fig. 24.51). Assume the pin diameter to be 6.35 mm and the coefficient of friction to be 0.22. Calculate the force required at the end of the handle.

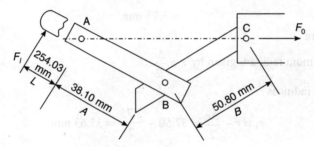

Fig. 24.51 Example 24.11.

Solution The force input is

$$F_i = 4\mu d F_0 \left(\frac{A+B}{LB}\right)$$

where A is the toggle distance from pin A to B;
B is the toggle distance from pin B to C;
μ is the coefficient of friction;
F_0 is the input force;
d is the pin diameter.

It is given that
$L = 254.00$ mm;
$A = 38.10$ mm;
$B = 50.80$ mm;
$\mu = 0.22$;
$d = 6.35$ mm;
$F_0 = 180$ kg.

Therefore, $F_i = 4 \times 0.22 \times 6.35 \times 180 \left(\dfrac{38.10 + 50.80}{254.00 \times 50.80}\right)$

$$= 7.00 \text{ kg}$$

24.7.4 Classification of Jig Bushes

Drill bushes may be classified as follows:

(i) Fixed bushes (plain or headed)
(ii) Liner bushes
(iii) Renewable bushes
(iv) Slip bushes
(v) Screw or clamp bushes
(vi) Special bushes.

Fixed bushes: Fixed bushes are pressed permanently into position and when washed out cannot be readily replaced. They are used when a hole is produced by one tool only. There are two types of this bush—plain and headed. One disadvantage of the plain bushes is that they are liable to be worn through the jig plate, either by the pressure of the drill or from the drilling machine spindle. Therefore, the headed bushes are commonly used.

While designing a jig bush, it should be kept in mind that all the sharp corners are chamfered adequately.

Liner bushes: Liner bushes are also press-fitted like the fixed bushes, but their main purpose is different. The linear bush is used to act as a hardened guide for both renewable and slip bushes. However, it can be used for guiding the tool—usually the largest in any combination.

Renewable bushes: These are special types of fixed bushes (see Fig. 24.52). When they need to be replaced due to wear, a retaining screw is removed and the worn bush is taken out. A new bush is then easily substituted. Renewable bushes are slipped over the linear bushes for easy removal.

Slip bushes: When a hole is to be drilled in stages and then subsequently too, the best practice is to use the slip bushes (see Fig. 24.53), rather than use separate jigs for each operation. Here the linear bush, in which the slip bush is placed, is also utilised for the last operation. Slip bushes are made in such a way that there is no need to open the retaining screw. The time required for slipping a bush is considerably minimised by properly designing the bush head. For taking a bush out, the head is given a back turn and then the bush is free of the retaining screw and can be easily taken out.

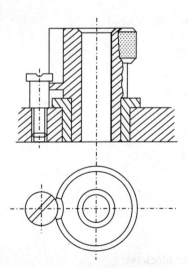

Fig. 24.52 Renewable bush arrangement.

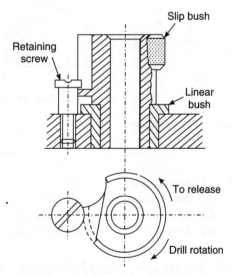

Fig. 24.53 Slip bush arrangement.

Screw or clamping bushes: The screw bushes are provided with threads on their outside diameters. For some type of light works, where a little metal is to be drilled, the screw bush is used to clamp the workpiece below, and also to guide the tool. Then it becomes a simple and inexpensive clamping device for the jig, but eccentricity between the thread of the bushing and the thread in the jig plate may result in the holes being drilled off the centre. Rapid wear of threads in the plate will cause inaccuracy in the exact location of the hole to be drilled. Due to these reasons, screw bushes are not preferred for accurate work.

Special bushes: While standard bushing should always be used wherever possible, sometimes it is necessary to design a special bushing. A typical case occurs when two holes are close together, and so the heads and walls of two standard bushes would interfere with each other. A special bush with two holes can be designed to meet these conditions.

EXAMPLE 24.12 A hole is to be drilled in the centre of a 200 mm diameter rod which has a tolerance of ± 0.25 mm. Assuming that the conditions are such that the vee block as a drill fixture shown in Fig. 24.54 must be used, what variation from the centreline of the drill bushing can be expected if

    (a) a 90° vee block is used;
    (b) a 70° vee block is used;
    (c) a 110° vee block is used?

What observations can you make from these answers?

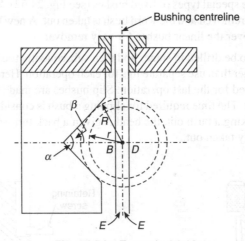

Fig. 24.54 Example 24.12.

Solution The variation

$$E = \operatorname{cosec} \beta \left(\frac{D-d}{4} \right)$$

where E is the variation;
        D is the maximum diameter of the work;
        d is the minimum diameter of the work;
        α is the included angle of the vee block;
        β is $\alpha/2$.

    (a) The variation from the bushing centreline for a 90° vee block is

$$E = \operatorname{cosec} \beta \left(\frac{D-d}{4} \right)$$

Since $D = 101.85$, $d = 101.35$, $\alpha = 90°$, and $\beta = 45°$, we have

$$E = \operatorname{cosec} 45° \left(\frac{101.85 - 101.35}{4} \right)$$

$$= 1.414 \times 0.125$$

$$= 0.177 \text{ mm}$$

(b) The variation from the bushing centreline for a 70° vee block is

$$E = \operatorname{cosec} \beta \left(\frac{D - d}{4} \right)$$

Since $\alpha = 70°$

$$E = \operatorname{cosec} \left(\frac{70°}{2} \right) \left(\frac{101.85 - 101.35}{4} \right)$$

$$= 1.7434 \times 0.125$$

$$= 0.218 \text{ mm}$$

(c) The variation from the bushing centreline for a 110° vee block is

$$E = \operatorname{cosec} \beta \left(\frac{D - d}{4} \right)$$

Since $\alpha = 110°$

$$E = \operatorname{cosec} \left(\frac{110°}{2} \right) \left(\frac{101.85 - 101.35}{4} \right)$$

$$= 1.2208 \times 0.125$$

$$= 0.153 \text{ mm}$$

(d) The observation is that a 70° vee block will confine a workpiece more than a 110° vee block. However, a comparable change in the workpiece diameter size will produce a greater variation in the location of the workpiece with reference to the drill bushing centreline when the 70° vee block is used than when the 110° vee block is used.

24.8 JIG BASE AND JIG FEET

28.8.1 Jig Base

A jig which is not bolted to the machine table must be provided with four jig feet fitted to a jig base. A base is necessarily to be provided whether or not feet are to be provided. Some design criteria may differ in the two cases.

24.8.2 Jig Feet

Drills (see Figs. 24.55 to 24.57) are provided feet at the jig base to ensure correct seating on the machine table. It must be understood that the purpose of the jig feet is entirely different from that of the tripod; hence four feet at the base of a jig are provided. The tripod must provide stability even when placed on an uneven seating plane, whereas the feet of the jig must ensure that the axes of the drill bushes are vertical to prevent

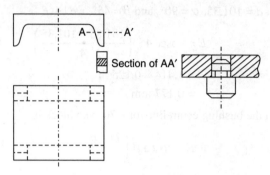

Fig. 24.55 Jig feet—Cast, welded or screwed.

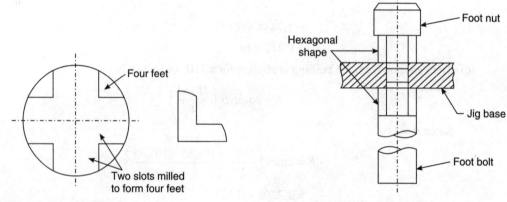

Fig. 24.56 Symmetrical feet of jig milled from metallic cross.   Fig. 24.57 Adjustable jig feet.

incorrect holes or the failure of the tools to prevent the pad from falling from the screw when released. The pad will adjust itself to accommodate any variation in the workpiece flatness.

24.9 CLASSIFICATION OF DRILL JIGS AND FIXTURES

24.9.1 Classification of Jigs

For a given component, different types of jigs can be made; each of them would be satisfactory for that component, but some would be more efficient than others. Again, they would vary considerably in manufacturing cost.

There cannot be drawn a strict line of demarcation while classifying jigs. Drilling jigs are broadly classified as follows:

   (i) Template jigs;
  (ii) Plate jigs;
 (iii) Channel jigs;
  (iv) Leaf jigs;
   (v) Solid jigs;
  (vi) Post jigs;
 (vii) Universal jigs;
(viii) Index jigs;
  (ix) Built-up jigs and welded jigs;
   (x) Trunnion jigs;
  (xi) Sandwich jigs;
 (xii) Nutcracker jigs;
(xiii) Table jigs;
 (xiv) Boring jigs.

Figure 24.58 shows the mechanism of depth control using a jig.

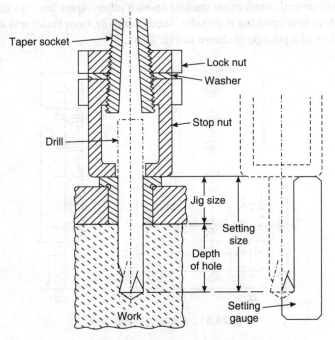

Fig. 24.58 Depth control.

Template jigs: The template jig (shown in Fig. 24.59) is the simplest type of jig, which can be held over the component while drilling small holes. As the jig is just positioned by the operator and does not contain location points, it is used where accuracy of holes relative to each other is not much important, and variations of positioning of the holes relative to workpiece can be tolerated. Such a template may or may not contain bushes to guide the drill. However, templates are sometimes used for marking off the holes to be drilled. The component and template are rested on a block, while centre-pops are made through the countershunk holes in the template. The template is then removed, and the component is set up for drilling to the centre pops. This method is not very accurate, but it saves the cost of marking.

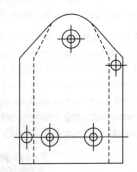

Fig. 24.59 Template jig.

Plate jigs: The main part of such a jig is the plate from which it takes its name. Other essential parts are the drill bushing and locating pins. A clamping device may be used, but for some jigs it is not needed and may consist of a C-clamp or a set of clamps. The plate jig becomes necessary where a whole pattern is to be drilled, and has a definite relation to the periphery of the component, whereas in the movement of the template the pins are used to locate the plate jig from the periphery to achieve an exact relationship of the hole and the periphery. Figure 24.60 shows the

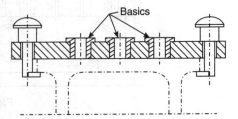

Fig. 24.60 Plate jig—side view.

component to be drilled. The centre line of the three holes from the periphery is important. The holes in other directions are exactly in the centre line. The tolerances have been mentioned for both the dimensions. The plate jig designed for this component has six location pins, and clamping screws are also provided on one side. Such a plate jig can drill several components stacked on each other. Apart from the clamping screws which must have floating pads to bear upon the plate sides, some clamping from above will also become necessary in that case. The top view of a plate jig is shown in Fig. 24.61.

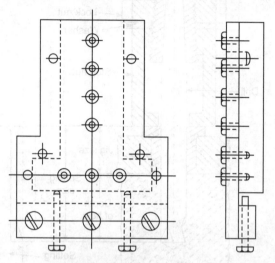

Fig. 24.61 Plate jig—Top view.

Plate jigs are widely used in different factories for drilling holes in long strips or angles. Several strips are stacked, and the plate jig is positioned above them, locating it separately from the strips.

Channel jigs: In this type, the workpiece is placed in a channel shaped like a trough and clamped by means of a screw. This type is limited to jobs having workpieces of simple symmetrical shape. A channel can also accommodate two angles in its corners. Clamping is done in the centre line of the channel at several places. For drilling machines used for drilling over a length of six to seven feet of angles held in the channel, the jig must move on its rails along the length of the channel. Such drilling machines are commonly used in a wagon manufacturing concern, where long angles, channels, plates and strips used in wagon construction are to be processed (see Fig. 24.62).

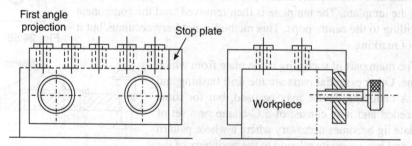

Fig. 24.62 Channel jig.

Leaf jigs or solid jigs: A leaf jig is generally a small jig incorporating a hinged leaf carrying the bushings and through which clamping pressure is applied. The leaf is held on to the jig body with the help of a cam. Leaf jigs are also called *latch jigs* (see Fig. 24.68). Most of the leaf jigs are easy to load, and their normally open design allows rapid and easy removal of chips and good visibility of the workpiece.

The clamping pressure on the workpiece should not be applied directly by the leaf part of the jig, but through a thumb screw in the leaf (see Fig. 24.63).

Leaf jig is also called *solid jig*, as its body is machined from a solid block of steel. It is suitable for drilling small workpiece.

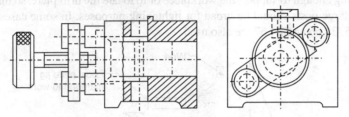

Fig. 24.63 Solid jig or leaf jig.

Box and tumble jigs: It looks like a closed box. It has drill bushes provided on two or more sides. Each side is brought under the drill spindle by tumbling the jig. In this way all jigs, whether of box type, leaf type or channel type, which have to be tumbled to present the desired work face to the drill spindle, are called *tumble jigs*. Feet are certainly needed, opposite to each bush plate in a box jig of headed bushes, when they are used. One side of a box jig is generally used for inserting the job and accommodating the clamping system. A workpiece in a box jig should be located by the 3–2–1 principle on supporting pins. Clamping screws placed opposite to the 2-pin and 1-pin planes force the workpiece into place. A clamping device opposite to the 3-pin plane finally constrains all degrees of freedom.

Post jigs: Circular components, which have both an external diameter and an internal diameter suitable for the purpose of location, are drilled in post jigs. The jig consists of two part (see Fig. 24.64). The body in the form of a post carries the workpiece, and the workpiece carries the bush plate. The bush plate is provided

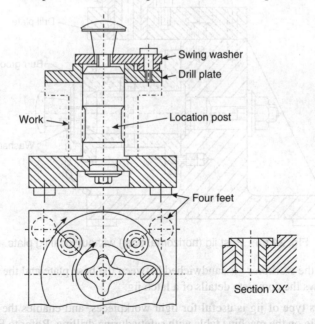

Fig. 24.64 Post jig.

with locating spigots, and can also be used without the post body. It must be locked in such a way that the face in which the hole is to be drilled is absolutely horizontal. The pin and locating hole for locking the jig should be wear-resistant. The drill plate also functions as the jig plate as shown in the top view in Fig. 24.64.

Post jigs are used for location from a bore. The post should be as short as possible to facilitate loading, but it must also be long enough to support the workpiece or to locate the drill plate, so that the location is only at its extreme ends; large posts should be broad for lightening purposes. In some cases, angular post jigs as shown in Figs. 24.65, 24.66 and 24.67 are also used.

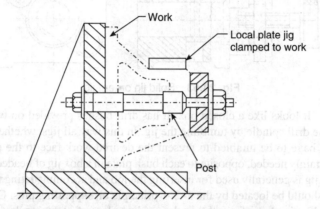

Fig. 24.65 Post jig (horizontal post) with local jig plate.

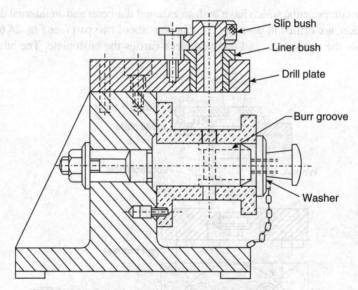

Fig. 24.66 Post jig (horizontal post) without local jig plate.

Sandwich jigs: Here the workpiece is sandwiched between the base plate and the drill plate by two swing bolts. Figure 24.68 shows the complete details of a latch jig.

Nutcracker jigs: This type of jig is useful for light workpieces, and enables the operator to load the jig rapidly and to hold the jig on the machine table with safety during drilling. Refer to Figs. 24.69(c) and 24.71.

Tumble (Turnover jigs or open jigs): These are used when it is necessary to locate the workpiece from the face that is to be drilled. The workpiece is located and clamped, and then inverted before positioning and machining. This type of jig presents no problems in swarf and cutting fluid disposal; loading and clamping is easy, but the workpiece is supported against the cutting forces only by the clamps. Refer to Fig. 24.70.

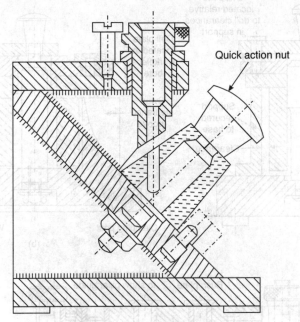

Fig. 24.67 Angular post jig.

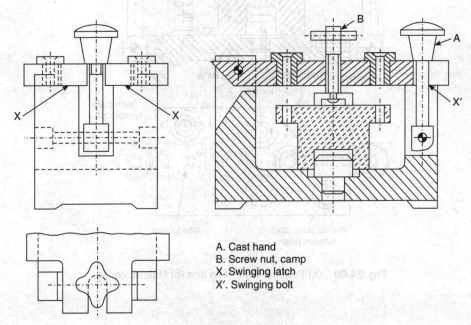

A. Cast hand
B. Screw nut, camp
X. Swinging latch
X'. Swinging bolt

Fig. 24.68 Latch jig.

Boring jigs: Boring jigs are commonly used for machining holes which must be aligned and sized with particular accuracy, or which are too large for drilling. They may also be used in finishing two holes in the same line.

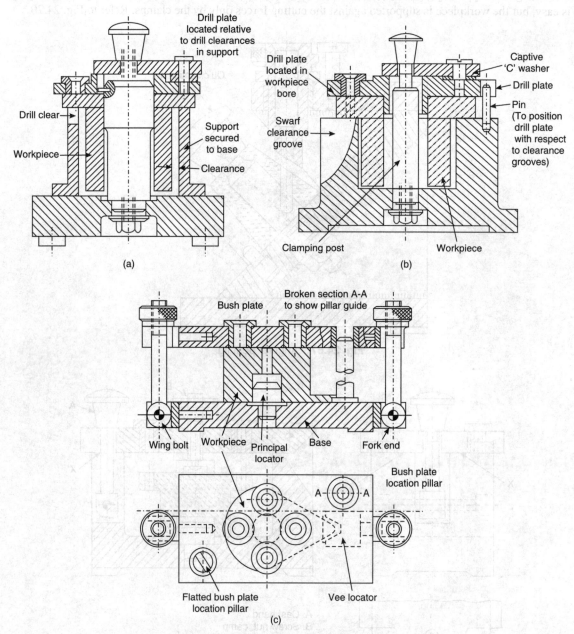

Fig. 24.69 (a) Post jig, (b) Pot jig and (c) Nutcracker jig.

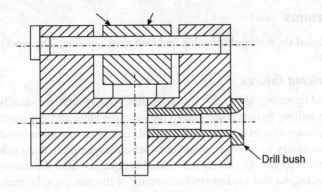

Fig. 24.70 Table jig.

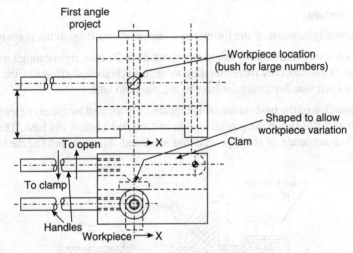

Fig. 24.71 Nutcracker jig mechanism.

24.10 CLASSIFICATION OF FIXTURES

There are many types of fixtures used in different industries for different components. But they may be classified on the basis of their working operations such as:

1. Milling fixtures
2. Turning fixtures
3. Grinding fixtures
4. Broaching fixtures
5. Assembly fixtures
6. Welding fixtures
7. Slotting fixtures
8. Boring fixtures
9. Miscellaneous fixtures.

24.10.1 Milling Fixtures

A milling fixture is located on the machine table and bolted in position. The workpiece is, in turn, located and clamped to the fixture.

General features of milling fixtures

A milling fixture should be strong and rigid. Unlike drilling jigs, which are usually moved about during a machining operation, a milling fixture is secured to the machine table and not moved till the completion of the batch; therefore, lightness is relatively unimportant. Milling is an operation that removes a large quantity of swarf, and so it is necessary to provide adequate swarf-clearance parts that enable swarf to be removed without having to invert the fixture. Generally the milling is done just after the casting; therefore an adequate clearance must be provided, so that loading and unloading of the piece can be made easily even when it is oversized. The supports of the fixture should be adjustable, because the casting surface is not even. The body of the milling fixture is generally made of gray cast iron due to its vibration-damping properties.

Elements of milling fixture

A milling fixture essentially consists of the following components built-up to the main base:

Base: The base of a fixture is sufficiently thick, say about 20 to 25 mm, rigid enough not to deflect upwards during up-cut milling. It also absorbs forces tending to set up vibrational effects of the chatter. The base is provided with lugs on each side for fixing the base to the machine table.

Tenon strip: The position of the base on the table is accurately located by means of tenon strips. The tenons are identical in width with the slot in the machine table and are fixed below the base. The length of the tenon is twice the width. They are made of steel, ground and hardened. They are held to the base with the help of screws. See Fig. 24.72.

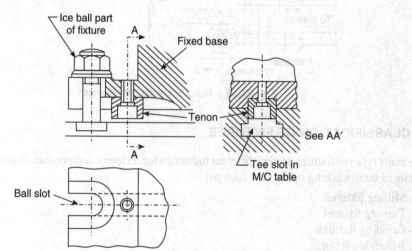

Fig. 24.72 Tenon and ball slot.

Setting block: Milling fixtures are provided with a setting block, so that a feeder gauge of 0.02 mm thickness may be used for setting the fixture relative to the cutters. It is made of steel, duly hardened and ground. The setting piece is fixed to the base of the fixture by means of screws and dowels. See Fig. 24.73.

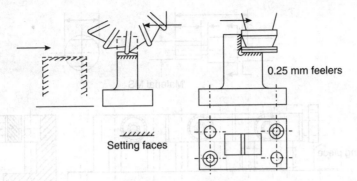

Fig. 24.73 Setting block.

Tee bolts: The fixture is bolted to the machine table with tee bolts suitable for the slots provided in the machine table. The shank of the bolt goes into the lug, and the nut is tightened on the screwed position.

Clamping device: A fixture may have any one of the clamping devices to clamp the workpieces.

Locating or positioning element: The excessive thrust of the cutter must be resisted by a fixed stop, because only one clamping device may not be sufficient.

Classification of milling fixtures

The fixture design depends upon the method of milling which is employed on machining. The following classification can be made on the basis of the milling method:

- (i) Plain milling fixtures
- (ii) String milling fixtures
- (iii) Gang milling fixtures
- (iv) Straddle milling fixtures
- (v) Profile milling fixtures
- (vi) Continuous rotary milling fixturess
- (vii) Indexing milling fixtures
- (viii) Pendulum milling fixtures
- (ix) Special vise jaw fixtures.

Milling requires special adjustment of the jack screws or the shaping of loaded rest pins individually for each component. Milling fixtures therefore hold only one component, because each component requires special attention. It is particularly true when the first operation is performed on a forging or casting when a locating surface is to be created for subsequent operations by the first operation fixture. The milled surface is to serve as a datum for all other dimensions in subsequent operations. Therefore, the first operation in milling fixtures requires great skill. Figure 24.74 shows the complete assembly of milling fixtures.

Plain milling fixtures: These fixtures are specially designed for components which are complicated in shape and are indicated in Figs. 24.75 and 24.76.

String milling fixtures: As the name suggests, this type occupies a considerable length of the machine table and has a number of components in a line or in tandem. The length of the row depends upon the size of the milling machine. This type of fixture is extensively used for milling forms on the heads of components having a shank. The pieces are held between hardened and ground vee blocks which are free to slide in a slot cut in the body of the fixture (see Figs. 24.77 and 24.78).

Gang milling fixtures: Three or more cutters can be mounted on the arbor, so that several faces can be machined at once. The operation is said to be gang milling. The machining table is positioned relative to one of the cutters in the gang. The heavy cutting load, which is generally associated with gang milling, makes it necessary to design very rigid fixtures, with plenty of metal to absorb the forces set up by cutting tools (see Figs. 24.79 and 24.80).

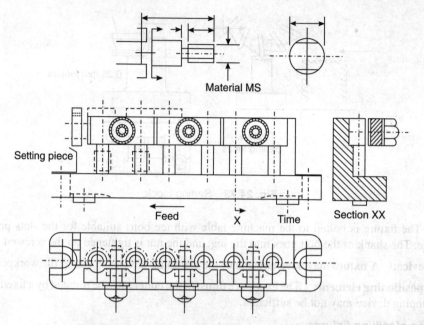

Fig. 24.74 Complete assembly of milling fixtures.

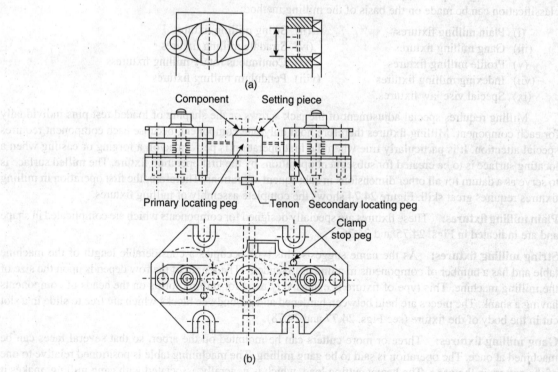

Fig. 24.75 Plain milling fixture.

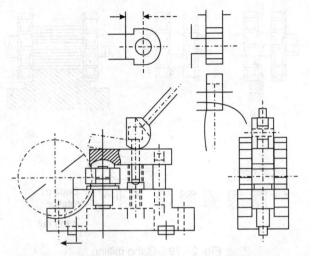

Fig. 24.76 Cam-operated clamping for plain milling fixture.

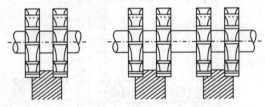

Fig. 24.77 String milling.

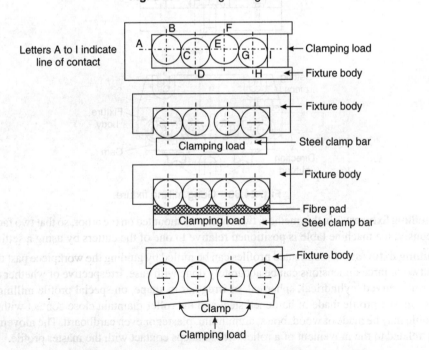

Fig. 24.78 Design of clamps for string milling.

660 TEXTBOOK OF PRODUCTION ENGINEERING

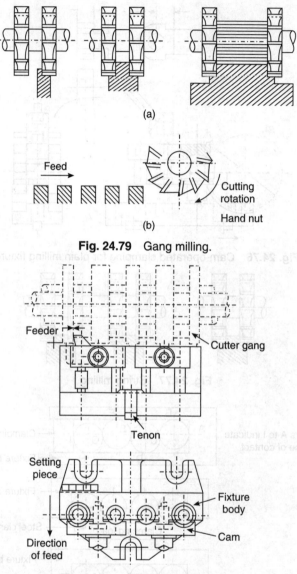

Fig. 24.79 Gang milling.

Fig. 24.80 Gang milling fixture.

Straddle milling fixtures: In this method, two cutters are mounted on the arbor, so that two faces are machined simultaneously; the machine table is positioned relative to one of the cutters by using a setting block.

Profile milling fixtures: Complicated profiles can be milled by guiding the workpiece past the cutter. Profile milling in two or three dimensions can be accurately done with ease, irrespective of whether the workpiece is flat, uniformly curved, cylindrical, spherical, or irregular in shape, on special profile milling machines. The template or master profile made of hardened steel and the roller maintain close contact with the profile. The master profile may be made of wood, brass, aluminium, plaster or even cardboard. The movement of the cutter is directly related to the movement of a roller, which keeps contact with the master profile.

Continuous rotary milling fixtures: The ideal arrangement for milling is done with the help of a continuous rotary milling fixture, in which the workpieces traverse past or under previously set cutters at a minimum feed consistent with the nature of the material, each one beginning correctly located and clamped. Such a fixture is designed to give minimum rigidity. For this type of fixture, the cutting time is usually less than that taken for clamping and unclamping by manual means, and some form of automatic clamping is devised. A rotary fixture for milling bosses of a connecting rod is common. The components are arranged radially round the fixture as closely as possible. The steel peg in the centre of the revolving table locates the fixture.

Indexing milling fixtures: A workpiece having a number of surfaces to be milled may be successively positioned by a single fixture provided with indexing arrangement. Indexing fixtures are used when necessary to move the workpiece, relative to the machine table or machine cutter spindle, between the machining processes of various kinds during an operation, but only when the fixture base must remain located relative to the machine. The dividing head used in milling practice is a universal indexing fixture. Indexing milling fixtures are simplified forms of the dividing head, designed to perform direct indexing (see Fig. 24.81).

Pendulum milling fixtures: Pendulum milling implies that cutting takes place when the table moves to the right and also when it moves to the left (see Fig. 24.82). In this method, one workpiece is machined at

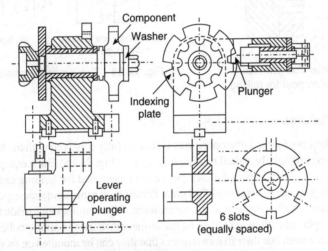

Fig. 24.81 Indexing milling fixture.

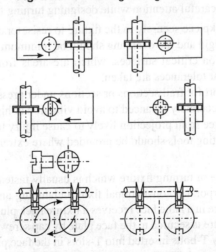

Fig. 24.82 Pendulum milling.

a time and is indexed between passes; indexing milling fixtures are also used in conjunction with pendulum milling.

Special vise jaw fixtures: A commonly used work holding device for milling is the plane or universal vise. Provision is made for attaching special jaw inserts to the fixed and movable vise jaw. Expenditure on special milling fixtures can often be avoided by careful arrangement of special vise jaws, so formed as to provide location and clamping for either one or a series of operations on a component. See Figs. 24.83 and 24.84.

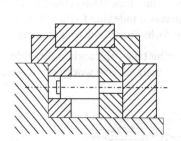

Fig. 24.83 Special vise jaws with location post for milling.

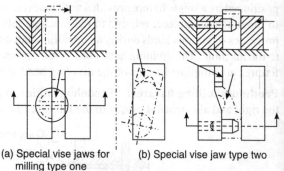

(a) Special vise jaws for milling type one

(b) Special vise jaw type two

Fig. 24.84 Two types of special vise jaws for milling.

24.10.2 Turning Fixtures

The holding of the workpieces for the lathe operations is successfully achieved with the help of numerous types of equipment available commercially, called *turning fixtures* (see Fig. 24.85). Such equipment can be classified into chucks, mandrels, and collets. Special chuck jaws can be designed for holding castings and forgings for first operation. Expanding mandrels and pegs find great favour with tool engineers, especially where the work affords a machined surface of sufficient accuracy to locate them. When components made from bars on automatic lathes require a second operation, they may often be finish-machined using a spring collet fitted with an internal stop. Some components are such that their fixture requires that they can be mounted on face plates with the help of dowels and screws. The workpieces are then located and clamped on these turning fixtures for further perations.

The following points need careful attention while designing turning fixtures:

1. Attach the rotating workpiece securely to the fixture to resist torsional forces.
2. The fixture should be rigid and the overhang should be minimum possible.
3. Locate the workpiece on critical surfaces, which are areas from which either all or the major dimensional and angular tolerances are taken.
4. Provide adequate support for frail sections or sections under pressure from lathe tools.
5. The fixture should be accurately balanced to avoid vibration at high spindle speeds.
6. The fixture should be free from projection likely to cause injury to the operator.
7. A pilot bush for supporting tools should be provided where extreme accuracy is required in boring operations.

Face plate: It is a common type of turning fixture which is usually fastened to the lathe face plate or back plate [see Fig. 24.85(b)]. It incorporates conventional fixture clamping and locating devices for holding a workpiece. A shallow counter bore in a lathe face receives a fixture back plug to locate the lathe fixture on the lathe spindle centre line. The fixture is secured to the face plate by cap screws inserted through the fixture into taped holes in the face plate, or by T-bolts inserted into T-slots in the face plate. A fixture used in high-speed lathes should be dynamically balanced. The face plate itself should be bolted to the spindle flange.

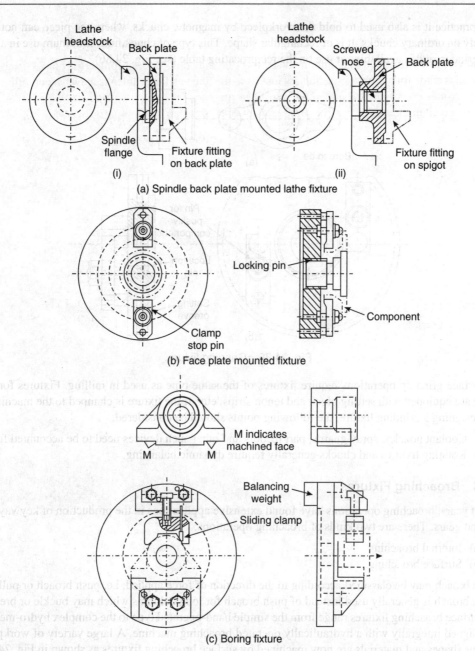

Fig. 24.85 Turning fixtures.

24.10.3 Grinding Fixtures

Every grinding operation needs some sort of fixture. Cylindrical grinding utilises mandrells for external surface grinding. The mandrell may be plain or tapered. For internal grinding operations the chuck is the most standard fixture. Special jaws as in lathes may be fixed here also for holding castings and forgings. In modern practice it is also used to hold the workpiece by magnetic chucks, where the piece can not be held adequately on ordinary chucks. It is of rectangular shape. This type of chuck finds minimum use in a vertical surface grinder, and is available for use on the reciprocating table (see Fig. 24.86).

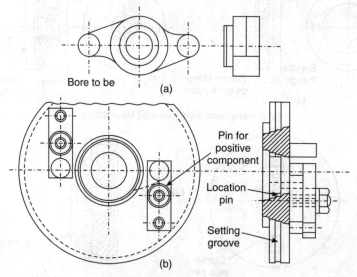

Fig. 24.86 Grinding fixture.

Surface grinding operations require fixtures of the same type as used in milling. Fixtures for surface grinding are equipped with setting block and tenon strips, etc., if the fixture is clamped to the machine table. While designing a grinding fixture, the following points should be considered:

1. Coolant nozzles, spray guards, part feeders and other such devices need to be accounted for.
2. Rotating fixtures and chucks generally require dynamic balancing.

24.10.4 Broaching Fixtures

In recent years, broaching operations have found extensive applications in the production of keyways, round holes, and gears. There are two kinds of broaching operations:

(a) Internal broaching;
(b) Surface broaching.

Again, a broach may be classified according to the direction of force applied, i.e. push broach or pull broach. The pull broach is generally used instead of push broach for long columns which may buckle or break.

Surface broaching fixtures range from the simple hand-operated type to the complex hydro-mechanical type designed integrally with a hydraulically operated broaching machine. A large variety of workpieces of numerous shapes and materials are now machined by surface broaching fixtures as shown in Fig. 24.87.

Many of the principles of milling fixtures may be applied to the design of surface broaching fixtures. A few essentials about surface broaching fixture design are:

1. All elements of the broaching surface must remain parallel with the broach axis.
2. The walls of the part being broached must be sufficiently heavy or adequately supported to withstand the pressures of the operation.

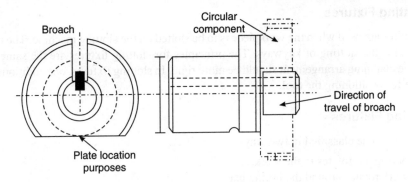

Fig. 24.87 Broaching fixture.

3. The amount of stock to be removed by the broach must be constant and controlled within reasonably close limits.
4. There must be no obstruction in the planes of the broached surfaces.

24.10.5 Assembly Fixtures

A work holding device will be used for an assembly operation if the operation is awkward. There are two types of assembly fixtures:

(i) Assembly fixtures for operations performed at ordinary temperature with mechanical means, which are riveting fixtures such as in wagon assembly fixtures. Two or more parts are held together in predetermined positions, and they are riveted by pneumatic riveters.
(ii) Fixtures for hot joining method of assembly work, where energy is used in the form of heat. All welding fixtures come under this category. Stresses resulting from thermal expansion of workpieces and for fixtures must be considered in the design of clamp and locators and on the proper positioning of the workpieces, before and during the assembly; fixtures for some operations may absorb considerable heat.

24.10.6 Welding Fixtures

A welding fixture can be regarded as a special assembly fixture. Welding fixtures may be grouped into three classes:

(a) Tacking jigs
(b) Welding fixtures
(c) Holding fixtures.

The following special requirements for designing by a welding fixture should be considered:

(i) The welder must have access to all the joints, so that he does not need to tack-weld and complete the welding when the assembly has been removed from the fixture.
(ii) The fixture must be designed so that it can be moved about during the welding operation. So, it should be light in weight.
(iii) The design must allow for the high concentrated heat associated with welding. The clamps must therefore hold the parts against the distortional forces. The positions of clamping and locating devices should be such that the heat does not damage them.
(iv) Screw threads and similar fixture parts must be protected against weld spatter.

24.10.7 Slotting Fixtures

The slotting fixtures are used when the workpiece is to be slotted on the slotting machine. The most important slotting operation is the slotting of keyways. The principles for slotting fixtures are the same as for milling fixtures. Here the clamping arrangement should be more rigid. In slotting fixtures, one additional arrangement should be made for positioning the tool.

24.10.8 Boring Fixtures

Boring operations can be classified in two ways:
   (a) The boring bar rotates in the work;
   (b) The work rotates around the boring bar.

Fixtures for category (a) are designed similar to drilling jigs. Vertical boring machines operate in the manner of category (b) and can utilise fixtures similar to lathes. Horizontal boring machines also work on the same principle sometimes.

Boring fixtures do not need to be as rigid as milling fixtures, because the load imposed by the boring tools rarely approaches the force exerted by a milling cutter.

24.10.9 Miscellaneous Fixtures

There are many fixtures which are used in practice. Some of these may be as follows:

   (a) Bending fixtures   (b) Varnishing fixtures
   (c) Brazing fixtures   (d) Stretching fixtures
   (e) Stamping fixtures   (f) Refilling fixtures
   (g) Tube flaring fixtures   (h) Gear checking fixtures, etc.

24.11 MATERIALS FOR MANUFACTURING OF JIGS AND FIXTURES

Since it is cheap and easy to work with, wood was probably the first material to be used in constructing jigs and fixtures. But now it is not considered good for most precession tooling purposes due to its lack of strength. Therefore, modem jigs and fixtures are made of metals. The most frequently used metals are iron and steel, although aluminium is also preferred sometimes for lightweight works.

In general, tooling metals should have the following physical characteristics:

   (i) Good formability
   (ii) Good strength properties
   (iii) Ability to retain close dimensions
   (iv) Resistance to handling impacts
   (v) Salvageability.

A widely available form, of iron called *cast iron*, should never be used for fabrication of jigs and fixtures, because cast iron is brittle and subject to breakage on handling. Wrought iron is an excellent material for fabrication on jigs and fixtures, but mild steel is used in modern practice because it has similar physical properties as wrought iron and it is even cheaper. The type of steel used in constructing a jig or fixture depends primarily upon such considerations as cost, the method to be used in fabricating or assembling the tool, and the load which the finished structure will be required to support.

Whenever a low-weight structure is required, aluminium and magnesium alloys may be advantageously used. Lightness in weight is required when the time and effort involved in the handling of the tool may seriously affect the cost of fabrication of assembly operations.

Materials for jig and fixture elements: In a general sense, mild steel can be used for the jig fixture body. But specifically, the following are the different materials for the various elements of jigs and fixtures:

Jig bodies: There are two common forms of construction for both jigs and fixtures: built-up bodies and cast iron bodies. In the built-up structure the component may be made of steel or cast iron. In the case of cast iron, the pieces are machined and joined with dowels and screws, because cast iron cannot be properly welded. Further, if the components are of steel, then it can either be built up using dowels and screws or fabricated by welding. Though cast iron is preferred for milling fixture due to its shock-absorbing properties, steel is taking its place due to case-in fabrication by welding. Welded bodies are lighter in weight by about 30% and they are also cheaper. The cost of pattern is also reduced by using mild steel structures. All important welded bodies must be heat-treated to relieve stresses before drilling holes etc. for bushes and locators.

When castings are used the feet are cast integral with the body. If the feet are to be added afterwards, then they are of steel and are hardened and finish-ground on assembly.

Jig bushes: In terms of minimum diameter the materials for drill bushes may be as follows:

1. Upto 15 mm outside diameter (O.D.)—Steel;
2. 15 to 30 mm O.D.—Cast steel;
3. Over 30 mm O.D.—Case-hardened mild steel.

The bore of a bush is generally ground and lapped to give a fine running fit with the tool with which it is to be used. A finish between 100 and 500 microns may be obtained without difficulty. Such a finish adds appreciably to the working life of the bushes fixed. Bushes are sometimes made from hard metal, but these are expensive due to difficulties in producing them, especially the bores. Bushes may be plated with hard chrome to prolong their lives.

Locating and clamping devices: Most of the locating and clamping devices are made from steel, but those components which come in contact with other components are hardened and working surfaces are ground. Silver steel is used for parts like dowel pins, bars for handles, etc. Threads should be left soft.

Commercial jig and fixture components: Except for jigs and bushes, the component parts for jigs and fixtures have not been standardised due to several difficulties. However, some important firms like Bharat Heavy Electricals Ltd. have prepared their own standards, which are made available for designers of jigs and fixtures. Details of these standards are available from the catalogues of the commercial firms. The main items in these catalogues comprise base plates, jig plates, hand nuts, thumb screws and nuts, clamping devices, locating pins, studs, tenons, locators, jacks, etc. The cost of these mass-produced components is usually considerably less than the cost of their fabrication in the plant.

Jig weight and strength: The designer must use his judgment in regard to the amount of metal put into the jig or fixture. It is desirable to make these tools as light as possible, in order that they may be easily handled, be of smaller size, and cost less in regard to the amount of material used for their making; but at the same time, it is poor economy to sacrifice any rigidity and stiffness to the tool, as this is one of the main considerations in obtaining efficient results. On large-sized jigs and fixtures it is possible to core out the metal in a number of places, without decreasing in the least the strength of the jig itself.

Providing jig feet: Ordinary drill jigs should always be provided with feet or legs on all sides which are opposite to the holes for the bushings, so that the jig can be placed level on the table of the machine. On the sides of the jig, where no feet are required, if the body is made from a casting, it is of advantage to have small projecting lugs for bearing surfaces when laying out and planning. While jigs are most commonly provided with four feet on each side, in some cases it is sufficient to provide the tool with only three feet; care should be taken in either case that all bushings and places where pressure will be applied to the tool are placed inside the geometrical figure obtained by connecting, using lines, the points of location of the feet.

Providing three feet means that the jig will obtain a bearing on all its legs, which will not be the case if four feet are provided if the machine table is not absolutely plane. But it is not quite safe to use the smaller number of supports, because a chip or some other object is liable to come under one foot and throw the jig and the piece out of the line, without this being noticed by the operator. If the same thing happens to a jig with four feet, it will rock invariably.

Methods of construction: Owing to the recent developments with welding, 'built-up' structures are now widely used. A built-up structure can be defined as a physical unit which is made by assembling two or more members. It cannot be constructed with the accuracy of a cast structure. Built-up structures normally comprise standard sections such as rounds, pipes, squares, etc. Further, to fabricate the steel structure, arc welding is preferred for economic reasons.

The basic structure of the average jig or fixture does not warrant extreme accuracy, because its only function is to provide a rigid support for a series of locating elements. The accuracy of the locating elements may be attained by machining the surfaces or members of the structure.

There are several methods followed in turning jig bushings. The most rapid method is to chuck out the hole and finish the outside setting, using the bar stock held in the chuck of a rigid engine lathe. But this method is not preferable in large bushings. For manufacturing the bushings, lapping and grinding allowances should be provided in design.

For hardening a tool steel bushing, it should be brought to an even red hot condition in a gas furnace. Then it should be dipped immediately in warm water. The bushes should then be heated to a dazzling heat, after which it is left in open air for cooling.

The hardness of a jig bush is generally kept at RC 58 to RC 60 (Rockwell Numbers).

Internal grinding followed by cast iron lapping for fine finishing is very suitable for finishing the holes.

24.12 ACCURACY

All operations of machining inevitably have variations in their final results. A jig or fixture designer has to determine what variations can be permitted in an operation. The variations produced can be attributed to the following causes:

(i) Variations in the dimensions of the workpiece caused by an operation
(ii) Variations in the material and its condition
(iii) Defects in tools and machines
(iv) Wear
(v) Deflection
(vi) Thermal expansion
(vii) Dirt chips and burrs
(viii) Errors of human judgement, limits, tolerances, and deficiencies of skill.

24.13 POSSIBLE WAYS OF AVOIDING INACCURACIES

During the process of design of any jig or fixture, it is better to go through the following list of important points, although in any jig design it is not necessary that all the points should be considered:

1. Can the component be inserted and withdrawn without difficulty?
2. Should the component be located to secure symmetry?
3. Have the best points of location been chosen with regard to the accuracy of location and the function of the component?

4. Are hardened location points provided where necessary?
5. Can locating points be adjusted, where required, to make allowance for the wear of forging dies or patterns?
6. Are locations clear of flashes and burrs?
7. Can jigs be easily cleared of swarf, particularly on the locating faces?
8. Are the clamps of sufficient strength?
9. Will any clamp-operating lever or nut be in a dangerous position, i.e. near the cutter?
10. Are the clamp and clamping screws in the most accessible and natural positions?
11. Can spanners be eliminated by the use of ball or eccentric levers?
12. Is the component well-supported against the action or pressure of the cutter?
13. Is the jig foolproof?
14. Has the operator an unobstructed view of the component, particularly at the points of location, clamping, and cut?
15. Is the jig as light as possible, consistent with strength?
16. Can the coolant, if used, reach the point of cut?
17. Have loose parts been used wherever possible?
18. Have standard parts been used where circumstances permit?
19. Where will burrs be formed, and is clearance for them arranged?
20. Are all corners and sharp edges likely to cut the operator (shown well reduced in the drawing)?
21. Are locating and other working faces and holes protected as far as possible from swarf?
22. Will the jig as designed produce components within the required degree of accuracy?
23. Have all slip bushes necessary for reaming, spot facing, tapping, counter-boring, seating, etc. been arranged?

24.14 ECONOMIC ASPECTS

The aim of industrial activity in general is to produce cheap goods, so that they are within the reach of the common man. Jigs and fixtures are utilised to increase production and reduce costs. Expenditure on the manufacturing of aiding equipment is justified only if the final account shows a profit. Therefore, before starting the design and construction of jigs or fixtures, economic analysis must be carried out. If the items to be produced are few, then a large investment on jigs or fixtures will not be justified.

A simple relationship between the initial investment and the production cost can be expressed in the following expression:

$$E - S = \frac{P}{N}$$

where E is the production cost of components by using the present method;
S is the production cost of the same component by using special jigs or fixtures;
P is the cost of the special equipment (jig or fixture);
N is the minimum number of components to be produced.

However, to have an exact analysis, many factors are accounted for in the relevant formulae. Due to the use of jigs and fixtures, we gain on account of the following savings:

Direct cost of labour (L)

$$L = Nl$$

where N is the number of components to be manufactured;
l is the direct labour saving per piece.

Savings in labour overhead (T)

$$T = L \times p$$

where p is the % of labour overhead saved.

Yearly expected profit

The expenses on jigs and fixtures are as follows:

(a) Interest cost: % yearly interest on investment (x)
(b) Fixed cost: % yearly expenses on insurance, tax, etc. (y)
(c) Maintenance cost: % yearly upkeep cost (z)
(d) Yearly % allowance for depreciation and obsolescence
(e) Initial cost (I).

Then, the savings will equal the total fixed charges per year when

$$N = \frac{L\left\{(x+y+z) + \dfrac{1}{w}\right\} + S}{I(l+p)}$$

or

$$l = \frac{NI(l+p)}{(x+y+2) + \dfrac{l}{w}}$$

or

$$-p = L\left\{(x+y+z) + \frac{1}{w}\right\} + S - NI \times l$$

or

$$w = \frac{L}{NI(l+p) - S - L(x+y+z) - P}$$

REVIEW QUESTIONS

1. Define the following terms:
   (a) Tool
   (b) Jig
   (c) Fixture.
2. Differentiate between jigs and fixtures.
3. What are the advantages of using jigs and fixtures?
4. What are the main principles of design of jigs and fixtures?
5. How are workpieces located? Discuss the various types of locators.
6. Write a short note on clamps and clamping.
7. What do you understand by the term *location of a piece*? What are the important factors of location to be considered while designing a locator?
8. Discuss the principle of least point and the principle of extreme position.
9. What is the best method of locating a rough surface?

10. How many degrees of freedom are restrained in a cylindrical location? What can be the problems in cylindrical locations? How are these overcome? How can a job be completely restrained using a cylindrical locator?
11. What are the advantages and limitations of a conical locator?
12. What is the principle of clamping? What factors govern the choice of a clamping device to achieve the purpose of clamping?
13. What are the advantages of hydraulic and pneumatic clamping over manual clamping?
14. Discuss the various elements of a milling fixture.
15. Distinguish between an air-indexing fixture and a continuous rotating fixture.
16. What is an indexing jig? What are the various types of commonly used indexing devices?
17. List the different types of jig bushes used in drilling jigs.
18. How does a template jig differ from a plate jig?
19. Why should a drill jig have four legs? Why not more or less?
20. What are the different types of material used for manufacturing of jigs and fixtures?
21. Write a short note on the methods of manufacturing jigs and fixtures.
22. What are the different types of drill jigs and fixtures?
23. What are the various types of jig bushes?

PART IV
METAL WORKING

CHAPTER 25

Metal Working Processes

25.1 DEFINITION

Metal working may be defined as plastic deformation performed to change dimensions, properties, and/or surface conditions by means of mechanical pressure. Non-cutting processes are generally referred to as *mechanical working processes*.

25.2 TYPES OF METAL WORKING PROCESSES

Depending on the type of metal, the shape desired, the relative cost of method, the following two types of mechanical working methods are employed:

(i) Cold working;
(ii) Hot working.

25.2.1 Cold Working

Cold working is defined as any form of mechanical deformation process carried out on the metal below its recrystallisation temperature. Most of the cold working processes are carried out at room temperature. Refer to Fig. 25.1 which shows the various types of cold working processes.

Effects of cold working

In general, cold working produces the following effects:

(i) Internal stresses are set up which remain in the metal unless they are removed by proper heat treatment.

676 TEXTBOOK OF PRODUCTION ENGINEERING

(ii) Distortion of the grain structure is created.
(iii) Strength and hardness of the metal are increased but ductility is decreased.
(iv) Smooth surface finish is produced.
(v) Accurate dimension of the parts can be maintained.

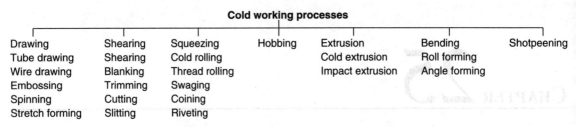

Fig. 25.1   Types of cold working processes.

Field of application

Cold working is chiefly employed as a finishing operation following the shaping of the metal by hot working. It is customary to produce cold worked products like strip and wire in different tempers, depending upon the degree of cold reduction following the last anneal. The cold worked condition is described variously as annealed (soft) temper, quarter hard, half hard, three quarters hard, full hard and spring temper. Each temper condition indicates a different percentage of cold reduction following the annealing treatment.

Advantages: The following are the main advantages of cold working:
   (i) It produces smooth surface finish.
   (ii) Cold working helps to attain better dimensional accuracy.
   (iii) Cold working increases the strength, elasticity and hardness of the metal parts worked.
   (iv) Small parts may be shaped easily, rapidly and at a lower cost than for most of the other methods.

Disadvantages
   (i) Some parts can not be cold worked because they are two brittle.
   (ii) It requires excessive energy to cold work large sections of most metals.
   (iii) The propagation of cracks is more possible in comparison to hot working.

The metals on which cold working is performed are as follows:
   (i) Mild steel of low carbon content
   (ii) Copper and its alloys, including brass and bronze
   (iii) Aluminium bronze having aluminium more than 7%
   (iv) Nickel bronze
   (v) Monel bronze
   (vi) Stainless steel
   (vii) Duralumin
   (viii) Aluminium alloys.

Types of cold working processes

Drawing operations involve the forming of metal through a die by means of a tensile force applied to the exit of the die. Most of the plastic flow is caused by the compressive force which arises from the reaction of the metal with the die. Rods, tubes and extrusions are often given cold finishing operations to reduce size, increase strength, improve finish, and provide better accuracy.

Tube drawing: The cold drawing of tubes, seamless or welded, is accomplished by drawing the tube over a mandrell and through the reduced die opening. Maximum reduction in one pass is limited to only 40%, since the material is heavily stressed. Before the second pass is given, it should be annealed. In general, it should be annealed after each pass. The draw bench requires a pulling power which ranges from 15,000 to 25,000 kg and may have a total length of 30 metres.

Wire drawing: Wires are manufactured by cold drawing. The raw material used for drawing is the rolled bar from hot rolling mill. Wire is pulled through the die, Fig. 25.2 shows a typical wire drawing die. Dies may be either carbide dies or diamond dies. Small diameter wires are drawn through a diamond die.

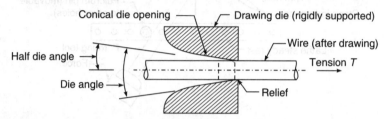

Fig. 25.2 Wire drawing die.

Embossing: Figure 25.3 shows the principle of embossing and an embossed component. The blank of sheet metal is kept in a suitable die and punched. The mating die conforms to the same configuration as the punch. The operation does not require much pressure. Embossing is a slow process.

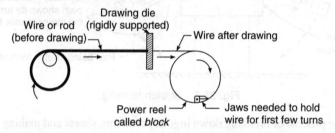

Fig. 25.3 Embossing and embossed component.

Cold spinning: This process is used to manufacture parts of circular cross-section from sheet metal. In this method, a former of the desired shape is fixed to the rotating chuck of the lathe. The circular blank of the sheet metal is pressed against the former with the help of a spinning total. The blank is held against the chuck of a lathe by the pressure of a freely rotating adapter on the lathe tail stock. Reflectors, funnels, aluminium teapots and other kitchenware are commonly produced by this method. Refer to Fig. 25.4.

This is a fast process and is mostly employed on aluminium brass, copper and silver.

Stretch forming: Figure 25.5 illustrates the principle of stretch forming. This process is adopted to provide double curvature on the same curved surface. The sheet metal is held by two jaws on two sides in a forming punch. The jaws are given horizontal motion and the forming punch is lifted up. A large force of 50–100 tonnes is required for the punch and the slides. The process is a stretching one, and the metal is stressed beyond its elastic limit while conforming to the die shape. The formed sheet is slightly thinner than the original sheet.

Shearing operations: These are dealt with in detail in Chapter 27.

Squeezing: It is a quick and widely used method of forming ductile metals. The main processes of squeezing are cold rolling, thread rolling, swaging, and coining.

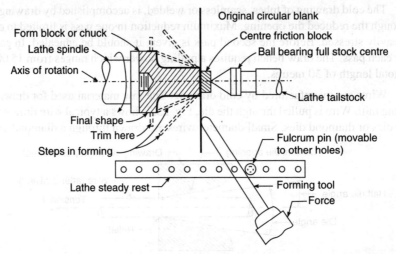

Fig. 25.4 Cold spinning.

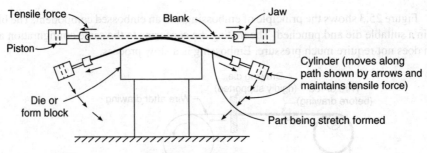

Fig. 25.5 Stretch forming.

Cold rolling: It is the process of breaking down ingots, producing sheets and making various sections such as rounds, hexagons, rails and girders. In this process, the section thickness is reduced by passing it between metal rollers revolving in the opposite directions. The simplest rolling mill consists of two rolls and is called the *two-high mill*. Similarly, a rolling machine consisting of three rollers is called the *three-high mill*.

The number of pairs of rollers needed depends upon the shape and ductility of the material. Plates may be rolled into tubes or rings if passed through three rollers (see Fig. 25.6). Two of the rollers are fixed and a third adjustable roller forms the piece.

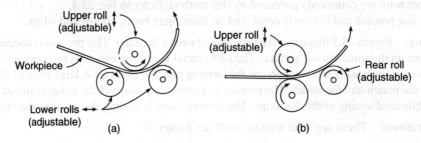

Fig. 25.6 Cold rolling.

Thread rolling: It is a mass production method for producing threads. The rolls have the thread form cut on their surfaces. The rolls rotate and are fed into the blank under pressure; metal flows into the die shape, forming a thread.

The threading dies may be in the form of flat plates. These have a reciprocating action, the blank being rolled between them to form the threads. Figure 25.7 shows the thread rolling process.

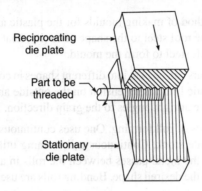

Fig. 25.7 Rolling thread using flat dies.

Advantages of thread rolling over thread cutting. These are:
  (i) The threads have smooth surfaces.
  (ii) Accuracy can be attained.
  (iii) Higher production rate can be obtained.
  (iv) Different threads can be easily formed.

Limitations of thread rolling. These are:
  (i) It is economical only when mass production is adopted.
  (ii) Only external threads can be produced.
  (iii) This can not be done on harder materials.
  (iv) The dies should have close tolerance.

Rotary swaging: This is used to reduce the cross-sectional area of rods and tubes. Swaging is often spoken of as a cold forging operation because the metal forming takes place under the hammering blows of die sections. The swaging machine consists mainly of a spindle which carries the die sections and rollers, as shown in Fig. 25.8.

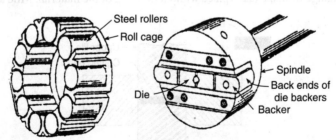

Fig. 25.8 Swaging machine.

Shotpeening: Shotpeening is done by blasting or hurling a rain of small shots at high velocity against workpieces to cause slight indentations.

Sizing, coining and hobbing: Parts of malleable iron, forged steel, powdered metals, aluminium and other ductile non-ferrous metals are commonly finished to thickness by using an operation called *sizing*. The special dies are needed for almost every job, but each piece can be sized in a fraction of the time of machining. Sizing is economical for mass production. Operations like sizing have been called *coining*, but coining more truely involves the impression and raising of images or charter from a punch and die into the metal. Hard money is the best example of coining.

Hobbing or hubbing is a method of making moulds for the plastic and die casting industries. A punch called the *hob* is machined from the tool steel to the shape of cavity, heat treated for hardness and polished. It is then pressed into a blank of soft steel to form the mould.

Cold bending: Tubes, rods and bars can be bent into different shapes in cold conditions. During cold bending the metal, stressed beyond the elastic limit, is in tension on the outside and in compression on the inside of the bend. Bending should take place at right angles to the grain direction.

Roll forming: There are two types of roll forming. One uses continuous strip material for high production work, the other uses sheet and plate stock. Continuous roll forming utilises a series of rolls to gradually change the shape of the metal. As the metal passes between the rolls in a fast-moving continuous strip, the cross-sectional shape is changed to the desired shape. Bending rolls are used for bending sheet and plate stock into cylindrical shape.

Cold extrusion: Cold extrusion is used to make small workpieces from the more ductile metals in cold conditions. The impact extrusion method is illustrated in Fig. 25.9. In this process, the blank is placed in the position into a blind die, and the punch is allowed to strike it. This results in the required component. Toothpaste tubes and collapsible medicine tubes are made by this process.

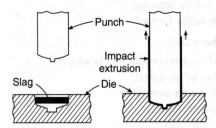

Fig. 25.9 Impact extrusion.

Cold pressing and deep drawing: In the cold pressing process shown in Fig. 25.10, the metal is shaped by pressing, and the change in shape takes place without thinning of the material. Mild steel is easily pressed.

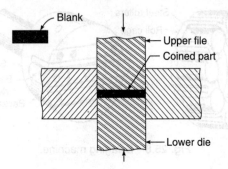

Fig. 25.10 Cold pressing.

Deep drawing requires very high ductility in the sheet stock, and this process can be carried out only on selected alloys such as 70–30 brass, cupro nickel, pure copper, pure aluminium and some of its alloys. Cold pressing and deep drawing are carried out on power presses consisting of die punches and blank holders.

25.2.2 Hot Working

In hot working of metals the shaping operations are conducted above the recrystallisation temperature. An increase in the temperature of the metal leads to an increase in the atom spacing and therefore the bond strength between the atoms of the metal grains is reduced; therefore it becomes easier to produce deformation and give the required shape to the metals.

Advantages: The following are the main advantages of hot working:
  (i) During hot working operations, the metal remains in plastic stage; hence large deformation is possible even with small amounts of force.
  (ii) The grains of the metal after hot working are refined, hence the mechanical properties such as toughness, ductility, and elongation, can be improved.
  (iii) By hot working process, resistance to impact gets increased.
  (iv) The blow holes present in the castings (ingots) are pressed together and hence eliminated by high working pressure used in hot working.
  (v) It welds up cracks and removes porosity in metals.
  (vi) Inclusions, if any, are broken down and distributed throughout the metal.
  (vii) The process is quick.
  (viii) The process is economical.
  (ix) The process is suitable for all commercial metals.
  (x) Thick sections can easily be shaped without damaging the structure.
  (xi) The power required for shaping the process is low.
  (xii) The strain-hardened grains are refined, resulting in better technological properties.

Distinction between hot working and cold working

Hot working	Cold working
It is done above the recrystallisation temperature.	It is done below and sometimes even at room temperature.
It is difficult to obtain good surface finish.	Good surface finish can be obtained.
It is a cheap process.	It is costlier than hot working.
Less power is required.	More power is required.
Lesser number of blows is required.	More blows are required.
Grain refinement is obtained.	Grain refinement can not be obtained.
Work hardening can not take place.	Work hardening takes place due to the shearing action of the metals.
Close tolerances can not be obtained.	Close tolerances can be obtained.
Tooling cost is high.	Tooling cost is less.
Handling is difficult and costly.	Handling is easy and cheap.
Porosity, cracks and impurities are eliminated.	These can not be eliminated.
The profiles are simple.	Intricate profiles can be made.

Disadvantages: These are:
  (i) It gives poor surface finish and close tolerances can not be maintained.
  (ii) Expensive tools are needed to work with metals at high temperature.
  (iii) Handling costs are high and the handling process is difficult.
  (iv) Tooling life is reduced due to working under high temperatures.

Hot working processes

The following are the principal hot working processes:

  (i) Hot rolling
  (ii) Hot extrusion
  (iii) Hot forging
  (iv) Roll piercing
  (v) Pipe welding
  (vi) Hot spinning
  (vii) Hot drawing

(i) Hot rolling: Hot rolling is the most rapid method of forming a metal into desired shape by plastic deformation between rolls. In deforming the metal between rolls, the work is subjected to high compressive stresses from the squeezing action of the rolls and to surface shear stresses, as a result of the friction between the rolls and the metal. The frictional forces are also responsible for drawing the metal into rolls.

Two-high rolling mill (*reversing*). Figure 25.11 shows a two-high mill. It has two heavy horizontal rolls, one over the other, i.e. in the same vertical plane. The space between the rolls can be adjusted by raising or lowering the upper roll. This rolling mill is normally used to roll the ingots. The metal in the hot plastic stage is passed between two rolls revolving at the same speed but in opposite directions. As the metal passes through the rolls, it is reduced in thickness and increased in length. The forming of bars, plates, sheets, rails and other structural sections is obtained by rolling processes.

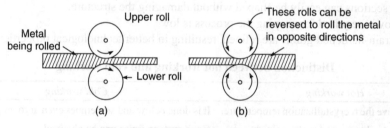

Fig. 25.11 Two-high rolling mill.

Three-high rolling mill. Figure 25.12 shows the arrangement of rollers in a three-high rolling mill. Three horizontal rolls are arranged one over the other. The upper and lower rolls rotate in the same direction, whereas the middle roll does so in the opposite direction. It is not as flexible as a two-high rolling mill because here, once the rolls are adjusted vertically, they remain fixed in their positions. In this process the production rate is high, since two or three pieces may be passing through the mill simultaneously.

Since the rolls' direction of rotation is not required to be reversed, smaller and less costly motive power is required. For big workpieces lifting tables are required, while small workpieces can be lifted manually. The three-high rolling mill can be used as blooming mill for billet rolling and finish rolling.

Four-high rolling mills. Figure 25.13 shows the arrangement of rollers in a four-high rolling mill. The working rolls which apply pressure on the metal are quite smaller in size. They are backed up by two big rolls. These backing rolls prevent deflections of the working rolls. These are commonly used for both hot and cold rolling of plates and sheets. Figure 25.14 shows a four-high rolling mill.

METAL WORKING PROCESSES **683**

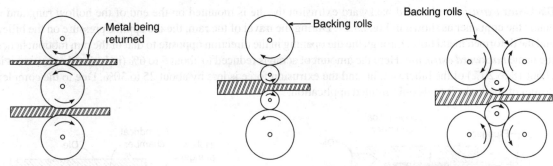

Fig. 25.12 Arrangement of rollers in three-high rolling mill.   **Fig. 25.13** Arrangement of rollers in four-high rolling mill.   **Fig. 25.14** Four-high rolling mill.

Planetary mill: This mill is used to reduce slabs to coiled hot-rolled strips in a single pass. Such a mill consists of the planetary assemblies and feed rolls, which push the slab through a guide into planetary rolls where the main reduction takes place as shown in Fig. 25.15. The mill is followed by a two- or four-high planishing mill and a coiler. The maximum width possible is 2000 mm. Here the synchronising of the speed of the rolls is very important and calls for a computer.

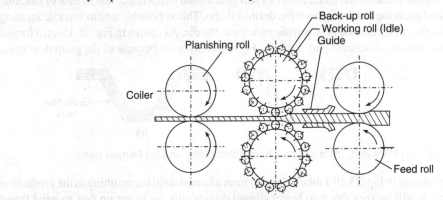

Fig. 25.15 Planetary mill.

(ii) Hot extrusion methods: Extrusion is defined as a process in which metal is reduced in cross-section by forcing it to flow through a die to produce cylindrical base or hollow tubes. Shapes of irregular cross-section can also be produced. The raw material is extruded in cast or rolled millets. The process of extrusion can be broadly classified into the following groups:

Forward or direct extrusion. In this process, the slug or the billet is kept in the container portion of the die. The required force is applied to the slug by means of the ram through the pressure plate, as shown in Fig. 25.16. In this process, the direction in which the material leaves the die is the same as that of the punch motion; hence the name *forward extrusion*. In the case of hollow forward extrusion (also called *Hooker extrusion*), the slug is a hollow piece. In this case, the punch has a shoulder as shown in Fig. 25.16, and acts as a mandrell. The bottom of the cup may be either closed or opened. At the beginning of the process, the mandrell should extend upto the level of the die shoulder and project a definite distance past the die. During the downward motion of the punch the metal is forced through the annular opening, forming a cup. The flange can not be avoided in this process. But it can be reduced to a very small thickness by having proper tool design. In case of flange, it has to be removed by machined or cut-off operation.

Backward extrusion. In solid backward extrusion the die is mounted on the end of the hollow ram, and it enters the container as shown in Fig. 25.17. During the travel of the ram, the die applies pressure on the billet, and the deformed metal flows through the die opening in the direction opposite to that of the ram motion; hence the name *backward extrusion*. Here the amount of scrap is reduced to about 5 to 6% (in forward extrusion it is about 18 to 20%) of the billet weight, and the extrusion force is less by about 25 to 30%. Due to the complex design of the tools, it finds only limited application.

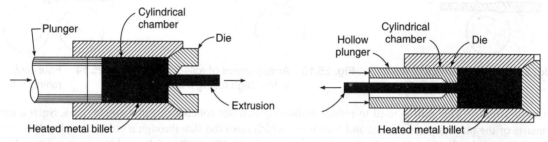

Fig. 25.16 Forward or direct extrusion.   **Fig. 25.17** Backward extrusion.

(iii) Hot forging: Comprises heating the metal above its recrystalization temperature and in case of machine forging pressure is applied upon the blank to give it the desired shape. This is possible due to suitable locating the heated blank between the cavity created between the punch and the die. As shown in Fig. 25.18(c), a forged part is stronger than a cast part or a machined part shown in Fig. 25.18(a) or (b) because of the grain flow lines.

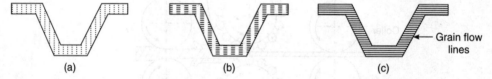

Fig. 25.18 (a) Cats part, (b) machined from rolled plate, and (c) Forged part.

(iv) Roll piercing: As shown in Fig. 25.19 a tube is formed from a heated solid bar resulting in the production of seamless tube or pipe. It will be seen that two barrel shaped driven rolls are so set up that an axial thrust is developed due to the rotation of the rolls in the solid bar. The solid bar also rotates due to friction with the two rolls. It is pushed against a mandrel having a conical front. This interaction between the solid bar and sizing mandrel generates a hole in the bar resulting in a seamless tube. The production rate of the seamless tube could be of the order of 24 metres per minute for tubes upto 150 mm diameter.

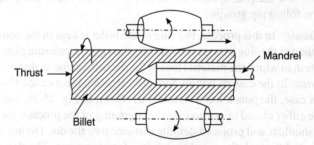

Fig. 25.19 Roll piercing operation (Mannesmaan process).

(v) Pipe welding: Pipe welding is categorized into two classes:

1. *Welding of pipes of shorter length:* In this process shown in Fig. 25.20(a), a measured length of the plate is first curled and then passed through a die as shown. The curled plate called skelp is first heated in a furnace and then pulled through the die so as to bring the bevelled ends together to be fusion welded. The pipe so formed has a short predetermined length and is finally straightened.

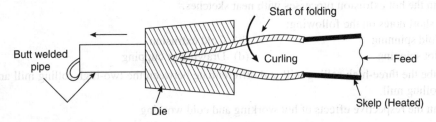

Fig. 25.20(a) Production of short butt welded pipe.

2. *Pipe welding of long pipes by continuous process:* In this process, the sheet to be converted into a pipe passes through a number of vertical and horizontal rolls which give the sheet the right diameter. As before, the sheet is subjected to heating in a furnace so that bending the sheet to a circular cross section becomes easier. In this process, electrical resistance heating is also done at the joint to ensure a strong joint. Figure 25.20(b) shows a schematic diagram of continuous resistance butt welding.

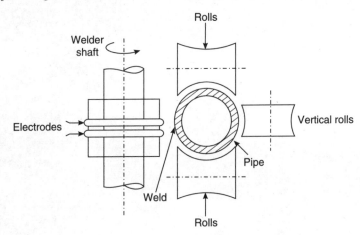

Fig. 25.20(b) Continuous pipe manufacture by resistence butt welding (Front view).

(vi) Hot spinning: Hot spinning is similar to cold spinning, except that in hot spinning the blank of the sheet metal is first heated and then held against the rotating chuck of the lathe. This process is quite commonly used to shape thicker metallic sheets.

(vii) Hot drawing: It is used to make cup-shaped components from sheet metal blanks. Usually drawing is performed when the metal is in cold condition, but when the metal thickness is more, hot drawing is carried out. In hot drawing, the heated blank is placed in position over the die, and the punch of the press is allowed to strike the blank and thus push the material into the die.

REVIEW QUESTIONS

1. What are the cold working and hot working processes?
2. Differentiate between hot working and cold working of metals.
3. Explain, with the help of neat sketches, any two cold working and any two hot working processes.
4. Explain the hot extrusion processes with neat sketches.
5. Write short notes on the following:
   (a) Cold spinning
   (b) Cold extrusion
   (c) Hot spinning
   (d) Drawing or cupping.
6. Describe the three-high rolling mill and differentiate between the two-high rolling mill and the three-high rolling mill.
7. Explain the respective effects of hot working and cold working.
8. List the respective advantages of hot working and cold working.

CHAPTER 26

Theory of Metal Working Processes

26.1 INTRODUCTION

Forming can be defined as a process in which the desired size and shape are obtained through the plastic deformation of the material. The stresses induced during the process are greater than the yield strength, but less than the fracture strength of the material. The type of loading may be tensile, compressive, bending, shearing, or a combination of these. This is a very economical process, as the desired shape, size and finish can be obtained without any significant loss of material. Moreover, a part of the input energy is utilized in improving the strength of the product through strain hardening.

26.2 METHODS OF PLASTICITY ANALYSIS OF MANUFACTURING PROCESSES

When a metal is deformed by a manufacturing process, the total work per unit volume done on the metal is given by

$$W_T = W_p + W_f + W_r$$

where W_p is the ideal work of deformation;
W_f is the work to overcome friction at the metal–tool interface;
W_r is the redundant work.

The redundant work is the work involved in the internal shearing process due to non-uniform deformation. It does not contribute to the change in the shape of the body. This concept is explained in Fig. 26.1, where a billet is shown before compression and after compression. If there is perfect lubrication at the metal–tool interface, the grid will remain undistorted. The square grid (before compression) will become rectangular (after compression). The height of each grid (before compression) will become rectangular (after compression). The height of each grid block will get shortened and its length will get increased. Such deformation is called *uniform deformation*. However, if friction is present at the metal–tool interface, then the grid will become distorted, accompanied by barrelling or bulging on the sides. The extra work which goes into distorting the grid and also into creating bulge is a waste and is called the *redundant work*.

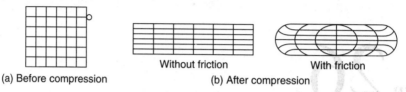

(a) Before compression   Without friction   With friction
(b) After compression

Figure 26.1 Mode of deformation.

The commonly used methods for the analysis of metal forming processes are:

1. Slab method
2. Upper-bond method
3. Slip-line method.

The analysis is basically the same for cold forming and hot forming, with difference only in flow and friction characteristics. The analysis of the metal forming processes by the above methods is only approximate, due to the many assumptions made regarding the behaviour of the work material and the mode of deformation. The approximation is better at low co-efficient of friction. Thus, the correlation is better with cold working, low-friction processes than with hot working, high-friction processes.

In this chapter, we shall deal with only the slab method. 'Slab-analysis technique' or 'elementary theory' makes the following assumptions:

1. The material is isotropic and incompressible, and the elastic strains are negligible.
2. Deformation is homogeneous throughout the deforming materials under study.
3. Stresses on a plane normal to the flow direction are principal stresses.

In this method, a 'slab' of infinitesimal thickness is considered and a force balance (equilibrium equation) is made on it. The resulting differential equation of static equilibrium is solved with the help of appropriate boundary conditions and yield criteria.

Analysis of metal forming mechanics

During the process of metal forming, a material is deformed inelastically. Considering a tensile test specimen, let l_1 be the initial gauge length. During the process of plastic flow and necking, the gauge length changes its value continuously. However, considering discrete steps of gauge length $l_1, l_2, l_3, \ldots$, we can write

Total strain $\quad \varepsilon = \left(\dfrac{dl_1}{l_1}\right) + \left(\dfrac{dl_2}{l_2}\right) + \left(\dfrac{dl_3}{l_3}\right) + \cdots$

Therefore $\quad \varepsilon = \displaystyle\int_{l_i}^{l_f} \left(\dfrac{dl}{l}\right) = \ln\left(\dfrac{l_f}{l_i}\right)$

where l_f is the final length and l_i is the initial length.

Since the volume V is constant, the equation can be written on volume constancy as

$$V = l_i A_i = l_f A_f \quad \text{or} \quad \dfrac{l_f}{l_i} = \dfrac{A_i}{A_f}$$

Thus

$$\text{Total strain} = \text{True strain} = \varepsilon = \ln\left(\frac{A_i}{A_f}\right)$$

The work done per unit volume in homogeneous deformation is

$$W = \text{True stress} \times \text{True strain} = \sigma_{ay} \times \ln\left(\frac{A_i}{A_f}\right)$$

If the effects of internal friction due to internal metallic shearing at shear or slip planes is not accounted for, and external friction is also not considered, the above formula can be applied to problems such as extrusion or wire drawing. In extrusion, considering the ram pressure p and the area l, the work done per unit volume of material for a displacement of the billet by unit distance is given by

$$p \times l = \sigma_{ay} \times \ln\left(\frac{A_i}{A_f}\right)$$

As a result of metal working research, Rowe (1985), Johnson and some others have found that

$$p = 2K\left[a + b \ln\left(\frac{A_1}{A_2}\right)\right]$$

where $a = 0.7$, a factor that accounts for the redundant work or the internal metallic shear, and $b = 1.5$, a factor that accounts for the friction between the billet and the container wall.

26.3 FORGING ANALYSIS USING SLAB METHOD

In this section, our analysis is mainly devoted to determining the maximum force required for forging a strip and a disc between two parallel dies. Obviously, it is a case of open die forging.

Forging of strip

Figure 26.2(a) shows a typical open die forging of a flat strip. To simplify our analysis, we shall make the following assumptions:

(i) The forging force F attains its maximum value at the end of the operation;
(ii) The coefficient of friction μ between the workpiece and the dies (platens) is constant;
(iii) The thickness of the workpiece is small as compared with its other dimensions, and the variation of the stress field along the y-direction is negligible;
(iv) The length of the strip is much more than the width and the problem is one of plane strain type;
(v) The entire workpiece is in the plastic state during the process.

At the instant shown in Fig. 26.2(a), the thickness of the workpiece is h and the width is $2l$. Let us consider an element of width dx at a distance x from the origin. [In our analysis, we take the length of the workpiece as unity (in the z-direction).] Figure 26.2(b) shows the same element with all the stresses acting on it. Considering the equilibrium of the element in the x-direction, we get

$$h d\sigma_x + 2\tau dx = 0 \tag{26.1}$$

where $d\sigma_x = (2/h) dx \cdot \tau$ is the frictional stress. To make the analysis simpler, p and σ_x are considered the principal stresses. The problem being of a plane-strain type, Eq. (26.1) may be used as the yield criterion. Thus

$$\sigma_x + p = 2K \quad \text{or} \quad d\sigma_x = -dp$$

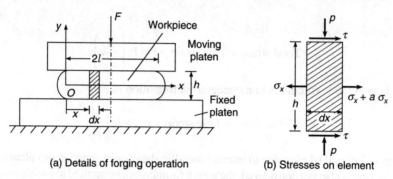

(a) Details of forging operation   (b) Stresses on element

Figure 26.2 Force and stresses during forging.

Substituting $d\sigma_x$ from the foregoing relation in Eq. (26.1), we get

$$dp = \frac{2\tau}{h} dx \qquad (26.2)$$

Near the free ends, i.e. when x is small (and also at $x \approx 2l$; the problem being symmetric about the midplane, we are considering only one-half in our analysis, i.e. $0 \leq x \leq l$), sliding between the workpiece and the dies must take place to allow for the required expansion of the workpiece. However, beyond a certain value of x (in the region $0 \leq x \leq 1$), say, x_s, there is no sliding between the workpiece and the dies. This is due to the increasing frictional stress which leads to the maximum value, equal to the shear yield stress, at $x = x_s$ and remains so in the rest of the zone, $x_s \leq x \leq l$. Hence, for $0 \leq x \leq x_s$

$$\tau = \mu p \qquad (26.3)$$

and for $x_s \leq x \leq l$

$$\tau = K \qquad (26.4)$$

However, it should be noted that this assumption is incorrect, as the stress τ acts on the planes on which p is acting (see Fig. 26.2).

For the sliding (nonsticking) zone, using Eq. (26.3) in Eq. (26.4) and integrating, we have

$$\int \frac{dp}{p} = \frac{2\mu}{h} \int dx + C_1 \qquad (0 \leq x \leq x_s)$$

or

$$\ln p = \frac{2\mu x}{h} + C_1$$

Now, at $x = 0$, $\sigma_x = 0$, i.e. $p = 2K$ (from the yield criterion). So

$$C_1 = \ln 2K$$

or

$$p = 2K \exp\left(\frac{2\mu x}{h}\right) \qquad (26.5)$$

For the sticking zone, using Eq. (26.4) in Eq. (26.2) and integrating, we have

$$\int dp = \frac{2K}{h} \int dx + C_2 \qquad (x_s \leq x \leq 1)$$

or

$$p = \frac{2Kx}{h} + C_2$$

If $p = p_s$ at $x = x_s$, then $C_2 = p_s - \dfrac{2Kx_s}{h}$. Thus

$$p - p_s = \frac{2K}{h}(x - x_s) \tag{26.6}$$

Again, from Eq. (26.5)

$$p_s = 2K \exp\left(\frac{2\mu x_s}{h}\right) \tag{26.7}$$

Therefore

$$p = 2K\left[\exp\left(\frac{2\mu x_s}{h}\right) + \frac{1}{h}(x - x_s)\right] \tag{26.8}$$

At $x = x_s$, $\mu p_s = K$. Using this along with the expression for p_s, we get

$$\mu 2K \exp\left(\frac{2\mu x_s}{h}\right) = K$$

or

$$\frac{2\mu x_s}{h} = \ln\left(\frac{1}{2\mu}\right)$$

or

$$x_s = \frac{h}{2\mu} \ln\left(\frac{1}{2\mu}\right) \tag{26.9}$$

Substituting this value of x_s in Eq. (26.8), we obtain

$$p = 2K\left[\frac{1}{2\mu}\left\{1 - \ln\left(\frac{1}{2\mu}\right)\right\} + \frac{x}{h}\right] \quad (x_s \leq x \leq 1) \tag{26.10}$$

The total forging force per unit length of the workpiece is given as

$$F = 2\left(\int_0^{x_s} p_1 \, dx + \int_{x_s}^{1} p_2 \, dx\right) \tag{26.11}$$

where p_1 and p_2 are the pressures given by Eq. (26.10).

EXAMPLE 26.1 A strip of lead with initial dimensions 24 mm × 24 mm × 150 mm is forged between two flat dies to a final size of 6 mm × 96 mm × 150 mm. If the coefficient of friction between the job and the dies is 0.25, determine the maximum forging force. The average yield stress of lead in tension is 7 N/mm².

Solution First, let us determine the shear yield stress K for lead as follows:

$$K = \frac{1}{\sqrt{3}} \sigma_y = 4.04 \text{ N/mm}^2$$

To use Eq. (26.8), the value of x_s is required. From Eq. (26.9)

$$x_s = \frac{6}{2 \times 0.25} \ln \frac{1}{2 \times 0.25} \text{ mm} = 8.3 \text{ mm}$$

Now, from Eq. (26.7), the expressions for the pressures p_1 and p_2 (for the nonsticking and the sticking zones respectively) can be found out. Thus

$$p_1 = 8.08 e^{0.083x} \text{ N/mm}^2 \quad (0 \leq x \leq 8.3 \text{ mm})$$

$$p_2 = 8.08(0.614 + 0.167x) \text{ N/mm}^2 \quad (8.3 \text{ mm} \leq x \leq 48 \text{ mm})$$

Using Eq. (26.11), the force per unit length we get is

$$F = 2\left[\int_0^{8.3} 8.08e^{0.083x}\,dx + \int_{8.3}^{48} 8.08(0.614 + 0.167x)\,dx\right] \text{N/mm}$$

$$= 3602.5 \text{ N/mm}$$

Since the length of the strip is 150 mm, the total forging force is 150×3602.5 N $= 0.54 \times 10^6$ N.

EXAMPLE 26.2 Solve Example 26.1 when the coefficient of friction $\mu = 0.08$.

Solution Using Eq. (26.9), we obtain

$$x_s = \frac{6}{0.16}\ln\frac{1}{0.16} = 37.5 \times 1.83 = 68.74 \text{ mm}$$

Since x_s is more than l, the entire zone is nonsticking.

Therefore, $\quad p = 8.08e^{0.027x}$ N/mm^2 $(0 \le z \le 48$ mm$)$

So $\quad F = 2\int_0^{48} 8.08e^{0.027x}\,dx = 1588.5$ N/mm^2

Total forging force $= 150 \times 1588.5$ N $= 0.238 \times 10^6$ N.

26.4 FLAT ROLLING ANALYSIS USING SLAB METHOD

The typical flat rolling of plates, sheets, or strips is essentially a plane-strain operation, since little widening of the workpiece results. This can be explained by examining the shape of the deformation zone as shown in Fig. 26.3. The length of contact L between the rolls and the workpiece is usually much smaller than the width w of the sheet. As the rolls induce a compressive stress σ_z, the plastic zone is thinned and is free to expand in the rolling direction x, but lateral expansion in the y direction is effectively restrained by the undeforming material on both sides of the roll gap. The net effect is a condition of plane strain, where $\varepsilon_y \approx 0$ and $\varepsilon_z \approx -\varepsilon_x$, except at the edges of the workpiece.

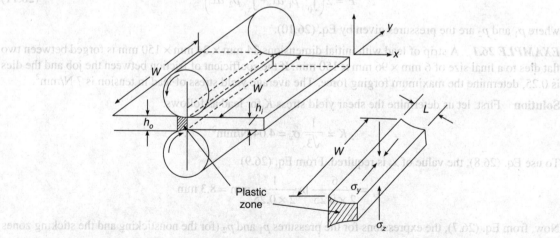

Fig. 26.3 Schematic diagram of deformation zone in flat rolling.

The effects of roll geometry and friction on the rolling process can be understood in terms of the friction-hill effect. Consider the rolling geometry in Fig. 26.4, where at some point N in the roll gap the surface velocities of the rolls and the work material are equal. To the left of this point, the surface velocity of the metal is lower, and the friction between the metal and the rolls tends to draw the metal into the gap. To the right of N the metal velocity is higher than the roll velocity; so the friction acts to the left on the metal. The net effect is to produce a friction hill.

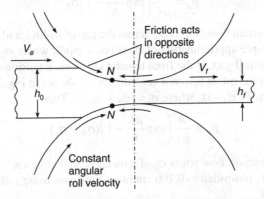

Fig. 26.4 Schematic diagram of flat rolling showing natural point N.

Consider the roll-gap geometry in Fig. 26.5, where R is the roll radius and $\Delta h = h_0 - h_f$ is the chord of the arc of contact. It is seen that

$$L^2 = Q^2 + \left(\frac{\Delta h}{2}\right)^2$$

$$Q^2 = R^2 - \left(R - \frac{\Delta h}{2}\right)^2 = R\Delta h - \left(\frac{\Delta h}{2}\right)^2 \tag{26.12}$$

$$L = \sqrt{R\Delta h} = \sqrt{Rrh_0} \tag{26.13}$$

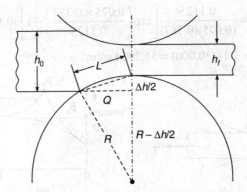

Fig. 26.5 Dimensional relations in roll gap.

The frictional effects are similar to those in plane-strain compression; in fact, if the roll curvature is neglected, the material in the roll gap can be considered as being under plane-strain compression. With sliding friction, Eq. (26.12) is applicable for the variation in the rolling pressure and Eq. (26.13) for the average roll pressure. For b in those equations, the roll contact length $L = \sqrt{R\Delta h}$ should be used, where h is the average height of the metal in the roll gap [i.e. $h = (h_0 + h_f)/2$]. Thus, in terms of these parameters:

$$p_a = \frac{h}{\mu L}\left(\exp\frac{\mu L}{h} - 1\right)\sigma_0 \qquad (26.14)$$

where σ_0 is the average plane-strain flow stress ($2k$) across the gap of width L and height h. If the material work hardens, a reasonable and simple approximation of σ_0 is $(\sigma_1 + \sigma_2)/2$, where σ_1 and σ_2 are the flow stresses at the entrance and exit of the roll gap. If the front tension σ_{ft} or back tension σ_{bt} is applied during rolling, the compressive stress p from the roll is lowered accordingly, as shown in Fig. 26.6(b). This can be handled approximately by replacing σ_0 by $\sigma_0 - \sigma_t$, where $\sigma_t = (\sigma_{ft} + \sigma_{bt})/2$. Then

$$p_a = \frac{h}{\mu L}\left(\exp\frac{\mu L}{h} - 1\right)(\sigma_0 - \sigma_t) \qquad (26.15)$$

EXAMPLE 26.3 The plane-strain flow stress σ_0 of a metal is 30,000 kg/cm². A sheet of this metal, 24 cm wide by 1/8 cm thick, is to be cold-rolled to 0.100 cm in a single pass using rolls of 12 cm diameter, and the coefficient of friction is 0.075.

(a) Compute the average pressure between the rolls and the sheet.
(b) If a front tension of 10,000 kg cm² were applied, what would be the average pressure?

Solution The theory developed in 26.4 is now applied to Rolling. Figure 26.6(a) assumes the material compressed between rolls as analogous to compression between platens. Figure 26.6(b) shows average pressure p_a and p_{max} at the peak of "friction hill".

(a) From Eqs. (26.13) and (26.14) as well as the accompanying discussion

$$h = \frac{0.125 + 0.100}{2} = 0.1125 \text{ cm}$$

$$L = \sqrt{6(0.025)} = 0.387 \text{ cm}$$

$$p_a = \frac{0.1125}{(0.075)(0.387)}\left[\exp\left(\frac{0.075 \times 0.387}{0.1125}\right) - 1\right]30,000$$

$$= 1.14(30,000) = 34,200 \text{ kg/cm}^2$$

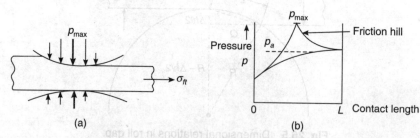

Fig. 26.6 Roll gap stresses.

(b) From Eq. (26.15) and with the same coefficient from part (a)

$$p_a = 1.14(30,000 - 5,000) = 28,500 \text{ kg/cm2}$$

Note the friction hill in Fig. 26.6(b). On increasing either the front or the back tension causes the hill to decrease and therefore lowers the average pressure. The neutral point is also affected by such tensions. As the back tension increases, the shift of this point is towards the exit; lowering the friction or increasing the reduction has the same tendency towards shifting the neutral point. Regarding an increase in the front tension, increasing the friction and decreasing reduction both tend to shift the neutral point towards the entry of the material.

If the sheet is thin (leading to small h) and the roll diameter is large (making L large), a large friction hill results and causes a high value of p_a. The roll-separating force per unit w, given by $F_s = p_a L$, increases even more rapidly. One consequence of high roll-separating forces is that they cause an elastic flattening of the rolls, much as an automobile tyre flattens under the weight of a car. The actual radius of curvature of the roll, R', of the contact area between the roll and the workpiece becomes larger than the unloaded radius R.

26.5 DEEP DRAWING ANALYSIS USING THEORY OF PLASTICITY

From the point of view of analysis, the process of deep drawing is very complex. In this process, various types of forces operate simultaneously. The annular portion of the sheet metal workpiece (see Fig. 26.7) between the blank holder and the die is subjected to a pure radial drawing, whereas the portions of the workpiece around the corners of the punch and the die are subjected to a bending operation. Further, the portion of the job between the punch and the die walls undergoes a longitudinal drawing. Though variation in the amount of thickening and thinning of the workpiece is unavoidable in this operation, we shall not take this into consideration in our analysis.

The major objectives of our analysis are (i) to correlate the initial and final dimensions of the job, and (ii) to estimate the drawing force F. Figure 26.7 shows the drawing operation with the important dimensions.

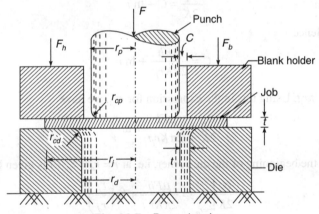

Fig. 26.7 Deep drawing.

The radii of the punch, the job, and the die are r_p, r_j, and r_d respectively. Obviously, without taking the thickening and thinning into account, the clearance between the die and the punch $(r_d - r_p)$ is equal to the job thickness t. The corners of the punch and the die are provided with radii r_{cp} and r_{cd} respectively. A clearance C is maintained between the punch and the blank holder.

To start with, let us consider the portion of the job between the blank holder and the die. Figure 26.8(a) shows the stresses acting on an element in this region. It should be noted that the maximum thickening (due to the decreasing circumference of the job causing a compressive hoop stress) takes place at the outer periphery, generating a line of contact between the holder and the job. As a result, the entire blank holder force F_h is assumed to act along the circumference [see Fig. 26.8(b)]. Thus, the radial stress due to friction can also be represented by an equivalent radial stress at the outer periphery.

Now, considering the radial equilibrium of the element shown in Fig. 26.8, we get

At equilibrium

$$r d\sigma_r + \sigma_r dr - \sigma_\theta dr = 0 \qquad (26.16)$$

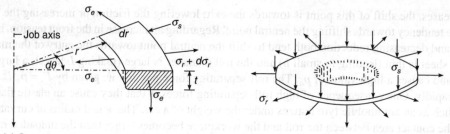

(a) Stresses acting on element during drawing   (b) Radial stress due to blank holding pressure

Fig. 26.8 Analysis of deep drawing operation.

and in plasticity condition

$$\sigma_r - \sigma_\theta = 2K \qquad (26.17)$$

Substituting σ_θ from Eq. (26.16) in Eq. (26.15), we get

$$\frac{dr}{r} + \frac{d\sigma_r}{2K} = 0$$

Integrating, we obtain

$$\frac{\sigma_r}{2K} = C - \ln r$$

as already mentioned. Hence

$$C = \frac{\mu F h}{2\pi K r_j t} + \ln r_j$$

Now, at $r = r_y$, $\sigma_r = \mu F_h/\pi r_j t$. Using this in the expression for σ_{rs}, we have

$$\frac{\sigma_r}{2K} = \frac{\mu F h}{2\pi K r_j t} + \ln \frac{r_j}{r}$$

Thus the radial stress at the beginning of the die corner, i.e. at $r = r_d = r_p + t$ is given by

$$\left.\frac{\sigma_r}{2K}\right|_{r=r_\alpha} = \frac{\mu F h}{2\pi K r_j t} + \ln\left(\frac{r_j}{r_d}\right) \qquad (26.18)$$

As the job slides along the die corner, the radial stress given by Eq. (26.18) increases to σ_z due to the frictional forces as shown in Fig. 26.9.

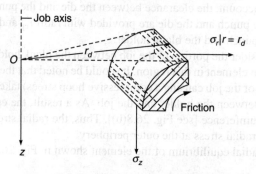

Fig. 26.9 Effect of friction at corners.

This increment can be roughly estimated by using a belt-pulley analogy. Thus

$$\frac{\sigma_z}{\sigma_r|r=r_d} = e^{\mu\pi/2} \qquad (26.19)$$

where μ is the coefficient of friction between the workpiece and the die.

There is a further increase in the stress level around the punch corner due to bending. As a result, the drawn cup normally tears around this region. However, to avoid this, an estimate of the maximum permissible value of r_j/r_d can be obtained by using Eqs. (26.18) and (26.19), with σ_z equal to the maximum allowable stress of the material. Since r_d is the final outside diameter of the product, it is easy to arrive at such an estimate. This estimate is based on the consideration of fracture of the material. However, to avoid buckling (due to the compressive hoop stress in the flange region), $r_j - r_p$ should not, for most materials, exceed $4t$.

The blank holder force is given by

$$F_h = \beta\pi r_j^2 K \qquad (26.20)$$

where β is between 0.02 and 0.08. An estimate of the drawing force F (neglecting the friction between the job and the die wall) can easily be obtained, using Eq. (26.19), as

$$F \approx \sigma_z 2\pi r_p t \qquad (26.21)$$

EXAMPLE 26.4 A cold rolled steel cup with an inside radius of 30 mm and a thickness of 3 mm is to be drawn from a blank of diameter 40 mm. The shear yield stress and the maximum allowable stress of the material can be taken as 210 N/mm² and 600 N/mm² respectively.

 (i) Determine the drawing force, assuming that the coefficient of friction $\mu = 0.1$ and $\beta = 0.05$.
 (ii) Determine the minimum possible radius of the cup which can be drawn from the given blank without causing a fracture.

Solution
 (i) We first calculate the blank holding force F_h from the given data as

$$F_h = 0.05\pi \times 40^2 \times 210 \text{ N} = 52{,}778 \text{ N}$$

Next, we find the value of σ_r at $r = r_d$ by using Eq. (26.18). Thus

$$\sigma_r \text{ at } r = r_d = r_\sigma = \frac{0.1 \times 52{,}778}{\pi \times 40 \times 3} + 2 \times 210 \times \ln\left(\frac{40}{30}\right)$$

$$= 14 + 80.8 \text{ N/mm}^2 = 94.8 \text{ N/mm}^2$$

Now, using Eq. (26.19), we get

$$\sigma_r = 94.8 \times e^{0.1\pi/2} \text{ N/mm}^2 = 110.9 \text{ N/mm}^2$$

It should be noted that this is much less than the fracture strength though $r_j - r_p = 10$ mm $= 3.33t$, i.e. very close to the limit set by the condition of plastic buckling. From Eq. (26.20), the drawing force is found to be s

$$F = 2\pi \times 30 \times 110.9 \text{ N} = 62{,}680 \text{ N}$$

In this case, $\sigma_z = 600$ N/mm². From Eq. (26.19)

$$\frac{\sigma_r}{r} = r_d = \frac{600}{e^{0.05\pi}} \text{ N/mm}^2 = 512.8 \text{ N/mm}^2$$

 (ii) Using Eq. (26.18) and taking $F_h = 52{,}778$ N, we get

$$\ln\left(\frac{40}{r_p + 3}\right) = \frac{512.8}{420} - \frac{0.1 \times 52{,}778}{2\pi \times 21{,}040 \times 3} = 1.221 - 0.033 = 1.188$$

or
$$r_p = \frac{40}{e^{1.188}} - 3 = 6.2 \text{ mm}$$

Again, it is interesting to note that $r_j - r_p = 30.8$ mm $\geq 4t$, and this goes much beyond the limit set by the plastic buckling condition.

26.6 EXTRUSION ANALYSIS USING SLAB METHOD

In the slab method, a thin slab of thickness dx of material being plastically deformed is considered. Two conditions are applied to the material in the slab.
1. Conditions of equilibrium are fulfilled. That is, the summation of all forces in the X-directon is zero; $\Sigma F_x = 0$.
2. The slab is in a fully plastic state. That is, the difference of principal stresses acting in two mutually perpendicular directions on the slab equals twice the yield shear stress of the slab material. This is also called the *condition of plasticity*.

Figure 26.10 indicates the problem of drawing of a sheet of half thickness h_b to a reduction resulting in half thickness h_a with a back tension σ_b applied to the sheet. Consider only the upper half (above the centreline).

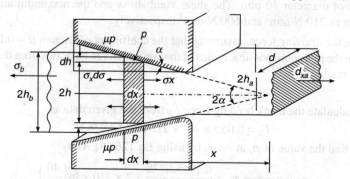

Fig. 26.10 Slab method of metal working.

Condition of equilibrium: For equilibrium along the x-axis

$$(\sigma_x + d\sigma_x)(h + dh) - \sigma_x \cdot h + p\left(\frac{dx}{\cos \alpha}\right)(\sin \alpha) + \mu p\left(\frac{d}{\cos \alpha}\right)\cos \alpha = 0$$

$$\therefore \quad \sigma_x\, dh + h d\sigma_x + p\, dh + \mu p\, dh \cot \sigma = 0$$

We use $B = \mu \cot \sigma$ in the subsequent calculations.

Condition of plasticity: Now, we can utilize the condition of plasticity to eliminate p from Eq. (26.21). Then

$$\sigma_x + p = 2k = S$$

where k is shear flow stress and S is direct yield stress.

$$\frac{d\sigma_x}{\sigma_x + (S - \sigma_x)(1 + B)} = -\frac{dh}{h}$$

$$\frac{d\sigma_x}{B\sigma_x - S(1+B)} = -\frac{dh}{h}$$

$$\frac{1}{B}\ln[B\sigma_x - S(1+B)] = \ln h + C$$

To evaluate the constant of integration, we apply the boundary condition $h = h_b$, $\sigma_x = \sigma_b$. Therefore

$$C = \frac{1}{B}\ln[B\sigma_b - S(1+B)] - \ln h_b \tag{26.22}$$

$$\frac{1}{B}\ln[B\sigma_x - S(1+B)] = \ln h + \frac{1}{B}\ln[B\sigma_b - S(1+B)] - \ln h_b \tag{26.23}$$

$$\ln[B\sigma_x - S(1+B)] = \ln\left(\frac{h}{h_b}\right)^B [B\sigma_b - S(1+B)] \tag{26.24}$$

Taking antilog, we get

$$[B\sigma_x - S(1+B)] = \ln\left(\frac{h}{h_b}\right)^B [B\sigma_b - S(1+B)] \tag{26.25}$$

Dividing throughout by SB, we get

$$\left(\frac{\sigma_x}{S} - \frac{1+B}{B}\right) = \left(\frac{h}{h_b}\right)^B \left(\frac{\sigma_b}{S} - \frac{1+B}{B}\right) \tag{26.26}$$

$$\frac{\sigma_x}{S} = \frac{1+B}{B}\left[1 - \left(\frac{h}{h_b}\right)^B\right] + \left(\frac{h}{h_b}\right)^B \left[\frac{\sigma_b}{S}\right] \tag{26.27}$$

To calculate the value of the front tension, put $\sigma_x = \sigma_{xa}$ and $h = h_a$ in Eq. (26.27). Then we have

$$\frac{\sigma_{xa}}{S} = \frac{1+B}{B}\left[1 - \left(\frac{h_a}{h_b}\right)^B\right] + \left(\frac{h_a}{h_b}\right)^B \frac{\sigma_b}{S} \tag{26.28}$$

Without backpull, putting $\sigma_b = 0$ Eq. (26.28) reduces to

$$\frac{\sigma_{xa}}{S} = \frac{1+B}{B}\left(1 - \frac{h_a}{h_b}\right)^B \tag{26.29}$$

26.6.1 Particular Cases

Equation (26.29) is very important and can be utilized to develop analogous expressions for rod drawings, tube drawing, etc., if (h_a/h_b) is regarded as an area ratio. For example, in the case of rod drawing or wire drawings:

$$\frac{\sigma_{xa}}{S} = \frac{1+B}{B}\left(1 - \frac{D_a}{D_b}\right)^{2B}$$

Note that the area of the rod is proportional to the square of the diameter. For tube drawing without mandrel

$$\frac{\sigma_{xa}}{S} = \frac{1+B}{B}\left(1 - \frac{D_a}{D_b}\right)^B$$

[Note that the area of the tube annulus is proportional to $D \times h$.]

Extrusion: In case of extrusion, Eq. (26.28) can be utilized. We put, at exit, front tensile stress $\sigma_{xa} = 0$, since there is no backpull but a pressure (p) with minus sign. Thus, substituting in Eq. (26.28) the above boundary conditions, we carry out manipulation with some algebraic expressions in Eq. (26.28). Thus, putting $\sigma_{xa} = 0$ and $\sigma_{xb} = -p$ in Eq. (26.28), we get

$$0 = \frac{1+B}{B}\left[1 - \left(\frac{h_a}{h_b}\right)^B\right] - \left(\frac{h_a}{h_b}\right)^B \frac{p}{S}$$

Thus, rearranging $\dfrac{p}{S_a} = \dfrac{1+B}{B}\left[1 - \left(\dfrac{h_a}{h_b}\right)^B\right]\left(\dfrac{h_b}{h_a}\right)^B$ where p is extrusion pressure.

EXAMPLE 26.5 Determine the power required to draw a wire from 12.5 mm to 10 mm in diameter at 100 m/min, when $\mu = 0.1$, $\alpha = 4°$ and $S = 2K = 30$ kg/mm^2. Also, calculate the maximum reduction possible.

Solution (a) We have $D_a = 10$, $D_b = 12.5$, $B = \mu \cot \alpha = 1.43$ and $2k = 30$ kg/mm^2. Then

$$\frac{\sigma_{xa}}{S} = \frac{1+B}{B}\left(1 - \frac{D_a}{D_b}\right)^{2B}$$

$$\frac{\sigma_{xa}}{30} = \frac{1+1.43}{1.43}\left[1 - \left(\frac{10}{12.5}\right)^{2.86}\right] = 0.80$$

$$\sigma_{xa} = 30 \times 0.8012 = 24.3 \text{ kg/mm}^2$$

Draw force $\left(\dfrac{\pi}{4}\right)D^2 \sigma_{xa} = \left(\dfrac{\pi}{4}\right) \times 100 \times 24.03 = 1886$ kg

Power required = 1886×100 kg m/min = 41.92 H.P.

To obtain maximum reduction, putting $\sigma_{xa} = S$, we have

$$\frac{\sigma_{xa}}{S} = 1 = \frac{2.43}{1.43}\left[1 - \left(\frac{D_a}{D_b}\right)^{2.86}\right]$$

$$\therefore \qquad \frac{D_a}{D_b} = 0.735$$

The initial magnitude of reduction was $\dfrac{10}{12.5} = 0.8$, when tensile stress was 24.03 kg/mm^2.

(b) Maximum reduction possible in tube sinkings: As discussed under wire drawing, the maximum reduction is limited by the mechanical strength of the exit end of the tube. That is $\sigma_{xa}/S = 1$ in the limiting case.

As a particular case of the equation, assuming $D_a = D_b$,

$$\left(\frac{\sigma_{xa}}{S}\right) = \frac{1+B}{B}\left[1-\left(\frac{h_a}{h_b}\right)^B\right]$$

Since $\sigma_{xa}/S = 1$ when stress $\sigma_{xa} = S$, we have

$$\frac{1+B}{B}\left[1-\left(\frac{h_a}{h_b}\right)^B\right]_{max} = 1$$

Let us take $\mu_1 = \mu_2 = 0.05$, $\alpha = 15°$ and $\beta = 0$. Then

$$B = \frac{\mu_1 + \mu_2}{\tan \alpha} = \frac{0.10}{0.268} = 0.373$$

$$\frac{1+B}{B} = 3.68$$

or

$$\left(\frac{h_a}{h_b}\right)_{max} = (0.7283)^{2.68} = 0.4275$$

Thus the maximum possible reduction is about 43%.

26.7 ROLLING

In rolling, the metal is plastically deformed by passing it between rolls. Rolling is done both hot and cold. The starting material is cast ingot, which is broken down by hot rolling into blooms, billets and slabs, which are further hot rolled into plate, sheet, rod, bar, pipe, rail or other structural shapes. Cold rolling is usually a finishing process in which products made by hot rolling are given a good surface finish, with increased mechanical strength of the material. The main objective in rolling is to decrease the thickness of the metal. Ordinarily, there is negligible increase in width, so that the decrease in thickness results in an increase in length.

Figure 26.11 shows the typical geometry for rolling. A metal sheet with a thickness h_0 enters the rolls, passes through the roll gap and leaves with a reduced thickness h_1. Since the volume rate of metal flow has to remain constant, velocity at exit V_1 will be more than velocity at entrance V_0. The roll has a constant surface velocity V. Thus, there is relative sliding between the roll and the workpiece. The direction of this relative velocity changes at a point along the contact area; this point is known as the *no slip point* or *neutral point N*, where the velocities of the roll and the workpiece will be equal. Therefore

$$V_0 h_0 b_0 = V_1 h_1 b_1$$

$$b \cong b_1$$

for plane-strain rolling.

At neutral plane, $V = V_0$ (actually V_0 is the component of V in horizontal direction, but as the angle of inclination is very small, we can take $V = V_0$).

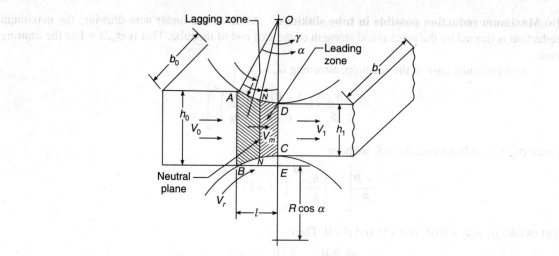

Fig. 26.11 Geometry of rolling process.

Because of the relative velocities involved, we have two velocities of rolled metal, V_0 and V_1 at entrance and exit respectively, V_m is the velocity of rolled metal at neutral point. It may be assumed without much error that V_m is equal to V_r which is the roll velocity at point N.

As
$$V_1 = \frac{V_0 h_0}{h_1}$$

∴
$$h_0 > h_1$$
$$V_1 > V_0$$
$$V_0 < V_r < V_t$$

$$\text{Backward slip} = \frac{V_r - V_0}{V_r}$$

$$\text{Forward slip} = \frac{V_1 - V_r}{V_r}$$

Zone NDCN is the *lagging zone* ($V_0 < V$), and zone ANNB is the *leading zone* ($V_0 > V$).

The angle a between the entrance plane and the centreline of rolls is called the *angle of contact* or *angle of bite*.

Let us consider the moment when the rolling process is just going to start. The roller contacts the entering material (may be the strip) at point A (see Fig. 26.11). This contact results in a normal force P between the roll and the workpiece, and the tangential frictional force F. If the resultant of P and F, that is T, is sloped to the right, then its component along the x-axis, $T_x > 0$, tends to push the workpiece into the roll opening and thus ensures *biting*. Thus the condition of biting, that is the condition for unaided entry of the workpiece into the rolls, is

$$T_x = F \cos \alpha - P_r \sin \alpha > 0$$

or
$$\frac{F}{P_r} > \tan \alpha$$

But
$$\mu > \tan \alpha$$

Let $\mu = \tan \beta$, where β is the angle of friction; therefore the limiting condition becomes

$$\tan \beta > \tan \alpha$$

or
$$\beta > \alpha$$

or
$$\alpha > \beta$$

which means that rolls 'bite' a workpiece if the angle of friction is greater than the angle of tangency or biting (see Fig. 26.12). If $\tan \alpha$ exceeds μ, then the workpiece cannot be drawn into the rolls. The values of biting angles are usually

$\alpha = 3°$ to $4°$ for cold rolling of steel and other metals, with lubrication or well-ground rolls;
$\quad = 6°$ to $8°$ for cold rolling of steel and other metals, with lubrication on rough rolls;
$\quad = 18°$ to $22°$ for hot rolling of steel sheets;
$\quad = 20°$ to $22°$ for hot rolling aluminium at $350°C$;
$\quad = 3°$ to $4°$ for cold rolling of steel and other metals, with lubrication or well-ground rolls;
$\quad = 28°$ for hot rolling of steel in ragged or well-roughed rolls.

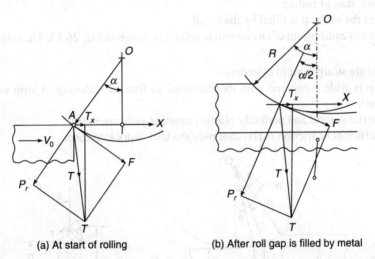

(a) At start of rolling   (b) After roll gap is filled by metal

Fig. 26.12 Biting of workpiece by rolls.

Ragging is the process of making certain fine grooves on the surface of the roll to increase the friction. In cold rolling the co-efficient of friction is usually of the order of 0.1 (due to the possibility of lubrication), but in hot rolling usually the sticking conditions exist.

Total reduction or *draft* taken in rolling

$$\Delta h = h_0 - h_1 = 2(R - R \cos \alpha) = D(1 - \cos \alpha)$$
$$l = R \sin \alpha$$

or

now
$$l = \sqrt{BC^2 - CE^2}$$
$$BC = \sqrt{R \cdot \Delta h} \text{ and } CE = R(1 - \cos \alpha) = 0.5 \Delta h$$
$$l \cong (R\Delta h)^{1/2}$$

The vertical component of P is known as the *rolling load*. This is the force with which the rolls press against the metal. Due to reaction, the metal tends to separate the rolls apart. Therefore, this force is also called the *separating force*.

The specific roll pressure

$$p = \frac{P}{\text{Contact area}} = \frac{P}{b \cdot l}$$

where b is the width of the sheet. Thus, the rolling load is given as $P = p \cdot l \cdot b$. The various methods to reduce the separating force are:

1. Smaller roll diameter (which reduces contact area);
2. Lower friction;
3. Higher workpiece temperature (even though friction will increase, p will be smaller);
4. Smaller bites, thereby reducing the contact area;
5. It is clear that the yield stress of the material in one direction is a function of the stress in the other principal directions. Therefore, if we apply tensile force to the workpiece in the horizontal direction, the compressive yield strength of the material in the vertical direction will be lower. Hence, the separating force will be smaller. Both "back tension" and "front tension" can be applied
    (a) At the start of rolling;
    (b) After the roll gap is filled by the metal.

Analysis: The stress equilibrium of an element in rolling is shown in Fig. 26.13. The following assumptions are made:

1. The rolls are straight, rigid cylinders.
2. The strip is wide compared with its thickness, so that no widening of strip occurs (plane-strain condition).
3. The material is rigid and perfectly plastic (constant yield strength).
4. The co-efficient of friction is constant over the tool–work interface.

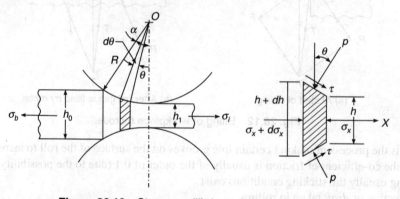

Figure 26.13 Stress equilibrium of element in rolling.

Considering the thickness of the element perpendicular to the plane of paper to be unity, we get the equilibrium equation in the x-direction as

$$-\sigma_x h + (\sigma_x + d\sigma_x)(h + dh) - 2PR d\theta \sin\theta + 2\tau_x R d\theta \cos\theta = 0$$

For sliding friction, $\tau_x = \mu p$. Simplifying and neglecting the second order terms, we get

$$\frac{d(\sigma_x h)}{d\theta} = 2pR(\sin\theta \pm \mu\cos\theta) \qquad (26.30)$$

The negative sign applies within the lagging zone and the positive sign applies within the leading zone, because the direction of the friction force changes at the neutral plane.

The mathematics for the solution of Eq. (26.30) is very complex, and various assumptions and approximations have to be made to get a tractable solution.

In rolling, the angle α is usually small; therefore $\sin\theta = \theta$ and $\cos\theta = 1$.
Thus, Eq. (26.30) reduces to

$$\frac{d(\sigma_x h)}{d\theta} = 2pR(\theta \pm \mu) \qquad (26.31)$$

Again, it is assumed that p and σ_x are principal stresses. Therefore, as the von Mises yield condition for plane strain states

$$p - \sigma_x = \frac{2}{\sqrt{3}}\sigma_0' \qquad (26.32)$$

where p is the greater of the two compressive stresses. In agreement with the literature on rolling, compressive stresses are taken as positive. In the general case, σ_0' varies along the deformation zone and depends upon the degree and rate of deformation of the metal. In cold rolling, it may be taken as the mean of the values at entry and exit.

With the help of Eq. (26.32), Eq. (26.31) can be written as

$$\frac{d}{d\theta}[h(p - \sigma_0)] = 2pR(\theta \pm \mu)$$

$$\frac{d}{d\theta}\left[\sigma_0 h\left(\frac{p}{\sigma_0} - 1\right)\right] = 2pR(\theta \pm \mu)$$

$$\therefore \quad \sigma_0 h \frac{d}{d\theta}\left(\frac{p}{\sigma_0}\right) + \left(\frac{p}{\sigma_0} - 1\right)\frac{d}{d\theta}(\sigma_0 h) = 2pR(\theta \pm \mu)$$

Now, as noted above, due to cold rolling, σ_0 increases as h decreases, thus making $\sigma_0 h$ nearly a constant and its derivative zero.

Dividing by ph, we get

$$\frac{d(p/\sigma_0)/d\theta}{p/\sigma_0} = \frac{2R}{h}(\theta \pm \mu) \qquad (26.33)$$

Now

$$h = h_1 + 2R(1 - \cos\theta)$$

or approximately

$$h = h_1 + R\theta^2$$

Equation (26.33) becomes

$$\frac{d(p/\sigma_0)/d\theta}{p/\sigma_0} = \frac{2R}{h_1 + R\theta^2}(\theta \pm \mu)$$

$$\frac{d(p/\sigma_0)}{p/\sigma_0} = \frac{2R}{h_1 + R\theta^2}(\theta \pm \mu)d\theta$$

Integrating this equation

$$\log_e(p/\sigma_0) = \int \frac{2R\theta d\theta}{h_1 + R\theta^2} \pm \int \frac{2R\mu}{h_1 + R\theta^2} \cdot d\theta \qquad (26.34)$$

Now the first term of the RHS is

$$\int \frac{2R\theta d\theta}{h_1 + R\theta^2} = \int \frac{2R\theta d\theta}{h}$$

$$= \int \frac{2\theta d\theta}{h/R}$$

Now $\dfrac{h}{R} = \dfrac{h_1}{R} + \theta^2$

$\dfrac{d}{d\theta}\left(\dfrac{h}{R}\right) = 2\theta$, because h and R are constant

Therefore, $\int \dfrac{2\theta d\theta}{h/R} = \log_e \dfrac{h}{R}$

The second term on the RHS is

$$\int \frac{2R\mu}{h_1 + R\theta^2} d\theta = \int \frac{2\mu}{h_1/R + \theta^2} d\theta$$

$$= 2\mu \sqrt{\frac{R}{h_1}} \tan^{-1}\left(\sqrt{\frac{R}{h_1}}\,\theta\right)$$

Equation (26.34) becomes

$$\log_e\left(\frac{p}{\sigma_0}\right) = \log_e\left(\frac{h}{R}\right) \pm 2\mu \sqrt{\frac{R}{h_1}} \tan^{-1}\left(\sqrt{\frac{R}{h_1}}\,\theta\right) + \log_e C$$

$$p = C\sigma_0 \left(\frac{h}{R}\right) e^{\mp \mu H} \quad (26.35)$$

where

$$H = 2\sqrt{\frac{R}{h_1}} \tan^{-1}\left(\sqrt{\frac{R}{h_1}}\,\theta\right) \quad (26.36)$$

Now at entry, $\theta = \alpha$,

Hence $H = H_0$ with θ replaced by α in Eq. (26.36).

Since $\theta = 0$ at exit

∴ $\quad H = H_1 = 0$

Also at entry and exit (free pounders) $\sigma_x = 0$

From Eq. (26.32), $p = \sigma_0'$

Therefore, in the lagging zone (entry zone)

$$\sigma_0' = C\sigma_0 \left(\frac{h_0}{R}\right) e^{-\mu H_0}$$

$$C = \left(\frac{R}{h_0}\right)e^{\mu H_0}$$

$$p = \sigma'_0 \frac{h_n}{h_0} e^{\mu(H_0-H)} \tag{26.37}$$

In the leading zone (exit zone), $C = \frac{R}{h_1}$.

Therefore,
$$p = \sigma'_0 \frac{h_n}{h_1} e^{\mu H} \tag{26.38}$$

The pressure rises from both entry and exit (where it is equal to σ'_0) to a maximum at an intermediate point, forming a friction hill, where the direction of the friction force changes. The neutral point which can be determined by using Eqs. (26.37) and (26.38) is given by

$$\frac{h_n}{h_0} \cdot e^{\mu(H_0-H_n)} = \frac{h_n}{h_1} e^{\mu H_n}; \quad \frac{h_0}{h_1} = e^{\mu(H_0-2H_n)}$$

$$H_n = \frac{1}{2}\left(H_0 - \frac{1}{\mu}\log_e \frac{h_0}{h_1}\right) \tag{26.39}$$

From Eq. (26.36) above

$$\theta_n = \sqrt{\frac{h_1}{R}} \tan\left(\sqrt{\frac{h_1}{R}} \cdot \frac{H_n}{2}\right) \tag{26.40}$$

or
$$h_n = h_1 + 2R(1 - \cos\theta_n)$$

As discussed above, the rolling load can be reduced by applying both "back tension" and "front tension", either individually or together. The back tension is applied by controlling the speed of the uncoiler relative to the roll speed, and the front tension may be created by controlling the coiler. The application of the back tension moves the neutral plane towards the exit. If its value is continuously increased, the neutral point will eventually reach the roll exit and the rolls will start slipping over the metal surface. The application of the front tension shifts the neutral plane towards the roll entry. The possible height reduction decreases with the back tension, since it will be difficult to roll the metal continuously. With the front tension, the maximum possible height reduction increases, since the pulling force increases. A study of the effect of sheet tensions has shown that the back tension is about twice as effective as the front tension in reducing the rolling load.

At entry
$$\sigma_x = -\sigma_b \text{ (back tension)}$$
$$\sigma_x = -\sigma_f \text{ (front tension)}$$

Equations (26.37) and (26.38) will become

$$p = (\sigma_0 - \sigma_b)\frac{h}{h_0}e^{\mu(H_0 - H)} \tag{26.41}$$

$$p = (\sigma_0 - \sigma_f)\frac{h}{h_1}e^{\mu H} \tag{26.42}$$

Therefore, for the same Δh, if R is less, l will decrease and so load will decrease, since the contact area is reduced. As the thickness of the sheet to be rolled goes on decreasing, the roll diameters go on decreasing. The 'spread' which is defined as

$$\Delta b = (b_1 - b_0)$$

increases with an increase in the roll diameter and the co-efficient of friction and a fall in the temperature of the metal during hot rolling. The forward slip, which is about 3 to 10%, increases with increased radius of the roll and the co-efficient of friction, and with a decrease in the thickness of the bar. The above analysis is only an approximate one and it tends to correlate better with cold-working, low-friction processes.

Maximum draft: It has already been proved that if the strip is to enter the rolls unaided, then the following relation has to be satisfied between the angle of bite and the co-efficient of friction between the roll and the material surfaces:

$$\mu > \tan \alpha$$

$$l = \sqrt{R \cdot \Delta h}$$

$$\tan \alpha = \frac{l}{R - \dfrac{\Delta h}{R}} = \frac{\sqrt{R\,\Delta h}}{R - 0.5\,\Delta h}$$

Since $R >> 0.5\,\Delta h$, it can be written that

$$\tan \alpha \approx \sqrt{\frac{\Delta h}{R}}$$

The maximum draft is obtained when

$$\mu \geq \sqrt{\frac{\Delta h}{R}}$$

$$(\Delta h)_{max} = \mu^2 R \qquad (26.43)$$

It is clear that large rolls and high friction allow heavy draft. It is however possible to roll with greater draft if the strip is pushed into the rolls or is accelerated prior to biting, in order to make use of the force of inertia. If the friction is too high, the load becomes excessive. Again, a compromise can be obtained with the help of theory.

Also, after the gap between the rolls is filled with metal, the rolling may proceed with an angle of tangency α greater than β. The limit for both these conditions may be found by considering the forces on the rolls. Let us assume that the resultant of the pressure force P_r is located in the middle of the arc of contact of the strip with the roll. The rolls will start slipping when the horizontal component of the average roll pressure P_r equals that of the surface friction. For small angles, the condition for no slipping of rolls or for continuous rolling will be

$$F \cos \frac{\alpha}{2} > P_r \sin \frac{\alpha}{2}$$

$$\frac{F}{P_r} > \tan \frac{\alpha}{2}$$

or

$$\tan \frac{\alpha}{2} = \mu$$

or

$$\alpha \approx 2 \tan^{-1} \mu$$

or

$$\alpha \approx 2\beta$$

It is thus clear that the angle of bite can be increased by a factor of 2 after the strip enters the rolls. This angle is called the *angle of nip*. Thus, the angle of nip is twice the angle of bite. This angle is of importance when very large drafts are required. The rolling can then be started by tapering the front end of the strip.

$$P = R \cdot b \int_0^a p \cdot d\theta$$

This is best evaluated by graphical integration.

Note. Here the flattening of the rolls and the effect of inclination have been neglected.

Roll torque and power: The total rolling load is distributed over the arc of contact. However, it can be assumed to be concentrated at a point along the arc of contact at a distance a from the centre line of the rolls. So, the torque on each roll is

$$T = p \cdot a$$

The moment arm a is generally expressed through the arm factor $\lambda = \dfrac{a}{l}$.

The values of λ can be determined experimentally. Its values can be taken as

$\lambda = 0.43$ for cold rolling with matte finished rolls
$\lambda = 0.48$ for cold rolling with a smooth surface

These values of λ neglect the effect of the elastic deformation of the rolls. Considering this, however, the value of λ can be taken as 0.3 to 0.4.

For hot rolling, for $l/h_m \geq 1.5$, its value range can be taken as 0.35 to 0.45 (approximately), with the values increasing with temperature. For $l/h_m < 1.5$, λ depends heavily on the l/h_m ratio, and its value should be taken from the appropriate curve. Hence

$$h_m = \text{Mean height} = 0.5(h_0 + h_1)$$

Assuming that the power consumed is equal on both rolls

$$\text{Power} = \frac{2T\omega}{100} \text{ in kW}$$

where T is in Nm and ω in rad/s.

Power loss bearings

The friction in the bearings supporting the rolls obviously causes some power loss. An exact analysis of the power loss in the bearings is too complicated. However, to estimate the power requirement of the rolling mill, it is sufficient to assume that the power loss in each bearing is given by

$$p_b = \frac{1}{2} \mu_b F_b d_b \omega$$

where μ_b is the coefficient of friction in the bearing (whose typical value is in the range of 0.002–0.01).

REVIEW QUESTIONS

1. An aluminium disc with 200 mm diameter and 25 mm thickness is forged to a final thickness of 15 mm. Estimate the maximum forging force when the coefficient of friction = 0.3 and tensile yield stress = 25 N/mm². Neglect strain hardening.
2. A cylindrical lead alloy billet of 50 mm diameter and 100 mm length is extruded to a final diameter of 25 mm by using a direct extrusion process. The average tensile yield stress for the alloy is 12 N/mm². Estimate the maximum force required and the fraction of the total power lost in friction for this operation.

3. If the operation in Exercise 2 is performed by a backward extrusion process, determine the maximum force required. Make suitable assumptions to first derive the expression for the force.
4. An aluminum cup of inside radius 40 mm and thickness 5 mm is to be drawn from a blank with a diameter of 50 mm. The shear yield stess and the maximum allowable stress in the material can be taken respectively as 14 N/mm² and 80 N/mm² respectively. Determine (i) the drawing force and (ii) the minimum possible radius of the cup which can be drawn from the given blank without causing a fracture. It is given that $\mu = 0.1$ and $\beta = 0.05$.
5. Circular aluminium blanks with 25 mm diameter and 3 mm thickness are to be produced as the starting material for toothpaste tube extrusion. The press has a capacity of 500 kN. The fracture strain and stress of the material are 2 and 80 N/mm² respectively. No shear is provided to the punch. Determine (i) the optimum clearance between the die and the punch, (ii) the maximum number of blanks which can be punched simultaneously per stroke, and (ii) the power required if the punch speed is 60 cycles/min.
6. A charge of TNT weighing 10 N is used in an unconfined explosive forming operation. Plot the peak pressure over the work surface with the stand-off distance.
7. When extruding 40 mm diameter billets through a die with a 28 mm opening and 45° half-cone angle, the centre-burst defect is found to be very frequent. Suggest a remedy along with its justification.
8. A wide strip is rolled to a final thickness of 0.5 cm with a reduction of 30%. The material of the strip has a curve of $\sigma = 7500\ \varepsilon^{0.4}$. The roll radius is 50 cm and the coefficient of friction is 0.2. Determine the total separating force per unit width of the strip.

Solution We have

Initial thickness of the strip
$$h v = 0.5 \times \frac{100}{100 - 30}$$
$$= 0.714 \text{ cm}$$

Change in thickness of strip = 0.214 cm

Contact angle
$$\alpha_2 = \cos^{-1}\left[\frac{50 - \left(\frac{0.214}{2}\right)}{50}\right] = \cos^{-1} 0.9979$$
$$= 4°12' \approx 0.0733 \text{ radians}$$

Now
$$K = \frac{2\mu}{\sqrt{h_f/R_0}} = \frac{2 \times 0.2}{0.1} = 4 \quad \text{and} \quad \tan \alpha_2 = 0.733$$

The determination of w at the neutral point (say, $W = W_n$) is made simply by equating Eqs. (26.32) and (26.27) (with the condition that $\sigma_{xb} = 0$, $\sigma_{xf} = 0$, and $A_0\varphi_0 = 0$) and hence the value of the neutral angle may be obtained. The pressure p in the entry and exit zones may be calculated using Eqs. (26.37) and (26.38) respectively. Divide the contact length into segments of 0.01 radian each. For each α, determine h, w and σ_0. Present a summary of the calculations in the tabular form and show that the value of the average contact area between the roll and the strip is simply $R_0\alpha_2$. The total separating force is $p_{avg}\ R_0\alpha_2$.

9. List and explain the methods you would use to reduce the separating force in strip rolling.
10. If the maximum angle of bite in rolling is given by the expression $\alpha = \tan^{-1}\mu$, show that the maximum draft, i.e. by $h_v - h_f$ is given by $\Delta_{max} = \mu^2 R_0$.
11. (i) What would be the approximate roll load necessary to reduce a 180 cm wide and 0.25 cm thick aluminium slab to 0.2 cm thickness in one pass, using 35 cm diameter rolls?

(ii) If the pass were increased from a draft of 0.05 cm to 0.075 cm, what influence would there be on the load?

Solution We have

(i) Strain imparted $\quad \varepsilon_1 = \ln \dfrac{0.25}{0.2} = 0.223$

Approximate $\quad \sigma_0 = 9.1$ kg/mm^2

Length of contact $\quad L = \sqrt{R_0 dh} = \sqrt{350 \times 0.5} = 13.21$

Load per unit width $\quad \dfrac{P}{b} = 9.1 \times 13.21 = 120$ kg/mm

Therefore $\quad P = 120 \times 180 \times 10 = 216000$ kg $= 260$ t

(ii) Strain imparted $\quad \varepsilon_1 = \ln \dfrac{0.25}{0.175} = 0.257$

Approximate $\quad \sigma_0 = 10.0$ kg/mm^2

Length of contact $\quad L = \sqrt{R_0 dh} = 16.2$

Load per unit width $\quad \dfrac{P}{b} = 10 \times 16.2 = 162$ kg/mm

$$P = 162 \times 180 \times 10 = 291600 \text{ kg}$$

Allowing 20% for friction

$$P = 349920 \text{ kg} \times 350 \text{ t}$$

12. (i) A 0.25 cm thick annealed mild steel strip is rolled to 0.2 cm in one pass and then to 0.15 cm. What is the mean yield stress in the second pass? What error would be involved in assuming the yield stress constant at its final value?
    (ii) If the strip is further rolled to 0.125 cm, what would be the mean yield stress for this pass, and by how much would it differ from the final yield stress?
13. In strip rolling, discuss if it would be possible for the neutral point to be towards the entry rather than the exit zone.
14. Deduce an expression for the normal roll pressure in the case of cold rolling of a strip, with no front and back tension. Enumerate the various assumptions made therein.
15. Explain the state of stresses in various parts of the workpiece for a deep drawing process.
16. In deep drawing of a cylindrical cup, is it necessary that there should always be tensile circumferential stresses in the formed cup? Explain.
17. How does the entropy of materials help in obtaining deeper draws in the deep drawing operations?
18. Obtain the expression for the radial drawing stress in a deep drawing operation, making suitable assumptions.
19. Discuss the importance of *drawing ratio*.
20. Sketch the cup drawing operation of a circular blank and explain the processes occurring in the different zones of the blank.
21. Solid discs must be forged from the *b/h* value range of 20 to 30. A viscous lubricant which will cause the coefficient of friction to be 0.06 and a solid foil lubricant which will cause a constant shear factor of 0.2 are available. Find which method is more effective in reducing forging force.

22. A disc is formed with pointed platens as shown in Fig. 26.14. The angle α is very small, and therefore one can assume that $\sin \alpha \approx \tan \alpha \approx \alpha \cos \alpha \approx 1$, $2r\alpha \approx 2b\alpha \approx h$, and the von Mises yield criterion is $p = \sigma_0 - \sigma_{rr}$.
   (i) Show the free body equilibrium of forces for sliding;
   (ii) Solve the resulting differential equation by putting $\sin \alpha \approx \tan \alpha \approx \mu$;
   (iii) What is the advantage of pointed platens? Discuss.

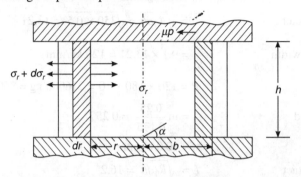

Fig. 26.14 Exercise 22.

23. For which of the following circular discs will the average pressure be higher?
   (i) $b = 5.0$ cm, $h = 1.2$ cm, $\mu = 0.04$;
   (ii) $b = 5.0$ cm, $h = 0.6$ cm, $\mu = 0.02$.

24. Approximately what forging load would be required to transform a 1 m long, 1 m diameter cylindrical bloom into a square section of equal area in a hydraulic press? $\sigma_0 = 450$ kg/cm².
   Solution The cross-sectional area of the bloom is $\pi(100/2)$ cm². So the squares would have sides equal to $50\sqrt{\pi} = 88.5$ cm.
   Assuming plane strain and full ticking friction ($m = 1$), from the equation
   $$\left(\frac{p}{\sigma_0}\right)_{max} = 1 + \frac{mb}{h} \quad \text{and} \quad \left(\frac{p}{\sigma_0}\right)_{avg} = 1 + \frac{mb}{2h}$$
   Now, $2b = 88.5$ cm and $h = 88.5$ cm.
   $$\sigma_0' = 2\sqrt{3}\,\sigma_0 = 1.155\,\sigma_0$$
   assuming von Mises yield criteria when a multiplier 1.155 enters the calculation. Therefore
   $$p_{avg} = 1.155 \times 450\,(1 + 0.25) = 650 \text{ kg/cm}^2$$
   The area of platen contact is 100×88.5; so the forging load
   $$= 8850 \times 650 = 5752500 \text{ kg} = 5752.5 \text{ t}.$$

25. Evaluate the press capacity necessary for forging a 1 m long cylindrical bloom to a hexagonal section with approximately 0.33 m side, if the yield stress is initially 400 kg/cm² but increases to 680 kg/cm² at the end of the operation. Assume (i) that the bloom is partially lubricated, so that $\mu = 0.3$ and (ii) there is no lubrication. What maximum pressures would be expected?

Solution A reasonable approximation is to assume that the load is equal to that necessary to cause yielding in rectangular billet with $2b = 33$ cm and $h = 66 \sin 60°$, under plane strain conditions. Therefore

$$\left(\frac{p}{\sigma_0}\right)_{max} = e^{2\mu b/h}$$

or

$$\left(\frac{p}{\sigma_0}\right)_{max} = 1.19$$

$$p_{max} = 680 \times 1.19 = 809.20 \text{ kg/cm}^2$$

(i) Again, the above equation may be written in approximate form (if μ is low) as

$$\left(\frac{p}{\sigma_0}\right)_{max} = \left(1 + \frac{2\mu b}{h}\right)$$

The mean pressure p on the platens is then

$$\bar{p} \approx \sigma_0 \left(1 + \frac{1}{2}\frac{2\mu b}{h}\right) = 680 \times 1.087 \times 740 \text{ kg/cm}^2$$

The contact area of the platen is 33×100 cm^2.
So the force is $740 \times 3300 = 2442000$ kg $= 2442$ t.

(ii) When there is no lubrication, there will be full sticking friction and $m = 1$. Hence

$$\left(\frac{p}{\sigma_0}\right)_{max} = 1 + \frac{mb}{h} = 1.29$$

$$p_{max} = 680 \times 1.29 = 877.2 \text{ kg/cm}^2$$

The mean pressure $\bar{p}$ on the platen is

$$\bar{p} = \sigma_0 \left(1 + \frac{mb}{2h}\right) = 680 \times 1.145 = 778.6 \text{ kg/cm}^2$$

Force = $778.6 \times 3300 = 2569380$ kg $\times 2569.38$ t.

26. Explain why there might be an increase in the density of a forged product.

CHAPTER 27

Press Tools and Their Design

27.1 INTRODUCTION

Press work has a very important position as a manufacturing process. The press is a metal forming machine tool designed to shape or cut metal by applying mechanical force or pressure. Power presses are used for producing large quantities of articles quickly, accurately and cheaply from the cold working of mild and other ductile materials.

27.2 ADVANTAGES

The following are the main advantages of press work:

Mass production: The presses are exclusively intended for mass production work.

Fast operation: Press work represents the fastest and most efficient way to form a sheet metal into finished products.

Accuracy: Articles produced by this process are quite accurate.

Economy: The components produced are in large quantity; as such the advantages of "scale of operation" result in economical operations.

Wide scope: The components produced range over an extremely wide field and are used throughout industry.

Ease in operation: Modern presses are designed for ease and speed in handling work into and out of them.

Low maintenance cost: The maintenance cost is very low and it is very easy to maintain, as there exists easy accessibility to operating mechanism, safety guards, as well as convenience in substituting repair parts.

Efficient lubrication: In press tools, the provision of efficient lubrication systems is made, resulting in higher efficiency.

Longer life: Press tools generally have longer life.

Less supervision: It requires very little supervision during operation.

27.3 TYPES OF PRESSES

Presses are classified on the basis of

    (i) Source of power   (ii) Design of frame
    (iii) Method of actuation of ram   (iv) Number of slides
    (v) Intended use.

These are described below.

27.3.1 Classification Based on Source of Power

This classification includes

    (a) Manual power or fly press
    (b) Power press of the following kinds:
        (i) Mechanical
        (ii) Hydraulic
        (iii) Pneumatic.

27.3.2 Classification Based on Design of Frame

This includes

    (a) Gap press   (b) Bench type press
    (c) Inclinable press   (d) Adjustable bed press
    (e) Horn press   (f) Straight side press
    (g) Pillar press   (h) Arch press.

27.3.3 Classification Based on Actuation of Ram

This consists of

    (a) Crank and connecting rod drive   (b) Eccentric drive
    (c) Friction disc drive   (d) Power screw drive
    (e) Cam and follower drive   (f) Rack and pinion drive
    (g) Knuckle joint drive   (h) Hydraulic or pneumatic drive.

27.3.4 Classification Based on Number of Slides

This includes

    (a) Single slide
    (b) Double slide.

27.3.5 Classification Based on Intended Use

This comprises

    (a) Shearing
    (b) Blanking

Presses based on power

Manual power or fly press: The fly press working details are shown in Fig. 27.1. The frame of the machine is a rigid C casting which carries all loads of shearing and pressing through the ram;

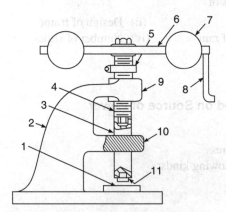

1. Die, 2. Frame, 3. Ram, 4. Screw, 5. Arrestor,
6. Arm, 7. Cast iron balls, 8. Handle, 9. Nut,
10. Guide, 11. Punch

Fig. 27.1 Fly press.

The typical shape of the frame leaves the front open and easy for the operators. The screw of the press operates through a nut in the top part of the frame. The two heavy walls on the left and right rotate and store energy, which is expanded while shearing or pressing. The handle on which the walls are mounted rotates in turn and moves the ram through the nut up and down, which is the required movement of the ram. The die is fitted in the lower ram of the C frame. Thus the punch and the die constitute the press tool. The sheet metal to be worked is placed on the die below the ram, and cast iron balls are rotated with a certain speed to store kinetic energy which is expanded during the operation performed on the sheet metal. A properly designed press of this type can be used for rapid work for small items, covering almost all operations that can be done on a power press.

Power press: Power press is a magnified version of the fly press and a power drive. Such type of press is designed for mechanical, hydraulic and pneumatic powers, and for the various mechanisms as listed above for transmitting power to the ram. In this press, the rotary motion obtained from the engine or motor is converted to reciprocating motion of the ram by any one of the mechanisms listed in Section 27.3.3. Hydraulic presses are slow and smooth in action but powerful and have small strokes. On the other hand, pneumatic presses are weak with long working strokes; the fluid is pumped to drive the ram or the piston, which is achieved by the air under pressure. The reciprocating movement of the piston is achieved by the exit of the fluid under pressure. In Fig. 27.2 the power press is shown; the driving mechanism is a crank and connecting rod, and a flywheel is used to minimise fluctuations in the operation of the press.

PRESS TOOLS AND THEIR DESIGN **717**

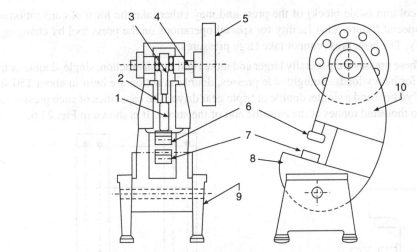

1. Ram, 2. Slide, 3. Driving mechanism, 4. Shaft, 5. Fly wheel,
6. Punch, 7. Die, 8. Bolster plate, 9. Base, 10. Frame

Fig. 27.2 Power press.

27.4 DESIGN OF PRESS FRAME

Gap press: The two presses described in the preceding section are similar to each other except for the power aspect; so, as the gap type press is adapted to all practical processes having a motor drive, it is used for light operations generally. Press tools of this class are generally used for making buttons, jewellery works, toys, etc. The capacity ranges from half a tonne to two tonnes. Roll feed magazine, sliding die, transfer feed and automatic feed may be employed to increase the production.

Bench type press: This is similar to the gap type press, except for having manual drive.

Inclinable Press: The inclinable press shown in Fig. 27.3 is extremely flexible in application, simple in construction and easy to operate. They are most commonly used in all industries for a variety of jobs. The principal identifying characteristic of such type of press is its ability to tilt back on its base, permitting the finished stampings to drop under gravity from the die, thus eliminating handling of the product.

Adjustable bed press: The adjustable bed press shown in Fig. 27.4 is incorporated with a screw, below the machine table bed on which the dies are mounted so as to facilitate lowering or raising of the bed. This facility enables the handling of various jobs of different sizes, but due to this facility it loses its rigidity.

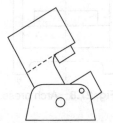

Fig. 27.3 Inclinable press.

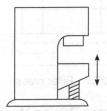

Fig. 27.4 Adjustable bed press.

Horn press: The horn press shown in Fig. 27.5 is identified by a detachable support—the horn. In all other types of pressure die, the support is integral with the frame as a fixed and rigid part of the press tool. The horn is bolted to the vertical column (slide block) of the press and may either take the form of conventional die support or feature as a special fixture. This facility for special operations on the press tool by changing the horn weakens the rigidity. The horn press cannot take large pressures.

Straight side press: These presses are generally larger and more rigid in construction, single, double or triple in action, and designed for heavy loads. Straight side presses, double in action are built in about 150 sizes. They may have a plane flywheel and a single, double or triple gear drive. The capacities of such presses range from a few tonnes to two thousand tonnes at the extreme end of the stroke. It is shown in Fig. 27.6.

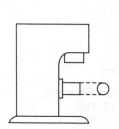

Fig. 27.5 Horn press.

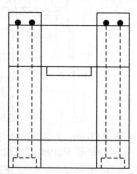

Fig. 27.6 Straight side press.

Pillar press: It is shown in Fig. 27.7. This press is similar in construction to a radial drilling machine and is available for various tonnage capacities.

Arch press: This is one of the smaller straight side presses. Its utility lies in the area of its bed and slide (large in proportion to its size). It has the advantage of straight sides and open front. Hence the work must be fed from the front to the back. It is slow in action and is used for blanking, trimming, bending, kitchenware making and embossing. Refer to Fig. 27.8.

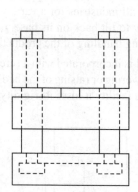

Fig. 27.7 Pillar press.

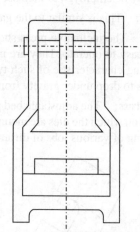

Fig. 27.8 Arch press.

27.4.1 Power Press Driving Mechanism

Crank and connecting rod: This mechanism has been illustrated in Fig. 27.9(a) to (c). It has its cutting stroke at the maximum velocity point, and it is used for large works; the length of the stroke can be changed in this system. Figure 27.9(d) shows a multiple reduction gear drive.

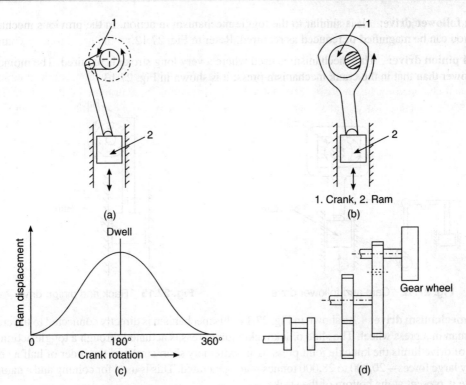

Fig. 27.9 (a) to (c) Crank and connecting rod and (d) Multiple reduction gear drive.

Eccentric drive: This drive mechanism is used for smaller strokes but heavier jobs. It has huge rigidity and has its cutting action at the middle of the strokes.

Friction and power drive: In this the slide is accelerated by means of a pair of friction discs engaging a flywheel at the bottom of the stroke; the entire energy is expended in the work (refer to Fig. 27.10).

Toggle mechanism: This is illustrated in Fig. 27.11 and is generally used for drawing operations. The principal aim for this mechanism is to obtain motion having a suitable dwell period, so that the blank may be effectively held. The dwell period is the period when some parts remain stationary.

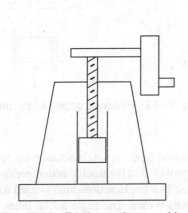

Fig. 27.10 Friction and power drive.

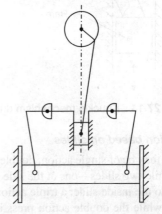

Fig. 27.11 Toggle mechanism.

Cam and follower drive: It is similar to the toggle mechanism in action. In the previous mechanism the dwell period can be magnified or reduced as required. Refer to Fig. 27.12.

Rack and pinion drive: This mechanism is used where a very long stroke is required. The moment of the slide is slower than that in the crank mechanism press; it is shown in Fig. 27.13.

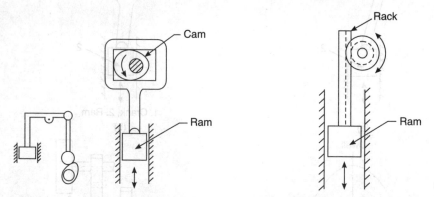

Fig. 27.12 Cam and follower drive.

Fig. 27.13 Rack and pinion drive.

Knuckle mechanism drive: As shown in Fig. 27.14, this mechanism is directly connected to the crank shaft by the pitman in a press wheel. The slide of a knuckle joint press is actuated through a toggle action linkage. This type of drive limits the knuckle joint press to an extremely short stroke of the order of half a centimetre or so. Very large forces—20,000 to 28,000 tonnes—are generated. This is used for coining and a natural dwell of the slide is present at the bottom of the stroke.

Hydraulic or pneumatic drive: It is shown in Fig. 27.15. This drive is used as an intensifier and generates very large pressures. This press is used for forming and drawing operations.

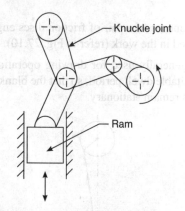

Fig. 27.14 Knuckle mechanism drive.

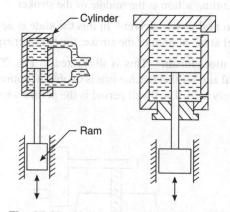

Fig. 27.15 Hydraulic or pneumatic drive.

Classification based on slides

The presses may be of single action, double action and triple action types. A single action press has one slide (or ram), while two slides—one at the side of the other—are provided in the double action press. The punch is attached to the inside slide; a triple action press has 3 slides. The single action press is used for stamping operations, while the double action press is used for combination dies. The triple action press is used for combination, embossing, blanking, and drawing operations.

Classification based on use

A press may be useful for punching, blanking, drawing, forming, coining, and embossing purposes, and the press may be named accordingly.

Transfer presses

These are high production presses and are suitable for light works, though they may also be designed for medium and heavy work.

The transfer press produces finished products from a single piece of stock, which is transferred from one station to another by means of transfer devices such as push bars, fingers and gripping mechanisms. It is often used for deep drawing operations. To avoid mechanical feeding, arrangements are used. Dangerous areas are railed around to avoid accidents.

Components of press

The major components of a press are as follows:

Bed: It is generally of rectangular shape and is a part of the frame. It is generally open in the centre to allow the scrap or blank to fall down. It also supports the bolster plate.

Bolster plate: It mounts press tools and accessories and is of sufficient thickness. It is generally fixed permanently on the bed and is never changed.

Ram (Slide): It moves through its fixed stroke. The position of the ram can be changed, but its stroke is generally fixed.

Knockout: It is a mechanism operating on the upstroke of a press, which ejects workpieces or blanks from the press tool.

Cushion: It is a press accessory located beneath or within the bolster for producing an upward motion of force and is actuated by air, oil or springs, or a combination thereof.

Flywheel: The flywheel is mounted at the end of the driving shaft and is connected through a clutch. The flywheel is directly coupled with the electric motor.

Clutch: It is used for connecting and disconnecting the driving shaft with the flywheel when it is necessary to start or stop the movement of the ram.

Brakes: These are used to stop the movement of the driving shaft immediately after it is disconnected from the flywheel.

27.5 METHODS OF PUNCH SUPPORT

The punches are usually held in steel punch plates of the punch holder which is clamped to the lower end of the ram. The various methods of securing punches in the punch plate are illustrated in Fig. 27.16.

27.6 METHODS OF DIE SUPPORT

The die is usually held in the die holder which is again clamped to the bolster plate mounted on the table. The different methods of securing the die blocks to the die holders are illustrated in Fig. 27.17.

View A: The die block is secured to the die holder by four set screws (only one is shown) and is located by dowels.

View B: The die block is secured by set screws from the bottom of the holder.

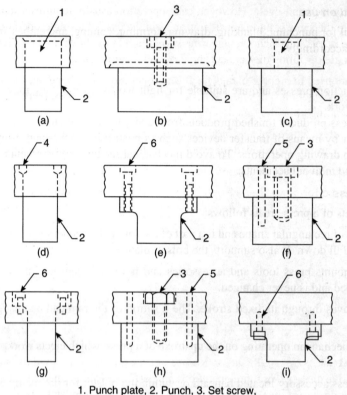

1. Punch plate, 2. Punch, 3. Set screw,
4. Grub screw, 5. Dowel pin, 6. Set screw

Fig. 27.16 Methods of securing punch holders.

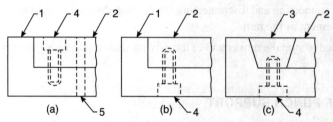

1. Die holder, 2. Die block, 3. Wedge, 4. Set screw, 5. Dowel

Fig. 27.17 Methods of securing die blocks.

Classification of dies

Basically, dies are of the following two types:

Single operation dies: The dies which are designed to perform only one operation with each stroke of the ram. They are of single design.

Multioperation dies: If several operations are completed in one stroke of the ram (or rams), then these are called *multi operation dies*. Multioperation dies are often used to bring two or more cutting and forming operations together into one work cycle. However, each operation usually requires a separate punch and die unit, which is subject to the same general consideration as in the case of a single-operation die that performs the same function. The broad classification under the two groups is as shown in the following lines.

The principal single operation dies are cutting dies and forming dies.

Cutting dies. They are used to prepare blanks for further operations like bending, forming and drawing by utilising cutting or shearing action. They are further classified as blanking dies, piercing dies, trimming dies, shaving dies, etc., depending upon the function performed.

Forming dies. These dies change the appearance of the blank without removing any stock. They add third dimension detail. In this group are included the bending, curling, forming and drawing dies.

The principal types of multioperation dies are compound dies, combination dies and progressive dies.

Compound dies. They perform two or more cutting operations such as blanking and piercing at one station of the press in every stroke of the ram. They are usually single action dies where all the operations are completed with one ram stroke at the same station.

Combination dies. They combine cutting with forming or drawing operation at one station of the press in every stroke of the ram. They are usually multiple action dies with one operation succeeding another. This is achieved by designing the die for use on a double action press which has two independent rams or slides, one moving on one side of another.

Progressive dies. The workpiece in progressive dies travels from one station to another with separate operations being performed at each station. Usually the workpiece is relative in the stroke until it reaches the final station which cuts off the finished piece. All stations work simultaneously but at different points along the work strip which advances one station at each stroke of the ram. These are used for blanking and piercing operations. In progressive dies, two or more operations are performed simultaneously at a single stroke of the ram by mounting separate sets of dies and punches at two or more different stations.

27.7 PRESS CAPACITIES

27.7.1 Tonnage Capacity

In rating presses, the term *tonnage* indicates the amount of pressure in tonnes that a press exerts on the work. There are four steps, rather separate and particularly important, in determining the capacity of a press for doing work. These are:

1. The strength of the body including side, pitman and other major parts;
2. The torsional strength of the shaft and clutch pans;
3. The amount of kinetic energy that can be stored in the flywheel system;
4. The horse power of the motor.

27.7.2 Catalogue Tonnage Rating

The conventional tonnage rating, as usually given in the manufacturers' catalogues, is determined by the strength of the press and is applicable only near the bottom of the stroke. It is the only limitation that is usually considered in a casual discussion of pan capacity. It has been customary to associate this rating with the shaft diameter at the main bearings to a "thumb rule" expressed by the equation

$$P = CD^2$$

where P is the tonnage capacity near the bottom of the stroke;
D is the shaft diameter in mm;

C is a constant (appropriate to the type and stroke of the shaft and available from the standard tables). In power press, the tonnage of the press is limited by a number of factors.

Tonnage limited by torsion

If a press load is applied at some appreciable distance above the bottom of the stroke, torque considerations limit the press capacity. This limitation depends on the shaft diameter, stroke and distance above the bottom, where the load is applied. The approximate tonnage for various working strokes can be determined from any press tool design monograph.

Tonnage limited by flywheel capacity

The energy expended on the work at each stroke of the press equals the tonnage required multiplied by the distance through which this tonnage must act. This energy is available from that stored in the flywheel and is given to the work as the flywheel slows down. This consideration is a particularly important factor in a continuously running press. This energy may be calculated by the equation

$$E_{max} \propto N^2 D^2 W$$

where E is the energy based on a radius of gyration of 0.375 D approximately;
N is the revolution per minute of the flywheel;
W is the weight of the flywheel;
D is the flywheel diameter.

If the tonnage is assumed as supplied uniformly from the point of contact with the work, the tonnage rating is limited by the flywheel energy T_f and can be expressed by the equation

$$T_f = \frac{E}{N}$$

where E is taken from the main equation.

Tonnage limited by motor

The motor must, during the non-working portion of the stroke, restore energy given up by the stroke. The tonnage T_m is limited by the motor and may be calculated by using the equation

$$T_m = \frac{150\ HP}{H \times Spm}$$

where HP is the horsepower of the motor;
Spm is the work stored per minute;
H is the distance above the bottom of the stroke whose work is indicated in the equation, and is based on the assumption that the tonnage capacity applies throughout the distance H and that 25% of the power will be lost due to friction.

Tonnage limited for safety

It is advisable to maintain a wide margin of safety between the calculated tonnage required for any job and the tonnage capacity of the press, because of the tendency of certain materials to work-harden and partly because of the conditions inherent in most dies. The tonnage requirements for cutting, shearing, piercing and blanking are customarily figured out on the basis of shaft dies with correct clearances. As the dies wear or get dull, the requirements may become double the original ones. This is particularly true when materials 0.500 mm thick or less are worked; drawing and forming operations are affected by die clearances, and worn or mismatched dies may make a noticeable difference in pressure requirements. Press failure or breakage is frequently the result of using dull dies.

27.7.3 Rating of Hydraulic Presses

The capacity of a hydraulic press is determined by the strength of the press body, the unit including the slide, the diameter and stroke of the hydraulic piston, and the hydraulic pressure.

27.8 PRESS OPERATIONS

There are a number of operations that can be performed on various presses. Some of the operations that can be performed on presses are described in the following sections.

The two major types of operations that can be done on presses are:
- (a) Cutting operations;
- (b) Non-cutting or non-shearing operations.

27.8.1 Cutting Operations

The following are the cutting operations:
- (i) Blanking
- (ii) Piercing
- (iii) Trimming
- (iv) Perforating
- (v) Notching
- (vi) Lancing
- (vii) Shaving
- (viii) Dinking.

Blanking: The term *blanking* indicates the operation of cutting flat shapes called *blanks*, which ordinarily can be further processed to give the desired shapes. This is illustrated in Fig. 27.18.

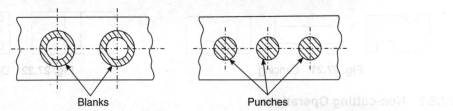

Fig. 27.18 Blanking.

Piercing: Holes of small sizes are pierced. When two or more piercing punches are employed together, their lengths should differ slightly in order to reduce the force and the impact at a time. This process is shown in Fig. 27.19.

Trimming: This operation is a finishing operation on completed jobs. Unwanted material projecting out from drawn or formed items is taken out in this process. Generally all operations are followed by a trimming operation.

Perforating: This is the production of a very large number of small holes on a sheet metal object.

Notching: Notching is the cutting of or taking out of small indentations from the edges of the workpieces. The edge of the sheet metal or of any other workpiece forms a portion of the periphery of the shape which has been punched out. The notching may have any desired shape. This is shown in Fig. 27.20.

Lancing: It is the combination of cutting and bending operations. This process is similar to some extent to the piercing operation which leads to the formation of either a slit in the metal sheet or an actual hole. The purpose of lancing is to permit the adjacent metal to flow more readily in subsequent drawing operations. This process is shown in Fig. 27.21.

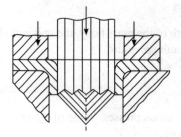

Fig. 27.19 Piercing.

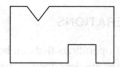

Fig. 27.20 Notching.

Shaving: When some change in the dimensions is necessitated due to some urgent requirements, e.g., when we want to reduce the blank size from 100 mm to 95 nm, then we use this process. A slight amount of metal, say, 10% of the metal dimensions, is removed or shaved from the blanked or pierced object. In this process we obtain closer dimensional tolerances with the required surface smoothness.

Dinking: It is a process generally used with paper, leather, plastics, etc. It is a modified shearing operation which is used to plank out shapes from low strength materials, primarily rubber, fibre and other materials as listed above. The punch may either be operated with hammer or mallet, or may be attached to the ram of a fly press. This is shown in Fig. 27.22.

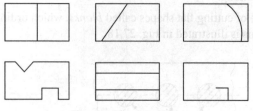

Fig. 27.21 Lancing.

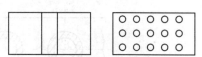

Fig. 27.22 Dinking.

27.8.2 Non-cutting Operations

The characteristic of non-cutting operations is that the plastic flow of the material will change the thickness of the object while no cutting action will take place. The following are the non-cutting operations:

- (i) Bending
- (ii) Forming
- (iii) Embossing
- (iv) Extrusing
- (v) Drawing
- (vi) Coining
- (vii) Curling
- (viii) Flattening or planishing.

Bending: In this process, plastic deformation takes place about a linear axis with or without change in the cross-sectional area. Successful bending depends on several factors, e.g. a minimum radius that can be bent without cracking, location of neutral axis, and the length that can be bent successfully. The process is illustrated in Fig. 27.23.

Forming: When two or more bends are made simultaneously with the use of a die, the process is generally referred to as forming.

Embossing: In this operation, we obtain impressions of desired size as shown in Fig. 27.24. The metal in this operation is stretched rather than compressed. It is used for stiffening of the bottom of pans or containers. Embossing boxes or big containers by cross ribs help make the section of a blank stronger. The embossing die may have engraving, or both the die and the punch may have impressions which are reproduced on the sheet metal by squeezing.

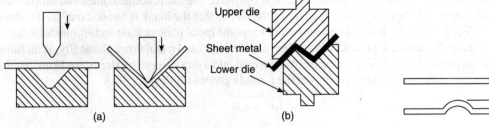

Fig. 27.23 (a) Bending and (b) angle bending.

Fig. 27.24 Embossing.

Extrusion: This is a perfect plastic flow operation which is covered in detail in the discussion of the forging operation. Therefore, it is not described in detail over here.

Drawing: When a metal blank is drawn into a die to obtain a cup-shaped pan, the flow of metal is in the plane parallel to the die face, such that the thickness and the surface area of the sheet remains about the same. The process of drawing is shown in Fig. 27.25(a) to (d).

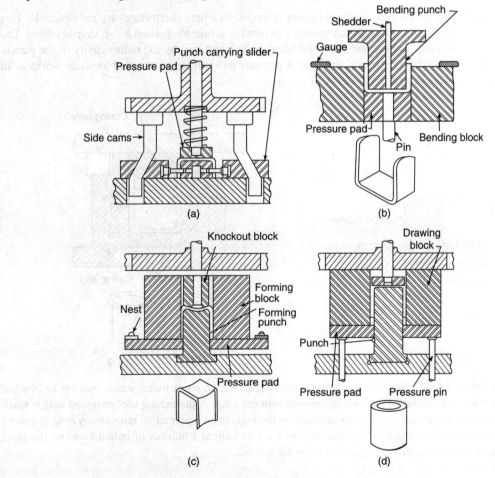

Fig. 27.25 Drawing.

The blank is placed on the die face and the press is operated. The blank holder comes into action before the punch and presses the blank tightly on the face of the die, so that the blank is "ironed out" as it is drawn over the reduced edge of the die hole. The punch holder keeps the metal plate straight and eliminates wrinkles. As the punch descends further, the blank is forced into the die cavity and metal flows plastically. With further movement of the punch in the die cavity, the blank is formed into a cup shape. The size of the blank required for a particular cup shape can be calculated from the formula given as follows:

$$D = \sqrt{d^2 + 4dh}$$

where D is the diameter of the blank
 d is the diameter of the shell
 h is the height of the shell.

Coining: In this operation, the plastic flow of the material takes place in the cavity of the die, or the punch and the die together forms a cavity. Very large squeezing pressures are applied on the material to be coined. It is required that good plasticity is possessed by the material. The material thus compressed will have engravings on both the punch and the die. It is shown in Fig. 27.26.

Curling: Curling is required to be done on certain components where sharp edges are not desirable. This is illustrated in Fig. 27.27. A curling punch operates to provide a suitable curl on a cup shaped object. The punch moves down towards the die and the metal blank rolls into a curl by the radius cavity of the punch. The curled edge is made of a wire to add strength. A pressure pad is fitted on the die block and works as an ejector for the curled component in the cavity of the die.

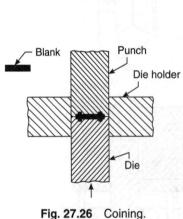

Fig. 27.26 Coining.

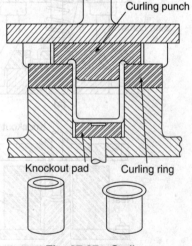

Fig. 27.27 Curling.

Flattening or planishing: The flattening operation is required on sheet metal, either blanker or blanked ones. This operation is similar to the ironing operation in drawing. A planishing tool provided with a small projection descends downwards and exerts pressure on the plate to be flattened, to remove any bent or curved surface present in the sheet metal. The planishing tool leaves behind a number of indentations on the sheet metal surface, which need a further operation to be eliminated.

27.9 CHOICE OF PRESS

For different types of operations, different press machines are in use, and therefore to make it economical and precise, the proper choice of press machine is essential. The following points need to be considered:

- (i) Tonnage capacity;
- (ii) Energy of flywheel;
- (iii) Crank pin speed;
- (iv) Capacity of motor;
- (v) Size and type of die;
- (vi) Stroke length;
- (vii) Clearance.

A large number of single action presses are direct driven, that is, the flywheel runs on the crank shaft and by means of a clutch causes the crank shaft to rotate. The presses are usually used for jobs which call for the work to be done at the bottom of the stroke. The usual operations of blanking, piercing and bending are typical of the work done, as geared presses run at a high speed.

General presses include both single and double action types. These machines, besides being capable of performing the above operations, are used for cupping, raising, forming and drawing operations, which in many cases call for the distribution of the load over a greater portion of the stroke. The geared press usually makes fewer stokes per minute than the direct driven type, as the gearing permits this, whilst the momentum in the quickly moving flywheel and the slower moving punch flywheel supplies the necessary power to carry the crank over. It should be noted that the slower movement of the geared press gives a larger time for the metal to flow.

The question of what type of press to choose is difficult to answer without a clear idea of the class of work to be done. When the work is of the type calling for simple blanking, or follow-on tools with automatic feeds, a short-stroke single action press is suitable. This type of press is also very useful for such tools as cropping and piercing. A short-stroke machine is advantageous for this class of the work, as the tools may be designed to prevent the punch leaving the stripper or guide plate. With a good design, the tools then slide from an effective guard.

When the work is of a general nature, such as blanking, piercing, clipping or forming, hand feeding is resorted to for a blank press with, say, a 100 mm stroke, and a pressure of 60 tonnes enables a great range of work to be done. The operator can see and handle the work easily, whilst the longer strokes permit the guards to work more effectively. A first class geared, general utility, single action press for work within its tonnage has a closed height of 200 mm between the load and the ram, and can handle, say, a 300 mm blank. As it is for short runs, an open fronted C frame machine is satisfactory. Thus, press selection is an important task that depends upon the experience of the production engineer.

27.10 DESIGN FUNDAMENTALS AND CONSTRUCTION FEATURES OF BLANKING, PIERCING AND CROPPING TOOLS

27.10.1 Design Analysis

In the field of design, the design activities are of three main types:

- (i) Improvement design
- (ii) Adopted design
- (iii) Creative design.

In improvement design type of work, the existing design is taken as a basis and certain improvements are made in the existing design.

In adoptive design, the existing design is adopted to new requirements by effecting some changes. The new requirements may be new working conditions, changes in materials, etc.

27.10.2 General Notes on Press Tool Design

The following points need to be kept in mind:
   (i) Aim for simplicity in design and operation.
   (ii) Provide adequate strength to all parts which have to withstand repeated blows.
   (iii) Study the relative methods of manufacture, i.e. expensive tools against several simple tools for each operation.
   (iv) Ensure that the tool cost bears a direct relation to the quantities to be produced.
   (v) If using a roll feed for follow-on tools, etc. check the feed of the rolls before deciding on the pitch of the blanks.
   (vi) Avoid change in the design for change's sake; keep to standard whenever possible.
   (vii) Watch carefully the limits you place on the tools; too light a limit means extra cost, and often greater trouble when setting and operating.
   (viii) Specify all gauges in decimals; it helps to avoid errors.
   (ix) When raising and cupping, an allowance beyond that required for the actual metal thickness gives better and quicker production.
   (x) When raising and cupping, ensure that the radii on the drawing edge on all beds is as large as possible.
   (xi) For drawing operations, use a form that will permit the metal to flow easily; avoid sharp corners; watch the stages from one operation to another.
   (xii) Give, wherever possible, stripping from both the bed and the punch. This applies to all press operations.
   (xiii) On the blanking, the die size determines the blank dimensions.
   (xiv) On piercing, the punch size determines the dimension of the hole. There may be a little springing in, and on fine work a margin of 0.25 to 50 mm is necessary to counteract this.
   (xv) It is usual to blank on the low limit.
   (xvi) When piercing, the high limit is chosen.
   (xvii) The die clearance varies from (½)° to 3° according to the job. On fine work, the first 3 to 5 mm may be left parallel, to ensure that the size is maintained for longer periods. Constant regrinding renders the die oversized.
   (xviii) The bridge between the blanks varies according to the thickness of the stock. The minimum bridge is roughly 0.50 mm.
   (xix) When bending, avoid sharp corners. The suitable radius is equal to twice the metal thickness.
   (xx) Bend, if possible, across the grain.
   (xxi) Keep an up-to-date record of machine capacities.
   (xxii) Shear on blanking or piercing tools reduces the load on the press.
   (xxiii) Shear on the punch gives curved blanks.
   (xxiv) Shear on the bed and punch flat gives a flat blank.
   (xxv) Burr from blanking is on the face of the metal in contact with the bed.
   (xxvi) The side clearance between a blanking or piercing bed is approximately one-twentieth the thickness of the stock; hence the difference between the diameters of the punch and the die is equal to approximately one-tenth the thickness of the stock.
   (xxvii) Design all tools with an eye on the safety of both operator and setter.

(xxviii) Keep all tools as solid as possible; on indexing tools, it is better for the indexing mechanism to rotate. Hence keep the bed fixed.
(xxix) Watch the limits given on pressing. Some of these are often too fine; get them altered, so that the job can become an economical one.
(xxx) When designing, think of the setter, and give him the best chance of setting the job.
(xxxi) See each location in a positive light and, unless completed, do not change your datum; you invite errors if you repeatedly change the datum.
(xxxii) When the screws have to be used repeatedly, see that the length of the thread is sufficient.
(xxxiii) See that the clamping legs are adequate for the job. Aim at keeping to the height of these standards. It prevents the setter from wasting time by looking for packing.
(xxxiv) Design all tools to suite as wide a range of presses as possible. By so doing, the production schedule is aided.
(xxxv) Don't forget that an air blast will remove small components quickly from the face of the tools, as well as dirt. If no air blast is available, spring often will do the trick.
(xxxvi) Metric threads on the small sizes may be too fine for use in cast iron. In this material, it is better to stick to the Whitworth standards, removing the fine pitch screws for mild steel, brass, etc.
(xxxvii) Ensure that there is adequate clearance on the tools to permit the work to drop; watch the clearances, holes, sharp edges, shoulders, etc. in this respect.

27.10.3 Press Tool and Its Parts

The blanking tools in general comprise the following items:

(i) Top and bottom bolster
(ii) Punch and die shoes
(iii) Punches
(iv) Die blocks
(v) Strippers and knock-cuts
(vi) Pillars and bushings.

The press is a universal type of machine tool which is used for many different operations and jobs. Special tooling is designed to adopt the press to a specific job or operation. This tooling constitutes the die assembly or simply a press tool die. The assembly consists of two holds, one mounted on the moving ram and the other bolted or clamped on the stationary press bed or bolster plate.

The assembly incorporates two main members—a male member and a female member. The male member is the punch and the female member is the die block or simply a die. The main components are as follows:

Punch: It is the male member of the unit and is kept as small as possible, consistent with the required strength and rigidity. The punch is made of a hard, wear resistant metal and is finally ground to the predetermined size, providing just optimum clearance between the punch and the die.

Punch retainer or punch plate: It fits closely over the body of the punch and holds it in the proper relative position. The retainer in turn is bolted to the punch holder.

Punch holder: It provides a wide flat surface which faces against the lower end of the punch and is anchored to the punch with the help of a shank, which is an integral part of the punch holder. The shank exactly fits into the ram opening, to help in properly positioning and aligning the punch. The punch holder is made of cast steel.

Backing plate: Whenever the punch is headless, a hardened steel backing plate is introduced between the back of the punch and the punch holder, so that the intensity of the pressure does not become excessive on the punch holder. The backing plate distributes the pressure over a wide area, and the intensity of pressure on the punch holder is reduced to avoid crushing.

Die block: It is the female working member and is kept as small as possible, consistent with the required strength. It is also made of a hard, wear resistant metal and is finish-ground to the predetermined size and tolerance.

Die retainer: Just like the punch retainer, the die retainer also holds the die block at the proper position with respect to the punch. The retainer is mounted on the die shoe or holder. In certain designs, the die shoe itself serves as a retainer for the die block. The die block is then mounted directly on the die shoe.

Die shoe: The die shoe assembly consisting of the die block and, die shoe is in turn bolted or clamped to the bolster plate.

Guide posts and bushings: The punch and die members, once properly located and aligned, are held in alignment by means of guide posts and bushings which resist movement or deflection of die members as operating pressure increases. The guide post and the bushings are part of the commercially available punch and die holders. The components are shown in Fig. 27.28.

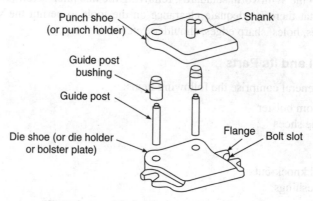

Fig. 27.28 Components of typical die set.

Stripper or stripper plate: When the punch completes its downward movement and starts returning, the scrap strip tries to go up along with it. The stripper plate prevents this upward movement of the scrap strip and positions the strip for the next stroke. The strippers are of many designs.

Stock stops and stock guides: Fixed types of strippers are sometimes also used to guide the stock, whereas stock stops locate the work material at a suitable position in relation to the previously blanked surface, in preparation for the next downward movement of the punch. Stock stops like strippers are always available in a variety of designs.

Cutting action in die: Blanking and piercing operations are performed to prepare the stock for further processing in forming die for bending, forming and drawing operations. In blanking operations, the piece to be used or further processed is punched from the strip stock, whereas in piercing operation a portion is further punched from piece to be used or further processed. Blanking and piercing operations are cutting operations performed in cutting die components.

The cutting of metal between die components is a shearing process, in which the metal is stressed in shear between two cutting edges to the point of fracture or beyond its ultimate strength. The metal is subjected to both tensile and compressive stresses. Stretching beyond the elastic limit occurs. Then there is plastic deformation and reduction in area. Finally, fracturing starts through clearance planes in the reduced area and becomes complete. In intensive testing of a long specimen of mild steel, the fracture takes place along the plane of maximum shear (see Fig. 27.29).

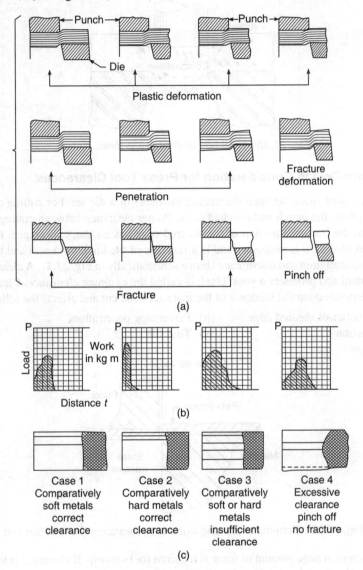

Fig. 27.29 Effect of certain clearance.

Shut height: This is distance from the bottom of the die shoe to the top of the punch holder when the die is in its closed position. The overall height of the guide posts must always be significantly less than the shut height, in order to ensure that the ram will not strike against the ends of the guide posts. For cutting dies, if possible the guide posts should be short enough to accommodate the total amount that the shut height will be lowered by because of sharpening. In general, the shut height of the press is the maximum die height that can be accommodated for normal operation, taking the bolster into consideration. It is fixed for a particular machine.

Figure 27.30 shows the direction of stresses in metal cutting.

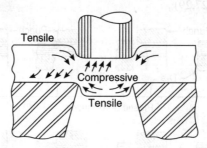

Fig. 27.30 Direction of stresses in metal cutting.

27.10.4 Important Design Consideration for Press Tool Clearances

Clearance is the measured space between the mating members of a die set. For cutting contour surfaces, it is the difference between the punch and die diameters. Proper clearance between cutting edges enables the fracture to meet, and the fractured portion of the sheared edge has a clean appearance. For optimum finish of a cut edge, proper clearance is necessary and is a function of the kind, thickness and temper of the work material. Clearance, penetration and fracture are shown schematically in Fig. 27.31. A clearance that is neither excessive nor too small and produces a good blank is called the *optimum clearance*. Clearance between the punch and the die depends upon the hardness of the material to be cut and affects the following:

- (i) Characteristics of sheared edge
- (ii) Percentage penetration
- (iii) Press exertion
- (iv) Tool life
- (v) Burr height.

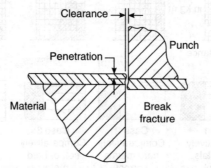

Fig. 27.31 Schematic drawing illustrating clearance, penetration and fracture.

If clearance is zero, a huge amount of force is required for blanking. If clearance is too much, there will be no cutting action. Hence optimum clearance is desired.

Figure 27.32 shows the various steps in metal shearing.

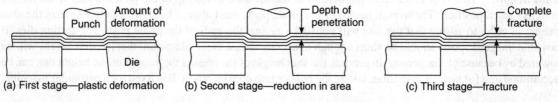

Fig. 27.32 Steps in shearing of metal.

Characteristics of sheared edge: In Fig. 27.33, the characteristics of the cut edge on the stock and blank, with normal clearance, are schematically shown. The upper corner of the cut edge of the stock (A) and lower corner of the blank (A′) will have a radius where the punch and die edges respectively make contact with the material. This is due to the plastic deformation taking place. This edge radius will be more pronounced when cutting soft metal. Excessive clearance will also cause a large radius at these corners, as well as burr on the opposite corners.

In ideal cutting operations, the punch penetrates the material to a depth equal to about one-third of its thickness, before fracture occurs and forces an equal portion of material into the die opening. That portion of the thickness so penetrated will be highly bruised, appearing on the cut edge as a bright bend around the entire contour of the cut adjacent to the edge radius indicated at B and B′ in Fig. 27.33. When the cutting clearance is not sufficient, additional bands of metal must be cut before complete separation is accomplished, as is shown at B in Fig. 27.33. When correct cutting clearance is used, the material below the cut will be rough on both the stock and the slug. With correct clearance, the edge of the fracture will permit a clean break below the cut band, because the upper and lower fractures extend towards one another. Excessive clearance will result in a tapered edge.

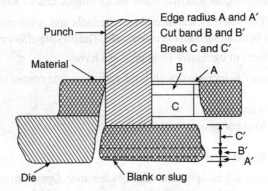

Fig. 27.33 Cut edge characteristics.

Figure 27.34 shows the effect of clearance on the shearing edge.

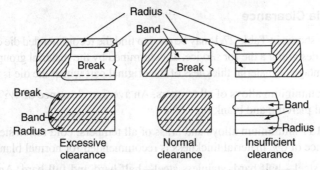

Fig. 27.34 Effect of clearance on shear edge.

Thus, the process of shearing gives rise to a sheared edge that has three characteristics:
(i) A corner radius;
(ii) A clean brushed surface;
(iii) A rough burred taper surface, dull gray in colour.

Percentage penetration: The depth of the white polished band formed in the second stage of the shearing action determines the percentage penetration, which is the ratio of this depth to the original thickness of the stock expressed in hundreds. The percentage penetration is about 33%, but it greatly depends upon three factors, i.e. the clearance, hardness and thickness of the material to be punched. As the hardness of the material increases, the material becomes more rigid, and the punch can travel only a shallow depth before the fracture starts. The thickness of the stock also has a similar effect. The greater the thickness, the more rigid the sheet becomes and lowers the percentage penetration. Thus, if the material is hard and thick the clearance will have to be increased accordingly, but it should also not become excessive.

Press exertion: Excessively small clearances require excessive forces to shear the material. Apart from increasing the clearances, the shear angle on punches and dies reduces the shearing force required at the peak point. The shearing force is thus small but uniform over a large part of the stroke.

Tool life: Both punches and dies become blunt if the clearance is too small. They have to be ground after short runs and set again. Blunt edges require excessive force to punch the material. The burrs on the work become predominant. Blunt cutting edges also cause excessive radii on the opposite work surfaces.

The fractured surfaces are rougher and may show many minute cracks which weaken the edges.

Burrs and bowing effects: Too small a clearance blunts the tools and is the cause of burrs after some time. Excessive clearance also causes burrs along with excessive radii. The bowing effect increases with excessive radii.

Summing up the bad effects of excessive clearance, we have:
(i) The sheared edge taper is increased with large radius on one corner of the sheared edge and burr effect on the other;
(ii) The bowing effect is increased;
(iii) The force required to shear becomes excessive;
(iv) Tool life decreases;
(v) The sides are smooth and vertical.

The aim of the tool engineer is to reach an optimum value of clearance. Optimum clearance should give smooth vertical sheared edges with minimum bowing effect and maximum tool life within regrinds. It is about 10% of the thickness of the material, but varies with hardness. Hard materials require more clearance than soft materials for the same thickness.

27.10.5 Practical Die Clearance

The die clearance chart as shown in Table 27.1 may be used to find the recommended die clearance to be allowed and to be provided for in designing a die for service, as determined by the material groups listed below, and for the pro-established percentage of material thickness of the original part which the die is designed to produce.

Group 1: 2S and 52S aluminium alloys of all tempers. An average clearance of 4½% material thickness is recommended for normal piercing and blanking.

Group 2: 24ST and 61ST aluminium alloys and brass of all tempers; cold rolled steel, dead soft, stainless steel. An average clearance of 6% material thickness is recommended for normal blanking and piercing.

Group 3: Cold rolled steel—half hard; stainless steel—half hard and full hard. An average clearance of 7½% is recommended for normal piercing and blanking.

Clearances for punching electrical steel laminations are listed in Table 27.1, arranged in the order of decreasing silicon content. The data indicate that the greater the silicon content, the greater is the required die clearance. Softer stock will require smaller die clearance but greater angular clearance to prevent scoring of die walls. The angular clearance per side is 38.0 mm length ground after hardening to 0.0254 to 0.0508 mm for hard stock and 0.0508 to 0.0762 mm for soft stock.

Table 27.1 Per side clearance (mm) lamination dies as per ISI

Grades of steel	0355 gauge (00155)	ISI (mm)	26 gauge = 0.5 mm (00.0186″)		24 gauge = (0.630 mm) (0.0249″)	
Transformer grades	0.0007	0.0178	0.00085	0.0216	0.0001	0.0254
Dynamo special	0.0007	0.0178	0.00085	0.0216	0.001	0.0254
Dynamo	0.0006	0.0152	0.00075	0.0191	0.0009	0.0254
Electrical	0.0006	0.0152	0.00075	0.0191	0.0009	0.0229
Armature	0.0005	0.0127	0.00065	0.0165	0.0008	0.0203

Control of hole and blank sizes by clearance location

A blank of a given size is to be prepared then the die is to be made to size and punched smaller by total clearance. The hole to be made is to be of exact size and its inner portion called *slug* is to be scrapped; then it is to be punched to size, with the die larger by the amount of total clearance 2C (see Fig. 27.35). The application of clearances on an irregularly shaped punch and die is shown in Fig. 27.36.

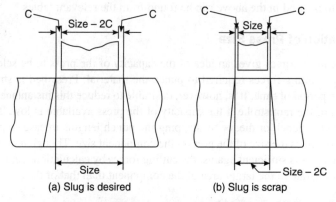

Fig. 27.35 Clearance location related to part punch and die dimensions.

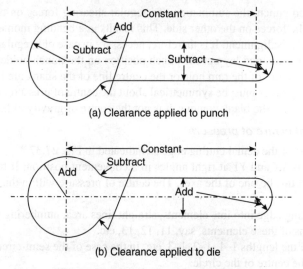

Fig. 27.36 Application of clearances.

Angular clearance: If the blank or the slug is to pass through the die in a blanking or piercing operation, the angular clearance is necessary in order to prevent the possibility of a blank or slug jamming in the passage. The angular clearance is usually ground from 6.25 to 37.5 mm per side. The die is usually machined straight for 3 mm, in order that the punch clearance is not increased when the die is sharpened by grinding the face.

27.10.6 Determination of Blanking Pressure

The pressure required to shear the material in blanking and shearing operations is given by the formulae

$$P = \pi D \times \sigma_s \times t \quad \text{for round holes}$$

and

$$P = L \times \sigma_s \times t \quad \text{for other contours}$$

where σ_s is the shear strength of the material in kg/mm^2;
  D is the hole diameter in mm;
  t is the material thickness in mm;
  L is the shear length or perimeter in mm;
  P is the punch force in kg.

The shearing strength to be used in the above can be found from the relevant tables.

27.10.7 Determination of Press Size

The determination of cutting forces gives an idea of the capacity of the press to be selected. Press capacity must be higher than the cutting forces required to punch the material. Press bed is stressed by the cutting forces over a very short period of time. It is, however, desirable to reduce this instantaneous force and spread it over a larger portion of the ram stroke if the capacity of the press available is low. This can be achieved by providing a shear to the punch or die, or by stepping the punch lengths in case of progressive dies. The other factor which also decides the size of the press is the component size. Though it may be possible that the tonnage capacity of the press is sufficient against the cutting forces by calculation, yet it may be possible that a bigger size press is used, due to the larger area of the component than that of the press bed.

27.10.8 Determination of Centre of Pressure

In case of irregular shaped punch, the summation of irregular shearing forces on one side of the centre of ram may greatly exceed the forces on the other side. This results in a bending moment in the press ram, and undesirable deflection and misalignment. It is therefore, necessary in case of irregular shaped punches to find out the exact centre of pressure, and lay out the punch position on the punch holder in such a way that the centre of pressure and the centre of the ram hole or the centre line of the shank are in the same straight line. The summation of shearing forces must be symmetrical about the centre of pressure. It is the centre of gravity of the line, i.e. the perimeter of the blank contour. It is not the centre of gravity of the area.

Method of calculation of centre of pressure

1. Draw an outline of the actual cutting edge as indicated in Fig. 27.37.
2. Divide the areas *XX* and *YY* at right angles in a convenient position. If the figure is symmetrical about a line, let this be one of the axes. The centre of pressure will in this case be the same as the axis.
3. Divide the cutting edge into line elements, straight lines area, numbering them 1, 2, 3, etc.
4. Find the lengths of these elements, say, 11, 12, 13, etc.
5. Find the CG of the lengths 1–1, 1–2, 1–3, etc. In the case of the semi-circular arc, the CG is given by $2r/\pi$ from the centre of the circle.

6. Find X_1, X_2, etc. and Y_1, Y_2, etc., i.e. the distance of the CG of each line element from the YY and XX axes respectively.
7. Calculate the distance X of the centre of pressure from the YY axis by the formula

$$X = \frac{11X_1 + 12X_2 + 13X_3 + 14Y_4 + \cdots}{L_1 + L_2 + L_3 + L_4 + \cdots}$$

8. Similarly, find

$$Y = \frac{L_1Y_1 + L_2Y_2 + L_3Y_3 + L_4Y_4 + \cdots}{L_1 + L_2 + L_3 + L_4 + \cdots}$$

The exact position of centre of pressure is thus determined.

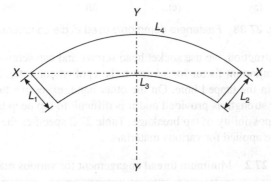

Fig. 27.37 Outline sketch.

27.10.9 Design of Screws and Dowels

In the designs of tools and dies, the fasteners are often the weakest link in the tool. If they are not selected and applied correctly, they can become the cause of failure of the entire tool; their size, location or number is fixed to avoid hardening of cracks and breaks, stripping of threads, distortion by release of internal stresses, and misalignment of holes.

Screws are used to hold components together, and dowels are meant to locate components sideways relative to each other. Clear holes for screws are drilled 0.3969 mm larger in diameter than the body diameter of the screw and counter bore holes for screw heads 0.7938 mm larger. These clearances will allow side shift if screws are loosened. Dowels are applied to effect accurate relative positioning.

Safe strength of screws

As many factors determine the strength of screws, we must know the amount of strength remaining in it after it has been tightened. A considerable proportion of the strength is expanded in tightening; what is left is the effective load carrying capacity.

The following are the types of fasteners commonly used in die construction as shown in the respective figures:

(i) Socket cap screws [Fig. 27.38(a)].
(ii) Dowels [Fig. 27.38(b)].
(iii) Socket flat head screws [Fig. 27.38(c)].
(iv) Socket bottom head screws [Fig. 27.38(d)].
(v) Stripper bolts [Fig. 27.38(e)].
(vi) Socket set head screws [Fig. 27.38(f)].

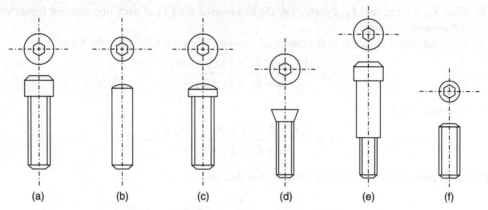

Fig. 27.38 Fasteners commonly used in die construction.

Generally for die construction, we use socket head screws and cap screws. The thread engagement as shown in Fig. 27.38 is an important factor for screws to be applied correctly. If too little is specified, it is possible to strip the thread in the tapped hole. On the other hand. excessive thread engagement should be avoided, because no greater strength is provided and it is difficult to tap deep holes. Also, as a deeper hole is tapped, the greater is the possibility of tap breakage. Table 27.2 specifies the minimum amount of thread engagement which should be applied for various materials.

Table 27.2 Minimum thread engagement for various materials

Material	Minimum thread engagement
Steel	$1\frac{1}{2}D$
Cast iron	$2D$
Magnesium	$2\frac{1}{2}D$
Aluminium	$2\frac{1}{2}D$
Fibre and plastics	$3D$ and more

For general purpose [see Fig. 27.38(d)],

$$I = 2D + 12.7 \text{ mm or } \frac{L}{2}$$

whichever is greater for coarse thread series, where L is the screw length, I is the thread length and D is the screw diameter.

For fine thread series

$$l = 1\frac{1}{2}D + 12.7 \text{ mm or } \frac{3}{8L}$$

whichever is greater.

Figure 27.39 shows the positioning and proportioning of screws. Here, t is the thickness of plate and d is the diameter of the dowel hole.

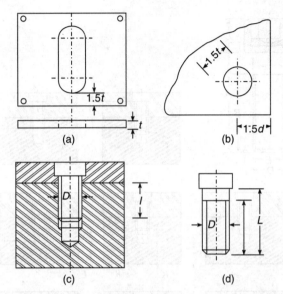

Fig. 27.39 Positioning and proportioning of screws.

Proportions of socket cap screws

Socket cap screws (see Fig. 27.40) are drawn according to the following factors:

A is the length of the screw;
B is the thread length;
C is the head height;
D is the body diameter ($C = D$);
E is the root diameter of the thread;
F is the thread diameter drawn;
G is the angle of the chamfer—45 degrees, 0.7938 to 2.3812 mm wide, depending on the screw size.

When the block to be fastened is made of hardened tool steel, the distance H should never be less than 1½D. For machine steel, it need only be thick enough for adequate strength. The length of thread engagement I is 1½ for steel, with a 2D for cast iron and same for non-ferrous metals.

The depth of the tapped hole J is applied in increments of 1.5875 mm. The distance K between the ends of the screw and the bottom of the thread, and between the bottom of the thread and the bottom of the top drill hole, is normally made equal to 3.175 mm. Bottoming taps have from 1 to 1½ imperfect threads, and this fact must be taken into consideration when drawing blind holes. The thread must extend at least 1½ times the pitch of the thread past the end of the screw. The length L of the thread chamfer is made 1.5875 mm; the chamfer M is 45°. At the bottom of the tap, drill holes are drawn. Two lines, at angles of 30° to each other and of dimension N, represent the conical depression produced by the end of the tap drill.

For hardened tool steel parts, the distance between the end of the tap drill hole and the lower surface of the block must be greater than 1D. When this distance is less than 1D, the tap drill runs completely through the pan. If this is not done, it is possible for the thin circular section of the steel to crack during hardening and fall out like a rough-edged disc.

The hole which a screw engages is tapped all the way through when the distance P (0 for blind holes) from the end of the screw to the lower surface of the block is $1D$ (one diameter of the screw or less).

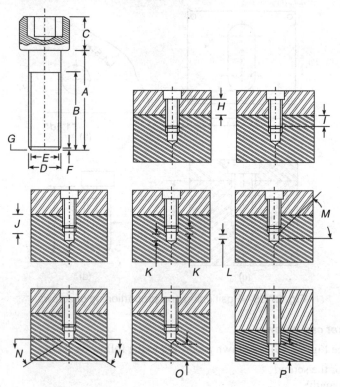

Fig. 27.40 Proportions of socket capscrews.

Dowels

Dowels are made of silver steel. These are heat treated to produce an extremely hard exterior surface with a somewhat soft and tough core to resist shear. Their surface hardness is R_c 60–64 and core hardness is R_c 50–54. Dowels are manufactured in two amounts of oversize. These are:

(i) Regular dowels employed for all new dies are made 0.00508 mm oversize to provide a secure press fit.
(ii) Oversize dowels are made 0.0254 mm oversize. They are used for repair work when the dowel holes have been enlarged through repeated pressing of dowels in and out, and when holes have been accidentally machined oversize.

Shear strength, the safe load that dowels can resist in shear, is determined by dividing the ultimate strength of the dowel. The ultimate shear strength is taken at 13358 kg/cm² and a safety factor of 12 has been applied. Therefore, there are safe loads for dowels in single shear under shock conditions, as accounted for in the design of dies. For double shear application values, appropriate values must be taken.

Proportions of dowels: The conditions under which dowels are employed determine the types of applications as selected. These are given below.

Through dowels. In this the hole is reamed all the way through the components, and the dowels can be pressed out from either side. When the dimension A is 50.8 mm or less, the dowel engagement B is between 1.5 to 2 times the diameter D, as shown in Fig. 27.41.

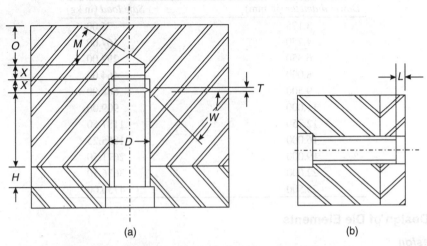

Fig. 27.41 Proportions of dowel screws for tool steel dies.

Semi blind dowels. In this, the dowel hole is drilled and reamed from one side at least 3.175 mm deeper than the dowel length. A smaller hole is drilled through the block, and the dowel can be pressed out from one side only. The dimension C of the knockout hole is made equal to $(1/2)D + 0.40$ mm as shown in Fig. 27.41.

Blind dowels. This dowel is applied in a blind hole—one not drilled and reamed completely through. The application should be avoided whenever possible, as blind dowels are more difficult to fit because of trapped air and their removal can be troublesome. The diameter E is made a press fit. The diameter F at the blind side should be a slip fit. Refer to Fig. 27.41.

Relieve dowels. When doweling blocks are over 50.8 mm thick (dimension G), standard length dowels are employed and the hole H is specified as 0.7938 mm larger than the diameter of the dowels for relief. This relief is applied to the portion of the hole not in actual contact with the dowel surface. The engagement length I is $1\frac{1}{2}$ to $2D$. Refer to Fig. 27.41.

Spacing of screws and dowels: For the spacing between the holes and edges of parts, see in Fig. 27.41(a). This is particularly important for tool steel to be hardened. If too little space is allowed, there is a possibility of the block cracking in the hardening process. But on the other hand, it is desirable to have the screw holes as close to the cutting edges as possible. For accurate positioning, the dowels are kept far enough apart. For tool steel, $C = 1 - \frac{1}{2}d$, and for machine steel, $C = 1\frac{1}{2}d$. For tool steel, $D = 1\frac{1}{2}d$ and for machine steel, $D = 1d$.

Normally, dowels are press fitted into both members. Holes are drilled and reamed after the fit fixture or die has been tried out. However, as for instance in sharpening, the dowels are press-fitted in one component and made a sliding fit in the other for ease of disassembly.

Table 27.3 shows the safe loads for various dowel diameters.

Table 27.3 Safe load for designing of dowel diameter

Dowel diameter (in mm)	Safe load (in kg)
3.175	75.00
4.750	167.00
6.350	297.00
8.000	464.00
9.500	668.00
11.000	909.00
12.750	1187.50
16.000	1855.25
19.000	2671.50
22.000	3636.50
25.500	4750.00

27.10.10 Design of Die Elements

Die block design

Four factors influence the design of the die block for any particular die. They are:

   (i) Part size   (ii) Part thickness
  (iii) Intricacy of part contour   (iv) Type of die.

Figure 27.42 shows the proportions of die screws.

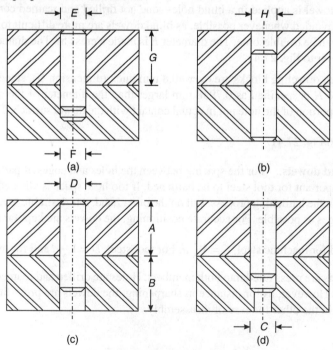

Fig. 27.42 Proportions of die screws.

(i) Part size: As the die block is made of superior tool steel, its overall dimensions should be such that they are at least consistent with the strength requirements. The overall dimensions should be obtained by having the maximum die wall thickness as the requirement for strength and by taking into account the space needed for mounting the screws, dowels and stripper plate.

Small dies, such as those used for producing weighing machine pans, usually have a solid die block. Only for intricate pan contours should the die block be sectioned to facilitate machining. Large die blocks are made in sections for easy machining, hardening and grinding.

(ii) Part thickness: Die block thickness is also governed by strength requirements of the materials to be punched. In fact it is the minimum area (i.e., wall thickness × block thickness) which really resists the forces involved in bursting the die. The minimum area provided should be effective in resisting these bursting forces. There are two methods followed in obtaining the dimensions of the die block. The first one is based on experience and the second one tries to base the dimensions on some calculations.

In the first method the dimensions and layout of the die block are obtained on the basis of the strip thickness (the thickness of the material to be punched), which is decided after consulting tables and charts prepared by some highly experienced researchers and accepted by the majority of engineers.

Strip thickness. Die block thickness is decided by considering the shut height of the selected press machine. As it is made of superior tool steel which is very costly, in no case is it made more than 44 to 55 mm thick.

There are some relationships between strip thickness, die plate height and angular clearance, which are presented in Tables 27.4 and 27.5. From these tables too, we can choose the proper die block height.

Table 27.4 Strip thickness versus die plate height

Strip thickness (in mm)	Die plate height (in mm)
0–1.600	24.00
1.60–3.15	28.00
3.15–4.50	35.00
4.5–6.30	41.25
6.30 and above	47.50

Table 27.5 Strip thickness versus angular clearance

Strip thickness (in mm)	Angular (N) clearance (in degrees)
0–1.60	0.25
1.60–3.15	0.50
3.15–4.50	0.75
4.50–6.30	1.00

Angular clearance or angular relief. The angular relief may also be selected from the tables relating strip thickness and angular relief, as shown in Table 27.5.

Positioning of dowels and screws. The spacing between the holes and edges of the parts of a die (see Fig. 27.45) is particularly important for tool steel to be hardened. If too little space is allowed, there is a possibility of the cracking of the die block during the hardening process. But on the other hand it is desirable to have the screw holes as close to the cutting edge as possible. In case of dowels, two and only two dowels should be provided in each block or element that requires accurate and permanent positioning. They should be located as far apart as possible for maximum locating effect, usually near diagonally opposite corners, while two or more screws should be used depending on the size of the die block. Mostly, the distance of a screw or dowel from the outer or inner edge is made equal to one and a half times its diameter. These different relations are shown in Fig. 27.41(a).

Figure 27.43 shows the various aspects of proportioning and positioning of holes in a die plate. *A*, *B*, *C* and *D* are distances of the die plates from the positions of holes as shown in Fig. 27.43(a) and 27.43(b).

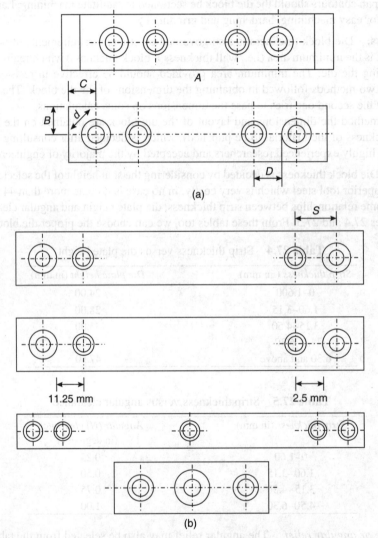

Fig. 27.43 (a) Proportion of distance of holes from edge of die plate and
(b) proportioning and positioning of holes in die plate.

In the second method the dimensions of the die block are found out first by calculation and the results so obtained are corrected using standard tables and past experience. The process of designing the die plates is divided into the following two steps:

(i) Determination of die plate surface;
(ii) Correction of die plate surface.

Die plate thickness. It is calculated on the basis of cutting load. It is given by the formula given below.
$$T = P^{0.015}$$
where T is the die plate thickness
       P is the cutting pressure in tonnes.
Here
$$P = L \times f_s \times t$$
where L is the cutting perimeter;
       t is the stock thickness;
       f_s is the shear strength of the stock.

Material to be cut is measured in kg/cm².

Corrections. After the preliminary calculations, the following corrections are accounted for:

(a) Under no circumstances should the die plate be thinner than 8.5 to 9.5 mm; even if the calculation gives the smaller values, we will increase them up to 7.5 mm for inferior die surfaces and 9.25 mm for superior die surfaces.
(b) The formulae given above are intended for comparatively small die, i.e. with cutting perimeter smaller than 51.00 mm. In the case of longer cutting edge length, the results obtained by the formula must be multiplied by the appropriate factor.
(c) In every case it is supposed that the die plate is made from first grade tool steel, correctly machined and perfectly heat treated. In the case of special alloy tool steel with very high tensile strength, the calculated values can be decreased and vice versa. If there is fear that heat treatment will not be adequate, it is advisable to increase the calculated values.
(d) Die plates must always rest on flat bases, i.e. die shoes or die holders, because if dies are directly put upon bolster plate openings, they are subjected to heavy flexible stresses, which they are not able to resist. If the dies are encased adequately in die shoes, the thinness can be decreased by up to 5% because of strong die support.
(e) In the case of high production tools, when it is foreseen that there will be a great number of regrindings, sufficient grinding allowance (0.254 to 0.508 mm per grinding) is added.
(f) In every case the final value must be rounded up to an available stock size; also, from each side, de-carburized layers of approximately 1.6 mm width should be machined away. This is termed *machining allowance*.

Layout. The critical distance between the blank opening edges and the die plate border must be at least 1.2 to 2 times the die plate thickness (see Fig. 27.44) in larger tools. In order to avoid development of cracks during heat treatment, any hole in a die plate corner must be located so that the maximum distance between its centre line and the die plate edge is at least 1.5 times the hole diameter, as shown in Fig. 27.45.

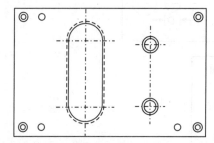

Fig. 27.44 Die plate layout.

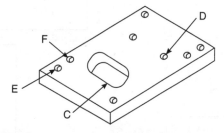

Fig. 27.45 Holes for positioning dowels sockets.

Figure 27.46 shows a progressive die for producing a washer. The punch assembly is not shown. The figure shows in detail the locating dowels and clamping serving for stripper, die and bolster.

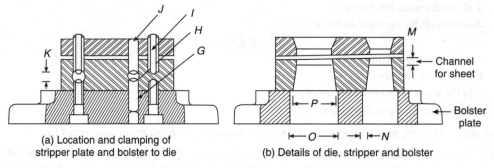

(a) Location and clamping of stripper plate and bolster to die

(b) Details of die, stripper and bolster

Fig. 27.46 Progressive die arrangement.

Figure 27.47 shows the proportions for hole spacings.

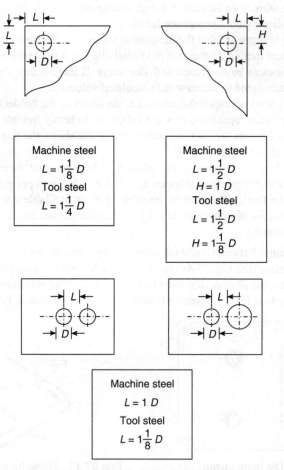

Fig. 27.47 Proportions for hole spacing.

The various hole sizes, the minimum distance between two holes having the same size and the minimum distance between two holes of different sizes are given in Fig. 27.48.

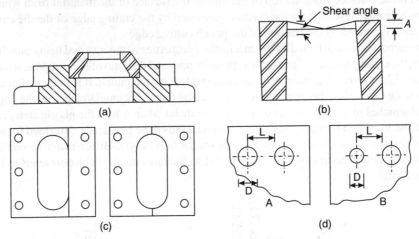

Fig. 27.48 Size of holes and distances between them.

Figure 27.49 shows the section view of the positions of the thrust plate, punch and punch plate.

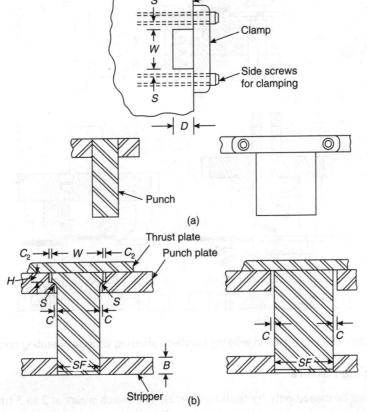

Fig. 27.49 Section view showing positions of thrust plate, punch and punch plate.

27.10.11 Burr Height

The burr height is the distance from the top of the burr to the surface of the material from which it projects. The burr on the hole edge results from the fracture generated by the cutting edge of the die cavity. The burr on the slug or bank is caused by the fracture at the punch cutting edge.

These characteristics are affected by the mechanical properties of the material being punched, the punch to die clearance, the cutting edge conditions, the type of stripper and the ratio of the component size to the stock thickness. Figure 27.50 shows the effect of worn cutting edges on the punch. It is evident that a considerable amount of smearing takes place because of the plastic metal that is accumulated below the corner radius on the punch. Dull punches of die cavities tend to smear the metal when it is in the plastic state, causing longer burnished land and increased burr height. Ductile material tends to be plastically deformed for a considerable distance before fracture occurs, and as a result increases the burr height. Brittle material, on the other hand, cannot be plastically deformed to any major degree and so fractures readily. The burr tends to be small. This is shown in the figure.

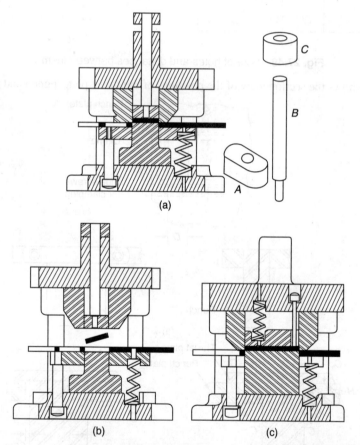

Fig. 27.50 Die and punch in working condition showing various operating conditions.

27.10.12 Stripping Factors

Component wear can be caused only by frictional forces. The punch wears at 2 to 3 times the rate of the die cavity. The basic explanation is that the punch goes through the stock twice, while the part or slug goes through the cavity only once. The stripping friction generates approximately twice the wear rate of the shearing friction, i.e. the burnished land is produced by plastic deformation of the material being punched. By the time the stripping cycle begins, the burnished land has become work-hardened. This results in an increase in the coefficient of friction.

Until very recently there was not enough knowledge relative to the factors that influence the force of friction. Consequently, performance improvement was limited to a change in the punch materials. With a fixed value of friction, the only course of action was to select a new tool steel with greater abrasion resistance. Unfortunately, the shock resistance of the tool is reduced as the abrasion resistance is increased. As a consequence, it was not unusual to exchange the tool failure potential from premature wear to that from breakage.

Punch to die clearance offers the greatest potential for increasing die performance. Since a larger than conventional clearance value generally results in the slug pulling problem, there must be a change in the slug size.

Figure 27.51 shows the die and punch assembly with the stripper plate in position.

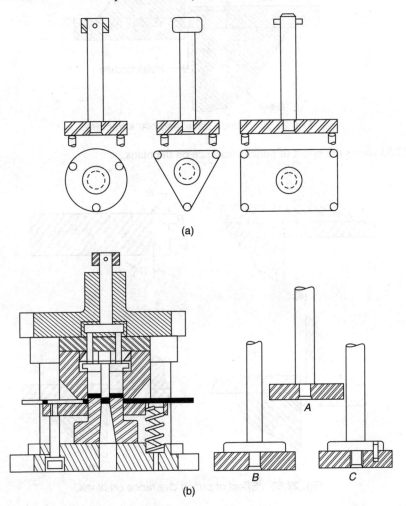

Fig. 27.51 Die and punch assembly with stripper plate in position.

27.10.13 Design of Punch

The exact dimensions of the punch on the diameter are determined by providing clearance between the punch and the die. If it is a blanking punch the size is less than normal, and if it is a piercing punch it is of the exact normal size. The punch is usually designed with a wide shoulder to facilitate mounting and to help prevent deflection under load.

Figure 27.52 shows punch travel and blank fracture.

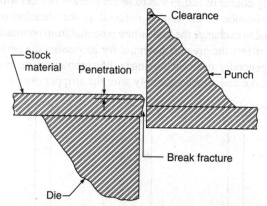

Fig. 27.52 Punch travel and blank fracture.

Figure 27.53 shows the effect of punch clearance on the blank.

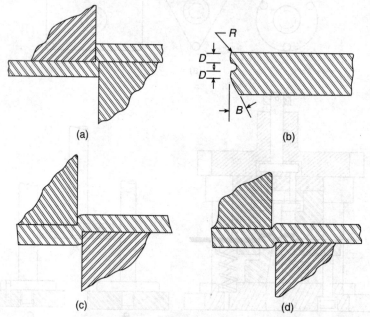

Fig. 27.53 Effect of punch clearance on blank.

Blanking punch: The blanking punches range from tiny punches for watches and instruments to large, multi-unit members for blanking large components. The size of the blank to be produced determines the type of punch to use. Design considerations include:

(i) Stability to prevent deflection;
(ii) Adequate screws to overcome stripping load;
(iii) Good dowelling practice for assurance location;
(iv) Sectioning, if required, for proper heat treatment.

The points relevant here are such considerations as keying the punch in order to keep it from turning, use of inserts for ease and economy of replacement, use of sectioning to facilitate heat treating and minimise distortion, use of shedders to prevent clinging of blanks to punch faces, and proper proportioning of and construction of blanking punches.

Larger blanking punches can be made in sections to facilitate heat treating and minimise distortion. These do not require flanges. They can be fastened individually to the punch holder with screws and dowels applied from the back.

The exact length of a punch can be found out only by laying out the whole assembly drawing, as the shut height is to be made up from the die block, die shoe, punch and punch holder, and sometimes the back-up plates. Only in rare cases are the punches over 101.5 mm in length. An allowance of 5.00 mm to 25.00 mm for sharpening should be made wherever possible in the length of the punches.

Figure 27.54 shows punch movement and blank sheet fracture.

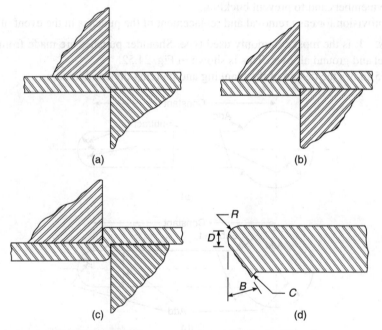

Fig. 27.54 Punch movement and blank sheet fracture.

To determine punch dimensions, practical experience counts a lot. However, some calculations are also required.

When the hole diameter equals the stock thickness, the unit compressive stress on the punch is four times the unit share stress on the cut area of the stock. Mathematically

$$\frac{4S_s t}{S_c d} = 1$$

where S_s is the unit shear stress on the stock in kg/cm²;
   S_c is the unit compressive stress on the punch in kg/cm²;
   t is the stock thickness in cm;
   d is the diameter of punched holes in cm. ($d/t \geq 1.1$ in all practical cases.)

The length of the punch can be calculated from the formula

$$L = \frac{\pi d}{8} \sqrt{\frac{ED}{S_s t}}$$

where E is Young's modulus of the punch material (for round punches).

For punching holes of $\frac{d}{t} \leq 1.1$, special steel punches are used.

Piercing punches.   Piercing punches are usually the weakest link in any die design. Therefore, the following factors must always be taken into consideration:
   (i) Make the punches strong enough, so that repeated shock in operation will not cause fracture.
   (ii) Slender punches must be sufficiently guided and supported to ensure alignment between the punch and die members and to prevent buckling.
   (iii) Make provision for easy removal and replacement of the punches in the event of breakage.

Shoulder punches.   It is the most commonly used type. Shoulder punches are made from a good grade of tool hardened steel and ground all over. This is shown in Fig. 24.52.

Figure 27.55 shows the design of the blanking and piercing die combination.

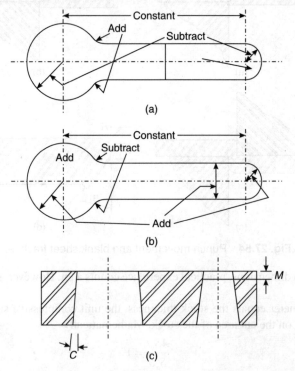

Fig. 27.55   Design of blanking and piercing die combination.

Figure 27.56 shows the effect of punch and die clearance.

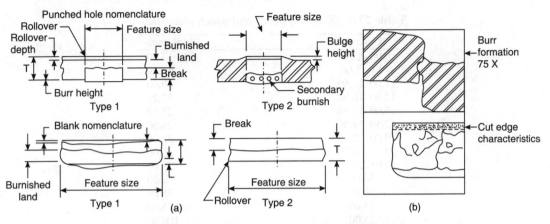

Fig. 27.56 Effect of punch and die clearance.

In Fig. 27.57, the diameter A is a press fit in the punch plate. The diameter B which extends to at least 3.175 mm is a slip fit for good alignment while pressing. The shoulder C is usually made 3.174 mm larger in diameter than A; the shoulder height D is 3.175 to 9.525 mm, depending upon size. The piercing diameter E is always on the high side of the tolerance. For example, if the hole is dimensioned 12.7254/0.1270 diameter on the part print, the punch diameter would be made 12.7254. The blending radius which connects the diameters B and E should be as large as possible, and the surfaces should be polished smooth because ridges would present focal points for fracture.

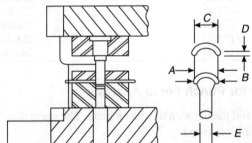

Fig. 27.57 Proportions of punch dimensions for design purpose.

Figure 27.58 shows the standard punch plate sizes.

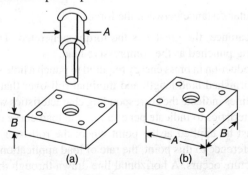

Fig. 27.58 Standard punch plate sizes.

Table 27.6 shows the relationship between the standard punch and punch plate sizes.

Table 27.6 Standard punch and punch plate sizes

A (mm)	B (mm)
0.00–8.00	12.50
8.00–11.00	16.00
11.00–12.50	19.00
12.50–16.00	22.25
16.00–17.50	25.50
17.50–19.00	28.50
19.00–22.25	32.00
22.25–24.00	35.00
24.00–25.50	38.00
50.00	50.00
50.00	75.00
75.00	75.00
75.00	100.00
75.00	128.00
100.00	100.00
100.00	128.00
100.00	152.00
100.00	128.00
128.00	152.00
128.00	178.00
152.00	152.00
152.00	178.00
152.00	203.00
178.00	254.00
178.00	178.00
178.00	228.00
178.00	280.00

27.10.14 Shear Diagram for Punch Force Analysis

Figure 27.59 shows the punch and the die with the punch plate. The shear diagram illustrates the effect of the punch to die clearance. By definition, (see Fig. 27.60),

$$\text{Torque} = F \times \frac{C}{2}$$

where F is the force;

$C/2$ is the perpendicular distance between the forces.

Obviously, the greater the clearance, the greater is the torque generated. Torque as a rolling action is as destructive to the material being punched as the compressive load.

Figure 27.60 shows the reduction in press energy required to punch a hole when the clearance is increased. The solid lines indicate the burnished land length and minimal roll-over that results from the conventional clearance values. The full end lines indicate the corresponding characteristics while the clearances are utilised. By using the hole edge characteristics to indicate penetration, the load curve can be drawn.

The maximum load must be applied at the point where the roll-over meets the burnished land. By drawing a line from the 0–0 reference to this point, the rate of load application will be shown. The maximum load must be applied until fracture occurs. A horizontal line drawn through the length of the burnished land

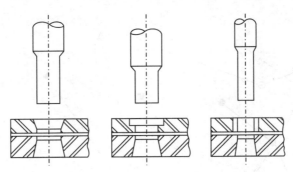

Fig. 27.59 Punch and die with punch plate.

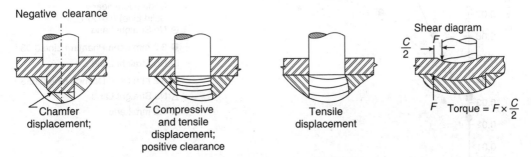

Fig. 27.60 Material deformation of sheet metal under punch.

represents the duration of the application of the maximum load. At fracture, the slug is separated from the stock and the only load necessary is that required to push the slugs out of the die cavity. By comparing the load curves, we can see that conventional clearances have generated a load curve that encompasses a very substantial area. The shaded areas illustrate the greater energy requirement for tight clearance. Obviously, the larger the clearance, the less will be the energy required.

The punching load formula does not provide for any effect of the various punch to die clearances, which indicates that it provides a large safety factor.

Figure 27.61 shows the relationship between the burr height and the number of strokes as charted for various clearances and die hole configurations.

27.10.15 Punch Plate Design

Punch plates hold and support piercing, notching and cut-off punches. They are usually made of machine steel, but can also be made of tool steel left too soft for high grade dies. Punch plates range from small simple blocks for holding single piercing punches to large, precision-machined plates for holding hundreds of perforators. Important design considerations include:

    (i) Adequate thickness for proper punch support;
    (ii) Good dowelling practice to ensure accurate location;
    (iii) Sufficient screws to overcome stripping load.

Thickness: The punch plate thickness B should be approximately $1\frac{1}{2}$ times the diameter A of the piercing punch. With the help of tables the die designer can establish the punch plate thickness quickly. While considering the thickness of the punch plate, the shut height of the press should also be kept in mind.

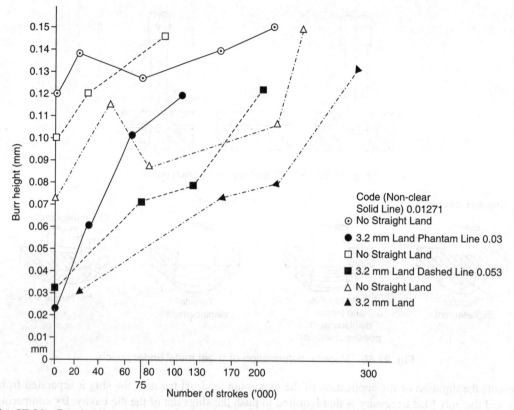

Fig. 27.61 Burr height versus number of strokes for variety of clearances and die hole configurations.

The punch plates used in compound dies are made larger so that they can act as spacers for the inverted die block A. The long socket cap screws, applied from the top of the die set, pass through clearance holes in the punch plates and are threaded into the die block.

27.10.16 Stripper Design

The primary purpose of a stripper is to remove the stock from the punch after blanking and piercing operations. However, the stripper also serves two other secondary functions. Firstly, it guides the strip if it is fixed to the die block surfaces. Secondly, it holds the blanks under pressure before the punch descends fully, if it is of the spring-loaded type.

Strippers are basically of two types:
 (i) Fixed stripper
 (ii) Spring stripper.

Fixed stripper: A simple fixed stripper may not guide the sheet, strip or workpiece, but with a little change in design the fixed stripper can be used to guide the workpiece. Then it is also called *channel stripper*. The simple fixed stripper and the channel fixed stripper are usually of the same width and length at the die block. The stripper is fastened with the same screws and dowels that fasten the die block, and the screw heads are counter bored into the stripper. However, if the die block is fastened to the die shoe from below, then the stripper screws and dowels are independent. The stripper thickness should be sufficient to withstand the force required to strip the stock from the punch. Additional thickness is provided at sides to provide the stock channel. The additional thickness is obtained by introducing small stripper blocks or by machining the original stripper plate, so as to provide a channel when fixed on the die. Screws of 10–15 mm size are used for fixing the stripper to the die block, but in either design the thickness of the stripper should be able to provide counterbores for the screw heads. The stripper thickness is generally taken on experience basis, though the shut height of the press is also a factor to be considered.

A 6 mm thick stripper is quite adequate for normal use. The height of the stock strip channel, formed either by separate blocks or by machining grooves in a thick plate, is 1½ times the stock's thickness. This should be increased if the stock is to be lifted over a fixed pin stop. The width of the channel should be at least 0.12 mm greater than the stock width. If a channel stripper is to serve as a guide properly, then apart from the provided variation of 0.12 mm in the channel width, its length should be at least twice the width of the stock or channel. This length of channel is designed by extending the stripper length on the feed end; then it guides the stock better. The block length below the stripper is also extended on the feed end side by fastening a sheet metal plate to the die block.

Spring strippers: Spring stripper plates may be made of cold rolled steel if they are not to be machined, except for holes. When machining must be applied to clear gauges, the plates should be made of machine steel, which is not subject to distortion.

Spring strippers, though complex, should be used when the following conditions are present:

(i) When perfectly flat and accurate blanks are required, because spring strippers flatten the sheet before cutting begins;
(ii) When blanking and piercing very thin material, to prevent uneven fracture and rounded blank edges;
(iii) When parts are to be pressed from waste strips left over from other operations, for good visibility to the operator for gauging purposes;
(iv) When stripping occurs immediately, as small punches are not subject to breakage;
(v) In secondary operations, such as in piercing dies, where the increased visibility provided by spring strippers allows fast loading of work and increased production.

Figures 27.62(a) and (b) illustrate the operation of the spring stripper plate. Springs, arranged around the blanking punch, provide stripping pressure. Four stripper bolts, located at the corners, clear die posts and guide

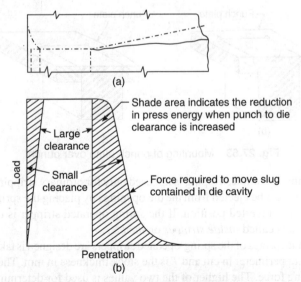

Fig. 27.62 Spring stripper plate operation.

bushings, thereby allowing the use of a smaller die set. The springs are retained in pockets counterbored in the punch holder and in the stripper plate. The holes are counter-sunk 3.2 mm by 45° in the punch holder and 1.6 mm by 45° in the stripper to guide the spring coils. A small hole is drilled completely through the punch holder and the stripper plate while the two are clamped together. This hole provides for engagement of the counterbored pilot.

Figure 27.63 shows the mounting of the punch plate over the punch.

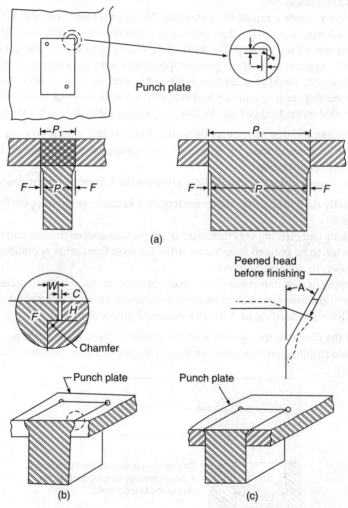

Fig. 27.63 Mounting of punch plate over punch.

The principle of spring operation stripping is not used only for ordinary stripping of the punch and work; blank or reject materials can also be ejected from the die openings by placing the spring inside the die, whether they are in upright position or inverted position. If the spring operated stripper is used to remove blanks or slug from an inverted die, it is called *inside stripper* or *shedder*.

The force for which the springs of the spring-operated strippers are designed is taken as $F = 1500 \times L \times T$ kg, where F is in kg, L is the cut perimeter in cm and T is the stock thickness in mm. The stripper force may be as high as 20% of the blanking force. The higher of the two values is used for determining the number and type of springs required. The required travel plus preload deflection will be the total deflection, and will determine the length of the spring required to stay within the allowable percentage of deflection limits. As the punch is re-sharpened, the deflections will increase and should also be allowed for.

Figure 27.64 shows the positioning of the punch stripper.

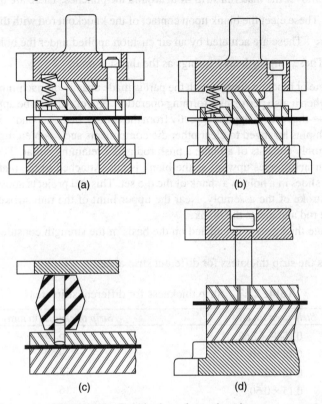

Fig. 27.64 Positioning of punch stripper.

New developments: The design of the stripper has undergone numerous changes in the past twenty years, perhaps more than any other part of the die. Formerly, when virtually all strippers were bridge type units, punch openings were hand-filled and the stripper material was soft steel. The limitations of this design were insufficient support of the punches and a tendency for the stock to tip during punch withdrawal. Both conditions led to excessive punch breakage. Today's stripper is an integral working member of the die. It moves with the upper die on guide pins that are always engaged with the top and bottom members. This design rules out the possibility of any cocking movement in the stripper movement that would easily snap the punches.

The importance of close control over stripper movement is seen in the tightness of clearance between punch and stripper. The clearance between punch and stripper must be 50% of the punch die clearance, which is far too little to sustain movement or peeking of the stripper. In some of the designs, the stripper is guided and supported by the die guide pins. It is preferable to have a smaller stripper containing its own guide pins, since the smaller the stripper the less the tendency to flexure. Moreover, the reduction in overall dimension permits greater stripper thickness and this too provides greater insurance against tilting and flexure.

Additionally, the fixing of the guide poles in the stripper provides greater bearing surface than can be obtained in the die set guide post. Such strippers are made of medium carbon steel hardened to 38 R_c. The stripper inserts which guide the punches are high carbon high chrome steel hardened to 60 to 62 R_c.

27.10.17 Knockout Design

Knockout removes or strips the completed blanks from the die members. They differ from stripper plates in that the stripper plate removes the material strip from around the punches. There are three types of knockouts.

Positive knockouts: These eject the blank upon contact of the knockout rod with the knockout of the press.

Pneumatic knockouts: These are actuated by an air cushion applied under the bolster plate of the press.

Spring knockouts: These employ heavy springs as the thrust source.

The knockout plate or block in contact with the part is made usually of machining steel; but it is made of heat treated tool steel when it also performs a forming operation. Knockouts can be applied in two ways. In the first—direct knockouts—the force is applied directly from the source. In the second—indirect knockouts—the force is applied through pins arranged to clear other die components such as piercing punches.

A knockout assembly consists of a plate, a push rod and a retaining collar. The plate is loose-fitted in the die opening contour and moves upward as the blank is cut. Attached to the plate usually by rivets is a heavy push rod, which slides in a hole in a shank of the die set. This rod projects above the shank and a collar retains and limits the stroke of the assembly. Near the upper limit of the ram stroke, a knockout bar in the press contacts the push rod and ejects the blank.

The knockout plate thickness is calculated on the basis of the strength consideration, i.e., the pressure applied.

Table 27.7 shows the strip thickness for different strip areas.

Table 27.7 Strip thickness for different strip areas

Strip area (in cm^2)	*Strip thickness* (in mm)
0.1588 × 7.62	6.25
0.15 × 7.50	9.50
0.15 × 23	12.75
0.15 × 0.50	16
0.32 × 7.50	9.50
0.32 × 15.25	12.70
0.32 × 23.00	16
0.32 × 30.50	19.00
0.5 × 7.50	12.70
0.5 × 15.25	16
0.5 × 23.00	19.00
0.5 × 30.50	22.25
0.05 × 7.50	16
0.65 × 15.25	22.25
0.65 × 23.00	19.00
0.65 × 30.50	25.50
0.80 × 7.50	19.00
0.798 × 15.25	25.25
0.798 × 23	25.50
0.8 × 30.50	28.50

Table 27.8 shows the various sizes of screws for holes.

Table 27.8 Various sizes of screws for holes

Screw size (in mm)	Socket cap screw	
	N.C.	N.F.
10.00	540	585
16.00	810	910
20.50	1250	1315
25.40	1565	1790
6.35	2850	3260
8.00	4620	5200
9.525	6950	7900
11.00	12400	10400
12.70	15500	14000
14.30	19150	17200
16	27500	21700
19.00	3050	30750
22.25	9250	41900
25.50	48500	54500

Table 27.9 shows the angular clearances for various stripper thicknesses.

Table 27.9 Angular clearances for various stripper thicknesses

Stripper thickness (in mm)	Angular clearance (in degrees)
0–1.60	0.25
1.60–4.75	0.50
4.75–8.00	0.75
8.00–0	1.00

Figure 27.65 shows the design of strippers for different punches.

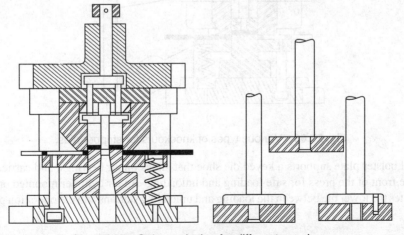

Fig. 27.65 Stripper design for different punches.

27.10.18 Design of Bushes

The pillar and bushings are provided in a press tool to maintain a constant relation between the top and bottom members. It does not allow variations in the die and punch clearance while the shearing operation is taking place.

The length of the pillar and the bushes is decided on the basis of the shut height of the press and the stroke of the press. The pillar diameter is decided on the basis of the side thrust.

27.10.19 Pneumatic Loading of Dies

Effective from 31 August 1974, any die in use that requires the operator to put his hands or fingers in the die area is in violation of DSHA standards. The simple, efficient and inexpensive means of avoiding this situation are as shown in Fig. 27.66.

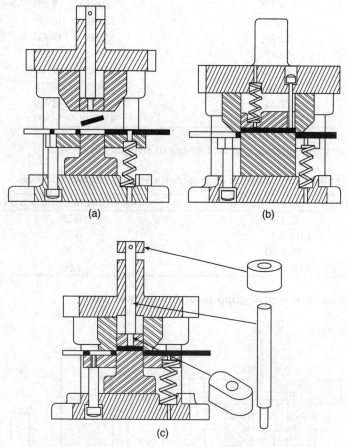

Fig. 27.66 Various types of knockouts for stripper dies.

A special bolster plate supports a keyed die shoe that slides on air bearings, and carries the lower half of the die to the front of the press for safe loading and unloading. An air cylinder mounted on the rear of the press reciprocates the lower die between the loading and unloading stations and the operation position beneath the press ram.

The standard bolster plate on the press is replaced with a mild steel plate that extends beyond the front of the press distance required for the die trend. This bolster plate is secured to the press bed, and a key-way is milled in it, front to back, on the centre line of the press. A hardened and ground key is secured in the key way with screws to guide the lower die shoe in its travel. The key strides in a mating key way milled in the bottom of the lower die.

Table 27.10 shows the various dimensions of punch and die holders.

Table 27.10 Various dimensions of punch and die holders

Die A (cm)	Space B (cm)	Punch holder thickness C (mm)	Die holder thickness D (mm)	Force E (in tonnes)
38.00	25.50	31.75	38.10	00.00 – 10.16
76.25	51.00	44.50	51.00	10.16 – 30.50
114.25	76.25	51.00	57.00	30.50 – 51.00
752.50	101.50	63.50	76.250	51.00 – 71.04
190.50	127.00	76.25	89.00	71.00 – 99.50
228.50	152.50	89.00	101.50	99.50 – 111.75
266.50	178.00	101.50	114.25	111.75 – 132.00
305.00	203.25	114.25	127.00	132.00 – 152.50
343.00	228.40	127.00	139.75	152.50 – 213.65
380.00	255.00	139.75	15.25	213.25 and more

Figure 27.67 shows the process of determining punch and die holder thickness.

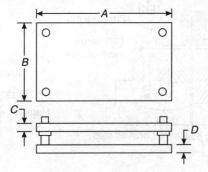

Fig. 27.67 Determining punch and die holder thickness.

Four air pockets (two on each side of the key way) are milled to a depth of 3.2 mm in the bottom of the die shoe, and holes are drilled to permit shop air to be ported to each pocket from one supply point. These pockets are spaced in a balanced pattern with relation to the die shoe centre line. When air pressure is applied, the die shoe laterally floats and one person can easily move it with finger pressure. To automate the reciprocation, a small air cylinder is attached to the press frame with the cylinder rod secured to the rear of the shoe. Simple air and electrical interlocking circuits are provided for automatic sequencing.

A dowel is pressed into the bolster plates and acts as positive rear stop for the die when it is in the press, and a hardened plate mounted on the back of the die shoe serves as a stop block when it contacts the dowel. Adjustments can be made by grinding this plate to the required thickness to assure precise location of the die under the press ram. A block mounted at the front of the bolster serves as a stop in the loading and unloading positions. Holes are drilled in the die shoe and mating holes tapped in the bolster plate to secure these two members together for transport and storage purposes.

When the operator depresses two flat buttons, air is ported to the bearing pockets, lifting the shoe and the die from the bolster by about 0.1 to 0.75 mm. At the same time, air ports to the cylinder push the die into the press. Flow control valves and cylinder cushions minimise shock at contact with the stop.

With the cylinder holding the die against the rear stock, a limit switch is automatically calculated to dump air from the pocket and initiate the press stroke.

27.10.20 Design of Press Tools with Ferrotic as Die Material

Ferrotic is a special die and tool material, far better than special die steels. It is widely used as die material (supported by special steels) and is widely used in large scale production of dies suitable for the manufacture of laminations of electric motor rotors. The die and punch design with ferrotic is different from conventional die-punch design with steel as a tool material. In the following four pages, the authors describe how design is done with ferrotic material.

Experience with steel bounded carbides has shown that sound design practice for tool steels can be readily followed when designing for the new tool material. In cases where modern rigid designs are employed, they can be utilized in connection with ferrotic with only minor modifications. On the other hand, where old designs are employed, e.g. those incorporating solid die plate instead of sectional design at L or T shaped punches, or where pressure pads and punch holders are inequitably guided, complete redesign of the tools is necessary to enable a high production material such as ferrotic to be successfully employed.

Sectional dies: Sectional design is recommended for ferrotic stamping dies. An important advantage of such design is the relative ease with which complex sections can be produced by plunge grinding with inexpensive aluminium oxide wheels (see Fig. 27.68).

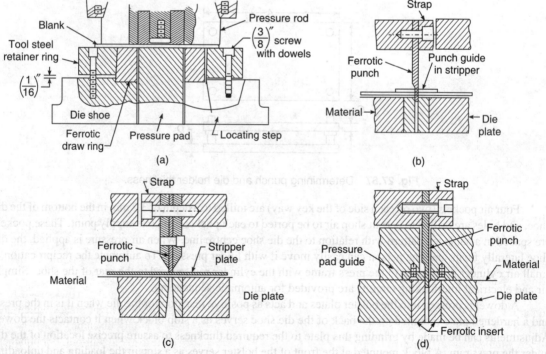

Fig. 27.68 (a) Partly sectional view of drawing tool with Ferrotic die supported by taper-bored tool steel ring, (b) Slender Ferrotic punch guided in strip per plate, (c) Double reinforcement for Slender Ferrotic punch and (d) Sectional view of tool with guided Ferrotic punch and heavy pressure pad.

Typical advantages of the plunge grinding area. The entire die section can be ground in one operation; grinding time may be as little as one hour—accuracy and reducibility are assumed; symmetrical die shapes are easily produced by placing two blanks in random and grinding both at one set-up; additional die sections for multiple cavity and couple lamination dies can be made at very little additional cost; templates are permanent and always available for security reproduction if replacement sections are needed in future; steel bounded carbide has high stability during the quench hardening process; and in most cases the need for regrinding the form after hardening is avoided. For high production dies, it is most economical to provide the section with tapped holes in the bottom faces, so that they can be clamped to the die from beneath in the conventional manner. This design creates the maximum possibility for tool regrinding and thus for long overall tool life.

Figure 27.69 shows the sectional view of a die with stripper plate and other elements.

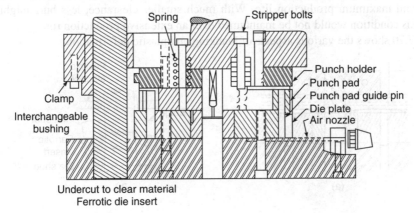

Fig. 27.69 Sectional view of die with stripper plate and other elements.

Drawing inserts: When steel bounded carbide inserts are used, it is necessary to provide suitable steel retainer rings capable of withstanding the high internal stresses which are exerted radially on the inside wall of the die during the operation. Actually, the initial compressive stress should be equal to or greater than the loop stress generated in the draw ring as a result of the drawing action. Depending upon the severity of the draw and the amount of ironing that takes place, a relatively heavy shrink fit may be used to advantage. Another design of supporting such inserts, tapered without employing the shrinking technique, is to retain a 3° taper which fits over a similarly tapered insert and is clamped down to the die support plate with a series of equally spaced bolts. This method ensures satisfactory functioning, as enough compression can be applied to the draw insert.

Punch design: In designing ferrotic C punches the need for high rigidity and economical production must both be considered. L or O shaped punches should be avoided where possible. Apart from the fact that the heavy base of an L or T shaped punch requires considerably more carbide material, the previously described formed grinding technique cannot be used, as the punch base prevents through grinding with the formed wheels.

A straight punch design with a 'nested base', as shown in Fig. 27.67, combines stiffness with low production cost. Thus the punch diameter can be obtained by dress grinding for straight pieces which are clamped in a grinding vice or held flat on the magnetic chuck of the surface grinding machine.

A small screw on the top of the punch serves to attach it to the punch plate to avoid pull-out. For punches of unusually thin cross-section, a reinforced design is recommended as shown in Fig. 27.68. The punch should be provided with a load, particularly when it is to be used for punching thick stock or in a notching operation, whereby a side thrust might be exerted on the punch.

This 5-stage tool for piercing and blanking rotor and stator laminations for electric motors incorporates slot punches made from Ferrotic machinable carbide and are typical of tools built by Brook Motors, which provide for longer usage between regrinds as compared with the tools incorporating high chromium components.

Clearance and edge preparation: For Ferrotic C, some users recommend that slightly larger clearances should be employed than with conventional alloy tool steels. 7 to 8% of stock thickness per side is suitable clearance for mild steel, alloy steel or non-ferrous material, and 9 to 10% per side of stock thickness is suitable for spring steel or stainless steel.

Some stamping specialists express concern that such a relatively wide clearance would cause excessive burrs. Actual experience in many cases has shown that these clearances in fact ensure minimum burr in combination with maximum production life. With much smaller clearance, less burr might be produced initially. But this condition would not be maintained over an extensive production run.

Figure 27.70 shows the various aspects of blanking and plastic deformation.

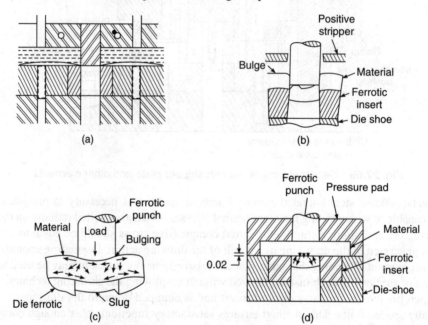

Fig. 27.70 (a) Blanking die with positive stripper plate, (b) Plastic deformation of punched sheet and slug, (c) Stress distribution in blank with punch movement and (d) Effect of pressure pad on plastic deformation of punched sheet and slug.

Figure 27.71 shows the various kinds of inserts used for different types of laminations.

Figure 27.72 shows the close-up view of a tool during the course of construction, consisting of one die ring and the associated punch assembly.

Unless precautions are taken when grinding the form on the punches, the sides of the punch face at the end and at the start of a grinding pass can have different properties, with the result that unequal rates of wear occur at different parts of the profile when the punch is in use (see Fig. 27.73).

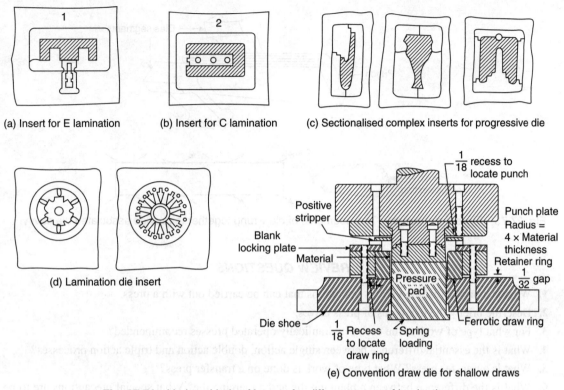

(a) Insert for E lamination
(b) Insert for C lamination
(c) Sectionalised complex inserts for progressive die
(d) Lamination die insert
(e) Convention draw die for shallow draws

Fig. 27.71 Various kinds of inserts for different types of laminations.

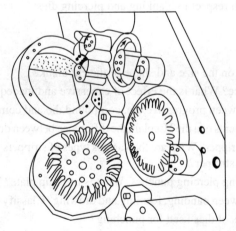

Fig. 27.72 Close-up view of tool during construction, showing one die ring and associated punch assembly.

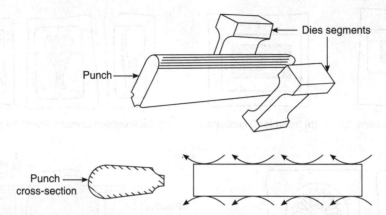

Fig. 27.73 Typical slot punch—Two segments of die wrung together to provide associated die cavity.

REVIEW QUESTIONS

1. What is a press? List the various operations that can be carried out with a press.
2. Describe briefly the five types of press drives.
3. For what type of work is the use of hydraulically operated presses recommended?
4. What is the essential difference between single action, double action and triple action processes?
5. What is a transfer press? What type of work is done on a transfer press?
6. What is the difference between a blanking die and a piercing die? What special precautions are to be taken while designing a piercing punch?
7. Explain the following with respect to blanking and piercing dies:
   (a) Clearance
   (b) Angular clearance
   (c) Shear

   What is the effect of each on the size and shape of the cut blank?
8. What is optimum clearance? What is the effect of excessive and inadequate clearance?
9. What is the difference between progressive die, compound die and combination die?
10. What are the essential elements of a die? Explain the difference between drop-through die and inverted die.
11. What is the function of a stripper? What are the various types of strippers in use in dies? How is knockout advantageous over spring stripper?
12. Sketch the ways of retaining piercing punches. What is back-up plate? Where is it most used?
13. What is the difference between banding, forming and drawing? Classify the die used for each operation.
14. What are the various types of dies used for punching?
15. How is the number of draws decided?
16. What are air bend dies? Where are they used?

PART V
Modern Methods of Manufacturing

Chapter 28

Unconventional Methods of Machining

28.1 INTRODUCTION

In recent years a number of new materials, which are harder, tough and have strong heat and shear resistance, have been developed. These materials are difficult to machine. Materials such as hastalloy, nitralloy, waspalloy, nominics, carbides, etc. are widely used in aeroplanes, nuclear engineering, space research and missile development. To machine these newer classes of materials, unconventional methods of machining are adopted.

28.2 DEFINITION OF UNCONVENTIONAL METHOD

The expression *unconventional method* is used in the sense that material removal in such cases does not occur due to plastic deformation and the formation of chips. These processes do not employ a conventional or traditional tool for metal removal; instead, they directly utilise some form of energy for metal machining.

28.3 MAJOR UNCONVENTIONAL MACHINING PROCESSES

The major unconventional machining processes include:

   (i) Ultrasonic Machining (USM) (ii) Abrasive Jet Machining (AJM)
   (iii) Electrochemical Machining (ECM) (iv) Electro Discharge Machining (EDM)
   (v) Laser Beam Machining (LBM) (vi) Electron Beam Machining (EBM)
   (vii) Plasma Arc Machining (viii) Ion Beam Machining.

28.4 PROCESS CAPABILITIES OF UNCONVENTIONAL MACHINING PROCESSES

The main machining characteristics and process capabilities are as shown in Table 28.1.

Table 28.1 Main machining characteristics and process capabilities

Process	Metal removal rate (mm/min)	Tolerance (micron)	Surface finish CLA (micron)	Depth of surface damaged (micron)	Power consumption (watts)	Work material
USM	300	7.5	0.2–0.5	2.5	2400	Refractories/ceramic glass
AJM	0.8	50	0.5–12	2.5	250	Super alloy/refractories/ceramics/glass
ECM	15,000	50	0.1–2.5	5	100,000	Steel/super alloy
EDM	800	15	0.2–1.2	125	2700	Steel/super alloy/titanium refractories
EBM	1.6	25	0.5–2.5	250	150	Refractories/ceramics
LBM	0.1	25	0.5–1.2	125	2 (average 2000)	Ceramics
PAM	75,000	125	Rough	500	50,000	Aluminium steel/super alloy
Conventional	50.0	50	0.5–5	25	3000	Ferro/organic/nonferrous metal lloys

The comparison of traditional and non-traditional machining processes is shown in Table 28.2.

Table 28.2 Comparison of traditional and non-traditional machining processes

Process	Max metal removal rate (mm^3/min)	Power consumed (kW/cm^3/min)	Cutting speed (m/min)	Penetration rate (mm/min)	Accuracy, mm Attainable	Accuracy, mm At max. metal removal rates	Typical machine input power (kW)
Conventional turning	33×10^2	0.045	75	—	0.002	0.1	22
Conventional grinding	82×10^1	0.45	1500	—	0.002	0.05	18
Ultrasonic machining (USM)	8×10^3	9.00	—	0.50	0.005	0.025	11
Electrochemical machining (ECM)	16×10^3	7.20	—	12	0.01	0.1	150
Chemical machining (CHM)	49×10^3	—	—	0.02	0.01	0.05	—
Electrical discharge machining (EDM)	49×10^3	1.80	—	12	0.01	0.1	11
Electron beam machining (EBM)	8	450	60	160	0.005	0.025	7.5
Laser beam machining (LBM)	5	2700	—	100	0.01	0.01	15
Plasma arc machining (PAM)	16×10^4	0.90	15	250	0.250	2.5	150

The typical surface finishes from non-traditional material removal processes is shown in Fig. 28.1.

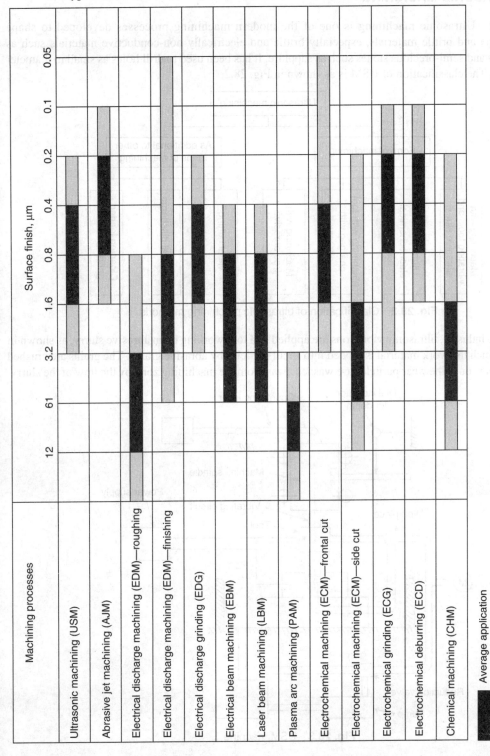

Fig. 28.1 Typical surface finishes from non-traditional material removal processes.

28.5 ULTRASONIC MACHINING

Introduction: Ultrasonic machining is one of the modern machining processes developed to shape hard-to-machine and brittle materials, especially brittle and electrically non-conductive materials such as glass, ceramics and semi-precious stones such as sapphire. It has been used to drill holes as small in diameter as 0.0125 mm. The classification of USM is as shown in Fig. 28.2.

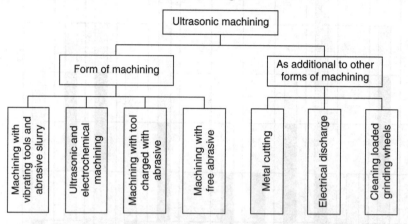

Fig. 28.2 Classification of ultrasonic machining methods.

Principle: In industry, ultrasonic vibrations are applied to a tool working in an abrasive slurry, as shown in Fig. 28.3, in which the work material is eroded into small particles by abrasive grains. The grains are crushed and the tool wears out. The wear particles are washed away from the machining zone by the flow of the slurry.

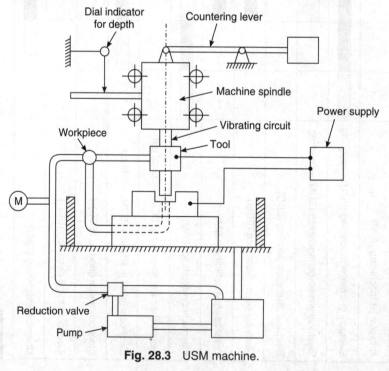

Fig. 28.3 USM machine.

The optimum working conditions, which are minimum energy consumption, low noise and a compact vibrator, are obtained by working near the low limit of the ultrasonic range (18 to 25 kHz). Hence the term *ultrasonic machining*, although the process is purely mechanical with energy imparted to the abrasive grains in the form of pulses.

Application: Ultrasonic machining is particularly effective for producing intricate shapes and openings in the workpieces of hard and brittle materials (see Table 28.3), which are most difficult or even impossible to machine by any other method. These materials are being increasingly used in electronics, instrumentation and other branches of industry, and their use has led to accelerated development of ultrasonic machining methods and exclusively ultrasonic type drilling machines. Ultrasonics may be used in turning, grinding and other metal cutting operations by adapting the feed mechanism to impart movement in the required direction. However, efficiency low, which greatly limits their application.

The relationship between machining rate and relative wear of steel tool for various machined materials is shown in Table 28.3.

Table 28.3 Relationship between machining rate and relative wear of steel tool for various machined materials

Machined material	Machining rate (%)	Relative wear of steel tool (%)
Glass*	100	0.5–1
Marble	300	—
Fluorite	280	—
Barium titanate	110	—
Germanium	100–200	0.5–1
Silicon	50–100	1–2
Jasper	95	—
Amazonite	80	—
Porcelain	60–70	—
Quartz	40–70	2
Agate	25–50	—
Topaz	40	—
Sapphire	20	—
Ruby	15–30	15
Corundum	9	—
Cemented carbide	2–5	40–80

*In normal conditions the tool penetration rate is at least 5 mm/min.

28.6 USM MACHINE

The diagram of a USM machine is shown in Fig. 28.3. The concentrator tool is connected to a vibrating circuit fitted to a machine spindle. The spindle is mounted on anti-friction slideways and is loaded against the workpiece with a force determined by a counterweight attached to a lever. The depth of penetration is shown by a dial indicator graduated in 0.01 mm divisions.

The slurry (a water suspension of abrasive powder or an electrolyte of sodium nitrate of 10 to 20% concentration) is pumped in by a special pump and the reduction valve, which provides a pumping pressure of 0 to 6 atm. The slurry is fed through a hole in the tool or workpiece. When ultrasonic and electrochemical machining are used in combination, a current of 200 A and 15 V potential is applied from the power supply consisting of a transformer, a two-half-wave type rectifier on silicon power diodes, and a current controller.

The ultrasonic head is mounted on a cylindrical column and has a vertical movement of 200 mm (power drive). The precision table, with an optical measuring system, ensures a positioning error of less than 10 mm.

1. Tool
2. Vibrating circuit
3. Machine spindle
4. Workpiece
5. Countering lever
6. Dial indicator for depth
7. Pump
8. Reduction valve
9. Power supply.

Ultrasonic machines are used in industry for drilling and reproducing cavities in cemented carbide dies; for blanking, heading and coining drawing dies and moulds for cutting optical lens blanks; and dies for cutting ceramics, ferrites and other brittle materials [see Fig. 28.4(a) to (h)]. When using an abrasive of 10.3 grain size, ultrasonic machining produces surface finishes within the 6.8 pm limit on cemented carbides, and the surface finish beneath the tool end face is one degree higher than that on the sidewalls. Surface finishes one class lower are obtained on glass. The optimum power requirements are 2.5 W/mm^2 and the slurry flow rate (boron carbide) is about 0.6 g/min cm^2.

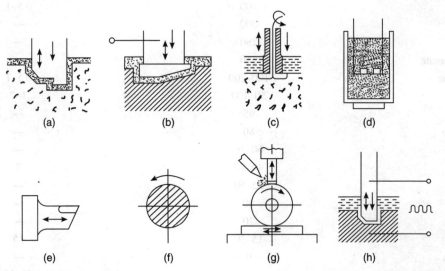

Fig. 28.4 Diagrams showing the use of ultrasonic vibrations in various methods of machining.

Machining with a free abrasive (in a slurry) and with randomly orientated ultrasonic vibrations [see Fig. 28.4(d)] is used to some extent in industry for debarring small components in mass production conditions. This method shows promise when applied with a high hydrostatic pressure, which greatly increases the output of the operation.

In certain cases, good results can be obtained by applying ultrasonic vibrations to a cutting tool, for instance a single point turning tool, as shown in Fig. 28.4(e), or to a drill, tap, milling cutter, etc. The surface finish is then improved by one or two classes, a built-up edge is prevented, chatter is eliminated, and the cutting temperature and forces are minimised due to the reduction in friction.

Slurry: The slurry can be considered a paste of abrasive grains in liquid. Both the abrasive grains and the liquid affect the performance of the process; we shall discuss them in detail.

The abrasives generally used are:

(a) Silicon carbide (SiC)
(b) Boron carbide (B_4C)
(c) Aluminium oxide (Al_2O_3).

Advantages of USM: Ultrasonics are very effective when used to clean grinding wheels. The extensive cavitation in the cutting fluid loosens and removes the particles from the loaded grinding wheel, thus improving the surface finish of the ground work by one to two classes and reducing wheel wear by more than 80%.

Ultrasonics are also used to intensify electrical discharge machining, for which purpose ultrasonic vibrations are applied either through the fluid or directly to the tool-electrode workpiece as shown in Fig. 28.4(g). The first method is recommended for machining with a wire electrode.

In general the advantages are:

(i) There is no appreciable rise in the temperature of the work, and hence no change in the physical properties of the workmaterial.
(ii) Cheap abrasives are used for the machining operations.
(iii) Brittle materials like glass, ceramic and Jew stones can be machined quite efficiently.
(iv) The photo-elastic examination of machined glass workpieces shows that there is no residual mechanical stress in the workpiece.
(v) The process is soundless and can be sensed by touching the equipment only.
(iv) The equipment is quite safe to handle and requires skilled labour only for setting the tool.

Accuracy: Dimensional accuracy up to ± 0.005 mm is possible and surface finish up to Re 0.1–0.125 micron can be obtained.

Limitations: The limitations are:

(i) Low metal removal rate (3 mm^3/s)
(ii) High power consumption.

Recent developments: Recently a new development in ultrasonic machining has taken place, in which a tool impregnated with diamond dust is used and no slurry is used. The tool is oscillated at ultrasonic frequencies as well as rotated. If it is not possible to rotate the tool the workpiece may be rotated.

This innovation has removed some of the drawbacks of the conventional process of drilling deep holes. For instance the hole dimension can be kept within ± 0.125 mm. Holes upto 75-mm depth have been drilled in ceramics without any fall in the rate of machining as is experienced in the conventional process.

$$\text{Metal Removal Rate} = 5.9 - V_0 \frac{R}{2} \frac{\sigma}{H} \text{ (in } mm^2/s)$$

when f is the frequency of the active grits striking the work surface in cps;
  R is the grit radius in mm;
  H is the surface hardness in kg/mm^2;
  V_0 is the amplitude of the vibration in mm;
  σ is the stress developed in the tool in kg/mm^2.

28.7 ABRASIVE JET MACHINING

Principle of operation: It involves the use of a high speed stream of abrasive particles carried by a high pressure gas on the work surface through a nozzle. The metal is removed due to the erosion caused by the abrasive particles impacting the work surface at high speed.

The schematic diagram of the working process is shown in Fig. 28.5. Filtered compressed air or gas is passed into the mixing chamber, where the abrasive particles are fed through a sieve which is made to vibrate at a suitable frequency. The mixing ratio is controlled by the amplitude of the vibration of the sieve. The stream of compressed air and abrasive particles is finally made to pass through a nozzle, so that it comes out in the form of a high velocity jet which impinges on the workpiece.

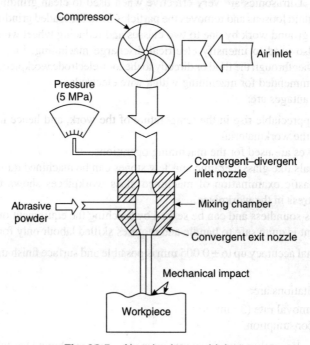

Fig. 28.5 Abrasive jet machining.

Data: The materials used are:
Abrasives:
    1. Al_2O_3 (Aluminium oxide)
    2. SiC_2 (Silicon carbide for cutting and grounding)
    3. $NaHCO_3$ (Sodium bicarbonate for high finishing)
    4. Dolomite (mineral for etching and polishing)
    5. Glass beads (for light polishing);

Carrier gas: Air, CO_2, N_2; O_2 is never used;

Size of abrasive particles: Up to 100 microns;

Pressure of nozzle: 2 to 8 kg/m^2;

Shape: Round, square, rectangular;

Stand-off distance: 0.7 mm to 15 mm;

Velocity of mixture emerging from nozzle: 150 to 300 m/min;

Accuracy: Up to tolerance ± 0.05 mm.

Advantages: These are:
    1. It can cut intricate holes and hard materials;
    2. Thin sections can be machined;

3. It has low capital cost;
4. High quality surface finish can be obtained;
5. No mechanical contact between work and tool is there.

Limitations: These are:
1. MRR is slow;
2. There is high nozzle wear rate;
3. The machining accuracy is poor;
4. Additional cleaning of work surface may be necessary due to the possibility of abrasive grains sticking in softer material.

Applications: Abrasive jet machining is used:
1. To cut slots;
2. To cut thin sections;
3. To cut confers;
4. For drilling, producing shallow crevices and deburring;
5. For producing intricate shapes in hard and brittle material;
6. For cleaning and polishing of plastics like nylon and teflon;
7. For frosting of interior surfaces of glass etching, marking of cylinder, etc.

28.8 ELECTRICAL DISCHARGE MACHINING (EDM)

EDM is a process of machining by "melting" large numbers of very small particles of metal by bombarding the metal with sparks from a shaped electrode. The voltage used is low (from 1 to 45 V). The amperage increases as the area covered increases; 5 to 60 A is commonly used. However, machines are made with capacities of several hundred amperes.

EDM has been used for small work for many years. As the power supplies improved, EDM machines began to be used for making punches and dies for presswork and cavities for forging dies. Today, a die-making shop can not compete without the use of electrical discharge machines.

In the last decade, the aerospace industries began to use EDM for certain difficult jobs in machining the tough metals they use. Since then the use of EDM in production lines for such jobs as drilling the holes in carburettors and many other jobs has rapidly increased.

28.8.1 Theory and Analysis of EDM

Electrical discharge machining comprises a number of physical processes which take place within the restricted space of the spark gap, and which share one energy source, namely an electrical discharge. This electrical discharge causes electrical energy to be converted into thermal and mechanical energy with a density attaining 30,000 J/mm^3 and power approaching hundreds of kW/mm^3.

The discharge occupies a certain volume in space, which is determined by the cross-sectional area of the discharge channel from microns to 1 mm^2 and by its length (from hundredths to tenths of a millimetre). The channel is divided into three sections characterised by voltage drops across the column and across the anode and cathode. In spite of the fact that the sections near the electrode can be measured in fractions of a micron, the voltage drop in these sections is over two-thirds of the total voltage, and consequently the greater part of the energy is released in these sections.

The energy distributed between the electrodes (tool and workpiece) and the spark gap (dielectric) gives rise to (in both electrodes) pyrolysis in the fluid as well as mechanical shock waves. Thus, an electrical discharge in an electric drive is a convened source of energy for the spark erosion process. A general representation of the phenomena occurring in spark erosion and the various relations between them is shown in the diagram in Fig. 28.6. The output characteristics (surface finish and accuracy) for a given workpiece and set of tool materials is a function of the discharge parameters and therefore of the pulse type and the current produced by the pulse generator.

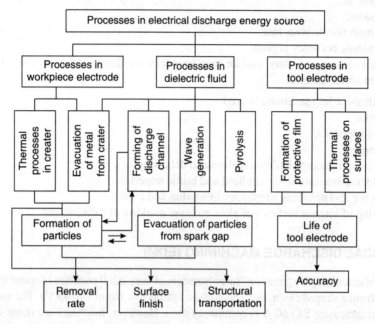

Fig. 28.6 Relationship between main physical processes in electrical discharge machining.

The EDM processes are all interconnected and consist of the heating and melting (or evaporation) of metal and its evacuation. When the heating temperature is below the melting point, there is no metal removed, nor will there be any if there are not sufficient forces to evacuate the molten metal. In an actual spark erosion process, the volume of material evacuated from a crater is determined by the relationships between these processes.

28.8.2 Basics of EDM

When two conductors (a properly shaped electrode and a workpiece) in an electric circuit are brought close together, a spark is generated between them. As shown schematically in Fig. 28.7(a), this spark (estimated to be at about 10,000 F/5500°C) melts and vaporises a spot in the metal. The direct current (DC) circuit is caused to cycle on and off (pulsed) from 250 to 500,000 times per second (250 Hz to 500 kHz), creating the condition shown in Fig. 28.7(b).

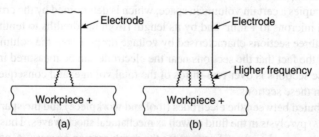

Fig. 28.7 Comparison of low- and high-frequency sparks in EDM.

According to the way electrons flow, work is normally positive and the electrode is the negative side of the circuit. Reverse polarity is used for some work.

The tank in which machining is done (or *burning* as some people call it) is filled with an electrolyte. This is a non-conducting liquid, often a petroleum distillate with additives. The vaporized metal, coming immediately in contact with the liquid, solidifies into a tiny hollow ball. Thus, the wet chips from EDM look like a fine black sludge.

28.8.3 Construction Features of EDM Machine

The electrical discharge machine itself is basically a vertical-spindle milling machine with a rectangular tank on the work table. The machine table may be moved along both the X and Y axes by hand-wheels or occasionally by numerical control.

The tank, which is fastened to the table, may be 330 × 500 mm in size; a small EDM machine of size up to about 1200 × 2400 mm costs about ₹ 70,000 to ₹ 1 lac, with larger sizes costing quite high. A typical schematic of the entire set-up is shown in Fig. 28.8.

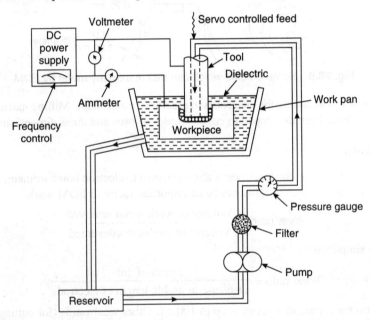

Fig. 28.8 Schematic drawing of EDM system. (The "tool" is the electrode.)

The spindle does not usually rotate but is fed downwards by a very sensitive servomechanism which keeps the electrode properly positioned as it cuts. The spindle is equipped with collies, v clamps, or a flat plate for holding various shapes of electrodes.

A flushing system is a part of the electrical discharge machine base and holds 30 to 75 gallons (114 to 284 litres). This is used to fill the work tank to about 25 mm above the work. A pump circulates this dielectric (non-conducting fluid) through a fine filter to remove the chips. The electrolyte is also used to flush the chips away from the electrode, as will be described later.

The electrical system, except in the small 10 to 15 A machines, is sold separately. These are rated at 25 to 400 A and cost between ₹ 6,000 and ₹ 50,000. Most units today are made so that split circuits can be used; that is, a 50 A service may be used to feed two electrodes using 25 A each.

Controls are used to vary the frequency, amperage, voltage, capacitance and the off-on time cycle.

Electrode: The electrodes are the cutting tools. Today they are usually made of fine grain carbon (graphite) or a carbon–copper mixture. These are easily machined to shape, are fairly inexpensive, will cut accurately, and can give good finish. A 25 × 25 × 150 mm electrode will cost about ₹ 50 to ₹100. Copper, brass, tungsten, and silver tungsten are also used. The last two give superb finish, but are difficult to cut to shape. The electrode and the workpiece should not be made of the same material.

The shape of the electrode is the shape of the cut. Thus, as shown in Fig. 28.9, any shape can be cut into metal. Electrical discharge machines which will cut 0.125 mm diameter holes in hardened steel using copper-tubing electrodes are made. The large machines finish-cut large forging, blanking, or press dies.

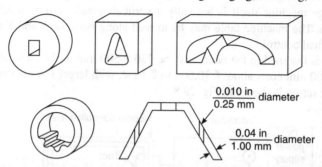

Fig. 28.9 Some shapes which can be cut into any metal by EDM.

Thus, the real skill is in cutting the electrodes to precise size and shape. Milling machines, tracer millers, and special machines have been developed to cut electrodes to two- and three-dimensional shapes.

28.8.4 Wear Ratio

As the spark bombards the work, the electrode is also subjected to electron bombardment. Thus, the electrode is eroded (worn away) during the cut. This can be an important factor in EDM work.

$$\text{Wear ratio} = \frac{\text{Volume of work metal removed}}{\text{Volume of electrode consumed}}$$

This is often simplified to

$$\text{Wear ratio} = \frac{\text{Depth of cut}}{\text{Decrease in usable length of electrode}}$$

The wear ratio for carbon electrodes is up to 100 : 1. Other wear ratios (for cutting steel) are copper, (2 : 1), brass (1 : 1), and copper tungsten (8 : 1).

Thus, a piece of copper cutting 25 mm deep into steel will wear 12.5 mm. Because of this, the spindle must travel 37.5 mm to make the cut. These ratios are approximate and will vary considerably.

Reverse polarity (electrode positive) practically eliminates electrode wear. The workpiece metal will "plate" onto the electrode, so that it will actually grow very slightly larger. This is called *no-wear EDM*. It is used mostly for roughing cuts and is slower than conventional EDM. However, it does save the expense of making a new electrode for each part made.

All electrodes will cut a hole or cavity larger than their own size. This overcut will be 0.025 to 0.2 mm on all surfaces.

This overcut increases with higher current and decreases with higher frequency. The amount of overcut must be known if the final work size is to be held. All manufacturers today can supply charts such as those shown in Fig. 28.10. These are quite accurate, and so the electrode size can be figured out ahead of time.

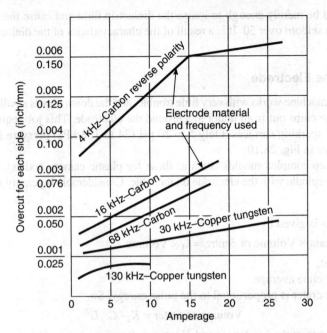

Fig. 28.10 Overcut versus amperage for several combinations of electrode and frequency.

28.8.5 Metal Removal Rate (MRR)

Electrical discharge machining is not a fast way to remove metal. The maximum rate is 0.045 in^3 (0.74 cm^3)/hr/A. Thus, a 25 A EDM machine will remove a maximum of 1.13 in.^{3} (18.5 cm^3) of material per hour. Fine finishing cuts may be made at one-tenth this rate.

Thus, if a large cavity (such a large forging or forming die) is to be made, it is often more economical to rough out most of the metal, in the annealed condition, by conventional milling. The die may then be hardened, and EDM will economically finish the shape to size, though some filing and polishing may still be required.

The material being cut will affect the MRR. Experiments indicate that the metal removal rate varies inversely as the melting point of the metal. The approximate value is given by

$$\text{MRR} = \frac{2.4}{(\text{Melting point in } °C)^{1.25}}$$

Thus EDM will cut aluminium much faster than steel.

Feeds and speeds for EDM cannot easily be taken from published tables, as they can be for either turning, drilling, or milling. The first run of a new job often requires some experimenting with the control settings. However, there are controls, and they do affect the metal removal rate. Some guidelines are:

Frequency: Lower frequencies give higher metal removal rates and poorer finishes. The top amperage is limited by the area of the electrode and by the gap between the electrode and the work. The current control knob is sometimes labelled the *removal rate selector*. The finish gets poorer as the amperage is increased. The maximum amperage with carbon electrodes is 50 A/in^2 (7.75 A/cm^2).

Duty cycle: This is the relative off-on time of each pulse of electricity. On most EDM machines this can be hand-controlled. A longer duty cycle tends to increase the MRR but causes a rougher finish.

Voltage: This should be merely enough to ionise the dielectric fluid and cause the spark to cross the gap. The working voltage is seldom over 50. It is a result of the characteristics of the dielectric, the work material, and the electrode.

28.8.6 Flushing the Electrode

An electrical discharge machine works with very little trouble, and the down time is small. The most troublesome job is getting the sludge chips out from under and around the electrode. This job requires a smooth constant flow of fluid across all sparking surfaces. Only 5 to 20 psi (34 to 138 kPa) pressure is needed. Some of the methods used are shown in Fig. 28.10.

In burring out deep complex moulds (such as those for plastic camera bodies), the work is sometimes put upside down on the spindle with the electrode on the table. Considerable ingenuity is needed in such cases.

Surface finish

According to Brash this is given by
$$(h_{av})^3 = \text{Constant} \times \text{Volume of centre} = k_1 \times \text{Volume}$$
where k_1 is a constant;
h_{av} is the centreline average.

Again, the volume of a crater is proportional to the pulse energy. So
$$\text{Volume of crater} = K_2 \cdot C \cdot U^2$$
where C is the capacitance of the condenser and U is the voltage.

Therefore $\quad (h_{av})^3 = K_1 \cdot K_2 C \cdot U^2$

or $\quad h_{av} = K_3 \cdot C^{1/3} \cdot U^{2/3}$

28.9 DETERMINATION OF METAL REMOVAL RATE IN RELAXATION CIRCUIT

The relaxation type of the circuit can be considered to consist of three parts:
  (i) Charging circuit;
  (ii) Discharging circuit;
  (iii) Sparking portion.

Let E_a be the applied voltage for charging the circuit in volts;
E_b be the charging voltage of the condenser in volts;
R_1 be the charging resistance (or resistance in the charging circuit) in ohms; and
C be the capacitance of the condenser in farads.

At any time t, the current flowing through the charging circuit i_1 is given by
$$i_1 = \frac{E_a - E_b}{R_1}$$

Again $\quad i_1 = C\dfrac{d}{dt}(E_b)$

Combining the two equations, we get
$$\frac{E_a - E_b}{R_1} = C\frac{d}{dt}(E_b)$$

or
$$\frac{dE_b}{E_b - E_a} = -\frac{1}{R_1 \cdot C} \cdot dt$$

Integrating both sides

$$\int_0^1 \frac{dE_b}{E_b - E_a} = -\int_0^1 \frac{1}{R_1 \cdot C} \cdot dt$$

or

$$\log(E_b - E_a) = \frac{t}{R_1 \cdot C} + K \quad \text{where } K \text{ is a constant}$$

$$E_b - E_a = e^{(-t/R_1C)+K}$$

$$= e^{(t/R_1C)} \cdot e^K$$

$$= e^{-t/R_1C} K_1 \quad \text{where } K_1 \text{ is a constant}$$

The boundary conditions are: $t = 0$, $E_b = 0$.
Substituting this in the above equation, we get

$$K_1 = -E_a$$

$$E_b - E_a = -E_a \cdot e^{-t/R_1C}$$

or

$$E_b = E_a(1 - e^{-t/R_1C}) \tag{28.1}$$

or

$$\frac{E_b}{E_a} = 1 - e^{-t/R_1C}$$

or

$$e^{-t/R_1C} = 1 - \frac{E_b}{E_a}$$

Taking log on both sides

$$\frac{t}{R_1 \cdot C} = -\log_e\left(1 - \frac{E_b}{E_a}\right)$$

$$t = R_1 \cdot C \cdot \log_e \frac{1}{\left(1 - \frac{E_b}{E_a}\right)}$$

The frequency of charging (or spark)

$$f_e = \frac{1}{t} = \frac{1}{R_1 \cdot C} \cdot \log\left(1 - \frac{E_b}{E_a}\right)$$

The energy delivered in each spark $= \frac{1}{2} \cdot C \cdot E_b^2$.
Total energy per second

$$= \frac{1}{2} \times \text{Frequency of sparking} \times \text{Energy per spark}$$

$$= \frac{1}{2} f_e \cdot C \cdot E_b^2$$

The metal removal rate is proportional to the energy.
Hence, metal removal rate

$$\text{MRR} \propto \frac{1}{2} f_e \cdot C \cdot E_b^2$$

$$\propto \frac{1}{2} \cdot \frac{1}{R_1} \frac{1}{\log_e\left(1 - \dfrac{E_b}{E_a}\right) E_b^2} \tag{28.2}$$

Conclusion: It is apparent from Eq. (28.2) that if the resistance R_1 decreases the metal removal rate increases. But if R_1 is made too small, then circuit arching will take place instead of sparking, which may damage the work material metallurgically. So an optimum value of resistance should be kept in the charging circuit.

This optimum resistance is also called *critical resistance*.

28.10 CRITICAL RESISTANCE

This is given by

$$R_1 C \geq 30 \sqrt{\frac{L}{C}}$$

where L is the equivalent inductance of the charging circuit;
C is the capacitance.

The frequency of cutting

$$f = \frac{1}{\text{Charging time} + \text{Sparking time}}$$

Since the sparking time is very small

$$f = \frac{1}{\text{Charging time}}$$

$$f = \frac{1}{R_1 E_{m-a} \times C} = \frac{1}{30\sqrt{\dfrac{L}{C}} \times C} = \frac{1}{30\sqrt{LC}} = \frac{0.03}{\sqrt{LC}} \tag{28.3}$$

28.11 CONDITION FOR MAXIMUM POWER

For maximum power delivery to the discharging circuit

$$E_b = E_a(1 - e^{-t/R_1 C})$$

The energy delivered at any period of time

$$dE = i_1 E_b \, dt$$

now

$$i_1 = C \frac{dE}{E_b \, dt} \tag{28.4}$$

Substituting the value of E_b from Eq. (28.1)

$$i_1 = C \cdot \frac{d}{dt}[E_a(1 - e^{-t/R_1 C})]$$

$$i_1 = C\left(\frac{d}{dt} E_a - \frac{d}{dt} E_a e^{-t/R_1 C}\right)$$

$$= 0 - C \cdot E_a \cdot e^{-t/R_1 C} \cdot \frac{d}{dt}\left(-\frac{t}{R_1 C}\right)$$

$$= C \cdot E_a \cdot e^{-t/R_1 C} \cdot \frac{1}{R_1 C}$$

$$= \frac{E_a}{R_1} \cdot e^{-t/R_1 C}$$

Substituting it in Eq. (28.4), we get

$$dE = \frac{E_a}{R_1} \cdot e^{-1/R_1 C} E_b \cdot dt$$

$$= \frac{E_a}{R_1} \cdot e^{-1/R_1 C} E_a (1 - e^{-1/R_1 C}) \cdot dt$$

$$= \frac{E_a^2}{R_1}(e^{-1/R_1 C} - e^{-2/R_1 C}) dt$$

Integrating both sides, we get the total energy stored, that is

$$\int dE = \int \frac{E_a^2}{R_1}(e^{-1/R_1 C} - e^{-2/R_1 C}) dt$$

$$E = \frac{E_a^2}{R_1}\left[e^{-1/R_1 C}\left(-\frac{1}{\frac{1}{R_1 C}}\right) - e^{\frac{2t}{R_1 C}}\left(-\frac{1}{\frac{2}{R_1 C}}\right)\right] + K_1$$

where K_1 is a constant

$$= \frac{E_a^2}{R_1}\left(-R_1 C \, e^{-1/R_1 C} + e^{2t} \frac{R_1 C}{2}\right) + K_1 \quad (28.5)$$

Putting the boundary conditions in the above equation to evaluate K_1 at $t = 0$, $E = 0$

$$0 = \frac{E_a^2}{R_1}\left(-RC_1 + \frac{R_1 C}{2}\right) + K_1$$

$$K_1 = \frac{R_1 C}{2} \frac{E_a^2}{R_1}$$

Substituting it in Eq. (28.5), we get

$$E = \frac{E_a^2}{R_1}\left(-R_1 C e^{-1/R_1 C} + e^{2t/R_1 C} \frac{R_1 C}{2}\right) + \frac{R_1 C}{2} \frac{E_a^2}{R_1}$$

or

$$= \frac{E_a^2}{R_1} R_1 C \left(-e^{-t/R_1 C} + \frac{1}{2} e^{2t/R_1 C} + \frac{1}{2}\right) \quad (28.6)$$

If this energy is delivered to the discharging circuit in time $t = t_1$, then the average power P_{av} delivered is obtained by dividing Eq. (28.6) by t_1.

$$P_{av} = \frac{E_a^2 \cdot R_1 \cdot C}{R_1 \cdot t_1}\left(-e^{-t/R_1C} + \frac{1}{2}e^{2t/R_1C} + \frac{1}{2}\right)$$

We have the equation

$$E_b = E_a(1 - e^{-1/R_1C})$$

If we know the value (numerical) of $\frac{1}{R_1C}$, we can get a simplified relation between E_b and E_a.

Now differentiating P_{av} with respect to $\frac{1}{R_1C}$, and putting it equal to zero, we can get the value of $\frac{1}{R_1C}$.

Therefore, for maximum power

$$\frac{dP_{av}}{d\left(\frac{t_1}{R_1C}\right)} = 0$$

(as $t = t_1$)

$$= \frac{d}{d\left(\frac{t_1}{R_1C}\right)}\left[\frac{E_a^2}{R_1} \cdot \frac{R_1 \cdot C}{t_1}\left(e^{-t_1/R_1C} + \frac{1}{2}e^{-2t_1/R_1C} + \frac{1}{2}\right)\right]$$

or

$$\frac{t}{R_1 \cdot C} = 1.26$$

Substituting for E_b, we get

$$E_b = E_a(1 - e^{-1.26}) = 0.76\, E_a$$

28.12 WIRE CUT EDM

It uses a very thin wire (0.02 to 0.3 mm in diameter) as an electrode and machines a workpiece with electrical discharge like a band saw by moving either the workpiece or the wire. The erosion of the metal utilising the phenomenon of spark discharge is the same as in conventional EDM. The prominent feature of a moving wire is that a complicated cut-out can easily be machined without using a forming electrode. Figure 28.11 shows the principle of wire-cut EDM. Basically it consists of a machine proper, composed of a workpiece, a contour movement controlled NC unit, a workpiece mounting table, a wire drive, a square power supply system, and a dielectric fluid unit with constant specific resistance.

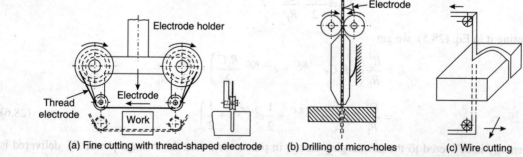

(a) Fine cutting with thread-shaped electrode (b) Drilling of micro-holes (c) Wire cutting

Fig. 28.11 Wire-cut EDM—NC machining (or optical copying) with wire electrode.

Advantages

These are:
1. The forming electrode adapted to the product shape is not required.
2. The electrode wear is negligible.
3. The machined surfaces are smooth.
4. The geometrical and dimensional tolerances are light.
5. The relative tolerances between punch and die is extremely high.
6. The die life is more.
7. The operating rate is high.
8. Low level of skill is required for operation.
9. Low level of maintenance and planning is sometimes needed to get efficient hushing around the electrode.

Application of EDM process and future developments: At present the following electrical discharge machining processes are being used in industry—(1) cavity sinking machines for forming steel forging dies and moulds, roughing turbine blade forgings, sinking openings in drawing dies and spinnerets, piercing nozzle holes and piercing channels in air and hydraulic equipment; (2) wire electrode machines for cutting out profiles in electronic components, for making templates and some types of bending dies, etc.; (3) cut-off machines for slitting blanks up to 750 mm diameter, particularly in materials which are difficult to machine; (4) grinding machines for removing casting cores and for marking, engraving and manufacturing of rubber sole moulds for shoe industry, tyre moulds and glass moulds.

The current technical knowhow is utilised to develop electrical discharge machining processes for intricate-shaped components and large components made from new materials.

The industrial development of electrical discharge machining is becoming increasingly important and includes the following features: the development of a general classification of components; an investigation into the stability of the process parameters; the determination of the spark gaps and the causes of oscillations in spark gaps; the development of fixtures and methods for compensating errors; combining electrical discharge machining with metal cutting, etc. Research may also be carried out on new methods of precision electrical discharge machining. In view of the marked reduction in the wear of metal electrodes, new methods may be developed for the precision manufacturing of electrodes. The ongoing research on generation and control equipment is fully based on theoretical investigations into the electrical discharge processes, and covers various conditions of precision machining within a wide range of frequencies.

The various methods of flushing chips from parts while cutting by EDM are shown in Fig. 28.12.

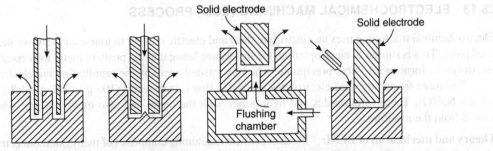

Fig. 28.12 Methods of flushing chips from parts while cutting by EDM.

In addition to developments which have improved machine and magnetic saturation generators, work is being carried out on the development of transistorised and thyristor pulse generators with a wide frequency range, which are the bases for the next stage in the development of electrical discharge machining.

An electrical discharge machine is a complex machine tool consisting of the following units: spark gap controller with manual or automatic setting; controller operating on the basis of metal removal rate and machined area, with predetermined programme or programme given automatically; controller of finishing condition motions with manual or automatic programming; controllers of auxiliary (vibrations and amplitudes of planetary motion); controllers in the system for the evacuation of erosion products (e.g., controllers of pressure, fluid temperature, viscosity, filtration, etc).

A great deal of work is being done on the development of new machines. At present, various types of universal and special purpose machines are manufactured by industry for various methods of electrical discharge machining and for carrying out various operations. The new models of universal machines are designed on the basis of the standard ranges of machines and modifications to them.

The machine tool industry should start manufacturing a group of electrical discharge machines for cavity sinking and wire electrode machines based on conventional jig boring machines.

Advantages and limitations of EDM process: Electrical discharge machining can be used to cut any shape into almost any conductive material (including the carbides), regardless of how hard it is. No burrs are formed, and no forces are exerted on the workpiece. Discs can be finished after hardening, thus avoiding the warpage which occurs if they are hardened after finish machining.

The sizes of cuts are from 0.010 in. diameter (0.25 mm); holes to press dies are over 48 in. (1200 mm) long, though most electrodes are from 2 to 12 in. (50 to 200 mm) on their largest sides.

Finishes of 10 μm to 0.25 μm can be obtained, and tolerances of ± 0.002 in. (0.05 mm) or even better can be held. The EDM machine, even on long duration cutting, needs very little attention. One operator can attend to two or three machines.

The major disadvantage of EDM is its slowness in removing metal. However, in die work and for narrow slots and small diameter holes, the time taken still may be less than half of that required by any other method.

The electrode must be accurately machined, which needs skilled work. However, cutting carbon is a lot faster and easier than cutting steel.

The EDM process, due to the heat involved, leaves 0.001 to 0.005 in. (0.025 to 0.125 mm) of resolidified metal around the surface of the cut, as well as an annealed layer under this. In die work this sometimes seems to be an advantage. However, if the work is used under stress, the hairline cracks in the resolidified areas may cause early failure.

Because of the wearing of the electrode, two or more electrodes may have to be made to complete one job. As many as six electrodes are needed to accurately complete complex moulding dies.

28.13 ELECTROCHEMICAL MACHINING (ECM) PROCESS

Electrochemical machining uses an electrolyte fluid and electric current to ionise and remove metal from the workpiece. This is similar to electroplating, the difference being that the positive metal ions combine with the electrolyte to form an insoluble precipitate, which is washed away by the rapidly flowing electrolyte.

The most frequently used electrolytes are NaCl, that is, salt water (100 g to 300 g salt/L) and sodium nitrate, $NaNO_3$. These form oxides and hydroxides with the metal and also hydrogen gas, which must be vented from the machine.

Theory and mechanism of ECM: Electrochemical machining comprises of many interconnected processes: electrode and hydrodynamic, mass and heat transfer, movements of charges in the electrical fields, and electrochemical reactions in the flow of the electrolyte.

Electrochemical reactions occur between the metal–electrolyte boundary surfaces. In oxidation reactions, electrons are transferred to the metal from the ions in the electrolyte solution in the regeneration reactions, electrons are transferred from the metal to the ions in the electrolyte. If there is no external current flowing through the boundary, then in this so-called *equilibrium condition* an equilibrium potential appears on the boundary at which the oxidation current is equal to the regeneration current. In electrochemical machining conditions, the potential drop is localised within the limits of the dense electrical layer of several Angstroms. This layer has considerable capacitance. Each electrochemical reaction has its own equilibrium potential depending on the temperature and the activity of the components in the reaction.

As a result of the electrode reactions a gradient of the concentrations of reacting particles is formed in the electrolyte. This means that when an external current flows through the boundary between the metal and the electrolyte, the concentration of the reacting particles near an electrode is greater than that of the particles in equilibrium conditions. The potential drop is the derivative of the concentration potential, which is proportional to the logarithm of the above-mentioned concentrations. The transfer of ions in the electrolyte may be either by diffusion, migration in an electrical field or convections in the flow. It is assumed that the magnetic field of the working current has practically no effect on the movement of the ions in electrochemical machining conditions. In these conditions, at distances less than 10^{-3} to 10^{-4} cm, diffusion transfer is predominant. Equalisation of the concentrations near the electrodes occurs at the depth of the diffusion layer.

In electrochemical machining, OH^- ions and electrolyte anions take part in anodic reactions, which at a specific electrode potential begin to detach oxygen (partly or completely) from the oxides, thus forming intermediate dissolved compounds. The problem of the machineability of metals is therefore reduced to that of the interaction between OH^- ions, activation anions and the passivating film. An oxide film several Angstroms thick is complex and has considerable capacitance.

The molecular and electrostatic action of the electrodes on the fluid extends into the electrolyte to a depth of several ion layers, that is, approximately 100 Å. If the gaps are not more than 0.1 to 0.2 mm, this interaction may be disregarded. In typical cases of electrochemical machining the Reynolds number is within 10^3 to 10^4, and the flow conditions may then be turbulent as well as laminar. The small gas bubbles (0.01 cm diameter) behave as solid spherical particles. In turbulent flow the small gas bubbles are distributed fairly uniformly and hydrogen bubbles are easily detached from the cathode by the hydrodynamic action of the flow. Gas bubbles (for example, air introduced with electrolyte) are split by the turbulent flow. If large bubbles enter the gap, these may lead to gas plugs. When the electrolyte flow is increased, cavitation occurs. This results in an increase in the localised electrical resistance in the gap, a reduction in the area of contact between the anode and the electrolyte, and a reduction in machining accuracy.

Mass transfer (ion movement) in forced convection conditions is described by the equation of convection diffusion.

The thermal phenomena have a considerable influence on the electrochemical process, mainly by altering the electrical conductivity, viscosity and electrode potentials. In power machines, heat is generated mainly by the current flow. In steady-state electrochemical machining conditions, the amount of heat generated by the process is equal to the amount of heat removed by the electrolyte flow. The distribution of heat in the gap can be determined by solving a partial differential equation.

The dissolution process on each element of the anode surface also depends on the local concentration of activating anions, the intermediate and final products of the anodic process, the hydrogen in the concentration index (pH) determined by the rate of the electrode processes, and the irreversible chemical reactions in the flow of OH^- ions with the products of the anodic dissolution. These reactions form insoluble hydroxides.

To calculate the metal removal rates from the anode, it is necessary to determine the current distribution, which is found from the field theory equations. The cathode and anode potentials are boundary conditions for the gap potential. In a general case, the field problems are characterised by the mixed (varying) boundary conditions of a potential, particularly that on the anode. The simplest field problems (Laplace equations) are derived on the assumption that the electrical conductivity in the gap is constant. The electrical conductivity is increased due to heating, with the result that the gap emission on the electrodes is reduced. It can be shown that on using a 25% solution of sodium chloride, a total machine current of 10000 A, a 10 V potential and an electrolyte flow rate of 200 L/mm, the electrical conductivity at the edges of the gap is increased by 20% due to the generated heat, and is reduced by 5% as a result of gas emission. In the case of non-uniform distribution of conductivity k in the gap, more complex field equations have to be used.

The quality of finish on the machined surface is determined by the statistical properties of the electrochemical machining process; the statistical character of the distribution of the atom energy in a polycrystalline metal; and the statistical properties of the anode. The quality of the anode surface also depends on the current density I_a on the anode and the polarisation of the anode ($\delta V_{an}/I_a$), where V_{an} is the anode potential.

From the point of view of analysis, all methods of electrochemical machining may be divided into the following two groups:

1. *Steady*, in which the anode form remains stable in time relative to the cathode. The method is only possible where the motion of the anode or cathode is uniform (for instance, the sinking of a cavity).
2. *Unsteady*, in which the anode form changes with time.

In any case, the rate of local removal on the anode is proportional to the local current density, chemical equivalent and current output. Analysis shows that there is a class of functions, $f_a(x, y)$, which describes the surface of the workpiece (anode) produced in steady-state electrochemical machining.

The steady-state methods may be divided into direct and reverse forms. In the direct method the tool form, which in specific conditions will be reproduced on the anode, should be determined from the given stable anode form. In the reverse method, the workpiece form must be determined from a given cathode form in specific steady conditions. The direct problem is reduced to that of determining the field of potential V near the anode, from the value of the potential on the anode surface and its first derivative. The local current density is determined on the basis of the steady-state condition on the given anode surface to be produced. The cathode form coincides with the equivalent potential corresponding to the cathode potential V_{ca}.

Experimental and theoretical results show that a steady electrochemical machining process can be carried out in a specific machine having a specific potential form and with a corresponding cathode and anode. Such a process is equally determined by ten parameters of the machine, which can be easily measured without disturbing the machining process. These parameters are as follows: supply source potential; cathode feed rate; temperature; concentration; pH value; electrical conductivity; the kinematic viscosity of the electrolyte at the entry to the working gap; electrolyte pressure (delivery rate); and the flow of the compressed gas supplied to the electrolyte.

Electrochemical machine: The principal parts of an electrochemical machine are shown in Fig. 28.13. The electrode holder is controlled in much the same way as in the electrical discharge machines. The worktable is moved along the X and Y axes by handwheels or automatic controls. The work-holding tank is usually made of stainless steel or plastic to resist the corrosive action of the electrolyte.

The power supply furnishes only 5 to 24 V but 500 to 25,000 A of direct current (DC), though most work can be done with 1000 to 5000 A. The power supply is kept separate from the machine to avoid corrosion from the electrolyte.

The electrolyte is circulated under pressure, usually 50 to 150 psi (345 to 1035 kN/m^2 or kPa) and occasionally higher. A 2500 A machine is supplied with a 200 gallon (750 L) tank, and a 5000 A machine may need an 800 gallons (3000 L) storage tank. Stainless steel circulating pumps supply up to 120 gallons per minute (450 L/min) at the pressure given above.

Filters, settling tanks or centrifuges are used to clean the electrolyte, and cooling coils may be needed for the larger machines when they are used near full capacity. The temperature should be about 120°F (49°C) maximum.

UNCONVENTIONAL METHODS OF MACHINING 795

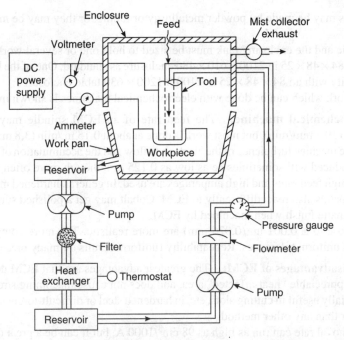

Fig. 28.13 Schematic drawing of ECM system.

The electrode is usually made of copper, stainless steel, or tungsten copper. As shown in Fig. 28.14, it is insulated except at the cutting tip. Insulation today is usually plastic, though back-on ceramic has been used. There is almost no wear on the electrodes, and so they last indefinitely on most types of ECM machines.

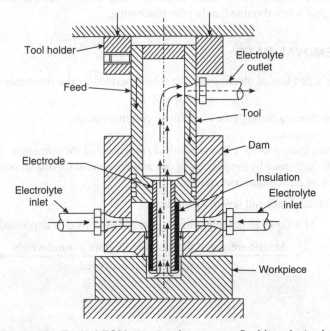

Fig. 28.14 Typical ECM set-up using reverse flushing electrode.

796 TEXTBOOK OF PRODUCTION ENGINEERING

The electrodes may be made by powder metallurgy or tubing, or they may be machined to shape as in EDM.

The work table and the enclosing tank must be sized to hold one or several workpieces. One company lists 41 × 24 × 19 in 84 × 48 × 25 in 2100 × 610 × 480 enclosure as standard. One of the largest ECM machines has 25,000 A capacity with an 84 × 48 × 25 in 2100 × 1200 × 630 table enclosure.

The typical work which can be done with electrochemical machines is shown in Fig. 28.14.

Control of electrochemical machine: The feed rate of an ECM spindle may be set from 0.05 to 0.50 in./min (1.25 to 19.7 mm/min), but most work is done at about 0.15 in./min (3.8 mm/min). This is limited by the capacity of the machine turbulence of the electrolyte flow and the accumulation of hydrogen gas bubbles.

The finish produced will sometimes be as low as 0.125 µm, though more often 0.5 to 3.75 µm. Better finish results when high feed rates and high amperages are used. In general, hardened metals machine to better finishes. Different metals also react differently to ECM. Cobalt may get a polished surface, while aluminium will always have a matte finish when machined by ECM.

Tolerances close to ± 0.001 in. (0.025 mm) are more realistic. The more complex the cut, the more difficult it is to hold uniform tolerances. Repeatability (uniform results on many pieces) is excellent.

Advantages and disadvantages of ECM: The greatest advantages are that ECM does not produce burrs, does not have any appreciable "heat-affected" area, and does not cause machining stress.

ECM is especially useful in cutting slots, etc. in hardened steel or difficult-to-cut alloys. On tough alloys, ECM may cut faster than any other method.

The metal removal rate can run as high as 98 cm^3/1000 A, but it can be a great deal less when ECM is used to cut small-diameter holes (sometimes 25 at a time) through tough metals.

The initial cost is the principal item slowing down the use of ECM. A 500 A electrochemical machine costs about ₹ 190,000, and a 5000 A machine may cost ₹ 450,000 plus accessories and tooling for each job.

The tooling (electrodes, flushing arrangement, etc.) is complex and thus more expensive. The use of a lot of electricity and the corrosive nature of the electrolyte must also be considered. The parts machined by ECM must be thoroughly washed immediately after machining.

28.14 METAL REMOVAL RATE

According to Faraday's first law of electrolysis, the mass liberated by the substance

$$M = Z \cdot I \cdot t \tag{28.7}$$

where I is the current flowing through the electrolytic cell in amperes;

t is the time in seconds;

Z is the constant known as the *electrochemical equivalent* of the substance. It is equal to the mass of the ions liberated by the substance by the passage of one ampere current for one second through the electrolytic solution or by the flow of one coloumb of charge.

According to Faraday's second law of electrolysis

M = Equivalent weight of a substance dissolved or deposited

$$= \frac{\text{Atomic weight of the material in gram atomic weight}}{\text{Valency of metal dissolved}} = \frac{A_w}{v}$$

Also

$$Z = \frac{1}{F} \cdot \frac{A_w}{v}$$

where F = Faraday's constant = 96,500 C = 26.8 A-h.

Thus the equation $M = ZIt$ becomes

$$M = \frac{1}{F} \cdot \frac{A_w}{v} \cdot I \cdot t$$

and metal removal rate

$$\text{MRR} = \frac{M}{a\rho t} \text{ cm/min (approx.)}$$

assuming 100% current efficiency; where

a is the machined area in cm^2;
ρ is the density of the workpiece in g/cm^3;
t is the time of current applied;

or

$$\text{MRR} = \frac{Ft}{F} \cdot \frac{A_w}{v} \cdot \frac{M}{a\rho t}$$

$$= \frac{1}{F} \cdot \frac{A_w}{v} \cdot \frac{It}{a\rho t}$$

$$= \frac{A_w}{v} \cdot \frac{1}{F} \cdot \frac{\rho_c}{\rho} \quad \text{where } \rho_c = \frac{I}{a}$$

EXAMPLE 28.1 In an ECM process for machining iron, it is desired to obtain a metal removal rate of 1 cm^3/min. Determine the amount of current required for the process, assuming that at. wt. of iron = 7.8 g/cm^3, valency at which dissolution occurs = 2, density of iron = 7.8 g/cm^3, and Faraday's constant = 1609 A-min.

Solution Given atomic weight of iron, $A_w = 56$ g
Valency of iron dissolution, $v = 2$
Density of iron, $\rho = 7.8$ g/cm^3
Faraday's constant, $F = 1609$ A-min
Metal removal rate (MRR) = 1 cm^3/min

$$1 = \frac{A_w}{Fv\rho} \times I$$

$$I = \frac{1609 \times 2 \times 7.8}{56} = 448 \text{ A}$$

EXAMPLE 28.2 In an ECM process machining iron by using copper tool and a saturated solution of NaCl in water as the electrolyte, the electrode area = 2 cm × 2 cm and the initial gap (h) for the electrolyte to pass is equal to 0.020 cm. For the electrolyte, specific heat = 0.997 cal per g per °C, density = 1 g/cm^3, and specific resistance = 3 ohm-cm. Calculate: (i) the permissible fluid flow velocity if the maximum permissible temperature of the electrolyte is the boiling point (95°C), the ambient temperature is 25°C and the applied voltage is 10 V; (ii) the maximum metal removal rate if the permissible current density is 150 A-cm^2. We are given that permissible fluid velocity for a rectangular electrode as per the following expression

$$\frac{V^2 l}{4.187 \, r_e \cdot h^2 \rho_e C_s (\theta_B - \theta_A)}$$

where V is the voltage applied;
   l is the length of the electrode;
   r_e is the specific resistance of the electrolyte;
   h is the gap length;
   ρ_e is the density of the electrolyte;
   C_s is the specific heat of the electrolyte;
   θ_B is the boiling temperature of the electrolyte; and
   θ_A is the ambient temperature.

Solution $V = 10$ V, $l = 2.0$ cm, $r_e = 3$ ohm-cm, $h = 0.020$ cm, $\rho_e = 1$ g/cm³, $C_s = 0.997$ cal/g °C, $\theta_B - \theta_A = 95° - 25° = 70°$.

$$\text{Velocity of fluid flow} = \frac{10^2 \times 2.0}{4.187 \times 3 \times (0.02)^2 \times 1 \times 0.997 \times 70}$$

$$= 558 \text{ cm/s} = 334.8 \text{ m/min}$$

$$\text{Metal removal rate (MRR)} = \frac{A_w}{v} \times \frac{1}{A} \times \frac{1}{F_\rho}$$

Putting the values, we have

$$\text{MRR} = \frac{56}{0.2} \times 150 \times \frac{1}{96540 \times 7.86}$$

$$= 0.0055 \text{ cm/s or } 0.33 \text{ cm/min}$$

28.15 ELECTROCHEMICAL OR ELECTROLYTIC GRINDING FOR TOOLS AND CUTTERS

Introduction: Electrolytic grinding is used to save time, and to greatly cut the cost of grinding wheels for grinding multiple-tooth milling cutters, carbide tipped tools, and tool bits. Also, because like all electrochemical processes it puts very little pressure on the work, it is used to grind slots, flats, and forms in thin-walled or fragile workpieces.

Theory and mechanism of ECG process: Electrolytic grinding (see Fig. 28.15) is used to grind cemented carbides and other current conducting materials which are difficult to machine. The wheels used in electrolyte grinding, namely, diamond wheels for cemented carbides and conventional abrasive metal bonded wheels for steel. In electrolytic grinding, metal is removed by a mechanical action, anodic dissolution and removal of oxide films.

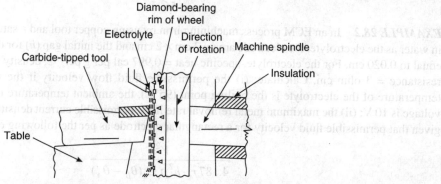

Fig. 28.15 Electrolytic grinding process.

The diamond grinding wheel is subjected to a low potential (6 V), and current is passed between the wheel and the workpiece through the medium of an electrolyte pumped continuously into the contact zone, used most frequently for grinding cemented carbide tools. The effectiveness of the process depends on the wheel specification, feeds and speeds, electrolyte properties, electrolyte concentration and method of delivery, wheel running accuracy, machine stiffness, etc.

From among the diamond wheels produced at present, the most suitable for electrolytic grinding are wheels on the cast metal bond, with 8 to 12 grain size and 100 to 150% concentration.

To ensure maximum output, the diamond coating should be at least 20 mm wide.

The following are the recommended conditions: wheel speed—18 to 22 m/s, longitudinal feed—1 to 2 m/min, wheel pressure for medium-sized tools—6 to 12 kgf/cm^2 of the contact area. The optimum potential (according to literature data, 5 to 6 V) is reduced. When the potential is increased above this value, wheel life is greatly reduced. In optimum conditions the current is 80 to 100 A. Numerous experiments carried out by many organisations have established that the best electrolyte is composed of 5% KNO_3 and 0.3% $NaNO_2$, with water as the remainder. The electrolyte density should be 1.035 g/cm^3; the recommended flow rate is 4 to 64 mm^3/min.

The metal removal rate may be varied within wide limits by a suitable selection of wheel specifications, speeds and feeds, etc. It is obvious that this will be accompanied by wide variations in diamond consumption. Because the in feed on this machine is provided with an elastic mounting of the tool, with all other conditions equal, the output rate and wheel wear will depend on the pressure of the wheel on the tool. By changing the wheel pressure, the output and wheel wear may be varied. The most economic pressure is that which ensures a metal removal of 300 to 450 mm/min, for a contact area of 1 cm^2 between the wheel and the cemented carbide tool. In these conditions, diamond wheel consumption is lower than in conventional diamond grinding.

When the wheel pressure is increased, the cemented carbide removal rate (according to some organisations) increases at the rate of 1.5, while the wheel wear increases at a rate of 1.7. This shows that with a further increase in pressure, the wheel wear is greater than in conventional diamond grinding. Calculations show, however, that when the diamond consumption is increased by 30%, the cost of grinding cemented carbides electrolytically is 1.2 to 1.5 times lower than in conventional grinding.

An important advantage of electrolytic grinding is the possibility of simultaneously grinding the cemented carbide tip and the steel shank, but in such a case the wheel wear is increased and output reduced by 1.5 to 2.5 times. For this reason, the ground area of the steel shank should be as small as possible and should not exceed the tip area of the cemented carbide. Electrolytic grinding produces a good surface finish.

Investigations have shown that electrolytic grinding produces a layer with a changed structure extending in depth to 10 to 25 µm, causing rapid tool wear after an initial wear of 0.1 mm.

To increase the tool life, this layer having an altered structure should be removed by sparking out for 1 to 3 seconds with the working current disconnected. This sparking out improves the surface finish.

ECG machines and process: An ECG machine looks much like a standard grinding machine. In fact, standard grinders can be converted to electrochemical grinders. The most used styles are surface grinders, tool grinders and form grinders, though OD and ID grinders are also used if the quantity of production makes them economical.

The total package uses the grinding machine, a 95–150 L tank, and a 250–1000 A power supply at 4–15 V. The electrolytes used are the same as in other electrochemical machining processors. The enclosure is usually made of plastic to resist the corrosive action of the salt.

ECG machines must have the spindle insulated from the rest of the machine. Brushes, or the newer mercury bearings, are used to conduct electricity to the grinding wheel. The grinding wheel is the cathode (negative) side of the DC circuit.

The flow of the electrolyte between the wheel (cathode) and the work (anode) dissolves the metal. The grits projecting from the wheel are set to just barely touch the work. Thus, they act as spacers and also "wipe" away the oxides formed on the workpiece. As the oxides are soft, there is very little grinding wheel wear (see Fig. 28.16).

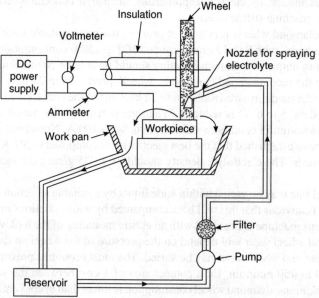

Fig. 28.16 Electrochemical surface grinder for flat or cortour work.

Grinding wheels: The grinding wheels have standard round, cup, or formed shapes; they must contain a conductive bond and nonconductive abrasive. The abrasives most often used are diamonds, Borazon, and aluminium oxide, often about 100 grit in size.

The bond is often of copper, copper–carbon, or copper–plastic mix. The wheel is trued or shaped with a diamond tool, the same as for conventional grinding wheels. However, the abrasive must project slightly beyond the bond; so the dressing is done by reversing the current and depleting a few thousandths of an inch (or hundredths of a millimetre) of the metal bond.

The schematic diagram of an electrolytic grinding machine is shown in Fig. 28.17.

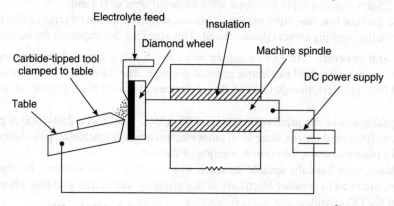

Wheel 300/1, D, 120, 68 concentration Carbide P 12.2 × 1 cm²

Fig. 28.17 Schematic diagram of electrolytic grinding machine.

Advantages and disadvantages of ECG: Reports indicate the following advantages:
  (i) An ECG machine requires only normal maintenance;
  (ii) Wheel wear is greatly reduced. An automotive company grinding multiple-tooth face mills reports using only one diamond wheel per year (previous use: over 50 per year at ₹ 5000 per wheel);
  (iii) The entire machine should be thoroughly washed occasionally, as the electrolyte is corrosive;
  (iv) The greatest advantage is that all work is completely free of burrs;
  (v) No heat is developed;
  (vi) No heat cracks or distortions are developed;
  (vii) Very little pressure is exerted on the work.

ECG, when grinding cutting tools, takes the full cut in one pass (no rough and finish cuts). In fact, the larger the wheel "contact" area, the better the ECG work.

The disadvantages are:
  (i) The major disadvantage is the cost of the ECG system. A 1000 A machine, with tank, pumps, and electrical controls, may cost ₹ 500,000. Because of this, ECG is most often used when long runs are made between set-ups, or when conventional grinding or milling of slots and forms creates excessive burrs (or, especially on tough aerospace materials, takes too long).

Figure 28.18 shows the plot of the work–wheel interface pressure test.

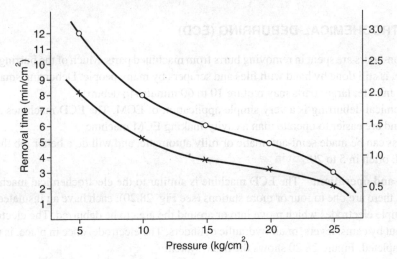

Fig. 28.18 Work–wheel interface pressure test: Oscillation length—15–18 mm; oscillation rate—65 double strokes/min.

  (ii) There is high power consumption;
  (iii) There is low metal removal rate;
  (iv) The work material must be a good conductor of electricity.

Figure 28.19 shows the machining time comparison when grinding various tool face sizes using the ECG process.

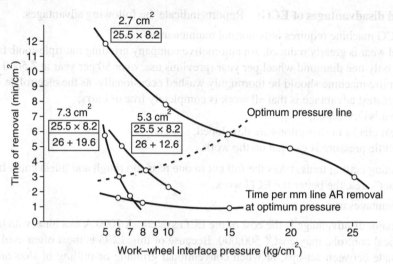

Fig. 29.19 Machining time comparison when grinding with different tool face sizes by electrolytic diamond grinding process.

28.16 ELECTROCHEMICAL DEBURRING (ECD)

A great many man-hours are spent in removing burrs from machined parts. Much of the burring, more properly called *deburring*, is still done by hand with files and scrapers by many people. Deburring small parts by hand takes less than a minute; larger parts may require 10 to 60 minutes to deburr.

Electrochemical deburring is a very simple application of ECM. The ECD machines are smaller, use less electricity and are easier to operate than a cavity-making ECM machine.

The process can be made semi-automatic or fully automatic, and will do a better job than can possibly be done by hand, often in 5 to 20.

ECD machines and operation: The ECD machine is similar to the electrochemical machine without the spindle. Instead, there are one to four or more stations (see Fig. 28.20); each have an insulated fixture to hold the work and simple electrodes which move into or around the area to be deburred. The electrode movement may be carried out by cams, levers, or air hydraulic cylinders. The electrode, once in place, is not moved until deburring is completed. Figure 28.20 shows the basic system.

The electrolyte is the same as for ECM, but needs only 10 to 50 psi (70 to 345 kPa) pressure, and often only 25 to 200 A of electricity per station will do the job. The electrolyte is stored in the reinforced fiber-glass tank in the base of the machine.

The copper or stainless steel electrode is positioned with a 0.005 to 0.025 in. (0.125 to 0.63 mm) clearance from the burr. It thus often cuts the burr off at the root. The length of deburring time and the amount of current used regulate the amount of radius or chamfer on the corners of the work. The parts are transferred to a washing tank immediately after deburring.

Doughnut-shaped electrodes are used for external burrs and gear shapes, and any desired irregularly shaped electrode can be made quite easily. A 2000 A, four-station ECD machine will cost about ₹ 250,000, and machines as small as 250 A are made, costing about ₹ 80,000. The tooling for both machines costs extra.

UNCONVENTIONAL METHODS OF MACHINING   **803**

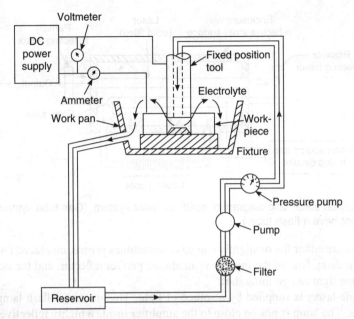

Fig. 28.20 Schematic drawing of electrochemical deburring machine.

28.17 LASER BEAM MACHINING APPLICATIONS AND PROBLEMS

Introduction: A laser beam, which is virtually a monochromatic light beam concentrated over an area, has a number of specific features which determine its application as a tool. Experimental results and modern developments in laser beam installation indicate that a laser beam may be economically used as a tool for producing holes or slots of less than 0.5 mm diameter (width), that is, when other mechanical drilling and electrical methods are either impossible to apply or are uneconomical. The maximum depth of hole formed in steel by a single pulse is approximately 4 to 5 mm.

The term *laser* is an acronym for "light amplification by stimulated emission of radiation". The most commonly used light-emitter is a cylindrical ruby rod which is "pumped" by light from a flash tube. When electrical energy is discharged into the lamp, there is a powerful flash of light that "pumps" ions within the ruby rod to higher-than-normal energy levels. When a few ions drop to the normal energy level, they emit photons that simulate other ions to do the same. The result is a chain reaction that causes all of the "pumped-up" ions to drop to the normal energy level almost simultaneously. The resulting pulse of light is of a single wavelength (6943 Å, within the region of visible radiation), and much of it travels parallel to the axis of the rod. A mirror at one end reflects the light emerging from that end of the rod back through the rod, so that light is added to the light emerging from the other end of the rod.

The resulting beam can be focused by a simple lens, so that its energy is concentrated on an extremely small area. This is what makes lasers of interest to manufacturing engineers—the focused laser beam can be made sufficiently powerful to vapourise any known material, or with somewhat less power, laser beams become excellent tools for spot type welding operations.

Laser system—Basic construction and operation: A laser operates on the principle that electrons in certain atoms oscillate when energy is supplied. A basic laser circuit (see Fig. 28.21) consists of three parts—a pair of mirrors, a source of energy, and an optical amplifier. This amplifier is popularly called the *laser*. A control system and a cooling system must be added to these basic parts.

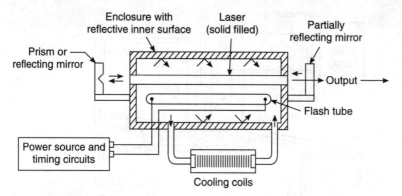

Fig. 28.21 Simplified schematic diagram of solid-rod laser system. (Gas-tube systems are similar but do not have a flash tube.)

The two mirrors are either flat or slightly concave. Sometimes prisms are placed facing each other with the amplifier between them. One of the mirrors is an almost perfect reflector, and the other is only partially reflective, that is, some light can go through it.

Energy for some lasers is supplied by a source of light. This may be a flash lamp filled with xenon, argon, or krypton gas. The lamp is placed close to the amplifier inside a highly reflective cylinder, so that as much energy as possible can be absorbed by the laser material. Other lasers, mentioned later, are supplied with energy by pulses of DC electricity.

The amplifier (laser) is a solid or gaseous material or, as developed recently, a liquid material which contains a small percentage of a particular kind of atom or molecule. This amplifier may either be a rod smaller than a human finger or a gas-filled tube several metres long.

When energy is pumped into the amplifier, light begins to bounce back and forth between the mirrors. A particular colour component (wavelength) of the light is amplified each time it oscillates. This light rapidly becomes very intense. At some point in this build-up of energy, light goes out through the partial reflector. This is the useful output. All of this happens in 0.000001 to 0.001 s; each time a short burst of energy (pumping) is supplied.

This pulse or beam of light can be focused to a small diameter by lenses and reflected around corners by mirrors. It can also weld parts which are inside transparent glass or plastic enclosures.

Within limits set by the materials involved, the frequency and the amount of energy supplied by each "pumping" can be varied.

The cooling of the laser and the lamp is necessary because lasers are very inefficient. Often less than 1% of the pumped energy is sent out as useful laser energy, although efficiencies range from 0.02 to 20%.

Thus, 80 to 99.98% of the input energy is converted to heat, and so all but the smallest lasers must have a cooling system. As in gasoline engines, both air and water cooling are used. Often cooling water is circulated around both the laser and the lamp (when lamps are used), and the water is cooled in a radiator just as in an automobile. The laser radiator may be a square body of only 4 in. (100 mm) sides, or it may be larger than the one used in a car.

Laser materials: The first operating laser used a ruby rod, and ruby still appears to be the best material for high-power lasers. Many other solid materials that can be induced to "laser" are known—neodymium-doped glass, neodymium-doped calcium tungstate, gallium arsenide and even some organic materials. The glass rods convert energy more efficiently than ruby, but they have somewhat shorter lives than ruby rods.

Neodymium-doped calcium tungstate is a reasonably efficient laser material and it can be pumped without break for continuous-wave laser action. At present, the highest output obtained is about 1 W—too small for metalworking applications. The semiconductor lasers based on gallium arsenide materials have quite high efficiency but their output is only in milliwatts.

Gas lasers can be continuously pumped to give a continuous beam, but the continuous beams are of such low strength (in terms of power, energy, or time) that they are of little interest for machining or welding applications. For these applications, ruby-rod lasers of the pulse type—capable of delivering large amounts of energy in an almost instantaneous pulse—are most satisfactory.

Nd glass amplifiers are made of glass to which a small percentage of neodymium has been added. This is less expensive than the ruby rod.

Nd–YAG: Neodymium-doped (added to) yttrium-aluminium-garnet crystals are in some ways superior to both of the above, though they are expensive. A ¼ in. diameter × 3 in. long (0.25 × 75 mm) Nd–YAG rod costs over ₹ 10,000. The cheapest laser using such a rod costs about ₹ 40,000 (Fig. 29.22 shows a commercial Nd–YAG laser with control by NC tape). A complete outfit costs ₹ 3,00,000 to ₹ 5,00,000.

Neodymium lasers, just like most commercial solid lasers, operate in bursts of energy. These pulses can (when a device called *Q switch* is used) be either as short as a few billionths of a second (nanoseconds or ns) or about 1/100 s (10 ms) long.

Depending on the use, pulse rates commonly range from 1 to 20 per second, though higher and lower frequencies can also be used.

CO_2 gas lasers consist of a glass tube filled with carbon dioxide. The tubes vary in length from 250 mm to 30.5 m. The necessity for long lengths is minimised today by methods such as circulating the gas and using transverse excited atmospheric (TEA) higher-pressure gas in the tubes, with many discharge points along the length of the tube. CO_2 lasers usually cost ₹ 50,000 and more.

CO_2 lasers may have a beam diameter from 1 to 100 mm, and produce power levels from less than 1 mW to over 1 kW.

Helium–neon (He-Ne) gas lasers are made in small, inexpensive sizes that are used in virtually every industry in the country. They are safe, emitting only 0.001 W (1 mW) or so in a visible continuous orange beam of light (though 50 mW He-Ne lasers are also made; however, these are not safe). Their most important use is in alignment and measurement of machinery. The modern surveyor's transits now use He-Ne lasers, and some new machine safety devices also use them.

Applications of lasers: The economic effectiveness of a laser beam as a tool is also determined by the life of the flash tubes. A single tube may be used for piercing approximately 50,000 holes; therefore, if the production programme embraces a batch of components in which the total number of holes to be pierced is two to three times greater than the tube capacity, the cost of tubes may nullify any savings obtained by the reduction in production time.

The operations at present carried out with laser beam may be classified as follows:

1. Piercing single holes and systems of holes of diameters ranging from a few microns to 0.5 mm. Typical examples of this operation are those of piercing holes in diaphragms used in optical and electronic devices; piercing ferrite rings; producing holes in diamond and cemented carbide drawing dies; piercing watch jewel bearings; and producing dies for drawing synthetic yarns.
2. Operations in this category include forming slots up to 10 μm wide and marking the division marks on graduated scales.
3. The preparation of precision resistances in which the error BVs maintained within 0.1 to 0.05%.
4. The static and dynamic balancing of high-speed components without the application of mechanical forces.

The operations in the first group are particularly interesting. Considering the requirements for hole geometry and depending on the thickness of the components, the operation can be regarded as drilling thin plate type components up to 0.1 mm thick, and as drilling holes with a conical form at the entry and exit ends; for example, the holes in diamond drawing dies, watch jewels, etc. are drilled in this way.

The key feature of the method of drilling thin plates (up to 0.1 mm thick) is that the process may be regarded as one of metal removal solely by evaporation from the zone subjected to laser radiation. It may be assumed with sufficient accuracy that the radiation concentration on the zone will be given by

$$D = \frac{\lambda f}{d}$$

where λ is the wavelength of the light beam;
f is the focal length of the light beam;
d is the minimum beam size on the resonator output.

CO_2 lasers are used extensively for cutting and welding almost any material from cloth to steel slabs. They have been used to cut 75 mm steel slabs, and a clothing manufacturer can use a numerically controlled CO_2 laser for cutting men's suits from cloth. It is faster and more accurate, and the edges of synthetic materials are sealed by the heat of the beam.

However, even though higher-powered lasers are being made and present tantalising possibilities, the commercial uses of lasers for drilling, cutting, welding, and scribing are still usually associated with small work.

Applied to drilling.   Drilling of almost any material is possible with lasers. However, the holes often have a 1° to 10° taper (larger at the top) and are not always smooth inside.

The advantages are that there is no drill breakage or dulling, no pressure on the part, and depth to diameter ratios of up to 40 : 1 can be drilled. The beam diameter and power can be varied to meet nearly any condition.

Swiss watchmakers drill 0.030 mm diameter (0.0012 in.) holes in 0.25 mm thick (0.010 in.) rubies using Nd–YAG lasers. It takes three to six pulses per hole. Also, lasers using about 250 pulses per side are used for drilling diamond dies used for fine wire drawing.

A 0.05 mm (0.002 in.) hole is "drilled" through a 0.05 mm thick (0.002 in.) nickel foil (using an Nd–glass laser) for use in an electron gun.

Applied to welding.   Welding requires close-fitting joints, and penetration of most commercial welding done is not over 1.0 mm (0.040 in.), though high-power CO_2 lasers can penetrate deeper. One automobile company is using up to 10 kW lasers to weld body seams.

Stainless steel of 0.18 mm (0.007 in.) thickness is welded, using overlapping spot welds with a ruby laser. An Nd–YAG laser is used to weld two 0.05 mm (0.002 in.) wires to 0.38 mm thick (0.015 in.) terminals, and 0.12 mm thick (0.0047 in.) leads are welded to a thin layer of gold on a silicon chip. A pulsed YAG weld can be made on a small stainless steel angle.

Applied to cutting.   Cutting, especially of thin materials, can often be done economically with lasers. When metal or ceramics are being cut, the melted metal sometimes gets back and re-melts on the work. Thus a jet of oxygen, nitrogen, or argon is often used with the laser. CO_2 lasers are most frequently used for cutting.

Titanium is cut with oxygen at rates of 2.5 to 15.2 m/min (8 to 50 fpm). Plastics can be cut with lasers using nitrogen gas, which clears the area and cools the work. Numerical control is also used by an aerospace company to guide a laser for cutting any shape in 2.65 mn thick (0.1 in.) titanium. A 50% saving in time is reported. An NC-guided 6 kW CO_2 laser is used to cut 9.6 mm steel at 45 ipm.

Other uses. Other uses of lasers include using N–C and Nd–YAG lasers to generate the complex patterns on films which are used to make integrated circuits (ICs), to cut patterns on thin gold film in electronic equipment, and to "trim" carbon resistors to produce accurate readings.

When making precision resistance films, one needs to ensure an error of not more than 0.05%; here the laser beam may be used to vaporise part of the film. For this operation, the resistance after scaling and ageing is supplied with a 0.5% allowance. An important feature of this method is that the resistance is supplied for this operation after sealing and stabilising its parameters; thus the resistance is not altered in any way.

Safety with lasers: Special care must be taken to protect the eyes, as laser beams are completely or partially reflected from many surfaces. Snug-fitting safety glasses with special lenses should be worn any time a laser of over 2 mW capacity is operated, though lasers below 10 mW may not be highly damaging.

Of course, one's body should never be in front of any high-power laser, as it is easier to cut flesh than steel.

Many lasers use high-voltage, high-power circuits; so it is wise not to touch or operate this equipment unless you are trained to use it.

When a laser is used, locked doors, warning signs, or barricades should prevent non-operating people from getting too close. In spite of some of the pictures, you cannot see most laser beams.

28.18 ELECTRON BEAM MACHINING—EBM

Introduction: In electron beam machining the workpiece is placed in a sealed vacuum chamber, in which the high vacuum is maintained by constantly working pumps. A special electron gun together with an electron-optical system produces a highly focused stream of electrons, which is limited by a cathode and accelerated in vacuum by a 150 kV potential. In these conditions the electrons attain a velocity of 240,000 km/s.

A particular feature of moving electrons is the possibility of focusing them over a small area to obtain high power densities. Accelerated electrons penetrating into a solid react with the electrons and ions of the lattice, with the result that their velocity is reduced and their direction altered. In practice the full kinetic energy of the electrons is converted into heat in the thin surface layer of the irradiated solid. The depth of this layer is determined by the accelerating potential and the specific gravity of the solid material. For instance, a beam of electrons with 10 kV energy will penetrate tungsten to a depth of 0.1 μm, and when the energy is increased to 160 kV it will penetrate to a depth of 28 μm; in other words, an increase in the potential causes a considerable increase in the depth of penetration.

Principles of working: The kinetic energy of the electrons is converted into heat due to repeated impacts with the atoms of the metal solid. After each impact the angle of the probable change of direction of an electron, caused by a decrease in its velocity, is increased. The result is that in the final section of its travel, an electron will dissipate the main portion of its energy. That is, at a distance approximately equal to the depth of penetration, the losses are at the maximum and the layers of the solid distributed in this point are heated to the maximum temperature. Thus, the conditions of heating with an electron beam differ from the usual ones, where energy conversion takes place on the surface and the heat penetrates into the solid due to heat conductivity. In electron beam treatment, this occurs in the reverse order, with the inner layers being heated and heat penetrating upwards to the outside. Figure 28.22 indicates the schematic layout of EBM machine.

Advantages and limitations: Electron beam machining has its limitations, connected with the necessity of working in a vacuum and at high potentials, and cannot noticeably replace electrical discharge, electrochemical and ultrasonic methods in mechanical engineering, where these methods are more economical and productive. Electron beam machining, however, will be very effective for machining materials that are difficult to machine, for the straight and profile cutting of thin films and plates, and for cutting narrow slots,

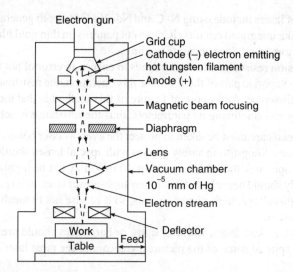

Fig. 28.22 Schematic layout of electron beam machine.

grooves, and small diameter holes. Thus, the main operations which can be performed by electron beam machining are as follows:

1. Piercing single holes below 0.1 mm diameter in corundum (jewel bearing), in cemented carbides (drawing dies and spinnerets), in various difficult-to-machine metals, in glass ceramics (spinnerets, injector holes), etc.;
2. Piercing of systems of holes and slots below 0.1 mm in diameter (grids, spinnerets, computer elements, micro diffraction gratings, etc.);
3. Profiling and slicing various elements (semiconductor materials, masks for integrated circuits), profiled slotting (electronic system elements and resistances), etc.

The widest application of electron beam machining should be in the piercing and milling of slots and grooves in micro-miniature components.

28.19 ELECTROLYTIC SAWING MECHANISM AND MACHINES

Electrolytic sawing uses a disc or band saws, as used for sawing metals, for difficult to machine materials on conventional machines.

Sawing by using this method is carried out in a water glass solution with a density of 1.23 to 1.32 g/cm³, which is pumped to the cutting zone. The tools are either circular saws or endless band saws made from mild steel. The machines are intended for use in mechanical engineering works, steel works, and steel mills.

An advantage of the electrolytic sawing machine is that very thin saws quickly convert metal in the cut into swarf. In the machines with circular saws the width of the cut is 1 to 3 mm and on the band saw machines it is 1 to 2 mm. The cut is square within 0.2 to 0.8 mm, and wandering of the cut is kept within 0.1 to 0.4 mm. The layer with structural changes varies from 0.05 to 0.6 mm, depending on the machining conditions. In all the machines mentioned, with the exception of some machines, the workpiece is held stationary and the tool is traversed. This principle is particularly convenient when sawing large workpieces weighing many tons, because the servo system for the moving tool in these machines may be composed of fairly light and simple mechanisms. The machines are supplied with hot water systems (for washing electrolytes from the mechanisms of the machines) and with fume extractors. A disadvantage is that it is necessary to use an aqueous solution of water-glass as an electrolyte, which has a tendency to solidify on the mechanisms of the machine.

28.20 MORE UNCONVENTIONAL PROCESSES

Chemical milling: This process was originally used on large panels for aerospace work, to make them a few pounds lighter by milling large areas a few thousandths of an inch below the normal surface. The panels may measure from 600 to 3000 mm in length. The panel is very thoroughly cleaned and then covered with a mask of rubbery plastic by spraying or dipping.

The mask is carefully cut and removed from the areas which are to be etched (milled). The plate is then dipped completely into a tank of chemicals which will dissolve (etch) away the exposed metal.

By controlling the time of immersion and the strength of the chemical bath, the depth can be controlled to ± 0.05 to 0.125 mm. By using the proper electrolyte, almost any metal can be chemically milled.

Chemical machining: Also, called chem-milling, chemical machining comprises photoforming, photofabrication, etching, and chemical blanking. This process, like chemical milling, uses a chemical to dissolve the metal, which may be 0.013 to 3.2 mm thick. Material as thin as 0.13 mm is almost impossible to run on a punch press. However, in chemical machining, the etching is allowed to eat completely through the metal, and so the result looks like a stamped-out part.

The chemical machining process starts with an accurately made drawing or artwork of the part. To reduce errors, this is usually made 10 or 20 times the actual size. The drawing is photographically reduced to the actual part size, and a negative or a positive is made.

In the meantime the metal (which is aluminium, steel, titanium, or super alloy) is coated with light sensitive photoresist. This surface is now exposed to light through the negative, just as in developing pictures. The exposed metal is then developed to remove unnecessary portions of the photoresist. This may be done on either one or both sides of the metal.

The treated metal is next put into a machine which sprays it with a chemical etchant, or it may be dipped into the solution. The etching chemical may be either sodium hydroxide (for aluminium), hydrofluoric acid (for titanium), or one of several other chemicals. Within 10 to 15 minutes, the unwanted metal is eaten away, and the finished part is then ready for immediate rinsing to remove the etchant.

Chemical machining is an excellent method of obtaining complex parts from very thin metals without the cost of a punch-press die. The advantages are that this process does not distort the workpiece, does not produce burrs, and can easily be used on the most difficult-to-machine materials. However, the process is slow, and thus it is not usually used to produce large quantities or to machine metal over 1/8 in. (3.2 mm) thick. Some small parts are made in lots of 10 to 100 at a time on a single plate, which speeds up production.

28.21 HOT MACHINING

Hot machining is employed for machining high strength, high hardness and high temperature resistant materials, which are difficult to machine at room temperature. Machining of hard metals at elevated temperatures is applied mainly to turning and milling operations. Since the shear strength of metal decreases at elevated temperature as compared to that at room temperature, the magnitude of cutting forces on the tool is lower. Further, as chip formation by plastic deformation in the shear plane ahead of the tool becomes easier at elevated temperature, and the cutting forces involved are less, the power requirements are low. However, at elevated temperature the property of the tool material is also changed due to its contact with high temperature material; therefore, tool life is also affected. It is found that tool life is maximum at a certain temperature and after that it decreases.

Also, there is a critical temperature of the workpiece (the work material and the tool material) at which the total metal removal rate per tool grinding will be maximum, irrespective of the speed.

Experiments in hot machining of austinic manganese steel with carbide tools, up to a maximum workpiece temperature of 6500°C, have produced the following findings:

(i) The work of a tool at high work temperature progresses in a way similar to that in machining at room temperature (see Fig. 28.23).
(ii) The tool wear rate decreases with increase in the workpiece temperature, thus increasing tool life.
(iii) There is an optimum value of the work temperature at which the tool life is maximum. For austinic manganese steel, it is about 6500°C for all cutting speeds.

·**Experimenting with electric current heating and carbide tools:** Baron has found that in hot machining, the flank wear decreases while the crater wear increases. For maximum tool life for any current value, the choice of feeds and speed is very critical. The depth of cut, however, does not affect the tool life to any appreciable extent.

An important aspect of hot machining is that at low cutting speed the tool life increases with increase in the cutting velocity. However, at high cutting speeds the tool life decreases with increase in the cutting speed. Break-back in the curves occurs at about 70 m/min. The tool life increases up to a certain value of workpiece temperature (see Fig. 28.23), after which it decreases.

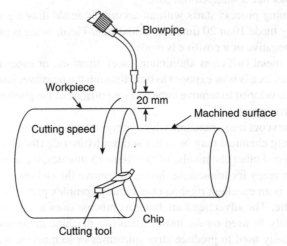

Fig. 28.23 Schematic diagram for hot machining.

As in the case of machining at room temperature, the cutting forces at high workpiece temperature are little affected by the variation in the cutting speed. The cutting force (F_c) in the machining of manganese steel at 5000°C varies directly with the depth of cut, but the thrust force (F_t) is little affected by a change in the depth of cut. The functional relationship of F_c with F_t with the feed and the depth of cut for machining manganese stainless steel at 5000°C is determined as

$$F_c = 229f \times 0.829d \qquad \text{in kgf}$$

$$F_t = 104f \times 0.16d \qquad \text{in kgf}$$

where f and d are measured in mm.

REVIEW QUESTIONS

1. (i) Describe the principle of operation of a relaxation generator used for electrical discharge machining. Explain how its disadvantages are overcome in other types of generators.
   (ii) Explain, with reference to the relaxation generator, why a servomechanism must be used to feed the tool. Show how an error signal may be obtained when using either this generator or a pulse generator.

2. (i) Draw a graph to show the relationship between the breakdown voltage of the dielectric fluid used in a spark erosion machine and the distance between the electrode and the workpiece. Discuss the influence of this relationship upon the design and operation of such machines.
   (ii) Show that for a spark machine operating on a relaxation circuit the breakdown voltage U is given by
   $$U = E(1 - e^{-t}RC)$$
   where E is the supply voltage;
   t is the charging time;
   R is the resistance of the circuit;
   C is the capacitance of the circuit.
   Hence determine for the above machine the average power output, given that resistance $R = 3.2\ \Omega$, capacitance $C = 150\ \mu F$, supply voltage $E = 200$ V and the breakdown voltage $U = 160$ V.

3. (i) Compare and contrast EDM and ECM with respect to
   (a) The principle of operation
   (b) The applications.
   (ii) When considering the use of electro-machining and the more traditional metal removal processes, what factors would influence the decision in deciding which techniques to apply?

4. (i) Explain the principle of metal removal by the EDM method and show by means of a well-proportioned diagram the general features of an ED machine.
   (ii) Compare the following processes in terms of the metal removal rate, surface finish and dimensional accuracy:
   (a) Electro-discharge machining;
   (b) Electro-chemical grinding;
   (c) Ultrasonic machining.
   For each process, indicate a typical application which would justify its use, giving reasons for your choice.

5. A component made of iron is to be machined by the ECM process. The valency of iron at which the dissolution takes place is 2. The density of iron is 7.8 g/cm². The atomic weight of iron is 56 g. Faraday's constant is 160 A-min.
   (i) Calcualte the amount of current required, if the rate of metal removal is 3 cm³/min.
   (ii) If the area of cross-section of the electrode is 1 cm², how much time would it take to drill a hole 6 cm deep? Take the current efficiency as 80%.

6. Tungsten was machined with the ECM process at 1000 A. The metal removal rate was observed to be 2 cm/min. Find out the valency of the metal at which it has been dissoluted electrochemically, given that
   Atomic weight of tungsten = 186 g
   Density of tungsten = 9.4 g/cm²
   Faraday's constant = 1609 A-min.

7. A chromium product which dissolves at valency 2 has been machined by the anodic dissolution process. The following data are given:
   Current density = 250 A/cm^2
   Atomic weight of chromium = 52 g
   Density of chromium = 7.2 g/cm^2
   Find out the metal removal rate.

8. A hole 2 cm in diameter and 10 cm deep is to be made in a product made of tungsten by the ECM process. The following data should be assumed:
   Current density = 200 A/cm^2
   Atomic weight of tungsten = 186 g.
   Calculate the:
   (i) Metal removal rate
   (ii) Time to make the hole.
   Assume $F = 96500$ C.

9. A nimonic alloy containing 20% cobalt, 60% nickel and 20% chromium is to be machined by ECM process with a current of 1000 A. The density of the alloy is 8.3 g cm^3. The additional data are as follows:
   (i) Nickel and cobalt dissolve at valency 3
   (ii) Chromium dissolves at valency 6.
   Evaluate the metal removal rate (in cm^2/s).

10. An iron workpiece was machined by the ECM process with copper tool. The electrolyte used was NaCl in water. The following properties of the electrolyte were recorded:
    Specific heat = 1.0 cal/g°C
    Specific resistance = 2 ohm-cm
    Density = 1 g/cm^3
    Electrode area = 1 × 1 cm^2
    Initial gap (h) = 0.01 cm
    Ambient temperature = 30°C
    Boiling temperature = 90°C
    Current density = 100 A/cm^2
    Voltage applied = 10 V.
    Calculate the permissible fluid velocity if the maximum temperature rise is limited to the boiling temperature of the electrolyte.

    Solution
    $$U = \frac{v^2}{4.187 r_s \cdot h^2} \frac{b}{\rho_e \cdot S_e (T_2 - T_1)}$$
    Substituting the numerical values in the above equation, we get
    $$U = \frac{10^2 \times 1.0}{4.187 \times 2 \times 0.01^2 \times 1.0 \times (90 - 30)} = 1990 \text{ cm/s} = 1194 \text{ m/min}$$

11. (i) Discuss the process of electro discharge machining with particular reference to the workpiece, surface finish, metal removal rate and tool wear. State the typical applications of the process.
    (ii) Compare the foregoing process with electrochemical machining and indicate graphically, and by description in the latter case, the general relationships which are known to exist between the working gap and the machining voltage. Indicate the effect of increasing the working gap on the permissible feed rate.

(iii) Calculate the amount of titanium removed in an electrochemical machining operation, given that
   Relative atomic mass = 47.9
   Machining time = 2 min
   Valency = 4
   Current density = 250 A/cm^2
   Faraday's constant = 96500 C.

12. Determine the metal removal rate in an ECM machining operation on an alloy of nickel, chromium and molybdenum at 10000 A. The density of the alloy is 8.91 g/cm^3 and its composition (%) is as follows: C 0.1, Cr 15, Fe 5.5, Mo 15.9, Ni 59.5, W 4.
   Neglecting carbon, which does not dissolve anodically, the atomic weights of the elements are 52.0, 55.86, 95.95, 58.69 and 183.92 g respectively. It is assumed that the following valences apply in the process: 2, 2, 3, 2 and 6.

13. The newly developed machining technology can be profitably employed for the following work materials compared to the conventional methods:
    (i) Soft materials like aluminium and mild steel
    (ii) Brittle materials like glass, ceramics, and carbides
    (iii) Any material

14. By new machining technology
    (i) The bulk of materials is removed
    (ii) A very small amount of material can be removed
    (iii) Any amount of material can be removed.

15. After machining by new technology
    (i) The product needs further finishing
    (ii) The product needs no further finishing
    (iii) A finer cut is essential for some products.

16. In ultrasonic machining
    (i) The workpiece is vibrated
    (ii) The tool is vibrated
    (iii) Both work and tool are vibrated
    (iv) No vibration is needed.

17. In ultrasonic machining, the tool is vibrated between the frequencies of
    (i) 1 and 5 kHz
    (ii) 5 and 10 kHz
    (iii) 10 and 15 kHz
    (iv) 15 kHz and higher.

18. During USM
    (i) There is tremendous noise
    (ii) There is no noise
    (iii) There is little noise
    (iv) No sound can be produced.

19. In USM the abrasives are fed as
    (i) Lumps
    (ii) Slurry
    (iii) Powder
    (iv) Small cubes.

20. In USM the purpose of the slurry of abrasive particles is
    (i) To cool the work
    (ii) To carry away the worn-out abrasives and eroded workpiece particles
    (iii) To cool the tool
    (iv) More than one of these.

21. The USM process is soundless because
    (i) No sound is generated
    (ii) The sound generated is within the audible range of the human ear
    (iii) The sound generated is above the audible range of the human ear.
22. The slurry is fed over the work
    (i) Continuously (ii) Intermittently
    (iii) After a predetermined time interval.
23. In USM process there is
    (i) Very little tool wear (ii) No tool wear
    (iii) Very high tool wear (iv) There is no question of wear.
24. For the USM process the tool material could be
    (i) Brittle (ii) Ductile
    (iii) Very hard (iv) Of any kind.
25. For the USM process the abrasive could be
    (i) Silicon carbide (ii) Boron carbide
    (iii) Aluminium oxide (iv) Any of these.
26. For any abrasive to be suitable for USM the following are important:
    (i) Grain size (ii) Durability of the edge of the grains
    (iii) Hardness of the grains (iv) All of these
    (v) Only (a) and (c).
27. The choice of a particular abrasive depends upon
    (i) Rate of material removal desired (ii) Surface finish and dimensional accuracy
    (iii) Hardness of work material (iv) All of these.
28. The liquid selected for making the abrasive slurry in USM should have
    (i) Very low viscosity
    (ii) Very high viscosity
    (iii) Viscosity which changes much with temperature.
29. Which of the following should be selected as the desired liquid for making a slurry of abrasives in USM?
    (i) Air (ii) Water
    (iii) Murcury (iv) Kerosene oil.
30. The metal removal rate in USM
    (i) Increases if the concentration of the slurry is very high
    (ii) Increases if the concentration of the slurry is very low
    (iii) Has no effect on the concentration of the slurry
    (iv) Does not increase if the concentration of the slurry is very high.
31. The metal removal rate in USM is
    (i) As high as that obtained in turning
    (ii) As low as that obtained in grinding and honing
    (iii) Lower than in these operations.
32. The finish of the work obtained in USM is
    (i) Better than in grinding (ii) Lower than in grinding
    (iii) Comparable to that in grinding (iv) Very dull on the surface.
33. The vibration of the tool in USM is obtained by
    (i) Cam system (ii) Eccentric mechanism
    (iii) Piezoelectric effect (iv) Magnetostriction effect.

34. The horn of the tool in USM can be
    - (i) Conical
    - (ii) Stepped
    - (iii) Exponential
    - (iv) Any of these.
35. The performance of the USM process
    - (i) Depends on the shape of the horn
    - (ii) Does not depend on the shape of the horn
    - (iii) The horn shape is not very important.
36. The performance of the USM process
    - (i) Improves by increasing the amplitude of vibration to a very high level
    - (ii) Deteriorates by increasing the amplitude of vibration to a very high level
    - (iii) Cannot be altered by changing the amplitude.
37. By the USM process the holes produced can be made
    - (i) Circular
    - (ii) Square
    - (iii) Hexagonal
    - (iv) Of any shape.
38. While drilling holes in a glass plate, there is
    - (i) No chance of any crack in the plate
    - (ii) Some chance of crack formation in the plate
    - (iii) Certainty of crack formation.
39. On the work machined by USM there is
    - (i) No question of stress
    - (ii) Virtually no residual stress in the work
    - (iii) Some residual stress in the work
    - (iv) Considerable residual stress in the work.
40. In the case of the refrigerated slurry used in USM
    - (i) The rate of metal cutting decreases
    - (ii) The rate of metal cutting increases
    - (iii) The rate of metal cutting remains the same as that of the slurry at room temperature.
41. If the slurry in USM is fed at elevated temperatures
    - (i) The metal removal rate does not change
    - (ii) The metal removal rate increases
    - (iii) The metal removal rate decreases.
42. The amplitude of vibration of the tool in USM varies from
    - (i) 0–0.5 mm
    - (ii) 0.1–0.5 mm
    - (iii) 0.5–1 mm
    - (iv) 1–2 mm.
43. By the USM process, holes of the following diameters could be drilled:
    - (i) As low as 0.002 mm
    - (ii) Not smaller than 0.1 mm
    - (iii) Not smaller than 0.5 mm
    - (iv) Not smaller than 1.0 mm.
44. In the EDM process
    - (i) The work is negative and the tool positive
    - (ii) The work is positive and the tool negative
    - (iii) Both are made positive
    - (iv) No electrical connections are made.
45. For proper application of the EDM process it is essential that
    - (i) The work is a conductor of heat
    - (ii) The work is a conductor of electricity
    - (iii) The work is magnetic
    - (iv) All of these.

CHAPTER 29

Grinding and Other Abrasive Metal Removal Processes

29.1 INTRODUCTION

Grinding as a metal removal process has been used by some research workers in metal cutting as micro-milling. However, engineers from well-known grinding wheel corporations like the Grindwell-Norton and Carborandum-Universal differ in their opinion and choose to place it as a unique process in a class by itself and feel that teachers as well as students study the grinding and other Abrasive processes from a different perspective. It has been stated emphatically that there is no surface of a product in engineering or in domestic use which is not touched by an abrasive grit.

Grinding as a metal removal process is associated with a large number of grinding variables (Fig. 29.1), which have a definite bearing on the quality of ground products. Wheel material, composition, dressing parameters and work material properties are the main variables in the grinding process. The intricacy of surfaces and their functional requirements have promoted research and development on this process which has made the higher metal removal rates possible. All this has resulted in the enhanced use of grinding process (by 30% to 40%) and it is likely to increase further in near future.

To 'Grind' means 'to abrade', to wear away by friction or to sharpen. In grinding, the material is removed by means of a rotating abrasive wheel. The action of grinding wheel is very similar to that of a milling cutter. The wheel is made up of a large number of cutting tools constituted by projected abrasive particles in the grinding wheel.

Grinding is done on the surfaces of almost all shapes and materials of all kinds and it is also able to produce accurate and fine surfaces.

GRINDING AND OTHER ABRASIVE METAL REMOVAL PROCESSES 817

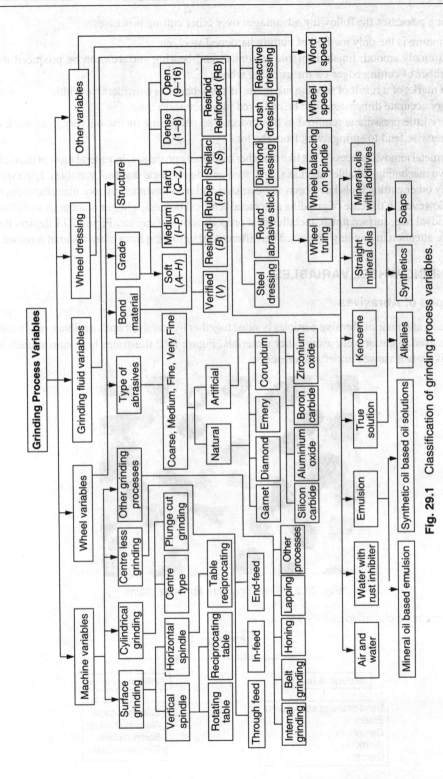

Fig. 29.1 Classification of grinding process variables.

Grinding possesses the following advantages over other cutting processes:
1. Grinding is the only method of cutting hardened steel, etc.
2. Extremely smooth finish desirable at contact and bearing surfaces can be produced due to large number of cutting edges on the grinding wheel.
3. No marks of a result of feeding are there, as the wheel has considerable width.
4. Very accurate dimensions can be achieved in a very short time.
5. Very little pressure is required in this process, so permitting its use on very light work that would otherwise tend to spring away from the tool.

Grinding as a metal removal process is not like the other machining processes. In general most of the metal removal processes have machining variables such as speed, feed and depth of cut and tool variables. High speed steel or carbide or any other cutting tools have been standardized and recommended by tool manufacturers to machine shop users. However, in the case of grinding as a metal removal process, a larger number of variables influence the metal removal rate, surface finish, metallurgical properties of surface, etc. Figure 29.1 depicts the variables which deserve attention in grinding process. This differentiates grinding from other material removal processes.

29.2 GRINDING WHEEL VARIABLES

29.2.1 Types of Abrasives

Grinding wheels are made of abrasive particles bonded together by suitable bond. An abrasive is a hard material which can be used to cut or wear away other materials. Figure 29.2 illustrates how silicon carbide abrasive grains look like when viewed under a microscope.

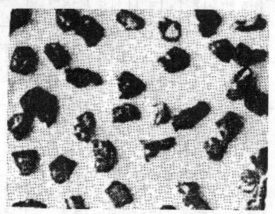

Fig. 29.2 Silicon carbide abrasive grains.

Abrasive can be classified as follows:

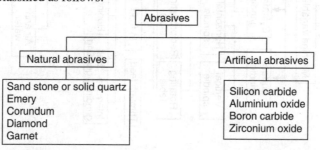

Grain Sizes

All varieties of abrasives are manufactured in (i) course (ii) medium (iii) fine and (iv) very fine grain sizes. However, the metal removal capability of abrasives depends on the following qualities.

  (i) Purity
 (ii) Uniformity in composition
(iii) Hardness
(iv) Toughness: If the abrasives are not tough, the wheel will fracture readily.
 (v) Sharpness of fracture: Better cutting action is obtained by sharp edged abrasives. Natural abrasives give rounded edges and are not efficient in cutting.

Abrasives must be refined and reduced to standard sizes before they are manufactured into wheels. For this, the abrasives are crushed and screened, that is, they are passed through screens containing meshes of standard sizes. These sizes are measured by the number of meshes per inch through which the grains pass. For example, a grain that passes a screen having 20 mesh opening per inch is called a 20 grit size and measures roughly 1/20 inch across. The grain size is indicated by grit number as shown below:

Coarse grain	–	4, 5, 6, 7, 8, 10, 12, 14, 16, 20, 24
Medium grain	–	30, 36, 46, 54, 60
Fine grain	–	70, 80, 100, 120, 150, 180
Very fine grain	–	220, 240, 280, 320, 400, 500, 600, 800, 1000, 1200

Sandstone

Sandstone wheels are rarely used. Although they are cut from high grade quartz or sandstone, yet they do not wear evenly because of variation in the natural bond.

Corrundum and emery

Both are composed of crystal aluminium oxide in combination with iron oxide and other impurities. These minerals lack in uniform bond and are seldom used in production.

Diamond

They are made with resinoid bond and are useful in sharpening cemented carbide tools. Due to their rapid cutting ability, slow wear and free cutting action, they are quite economical. Very little heat is generated with their use.

Silicon carbide

It was discovered during an attempt to manufacture precious gems in an electric furnace. The hardness of this material according to Moh's scale is slightly over 9.5, which approaches the hardness of a diamond. Raw materials consisting sand, petroleum, coke, saw dust and salt are heated in a furnace to around 2300°C and held there for a considerable period of time. The product consists of a mass of crystals surrounded by partially unconverted raw material. After cooling, the material is broken up, ground and crushed to grain size. Silicon carbide crystals are very sharp and extremely hard.

Aluminium oxide

It is made from the clay-like mineral called 'bauxite' which is the main source of aluminium. Aluminium oxide is slightly softer than silicon carbide, but is more friable and is preferred in grinding steel when a sustained cutting action is desired. Most manufactured wheels are made from aluminium oxide.

29.2.2 Bond Material

A bond is a material that holds the abrasive grains together enabling the mixture to be kept in desired shape. Various types of bonds are designed as follows:

(i) Vitrified bond   –   V
(ii) Silicate bond   –   S
(iii) Shellac bond   –   E
(iv) Rubber bond   –   R
(v) Resinoid bond (Bakelite) – B
(vi) Oxychloride bond   –   BF

The bond material should have the following properties:
  (i) The bonding material should withstand the grinding temperature.
  (ii) It should be able to withstand the spindle speed without disintegrating.
  (iii) The bonding material should be capable of retaining the abrasive grains during the cutting action.
  (iv) It should be friable enough to expose more sharp edges when one particular sharp edge undergoes breakage.
  (v) The bond should be rigid to enable the grit to penetrate into the work.

Various types of bonds are as follows:

Vitrified bond

The abrasive grains are mixed with clay-like ingredients that are changed to glass-like material upon being burnt at high temperature. The addition of water is required and the wheels are shaped in metal moulds in a hydraulic press. The time for burning, varies with wheel size, being anywhere from 1 to 14 days. These wheels are porous, strong, unaffected by water, acid, oils and climatic or temperature conditions. Vitrified bonds are used for most grinding operation. The maximum standard speed generally used for vitrified bond wheels is 33 m/sec.

Silicate bond

Silicate bonds require much lower temperature to harden the wheel than to vitrified bonds. In this case, sodium silicate is mixed with abrasive grains and mixture is tapped in metal moulds. After drying for several hours, the wheels are baked at 260°C for 1 to 3 days.

Silicate binders release the grains more readily giving a milder (or cooler) grinding action and are suitable for grinding edges of tools. The process is recommended for large wheels, as they tend not to crack or wrap in the baking process.

Shellac bond

Shellac bond, initially used for cut-off operations, are also used for hardened materials. The abrasive grains are first coated with shellac by being mixed in a steam-heated mixture. The material is then placed in heated moulds and rolled or pressed. The wheels are finally baked for few hours at 150°C. This bond is adopted for thin wheels as it is strong and elastic. Alkaline cutting fluids react with this bond.

Rubber bond

Pure rubber with sulphur as a vulcanizing agent is mixed with abrasives by feeding the material between heated mixing rolls. After it is rolled to thickness, the wheels are cut out with dies and vulcanized under pressure.
- Very thin wheels can be made by this process.
- Rubber bonded wheels are used for high speed grinding (45–80 m/s).

As they afford rapid removal of the stock, these wheels are used for cut-off operations. When subjected to heat, the bond softens and releases the grains. Thus, cutting fluid should normally be used with this bond.

Bakelite or resinoid bond

Abrasive grains in this process are first mixed with thermosetting synthetic resin powder and liquid solvent, the moulded and baked. The bond is very hard and strong and can be operated at speed of 45–80 m/s. It is used for general purpose grinding and for snagging purposes.

Oxychloride bond

This bond contains oxides and chlorides of magnesium in contrast with other bonding materials. Oxychloride is a cold setting cement requiring an aging period. As most cutting fluids affect this bond, it is normally used for dry grinding.

29.2.3 Grade

Grade of a wheel is a measure of how strongly the grains are held together by the bond. The bonding material in a wheel surrounds the individual grains and links them together. Different types of grades are represented by English letters from A to Z, as shown below.

A	to	H	–	Soft	
I	to	P	–	Medium	
Q	to	Z	–	Hard	

Different types of wheel grade are shown in Fig. 29.3.

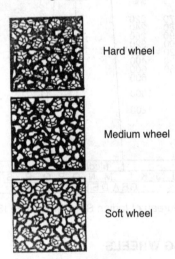

Fig. 29.3 Wheel grades.

29.2.4 Structure

The structure of grinding wheel refers to the relationship of abrasive grain and bonding material and of the spacing between the abrasive grits due to the air gap or the void between them. Different types of structures are represented by a range of numbers.

1	to	8	–	Dense
9	to	16	–	Open

29.3 STANDARD CODIFICATION OF A GRINDING WHEEL

To the diversity of methods employed by the various grinding wheel manufacturers for making grinding wheels, a new marking system has been accepted as standard. It is divided into six parts, with the marking placed in the following sequence:

Position 1	Position 2	Position 3	Position 4	Position 5	Position 6
Kind of abrasive and manufacture's prefix	Grain size	Grade	Structure	Bond type	Manufacturer record

Figure 29.4 shows how a specific wheel is marked and how it should be ordered from the manufacturer. The chart shows that a grinding wheel marked 51 A-36-L-5-V-93 has been chosen. The letters and numbers contained in this identification are explained in Fig. 29.4. This codification is as per the Bureau of Indian Standard Code IS 551:1989.

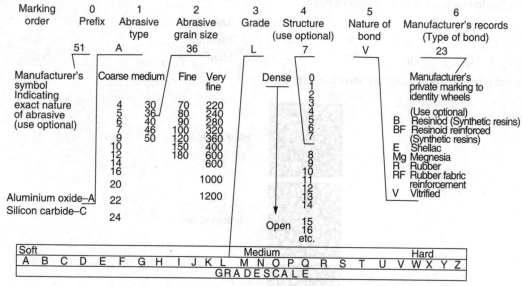

Fig. 29.4 Bureau of Indian Standards Code IS 551:1989.

29.4 SELECTION OF GRINDING WHEELS

Fixed factors

The four fixed factors for wheel selection are determined by work material, accuracy desired which is stated on the drawing of the component, type of grinding operation and area of contact between the wheel and the workpiece.

Some guidelines for wheel selection are given in the paragraphs below.

Material to be ground

Use a soft wheel, i.e., a wheel with low bond strength if the work material is hard and a hard wheel if the work material is soft. While needing a sustained cutting action, as in the case of ductile materials like steels, the use of Aluminium oxide grit wheels is recommended for cast iron and for similar hard materials, the use of silicon carbide grit wheels is recommended. For grinding ultra-hard materials like tungsten carbide, diamond grit wheels are required. Some other specific considerations are given below:

Grade range

H to K is suitable for the grinding of soft steels while grades F to J are suitable for grinding of hard steels.

Area of contact

If the type of operation is such that the area of contact is more as in case of internal grinding or surface grinding, the use of a soft wheel is recommended because the force per grit will be lower in these cases and self dressing action will be encouraged by using a wheel with lower bond strength. In case of cylindrical grinding on the periphery of the wheel, there is a line contact and a hard wheel is recommended. Where forms, fillets and small radii are required to be ground, fine grained hard wheels are recommended.

Accuracy and surface finish

It is desirable to use grain size ranging from 36 to 80 for commercial roughing, whereas for fine finishing grain, the sizes of 80 to 320 are preferred. On an average a grain size of 46 is utilized. The grain size indicates the number of perforations per inch in a mesh through which the grain will just pass. For example a grain size of 46 represents an average value while a grain size of more number like above 320 represents very fine grain suited for micro finishing like lapping. Where high stock removal is a requirement, softer than average grades and grain sizes even less than 46 are recommended.

Types of operation

The most commonly used grades for commercial surface grinding ranges from F to J. Hard, dense, fine-grained metals of low ductility require softer than average grades. Cylindrical grinding grades of M or above are preferred. The grain size is decided depending upon the stock removal rate.

Variable factors for wheel selection

In this category of wheel selection criteria, the following four factors which are within the control of the production engineer are considered:

    (i) Work speed v
   (ii) Wheel speed V
  (iii) Condition of the machine, and
  (iv) Skill of the operator.

The selection of grinding wheels is based on the recommendations of the grinding wheel manufacturers. The product application engineer of the wheel manufacturing firm like Grinder—Norton or Carborandum Universal advises the user about the appropriate wheel to be used for a specific application. However, for a logical explanation of rational wheel speed V and work speed v is possible only after inspecting a mathematical formula for mean undeformed chip thickness (t_m) in grinding as shown in Fig. 29.5.

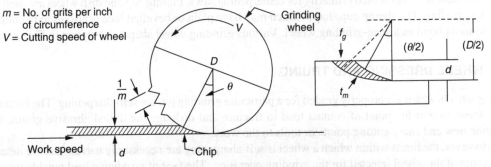

Fig. 29.5 Grinding process and its geometry of chip formation.

The formula for force acting on a single grit is given by treating grinding as a micro-milling operation. Thus, the mean chip thickness 'A' which determines the force on grit can be stated by suitably adapting Schlesinger's formula for milling, given in Chapter 6. Here,

$$t_m = f_g \left(\frac{d}{D}\right)^{0.5} = \frac{v}{mV}\left(\frac{d}{D}\right)^{0.5}$$

where f_g = Feed per grit
d = Wheel depth of cut which depends upon stock allowance to be removed
D = Wheel diameter
v = Table feed rate in surface grinding with a cylindrical wheel
m = Number of grits per inch of wheel circumference
V = Wheel speed

This depends on the openness or closeness of the wheel, i.e., wheel structure which ranges from 1 to 15, indicating the range from a close packed structure to a less dense structure.

If a wheel is acting hard, i.e. if the grains are not leaving the bond even after they have become blunt, then they will cause merely a rubbing action against the work surface even after their life is over. Thus, there will be no metal removal but only heat generation which will cause thermal and metallurgical damage to the work surface. In such a case, increase in v, i.e. table feed rate and decrease in wheel speed V will increase the chip load and result in self-dressing action. On the other hand, excessive wheel wear and loss of abrasive material result in under-utilization of the wheel, and the chip load can be reduced by decreasing v and increasing V which reduces the chip load. Another parameter which can be controlled is d, the stock removed per pass. It may be pointed out that for optimum wheel utilization, the ratio of amount of metal ground to the amount of wheel wear termed the grinding ratio should be at least 100 for rough grinding and from 200 to 600 for finish grinding.

Regarding the other two variable factors, it has been found that in the case of a new or semi-skilled operator or an old grinding machine, a hard wheel should be used.

29.5 GRINDING WHEEL SHAPES

Straight wheel, wheel recessed on one side and wheel recessed both sides are classified as straight wheel types and are used for cylindrical grinding, surface grinding, off-hand grinding and snag-grinding. The recess is provided to give clearance for mounting flanges. Tapered wheel is used for snag-grinding operations. Cylinder wheel type is used on either horizontal or vertical spindle surface grinding machine. Either the peripheral surface or the face of the wheel is used as a grinding surface. A straight cup wheel is used on horizontal or vertical spindle surface grinding machine. It is also used for off-hand grinding. The wheel face may be plain or bevelled. Saucer wheel is used primarily for resharpening saws. Flaring cup and dish wheel are used for tool and cutter grinding. The flaring cup shape, which may have plain or bevelled face is also used in conjunction with a resinoid bond as a snag-grinding wheel. Various grinding wheel shapes are shown in Fig. 29.6.

29.6 WHEEL DRESSING AND TRUING

Dressing wheels which are properly graded for a particular grinding job are self-sharpening. The forces acting on the wheel face at the point of contact tend to fracture and dislodge the dulled abrasive grains, thereby presenting new and sharp cutting points or tools to the work.

However, the limits within which a wheel is self-sharpening are necessarily narrow and are determined by the grade of the wheel selected for the grinding operation. The face of grinding wheel outside these limits must be sharpened by what is commonly called a dressing and truing tool.

GRINDING AND OTHER ABRASIVE METAL REMOVAL PROCESSES

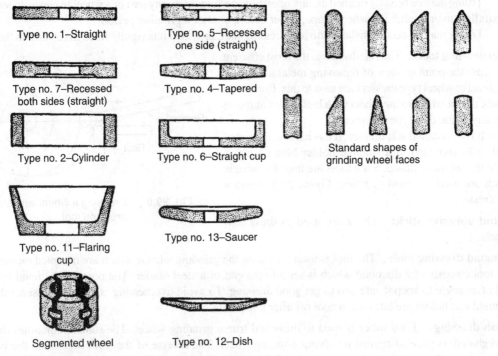

Fig. 29.6 Grinding wheel shapes.

Wheels that do not require dressing are usually too soft to hold size and profile of face or to produce good finishes. Wheels that require continuous or excessive dressing are usually too hard and the dressing operating frequently is wasteful. In either case, the wheel economy is reduced. The ideal wheel should be as self-sharpening as possible.

Dressing may be broadly defined as any operation performed on a wheel face changes the nature of its cutting action.

Dressing is necessary when the individual grain crystal in the wheel face becomes dulled and the rate of stock removal decreases constantly as the grinding operation prolongs without dressing (Fig. 29.7). Dressing is also necessary when the abrasive crystals or bond pores become loaded with metal or foreign matter, since this loading seriously affects the grinding action of the wheel.

Loaded wheel face

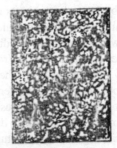
Glazed wheel face

Fig. 29.7 Wheel faces need dressing.

Truing may be broadly defined as, any operation, performed on any part of a wheel to create concentricity or parallelism or to alter the wheel shape, either before or after a grinding period.

Truing may be accomplished with any dressing tool, provided it is rigidly fixed in relation to the wheel.

Steel dressing tool: The star dresser is the most efficient tool from the point of view of removing metal filling and the abrasive wheel type dressers are next to this. Both serve to pick the metal out to the wheel with less loss of abrasive than any of the disc type dressers.

It is also called star dresser. It has hardened pointed steel disks mounted in a suitable holder held on a rest while the handle is raised. Such tools are used for wheels which are used for hand grinding. Figure 29.8 shows a star dresser.

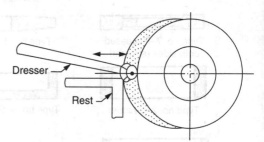

Fig. 29.8 Dressing a wheel with star dressing tool.

Round abrasive stick: These are used to dress thin wheels.

Diamond dressing tool: The tool is used for truing the grinding wheels which are mounted on machines. The tool consists of a diamond which is set into the end of a steel holder. The point of diamond is always kept at an angle to keep it safe and to get good dressing. To avoid overheating, a coolant is used, otherwise diamond and holder are allowed to cool off after a few cuts.

Crush dressing: Hard roller is used to dress and true a grinding wheel. The roller, sometimes diamond impregnated, is pressed against revolving grinding wheel. The reverse of the form needed on the wheel is given to the roller which displaces the grains and imprints the form on the wheel. For intricate forms, crush dressing is very economical.

Reactive dressing: It is a routine followed in grinding operations to restore cutting efficiency of wheel. When the wheel becomes dull and needs more power; the work speed is increased automatically. It causes the wheel to break down and resharpen itself, due to increase in grain depth of cut.

29.6.1 Dressing and Truing Procedures

Correct speed of dressing

Dressing and truing should be done at operating speeds or slower and never at higher speeds so that wheels are in satisfactory balance. Operating speed may be out of balance at higher speed and, therefore, any dressing or truing action must take place as near the normal operating conditions as possible.

Wheels may be dressed either wet or dry, but the operation should always be carried out under the same conditions as prevailing during the actual grinding operation. That is, if it is a wet grinding operation, wheel should be dressed wet, and in a dry grinding operation, it should be dressed dry to achieve the best grinding results. Wheel edges should always be rounded off with a hand stone or a precision tool before and after dressing, especially on fine finishing wheels. This prevents chipping the wheel edges and minimizes 'feed lines'. This ruler does not apply, however, when grinding to a shoulder where sharp corner on the work are required. The latest developments in dressing techniques are crush form dressing and continuous dressing. With special equipment, it is possible to dress intricate sharp, combining angles, radii and straight diameters into the face of a grinding wheel and in turn, grind this shape into the workpiece in a single operation.

In crush form dressing, a special roll, usually made of hardened high speed steel, tungsten carbide or boron carbide, four to six inches (100.0 to 150.0 mm) in diameter, is made in the profile of the work part. This is usually driven at a speed of between 150 and 300 surface feet per minute (0.7 and 1.5 metres per second) and brought into contact with the grinding wheel, which is not under power in this method. Pressure between the wheel and the roll increased through small increments of infeed to gradually crush the wheel face into the desire profile.

The profile is ready for the grinding operation and can be redressed at any time by bringing the wheel and the form roll into contact again. Crush form dressing techniques produce an open face on the grinding wheel so that usually finer grit sizes and harder grades are used.

The more recent development of continuous dressing is by using a metal bonded diamond wheel and the reverse of the shape of the part is ground into the face of the grinding wheel.

At fixed intervals, this dresser is brought into contact briefly with the grinding wheel for the reconditioning of the abrasive wheel face. The diamond wheel is driven by a flexible shaft and is usually mounted behind or above the grinding wheel.

Grit size of the grinding wheel has normal relationship to finish as when single point diamonds are used and grades are usually slightly softer.

Dressing of snagging wheels

Generally, the wheel face should be open and rough. Use the dresser, hold the tool to the wheel in one location until the wheel is sufficiently dressed. Then, move the tool to a new position three-fourths of the toolwidth away. Repeat this operation until the wheel face is fully dressed.

If the wheel is too hard or too fine for the job at hand, correction may sometimes be made by corrugating or grooving the wheel face.

This can be done by arranging the dresser discs in a pattern and looking them to prevent individual rotation on their supporting pin. This is easily accomplished with the locked type of dressing tool. In all cases, a rigid dresser support is essential. Dressing of snagging wheel is shown in Fig. 29.9.

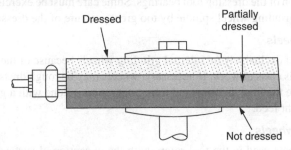

Fig. 29.9 Dressing snagging wheel.

Dressing of surface wheels

First type of wheel is mounted on horizontal spindle surface grinders. Roughing cuts may be satisfactorily made with a silicon carbide dresser stick mounted in an iron holder. The holder is held in place by a chuck, while the wheel is passed over it at a medium infeed allowing '0.003 to 0.005' (0.075 to 0.125 mm) for finishing, if required.

Finishing cuts, particularly for tool room grinding, are made with a diamond dressing tool at a slow traverse and light cut. Since many surface grinders do not use coolants, great care should be exercised by allowing frequent cooling periods to prevent undue heating of the diamond.

Second type known as segmental wheels are mounted on vertical spindle surface grinders. These wheels are best dressed with the star dresser and should be well supported.

Dressing of cylindrical wheels with diamond tools

(i) **Rough dressing:** Pass the diamond rapidly across the face of the wheel with a tool feed of 0.025 mm maximum. For most work, one pass of the dressing tool will be found to be sufficient.

If the wheel loads during a roughing cut, the wheel face may be opened by corrugating or grooving it with deep diamond cuts at a very rapid traverse, thus, allowing a rapid breaking down of the bond.

(ii) **Finish dressing:** If finishing cuts follow roughing cuts on the same wheel, make one pass of the dressing tool at a rapid traverse and a tool feed of 0.00015" (0.003800 mm) if the wheel is soft, or 0.005" (0.1270 mm) if it is hard. After marking this pass, reduce traverse to about one half and make one or two passes at a tool feed of 0.0003" (0.00762 mm) to 0.0005" (0.01270 mm) and repeat the traverse without dresser feed. With this type of dressing, slight spiral marks may appear on the first few pieces, but the cutting action of the wheel will be rapid and the finish good.

For finer finishes, follow the same procedure with several passes of the dressing tool across the wheel face at diminished feeds and traverse rate. This latter procedure reduces the cutting action of the wheel, therefore, wheel infeeds should be reduced accordingly.

Dressing of cylindrical wheels with abrasive wheel dresser

Dressing and truing with the abrasive wheel dresser on miscellaneous cylindrical grinding is an inexpensive method. Do not allow the tool to pass off the face of the wheel and confine the angle setting to between 5 and 7 degree from the face of the grinding wheel. To avoid glazing, both the dressing wheel and the grinding wheel, heavy infeeds of the dressing tool are necessary to a depth ranging from 0.025 mm to 0.05 mm.

Like all other types of rotary dressing tools, the abrasive wheel dresser will true a wheel only in proportion to the mechanical condition of the dressing tool bearings. Some care must be exercised with this type of dresser to prevent springing the grinding wheel spindle by too great pressure of the dressing tool against the wheel.

Dressing of internal wheels

Diamonds are used almost exclusively on internal grinding wheels because of the accuracy demanded of the dressing and truing for this class of work. If necessary, wheels may be rough dressed and trued with a dresser stick after mounting while the final dressing and truing operation is done with a good diamond dressing tool at a rapid traverse and light feed.

Dressing of tool room wheels

Saucer, dish and cup wheels used in the tool room, with the exception of surface grinding wheels, are best rough formed with a silicon carbide dresser stick which is hand controlled because of the varied shapes required. For sharp corners, small radii and finish forming, a hand diamond tool is required. Cuts must be light and intermittent to prevent excessive heating of the diamond. Surface grinding wheels in the tool room follow the procedure previously described.

Dressing of thin-edged wheels

Wheels trued to thin edges or sharp corners need sharp dressing tools. Dull tools require too much pressure to penetrate into the wheel body. Excessive pressure may cause chipping of the wheel edge or even breaking of the wheel itself. The dressing or truing cut should always start at the weakest point of the wheel face and, whenever possible, the cut should be from the extreme periphery toward the arbour.

29.6.2 Rules for Using Diamond Tools

Diamond dressing tools should always be used at an acute angle to the wheel face and to the direction of wheel rotation. Either of these two angles may be used, depending on the type of tool and its application.

Canting to prevent chatter

The first and the most important angle is one intended to prevent chatter in the dressing tool, gouging the wheel face and damaging to the diamond. This is usually referred to as 'canting'.

Assuming that the diamond point is in contact with the wheel on this line, raise the rear end of the tool (C), in the direction opposite the direction of wheel rotation (B), making an angle (A) between the radial line and the tool axis of from 3° to 15°. If in doubt as to the exact radial line, drop the point of the dressing tool to position not more than 6 mm below the supposed radial line. This contact point should never be above the radial line. With this setting, it is necessary to rotate the tool point about its axis to maintain a point on diamond.

Canting to sharpen diamond

Refer to Fig. 29.10. The second angle position C is intended to make the diamond self-sharpening and to avoid diamond marks on the wheel face. This angle constitutes a canting of the dressing tools to one side of the wheel. Again assuming that the diamond point is in contact with the wheel face, move the rear end of the tool holder to the side, making an angle of 30° between the tool axis and the plain of the wheel, or 60° from the grinding wheel face. With this angle, the diamond assumes and maintains a 60° included angle. As the diamond is rotated, it presents a sharp edge to the wheel, which will give an open dressing free of diamond called 'threads'. Here the diamond should set be slightly below rather than above the radial line.

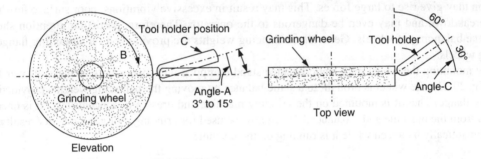

Fig. 29.10 Canting of diamond dressing tool.

Frequently, the wheel face is worn convex or tapered. It is a good practice to bring the diamond into contact with the wheel at the highest point on the wheel face before traversing, with any in-feed. This prevents excessive penetration with consequent possible damage to the diamond. To produce high finishes, this practice is not advisable, as the preliminary contact of the diamond with the wheel may leave a mark which may be transferred to the work. In this case, it is best to place the diamond slightly out of contact with the wheel, start the traverse and in-feed of 0.0254 mm at the end of each pass until contact is made and then proceed with the traverse.

A diamond is usually mounted so that the greatest practical amount of it is exposed for use. For this reason, it is poor practice to grind back into the diamond setting, unless the tool is actually designed to permit it, as in the case of certain trade marked tools, wherein the diamond is mounted on a 'cold setting'.

Keep diamond sharp

For dressing purpose it is quite essential that the diamond be sharp. The diamond dresser is a cutting tool, and like all cutting tools, its usefulness is in direct proportion to its sharpness. Dressing is not merely a matter of removing imbedded metal from the face of the wheel, in which case, the star dresser could be universally used. However, dressing is also a matter of sharpening and for that purpose diamonds are needed. It is false economy to allow a diamond to become dull. Frequent turning and occasional resetting of the diamond point may save many rupees in increased wheel life and better wheel efficiency.

To function properly, a grinding wheel must be dressed in such a manner that the abrasive grains are completely fractured and project above the bond. This can only be accomplished with a sharp dressing tool, since a dull tool merely crushes the grains down to the level of the bond and leaves some weakened by partial fracture. These weakened grains break in contact with the work and invariable cause scratching.

Then too, much a dull diamond frequently presses the wheel cuttings into the bond pores or voids in the wheel face, giving the effect of a loaded wheel. In every case of dull dressing, the wheel acts harder than normal and frequently causes failure of a properly graded wheel.

A decided advantage in maintaining a uniformly sharp diamond is to be found in the corresponding uniformity of grade effect. As diamond becomes dull in service, the wheel also gradually becomes duller at each dressing but the change is too gradual in most cases to be easily noticed. When the diamond is changed, however, the wheel acts softer, thus, changing the character of the operation and the finish.

29.7 BALANCING OF GRINDING WHEEL

From the point of view of wheel life and surface finish on the job, variation in grinding operation is critical. Assuming that the machine is rigid and bearing are in good condition; vibrations to a large extent are caused by out-of-balance and out-of-round wheels. Since the grinding wheel speeds are high, slight out-of-balance condition may give rise to large forces. This may result in excessive vibrations, poor surface finish, faster wheel breakdown and may even be dangerous to the operator. Therefore, particular attention should be paid to the balancing of wheels. Generally, balancing weights are provided on the mounting flange of the grinding wheels.

By mounting the wheel on a static balance stand equipped with two rollers or pairs of over-lapping discs (Fig. 29.11), the wheel is brought to a static balance by moving the balancing weights provided on the mounting flange. Then it is mounted on the grinding machine and dresses to concentricity. It is once again removed from the machine and rebalanced. It can now be used for grinding. For getting better results, wheel can be dynamically balanced while it is running on the machine.

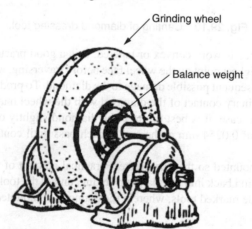

Fig. 29.11 Revolving wheel balancing stand.

Mounting of wheel on the spindle

The method of mounting a wheel on the spindle is shown in Fig. 29.12. Some important points to be considered are given as follows:

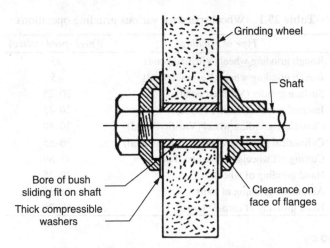

Fig. 29.12 Mounting of a grinding wheel.

(i) Make sure that the wheel is not cracked. Although the cracks are not visible; yet the crack in the wheel can be found. To test for cracks, hold the wheel through its hole and lightly tap it with an object like a wrench handle. A wheel in good condition will give a very clear ringing sound, while a cracked one will give a dull or flat sound.
(ii) Ensure that the wheel has proper packing on both sides around the hole.
(iii) The wheel should slide into the spindle or arbour having clearance up to 0.1 mm. If hole is small, it should be reamed.
(iv) The wheel flanges should be of safety type. In case of big stones, it is very important. Washers made of thick, soft pulp or rubber should be placed between flanges and wheel sides.
(v) Wheel should never be excessively tightened.
(vi) Wheel should never be operated without guard. Also, ensure that wheel moves freely in its guard and housing by rotating the wheel with hand.
(vii) See that wheel is balanced and dressed properly.
(viii) While starting the machine, keep yourself away to one side for a minute or so.

29.8 RECOMMENDED WHEEL AND WORK SPEEDS FOR GRINDING

It this section, we study the recommended wheel speed and work speed used for grinding processes.

29.8.1 Recommended Wheel Speeds

If the wheel speed is increased at a constant longitudinal or rotary feed rate, the size of the chip, removed by a single abrasive grain is reduced. This reduces the wear of the wheel. If the wheel speed is reduced, the wear is increased. From this, it is clear that from the view point of wear, it is better to operate at higher wheel speeds. However, it is limited by the allowable speeds at which the wheel can be worked as well as the power and rigidity of grinding machine. Normally, the grinding wheel speed ranges from 20 m/sec to 40 m/sec. The wheel speed also depends on the type of grinding operation and the bond of grinding wheel, e.g. resinoid bonded wheels can be generally used at higher peripheral speeds than vitrified bond wheels. Table 29.1 gives various speeds for various grinding operations and types of bonds.

Table 29.1 Wheel speed for various grinding operations

Type of grinding	Wheel speed (m/sec)
Rough grinding wheel (Vitrified bond)	25
Rough grinding wheel (Resinoid bond)	45
Surface grinder (Vitrified bond)	20–25
Internal grinder (Vitrified bond)	20–35
Centreless grinding wheel (Vitrified bond)	30–80
Cylindrical grinding wheel (Vitrified bond)	30–35
Cutting off wheels with resinoid bond	45–80
Hand grinding of wheels	20–25
Automatic grinding of tools	25–35
Hand grinding of carbide tools	18–25

29.8.2 Work Speeds

Work speed can be defined as the speed at which the workpiece traverses across the wheel face or rotates around between centres. If work speed is high; wheel wear is increased, but the heat produced is decreased. On the other hand, if the work speed is low, the wheel wear decreases, but heat produced is more. The ratio of wheel speed to work speed is of importance and should be maintained at proper value. Low work speed results in local overheating and bring about the deformation of hardened workpiece. It affects the mechanical properties of the workpiece and very often, micro cracks will appear on the workpiece. The increase in work speed is limited by premature wheel wear and vibrations induced by wear. The approximate work speed (in m/sec) for various materials are given in Table 29.2.

Table 29.2 Work speed for various materials

| Material | Cylinder grinding | | Internal grinding | Surface grinding |
	Roughing	Finishing		
Soft steel	11–15	6–8	15–20	
Hard steel	14–16	6–10	18–22	
C.I.	12–15	6–10	18–22	
Brass	18–20	14–16	28–32	8–15
Aluminium	50–70	30–40	32–35	

29.9 TYPES OF GRINDING MACHINES

Grinding machines may be of roughing type (such as grind stones, bench grinders, flexible shaft grinders, and cut-off wheels) or they may be of precision type.

Precision type grinding machines can be classified as:

1. **Surface grinding machine:**
    (i) Planer type
    (ii) Rotary type
2. **Cylindrical grinding machine:**
    (i) External type
       (a) Axial feed
       (b) Plunge cut
    (ii) Internal type

3. **Centreless grinding machine:** Through feed, infield, end feed type
4. **Other grinding machines:**
   (i) Tool and cutter grinder
   (ii) Special grinding machines, e.g. form grinder, thread grinder, gear grinder,
   (iii) Jig grinding machines
   (iv) Recently developed machines
      (a) Centreless grinding machines with profile dressing
      (b Roll chamber grinding machines,
      (c) Duplex grinder,
      (d) Angular wheel head grinder.

Each of these grinding machines will now be discussed in detail.

29.9.1 Surface Grinding Machines

The grinding of plane or flat surfaces is called surface grinding. There are two types of machines on which this operation is performed. These two types of machines are identified by the positions of the grinding wheel spindle in its relation to the table which holds the work. The spindle position is either horizontal or vertical. The types of wheels used are determined by the spindle position. Figure 29.13(a) illustrates the operation of surface grinding with the periphery of the grinding wheel. Figure 29.13(b) illustrates the operation of surface grinding on the face of a grinding wheel.

The choice of wheel shape and the method of mounting it provides the basis of classifying the two types of surface grinding operations.

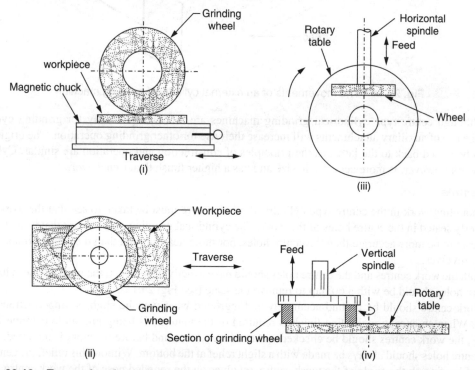

Fig. 29.13 Types of surface grinders (i) horizontal spindle reciprocating table (ii) vertical spindle reciprocating table (iii) horizontal spindle rotating table (iv) vertical spindle rotating table.

29.9.2 Cylindrical Grinders

Here, we discuss the various types of cylindrical grinders.

Centre type cylinder grinders

In these grinders, the workpiece is rotated between centres. These grinders are primarily used for grinding contoured cylinders, tapers, faces and shoulder, fillets and even cams and camshafts. There are two types of grinding operations where the workpiece is held between fixed centres (i) Traverse grinding and (ii) Plunge grinding.

In traverse grinding, the work reciprocates past the grinding wheel. As the workpiece passes, the wheel removes fixed amount of metal from the diameter of the workpiece. At the end of the pass, the wheel is advanced another increment of distance (just as in lathe) for removing further metal layer (Fig. 29.14). This grinding is employed when the workpiece is longer than the maximum width of the wheel.

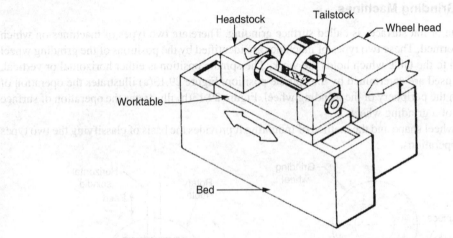

Fig. 29.14 The movements of an external cylindrical grinding machine.

Although centre type cylindrical grinding machines are intended primarily for grinding cylindrical pieces, the use of auxiliary attachments will increase their use for other grinding operation. The origin of this type can be traced back to the lathe, as the principles of construction and operation are similar. Cylindrical ground work however, is more accurate in size and has a higher finish than trench work.

Work centres

When mounting work in the centre type cylindrical grinder, care must be taken to see that the work centres are properly seated in the centre holes of the work. The cylindrical grinding of any work that is held between centres cannot be more accurate than the centre holes, nor more nearly round than the centre points in which the work revolves.

Both the work centres and the centre holes should have exactly 0° angles. If the work has been hardened, the centre holes should be with a 60° lap to remove the scale (see Fig. 29.15).

Work centre should be ground accurately and reground whenever the slightest imperfection in them is noted. When an exceptionally fine finish is required or the material is being ground to extreme limits of accuracy, the work centres should be checked and necessary, reground before the work is mounted.

Centre holes should always be made with a slight relief at the bottom. Without this relief, the centre hole, instead of bearing on the angle of the work centre, revolves on the rounded nose of the work centre, causing the work to wobble about on the work centre.

GRINDING AND OTHER ABRASIVE METAL REMOVAL PROCESSES 835

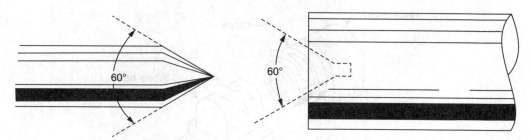

Fig. 29.15 Centering of job for cylindrical grinding.

Even though the work centres and centre holes are perfect, a great deal of trouble may be caused if foreign material is introduced between the bearing surfaces. This condition may develop because of improper cleaning of the centres and centre holes or because the lubricant used is contaminated with foreign substances.

Plunge-cut grinding (See Fig. 29.16)

There are, however, a large number of cylindrical grinding operations that are performed by the plunge-cut or straight infeed method where no traverse is involved. These plunge-cut operations may involve the use of plain or formed-face wheels to generate form on the cylindrical piece ground. When grinding form, the ability of the wheel face to hold its form operation is to be performed economically with respect to production requirements and dressing costs.

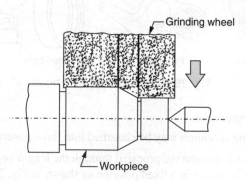

Fig. 29.16 Plunge grinding.

On all of these special operations performed on cylindrical grinding machines, many seeming contradictions in wheel action are apparent. This is because the operation has ceased to be simple cylindrical grinding of plain, straight cylinders.

Internal grinding machines

The grinding of internal surfaces of holes (inside diameters) is called internal grinding.

Figure 29.17 shows the simplest form of the internal grinding machine.

The application of internal grinding is extensive and the range of hole sizes and types of pieces are limited only by the capacity of the grinding machine. Both general purpose and special purpose grinders have been developed so that internal grinding can be performed rapidly and economically on a wide range of hole sizes and configurations as are shown in Fig. 29.18.

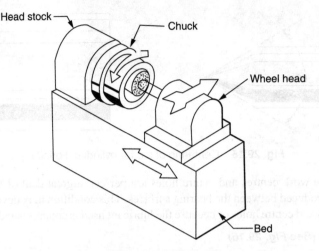

Fig. 29.17 Internal grinding.

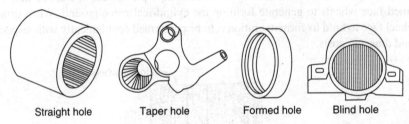

Straight hole   Taper hole   Formed hole   Blind hole

Fig. 29.18 Various types of holes.

Types of internal grinding machines

Single purpose internal grinding machines may be classified into three general types:

1. The wheel spindle is rotated and reciprocated through the length or depth of the hole being ground, while the workpiece is rotated in a fixed position as shown in Fig. 29.17.
2. The wheel spindle is rotated but is held in a fixed position, while the workpiece is rotated and reciprocated to obtain traverse.
3. Cylinder grinder is the oldest of the three types, however, it is not as commonly used as the other two. This set-up consists of the wheel spindle rotating while the workpiece is reciprocated to obtain traverse movement but is not rotated.

Internal grinding is also performed with an internal grinding attachment on a universal tool and cutter grinder. Tool post grinders attached to lathes, are also satisfactorily used.

Mounting the work

On the first class of grinders, the workpiece is either chucked or mounted on a face plate. The centre of the work and the centre of the wheel spindle are on a common vertical and horizontal centre lines. A cross slide adjustment is provided under work head or the wheel head to bring the wheel into contact with the work and allows the movement of either the wheel or the work along a common horizontal centre line.

    The general conditions applicable to the second class of machines are similar to those surrounding the first, except that the workpiece, instead of the wheel spindle, is reciprocated to obtain wheel traverse.

On the third class of cylinder grinders, the wheel spindle is mounted on a table in a fixed position with reference to the table. Vertical and horizontal table adjustments are provided to bring the wheel into the desired alignment with the wheel spindle.

Machine condition

Spindle condition: The heart of any precision grinding machine is the grinding wheel spindle and its bearing. Any wear in the bearing that permits the spindle to vibrate will render the machine useless for precision work.

Balance of both work and wheel is essential to good grinding. An out-balance condition of the work, particularly, when the work is not rigidly held in the machine, also creates vibration that varies the pressure (stresses) between the wheel and the work.

True running of wheel is also essential to the maintenance of steady pressure between wheel and work. True running is obtained by precision dressing and truing with a diamond.

Due to the great overhang of the wheel spindle on internal grinding machines, every precaution must be taken to keep wheel pressure low and vibration at a minimum for satisfactory operation. Low wheel pressures and minimum vibration are of much greater importance to the satisfactory operation of internal machines than on cylindrical machines.

29.9.3 Centreless Grinding

We study here the various types of centreless grinding in detail.

Centreless cylindrical grinding

It is accomplished by grinding cylindrical surfaces without rotating the workpiece between fixed centres. The workpiece is supported between three components as seen in Fig. 29.19. The three components are (i) Grinding wheel (ii) Regulating wheel and (iii) Work-rest.

The grinding wheel does the actual grinding while the work-rest positions the workpiece for grinding. The functions of the regulating wheel are to govern the speed of the rotation of the workpiece, to govern the sizing of the workpiece, and to govern the rate of workpiece travel through the grinder through feed grinding. The grinding and regulating wheels rotate in the same direction, while the workpiece rotates in the opposite direction. The three principal classes of centreless grinding are: (i) Through feed, (ii) In-feed and (iii) End feed.

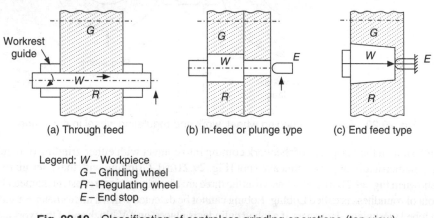

Legend: W – Workpiece
G – Grinding wheel
R – Regulating wheel
E – End stop

Fig. 29.19 Classification of centreless grinding operations (top view).

In through feed grinding shown in 29.19(a) the workpiece is passed between the grinding wheel and regulating wheel. The actual movement of the workpiece past the face of the grinding wheel is imparted by tilting the regulating wheel at slight angle about a horizontal axis from 0° to 7° or 10°.

The in-feed method shown in 29.19(b) is used on the work which has a shoulder, head or some portion larger than the ground diameter, so that it does not pass completely through the wheels.

The end feed method shown in 29.19(c) is used primarily on taper work. Either the grinding wheel or regulating wheel, or both are dressed to the desired taper. This is explained in detail in the foregoing descriptions.

The principles of centreless type of cylindrical grinding are different from those employed in centre type operations. The machine elements of a centreless grinding machine are; the grinding wheel, the regulating wheel and the work-rest blade, all three of which support the workpiece in the grinding position.

The regulating wheel usually consists of either a rubber bond or a resinoid bond abrasive wheel with frictional characteristics which impart a constant and uniform rotation to the workpiece at the same surface speed (surface meters per minute) as the regulating wheel itself. In other words, the workpiece, to all intents and purposes, is "geared" to the regulating wheel which, therefore, determines the speed of the workpiece.

The work-rest blade also acts as a means of supporting the workpiece during the grinding operation. It also incorporates suitable guides for directing the work to the wheels (grinding and regulating) and discharging it from the zone of grinding action.

Principles of centreless grinding—Rounding up action of the process

Whereas it is more or less obvious how workpieces are ground on a cylindrical centre type machine, the principles are not as apparent in centreless grinding.

In centreless type cylindrical grinding, the centre of the work is in line with the centres of the grinding and regulating wheels (Fig. 29.20). The flat top work supporting blade is used. The surfaces of the wheels together with the blade form three sides of a square.

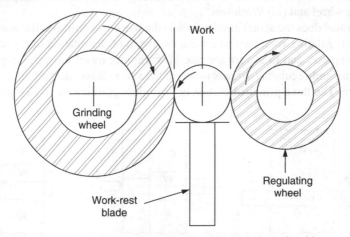

Fig. 29.20 Centre of grinding wheel, work and regulating wheel in one plane.

Any high spot on the periphery of the work coming into contact with either grinding or regulating wheel will produce a diametrically opposite concave spot [Fig. 29.21(a)]. Grinding with these set-up results in work generated is shown in Fig. 29.21(b), and is known as the three arc triangle, having a constant diameter but not round. This type of out of roundness is called Lobing. Lobing cannot be detected when a micrometer is used to check the accuracy of a lobed workpiece. However, if a lobed cylinder is placed on a V block and if the workpiece is rotated under the tip of a dial indicator, the lobing or out of roundness can be detected. It is obvious that a horizontal line of centres as stated above is undesirable in centreless grinding because of the lobing defect produced by such a setting.

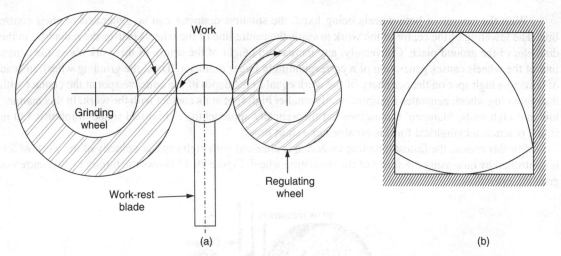

Fig. 29.21 (a) High spot on the work results in presentation of regulating wheel in this work and (b) A three arc triangle with constant diameter but not round.

Now, as a corrective measure, if the centre of the work is elevated above the centres of the wheels by raising the supporting blade, a low spot coming in contact with the regulating wheel will cause a high spot to be generated at the contact with the grinding wheel, but not diametrically opposite. As the piece being ground is rotated, the high and low spots will not come opposite to each other, as is shown in Figs. 29.20 and 29.21, and a gradual rounding up effect is thus obtained.

To attain maximum corrective rounding up action, use has been made of a blade with angular top as shown in Fig. 29.22(a). This produces distinct corrective phenomena which is diagrammatically illustrated in Fig. 29.22(b). Two lines, A–A and B–B are tangents drawn at the points of contact of the work with the wheels, and another line, C–C shows the plane of the angular top of the blade. If a low spot on the work comes in contact with either the blade or the regulating wheel, as in the case of the piece shown in dotted lines, the approximate centre of the work will be lowered.

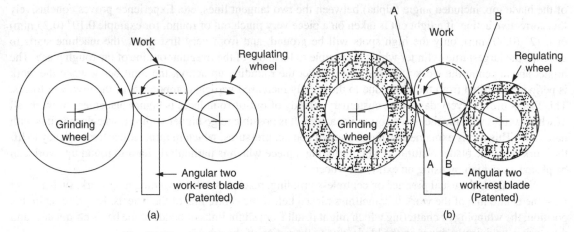

Fig. 29.22 Angular blade used for corrective rounding up action.

With the centres of both wheels being fixed, the smallest diameter can be generated on their centre line. The lowering of the centre of the work towards the centre line of the wheels causes the reduction in the diameter of the ground piece. Conversely, any increase in height of the centre of the work above the centre line of the wheels causes generation of a correspondingly larger diameter. Thus, the grinding wheel, instead of leaving a high spot on the periphery of the work equal to the depth of the concave spot at the contact with the regulating wheel, generates a proportionally smaller high spot at its contact with the work. In this manner, low and high spots, 'dampen' themselves but theoretically approaching cylindrical shape in infinity, but in actual practice, a cylindrical form is obtained in a short time.

For this reason, the fastest rounding up action is obtained with high angular velocity of the work which is controlled by increasing the speed of the regulating wheel. Figure 29.23 shows the through feed centreless grinding operation.

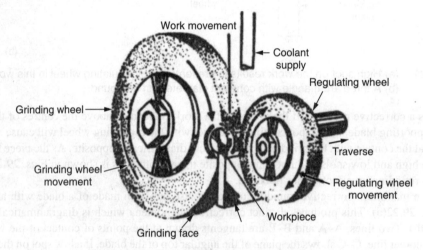

Fig. 29.23 Through feed centreless grinding.

This corrective action is very complex and depends upon many involved variables, such as the inclination of the blade top, included angle (alpha) between the two tangent lines, etc. Experience proves conclusively this corrective action, if a light cut is taken on a piece very much out of round, for example 0.10″ (0.25 mm) or 0.12″ (0.30 mm), only the high spots will be ground, and from very first spark, the machine starts to generate the largest true cylinder which is possible to form out of the irregular outline of the rough piece. The higher above centre the work is placed, the quicker the rounding up action, the limit being when the work is periodically lifted from the blade due to the greatly increased vertical components of the involved forces. This fact is proved every day as the customary remedy of out-of-roundness is higher placement of work, all of which agrees with the theory of corrective action. It is possible to go higher with soft wheels than it is with hard ones. This fact is explained by the decreased contact pressures, which in turn, reduce the tendency to lift the work from the blade. Naturally, when grinding a piece which is particularly hard to round up, work can be placed extra high by using an extra soft wheel.

Contrary to the usual practice on centreless grinding, machines, when grinding long work, such as steel bars, the centre line of the work is sometimes placed below the centre line of the wheels. By grinding in this position, the whipping or chattering which might result from slight links or bends of the bar is eliminated, and the work is held firmly down on the blade, due to the action of the wheels.

All centreless grinding machines employ so called negative work speed, or in other words, the angular rotation of the work and that of the grinding wheel, are in opposite directions. The grinding wheel revolves at standard grinding speeds approximating 6000 feet per minute (30 mps). The operating speeds of the regulating wheel can be varied at will, and have the range from 50 to 200 feet per minute (0.25 mps to 1 mps) peripheral speed.

The preceding discussion can be summarized in the following general rules:

For quick rounding up action, the work centre should be placed as high as possible, the angular velocity of the work should be high, and the rate of traverse (if any) across the grinding wheel face should be small.

For best straightening out effect of long single diameter work, the centre of the piece should be placed below the centre line of the wheels and the rate of traverse should be high. Grinding in this position is primarily for straightening the work. Subsequent passes can be made with normal set-up for corrective rounding action. It may be pointed out here that to achieve a line contact between the workpiece and the regulating or control wheel in through-feed centreless grinding, the latter should be a hyperboloid in shape. A cylindrical regulating wheel shall result in a point contact with the workpiece which is undesirable because there will be a slip between the work and the regulating wheel.

Classes of centreless grinding

The centreless grinding principles lend themselves to almost unlimited applications through the use of machine set-ups involving special relationships between the grinding wheel, the regulating wheel and the work-rest blade, combined with various types of work guides and feeding mechanisms.

The three principal classes of centreless grinding are:

(a) Through feed
(b) Infeed
(c) End feed.

(a) *Through feed* centreless grinding is accomplished by passing the workpiece between the grinding wheel and the regulating wheel. The longitudinal or actual movement of the work past the face of the grinding wheel (a movement corresponding to traverse in centre type grinding) is imparted by the regulating wheel.

To do this, the machine is arranged in such a manner that the regulating wheel can be swung about a horizontal axis from 0° to 7° or 10° relative to the axis of the grinding wheel spindle.

Moreover, the speed of the regulating wheel, which can be changed by means of a gear box, and the diameter of the regulating wheel also influence the feeding rate of the work. The interdependence of these factors is closely approximated by the following mathematical expression:

$$F = d\pi N \sin a$$

where F = Feed of the work (in mm per minute)
d = Diameter of the regulating wheel (in mm)
N = Speed of the regulating wheel, in revolutions per minute,
a = Angle of the inclination of the regulating wheel.

Figure 29.24 shows diagrammatically these conditions when viewing the arrangement at a right angle to the axis of the grinding wheel spindle. The theoretical feed is based on the assumption that there is no slippage of the work whatsoever in its contact with the regulating wheel. In actual practice, the error in this connection rarely exceeds two per cent. It is often necessary to pass work between the wheels more than once. The number of passes is determined by the amount of stock to be removed, the condition of the work as to roundness and straightness, the quality of the material, and the limits of accuracy required.

With this method, there is a fixed relationship between the grinding wheel, regulating wheel, and the work supporting blade. The wheels are adjusted so that the distance between their active

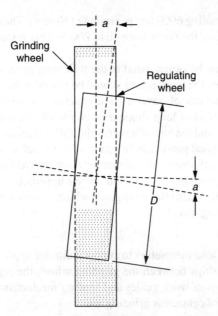

Fig. 29.24 Arrangement for through feed in centreless grinding.

surfaces, together with the height of the work blade, determines the diameter of the ground piece. The centres of the wheels are stationary during grinding operations and require slight re-adjustments from time to time to compensate for the wear of the grinding wheel.

A work-rest or fixture which provides means for holding the blade incorporates adjustable guides, both to the front and rear of the wheels. These guides must be accurately aligned with the regulating wheel face to ensure the work travelling in a straight line.

(b) *Infeed* method is usually employed when grinding work which has a shoulder, head, or some portion larger than the ground diameter. The same method is used for the simultaneous grinding of several diameters of the work as well as for finishing pieces with taper, spherical, or any other irregular profile.

In general, this method corresponds to the plunge cut or form grinding on the centre type grinder.

The length of the section or sections to be ground in any one operation is limited by the width of the grinding wheel. As there is no relative axial movement of the work, the regulating wheel is set with its axis approximately parallel to that of the grinding wheel. Only a slight angle is maintained to keep work tight against an end stop.

With the infeed method, there is a fixed relationship between the work support blade and the regulating wheel. These two units clamped together carry the work to and from the grinding wheel. This movement is performed by turning the infeed lever, 90°. As the lever is brought down, the desired size being secured when the lever has made its full swing. On reverse movement of this lever, the gap between the wheels is increased, and either manually or automatically operated ejector kicks the work out from between the wheels, and another piece is placed in position by the operator.

If the piece to be ground is longer than the width of the wheels and must be ground only a short distance from each end, a variation of infeed grinding is used, which is called the outboard roller support method. One end of the work is supported by the blade between the wheels and the other end rests on rollers which are usually outside of the machine, but form the part of the fixture which is called the infeed roller work-rest.

The same cycle of operations as described for infeed grinding is used, the only exception being that it is not always possible to use the ejector because of the size of work, which is usually larger than that of the standard infeed class.

(c) *End feed* method is used only on taper work. The grinding wheel, the regulating wheel, and the blade are set in a fixed relationship to each other, and the work is fed in from the front, manually or mechanically, to a fixed end stop. Either the grinding or the regulating wheel, or both, are dresses to the proper taper.

Special applications

Figure 29.25 illustrates the special applications of centreless grinding.

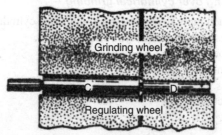
(a) Multiple diameter centreless grinding

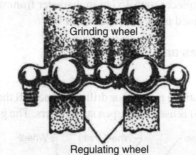
(b) Centreless profile end feed grinding

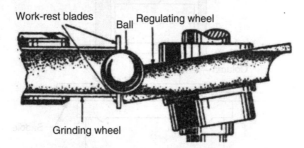
(c) Centreless grinding of spherical balls

Fig. 29.25 Centreless grinding operations—some typical applications.

All that has been said in regard to finish requirements and wheel selection on centre-type cylindrical grinding is applicable to centres type grinding too.

Advantages and limitations of different centreless grinding methods
Advantages:
   (i) The operation becomes automatic by employing continuous magazine feed for longer bars and hopper for small jobs.
   (ii) No holding of the workpiece is required except the work support and hence the long bars can be ground easily without any deflection.

Limitations:
   (i) Useful only for straight cylindrical parts.
   (ii) The form grinding operation cannot be carried out by this process.

Advantages of centreless grinding over cylindrical grinding
   (i) Rate of production is much more in centreless grinding than cylindrical grinding.
   (ii) Better stability as work is supported rigidly.
   (iii) Highly useful for long jobs.
   (iv) Job setting time is minimum.
   (v) Cost of production is less.
   (vi) Wear and tear of the machine is less.
   (vii) Maintenance cost is low.

Scope of centreless grinding
This process is most suitable for workpieces upto 15 cm in diameter from washer to bars of 7 to 8 metres in length. The accuracy that can be obtained is the order 0.25 mm.

29.9.4 Other Grinding Machines and Processes

Tool and cutter grinder
When cutting tools, such as milling cutters, reamers or drills become dull they require resharpening on a tool and cutter grinder machine (Fig. 29.26) is used to sharpen such cutters. The grinder can grind plain cylindrical

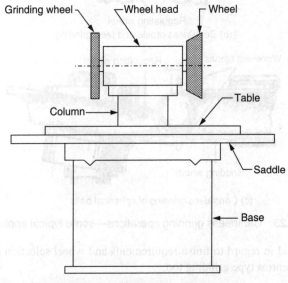

Fig. 29.26 Tool and cutter grinder.

cutters, angular cutters, end mills, formed cutters, reamers, taps and other cemented carbide tools. They are classified as:

(i) Universal tool and cutter grinder
(ii) Single purpose tool and cutter grinder.

Universal tool and cutter grinder is intended for sharpening of miscellaneous cutters, whereas single purpose grinder is used for grinding tools, such as drills, tool bits, etc. for production plants. A universal tool and cutter grinder has a heavy, rugged base with a saddle mounted on the top of the base. The column supporting the wheel head is mounted on the saddle and it can be moved up and down and swivelled. The table moves on a top base which is mounted over the saddle. The top of the base contains the gears and mechanism which controls the table movement. The head stock and tail stock are mounted on either side of column on the base of the machine. It can be swivelled and positioned on the base for varied setups.

Special grinding machines

In special grinding machines, the form grinder, thread grinder and gear grinder are very important.

(i) **Form grinder:** In form grinding, a formed wheel feeds towards the revolving work, which does not traverse. The wheel has the exact shape and form of the profile to be ground. Figure 29.27 shows the form grinding.

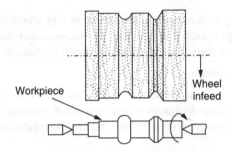

Fig. 29.27 Form grinding.

(ii) **Thread grinder:** In thread grinding, the profile of the grinding wheel depends on the shape of the thread to be ground. The wheel is so formed as to have the exact shape of the thread. Figure 29.28 shows thread grinding. Thread of any form can be ground on these machines, either from a blank or as finishing operation after milling.

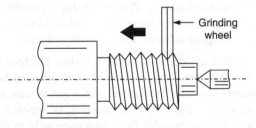

Fig. 29.28 Thread grinding.

(iii) **Gear grinder:** As shown in Fig. 29.29, this grinder makes use of a formed wheel, shaped exactly to the profile of the shape between two adjacent teeth. This curved profile varies for each change in gear diameter, or for each change in the number of teeth. Such a profile is difficult to form and

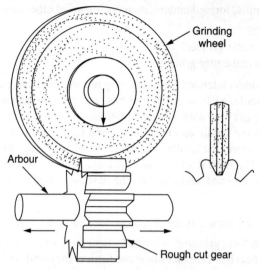

Fig. 29.29 Gear grinding.

maintain. A rough cut gear is mounted on an arbour and the arbour is held in an indexing and locking work head. With the gear locked in position, it is reciprocated slowly under the wheel. At proper intervals, indexing head shifts the gear from one tooth position to the next.

Newly developed grinding machines

Automation has been associated with advancement in technology. In the process of automation for small batch production and mass production, hydraulic tracer controlled machine tool and programmed operating cycle machine tools have been evolved. However, this requires templates, cams, stops, electrical trip dogs, timers, etc.

Keeping automation in mind different types of machines have been developed for the operations which were being done before, using some attachments or fixtures on cylindrical, internal, surface and centreless grinding machines. Thus, required jobs are being done achieving better results in minimum time, without fatigue.

(i) **Centreless grinding machine with profile dressing:** Both the grinding and the regulating wheel heads are provided with hydraulically operated wheel dressers having stepless variable dressing speeds and are controlled independently. Profile dressing is possible with the interchangeable copy template. Top slide of dressing units is mounted on preloaded antifriction ball guide ways. A spring loaded stylus against a template achieves high copying accuracy.

(ii) **Duplex grinder:** Duplex grinder is a surface grinding machine used to generate two parallel surfaces at high production rate. The machine has two grinding heads and utilizes the faces of bonded grinding wheels fastened to steel plates which are, in turn, attached to spindle face plate.

The two grinding wheels (Fig. 29.30) are running in opposite direction and workpieces are fed between them to grind two parallel faces simultaneously to close tolerance and fine finish. The production rate is quite high in this machine. Piston rings can be produced at a rate of about 1500 pieces/hour attaining flatness and parallelism within 0.01 mm.

The finish and flatness obtained by this process is better than any other grinding process using periphery of grinding wheel.

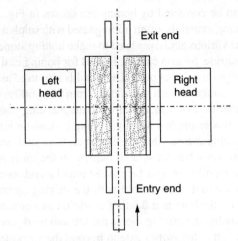

Fig. 29.30 Duplex grinder.

(iii) **Angular wheel head grinder with profile dressing:** For grinding multiple diameters, shoulders and profiles in one plunge cut, the wheel is trued with a hydraulic copy truing device mounted on the wheel head.

In this machine, wheel head is swivelled and fixed at 30° to the work axis (Fig. 29.31). By doing so, wheel width is increased and facing on job becomes easier and has been found to attain uniform wheel wear. This is because in case of grinding a shoulder along with a step parallel to the axis, a swivelled grinder head ensures a line contact along the shoulder as well as the step. Thus, the condition of contact between the wheel and the workpiece ensures equal load on the grits all along the profile which results in uniform wear and maintenance of wheel profile accuracy. This is shown in Fig. 29.31 where the actual length of contact (w) is more than wheel width (w_1).

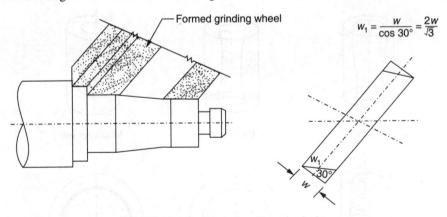

$$w_1 = \frac{w}{\cos 30°} = \frac{2w}{\sqrt{3}}$$

Fig. 29.31 Angular wheel head grinder.

29.10 HONING OPERATION

Honing is a grinding or abrading process used mainly for finishing round holes by means of bonded abrasive stones called hones. Honing is primarily used to correct out of roundness, taper, tool marks and axial distortion.

Various hole geometries that can be corrected by honing are shown in Fig. 29.32. Honing stones are made from common abrasive and bonding materials, often impregnated with sulphur, resin or wax to improve cutting action and lengthen tool life. The various abrasives used to make honing stones are silicon carbide, aluminum oxide, diamond or cubic boron nitride. Silicon carbide is used for honing cast iron and non-ferrous materials, whereas aluminium oxide is used to hone steel parts. Diamond is also used as an abrasive to hone parts made of ceramics or hard carbides. The abrasive grain size ranges from 80 to 600 grit. Almost every material can be efficiently honed. Steel of all varieties, cast iron, aluminium, magnesium, brass, bronze, glass, ceramics, hard rubber, graphite and silver are a few examples. Mostly, honing is done on internal cylindrical surfaces, such as automobile cylinder walls. Honing corrects defects like taper, bow, bellmouth, waviness out of roundness, etc. of cylinder liner bores. It creates a typical scratch pattern on the inner surface of cylinder liners which facilitate the retention of lubricating oil on this surface. Most popular and successful grinding fluid for honing is Kerosene. In honing, the honing tool is connected with the driving spindle through a universal joint so that the honing tool reciprocates in the bore in a floating condition and generates an almost perfect cylinder.

When honing is done manually; the honing tool is rotated and workpiece is passed back and forth over the tool. Length of motion is such that the stones extend beyond the workpiece surface at the end of stroke. For precision honing, shown in Fig 29.32 the work is usually held in a fixture and the tool is given a slow reciprocating motion as it rotates. The stones are thus given a complex motion as rotation is combined with

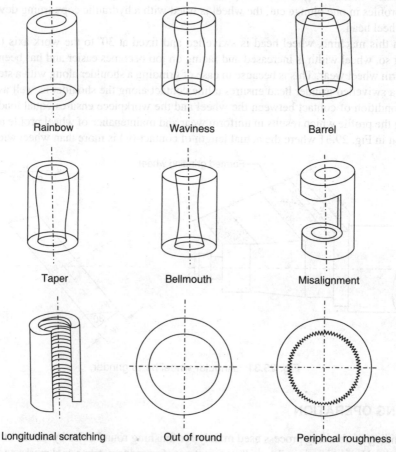

Fig. 29.32   Different hole geometries that can be corrected by honing.

oscillatory axial motion. These two motions combine to give a resulting cross-hatch lay pattern. Honing stones may be held in the honing head by cementing them into metal shells, which are clamped into holder or they are cemented directly into holders. During honing operation, the honing head is not guided externally but floats in the hole being guided by the work surface. Coolants are essential to the operation of this process, to flush away small chips and to keep the workpiece cool, kerosene is preferred to keep the temperatures uniform. Sulphurized mineral oil or lard oil is sometimes used for this purpose.

Honing differs from grinding due to the reason that honing requires the use of low speed and low pressure, which keeps the surface temperature relatively low, whereas grinding machines are run at high speed and high wheel pressure. Also, abrasive stones used in the honing machine have a relatively large area of abrasive in contact with the job. Due to large area of contact the honing sticks are subjected to less pressure and a softer grade of abrasive is preferred. As surface temperature is low, surface damage is kept to a minimum. The feature of honing, which allows it to develop round and straight bore is the relationship of cutting faces of the honing stones to the surface being honed. Either the tool or the fixture holding the work part floats. Due to this, the honing tool exerts equal pressure on all sides of bore. With honing process, it is possible to handle work parts having internal diameters from 1.5 mm to 15 mm. In honing, dimensional accuracy up to 0.00025 cm is common. Surface finish up to 0.1 μm can be expected. One limitation of honing is that honing produces a rather long chip with ductile materials and is unsuitable for certain composite materials. Clogging creates problem in such cases.

29.10.1 Type of Honing Machines

Honing can be done on general purpose machine such as lathe, the drill press, etc. but more economical results can be obtained by using machines for production work. The most common out of these are as follows:

Horizontal honing machines

These machines are mostly used for honing comparatively longer jobs, such as gun barrels and long tubular parts. All such machines carry a horizontal spindle, on which honing tool in mounted. In some machines, the workpiece is mounted on a table which can move the work to and for hydraulically. The hone reciprocates about its own axis and also simultaneously rotates. The motion of the honing tool may be controlled by an electric motor through the speed gearbox, whereas for reciprocation electro-mechanical, rope or hydraulic drive is employed. In some machines, the work is held in a horizontal position and rotated about its own axis. No reciprocating motion is given to it. Against this, the honing tool, which is mounted on a travelling head, is rotated and reciprocated to give the same results as above.

Vertical honing machines

These machines hold the work as well as the tool in vertical positions. Usually the spindle head and hence the tool reciprocates and not the workpiece. A vertical honing machine is shown in Fig. 29.33. The rotation of the spindle and hence the tool, is accomplished by means of hydraulic drive in vertical honing machine. The conditions for cooling the honing tool and carrying away the chips are more favourable in this machine, as the coolant is more uniformly distributed over the work surface. These machines are best suited for small jobs.

In the latest honing machines, inprocess gauging equipment is incorporated to gauge the bore diameter automatically, throughout the honing cycle. When the desired diameter is obtained, a signal is generated which stop the expansion of the honing tool. After this, the pressure on the honing sticks gets reduced gradually and the tool is with from the bore. For controlling the bore size, two types of devices are used, depending on whether the tool or bore is gauged. In tool gauging, a tool having a gauging ring is used. The honing sticks of the tool have plastic tabs molded on each and as shown in Fig. 29.33. These tabs wear down along with abrasives, therefore, their outer diameter always remains equal to bore diameter. The gauge ring is positioned

850 TEXTBOOK OF PRODUCTION ENGINEERING

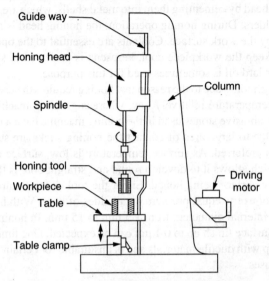

Fig. 29.33 Vertical spindle honing machine.

above the workpiece, so that the upper tabs enter the gauge ring at the top of each stroke. The inner diameter of ring is equal to the lower limit of the desired bore size. When the bore and the tabs acquire the same size as that of gauge ring, the friction between the ring and the tabs causes the gauge ring to turn, which subsequently generates a signal to stop the cycle.

Advantages of honing
  (i) The honing process enables highly accurate holes, as the possibility of variation is very less.
  (ii) Many holes can be honed simultaneously on multiple spindle machines.
  (iii) Hole of any dimension can be honed.
  (iv) As compared to other hole finishing methods, high productivity at low cost is obtained.

Disadvantages of honing
  (i) It is impossible to improve lack of straightness in holes.
  (ii) It is difficult to hone tough non-ferrous metals due to glazing or clogging of the pores of the abrasive sticks.

Various defects that can occur during honing process and the possible types of these defects are shown in Fig. 29.32.

29.10.2 Lapping

Lapping is basically an abrasive process in which loose abrasive function as cutting points find support of the lap.
  The process has the following features:
  (i) Use of loose abrasives between the lap and the work.
  (ii) The lap and workpiece are not positively driven, but are guided in contact with each other.
  (iii) Relative motion between the lap and work surface should be constantly changing. The most effective path is of non-repeating in figure of eight or cycloidal in nature.

The lubricant for lapping is liquid in which fine abrasive grains are suspended. Machine oil, soluble oil, grease, etc. are used as lapping vehicle. Figure 29.34 shows the principle of lapping process.

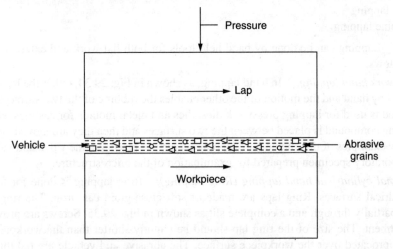

Fig. 29.34 Principle of lapping.

Lap material
Cast iron is the best lap material, but brass, bronze, lead, soft steel are also used. In any event, lap should be softer than the workpiece, so that the abrasive gets embedded in the lap.

Abrasive used in lapping
Silicon carbide is used for rapid stock removal and aluminium oxide for improved surface finish. Speeds between 1.50 m/sec and 4 m/sec are used.

Lapping speed
The speed of the lap relative to workpiece surface is chosen from Table 29.3 taking into account loss of work material and surface roughness.

Table 29.3 Lapping speed for various jobs

	Roughness value (Ra in microns)	Lapping speed (metre/min)
For medium accuracy	< 10	200–400
For accurate jobs	$10 \leq Ra < 12$	100–250
For very accurate jobs	$12 \leq Ra < 14$	10–100

Lapping is a precision finishing process done on precision tools, gauges, valves and on other similar places where resistance to wear of moving parts, better sealing characteristics and longer life of cutting edges are prominent factors. A very thin layer of metal from 0.005 to 0.01 mm is usually removed by lapping.

As lapping is not primarily meant for removing metal, so, it should be kept in mind that the material left on the work surface is minimum. Keeping in view the above discussions, the recommended range of lapping allowance to be left is as follows:

| Allowance on surface | 0.0075 mm to 0.0125 mm |
| Allowance on diameter or thickness | 0.015 mm to 0.05 mm |

Lapping methods and machines

Lapping is done in the following two ways:

 (i) Hand lapping
 (ii) Machine lapping.

Hand lapping: Lapping can be done by hand held tools for both flat work and external cylindrical work explained as follows:

 (a) *Flat work hand lapping:* In hand lapping, as shown in Fig. 29.34, either the lap or the workpiece is held by hand and the motion of the other enables the rubbing of the two surfaces in contact. This method is used for lapping presswork dies, dies and metal moulds for castings, etc. Sometimes, a lapping compound is placed between the two surfaces and then they are moved against each other. A few examples of this method are lapping of surface place, engine valve and valve seat, lapping of laboratory specimen prepared for examination of the microstructure.

 (b) *External cylindrical hand lapping (Ring lapping):* Ring lapping is done for finishing external cylindrical surfaces. Ring laps are made of soft close-grain cast iron. The ring lap has several cuts partially through and a complete slit as shown in Fig. 29.35. Screws are provided for precise adjustment. The size of the ring lap should be slightly shorter than the workpiece. The ring lap is reciprocated over the workpiece surface. The abrasive and vehicle are fed through the slot to maintain a straight round hole in the lap. This type of lapping is recommended for stepped plug gauges or gauges made in small quantities.

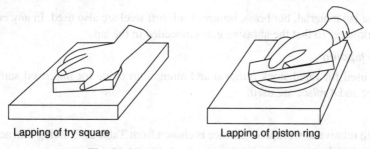

Lapping of try square   Lapping of piston ring

Fig. 29.35 Methods of hand lapping.

Machine lapping: This is performed for obtaining highly finished surface on many articles like races of ball and roller bearings, worm and worm gears, crankshafts, camshafts and various automobile engine parts like injector pump parts, spray nozzle, etc. Various types of machine lapping processes are explained as follows:

 (a) *External cylindrical machine lapping:* It is a vertical spindle machine carrying one upper stationary lap and one rotating lower lap. The upper lap is free to float and rest on the work which rides on the face of lower lap. Pressure is applied by gravity. The work is held loosely in the work guide, ring which carries the workpieces in slots made in the work ring so it follows a random path between the lap faces. An external cylindrical lapping machine is shown in Fig. 29.36. In this type of lapping, the cylindrical workpieces are placed horizontally in the work guide ring as lapping progresses. Parts like the rollers and needles of roller and needle bearing are lapped to a high degree of parallelism and accuracy.

 (b) *Flat machine lapping:* The work holder propels the work in this case. The rotating driving spindle may give a friction drive to the work holder, or a positive drive may be given through gear teeth on the periphery of spindle and work holder. The driving spindle rotates at a different speed than the lower lap and the motion given to the work holder causes the work to cover the entire lap surface.

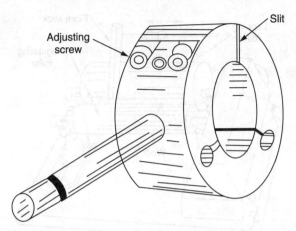

Fig. 29.36 Adjustable ring lap.

(c) *Centreless roll lapping machine:* The machine consists of two cast iron rollers; lapping roller and regulating roller. Lapping roller is twice in diameter as compared to that of regulating roller and both revolve in the same direction and at the same speed. The abrasive compound is applied to the rollers and the workpiece is laid between the two rollers as shown in Fig. 29.37. Lapping roller creates a rapid lapping action due to its increased surface speed. The workpiece is moved evenly over the entire surface of the roller by a fibre stick which is uniformly reciprocated.

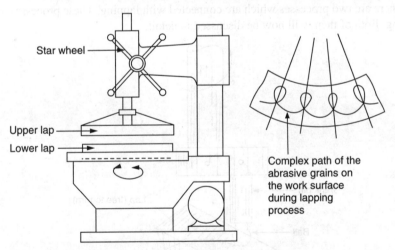

Fig. 29.37 Lapping machine for cylindrical work.

The machine is used for lapping one piece at a time and is designed for lapping plug gauges, measuring wires and cylindrical objects used in metrology.

(d) *Centreless lapping machine:* The machine is similar to centreless grinding machine except that extra-long grinding wheel and regulating wheels are used to allow the workpiece to remain in abrading contact for a longer time (Fig. 29.38). The lapping wheel speed falls in the range of 175–650 m/min, whereas regulating wheel has a speed of 70–175 m/min.

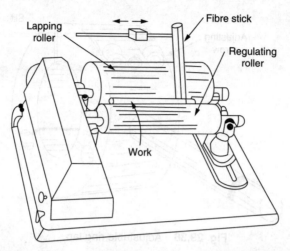

Fig. 29.38 Roller type machine for centreless cylindrical lapping.

(e) *Spherical lapping:* Spherical surfaces are lapped on a machine similar to a drill press. A cast iron lap is used which is the counterpart of the work surface to be lapped. A crank is held in the spindle and crankpin is provided with a ball that enters freely into a blind hole in the back of the lap as depicted in Fig. 29.39. The workpiece axis is aligned with spindle axis and the spindle is then rotated which gyrates the lap.

There are two processes which are connected with lapping. These processes are polishing and buffing. Both of them will now be discussed in detail.

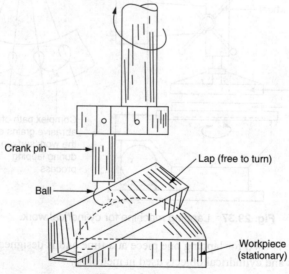

Fig. 29.39 Spherical lapping.

29.10.3 Superfinishing Operation

Super finishing is a micro-finishing operation in which a bonded abrasive sector or flat disc presses upon the surface to be micro-finished. The work surface and the bonded abrasive sector are in light contact under some pressure applied on the top of the abrasive sector. There is a film of viscous fluid in between the sector and the workpiece. In the beginning, the high spots on the surface of the rotating workpiece come in contact with the abrasive sector by rupturing the viscous oil film. Thus, the high spots on the workpiece are removed and the surface finish of the shaft type workpiece is improved. A gap is created between the shaft type workpiece and the abrasive sector separated again by the viscous oil film. On further increasing the pressure on the top of the abrasive sector, metallic contact between the shaft type workpiece and the abrasive sector is again established causing further metal removal. Such a process repeats itself a number of times until a very high degree of surface finish and close tolerance of the order of IT1 is achieved. Super finishing differs from lapping in the basic principle of working that in this case, the principle of hydrodynamic pressure build up by the rotating shaft creates high pressure by shearing of the oil film separating the workpiece and the grinding sector.

29.11 SNAGGING AND OFF-HAND GRINDING

Those grinding operations are termed 'snagging' that are ground in the field of grinding where the removal of metal is the prime function of the wheel. Finish requirements on the 'ground product' of little or no consideration. Snagging operations are performed in the cleaning rooms of (steel, malleable iron, cast iron, non-ferrous alloys, aluminium, brass) where gates, risers and headers are removed and parting line fins are reduced or ground off. They are also performed steel mills involving ingot, billet, bar, stab, and grinding. Those snagging operations are closely associated and found here the chipping hammer is used.

The reduction or removal of flash on forgings and the removal of excess material from welds also falls within this field of grinding.

29.11.1 Types of Snagging

The grinding machines used for snagging are varied and are differentiated from precision grinding machines (cylindrical, centreless, surface and internal grinders) by the fact that fixtures for holding the workpieces are not generally used or, wherever used, are not designed to give precise limits of accuracy to the work being ground.

On snagging operations, the machine is either brought to the work or the machine. In either case, the wheel and the workpiece are not fed into contact with each other by machine movements but rather by the movement of the entire grinding machine to the work or by the independent movement of the workpiece to and across the face of the wheel by the free or off-hand method grinding.

The grinding machines used on these operations are:

(i) Swing frame grinders
(ii) Floor stand grinders
(iii) Portable grinders.

Swing frame grinders are usually huge from an overhead suspension of mono-rail trolley and consist of a horizontal frame from 1.8 to 3.0 metres along with a motor mounted at the rear which actives, through belts, the grinding wheel spindle at the front end of the frame. The front end is equipped with handles by which the operator controls the movements of the machines. Since the suspension point is at the centre of gravity of the machine and is of a non-rigid type (easily turned and swivelled), the operator can move the machine and wheel in any direction.

The work has no movement during the grinding operation. It moves only to present new areas to be ground and usually is so heavy as to require service to handle.

Floor stand grinders consist of a fixed upright base or pedestal supporting wheels mounted on a horizontal spindle because most machines are designed to mount type wheels on both ends of the spindle. They are referred to as double-ended grinders. The wheel spindle is driven either by the direct drive method or by belts from a motor in the base or from an overhead drive shaft.

No machine movements are involved in the machine except the rotation of the wheel.

As the machine is firmly fixed to the floor, the casting or workpieces to be ground have to be brought to the machine by hand or by chain hoists. Since no movements are incorporated in the machine, the various surfaces to be ground on the work have to be presented to the grinding wheel by hand.

The work, therefore, is usually much lighter than that ground on swing frame grinders and is limited (except where chain hoists are used) to a weight and size which can be conveniently handled by the operator.

Portable grinders are self-contained machines which can be moved from place as is implied in the name. They are of a light and compact construction, carrying self-contained power units either electric motors or air turbines).

Their uses are varied, but their chief value lies in their ability to:

(i) Present the grinding wheel to surfaces that are inaccessible to other types of machines.
(ii) Remove small amount of stock from large workpieces at widely separated points. Here, the portability saves time and labour.
(iii) Reduce welds and remove small surface imperfections and such finishing operations as the blending of curved surfaces, etc.

Cumiflex Reinforced depressed centre wheels

Recent years have seen the development of the reinforced depressed centre wheel which represents a great step forward in the abrasive field.

Cumiflex: Reinforced depressed centre wheels have a wide field of application, combining a remarkable versatility, economy and fast rate of cut with a greater measure of safety than has previously been possible.

Their great versatility is shown by the fact that the same wheel has been successfully used for roughing, smoothing, grooving and cutting such widely differing materials as:

(i) Cast iron (ii) Marble
(iii) Steel (iv) Bronze
(v) Phosphor bronze (vi) Copper
(vii) Manganese bronze (viii) Magnesium bronze
(ix) Aluminium (x) Gunmetal

Wheels of this type are extremely economical, and can be used right down to the hub. Owing to their construction, they have an extraordinary capacity to withstand mechanical shock to the reinforcing layers giving greater tensile strength than was previously possible.

Operations with this type of wheel include cutting-off risers on ferrous and non-ferrous casting, smoothing jagged edges on flame cut plates, cutting sheet metal, removing rivet and bolt heads, grooving seams prior to welding, smoothing welds, cleaning up heavy castings and descaling and deburring.

Speeds

These three types of grinding machines for snagging operations (swing frame, floor stand and portable) are operated at both 'low' and 'high' speeds.

Those machines on which vitrified wheels are to be used must not be operated at speeds exceeding 6500 surface feet per minute (33 mps).

Those machines on which hard or strong organic bond wheels are to be used must not be operated at speeds exceeding 9500 surface feet per minute (45 mps).

Higher than normal speeds

Extensive testing has proved that the proper utilization of higher surface speeds has produced higher rates of metal removal, longer wheel life and definite increment in productivity. Therefore, there is a trend toward machines that are specifically engineered to withstand the stresses generated at 12,500 and 16,000 sfpm (60 mps and 80 mps). They are rigid, well guarded and equipped with proper spindle and bearings. Wheels used on these machines are, in turn, especially engineered and tested.

It is important for grinding wheel users to recognize that the benefits derived from these higher speeds are the result of an integrated system of wheel and machine, both designed for the purpose.

In no case should a machine be modified by the user to operate at a higher speed than that for which the machine was originally designed. It is even more important that a grinding wheel should never be operated at a higher speed than the maximum operating speed marked on the wheel.

Stresses involved in snagging operations

Since on all snagging operations, the machine operator controls the pressure built up between the wheel face and the work and is determining in the nature of the movements between wheel and work, it is most important that he adapts himself and his grinding techniques to the wheel.

It has been pointed out earlier that as the stresses on the face of a grinding wheel of a specific grain size, grade, bond and structure are increased, the wheel will act softer. Likewise, as the stresses on the wheel face are decreased, the wheel will act harder.

To keep these stresses constant in precision grinding operations (cylindrical, centreless, surface and internal), machine movements and speeds are relatively easier to control (since they are mechanical) than in snagging operations. And on these type of grinders, operations can be standardized by definite procedures being set up and followed which are not possible in snagging operations.

In snagging operations and in all off-hand grinding, the standardization of operation or equalization of stresses as developed on the wheel face can be attained only by operator education.

To obtain maximum efficiency, the machine operator must develop a 'grinding sense'. If the operator observes the flow of sparks (commonly called 'spark flow') from his wheel in relation to the pressure applied and related this to the wheel wear, he will soon sense the most efficient pressures to use to obtain maximum stock removal with reasonable wheel life.

Experience will teach the operator that a grinding wheel will remove stock more economically if there is a continuous relative movement between the wheel and workpiece at low pressures, as compared with little or no relative movement at high pressures. This is because large areas of contact are built up between the wheel and the work when the two are in contact without movement. The faster and more continuous the movement, the less the area of contact that is built up. Long arcs of contact and wide areas of contact cause interference with the disposal of grinding chips and result in a retarded cut.

29.11.2 Off-hand Grinding

In addition to the field of snagging operations previously described, there is a wide field of operations which is termed 'off-hand' grinding which cannot be clearly defined as snagging.

The types of equipment used for off-hand grinding are floor stand grinders and bench grinders, the latter being a modified floor stand grinder but with a shorter base enabling it to be easily mounted on a work bench or table. The bench grinders mount smaller wheels and are used as general purpose machines rather than as production snagging grinders.

Off-hand grinding machines

These machines are used for various grinding operations ranging from rough shaping of miscellaneous pieces to the shaping of small tools.

The off-hand grinding or sharpening of tools chipping of tools, gauge, chiesels, etc. is similar to snagging operations. However, in these cases, closer tolerances of size and shape are desired and obtained.

The wheels used on these machines for off-hand grinding are of finer grain sizes and sifter grades than those generally used in the snagging field as they are used to generate accurate cutting points and edges. Likewise, the finish on the cutting edges of the tools must be considered.

29.12 COATED ABRASIVES FOR OTHER INDUSTRIAL APPLICATIONS

In recent years, there has been an increasing application of the coated abrasives especially when grinding very hard materials using diamond grit or cubic boron nitride abrasive grit. The abrasive grit is coated on wheels made of metals or some other material like fibre reinforced plastics. These wheels have been developed by leading abrasive manufacturing companies like Norton or John Oakes and Mohan near Ghaziabad. Leading companies advertise the use of grinding abrasive discs. So far, industries prefer the use of bonded abrasive wheels because of their low cost due to their use of clay based vitrified bonds. However, in case of cloth or paper backed abrasives, the coated abrasives are preferred for economic reasons. In industry, coated abrasives play a vital role. Their versatility and the range of applications is much greater than most people realize, coated abrasives are employed in the manufacture of almost every product used in the factory, office and home and in air, marine, railway and highway vehicles. They are used in obtaining fine finishes for grinding and shaping products into various forms.

Coated abrasive production methods are indirect economic competition with various other tools and cutting mediums, including grinding wheels, polishing wheels or set-up wheels, lathe tools, milling machines and chipping hammers. In order to complete successfully against these long established tools, it has become important that proper abrasives and machines are used and the economics of each operation studied fully.

When selecting coated abrasives, the following points must be considered:

  (i) Type of machine used contact belt, platen, overhead, disc sander, etc. Speed in metres per minute must be specified.
  (ii) Type of job-grinding, finishing, etc.
  (iii) Nature of workpiece—ferrous, non-ferrous, wood, etc.
  (iv) Amount of stock removal—heavy, medium or light.
  (v) Type of finish required—coarse, medium, fine or superfine.
  (vi) Accuracy of finish.
  (vii) Whether coolant used or not—if coolant is used, then what type of it is?
  (viii) Details on jigs and fixtures.
  (ix) Any other information concerning the process of grinding.

Industry recognizes coated abrasives as efficient production tools for metal working. Various forms of coated abrasives are used extensively from first roughing to final finishing for both of hand and precision grinding.

The best results are obtained only by combining correct abrasives with proper methods. In the metal working industry, two main types of coated abrasives are used—Silicon carbide and Aloxite (Aluminium oxide).

Both silicon carbide and aloxite are products of the electric furnace. Silicon carbide is used for making abrasive cloth and paper for work on metal of low tensile strength and for finishing stainless steel. It is used on water-proof paper for wet sanding of primers, undercoats and colour coats in automotive and sheet metal industries.

Aluminium oxide is used for manufacture of abrasive cloth and paper-glue and resin bond-for grinding and finishing metals of high tensile strength. For heavy stock removal and severe grinding operations, aluminium oxide abrasives are ideal.

Coated abrasives for the wood working industry

For the wood working industry, coated abrasives help to get the best surface finishes. Flint and Garnet coated abrasives are available in a wide range of grades, sizes and shapes. They are used for hand application and also as belts, discs and rolls for drum sanders at speeds less than 1500 mpm. For high speed sanding aloxite abrasives work more efficiently.

The coarser grades help in fast and easy removal of material, whereas the finer grades help to produce sooth finishes.

Coated abrasives for the leather industry

On leather and leather products, silicon carbide and aloxite coated abrasives help to obtain a superfine finish. The surfaces of these abrasives are so sharp, that they cut the soft yielding surface of leather without tearing or pulling. Aloxite coated abrasives are particularly suited for high speed machines.

Resin sander discs

Resin sander discs are recommended for use with high speed portable tools for dressing heavy welds and for applications generating intensive heat and also where toughness of bond is required.

Whatever the finishing or sanding job may be, the best abrasive is the one that cuts fast, lasts long and produces the desired finish. With the wide range of coated abrasives manufactured to consistent quality, it is easy to select the correct abrasive for any type of job.

Abrasive company gives you complete abrasive service in following ways:

(i) Manufactures coated abrasive products of every type for every industrial use.
(ii) Employs most modern job-engineered to suit your needs.
(ii) Offers complete stocks and immediate service through an organized country-wide network of distributors.
(iv) Constantly pioneers new developments to keep your costs down.

Coated products-backing

Paper: Furnished in four weights. Weight of 'A' is the lightest and most pliable and is used mainly for finishing operations.

The weights of 'C' and 'D' are of moderate strength and pliability, and they are mainly used for hand standing and for light mechanical applications, such as vibrating sanders.

Weight of 'E' is heavy, tough and durable paper supplied mainly in roll, disc, or belt form for mechanical sanding.

Cloth: Drill cloth is used for applications where a backing stronger and more flexible than paper is required.

Comlcination (PAPER and CLOTH): Paper of weight 'E' is combined with light weight cloth. This product is stronger than paper and less expensive than drill cloth and is used where the stress and strain is greater than the paper is capable of standing. It will resist stretch better than cloth, but has little flexibility.

Cloth and fibre: Drill cloth combined with vulcanized fibre, mostly used for manufacture of abrasive discs for use on high speed sanders.

Fibre: Heavy and tough vulcanized fibre used for manufacturing of discs for use on high speed sanders.

Bond

Glue: Specially manufactured strong and flexible glue is used as bond for abrasives for dry mechanical and hand applications.

Modified glue: An inert filler is added to glue to provide a more heat resistant binding agent. Suitable for severe dry mechanized operations particularly with high speed machines.

Varnish: It is used for manufacture of waterproof abrasive papers for wet sanding of undercoats, paints, lacquers, etc.

Resin: Synthetic resin used for manufacture of discs and belts for heavy duty, high speed work.

Types of coating

Closed coating: As the name itself implies, the entire surface of the backing is covered with grain. This is the normal type of coating used for operations where a high rate of stock removal and a heavy working pressure is required.

Open coating: 50% to 70% of the backing is covered with grain. With greater spacing between grains, open grain coating allows the material sanded to free itself thereby reducing the loading or clogging of the surface.

Methods of coating

Electrocoating (E.C.): It is process where the abrasive grains are coated in an upright position on the backing surface in an electrostatic field created by high voltage equipment. This result in an absolutely uniform coating with sharp edges of the grains sticking out and broad bases firmly anchored in the adhesive layer. Hence electrocoated abrasives cut faster and cooler, last longer, clog less and remove much greater stock.

Gravity coating: In this process the spreading of gain is done by gravity by means of a controlled hopper feeding the grits on the adhesive coated backing surface. This method is used for abrasives for hand application, such as flint and emery sheets used in clearing of surfaces.

Flexing

Nonelex: The backing material furnished in these cases is fairly stiff and most suited to hand applications in sheet form.

90° flex-material has unidirectional strength and is flexible in lengthwise direction. Most suited for belt manufacture as it runs true over pulleys and has been pre-stretched.

45° flex-material flexible and has isotropic property in all directions it is particularly suitable where profile cleaning and polishing of work is required.

29.13 CENTRELESS BELT GRINDING (Refer to Fig. 29.40)

Centreless belt grinding, as the name itself implies, is a process where workpieces are ground without being held between centres. The principle of centreless grinding with abrasive belts is similar to that using grinding wheel with the exception that in place of grinding wheels, contact wheels are used and the abrasive belts mounted on them with idler pulleys holding them under tension.

The common method of centreless grinding is to pass the cylindrical workpiece such as tube or rod between the contact wheel and a control wheel allowing it to rest on a rest blade. The angle of inclination of the control wheel determines the rate of through feed or traverse of the workpiece. The control wheel is usually motorized to rotate at a speed lower than the contact wheel. This enables an easy and accurate control of rate of traverse, depth of cut and finish of the workpiece change figure.

A slight variation of the above method is to use a short conveyor belt in place of the control wheel as shown in Fig. 29.40. As with a control wheel, the angle of inclination of the belt determines the rate of traverse. This arrangement is particularly useful where long lengths of cylindrical workpiece are to be ground as the conveyor belt has better grip and exerts higher and more uniform pressure on the workpiece, thus speeding up the grinding process.

The scope of centreless grinding with abrasive belts is wide. Though very high precision finishing can only be obtained by using grinding wheels, abrasive belts can be employed to obtain a tolerance of about 0.25 mm

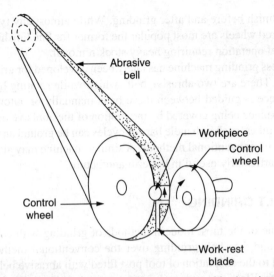

Fig. 29.40 Principle of centreless grinding with abrasive belt.

but the main advantage of using an abrasive belt instead of a grinding wheel lies in that the belt is faster and gives cooler cutting and far less time is consumed in changing belts of different grits. Moreover, with grinding wheels, the diameter gets gradually reduced due to wear and hence, to maintain the correct cutting speed, expensive attachments are required to progressively wheel. With the centreless machines employing two belt grinding units, abrasive belts of coarse and fine grits may be used so that in a single pass, required finish can be obtained. In special cases where fine finish is required, the abrasive belt can be replaced by a polishing belt of cotton or lamb's wool and finishing carried out with a polishing compound or the contact wheel with the abrasive belt or polishing belt may be replaced with a buffing mop and polishing done with suitable compounds.

These centreless units can also be arranged in tandem and have them mounted with grinding wheels belts of different grit sizes and polishing belts or mops so that a rough workpiece comes out of the production line completely ground and polished. Many of the centreless grinding operations now being carried out with set-up wheels (cloth or felt bobs coated with glue and grain) can be advantageously replaced with contact wheels mounted with abrasive belts. Set up wheels, being difficult to prepare, are prone to grain shedding and heating up of the workpiece. With the uniform coating of the modern abrasive belts, cutting action is faster and cooler and the life of the belts is much longer.

While centreless belt grinding is particularly useful for grinding cylindrical pieces such as tubes and rods, stepped down, tapered and shouldered workpieces can also be ground and polished. In such cases, the contact wheel and control wheel have to be profile to suit the workpiece and special fixtures may have to be made to bring the workpiece in contact with the abrasive.

Centreless grinding is generally carried out dry or with grease or grinding pastes. But where required, the operation can be performed wet with suitable cutting fluids for which waterproof abrasives will, of course, be necessary. However, if oil or kerosene are used, normal glue bond belts are quite satisfactory.

In many cases, a standard contact belt grinding unit can be adopted for centreless grinding by mounting a separate unit containing the motorized control wheel. This unit can be attached when carrying out general off hand grinding on the contact wheel.

The speed of the belts used in centreless grinding is nearly the same as for contact belt grinding and varies from 650–2700 mpm, depending upon the type of material being ground. The speed of traverse of the workpiece through the unit varies from 5 to 6.0 metres per minute and depends on the diameter of the workpiece and the surface finish before and after grinding. While almost any type of contact wheel may be used, cloth and rubber covered wheels are most popular the former for fine finishing and polishing operations and the later for rough initial operation requiring heavy stock removal.

A new type of centreless grinding machine has lately been developed for grinding and polishing of tubes of straight or curved shape. There are two abrasive belts which while rotating by themselves revolve round the workpiece. The workpiece is guided between these belts manually or automatically but does not itself rotate, the entire of the workpiece being covered by the motion of the abrasive belts themselves. In this way, quite complicated shapes of tubes, such as handle bars of cycles can be ground and polished very efficiently in fraction of the time required by conventional methods. A similar machine may also be developed for grinding the ends of tubes which is particularly useful in boiler manufacture.

29.14 TOOL POST BELT GRINDING

Tool post belt grinding is one of the most modern methods of grinding with coated abrasive belts. Certain advantages of contact belt and free belt grinding over the conventional methods, such as grinding with wheels or abrasive stripe led to the adaptation of tool post fitted with abrasive belts on turning, lathes, boring mills, planning machines, etc. Elasticity and cool cutting action of abrasive belts with little or on clogging of the abrasive surface has been responsible for the success of this method of grinding. With the knowledge gained during the years in manufacturing precision grinding attachments mounted with grinding wheels, the machinery manufacturers have now developed suitable attachments to take up abrasive belts for obtaining very high finish on cylindrical pieces such as axles, spindles, shafts, rolls and calendars and on flat parts such as machine beds and rails.

Surfaces with very fine finish are of great importance for rolls and calendars required in chemical, textile, paper, tinplate, photographic industries, etc. as well as for spindles and shafts in textile, machine tool and allied industries. The importance of high grade finish for machine beds or rails is well-known and in many cases, accuracy of movements of machines depends to a large extent on the finish. The same principle is now used for giving super-finish to new or overhauled crank shafts. The breakdown of many crankshafts has been detected to bad lubrication which in turn has been the result of insufficient surface finish on the crank shafts. By grinding and polishing with belts life of the crankshafts has been considerably improved.

The principle of the tool post grinder is to mount a motorized belt grinding attachment as on the tool post of lathes, planning machines, boring mills, milling machines, etc. The attachment is very similar to the belt grinding arm of a polishing lathe for contact wheel grinding and consists of a contact wheel mounted on the shaft of the motor with an idler pulley at the end of the arm to take the abrasive belts. These attachments can also be used on simple devices that take cylindrical workpieces between centres or give reciprocatory motion to flat surfaces. With these mechanisms, huge containers may also be ground or polished both on the exterior and interior surfaces.

For lathe attachments, while the workpiece is rotated on the lathe, the contact wheel with the belt is brought in contact with the work by moving the tool post. The lead screw of the lathe moves the tool post laterally along the entire surface of the workpiece. In effect, this is similar to the cylindrical grinding operation of the tool post grinding mounted with grinding wheel. In the case of flat grinding, the belt attachment is mounted on the tool post of the planning machine and the workpiece is given a reciprocator motion in the normal way.

Tool post belt grinder has proved to be a very economical tool especially for calendar and roll grinding in western countries. The rolls and calendars are required to be overhauled at frequent intervals and grinding them with grinding wheels is usually left to the specialists since considerable skill is required to get a good finish. Without sturdy machines and adequate knowledge of the grinding process, defects, such as chatter marks occur which mark the work. But with the abrasive belt, due to cushioning effect of the contact wheel and elasticity in the free belt, even semiskilled operations are able to finish the rolls satisfactorily. Hence roll and calendar grinding can now be undertaken in their own shops and charges for transporting the rolls to the shops of the specialists and the risk of damage to the rolls during transit are eliminated. In addition, it is not necessary to retain a great quantity of rolls and calendars in reserve with consequent economy in capital outlay.

However, no claim is made that tool post grinding with belt is superior in all cases to grinding with grinding wheels. Where precision is of greater importance, wheels are required to be used but when superior surface finish is wanted, belt grinding offers greater advantages. In many cases, the work may initially be ground with wheels and finished with abrasive belts.

These tool post units can also be used as normal grinding posts by clamping them on bench vices and grinding the workpiece on the contact wheel or on the free belt.

Usually, in tool post grinding, if the surface of the workpiece is fairly rough, an abrasive belt of medium grit is initially used and contact wheel brought in contact with the work. For the next operation a fine grit abrasive is used and finished by grinding with the free belt which due to lower rate of stock removal gives a finer surface than a contact wheel. Grease or grinding pastes can be applied to the surface of the abrasive to improve the finish. If a superfine finish is required, special polishing belts of felt, cotton or lambs wool may be used with or without the addition of polishing aids such as luster or compositions that are usually available for the polishing trade. Results obtained by this method have been extra-ordinary.

Speeds of the abrasive belts used depend on the nature of the material, its thickness or diameter, rate of stock removal and the finish required. In most cases, the speeds normally employed for contact belt grinding as given in our earlier pamphlet on this subject are used. The rate of tracers of the abrasive along the surface of the workpiece varies from 3 to 10 metres per minute. By regulating the revolutions of the workpiece mounted on the lathe, grinding effect can be varied while maintaining the same speed for the abrasive belt.

For initial grinding, abrasive belts of grits 80 to 120 may be used and for final finish grits 240 and 320 are recommended.

29.15 COATED ABRASIVES

Apart from bonded abrasives, coated abrasives find wide industrial application. Table 29.4 indicates a concise summary of coated abrasives and their area of application.

Table 29.4 Varities of coated abrasives products

Material	Grit range	Backing	Coating	Belts	Discs	Rolls	Sheets	Application
Flint paper	G. S/0-3½	C	Non EC				X	Hand sanding wood.
Flint paper	G.4/0-2½	E	EC	X	X	X		Mechanical application-wood and leather.
Flint paper	G.4/4-1	X	Non EC				X	Hand sanding wood.
Flint cloth	G.4/0-3	X	EC	X	X	X		Mechanical application-wood and leather.
Emery paper	G.5/0-3	C	Non EC				X	Hand sanding all types of metals.
Emery paper	G.4/0-1	E	EC	X	X	X		Belts and disc sanding of metals.
Emery cloth	G.5/0-3	X	Non EC				X	Hand sanding all types of metals.
Emery cloth	G.5/0-2½	X	EC	G	X	X		Belt and disc sanding of metals and few general workshop uses.

(Contd.)

Table 29.4 Varities of coated abrasives products (*Contd.*)

Material	Grit range	Backing	Coating	Belts	Discs	Rolls	Sheets	Application
Garnet paper	G.4/0-2/0	C	EC OP				X	Hand and vibrating sanding of wood.
Garnet paper	G.4/0-3	E	EC	X	X	X		Belt, drum and disc sanding of wood at speeds less than 1500 mpm.
Garnet cloth	G.4/0-3	C	EC	X	X	X	X	Belt, drum and disc sanding of wood at speeds less than 1500 mpm where greater backing strength is required.
Special garnet paper	150-30	E	EC	X	X	X		Belt, drum and disc sanding of wood at speeds less than 1500 mpm where greater backing strength is required.
Special garnet paper	120, 80, 60, 50 and 36	E	EC OP	X	X	X		Belt, drum and disc sanding of wood at speeds less than 1500 mpm where greater backing strength is required but where severe clogging is experienced.
Special garnet cloth	120-3	X	EC	X	X	X	X	Belt, drum and disc sanding of wood at speeds less than 1500 mpm where greater backing strength is required.
Aloxite paper	130-36	D	EC OP				X	Used on vibrating sander for wood.
Aloxite paper	S00-36	E	EC	X	X	X	X	Belt, drum and disc sanding of wood at speeds higher than 1500 mpm. Also for high tensile metal and high speed buffing of leather.
Aloxite cloth	400-24	X	EC	X	X	X	X	Mechanical application on high tensile metals & severe operation on all metals.
Aloxite cloth	120, 80, 50, 36, 30 and 24	X	ECM	X	X	X		Mechanical application on high tensile metals & severe operation on all metals for severe operations.
Silicon carbide paper	500-24	X	EC	X	X	X	X	Buffing leather and finishing low tensile metals.
Silicon carbide cloth	400-24	E	EC	X	X	X	X	Mechanical application on low tensile metals and in leather and shoe industry.
Aloxite sander discs	400-13	FC	EC		X			Portable disc sanding of all metals and hard wood.
Aloxite sander discs	150-16	FC	EC M		X			Portable disc sanding of all metals and hard wood for severe application.
Aloxite resin sander discs	80-24	F	EC M		X			For severe application at high speeds (2300 mpm and above).
Waterproof paper	600-400	A	EC				X	Welt and dry rubbing down of under-coats, colour coats, lacquers and varnishes,
Waterproof paper	S00-150	B	EC				X	
Waterproof paper	120-80	D	EC				X	Both water and kerosene can be used as coolants.

Note: A few types of abrasives are available in special shapes tike naumkegs, heel, breasters, websters, sleeves, shredded strips, etc.

Symbols

X	Denotes products available.	FC	Fiber and cloth combination
A	Light weight paper for hand application.	X	Drill cloth
C	Weight of moderate strength and pliability	EC	Electrocoat
D	Mainly used for sanding light mechanical applications,	OP	Open coat
E	Heavy, tough and durable paper for mechanical sanding.	ECM	Electrocoat modified glue
F	Fiber		

29.16 TROUBLE SHOOTING

Common problems, reasons and correction in grinding process are illustrated here. Table 29.5 shows common troubles, causes and their remedies in grinding process.

Table 29.5 Common troubles and their remedies

Common troubles	Causes	Remedies
Glazing of wheel.	Grade too hard, grain too small, wheel speed too high.	Wheel needs dressing.
Rounding of wheel.	Grain too small, grade too hard, wheel speed too high.	Wheel needs dressing.
Cutting action of seizer.	Grade too hard and wheel glazed.	Wheel needs dressing.
Rapid wheel wear.	Grade too soft, grain too large, wheel speed too low.	Give less feed and less traverse.
Wheel or work chatters or vibrates.	Wheel loaded, grade too hard, wheel not balanced.	Use centre rest or back rest.
Work surface gets overheated.	Wheel glazed or loaded.	Use coolant.
Work surface is ground inaccurate	Lack of coolant, work out of balance or failure to use centre rest or back rest.	Remove all defects.

29.17 GRINDING ERRORS

Following tables (Tables 29.6 to 29.10) show some of the common grinding errors and the methods of correction.

Table 29.6 Ground surface errors due to chatter

Indication	Cause	Methods of correction
Chatter marks may take any of several forms and may be the result of any of the causes listed.	Wheel out of balance.	Rebalance carefully on own mounting.
		Rebalance after truing operation.
		Run wheel without coolant to throw off excess water.
		When wheel is removed from machine, store on side to prevent water from setting at lower edge of wheel.
	Wheel out of round.	True before and after balancing.
		True sides to face.
	Wheel grading too hard.	Select softer grade, more open bond or coarser grit.
	Work centres or work rests not true, or improperly lubricated.	Check fit of centres and rests. Provide constant and even lubrication.
	Dressing.	Use sharp diamond—rigidly held close to wheel.

Table 29.7 Ground surface errors on work as scratches

Indication	Cause	Methods of Correction
Narrow and deep regular marks.	Wheel to coarse.	Use finer grain size.
Wide irregular marks of varying depth.	Wheel too soft.	Use harder grading.
Widely spaced spots on work.	Oil spots or glazed areas on wheel face.	Balance and true wheel. Avoid getting oil on wheel face.
Fine spiral or thread on work.	Faulty wheel dresser.	Replace cracked or broken diamonds.
		Use slower dressing traverse. Set dressing tool at angles of 5° down and 30° side.
		Turn diamond every third dressing.
		Tighten holder or diamond. Dress with less penetration.
		Do not allow tool to dwell in contact with wheel. Do not start dressing cuts on face—locate tool on face, but start cuts from edge. Make final pass in dressing in opposite direction to grinding traverse.
		Traverse diamond across face of wheel evenly. Round off wheel edges—chamfering or dressing back is not sufficient.
	Faulty operation.	Prevent penetration of advancing or following edge of wheel by being careful to dress wheel face parallel to work. Reduce wheel pressure. Provide additional steady—rests.
		Reduce traverse in relation to work rotation.
		When making numerous passes, make slight change in traverse rate at each pass to break up pattern.
Wave traverse lines.	Ragged wheel edges.	Round of wheel edges.
Isolated deep marks.	Improper wheel dressing.	Use sharper dressing tools. Brush wheel after dressing, preferably with a stiff bristle brush.
	Coarse grains foreign matter in wheel face.	Dress out.
	Bond disintegrates, grain pulls out.	Coolant too strong for some organic bonds; decrease soda content.
Irregular marks.	Loose dirt.	Keep machine clean.
Irregular marks of various length and width, scratches usually 'fishtail'.	Dirty coolant.	Clean tank frequently. Flush guards, etc. after dressing and when changing to finer wheels.
Deep irregular marks.	Loose wheel flanges.	Tighten flanges, using blotters.
Grain marks.	Wheel to coarse or too soft.	Select finer grain size of harder wheel.
	To much difference in grain size between roughing and finishing wheels.	Use finer roughing wheel or finish out better with roughing wheel.
	Dressing too coarse.	Less dresser penetration and slower dresser traverse.
	Improper cut from finishing wheel.	Start with high work and traverse speeds, to cut away previous wheel marks; finish out with high work and slow traverse speeds, all wing wheel to spark out entirely.

Table 29.8 Ground surface errors on work as spirals

Indication	Cause	Methods of correction
Spirals (traverse lines) same lead on work as rate of traverse.	Misalignment	Check alignment of head and tail stocks, also wheel head to work.
	Truing	Have truing tool set on work wheel contact line, but pointed down 3°. Round off edges of wheel face.
Lack of cut, glazing, some loading, burning of work, chatter.	Wheel too hard in effect.	Increase work and traverse speeds wheel pressure (infeed). Decrease spindle speed, wheel diameter and width of wheel face. Open up wheel by sharper dressing. Use thinner coolant. Avoid dwelling at end of traverse. Avoid gummy coolants. Use coarser grain size and softer grade.
Wheel marks, short wheel life, not holding cut, tapered work.	Wheel too soft in effect.	Decrease work and traverse speeds and wheel pressure (infeed). Increase spindle speed, wheel diameter and width of wheel face. Dress with slow traverse and slight penetration. Use heavier coolants. Do not pass off work at end of traverse.

Table 29.9 Ground surface errors on work due wheel loading

Indication	Cause	Methods of correction
Metal lodged on grains or in wheel pores.	Incorrect wheel	Use coarser grain size, or more open bond, to provide chip clearance. Use more coolant.
	Faulty dressing	Use sharper dresser. Dress faster. Clean wheel after dressing.
	Faulty coolant	Use more, cleaner and thinner coolant.
	Faulty operation	Manipulate operation to soften effect of wheel. Use more infeed.

Table 29.10 Ground surface errors on work due to wheel glazing

Indication	Cause	Methods of correction
Shiny appearance smooth feel.	Improper wheel	Use coarser grain size, softer grade. Manipulate operation to soften effect.
	Improper dressing	Keep wheel sharp with sharp dresser. Use faster dressing tool traverse. Use more dressing tool penetration.
	Faulty coolant	Useless oily coolant. Use more coolant.
	Gummy coolant	Increase soda content if water is hard. Do not use soluble oils in hard water.
	Faulty operations.	Use greater infeed.

REVIEW QUESTIONS

Tick the most correct response:
1. Which of the following processes remove maximum amount of material out of a workpiece.
   (a) Honing (b) Lapping (c) Grinding (d) Superfinishing
2. For achieving extremely smooth surface finish at bearing surfaces, the process of machining generally employed is:
   (a) Milling (b) Shaping (c) Drilling (d) Grinding
3. Grinding operation is used for
   (a) Forming (b) Shaping (c) Dressing (d) Finishing
4. For grinding shafts, spindles and bolts use:
   (a) Tool and cutter grinding (b) Cylindrical grinding
   (c) Thread grinding (d) Surface grinding
5. For grinding flat surfaces use:
   (a) Internal grinding (b) Thread grinding
   (c) Cylindrical grinding (d) Surface grinding
6. The highest cutting speed is used in:
   (a) Surface grinding (b) Centreless grinding
   (c) Internal grinding (d) Cylindrical grinding
7. Grinding ratio may be defined as:
   (a) Wear of grinding wheel/Volume of metal removed
   (b) Volume of metal removed/Wear of grinding wheel
   (c) Wear of grinding × Volume of metal removed
   (d) None of the above.
8. Surface speed (in m/min) of the grinding wheel in centreless grinding is:
   (a) 15–16 (b) 100–500 (c) 1000–1500 (d) 1500–1800
9. Workpiece is supported as follows in centreless grinding:
   (a) On magnetic chuck (b) In universal chuck
   (c) Incollet chuck (d) None of the above
10. The workpiece advances in centreless grinding due to:
    (a) Machine drive (b) Effort applied by operator
    (c) Force exerted by regulating wheel (d) Force exerted by grinding wheel
11. Artificial abrasive are:
    (a) Sandstone, emery, diamond, quartz (b) Silicon carbide, Aluminium oxide, Boron carbide
    (c) Garnet (d) Corundum
12. For grinding, low tensile strength materials like bronze, brass, copper, aluminium, etc. the abrasive used is:
    (a) Silicon carbide (b) Garnet
    (c) Aluminium oxide (d) Corundum
13. For softer material, the grain of abrasive used is:
    (a) Coarse grain (b) Fine grain
    (c) Medium grain (d) Both (b) and (c)
14. If condition of machine is such that it produces vibrations, then use:
    (a) Hard grade abrasive (b) Soft grade abrasive
    (c) Both (a) and (b) (d) Something else

15. For harder material, the grain of abrasive used is:
    (a) Coarse grain   (b) Fine grain
    (c) Medium grain   (d) Both (a) and (c)
16. Majority of the grinding wheels use the following types of bond:
    (a) Resinoid   (b) Silicate   (c) Rubber   (d) Vitrified
17. In finish grinding, the grinding ratio is kept as:
    (a) 5–10   (b) 10–25   (c) 25–50   (d) 50–100
18. Grade of a wheel, i.e. strength of a bond is graded from A to Z, where
    (a) A is very soft   (b) Z is very hard
    (c) Z is very soft   (d) Both (a) and (b) are true
19. If a grinding wheel is designed as 30 A 36 H 6 VB, then the letter H indicates:
    (a) Grade of bond   (b) Type of bond
    (c) Structure   (d) Abrasive
20. In the above question, the letter 'V' indicates:
    (a) Type of abrasive   (b) Type of bond
    (c) Type of structure   (d) Grade of bond
21. A grinding wheel is said to be glazed:
    (a) When it becomes unbalanced.
    (b) When the abrasive grains have been replaced.
    (c) When the abrasive grains become dull and stop cutting.
    (d) When some lubricant is added to the abrasive grains.
22. Truing of grinding wheel is done by:
    (a) Glazing the wheel   (b) Dressing the wheel
    (c) Loading the wheel   (d) Balancing the wheel
23. Grinding wheels are balanced _____ by shifting weights on one flange of the wheel mount.
    (a) Statically   (b) Dynamically
    (c) Both statically and dynamically
24. Which operation should be performed in the last in connection with grinding wheel?
    (a) Balancing   (b) Dressing
    (c) Truing   (d) Glazing
25. Which of the following grinding processes has highest possible speeds?
    (a) Internal grinding   (b) Surface grinding
    (c) Cylindrical grinding (d) Rubber shellac cutting off wheels.
26. What are natural and artificial abrasives? Why are the latter preferred over the former?
27. What is meant by grain, grit, structure and grade of a grinding wheel?
28. Why truing and dressing are necessary in grinding wheel?
29. Why balancing is done? Suggest some methods for balancing a grinding wheel.
30. What is surface grinding? Explain with suitable sketches, the relative work, wheel and table movements on different types of surface grinders.
31. What are the different types of internal grinders? Describe them.
32. Explain the principle of centreless grinding. How do 'Through feed' and 'Infeed' methods differ in centreless grinding?
33. Explain the factor to be kept in mind while selecting a grinding wheel?

References

Abuladze, N.G., *Character and Length of the Plastic Contact between a Chip and the Tool Rake Face: Collinear Machinability of Heat-Resisting and Titanium Alloys*, Kuibyshev Publishing House, 1962.

Acherkan, N., *Machine Tool Design*, Vols. I–IV, Mir Publishers, Moscow, 1968.

Acherkan, N., *Metal Cutting Machine Tools* (in Russian), State Scientific and Technical Publishing House of Machine Building Literature, Moscow, 1958.

Alden, G.I., 'Operation of Grinding Wheels in Machine Grinding', *Trans. ASME*, Vol. 36, 1914, p. 451.

American Society of Tool and Manufacturing Engineers, *Die Design Handbook*, McGraw–Hill, New York, 1955.

Anelchik, D.E., et al., 'An Investigation into the Instantaneous Temperature in the Grinding of Alloy Steel', *Industrial Diamond Review*, 1968, p. 539.

Archibald, F.R., 'Analysis of the Stresses in a Tool Cutting Edge', *Trans. ASME*, Vol. 78, No. 6, August 1956.

Armargeo EJ and G.Brown, "*Machining of Metals*", PHI Learning, 1985.

Artobolevskii, I.I., *Theory of Machines*, Nauka Publishers, Moscow, 1967.

Backer, W.R., E.R. Marshall, and M.C. Shaw, 'The Size Effect in Metal Cutting', *Trans. ASME*, Vol. 74, 1952, p.61.

Baranov, G.G., *A Course on Theory of Mechanism and Machines* (in Russian), Mashinostroenie Publishers, Moscow, 1967.

Baron, C.H., *Numerical Control for Machine Tools*, McGraw-Hill, New York, 1971.

Bartsch, W., *Numerical Control of Machine Tools*, Wiley Eastern, New Delhi, 1980.

Basu, S.K., *Design of Machine Tools*, Allied Publishers, New Delhi, 1965.

Belyaev, N.M., *Strength of Materials*, Mir Publishers, Moscow, 1979.

Bezier, P., *Numerical Control*, John Wiley, London, 1970.

Bhattacharya, A., and G.C. Sen, *Principles of Machine Tools*, New Central Book Agency, Calcutta, 1973.

Block, H., 'Theoretical Study of Temperature Rise at Surfaces of Actual Contact Under Oiliness Lubricating Conditions', *Proceedings of the General Discussion on Lubrication and Lubricants of The Institution of Mechanical Engineers*, London, Vol. 2, 1937, p. 222.

Boguslavsky, B.L., *Automatic and Semi Automatic Lathes*, Peace Publishers, Moscow.

Busch, G., P. Sehnud, and R. Spondin, 'Thermo-Voltage between Silicon Carbide, Copper and Platinum', *Helv. Phys. Acta.*, Vol. 20, No. 6, 1947.

Campbell, James S., Jr., *Principles of Manufacturing Materials and Processes*, McGraw-Hill, New York, 1961.

Carslaw, H.S., and J.C. Jaeger, *Conduction of Heat in Solids*, Clarendon Press, Oxford, England, 1947.

Carslaw, H.S., and J.C. Jaeger, *Conduction of Heat in Solids*, Oxford University Press, 1959, pp. 255–266.

Cates, P.D., 'How to Choose Coolants for Diamond Grinding Operations', *Industrial Diamond Review*, September 1976.

Chalkey, J.R., 'Some Further Experiments on the Use of Grinding Fluid', *University Review*, Vol. 4, 1968.

Chalkey, J.R., 'The Use of Grinding Fluids', *University Review*, Vol. 3, 1967, pp. 18–21.

Chao, B.T., and K.J. Trigger, 'Temperature Distribution at the Tool–Chip Interface in Metal Cutting', *Trans. ASME*, Vol. 77, 1955, pp. 1107–1121.

Chao, B.T., and K.J. Trigger, 'Temperature Distribution at the Tool–Chip and Tool–Work Interface in Metal Cutting', *Trans. ASME*, Vol. 80, 1958, pp. 311–329.

Chapman, W.H., 'Cylindrical Grinding in 1920', *Trans, ASME*, Vol. 42, 1920, p. 595.

'Checklist of Jig and Fixture Design', *American Machinist*, 12 September 1955 and 26 September 1955.

Childs, J.J., *Principles of Numerical Control*, Industrial Press, New York, 1969.

Coes, L., Jr., 'Knowledge of the Scientific Principles of Grinding is a Basis of Recent Progress in Abrasives', *Industrial and Engineering Chemistry*, Vol. 2493, December 1955.

Cook, N.H., E.G. Loewen, and M.C. Shaw, 'Machine Tool Dynamometers', *American Machinist*, 10 May 1954.

Cook, N.H., *Manufacturing Analysis*, Addison-Wesley, 1966.

Course on Computer Techniques for Design of Machine Tool Structures, CMTI, Bangalore, 1973.

Davis, A., 'How to Get the Best Out of Cutting Fluids', *Metal Working Production*, Vol. 124, February 1981, pp. 127–128.

DeVries, M.F., S.M. Wu, and J.W. Mitchell, 'Measurement of Drilling Temperature by the Garter Spring Thermocouple Method', *Microtechnic*, Vol. 53, No. 6, 1967.

Dobrovoskii, V.A., et al., *Machine Design* (in Russian), State Scientific and Technical Publishing House of Machine Building Literature, Moscow, 1962.

Duwell, E.J., E.S. Hong, and W.J. McDonal, 'The Role of Chemical Reactions in the Preparation of Metal Surfaces by Abrasions', *Wear*, Vol. 9, 1966, pp. 417–424.

Duwell, E.J., I.S. Hong, and W.J. McDonald, 'The Effect of Oxygen and Water on the Dynamics of Chip Formation During Grinding', *ASLE Trans.*, Vol. 12, 1969, pp. 86–93.

Egorov, N.K., and V.E. Barinov, 'Proximate Evaluation for The Effectiveness of Coolants During Grinding Operations', *Russian Engineering Journal*, Vol. 55, No. 4, 1975, pp. 55–56.

Elyashev, A., *Fundamentals of Machine Tool Design*, Lecture Notes for UNIDO Course, Moscow.

Epifanov, G.I., and P.A. Rebinder, 'Energy Balance of the Metal Cutting Process', Dokladi Akademii Nauk, USSR, Vol. 66, 1949, p. 653.

Feodosyev, V., *Strength of Materials*, Mir Publishers, Moscow, 1968.

Filonenko, S.N., et al., 'Effect of Grinding Speeds and Feeds on Cutting Zone Temperature', *Machs Tool*, XL, 56.

Filonenko–Borodich, M.M., *Mechanical Theory of Strength*, Moscow University, 1961.

Fisher, R.C., 'Grinding Dry with Water', *Grinding and Finishing*, March 1965, p. 32–34.

Fisher, R.C., 'How Wet is your Wet Grinding', *American Machinist, Metal Working Manufacturing*, Vol. 107, No. 7, April 1963, pp. 114–115.

Frokht, M., *Photoelasticity*, Vol. II, GONTI Publishers, Moscow, 1950.

Gassan, P., *Theory of Design*, B.T. Bansford, London, 1974.

Gaylor J.F. and C.R., *Metrology for Engineers*, Cassell, 1964.

Gettelman, K., 'New Nozzle Makes Grinding Coolant Work', *Modern Machine-Shop*, July 1970, p. 100–104.

Grandjean, E., *Fitting the Task to the Man*, Taylor and Francis, London, 1975.

Grover, G.K., *Mechanical Vibrations*, Nem Chand and Brothers, Roorkee, 1977.

Grunzweg et al., 'Calculation and Measurement of Wedge Indentation', *Journal of Mechanics and Physics of Solids*, Vol. 2, 1954, p. 81.

Guest, J.J., 'The Theory of Grinding with Reference to the Selection of Speeds in Plain and Internal Work', *Proceedings of the Institution of Mechanical Engineers*, London, 1915, p. 543.

Hahn, R., 'Grinding Wheel Wear, Fluid and Chatter in Limited Grinding Time: A Leverage For Reduction in Grinding Costs', Internal Communication to CIRP, STCG, 19 January 1976.

Hahn, R.S., 'The Effect of Wheel–Work Conformity in Precision Grinding', *Trans. ASME*, Vol. 77, 1955, pp. 1325–1329.

Halverstadt, R.D., 'The Development of a Test for Evaluating Grinding Fluids', *ASLE Annual Meeting*, Cincinnati, Paper No. 58-AM 4B, 1958.

Heinz, W.B., 'Metal Cutting with Abrasive Wheels', *Trans. ASME*, Vol. 65, 1943, p. 21.

Hervey, R.P., and N.H. Cook, 'Thermal Parameters in Drill Tool Life', ASME Paper No. 65-Prod-15, 1965.

Hoffman, A., Communication, *Maschinenhau*, January 1933.

Hughes, F.H., 'Coolant Additives and Diamond Wheel Efficiency', *The Engineer*, 1968, pp. 786–787.

Hughes, F.H., 'Low Cost Mods Boost Grinder Performance', *Engineering Production*, 24 February 1972.

Hughes, F.H., 'Making Your Coolant Work For You', *Talking Diamond Grinding*, Pt-7.

Hughes. F.H., *Using Ultrahard Abrasives for Grinding – A Primer*, unpublished book.

Hutchinson, R.V., 'Do We Understand the Grinding Process?', *SAE Journal*, Vol. 42, 1938, p. 89.

Jaeger, J.C., 'Moving Sources of Heat and Temperature at Sliding Contacts', *Proceedings of the Royal Society of New South Wales*, Vol. 76, 1942, p. 222.

Jeffery, W., 'Alternative to Oil Base Cutting Fluids for Metal Cutting, Metal Removal and Abrasive Machining', *Cutting Tool Engineering*, Vol. 33, No. 9–10, 1981.

Jeffery, W.O., 'The Effect of Hard Water on Cutting Fluids', *Cutting Tools Engineering*, November–December 1980.

Joshi, P.H., *Jigs and Fixtures*, A.H. Wheeler, 1996.

Kaminskaya, V.V., et al., *Beds and Housings of Machine Tools* (in Russian), Mashgiz Publishers, Moscow, 1960.

Karaim, I.P., 'Grinding Temperature Measurement Using a Thermal-Electrode', *Russ. Engg. Jl.*, Vol. 8, No. 51, 1964.

Karaim, I.P., 'Influence of Cooling Method on Grinding Zone Temperature', *Machs Tool*, XL, 42.

Kasatkin, A.S., *Fundamentals of Electrical Engineering* (in Russian), Energiya Publishers, Moscow, 1966.

Kececioglu and Dimitri, 'Shear Strain Rate in Metal Cutting Effects on Shear Flow Stress', *Trans. ASME*, Vol. 80, No. 1, January 1958, p. 158.

Kedrov, S.S., *Vibrations of Machine Tools* (in Russian), Mashinostroenie Publishers, Moscow, 1978.

Kevshinskeii, V.V., *Milling Operation* (in Russian), Mashinostroenie Publishers, Moscow, 1977.

Khudobin, I.L., 'A New Staged Method for Delivering Cutting Fluids in Grinding', *Russian Engineering Journal*, 1982.

Khudobin, I.V., et al., 'New Way of Applying Cutting Fluids in Grinding Operations', *Machines and Tooling*, Vol. KLI, 1970, p. 52–55.

Khudobin, L.V., and A.N. Samsonov, 'High-Pressure Jet Cooling in Grinding', *Machines and Tooling*, Vol. 44, No. 8, 1966.

Khudobin, L.V., and A.N. Samsonov, 'Optimum Grinding Conditions with High-Pressure Jet Cooling', *Machines and Tooling*, Vol. XXXVIII, No. 4, 1967.

Khudobin, L.V., and E.P. Gulnov, 'Selecting the Degree to which Cutting Fluids Should be Rid of Mechanical Impurities when Grinding,' *Machines and Tooling*, No. 12, 1976, pp. 31–32.

Khudobin, L.V., *Cutting Fluid Applied in Grinding*, Mashinostroenie Publishers, Moscow, 1971.

Klyuchnikov, A.V., 'Numerically Controlled Machine Tools', Lecture Notes for UNIDO Course, Moscow.

Koenigsberger, F., and J. Tlusty, *Machine Tool Structures*, Vol. I, Pergamon Press, Oxford, 1970.

Koenigsberger, F., *Design Principles of Metal Cutting Machine Tools*, Pergamon Press, Oxford, 1964.

Kovan, V. *Fundamentals of Process Engineering*, Foreign Languages Publishing House, Moscow.

Kudinov, V.A., *Dynamics of Machine Tools* (in Russian), Mashinostroenie Publishers, Moscow, 1977.

Kuklin, L.G., 'The Fatigue Strength of Carbide T5K10', *Machine and Tooling*, No. 4, 1961, pp. 33–34.

Kuklin, L.G., V.I. Sagalov, V.B. Serebrovskii and S.P. Shabashov, 'Increasing the Strength and Wear Resistance of Carbide Tools', *Mashgiz*, Vol. 24, 1960.

Kusher, A.M., *An Atlas of Kinematic Diagrams of Machine Tools* (in Russian), Mashinostroenie Publishers, Moscow, 1977.

Lebedev, V.G. et al., 'Micro-Thermocouple Using Diamond and Cubic Box on Nitride Wheels', *Machs Tool*, XL, 42.

Lebedev, V.G., et al., 'Assessing Ground–Surface Contact Temperature Using Grindable Thermocouples', *Machs Tool*, 1973, p. 45.

Leslie, W.H.P., *Numerical Control Users Handbook*, McGraw-Hill, New York, 1970.

Letner, H.R., 'A Modern Perspective of the Grinding Process', *Grinding and Finishing*, 1960, p. 36–41.

Letner, H.R., 'Influence of the Grinding Fluids upon Residual Stresses in Hardened Steel', ASME Paper No. 55-A-123, 1956.

Lindert, A.W., 'Some Problems Encountered in the Use of Soluble Oils', *Lubrication Engineering*, Vol. 7, 1951, p. 223–227.

Lissaman, A.J., and S.J. Martin, *Principles of Engineering Production*, ELBS/Hodder and Stoughton, London, 1983.

Littmann, W.E., and J. Wulff, 'The Influence of the Grinding Process on the Structure of Hardened Steels', *Trans. ASME*, Vol. 47, 1955, p. 692.

Loewen, E.G., and Shaw, M.C., 'On the Analysis of Cutting Tool Temperatures', *Trans. ASME*, Vol. 76, 1954, pp. 217–231.

Loktev, D.A., *Metal Cutting Machine Tools*, Mashinostroenie Publishers, Moscow, 1968 (in Russian).

Loladze, T.N., 'Cutting Tool Wear', *Mashgiz*, Vol. 5, 1958.

Machinery's Screw Thread Book, 20th edition, Machinery Publishing Company, 1972.

Malov, A., and Y. Ivanov, *Principles of Automation and Automated Production Processes*, Mir Publishers, Moscow, 1976.

Mankani, L.B., and H.H. Alvord, *Manual of Applied Machinery Design*, Roorkee University, Roorkee, 1962.

Marshal, W.B., and M.C. Shaw, 'Forces in Dry Surface Grinding', *Trans. ASME*, Vol. 74, 1952, pp. 51–59.

Martin, S.J., *Numerical Control of Machine Tools*, English University Press, London, 1970.

Maslov, D., et al., *Engineering Manufacturing Processes in Machine and Assembly Shops*, Mir Publishers, Moscow, 1967.

McCormick, E.J., *Human Factors in Engineering and Design*, Tata McGraw-Hill, New Delhi, 1976.

Merchant, M.E., 'Mechanics of the Metal Cutting Process', *Journal of Applied Physics*, Vol. 16, 1945, pp. 267 and 318.

Metals Handbook, American Society for Metals, Cleveland, 1948, pp. 313, 314.

Michelon, Leno C., *Industrial Inspection Methods*, Harper and Brothers, New York, 1950.

Mittal, R.N., T.M. Porter, and G.W. Rowe, 'Lubrication With Cutting and Grinding', Unpublished Report, University of Birmingham, England, 1981.

Modern Machine Shop, NC Guide Book and Directory Issue, Gardner Publications, Cincinnati, 1974.

Mokee, R.E., R.S. Moore, and O.W. Boston, 'A Study of Heat Developed in Cylindrical Grinding', *Trans. ASME*, Vol. 73, 1951, pp. 21–34.

Movnin, M., and D. Goltzker, *Machine Design*, Mir Publishers, Moscow, 1969.

Muller, J.A., 'Abrasive Report', Private Communication, Carborundum Company.

Muller, J.A., 'How Much Do Grinding Fluids Affect Wheel Performance', *Iron Age*, 1957, pp. 79–41 and 134–135.

Murrel, K.F.H., *Ergonomics*, Chapman Hall, London, 1962.

Naerman, M.S., and L.E. Pekarev, 'Increasing the Efficiency of Superfinishing by Adding Solid Lubricants to the Pores of Abrasive Stones', *Machines and Tooling*, Vol. 45, No. 81, 1975, pp. 448–50.

Nee, Y.C., 'The Effect of Grinding Fluid Additives on Diamond Wheel Efficiency', *International Jl. MTDR*, Vol. 19, 1979, p. 21–31.

Niebusch, R.B., and E.H. Strieder, 'The Application of Cutting Fluids to Machining Operations', *Mechanical Engineering*, Vol. 74, 1952, p. 203–207.

Olesten, N.O., *Numerical Control*, Wiley Interscience, New York, 1970.

Osman, M., and S. Malkin, 'Coolants for Cutting and Grinding With Diamond Impregnated Wheels', Burmah-Castrol Industries Industrial Lubrication Division Pub. No. IND/215/738, August 1973.

Osman, M., and S. Malkin, 'Lubrication by Grinding Fluids at Normal and High Wheel Speeds', *ASLE Trans.*, Vol. 15, 1972, pp. 261–268.

Outwater, J.O., 'The Mechanism of Grinding and the Function of the Lubricant', *Lubrication Engineering*, Vol. 7, 1952, pp. 123, 124, 144.

Outwater, J.O., and M.C. Shaw, 'Surface Temperature in Grinding', *Trans. ASME*, Vol. 74, 1952, p. 73.

Oxford, C.J., 'On the Drilling of Metals: I. Basic Mechanics of the Process', *Trans. ASME*, Vol. 77, 1955, pp. 103–114.

Oxley, P.L.B., 'Mechanics of Metal Cutting', *Proceedings of the International Production Engineering Conference*, 1963.

Parsegov, S.V., et al., 'The Use of Cutting Fluids and Solid Lubrications When Grinding Floor Cutters', *Soviet Engineering Research*, No. 11, 1981, pp. 86–88.

Patton, W.J., *Numerical Control — Practice and Application*, Reston Publishing, Reston, Virginia, 1972.

Perry C.C. "Data Reduction Algorithms for strain gauge rosetta measurements", Experimental techniques May 1989 (pp 13–18)

Peter, J., and R. Aerens, 'An Objective Method For Evaluating Grinding Fluids', *Annals of CIRP*, Vol. 25, No. 1, 1976, pp. 247–250.

Phelan, R.M., *Fundamentals of Mechanical Design*, McGraw-Hill, New York, 1957.

Pike, A.R., 'Fluids for Grinding Only', *Grinding and Finishing*, June 1965, pp. 30–33.

Pippenger, J.J., and T.G. Hicks, *Industrial Hydraulics*, McGraw-Hill, New York, 1962.

Pitts, G., *Techniques in Engineering Design*, Butterworth, London, 1973.

Preferred Limits and Fits for Cylindrical Parts, American Standard ASA B46.1-1955, American Standards Association, New York, 1955.

Push, V.E., *Design of Machine Tools* (in Russian), Mashinostroenie Publishers, Moscow, 1977.

Quadt, R., 'When to Use Cold Extrusion of Aluminium', *Design Engineering*, November 1956.

Rabinovich, A., *Speed Boxes of Machine Tools* (in Russian), Lvov University, Lvov, 1968.

Razumov, I.M., *Scientific Organization of Labour* (in Russian), Vyschaya Shkola Publishers, Moscow, 1978.

Reshetov, D.N., *Machine Design*, Mashinostroenie Publishers, Moscow, 1975.

Rippel, H.C., *Cast Bronze Hydrostatic Bearing Design Manual*, Cast Bronze Bearing Institute, USA, 1963.

Roberts, A.D., and R.C. Prentice, *Programming for Numerical Control Machines*, McGraw-Hill, New York, 1978.

Rowe, G.W., 'Control of Conditions and Fluids in Grinding', *Proceedings of the International Conference on Optimum Resources Utilization Through Tribology and Maintenance Management*, IIT Delhi, November/December 1981.

Rowe, G.W., 'Lubricant Testing For Grinding Operation', *Wear*, Vol. 77, 1982, pp. 73–80.

Rowe, G.W., *An Introductin to Industrial Metalworking Processes*, Edward Arnold, London, 1985.

Rowe, G.W., E. Smart, and K.C. Tripathi, 'Surface Adsorption Effects in Cutting and Grinding', *Trans. ASME*, No. 20, 1977, pp. 247–253.

Rozenberg, Y.A., 'Performance Properties of Lubricating Oils and Their Evaluation', *Russian Engineering Journal*, Vol. 55, No. 8, 1975, pp. 5–46.

Sagarda, A.A., and I.A. Khudobin, 'The Effectiveness of Coolants When Grinding with Diamond Wheels', *Sintetcheskiv Almazy*, No. 2, 1976, pp. 35–39.

Sakmann, B.W., 'Geometrical and Metallurgical Changes in Steel Surfaces under Conditions of Boundary Lubrication', *Journal of Applied Mechanics, Trans. ASME*, Vol. 69, 1947, p. A-43.

Sato, T., and M. Mizuna, 'Features and Principles of Through Wheel Coolant Grinding', Report 28, No. 1, Faculty of Engg. and Tech., Tohoku University, 1963, pp. 163–174.

Sharpe, K.W.B., *Practical Engineering Metrology*, Pitman, London, 1982.

Shaw, M.C., and C.J. Oxford, Jr., 'On the Drillings of Metals: II. The Torque and Thrust in Drilling'. *Trans. ASME*, Vol. 79, 1957, pp. 139–148.

Shaw, M.C., and C.T. Yang, 'Inorganic Grinding Fluids for Titanium Alloys', *Trans. ASME*, Vol. 78, 1956, p. 861–868.

Shaw, M.C., *Metal Cutting Principles*, Cambridge Technology Press, 1957.

Shaw, M.C., N.H. Cook, and P.A. Smith, 'The Mechanics of 3-Dimensional Cutting Operation', *Trans. ASME*, Vol. 74, 1952.

Shigley, J.E., *Kinematic Analysis of Mechanisms*, McGraw-Hill, New York, 1959.

Shore, H., 'Tool and Chip Temperature in Machine Shop Practice', Massachusetts Institute of Technology Thesis, 1924.

Shvartz, V.V., A.V. Elyashev, and N.N. Gudimenko, *Developing the Kinematic Diagrams of Machine Tool Speed Boxes*, Peoples Friendship University, Moscow, 1978.

Simon, W., *The Numerical Control of Machine Tools*, Edward Arnold, London, 1973.

Singh, D.V., et. al., *Bearings, Lubricants and Lubrication*, Roorkee University, Roorkee, 1975.

Sluhan, C., 'Grinding Mods', Industrial Diamond Conference, Chicago, November 1969.

Sluhan, C.A. 'Some Considerations in the Selection and Use of Water Soluble Grinding Fluids', *Lubrication Engineering*, Vol. 110, March 1960.

Sluhan, C.A., 'Grinding With Water Miscible Grinding Fluids'. *Lub. Fngg.*, Vol. 26, 1970, pp. 352–374.

Sluhan, W.A., 'Equipment for Control and Maintenance of Water Miscible Cutting and Grinding Fluids', *Cutting Tool Engineering*, May/June 1975, pp. 5–8.

Sluhan, W.A., 'Extending The Life of Water Miscible Cutting and Grinding Fluids', *ASLE/JSLE Joint Lubrication Conference*, Tokyo, 1975.

Stewart, D.A., and H.R. Serierstram, 'Residual Stresses When Grinding High Temperature Alloys: Cutting Fluid Selection', *Lubrication Engineering*, June 1961, pp. 286–290.

Strasser, F., 'Should Die Thickness Be Calculated?', *American Machinist*, 19 February 1951.

Surface Roughness, Waviness and Lay, American Standard ASA B46.1-1955, American Standards Association, New York, 1955.

Sweeney, G., *Vibration of Machine Tools*, Machinery Publishing, London, 1971.

Tarasov, L.P., 'Grinding Fluids', *Tool and Manufacturing Engineer*, June 1961, pp. 67–73.

Tatrenko, V.V., et al., 'Micro-thermocouple for Measuring the Temperature Field in Grinding', *Russ. Engg. Jl.*, XLIX, 53, 1969.

Tatrenko, V.V., et al., 'Universal Thermocouple for Measuring Grinding Temperature', *Measmt. Tech.*, Vol. 2, 1970, p. 265–266.

The Science of Precision Measurement, the DoALL Company, Des Plaines, Illinois, 1953.

Thomson, W.T., *Theory of Vibration*, Prentice-Hall of India, New Delhi, 1975.

Tobias, S.A., *Machine Tool Vibration*, Blackie, London, 1965.

Treatise on Milling and Milling Machines, 3rd edition, Cincinnati Milling Machine Company, Cincinnati, 1951.

Trigger, K.J., and B.T. Chao, 'An Analytical Evaluation of Metal-Cutting Temperatures', *Trans. ASME*, Vol. 73, 1951, p.57.

Trimal, G., and H. Kaliszer, "*Delivery of Cutting Fluids in Grinding*", Grinding Seminar of the 16th International MTDR at UMIST, September 11, 1975.

Tripathi, K.C., and G.W. Rowe, 'Grinding Fluid, Wheel Wear and Surface Generation', *Proceedings of the 17th International MTDR Conference*, Birmingham, September 1976, pp. 24–26.

Truckenmiller, W.C., 'Quenching Properties of Cutting Fluids', Private Communication, June 1953.

Tsueda, M., and Y. Hasegawa, 'The Study of Cutting Temperature in Drilling', *Journal of Precision Mechanics of Japan*, Vol. 25, No. 1, 1959, pp. 9–14.

Tsueda, M., Y. Hasegawa, and Y. Nisina, 'The Study of Cutting Temperature in Drilling: 1. On the Measuring Method of Cutting Temperatures', *Transactions of the Japanese Society of Mechanical Engineers*, Vol. 27, No. 181, 1961, pp. 1423–1430.

Vallance, A., and V.L. Doughtie, *Design of Machine Members*, McGraw-Hill, New York, 1951.

Vershinin, V.D., *Electrical Control Circuits of Machine Tools* (in Russian), Peoples Friendship University, Moscow, 1969.

Verson, J., 'Tooling for Cold Extrusion', *American Machinist*, 7 October 1957.

Wagner, H.W., 'Grinding Fluids: Characteristics and Applications', *Mechanical Journal*, 1982.

Watanabe, S., et al., 'Antirest and Lubricity Characteristics of Cutting Fluids and Additives', *Lubrication Engineering*, Vol. 38, No. 7, July 1982, pp. 412–415.

Welbourn, D.B., and J.D. Smith, *Machine Tool Dynamics: An Introduction*, Cambridge University Press, Cambridge, 1970.

Wilson, F.W., *Numerical Control in Manufacturing*, McGraw-Hill, New York, 1963.

Wu, S.M., and R.N. Meyer, 'Cutting Tool Temperature in Prediction Equation by Response Surface Methodology', *Journal of Engineering for Industry, Trans. ASME*, Series B, Vol. 86, No. 2, May 1964, pp. 150–156.

Yakimov, A.V., et. al., 'Investigation of Temperature in the Grinding Zone', *Russ. Engg. Jl.*, Vol. 8, No. 51, 1964.

Yang, C.T., and M.C. Shaw, 'The Grinding of Titanium Alloys', *Trans. ASME*, Vol. 77, 1955, p. 645.

Zorev, N.N., 'Mechanics of Metal Cutting', *Mashgiz*, Vol. 53, 1956.

Index

Abrasion wear, 225
Accuracy in jigs and fixtures, 668
Action of cutting fluids as lubricants, 263
Adjustable plate clamp, 632
Advantages
    of cutting fluid applications, 262
    and disadvantages of hydraulic drives, 372
    of using jigs and fixtures, 617
    of using press tools, 714
Air-operated clamp, 638
Alloy C.I., 8
American system, 35
Angle slip gauges, 556
Angular measurement, 556
    mechanical, 557
    optical, 558
    pneumatic, 560
Antifriction bearings, 324
Application
    of clearances, 734
    of metal working fluids, 264
ASA coordinate system (*see* American system), 35
Assembly fixtures, 665
Automation, 434

Belt grinding, 860–862
Best size wire method, 585
Bond material, 819

Boring
    fixtures, 666
    tools, 73
Braking, 406, 407
Brazed
    insert, 257
    tools
        diamond, 74
        tipped, 73
British maximum rake system, 34
Broaching, 531
    fixtures, 664
    methods, 531, 533–535
Built-up edge and tool failure, 221, 222

Calculation for design of beds, 292
Calibration of length standards, 553, 555
Cam-actuated clamp, 636
Capstan and turret lathes, 436, 439
    tooling layout for, 442
Cast iron (C.I.), 7
Cemented carbides, 53–56
Centre of pressure, 738
Centreless grinding, 837–844
Ceramics, 57
Cermets, 58
Channel jig, 650
Characteristics of good cutting fluid, 262

Chatter vibration, 422
Chemical wear, 228
Chilled C.I., 8
Chip
   breaker, 88–90
      clamp-type, 87
   control, 249
      through chip breakers, 87
      through tool grinding, 86
   formation, 82
      effect of various factors on, 85
      geometry of, 82
      mechanism of, 81
      Merchant's model of, 104
   thickness in milling, 132
Choice of press, 729
Clamping devices, 631
Classification
   of gears, 523
   of machine tools, 280
   of press
      by actuation of ram, 715
      by design of frame, 715
      by intended use, 716
      by number of slides, 715
      by slides, 720
      by source of power, 715
      by use, 721
Clutch control, 417
Coated abrasives, 858–860, 863, 664
Cold drawing, 675
Common features of machine tools, 282
Comparator, 557–560
Comparison of cutting tool materials, 61
Composites, 6
Conical locators, 622, 624
Constructional features of basic machines, 19
Continuous
   broaching, 534
   chips, 84
Cost analysis, 253
Crater wear, 224
Cut-off tools, 72
Cutting
   fluid, 261, 263
      and tool life, 264
   power in milling, 136
   speed, 221, 261

   tool, 68
      angles, 30–34
      design, 67
      materials, 50
      temperature by dimensional analysis, 167
Cylindrical location, 624

Damping of vibration, 290
Deep drawing analysis, 695
Definition
   of machinability, 247
   of machine tool, 279
Depth of cut and cutter wear, 132
Design
   of brazed tool seats, 72
   of die elements, 744
   of feed power, 310
   of frame, 717
      of machine tool
         beds, 284, 285
         gear box, 328
         guides, 294
   of jigs and fixtures
      principles, 618
      procedure, 619
   of nut and screw, 310
   of screws and dowels, 739
   of single-point cutting tool, 70
Desirable characteristics of dynamometer, 194
Development of series of numbers, 332
Dial gauge, 595
Diameter of screw thread, 585
Diamond tools, 58
Die support methods, 721
Difference
   between hot and cold working, 681
   between jig and fixture, 617
Diffusion wear, 227, 228
DIN system, 35
Disadvantages of using cutting fluids, 271
Discontinuous chips, 83
Down-cut milling, 130
Drilling, 13
   jig and fixtures classification, 645
   machine, 13
   mechanism of, 123

INDEX

Dynamometry, 193
   various types of, 198

Eccentric clamp, 634
Economic aspects of jigs and fixtures, 669
Economics of metal machining, 253
Effect
   of alloying elements, 251
   of cutting fluid, 264
   of error
      angle, 583, 584
      pitch, 583
   of shape and tool angles on tool life, 233
Electrical
   discharge, 783–785
   and electronic regulation, 398
   principles, 562
Electrochemical machining, 792
Electrolytic grinding, 798
Embossing, 677
Energy consideration, 100, 178
EP type lubrication mechanism, 270
Equalising clamp, 634
Equation on size of cut, 230
Experimental techniques of temperature measurement, 182
External micrometer, 545, 546

Factors
   affecting tool life, 230
   in design of beds, 285
Feed rate and cutter wear in milling, 132
Fits and allowances, 567
Fixtures
   manufacturing and materials aspects, 666
   milling of, 656
Flank wear, 223
Flat rolling, 692
Fly press, 716
Forces acting on a cutting tool, 68, 94
Forged tools, 25
Forging analysis, 689
Form tools, 27
Forming processes, 10
   non-ferrous, 61
   non-metallic, 63
   stretch, 677

Fractured chips, 85
Functions of cutting fluids, 261
Fundamental
   deviations, 570
   of mechanical regulation, 329

Gang milling fixtures, 657
Gaseous cutting fluids, 270
Gauge tolerances, 577
Gear
   bevel, 528
   cutting methods, 522–524, 526–530
   hobbing, 526, 528
   milling, 525, 527
   shaping, 528
   shaving, 530
   tooth proportions, 523
Generating methods, 525
Geometrical tolerances, 571
Grey C.I., 7
Grinding, 17
   and abrasive processes 816–820
   fixtures, 663
   kinetics, 143
   machines types, 832–837, 844–847
   mechanism of, 137
   undeformed chip
      thickness in, 141
   wheel codification, 822
      balancing, 830
      dressing and truing, 824, 826
      selection, 822
      shapes, 824
      speeds, 831, 832
Ground-in type chip breaker, 87, 89

Heat
   in grinding, 176
   in metal cutting and temperature measurement, 155
Heat and temperature
   in drilling, 173
   in milling, 173
High speed steels, 52
Honing, 847–850
Horizontal broaching machines, 532

INDEX

Hot
  extrusion, 683
  rolling, 682
  working, 682
Hydraulic drive, 370, 374, 379, 382
  in grinding machines, 393

Important features of simple broach, 531
Indexable inserts system, 257
Indexing milling fixtures, 661
Inhomogeneous chips, 84
Interferometry, 551
  optical flats in, 551
Internal broaching, 532
ISO system, 36

Jig base and jig feet, 647
Jigs and fixtures, 617

Knockout design, 762

Lapping, 18, 850–854
Laser beam machining, 803
Latch clamps, 634
Lathe, 19
  dynamometer, 212, 215
Lee and Shaffer's model of shear angle relation, 102, 104
Length standards, 548
Limit gauging, 576
Limits, fits and tolerance, 565
Line standard, 548
Linear measurement, 545
Link mechanisms in kinematic pair, 315
Liquid cutting fluids, 265
Locations, 620
Locators, 621
Low alloy steels, 52
Lubrication, 303

Machinability, 247
  evaluation of, 247
  index of, 252
Machine tools, 19, 279
  design, 279
  industry, 281

  testing, 595–601
  vibrations, 420
Machines of milling process, 130
Machining, 781
  processes, 12
  major unconventional, 773
Malleable C.I., 7
Manufacturing processes, 9
Material
  of beds, 292
  of guideways, 299
Maximum draft, 708
Measurement of temperature in milling, 188
Mechanical
  chipping, 222
  regulation of
    drives, 352
    speeds, 366
Mechanically
  clamped diamond tool, 75
  held tipped tools, 74
Mechanics of metal cutting, 94
Meehanite, 288
Merchant's model of shear angle relation, 104
Metal
  cutting, 100, 261
    fluid, 263
    process, 79
    theory of, 78
    tool, 24
  working, 10
    analysis, 688
Metallics, 4
Metric threads, 573
Microstructure of workpiece and tool life, 232
Milling, 12, 14, 16, 130
  work done in, 136
Milling and grinding dynamometers, 216
MRR in EDM, 785–787

Nature of cutting and tool life, 233
Nodular C.I., 8
Nomenclature systems, 34
Non-ferrous cast alloys, 52
Normal rake system, 35
NPL
  interferometer, 553
  method, 585
Nutcracker jig, 652, 655

INDEX

Optical projection, 564
Orthogonal and oblique cutting, 79

Pad locator, 622, 623
Particular cases for extrusion analysis, 698, 699
Pin locator, 622
Plain
    carbon steels, 51
    milling fixtures, 657
Planer, 22, 24
Plasticity analysis, 695
Plate
    clamp, 632
    jig, 649, 650
Pneumatic clamping, 637
Points of support, 550
Polymers, 5
Post
    jig, 651
    type locator, 623
Power
    in broaching, 144
    consumption, 247
    press, 714, 716, 718
    source of, 715
Practical die clearance, 736
Press
    operations, 725
    tools
        clearances, 734
        design, 729
        parts, 731
Process capabilities of unconventional machining processes, 774
Profile locator, 622
Properties of C.I., 7
Protection of guideways, 303
Punch support methods, 721

Rake system, 28
Ram
    actuation, 715
    driving mechanism, 718
Ray diagram, 340
Relationship between ASA and ISO systems, 37
Requirements of cutting, 50
Rigidity of workpiece and tool life, 233

Roll torque and power, 709
Rolling
    slab method, 698
    velocity analysis, 701
Rotary swaging, 679

Schlesinger's formula, 133
Screw
    clamps, 634
    thread, 572
        caliper gauge, 583
        gauging, 582
        Not Go gauge, 577
        rolling, 538–540
Screw-threads production, 535–540
    chasing, 536
    trapping, 538
Selection
    criteria for metal cutting and metal working fluids, 271, 273
    of machine tools, 282
Semi-automatic multi-tool lathes, 434
Shafts in machine, 319
Shape of beds, 293
Shaper, 21
Shaping, 13
Sheared edges in piercing, 735
Sheet drawing
    particular cases, 699
    by slab method, 698
Simple
    adjustable locator, 623
    approach to forging analysis, 689
Single-point tools, 24
Single-spindle automatic lathes, 447
Six-point location principles, 620
Sliding plate clamp in, 632
Slip gauges
    BS standards, 555
Slotter, 22
Snagging, 855–857
Solid
    cutting fluids, 265
    jigs, 650
Sources of error in linear measurement, 555
Speed control, 388
Spherical washer for small adjustment, 633
Spindle nose, 323

Spring-loaded adjustable locator, 623
Squeezing, 677
Stationary heat source, 156
Stepless regulation of speeds, 365
Strain, 101
    measurement, 195
Strength
    in bed design, 285
    of materials, 250
Stress and strain in chip, 99, 101
String milling fixture clamp, 659
Stripper plate, 751
Superfinishing machines, 18
Surface
    broaching, 533
    finish and its measurement, 586, 588–594
Surface finish terminology, 586
    evaluation, 588
    measurement, 590–594
Swing washer clamp, 635
Synthetic or chemical cutting fluids, 268

Temperatures in orthogonal cutting, 156
Theories of shear angle relationship, 103, 104
Thickness ratio, 92, 103
Thread rolling, 679
Three-point strap clamp, 633
Toggle clamps, 637
Tolerance, 566
    grades, 568
    for ISO, 573
Tool
    bit in holder, 26
    design, 72, 730
    face temperature, 162
    failure, 85, 184, 220
        criterion for, 220
    life, 229, 248
        measurement, 229
        specifications, 228
    material, 233
        and tool life, 232
        types, 51
    seats, 72
    tipped, 25, 26
    wear, 223
        measurement, 228

Torque and thrust in drilling processes, 124
Total force acting on a cutter, 135
Troubleshooting in grinding, 865–867
Tube
    drawing, 677
    sinkings, 701
Tumble jigs, 651
Turning, 12, 13
    fixtures, 616, 618
Two-point strap clamp, 633
Two-way latch type clamp, 634
Types
    of abrasives, 818
    of chips, 83
    of C.I, 7
    of cutting fluids, 265
    of guideways, 295
    of machine tool vibration, 420
    of materials, 4
    of presses, 715
    of tool failure, 221

Undeformed chip length, 130
Up-cut milling, 130
Uses of strain gauges, 197

Vee locator, 625
Velocity relationships
    in metal cutting, 91
    in orthogonal cutting, 91
Vernier
    calipers, 547
    scale, 548
Vibration, 290
    elimination of, 432
    response, 289
    of spindles, 322
Volume to weight ratio, 286

Wear mechanism of tools, 225
Wedge clamps, 633
Welding fixtures, 665
White C.I., 7
Wire cut EDM, 790
Wire drawing, 677
Wrought iron, 8